Basic Concepts of Iron and Steel Making

Sujay Kumar Dutta · Yakshil B. Chokshi

Basic Concepts of Iron and Steel Making

Sujay Kumar Dutta
Metallurgical and Materials
Engineering Department
Faculty of Technology and Engineering
The Maharaja Sayajirao University
of Baroda
Vadodara, Gujarat, India

Yakshil B. Chokshi
Metallurgical Engineering Department
Government Polytechnic Rajkot
Rajkot, Gujarat, India

ISBN 978-981-15-2439-4 ISBN 978-981-15-2437-0 (eBook)
https://doi.org/10.1007/978-981-15-2437-0

This Springer imprint is published by the registered company Springer Nature Singapore Pte Ltd.
The registered company address is: 152 Beach Road, #21-01/04 Gateway East, Singapore 189721, Singapore

Preface

The steel is the most widely used metallic (i.e. iron) alloy, and its production (in world) is more than fifty times total production of combined all other metals. That is why iron and steel form one group (i.e. *ferrous metals*), and other metal combinations (except iron) form other group (i.e. *non-ferrous metals*). India occupies third position in world steel's production and will become second in coming decade. All Metallurgical and Materials Engineering Departments in technical colleges/institutes, particularly in India, have teaching programmes in *ironmaking and steelmaking*. During the past three and half decades, many advances have been made in the implementation of new technologies in the *ironmaking and steelmaking*. These technical advances have been reflected also on the curriculum given at the technical colleges/institutes. This textbook covers almost all the important basic concepts, derivations and numerical for undergraduate and graduate engineering students in simple and easy to understand formats. Even plant's engineers or operators and research scientists can brush up their understanding of *ironmaking and steelmaking* and use as a source of reference.

This textbook is divided into six parts: Part I (**Ironmaking**) of the book covers **Raw Materials and Blast Furnace Ironmaking**, Part II covers **Alternate Methods of Ironmaking** including sponge iron, smelting reduction processes, etc., Part III covers **Physical Chemistry of Ironmaking** including thermodynamics and kinetics, Part IV covers all **Steelmaking Processes**, Part V covers **Thermodynamics** and **Physical Chemistry of Steelmaking**, and Part VI discusses **Pollution** in iron and steel plants.

Sujay Kumar Dutta taught process metallurgy, in general, and ironmaking and steelmaking in particular, at M. S. University of Baroda, India, for thirty-six years. He visited several iron and steel plants; interactions with professionals have considerably enhanced his knowledge of the subject. He wishes to gratefully acknowledge all of them. He expresses his gratitude to his teachers Prof. A. K. Chakrabarti (earlier *Bengal Engineering College*, Shibpur; presently *Indian Institute of Engineering Science and Technology*, Shibpur, India) for introducing him to the metallurgy of ironmaking and steelmaking and Prof. A. Ghosh (*Indian Institute of Technology* Kanpur, India) for advance ironmaking and steelmaking, which have inspired him throughout his career.

Despite taking all the possible care, there may be some errors or mistakes left out unnoticed. If so, please feel free to interact with us. In moulding and casting this textbook, we poured our long experience in it and also collected the information from several sources. We are indebted to one and all, from whose valuable knowledge we have been benefited. We thank our family members for their cooperation during the preparation of the book.

Vadodara, India Sujay Kumar Dutta
Rajkot, India Yakshil B. Chokshi

Contents

Part IV Steelmaking

About the Authors

Sujay Kumar Dutta is a former Professor and Head of the Department of Metallurgical & Materials Engineering at Maharaja Sayajirao University of Baroda, India. He received his Bachelor of Engineering (Metallurgy) from Calcutta University in 1975 and his Master of Engineering (Industrial Metallurgy) from MS University of Baroda in 1980. He completed his PhD at the Indian Institute of Technology, Kanpur, India, in 1992. He joined MS University of Baroda as a Lecturer in 1981 and was promoted to Professor in 2001. Prof. Dutta has received several awards, including an Essar Gold Medal (2006), a Fellowship (2014) and a Distinguished Educator Award (2015), all from the Indian Institute of Metals (IIM), Kolkata, in recognition of his distinguished service to the field of Metallurgical Education and to the IIM. He has authored of five books, two chapters of "Encyclopedia of Iron, Steel, and Their Alloys", and published more than 120 research papers in national/international journals and conference proceedings. Very recently Prof. Dutta was awarded "*SAIL Award 2019*" by The Institution of Engineers (India).

Yakshil B. Chokshi is a Lecturer at the Department of Metallurgy, Government Polytechnic Rajkot, India. He received his BE in Metallurgical & Materials Engineering from Gujarat University in 2010 and his ME in Industrial Metallurgy from MS University of Baroda in 2012.

Part I
Ironmaking

1 Raw Materials

Steel (i.e. refined impure iron) is the most used iron alloy; its production is more than 50 times total production of combined all other metals. That is why iron and iron alloy form one group (i.e. *ferrous metals*) by its own right. Iron ore/sinter/pellets, coke and limestone are used as raw materials for blast furnace. Impure liquid iron is produced in a blast furnace from reduction of iron ore by carbon of coke. Sinter making, pellets formation as well as coke making are discussed in detail in this chapter. Testing methods of room temperature physical properties of raw materials are also described.

1.1 Introduction of Ferrous Extractive Metallurgy

Looking at the periodic table (Fig. 1.1), it is found that more than 70 elements are metals out of 118 elements. Not all the metals are equally important according to their uses. The steel (i.e. refined impure iron) is the most used iron alloy; its production is more than 50 times total production of combined all other metals. That is why iron and iron alloy form one group (i.e. *ferrous metals*) by its own right, and other metals combination (except iron) forms other group (i.e. *non-ferrous metals*). General flow diagram for extraction of metal from ore is shown in Fig. 1.2. Extractive metallurgy for iron and steel is known as *Ferrous Extractive Metallurgy*.

Since extractive metallurgy is a science of separating metals from their ores (by chemically) and subsequently refining them to make pure form (to some extent), the production of hot metal/pig iron is a partial step involved for extractive metallurgy of iron, and it is not completed until it (i.e. hot metal) refines. Therefore, *steelmaking* is a refining process in which impurity elements are oxidized from hot metal and steel scrap that are charged to the process.

Hence, ferrous extractive metallurgy has two-stage operations:

- Reduction stage, i.e. ironmaking stage (iron ore reduced to hot metal),
- Oxidation stage, i.e. steelmaking stage (hot metal refined to steel).

S. K. Dutta and Y. B. Chokshi, *Basic Concepts of Iron and Steel Making*,
https://doi.org/10.1007/978-981-15-2437-0_1

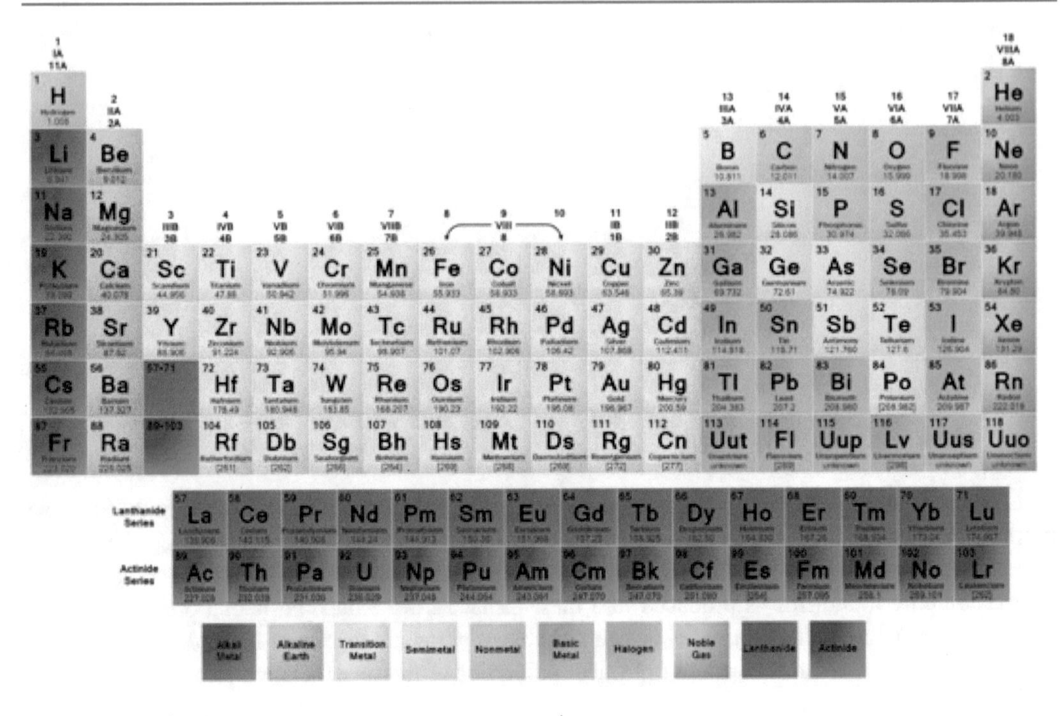

Fig. 1.1 Periodic table of elements

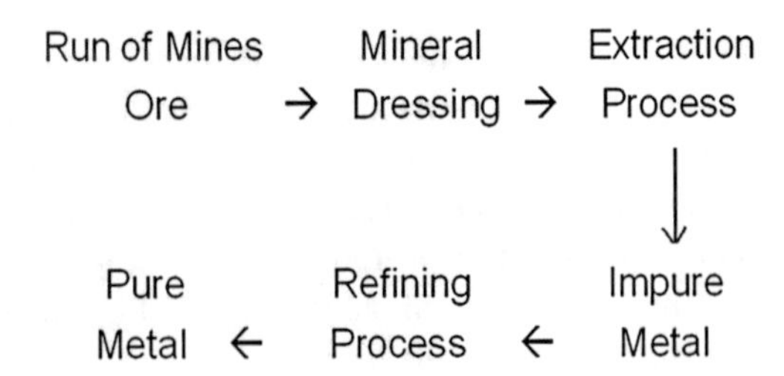

Fig. 1.2 General flow diagram for extraction of metal from ore

1.2 Raw Materials

Impure iron is produced from reduction of iron ore by means of coke in a blast furnace. The product is a liquid iron, known as hot metal (3.5–4.5% C, 0.5–1.25% Si, 0.5–0.75% Mn, etc.) which is raw material for production of steel. Hot metal is refined by oxidation process to form steel. Hot metal is solidified into a small sized ingots which are known as *pig iron*. From a bird's eye view, the pig iron casting looks like piglets which are feeding by their mother. That is why it was named pig iron for solid impure iron. The pig iron is very brittle in nature due to high carbon content, which makes it fit only as a casting of some components by adjusting composition. A schematic view of a blast furnace is shown in Fig. 1.3.

The raw materials used for blast furnace ironmaking are mainly (i) iron ore, (ii) metallurgical coke, (iii) limestone, (iv) air and (v) water (for cooling).

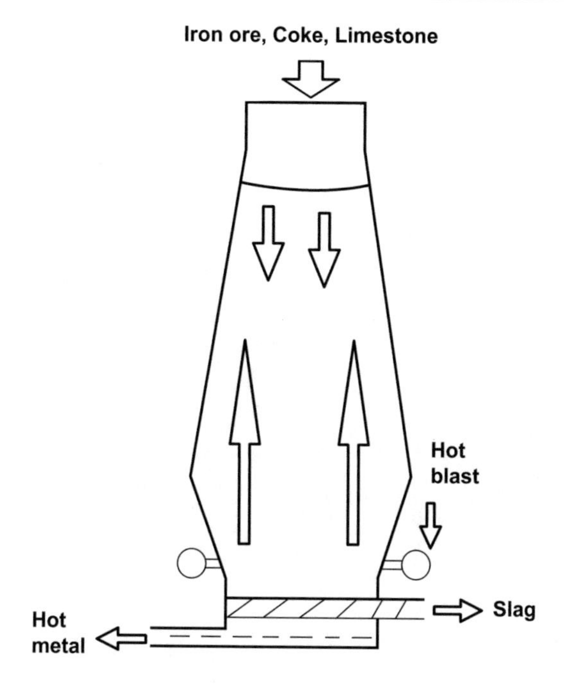

Fig. 1.3 Schematic representation of a blast furnace

1.2.1 Iron Ore

A wide variety of iron ore deposits have been found all around the world. These total over 350 billion tonnes at an average iron (Fe) content of 47%. A minor fraction of these deposits is currently commercially mined as iron ore with Fe contents ranging from as low as 30 up to 65%. The world's iron ore production and reserves are shown in Table 1.1. World consumption of iron ore grows on average 10% per annum with the main consumers China, Japan, India, Korea, USA and the European Union [1]. China is currently the largest consumer of iron ore in world and establishes as the world's highest steel producer.

Types of iron ores are: (a) haematite (Fe_2O_3), (b) magnetite (Fe_3O_4), (c) siderite ($FeCO_3$) and (d) limonite ($Fe_2O_3 \cdot 3H_2O$).

(a) **Haematite (Fe_2O_3)**: It is also called *red ore of iron*. The maximum theoretical iron contains 70% and 30% oxygen (for pure Fe_2O_3). In actual deposits, the iron content varies 50–65%. Its specific gravity is about 4.9–5.25. It is non-magnetic in nature and is more reducible compared to other minerals of iron. Figure 1.4a shows the haematite iron ore.
(b) **Magnetite (Fe_3O_4)**: It is called *black ore of iron*. Magnetite, combination of ferrous and ferric oxides (i.e. $Fe_3O_4 = FeO + Fe_2O_3$), is an isometric crystal system. Magnetite contains 31.03% FeO and 68.97% Fe_2O_3. The maximum theoretical iron contains 72.4% and 27.6% oxygen (for pure Fe_3O_4). In actual deposits, the iron content varies from 25 to 70%. This varies over a wide range depending upon the impurity (i.e. gangue) content of the ore. A grey metallic lustre is also observed. Its specific gravity is about 4.9–5.2. It is magnetic in nature, so it can be easily separated from the gangue minerals by magnet. The main disadvantage of the ore is that it has very poor reducibility. Many often magnetite ores required to be agglomerated by sintering or

Table 1.1 World iron ore production and reserves (Mt) [1]

Country	Mine production				Reserves	
	Useable ore (2016–2017)		Iron content (2016–2017)		Crude ore	Iron content
Australia	858	880	531	545	50,000	24,000
Russia	101	100	60	60	25,000	14,000
Brazil	430	440	275	280	23,000	12,000
China	348	340	216	210	21,000	7200
India	185	190	114	120	8100	5200
Ukraine	63	63	39	39	6500	2300
Canada	47	47	29	29	6000	2300
Sweden	27	27	16	16	3500	2200
Iran	35	35	23	23	2700	1500
USA	42	46	26	28	2900	760
Kazakhstan	34	34	10	10	2500	900
South Africa	66	68	42	42	1200	770
Other countries	116	110	72	68	18,000	9500
World (total)	2352	2380	1453	1470	170,400	82,630

Fig. 1.4 Iron ores: **a** haematite, **b** magnetite, **c** siderite and **d** limonite. (**a**, **b** Courtesy from MS University of Baroda, India; **c**, **d** reproduced with permission from 911 Metallurgist [23])

pelletizing before charge into the blast furnace. Due to high-temperature firing, Fe_3O_4 get converted into Fe_2O_3 and hence the poor reducibility of magnetite ore is indirectly taken care of it. Figure 1.4b shows the magnetite iron ore.

(c) **Siderite ($FeCO_3$)**: It is known as *spathic iron ore*. It contains 48.2% iron (for pure mineral) with various colours mainly from ash grey to brown. It requires calcination before charging into the blast furnace to eliminate CO_2 from the ore. It is often mixed with other carbonates (like $CaCO_3$, $MgCO_3$). Its specific gravity is about 3.7 to 3.9. Figure 1.4c shows siderite ore.

(d) **Limonite ($FeO(OH) \cdot nH_2O$)**: It is called *yellowish-brown ore*. The ore contains actually very low iron. The dehydration occurs by the pre-heating of the ore. Limonite is relatively dense with a specific gravity varying from 2.7 to 4.3. Figure 1.4d shows limonite ore.

1.2.1.1 Evaluation of Iron Ore

(i) Richness of deposit (i.e. % Fe in the ore): Indian iron ores are mostly rich in iron; generally, ores contain more than 50% Fe. Hence, those ores contain less than 50% Fe, that are called *low-grade ores* in Indian condition.
The metallurgical advantage of using iron ore rich in iron content is well known that for every 1% increase in Fe content, there is 1–3% increase in production and a similar decrease in coke rate.

(ii) Extent of deposit: Deposit of ore must be in good amount to run the plant for several years.

(ii) Location of deposit: Location of deposit must be as far as nearer to the plant; otherwise, transportation cost will be increased.

(iv) Composition of the gangue minerals: The gangue content in the ore should be as low as possible. The silica in the ore goes as waste and increases the slag volume and coke consumption. Disproportionately, high alumina content in the ore may also increase the slag volume and fluidity. The presence of 12–19% Al_2O_3 in the slag gives the desired fluidity for removal of sulphur from hot metal. With high phosphorous and sulphur content in the ore, production of quality hot metal becomes difficult. 1% increase in gangue content of the ore results in 1.75% increase in coke rate. It is obvious, therefore, that high-grade ore is most welcome by the blast furnace operator provided Al_2O_3/SiO_2 ratio is low. High ratio will lead to high Al_2O_3 slag with its associated smelting difficulties. Very high Al_2O_3 in slag has a lower de-sulphurizing capacity, and that may lead to undesirable high sulphur in the hot metal.

(v) Size of deposit: Permeability for gas flow in blast furnace depends on size of ore and void fraction. Void fraction again depends on the ratio of smaller and larger particles of charge materials.

1.2.2 Metallurgical Coke

The specifications of metallurgical coal, for making metallurgical coke, should be as follows:

Volatile matter: 26% in coal having 15% ash or about 30.59% on ash free basis,
Ash: not more than 15% (it can be acceptable coal with up to 17% ash content),
Phosphorous: not more than 0.15%,
Sulphur: not more than 0.6%.

Fig. 1.5 Coke. (Courtesy from MS University of Baroda, India)

Coking coals in coherent beds, when heated in absence of air, release volatiles and form strong and porous mass that are known as *coke* (as shown in Fig. 1.5). Coke, produced out of coal, contains 21.12% ash, 78% fixed carbon, 0.88% volatile matter, 0.18% P, 0.57% S and 1.4% moisture. The coke with these specifications may be considered as good coke. Coke is basically a strong solid material which forms lumps based on a structure of carbonaceous material internally binding together. The average size of the coke is much bigger than the iron ore, and the coke will remain solid throughout the blast furnace process. Coke is a solid and permeable material up to very high temperatures (>2000 °C), which is of importance in the hearth, melting and softening zones. Below the melting zone, coke is the only solid material; so, the total weight of the blast furnace charge is supported by the coke. The coke bed must be permeable, so that hot reducing gases going upward direction through it, and molten slag and iron can flow down easily through coke bed to accumulate at the hearth.

The coking coal used in India is of poor quality with high ash content (21–24%) even after washing as against ash content of 9–10% in technologically advanced countries. As a result, the high ash coke creates operational problems such as high coke rate and low productivity in the Indian's blast furnace (BF). Thus, high ash content also increases phosphorus content in the hot metal.

Due to nature of deposition, Indian coals, in general, are of inferior quality (i.e. high ash content) compared with coals available in the international market. Despite of this, Indian coals act as environmentally friendly because of the following:

- Low sulphur content,
- Low chorine content and
- Low toxic trace elements content.

1.2.2.1 Functions of Coke

Main functions of coke in a blast furnace are as follows:

(i) It acts as a fuel to provide heat for (a) requirement of endothermic chemical reactions and (b) melting of metal and slag.
(ii) It acts as a reducer by producing of reducing gases (CO, H_2) for the reduction of iron oxides in ore.

(iii) It provides the support the whole burden (in bosh region) during melting of iron-bearing burden.
(iv) It provides a permeable bed through which molten slag and metal droplets come down into the hearth and help hot reducing gases flow upward direction for heating raw materials as well as reduction of iron ore.
(v) It provides the carbon for proper carburization of the hot metal.

To fulfil above functions, the coke should satisfy the following requirements:

- To be a good fuel coke, it should have maximum carbon content and minimum ash content;
- To regenerate reducing gases and to produce heat, coke should have a high reactivity with oxygen, carbon dioxide and water vapour;
- In bosh region, permeability of the charge is maintained by the coke alone because other materials, except coke, are either semi-fused or molten stage. Hence, coke should be remained solid until it burns at the tuyere level.

The role of coke as a fuel as well as a generator of reducing gas can be considerably reduced using higher blast temperature and injection of auxiliary fuels in the blast furnace. This reduces the coke rate considerably from 700 to 900 kg/tonne of hot metal (kg/thm) to 400–500 kg/thm.

Combustion efficiency of coke is evaluated through the off-gas analysis (Eq. 1.1). The combustion efficiency would be 100% if off-gas of BF contained only CO_2 gas.

$$\text{Combustion efficiency} = \frac{(CO_2)}{(CO + CO_2)} \tag{1.1}$$

1.2.2.2 Quality of Coke

The efficiency of a blast furnace is directly dependent on the permeability of the charge in the furnace which is again directly depends on the quality of the coke. Therefore, it is expected that the quality of coke should be such that it gives minimum of operational difficulties and maximum of production rate at minimum of coke consumption.

The value of coke (as a BF fuel) depends on the following properties:

(i) chemical composition,
(ii) reactivity,
(iii) size range,
(iv) thermal stability at high temperatures and
(v) strength and abrasion resistance.

(i) Chemical composition of the coal depends on its proximate and ultimate (elemental) analyses. The proximate analysis includes content of moisture, volatile matter, fixed carbon and ash in coal. Elemental or ultimate analysis is the quantitative determination of carbon, hydrogen, nitrogen, sulphur and oxygen in the coal. Generally, chemical composition of coke means fixed carbon, volatile matter, ash, sulphur, phosphorus and other impurities. Fixed carbon in coke acts as a fuel as well as reducing agent in the BF. As the ash content increases, the available carbon of coke decreases, and subsequently, the heat supply to the furnace per unit weight of coke decreases. Very good quality coke should have >85% fixed carbon, <10% ash, <2% volatile matter, 0.6–1.5% S and 0.04% P. The ash is the inorganic residue after burning; ash contains refractory oxides, e.g. 55–60% SiO_2, 15–30% Al_2O_3, 4–10% Fe_2O_3, 2–3% CaO, 1–4% MgO, 2–8% Na_2O + K_2O and trace of TiO_2.

Due to the presence of high ash content in coke, (a) the coke rate in BF will be high, 1% increasing of ash, coke rate increases 1–2%; (b) slag volume will be more; (c) productivity of BF will be low, due high volume of slag to low utilization of useful volume of BF, 1% increasing of ash, productivity decreases 2–3%; and (d) phosphorus in hot metal will be increased.

Sulphur and phosphorus, usually present in coke in form of inorganic compounds, are later transferred into hot metal. Most of the sulphur and part of the phosphorus in hot metal comes from coke, and remaining phosphorus comes from the ore. North-eastern parts of India have coals of good coking quality, but it contains 2–7% S (major portion as inorganic); it cannot be economically used for making coke for BF. Increasing sulphur and phosphorus in coke means increased sulphur and phosphorus contents in hot metal and as well as more consumption of flux during steelmaking are likely to increase costs in the steelmaking. If ash of coke contains excess sulphur, that coke should be avoided in BF. Increasing phosphorus in coke means increased phosphorus content in hot metal.

The calorific value (Q) of coal is the heat generated by its complete combustion with oxygen. The calorific value is a complex function of the elemental composition of the coal. The calorific value can be determined experimentally using calorimeters. Dulong suggests the following approximate formula [2] for the calorific value (Q, kJ/kg) of coal, when the oxygen content is less than 10%:

$$Q\,(\mathrm{kJ/kg}) = 337\mathrm{C} + 1442\left\{\mathrm{H}-\left(\frac{\mathrm{O}}{8}\right)\right\} + 93\mathrm{S} \tag{1.2}$$

where C, H, O and S are the weight per cent of carbon, hydrogen, oxygen and sulphur in the coal, respectively.

The *heat of combustion* of a product measures the energy released (in kJ/mol) when that substance is burned in air. The *specific energy* of a fuel provides practical measures of the energy content of a fuel more commonly used in the storage and handling of this substance (energy per weight).

Specific energy (E_S) is the energy (Q) per unit mass (m) of a fuel as follows:

$$E_S = \left(\frac{Q}{m}\right) \tag{1.3}$$

(ii) Reactivity of coke may be defined as the rate of reaction between coke and oxygen or any other gaseous phase (e.g. CO_2, H_2O). The reactivity of a coke is a measure of its ability to react with CO_2 to form CO, according to *Boudouard reaction*:

$$\mathrm{C(s)} + \{\mathrm{CO_2}\} = 2\{\mathrm{CO}\}, \Delta\mathrm{H}^0_{298} = 85.35\,\mathrm{kJ/mol\ of\ CO} \tag{1.4}$$

i.e. for conversion of $CO_2 \rightarrow CO$, this is also known as *gasification of carbon* by carbon dioxide or *solution loss reaction* (because by this reaction not only loss of the heat, but also loss carbon). The reactivity is particularly important from the point of view of rate of its combustion at the tuyere level and its reaction with the ascending gases. Different cokes have different reactivities.

The degree of gasification of carbon (F_C) is defined as [3]:

$$F_C = \left(\frac{\text{Weight loss of carbon in the sample}}{\text{Total weight of carbon in the sample}}\right) \quad (1.5)$$

The term *reactivity* (R_C) is commonly used to denote the rate of the gasification reaction.

$$\text{i.e.}\, R_C = \left(\frac{dF_C}{dt}\right) \quad (1.6)$$

The higher the reactivity of the coke, the faster is the gasification reaction. In iron ore reduction, F_C has been seen to vary linearly with time (t), up to about F_C = 0.3–0.4. In this region, reactivity R_C is a constant.
Reactivity has a significant influence on the reduction process; hence, coke with high reactivity values is preferred. The rate of coke burning of overall controls the rate of production of a BF. The rate of coke burning is directly proportional to:

(a) coke area exposed to the blast,
(b) temperature and pressure of the blast and
(c) affinity of the carbon for oxygen.

More open and well-developed cellular structure of coke will expose maximum area of carbon to the blast. Coke, with higher porosity and adequate strength, is a better fuel. By increasing 7% of porosity, coke rate decreases 100 kg/thm.

(iii) Size range of coke affects the distribution of materials inside the furnace and consequently the gas flow rate which control production rate directly. The size of coke is generally chosen to match the size of other raw materials and is much bigger than the size of the iron-bearing materials to ensure maximum bed permeability for smooth furnace operation.
To ensure a sufficiently high permeability of the charge, the charge coke should be at least 80% of +40 mm size. If the size of iron-bearing material is 13 mm, size of coke should be 52 mm. Size ratio of coke: iron-bearing materials is 4 : 1, i.e. minimum size of coke should be 3–5 times more than of the iron-bearing material. +80 mm fraction of coke should be removed as far as possible.

(iv) Thermal stability at high temperatures: Due to descending of coke, the BF gets gradually heated from 200 to 1600 °C before it burns in front of the tuyeres. The temperature gradient increases with increasing in size and degradation of coke. The thermal stresses are greater for the larger size of charged coke; in other words, smaller size of coke is better for better thermal stability as follows:

(a) absence of large lumps in charged coke,
(b) uniformity of coke texture,
(c) high carbonization temperature (i.e. coke making),
(d) low chemical reactivity and
(e) minimum ash content.

(v) Strength and abrasion resistance: Coke should have sufficient strength to withstand handling and charging action before landing in the BF. Coke should stand high temperature and load of 23–25 m tall burden material standing above it. Height of the furnace is controlled by the strength of coke. If coke breaks down into fines under the conditions, that will affect the

furnace permeability. Shatter test gives resistance of coke to impact, and tumbler test to measure the resistance of coke to degradation by combination of impact and abrasive forces. Height of modern BF is controlled by the strength of the available coke.
The coke travelling from stack region to the dead man zone is exposed from low to high temperatures; high alkalis during long periods of time along with additional reactions (reduction of slag, carburization) mostly affect the surface of the coke lumps. Coke of dead man zone sampled by core drilling corresponds to the unreacted core of the initial lumps, and it is not surprising that it exhibits similar strength to the coke that is charged at the top.

1.2.2.3 Preparation of Coke

The coal selection to make coke is the most important variable that controls the coke properties. The rank and type of coal selected impact on coke strength, while coal chemistry largely determines chemistry of coke. In general, bituminous coals are selected for blending to make blast furnace coke of high strength with acceptable reactivity and at competitive cost. Table 1.2 shows the typical chemical composition of coke that may be of good quality. Ash directly replaces fixed carbon. The increased amount of ash increases volume of the slag which requires heat energy to melt and as well as more fluxes to make a liquid slag. Ash, sulphur, phosphorus and alkalis can be best controlled by careful selection of the coal, coke and burden materials.

Coke, which is used as fuel in blast furnaces during the ironmaking process, is made by heating coking coal at a high temperature in a series of ovens known as a *coke oven battery*. A coke oven battery is a group of ovens in which coals are heated to high temperature (1100 °C) in the absence of air. The carbon is concentrated, and the volatile mater is removed from the coal to produce coke by this process. Metallurgical coke is produced by heating coking coal (25–30 wt% volatiles content) in the absence of air. This causes the volatiles to be distilled from the coal to give a porous coke which is (i) reactive at high temperatures and (ii) strong enough to take whole load of materials within the blast furnace.

Figure 1.6 shows the coke oven battery. In general, ovens are made from silica bricks and a single battery can contain as many as 100 rows, vertical retort chambers, each separated from its neighbours by hot gases. Coal is heated by conduction through the walls, so that oven width is restricted to a maximum of about 0.5 m. The height of ovens can be as much as 7 m and their length 15 m.

>3 mm (80–90%) particle sizes are charged to the oven by pipe line. The carbonization takes place between 12 and 20 h to complete. The aim is to achieve a uniform surface temperature in the range 950–1050 °C. The doors of the oven are removed, and the whole mass is pushed by inserting a ram from the rear side of the battery and collecting coke in a quenching car at the front side. The red-hot coke is then quickly cooled by water spraying.

Stamp-Charging Battery

Stamp-charging is a pre-carbonization technique for coke making. Pre-carbonization technologies, such as selective crushing, pre-heating, partial briquette-blending and stamp-charging, are being utilized in different steel plants all over the world. In stamp-charging, the bulk density of the coal is increased by stamping the charge into a solid cake, outside the oven. This cake, which is almost similar in size to the oven, is then inserted into the oven.

Table 1.2 Typical coke analysis

Fixed carbon (%)	Volatile matter (%)	Ash (%)	Nitrogen (%)	Sulphur (%)
87–92	0.2–0.5	8–11	1.2–1.5	0.6–0.8

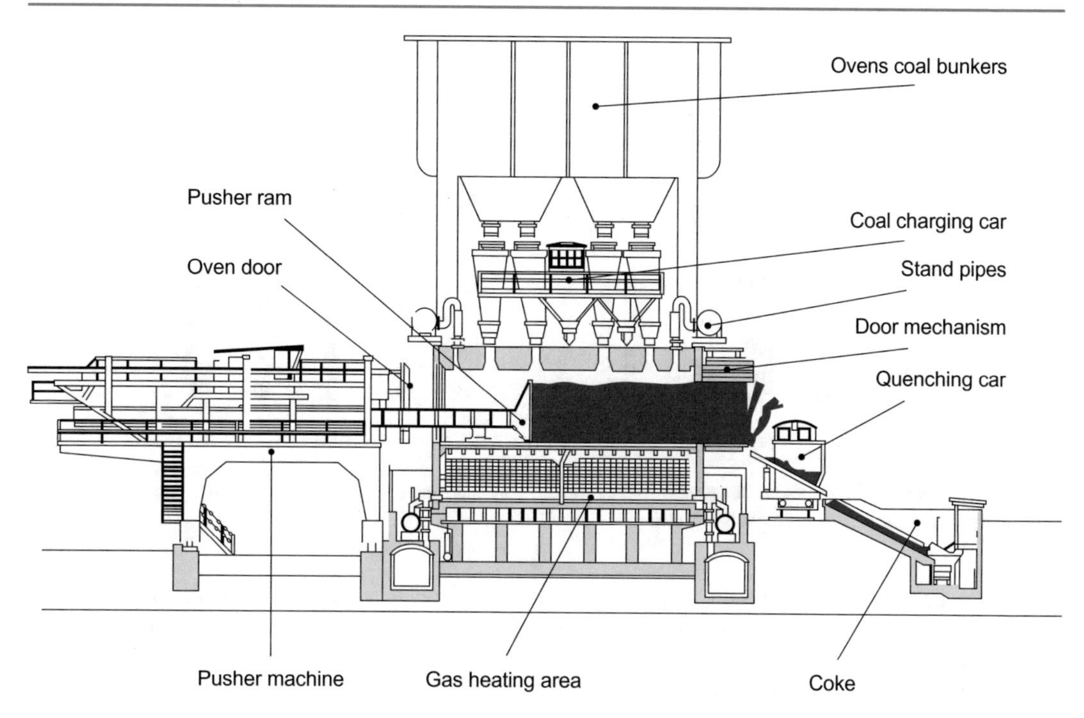

Fig. 1.6 General layout of a typical coke oven battery [24]

Any improvement in coke quality depends on the blend constituents, homogeneous distribution of the individual components of coals, as well as an increase in the bulk density of the blend. Compaction along with finer crushing leads to the maximum improvement in the bulk density, so stamp-charging has improved coke quality.

Low-rank, weakly coking and high-volatile coals can also be used, but because the charge is compacted to very high bulk densities, increased wall pressures can be a problem. To control the wall pressure within acceptable limits so that the refractory walls of oven are not damaged, so carefully chosen the coal blend. An optimum balance between high- and low-volatile coals is necessary to ensure that high-quality coke is produced without causing permanent damage to the oven walls and the oven supporting structure.

Indian coking coals are in poor quality. The ash content has steadily increased from 16 to 24% and is still increasing. To conserve the prime coking coal to optimize its use, Central Fuel Research Institute (CFRI), Dhanbad, India, has developed *blending of coking coals* using prime (40–50%), medium (30–40%) and semi- or weakly (10–20%) coking coals. The ash of blending coal is about 15%, finally produced coke of 22% ash.

In stamp-charging, coal finely ground to 90 ± 2% below 3.2 mm, with the moisture content of coal adjusted to 10 ± 1%, is compacted into a solid cake outside the oven, and the cake is inserted into the oven through the ram side door. The stamping operation is carried out by a special machine known as the *stamp-charging-cum-pushing machine*. The coal blend is gradually discharged from the bunker into the stamping box, and a series of hammers compact the coal mass into a solid cake and charge into the oven as a single piece (as shown in Fig. 1.7).

The first plant was started up in 1984, at ZKS Zentralkokerei Saar, Volklingen, Germany. Then, stamp-charging was adopted by Tata Steel in 1989 and other plants in India. During this century, an impressive capacity was built in China, with more than 100 Mtpy. The first modern stamp-charging plant in the Americas was built in 2010 in Thyssen Krupp Steel CSA, Santa Cruz, Brazil [4].

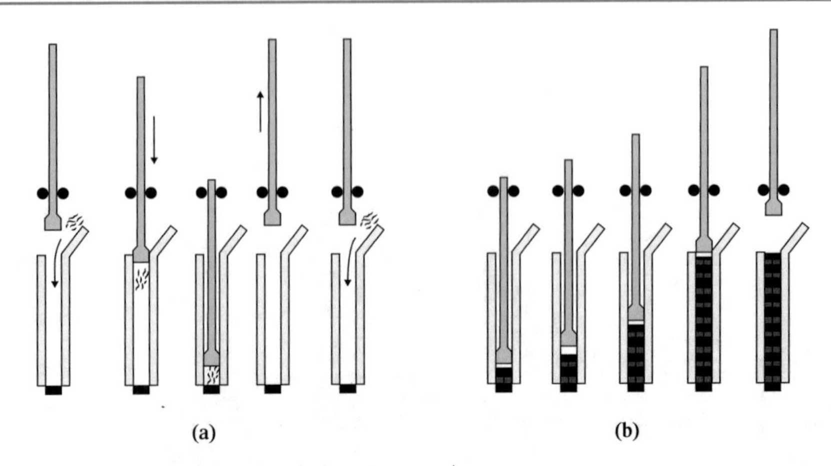

Fig. 1.7 Schematic representation of vertical coal stamping [25]. **a** Sequence of charging, pressing and new charging, **b** growing of the cake after each charging step

The success of stamp-charging depends on the formation of a stable coal-cake which can be charged into the oven without any breakage. Production of a stable cake depends on three basic factors: (i) correct shape of feed coal, (ii) precise moisture control and (iii) adequacy of stamping energy imported to the coal mass.

The strength of the cake and hence the stability is mainly due to mechanical interlocking of particles which is a function of surface area. For the best results, the size distribution is as: 90% <3.15 mm, 35–50% <0.5 mm, and average size is 1 mm.

The compressive strength of the cake and its stability depends on greatly the specific stamping energy imported to the coal mass.

$$\text{Stamping energy (Nm)} = G \cdot Z \cdot H \cdot t \cdot n \cdot g \tag{1.7}$$

where

G Weight of drop hammer, kg,
Z Number of drop hammers,
H Height of drop, m,
t Stamping time, s,
n Frequency/s,
g Gravitational force, m/s^2 = 9.81 m/s^2.

$$\text{Specific stamping energy (Nm/kg)} = \frac{\text{Stamping energy}}{\text{Weight of coal}} \tag{1.8}$$

The specific stamping energy of 420–450 Nm/kg is imported to the coal mass to achieve a typical bulk density of over 1050 kg/m^3 for requirement of cake stability. By using 20–30% imported coal reduced the requirement of prime coking coal from 25 to 50% in the case of stamp-charging without affecting the coke quality.

The conversion of coal to coke takes place by the process of carbonization, which consists of heating the coal in a closed chamber in absence of air. As the temperature rises, the coal blend becomes plastic (i.e. semi-fused) and distillation of the volatiles occurs. This can be divided into three stages as follows:

(i) Pre-plastic stage → evolution of moisture and a little of volatile matter at temperature up to 300 °C,
(ii) Plastic stage → extensive molecular disruption and evolution of large amounts of volatiles. Volume changes occur due to temperature raised from 300 to 700 °C,
(iii) Post-plastic stage → coal solidifies as a hard porous residue known as coke. Temperatures increased from 900 to 1200 °C, and high-temperature carbonization produces the strong coke. Cokes, carbonized at lower temperature, are faster rate of reaction with oxygen. Since strength is generally considered to be more important than reactivity, high-temperature coke is preferred by most BF operators.

During carbonization in a coke oven, fissures in the coke are generated due to stresses that arise from the differential contraction rates in adjacent layers of coke due to different temperatures. Typically, they are longitudinal, which are perpendicular to the oven walls. Additionally, many transverse fissures are also formed during pushing. These fissures determine the size distribution of the product coke by breakage along with their lines during subsequent handling.

Stamp-charging has following advantages [3]:

- Significantly higher bulk density (1150 kg/m^3),
- Increased throughput (8–10%) owing to higher bulk density,
- Improved coke strength due to compact of the individual coal particles during carbonization,
- Product coke is denser, smaller and more uniform in size,
- Since fine coal is not charged from the top of the ovens, less pollution is occurred.

The primary disadvantage is that the preparation and handling of the stamped cake increased capital cost. Stamp-charging has been successfully used in France, Poland, Republic of Czechoslovakia, Brazil, Germany, China and India.

Dry Cooling of Coke [5]

Modern battery uses coke dry cooling system to produce coke of moisture free. The coke dry cooling system is based on the coke dry quenching media, i.e. cooling of hot coke with inert gas circulating in close loop between the chamber of hot coke and the waste heat. It is also known as *coke dry quenching* (CDQ), which is an alternative to the traditional wet quenching. During wet quenching of oven coke, sensible heat of the hot coke is dissipated into the atmosphere and lost. In addition, there are air-borne emissions (0.5 tonne of steam per tonne of coke laden with phenol, cyanide, sulphide and dust), and a large quantity of water (around 0.6 m^3/t of coke) is needed for wet quenching. The contaminated water is also discharged in the environment. In a *coke dry cooling plant* (CDCP), red-hot coke is cooled by inert gases. The heat energy from the hot coke is recovered in a waste heat boiler for use as steam, resulting in energy conservation as well as a reduction in coke particle emissions. Around 80% of sensible heat is recovered.

Hot coke is brought from the battery to the CDCP in bottom opening bucket kept on the quenching car. This bucket is lifted at the CDCP by a hoisting/charging device to the top of the CDCP chamber, and red-hot coke is discharged in the cooling chamber. Hot coke (temperature around 1000–1100 °C) is cooled in the chamber by the inert gas. The inert gas is a mixed gas which consists of mainly nitrogen (70–75%) along with small amounts of CO_2 (0–15%), CO (8–10%) and H_2 (2–3%). In the cooling chamber, inert gas moves upwards, while the coke moves downwards by the gravity. The coke is discharged from the bottom. The passage time of the coke through the cooling chamber is around 5–6 h, and the flow rate of inert gas is around 82,000 Nm^3/h. The temperature of the coke at the time of discharge from the cooling chamber is below 200 °C. The hot inert gas, after picking heat from the hot coke, comes out from the top of the cooling chamber at around 800 °C and is passed to the waste heat recovery boiler where the sensible heat of the inert gas is used to produce steam (25 t/h) at high pressure (around 40 kg/cm^2) and temperature (around 440 °C). This steam produces electric power in a steam pressure turbine.

Primary and secondary dust collection systems are installed at the coke charging and the discharging area as well as in the boiler section. Dust emissions are reduced to 50 mg/Nm^3 in a bag filter house which is less than 0.09 kg of dust per tonne of dry cooled coke. Since cooling gas is recirculated in a closed system, there is no air-borne coke emissions from gas cooling. Also, no dust laden steam clouds are released by the CDCP. This helps in the improvement in the working environment in the area.

1.2.2.4 Solution for Long-Range Coking Coal Shortage

Solution for long-range coking coal shortage is as follows:

1. Blending with other coals,
2. Importing low ash coking coal,
3. Improving recovery from coal washeries,
4. Reducing coke rate in BF,
5. Using formed coke in BF,
6. Injecting of non-coking coal through tuyeres and
7. Alternative ironmaking, not based on coking coal.

(i) **Blending with imported low ash coking coals**: Imported coking coal (low ash content) used 30–40% maximum in blending mixture, because: (a) use of 100% imported coking coal needs addition of SiO_2 to lower down Al_2O_3/SiO_2 ratio in Indian iron ore, (b) cost of coke will be high by using 100% imported coal. Effect of imported coking coal on productivity and coke rate is shown in Fig. 1.8. Overall ash content in coke as well as coke rate decreases by blending with imported coking coal.

(ii) **Washing coking coal**: Coals have undergone the process of coal washing or coal beneficiation, resulting in value addition (i.e. carbon) to coal due to reduction in ash percentage. These

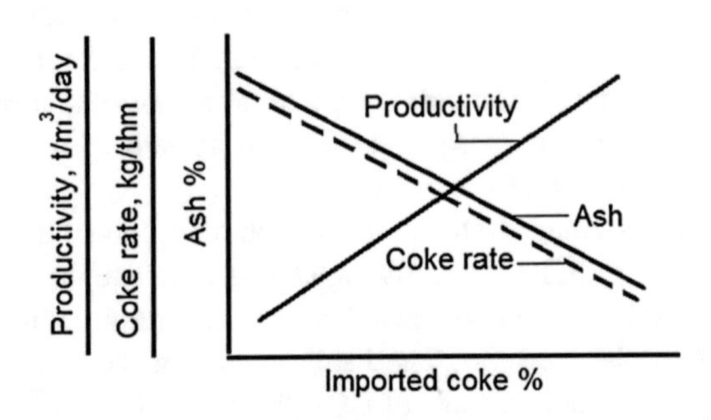

Fig. 1.8 Effect of productivity, coke rate and ash content in coke by imported coke

washed coals are used in production of coke for ironmaking. By washing coking coal, the ash content decreases and there are also some losses of coal.

iii) **Formed coke**: It is synthetically prepared coke. It is made from non-coking coal. Flow diagram of formed coke formation is shown in Fig. 1.9. Non-coking coal is pyrolysis at inert atmosphere, getting char and volatile matters; again, from volatile matters, liquid tar and gases are obtained. Char is mixed with tar (as binder), fluxes may be added, then briquetting the mixer by addition of catalyst and applying the pressure. The briquettes are cured at 100–200 °C.
Advantages of formed coke are as follows:

(i) Main raw material is non-coking coal,
(ii) Uniform size and shape,
(iii) Good permeability,
(iv) Addition of fluxes in formed coke making help in BF slag-making reactions.

Disadvantages of formed coke:

(i) It is costly material,
(ii) Cost is one and half times more than coke.

(iv) **Injection of non-coking coal through tuyeres**:
Coal types are discriminated according to their volatile matter content. The volatile matter is determined by weighing coal before and after heating for three minutes at 900 °C. Coals that have between 6 and 12% volatile mater are classified as low-volatile coal; those between 12

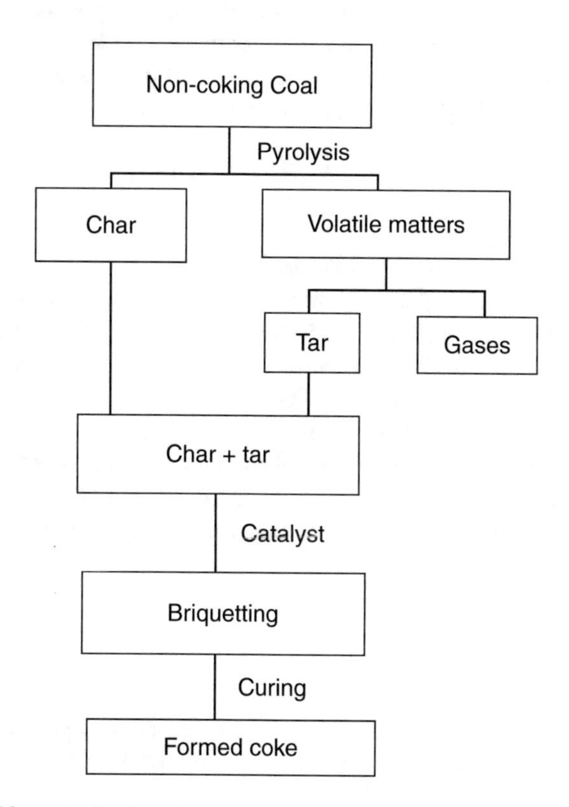

Fig. 1.9 Flow diagram of formed coke formation

and 30% are medium-volatile coal and anything over 30% are high-volatile coal. All types of coal have successfully been used. The most important property of the injection coal is the *replacement ratio* (RR) of coke. The composition and moisture content of the coals determine the amount of coke replaced by a certain type of coal. The replacement ratio of coal can be calculated with a mass and heat balance of the furnace. The chemical composition of the coal (i.e. carbon percentage, hydrogen percentage, ash content), the remaining moisture and the heat required to crack the coal chemical structure (especially the C–H bonds) must be considered.

A formula for the replacement ratio (RR, compared with coke with 87.5% carbon) is:

$$RR = \{(2 \times \%C) + (2.5 \times \%H) - (2 \times \%\text{moisture}) - 86 + (0.9 \times \%\text{ash})\} \quad (1.9)$$

where %C, %H, %moisture and %ash present in coal, respectively.

This formula shows that the coke replacement depends on carbon and hydrogen content of the coal. Any remaining moisture in the coal consumes energy introduced with the coal. The positive factor of the ash content comes from a correction for heat balance effects. High-volatile coals are easily gasified in the raceway but have lower replacement ratio in the process.

1.2.3 Fluxes

A flux is a substance which is added during smelting to help lower down the softening point of the gangue materials of ore, as well as ash of coke; to reduce the viscosity to the slag and to decrease the activity of some of its components for making stable in the slag phase. Slag is formed by the combination of gangue and flux, which is product (i.e. molten oxides) of smelting operation.

Limestone and dolomite act as fluxing material in blast furnace. It should contain less than 5% insoluble (i.e. $\sum SiO_2$, Al_2O_3, Fe_2O_3). The value of a flux is expressed in terms of available base.

$$\text{Therefore, available base} = [\%(CaO + MgO) - (\%SiO_2) \cdot B] \quad (1.10)$$

where B is the basicity of slag.

The chemical formula of limestone is $CaCO_3$. Limestone decomposes at 900–950 °C; marble particles expand first and then shrink during the heating stage, microfractures are generated, and the lump particle breaks into small particles. The chemical formula of dolomite is $CaMg(CO_3)_2$; it provides the MgO for blast furnace slag formation. Dolomite decomposes at two temperatures (~700 and ~950 °C), so the breakage is higher than limestone.

Calcined limestone ($CaCO_3$) and dolomite ($CaCO_3 \cdot MgCO_3$) fluxed the gangue materials (i.e. silica and alumina impurities) of ore and ash of coke to produce a low melting point (1300 °C) fluid slag. Lime (CaO) has the secondary effect of adding part of the sulphur in the charge, and sulphur is also introduced mainly as an impurity in the coke. Sulphur should be removed in the slag phase rather than in the product iron.

1.2.4 Air Supply

One of the most important raw materials in BF practice is the air supply. Steam and gas are operated by turbo-blower. The blast pressure normally ranges from 1 to 4 kg/cm^2 at the tuyeres, so that the turbo-blowers are designed to withstand pressure of 6–7 kg/cm^2. The advantages of the turbo-blowers appear to be:

(i) Smaller foundations and buildings are required,
(ii) Less space is necessary,
(iii) Maintenance costs are generally low,
(iv) Higher blast pressures can be obtained,
(v) A more regular and smoother blast results,
(vi) Blowers can be driven electrically.

There are three classes of turbo-blowers:

1. Vertical type in which the air cylinder is immediately above the steam cylinder, with both pistons on the same rod.
2. Horizontal type in which the air and steam cylinders are in a horizontal line. Again, both pistons operate on the same rod.
3. Vertical–horizontal type, in which the air cylinders are usually placed vertical and the steam cylinder horizontally.

1.2.5 Problem of Indian's Raw Materials

Blast furnace technology in India is more than 100 years old, but the performance indices like coke rate, productivity, etc., have hardly shown much improvement (as shown in Table 1.3). It is revealing to note that the productivity of Japanese blast furnace has increased from 1.0 to 2.5 tonne/m^3/day, and coke rate has gone down from 900 to 400 kg/thm.

The poor quality of raw materials is creating difficulties to achieve higher productivity. Indian iron ores are mostly rich in iron (52.2–69.4%); generally, ores contain average more than 60% Fe. But they have: (i) a high alumina-to-silica ratio (1.8–2.5), resulting in: (a) low blowing rates, (b) higher coke rate (due to high melting point of alumina, 2050 °C), (c) higher alumina in slag, so viscosity of slag increases too much (as shown in Fig. 1.10), de-sulphurization would be difficult (i.e. high sulphur in hot metal), and (d) other related operational problems in the blast furnace (i.e. (i) low productivity, (ii) the softness of ore, which leads to the problems of generation of fines in the furnace).

Due to high ash content in Indian coking coal, coke rate and slag volume are relatively high in Indian blast furnace. Reserve of coking coals is not enough for long lasting time.

1.3 Agglomeration Processes

The blast furnace is a counter-current gas–solid reactor in which the solid charge materials are moving downwards, while the hot reducing gases are flowing upwards. The best possible contact between the solids and the reducing gas is obtained with permeable burden which permits not only a high rate of gas flow but also a uniform gas flow rate with a minimum of channelling of the gas. 25–50 mm fraction of solid materials is considered most favourable sizes.

Agglomeration is the process for aggregate (i.e. bonding) of the small or fine particles to a useful size. The primary purpose of agglomeration is to improve burden permeability and gas–solid contact, thereby reducing blast furnace coke rates and increasing the rate of reduction, i.e. productivity. Good

Table 1.3 Typical performance of Indian blast furnace

Input	Output
1. Ore and other iron-bearing material: 1.1–1.5 tonne (t)	1. Hot metal: 1.0 tonne (t)
2. Coke: 0.6–0.8 t	2. Slag: 0.6–0.75 t
3. Limestone: 0.2–0.3 t	3. Flue dust: 25–60 kg
4. Air: 2500–3000 Nm^3	4. BF gas: 2500–3000 Nm^3
5. Water: 57 t	

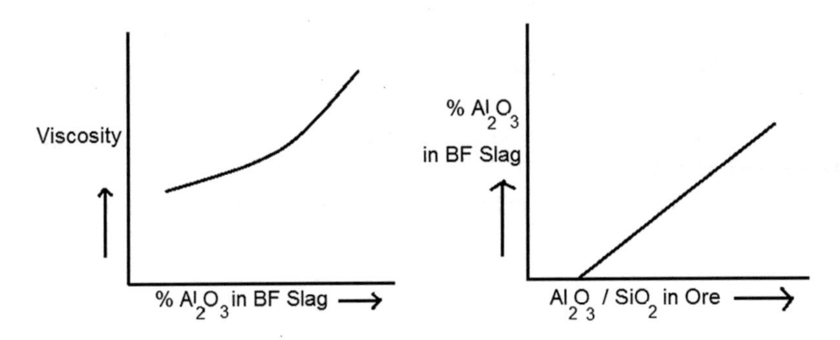

Fig. 1.10 Effect of Al_2O_3 in ore to the BF slag

agglomerate using for blast furnace should be contained 60% or more of iron, a minimum of undesirable constituents, and size range of material 5–12.5 mm. The agglomerate should be strong enough to withstand degradation during stock-piling, handling and transportation; to arrive at the furnace skip with a minimum of approximate 85–90% of plus 5 mm materials. In addition, the agglomerate must be able to withstand the high temperature and the degradation forces within the furnace without slumping or decrepitating. The agglomerate should also be reasonable reducible, so that it can reduce at a satisfactorily high rate in the blast furnace.

Agglomeration can be classified into four types as follows:

1. Sintering,
2. Pelletizing,
3. Briquetting and
4. Nodulizing.

1.3.1 Sintering

Sintering is an important process of agglomeration. Sintering is a process of heating fine materials (+3 mm sizes) to an elevated temperature without complete fusion such that the small solid particles are in contact with one another adhere and agglomerate into larger, more useful sizes. Sinter is an agglomerate made from small particles of iron ore or iron-bearing materials that are semi-fused or fritted together at a high temperature that is produced by combustion of coal. Now flux of small size is also added to the sinter mixture to flux partially of the gangue materials of iron ore to form slag before it is smelted in the blast furnace.

Small particles (+3 mm) of iron ore are mixed with 5–10% coal (fuel), 5–10% limestone (flux, will be more for self-fluxed and super-fluxed sinters) and 6–8% moisture. The sinter process is initiated by igniting the fuel in the top layer of the sinter mixture and is maintained by sucking air (down-draft) through the charge. The small ore particles in the bed are bonded together to a fused aggregate by the action of high temperature; combustion zone is moving in the direction of air flow. The incoming air is pre-heated by hot sinter before reaching the flame front and the waste gases pre-heat the green charge. Combustion air and charge meet each other in a pre-heated state in the combustion zone and produce a heat accumulation leading to incipient fusion. Figure 1.11 shows a cross-sectional profile of sinter bed.

The whole sintering process based on the heating stages can be split into five steps:

Step 1 Haematite (Fe_2O_3) reacts with CaO to generate the first calcium ferrites at around 1100 °C (solid–solid reaction). Around 1200 °C, these calcium ferrites start to convert into the liquid phase. The liquid is rich in CaO and Fe_2O_3 and begins to combine with ultra-fine iron oxides, SiO_2, Al_2O_3, and MgO.

Step 2 Liquid phase starts to generate while temperature keeps on rising, and the superficial disintegration of haematite starts.

Step 3 The liquid melt assimilates with CaO and Al_2O_3, and the evolution of the reaction between liquid and haematite forms acicular calcium ferrites in the solid state (needle form), which are rich in Al_2O_3 and SiO_2.

Step 4 If the process temperature does not reach above 1300 °C or the residence time above 1300 °C is very short, the microstructure after cooling will be rich in acicular calcium ferrites in a bulk of crystalline silicates and granular haematite. That is the heterogeneous sinter.

Step 5 If the process temperature exceeds 1300 °C and the residence time above 1300 °C is long, columnar calcium ferrites and coarser recrystallized particles precipitate from melt. During the cooling step, skeletal rhombohedral haematite precipitates from the liquid phase, and calcium ferrites crystallize as long columnar form flakes, i.e. the homogeneous sinter.

Many efforts have been made to develop new technologies aiming at decreasing the fossil fuels utilization due to the environmental restrictions and decrease the process carbon intensity. The sintering process is complex and involves various physical and chemical phenomena such as heat, mass and momentum transfer coupled with chemical reactions. These phenomena take place

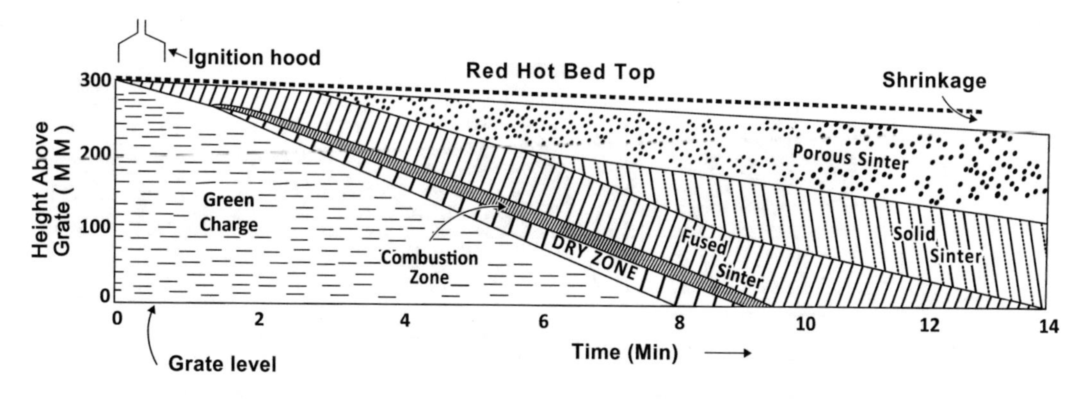

Fig. 1.11 A cross-sectional profile of sinter bed

simultaneously increasing considerably the complexity of process analysis. Thus, an effective way of developing new concepts and their quantification is to develop comprehensive mathematical models capable of simultaneously considering the mass transfer using reliable rate equations for the chemical reactions, momentum transfer for complex bed structure and interphase heat transfer considering simultaneously convective, radiation and chemical reactions heat transfer [6]. Actual sintering machine is developed to improve the flexibility of the process and allow simultaneously operation with gas recycling, fuel gas utilization, operation with partial operation of mill scale and biomass together with fossil fuels as coke breeze or anthracite. Figure 1.12 shows the new concept to attain these principles.

Factors affecting sinter quality:

(i) Particle size: Particle size of the charge is one of the most important factors affecting permeability of sinter bed. On the other hand, if the proportion of course grains are considerable, the heat content of waste gas cannot be sufficiently absorbed by the solid. In general, the mix consists of 8–10 mm, coke breeze of 4–5 mm, and fluxes of minimum grain size and return fines of 7–10 mm.

(ii) Moisture: The most important factor affecting bed permeability is the water content. Proper control of water should give maximum porosity and permeability. Water level is also important as a regulator of the speed of combustion front. Too little water increases speed and causes separation of combustion and heat fronts. Too much water requires high amounts of fuel and affects permeability.

(iii) Fuel: Generally, large amounts of fuel in the sinter mix increases sintering rates by giving higher flame front speed, if the air flow can be increased. But the maximum temperature increases with more fuel, this influences the nature and extent of slagging, and the sinter quality is adversely affected. Sinter contains 5–20% metallic iron due to present of solid fuel.

(iv) Flux: Production rates have been found to increase in the addition of limestone.

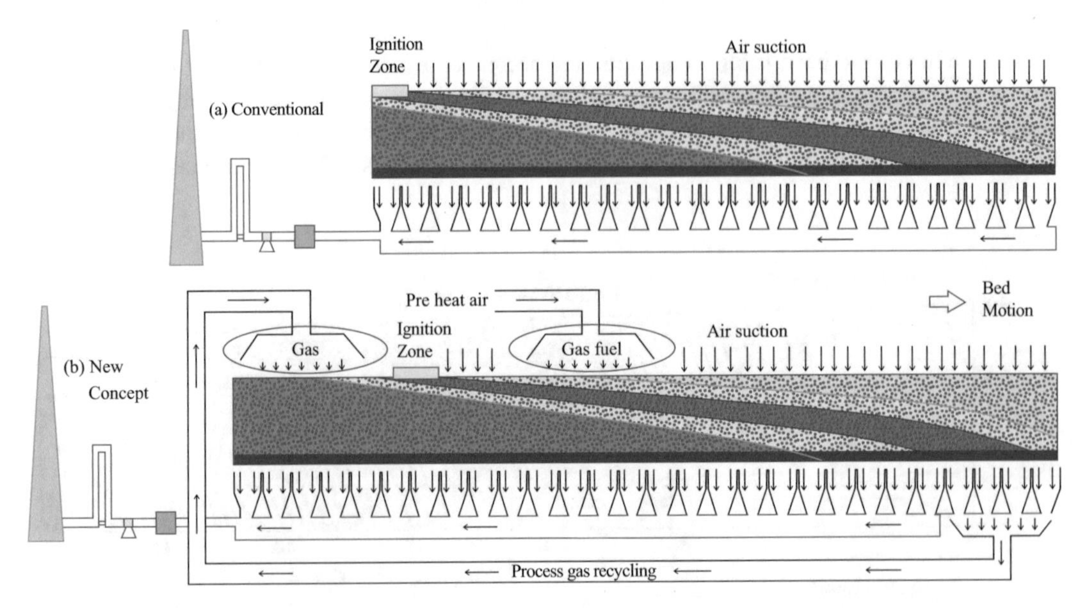

Fig. 1.12 New technology for sintering using recycling gas and gaseous fuels [6] (reproduced with permission from Springer Nature)

Fig. 1.13 Agglomerate product of iron ore **a** sinter and **b** pellets (Ø = 10–18 mm). (Courtesy from MS University of Baroda, India)

Additionally, limestone is added to the sinter mix to produce self-fluxing sinter with respect to the ordinary sinter. This is attributed to the correct slag formation resulting in higher bed permeability in the high-temperature zone. Additions of lime to the charge will increase the permeability of green bed. Figure 1.13a shows the sinter product.

Types of sinters are as follows:

(a) Acid sinter: when the basicity ratio (CaO/SiO_2) is less than 1.0, i.e. sinter without flux,
(b) Self-fluxed sinter: when the basicity ratio is more than 1.0,
(c) Super-fluxed sinter: when the basicity ratio is more than 2.0.

Sinter quality can be improved by the changing chemistry of sinter:

(i) increase the CaO content of sinter,
(ii) increase of MgO content,
(iii) a substantial reduction of FeO content.

Advantages of using sinter in blast furnace:

1. Any free or combined moisture in the ore is removed and the carbonates are decomposed,
2. A part of the sulphur present in the charge is removed during the sintering process,
3. The gangue constituents of the ore are pre-slagged in fluxed sinter,
4. Some amount of heat in the blast furnace is saved as dissociation of carbonates takes place during sintering,
5. Using super-fluxed sinter, the amount of limestone charged in the BF is reduced, and thus, the solution loss reaction (i.e. $C(s) + \{CO_2\} = 2\{CO\}$) in BF is decreased,
6. The evolution of CO_2 gas in BF, due to dissociation of carbonates, is much reduced, therefore favouring the indirect reduction of iron ore,
7. The productivity increases as the volume of material charged per tonne of hot metal is reduced by use of super-fluxed sinter,
8. The irregular surface of sinter helps to maintain an adequate permeability and permits high blowing rate,
9. It is generally possible to use higher top pressure and carry high blast temperature.

1.3.2 Pelletization

Pellets are agglomerated from very finely (minus 200 mesh size) iron ore concentrates to which a small quantity of binder has been added. The process of pelletization consists of formation of green balls (5–20 mm size) by rolling of moist iron ore fines with binder.

There are two types of pelletizers: (a) drum pelletizer and (b) disc pelletizer.

(a) Drum pelletizer: Drum pelletizer (Fig. 1.14) has the length and diameter ratio of 2.5–3.5. Length and diameter are 6–11 m and 2.0–3.5 m, respectively [7]; drum is rotated at the speed of 10–15 rpm. The materials roll over the surface of the rotating drum and slide downwards due to inclination of the drum. This motion of the material is called cascading. Drum does not act as a classifier.

(b) Disc pelletizer: Disc pelletizer (Fig. 1.15a) has diameter and height ratio of 10. Diameter and lip height are 3.5–7.5 m and 0.35–0.75 m, respectively; disc is rotated at the speed of 15–30 rpm, and the angle of inclination is 40–45°. As the speed is increased, the rolling improves and consequently the rate of growth increases. Excessive speed, however, may result in breaking the pellets by impact. The correct speed of rotation is 25–35% of the critical speed. The critical speed is the speed at which material will be centrifuged (i.e. materials will be rotating along with periphery of the disc). As the angle of inclination is increased, compaction will be more, and pellets of better strength will be obtained. When the disc is rotated, the material (iron ore mixed with binder) moves up due to frictional force and tends to slide down due to inclination of the disc and falls down by the gravity. The materials are also rolled over the surface of the rotating disc and slide down. By this way, small balls are formed. This motion of the balls (i.e. lifted and fall down) is known as *cascading* (as shown in Fig. 1.15b).

The rate of production of pellets depends on:

(i) Disc diameter,
(ii) Lip height,
(iii) Inclination angle of disc with horizontal,
(iv) Speed of rotation,
(v) Rate of charging,
(vi) Rate of moisture addition,
(vii) Nature and size of charge,

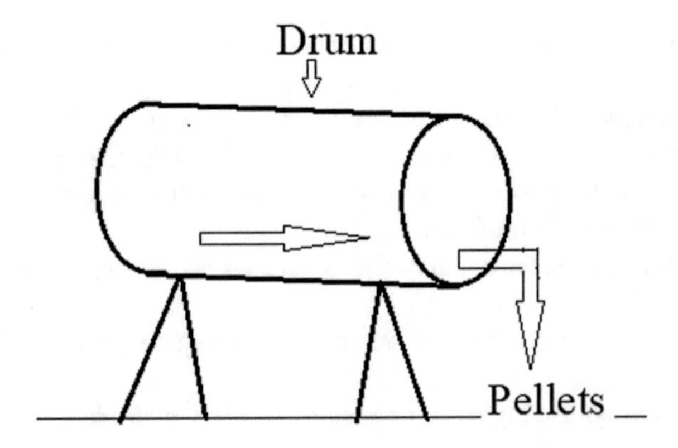

Fig. 1.14 Drum pelletizer

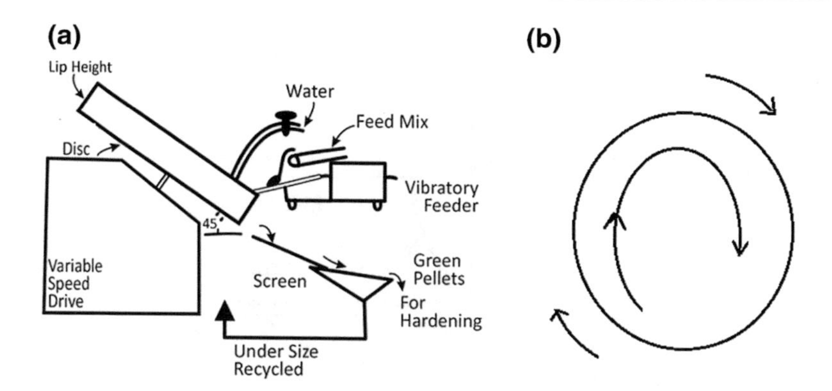

Fig. 1.15 **a** Disc pelletizer, **b** feed mix lifted and fall in disc due to rotation (i.e. cascading)

(viii) Desired size range of pellets,
(ix) Rate of withdrawal of the pellets,
(x) Any other additions like binder, flux, etc.

The pelletization process consists of the following steps:

1. Charge materials' preparation,
2. Formation of green ball and sizing,
3. Indurations or hardening of green ball: (i) drying, (ii) pre-heating and (iii) firing,
4. Cooling of hardened pellets.

1.3.2.1 Charge Materials' Preparation

To achieve good strength and porosity (for good reducibility) of pellets, the particle size should be −100 mesh (i.e. −200 μm) in size with more than 60% as −350 mesh (i.e. −44 μm). Right combination of sizes and shape of charge particles leads to max mechanical interlocking and surface tension to obtain strong green pellets. Surface roughness increases the surface area for the same size particle and improves green strength. Smaller particle fills up pores left in between the bigger particles and reduces the porosity of the pellets (20–30% porosity is alright). Iron-bearing particles are properly mixed with critical amount of moisture and binder to develop strength. Figure 1.13b shows the product of pellets.

1.3.2.2 Functions of a Binder

Binders play a very important role in the pelletization process. Binder is the material which serves as a bridge between the particles and thus increases the green or dry strength of the bonded particles. Binders act as two very important functions [8] in iron ore pelletization:

- The binder makes the moist ore plastic, so that it will nucleate seeds that grow at a controlled rate into well-formed pellets.
- During drying, the binder holds the particles in the agglomerates together, while the water is removed and continues to bind them together until the pellet is heated sufficiently to sinter the grains together.

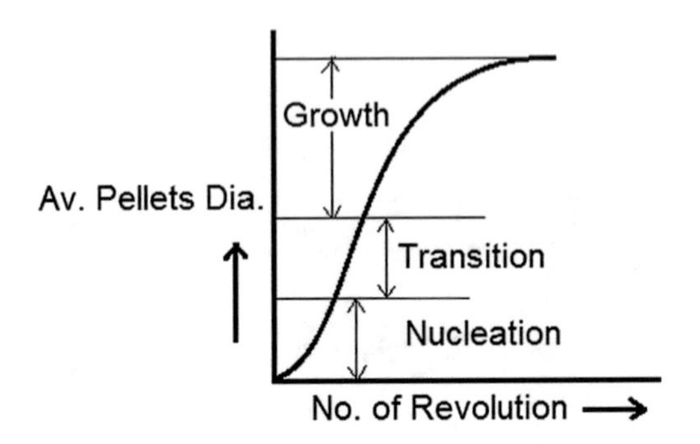

Fig. 1.16 Pellets' diameter versus number of revolutions

Selection of Binder

There are different types of binders available for pellet making. Binder (e.g. bentonite, cement, lime, $Ca(OH)_2$, dextin, dextrose, thermosetting resin) is the material which increases bonding strength to hold the agglomerate particles; it improves green ball formation and strength of ball.

Before selecting the binder, binder must satisfy the following requirements [9]:

1. Mixing behaviour: It should be spread out uniformly over the surface of particles.
2. Mechanical properties: A good binder should maintain good mechanical properties of pellets, including green, dry and fired pellets, e.g. deformation under load, resistance to fracture by impact and by compression, resistance to abrasion.
3. Chemical composition: A good binder should not bring any environmentally and metallurgically harmful elements such as S, P, As etc., into the product pellets. It should not contain impurities such as silica, alumina, etc.
4. Metallurgical behaviour: A good binder should maintain pellet's excellent metallurgical properties such as reducibility.
5. Processing behaviour: Adding, mixing, dispersion of binder, green ball preparation, pellet drying, etc., should not be complicated or essentially change conventional pellet production circuit.
6. Toxic factor: It should be harmless to the operating personals.
7. Cost factor: Price of binder should be acceptable for iron ore pellets production. It should be cheap and easily available in the market.

1.3.2.3 Green Balls Formation

Size of balls produced in a pelletizer, from a charge mixer containing right amount of moisture and binder, depends on the time and speed of pelletizer, i.e. numbers of revolution. Figure 1.16 shows pellets diameter vs number of revolutions.

The ball formation is a three-stage process: (i) nucleation or seed formation, (ii) transition region and (iii) ball growth region.

(i) Nucleation formation: Nucleation is formed only if critical amount of moisture level is maintained. Without these, process cannot proceed properly. The formation of balls on a pelletizer depends primarily on the moisture content. There must be sufficient moisture to cover the surface of the particle, so that when two of them come into contact, a liquid bridge is formed under the influence of surface tension; nucleation is formed. (i) If moisture is less than the critical amount, its distribution tends to be non-uniform; some of the particles remain dry.

(ii) If the moisture level is more than the critical value, growth rate is more, but the balls are deformed due to high plasticity.

When a wet particle meets with another wet or dry particle, bonding is immediately formed between the two particles during rotation. Surface tension of water alone is the binding force in these nuclei.

(ii) Transition region: After nuclei are formed, they pass through a transition period where the plastic nuclei further rearrange themselves and get compacted by eliminating the air voids. Whole thing moves from pendular state via funicular state to the capillary state of bonding. Rolling action causes the granules to increase density further; granules are still plastic with a water film on the surface and capable of bonding with other granules.

(iii) Ball growth region: Growth takes place by (i) layering and (ii) assimilation.

(i) Growth by layering is done when fresh dry charging material added. Amount of material picked up by the balls is directly proportional to the exposed surface. This is more predominant in the disc pelletizer.

(ii) Growth by assimilation is possible when balling proceeds without addition of new charging material. By rolling action, some of the granules may break, particularly the smaller ones. The bigger the ball, the larger it will grow under these conditions. Balls with higher moisture content grow to larger size, but the strength of the resulting balls is low. This is more predominant in the drum pelletizer.

1.3.2.4 Indurations or Hardening

(i) Drying: Green balls are dried by passing hot air through the bed of the pellets. Temperature depends on moisture content in pellets; haematite pellets are disintegrated at temperature 300 °C, so pellets are dried at <300 °C and magnetite pellets dried safely at even 475 °C.

(ii) Pre-heating is done at 600–700 °C: (a) to avoid cracking on pellets due to thermal stresses, (b) calcinations of carbonates take place, (c) partial oxidation of magnetite ore and (d) bonding of oxide particles.

(iii) Final firing is done at 1200–1300 °C to strengthen the pellets.

The main methods for firing of pellets are as follows [10]:

(1) Travelling grate process and

(2) Grate-kiln process.

(1) **Travelling Grate process** uses for firing at the two-thirds of the world's installed pelletizing processes. A typical flow diagram of the process is shown in Fig. 1.17.

It has three steps as follows:

- Raw material preparation,
- Forming green pellets,
- Pellet hardening.

The advantages of the travelling grate process are as follows:

1. Pellets remain undisturbed throughout the entire process (including cooling),
2. Uniform heating of pellets, leading to uniform product quality,
3. Minimized dust and fines generation,

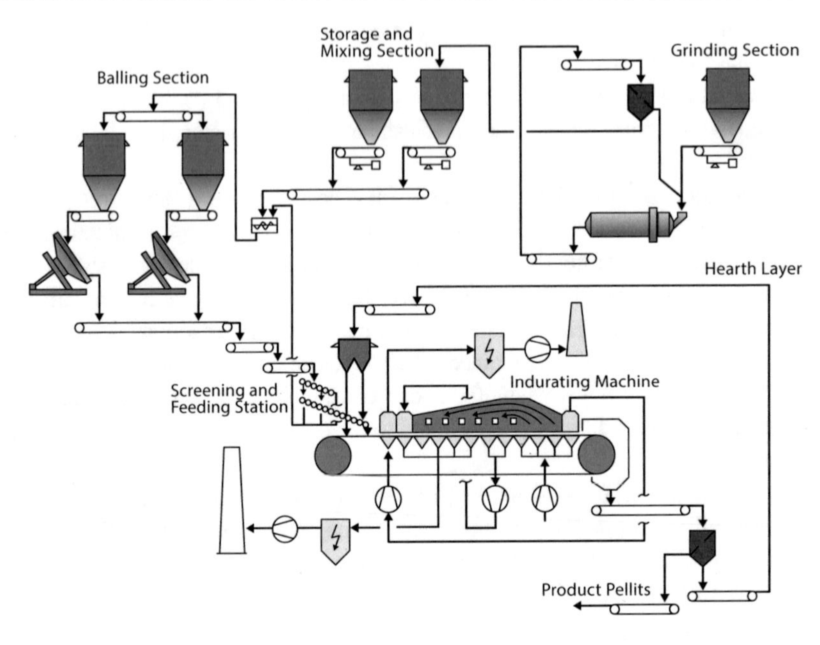

Fig. 1.17 Flow diagram of the travelling grate process [26]

4. Superior process flexibility and
5. Low specific heat consumption.

(2) The first ***grate-kiln system*** pellet plant was installed in 1960. The plant took iron ore concentrate and produced superior iron ore pellets for blast furnace and sponge iron unit. There are two main processes for producing iron ore pellets: (a) the grate-kiln system and (b) the straight grate system.

In the *straight grate system*, a continuous parade of grate cars moves at the same speed though the drying, hardening and cooling zones. Any change in one section affects the retention time in another. In the grate-kiln system (Fig. 1.18), independent speed control of the grate, kiln and cooler is available to the operator. This provides process flexibility to adjust to changes in concentrate feed.

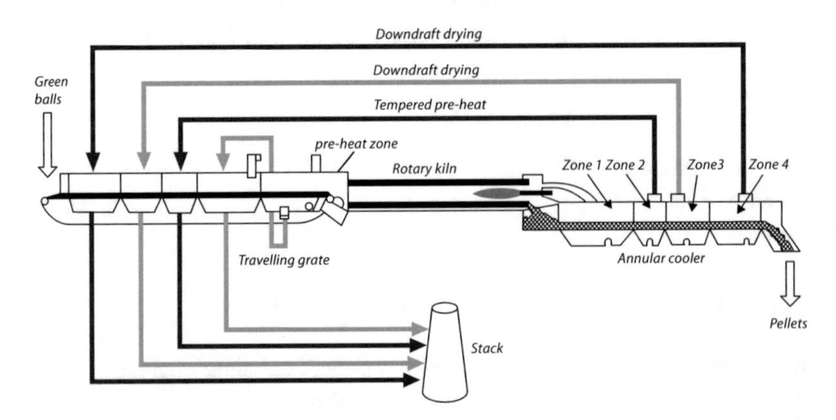

Fig. 1.18 Flow diagram of the grate-kiln process [26]

In a grate-kiln system, the travelling grate is used to dry and pre-heat the pellets. A refractory lined rotary kiln is used for hardening. In a straight grate, the grate cars should go through the drying and hardening zones. So, a deep bed of pellets with a hearth layer is required. Power requirement of a modern grate-kiln system will be less than 20 kwh/t, while a straight grate system will use over 35 kwh/t.

The grate-kiln system has flexibility due to one burner. The burner of kiln can use liquid, gas or solid (coal or wood) fuels separately or in combination. One burner reduces maintenance costs and improves fuel efficiency. In the grate-kiln system, 95% of the air is used for combustion (at 1000 °C) from the cooler. A straight grate, with up to 50 burners, cannot match this level of energy recovery.

1.3.2.5 Theory of Bonding

Green balls should have adequate strength and to stand overlying load and abrasion during screening and handling until finally hardened by induration. Theory of balling: (i) dry material does not pelletize, and presence of moisture is essential for conversion the fines into balls by rolling, (ii) surface tension of water in contact with the particles plays a dominant role in binding the particles together, (iii) rolling of moist material leads to the formation of very high densities balls and (iv) material can be rolled into balls which is directly proportional to the surface area of particles, i.e. its fineness.

There are three types of water–particle interaction: (a) pendular state, when water is present just at the point of contact of the particles and surface tension holds the particles together; (b) funicular state, when some pores are fully occupied by water. A film is covering the entire surface of the particles; (c) capillary state, when all the pores are filled with water and there is no coherent film covering the entire surface of the particles (as shown in Fig. 1.19).

Maximum strength of a green ball produced by compacting the material having low porosity and just enough water saturates the pores. The rolling action during pelletization is beneficial in reducing the internal pore space by effecting compaction and mechanical interlocking of the particles.

Generally, bentonite uses as binder for harden pellets; bentonite gives the slag bonding at high temperature (1200–1300 °C). Iron silicate ($FeO \cdot SiO_2$) is formed, which has low melting point, melts and spade in between capillary spaces.

1.3.2.6 Advantages of Pellets as Burden Material [10]

1. **Good reducibility**: Because of their high porosity that is (25–30%), pellets are usually reduced considerably faster than hard burden i.e. sinter or hard natural ores/lump ores.
2. **Good bed permeability**: Their spherical shapes containing open pores give them good bed permeability. Low angle of repose, however, is a drawback for pellet and creates uneven binder distribution.
3. **High uniform porosity** (25–30%): Because of high uniform porosity of pellets, faster reduction and high metallization takes place.

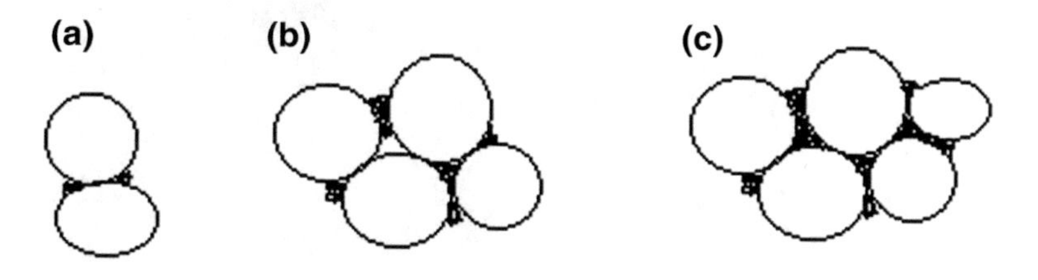

Fig. 1.19 Water–particle interaction: **a** pendular state, **b** funicular state and **c** capillary state

4. **Less heat consumption than sintering**: Approximately 35–40% less heat required than sintering.
5. **Uniform chemical composition and very-low LOI**: The chemical analysis is to a degree controllable in the concentration processing within limits dictated by economics. Very low or no loss of ignition (LOI) makes them cost-effective.
6. **Easy handling and transportation**: Unlike sinter, pellets have high strength and can be transported to long distances without much fine generation. It has also good resistance to disintegration.
7. **Utilization of waste**: Pellets also help in utilization of waste ore and lean ore as cheaper and abundant raw material.

1.3.2.7 Cold-Bonded Pelletizing [9]

Conventional pelletization process requires hardening of the pellets by firing at near fusion temperature (1200–1300 °C) using mostly oil-fired furnaces. Therefore, production costs of these indurated pellets are increasing day by day due to the increasing oil price. Alternative energy-saving process is the cold bonding technology for pellet making which is becoming more and more popular nowadays all over the world. The pellets are hardened in cold bonding processes, the hardening is achieved by additives with high bonding properties, due to physical–chemical changes of the binder at low temperatures; the free ore grains remaining intact. This benefits the reducibility of pellets as well.

The development of cold-bonded pellets at lower cost has brightened the prospects. Traditional pellet making is costlier process due to need of hardening in furnace at a temperature of 1200–1300 ° C. Cold-bonded pellets do not require hardening treatment and hence constitute a potential alternative route for utilization of ore fines. Binders play a very important role in the cold bonding pelletization process. The pellets in cold bonding are hardened due to the physico-chemical changes of the binders in ambient conditions or at slightly elevated temperatures (maximum 500 °C or so).

Bonding mechanisms of cold bonding processes are hydraulic or hydrothermal reactions from normal temperatures up to about 250 °C. The binders used should have such a chemical composition that no negative influence on the pellet properties is to be expected.

1.3.3 Briquetting

Briquetting is an old art that has been used to agglomerate or form small/large lumps of regular shape by pressing of iron ore fines, coal/coke fines with or without a binder. Briquette has some suitable size and shape (as shown in Fig. 1.20).

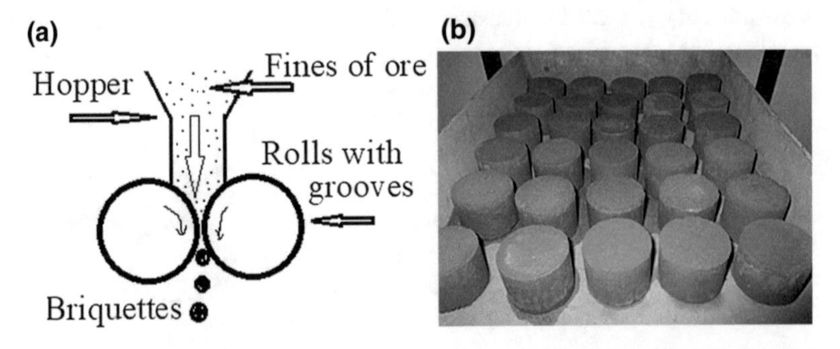

Fig. 1.20 **a** Briquetting machine and **b** briquettes (Ø = 50 mm). (Courtesy from MS University of Baroda, India)

In the early 1960s, briquettes of fine ore, produced in punch, drop, or toggle presses and fired at high temperature before charging to the BF were believed to increase furnace capacity and lower coke rate.

However, cold bonding briquettes can be produced using a binder such as lime, cement (8–10%), molasses and without firing; harden by ageing (for cement binder with spraying water) for several days. Composite briquettes are also produced by addition of 10–15% coal/coke fines and binder.

The hot-ore-briquetting process was developed by U.S. Steel, a suitable agglomerate for use as a BF charge. This process requires no additional binder. The hot-ore-briquetting process: iron ore fines are heated to 800–1000 °C and then briquetted in hot condition in a double-roll briquetting press at loads of 50–60 tonnes. As the temperature increases, the strength and density of briquettes are also increased. The hot-ore-briquettes have good cold strength and exhibit satisfactory hot strength under simulated BF conditions. The hot-ore-briquetting process requires only about half fuel per tonne of product as sintering and about the same as that for pelletizing.

1.3.4 Nodulizing

Nodulizing means conversion of finely divided iron ore into a nodular form. Aggregation of finely divided material such as mineral concentrates, by aiding of binder and perhaps kilning, into nodules sufficiently strong and heavy to facilitate subsequent use, such as charging into blast furnaces. Nodulizing improves the metallurgic properties of the ore by altering the chemical composition:

- It removes harmful impurities, such as sulphur and volatile gases, and
- Increases reducibility. As a result of nodulizing, mechanical strength is increased, and the ore acquires a porous structure.

Table 1.4 Difference between fired pellets and cold-bonded composite pellets [27]

Fired pellets	Cold-bonded composite pellets
Pellets are made from the iron ore fines only	Pellets can be made of iron ore and carbonaceous material
Reduction takes long time	Reduction takes within short time
If pellets used in blast furnace, coke is used as a reducing agent	This is coke-free technology; coke is replaced by non-coking coal
Pellets are used only after firing at high temperature indurations	In this, pellets are used in green/dried state; there are no indurations required
Fuel consumption is higher	Usage of furnace oil is either fully eliminated or reduced considerably
Energy is expensive, so production cost of pellet making is higher	Energy conservative process and less energy is required for pellet making
Chemical composition of raw material may change during induration in an oxidizing atmosphere	Chemical composition of raw material does not change
It consists of more steps to produce	It consists of less steps to produce
Strength of pellets is more	Strength of pellets is less
Capital and operating costs of fired pellets are more	Capital and operating costs of cold-bonded pellets are estimated to be only two-thirds of fired pellets

Fine ore along with some carbonaceous material like tar are charged to the rotary kiln (2 m diameter and 30–60 m length, rotate at 1–2 rpm) heated by gas or oil. Nodules or lumps are formed by the rolling of the charge heated to incipient fusion temperatures. The feed material travels counter-current to the gases. The temperature inside the kiln is just enough to soften the ore but not enough to fuse the ore. It takes about 1.5–2.0 h for charging to travel through the kiln. Size of nodules produced considerably depending upon the tar content and the temperature of the kiln.

It is costlier than sintering, due to high fuel consumption. Nodulizing has the advantage that feed moisture and particle size are not critical as they are in pelletizing. The formation of rings and large balls has been a serious problem at all installations at some time in their operation and seems to be most frequent cause of shutdown.

1.3.5 New Feed Material (Iron Ore–Coal Composite Pellet)

Iron ore–coal composite pellet technology has been under development for last many years without its significant contribution in ironmaking. The principle technological problem is to produce such composite pellet at comparatively lower cost. Of late, the development of cold-bonded composite pellets at lower cost has brightened the prospects. Traditional pellet making is costlier process due to need of hardening in furnace at a temperature of 1200–1300 °C. Cold-bonded pellets do not require hardening treatment and hence constitute a potential alternative route for utilization of ore fines. Cold-bonded iron ore–coal composite pellets also utilize cheap and readily available reductant such as coal fines, coke breeze and char fines. Table 1.4 shows the difference between fired pellets and cold-bonded composite pellets.

Iron ore–coal composite pellets/briquettes and self-reducing agglomerates all refer to carbon-bearing materials mixed with iron-bearing materials into agglomerates [11]. It can be mixtures of fine iron ore (haematite, magnetite, dust and pre-reduced iron-bearing ore fine, etc.) and fine carbonaceous materials (coke breeze, fine coal, charcoal, char, etc.) adding some binder. Therefore, in the context of process optimization, energy saving, waste recycling and environmental concerns, these agglomerates can be used in blast furnace as well as in other ironmaking processes as feed materials. Briquettes with high amount of coal tend to melt at higher temperatures, while those having less amount of coal tend to melt earlier due to the melting of slag containing unreduced iron oxide. Higher coal content briquettes will be preferred in blast furnace to avoid forming low melting slag and then maintaining the furnace gas permeability.

Composite pellets, containing iron oxide and carbon fines, should be prepared only by cold bonding technology. Cold bonding technology is preferred because:

1. The pellets are hardened in cold bonding process due to physico-chemical change of the binder at low temperature, the free iron ore grains remaining intact. This benefits the reducibility of the pellets.
2. It is an energy-saving process for pellets production and hence is becoming more popular.
3. If composite pellets are fired or hardened, simultaneously its reduction also occurs due to presence of carbon. Hence, instead of getting fired oxide pellets, some metals form due to reduction within the composite pellet and hardening of pellet does not take place; as well as carbon also loss from composite pellet.

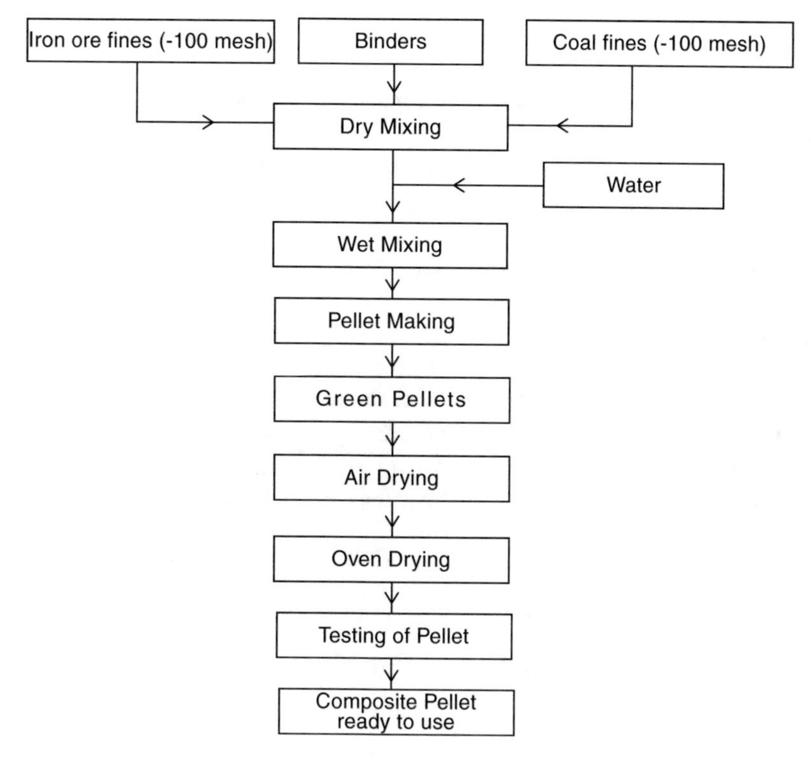

Fig. 1.21 Flow diagram of composite pellet making [11]

1.3.5.1 Definition of Composite Pellets

The term ***iron ore–coal composite pellet*** is being employed here to mean pellet containing mixture of fines of iron-bearing oxide and carbonaceous material (coal/coke/char) which has been imported sufficient green strength for subsequent handling by cold bonding techniques [11]. The pellet should have sufficient strength to withstand high temperatures and stresses during reduction in a furnace.

Interest in iron ore–coal composite pellets has grown from the 1980s because of the following advantages [12]:

1. Utilization of cheaper resources (such as iron-bearing fines, coal fines and coke freeze),
2. Fast reduction due to intimate contact between reductant and iron oxide particles,
3. Reduction in energy consumption, as cold-bonded composite pellets do not require induration (high temperature heating),
4. Promising prospects for ironmaking at a small scale with less capital investment,
5. Because of their uniform size and convenient form, pellets can be continuously charged into the furnace, leading to higher productivity and
6. Consistent product quality as the chemical composition of composite pellets does not change.

1.3.5.2 Binders for Composite Pellets

Research workers had tried to prepare iron oxide/ore and coke/coal/charcoal composite pellets, using various organic and inorganic binders. Binders used to prepare composite pellets are classified as:

1. Inorganic binders: Bentonite, cement, sponge iron powder, lime, $Ca(OH)_2$, silica, water glass, fly ash, fly ash and lime, boron compounds, sodium poly-acrylate (SPA), etc.,

2. Organic binders: Dextrin, dextrose, thermosetting resin (TSR), starch-based binder, molasses, poly-vinyl alcohol (PVA), etc.,
3. Combined binders: Different combination of organic and inorganic binders, such as dextrin, TSR, molasses, lime, and $Ca(OH)_2$.

Figure 1.21 shows the flow diagram for composite pellet making. Iron ore–coal composite pellets were prepared using different combination of organic/inorganic as binder.

1.3.5.3 Binding Mechanism for Composite Pellets

1. **Formation of Calcite Structure**

Based on addition of lime method, the resultant strengthening is being due to the formation of a carbonate network inside the ball. Blending of calcium hydroxide with the concentrate and wetting of mixture is followed by pelletizing. The pellets are pre-dried and hardened by action of CO_2. The catalytic action is given by the sugar, and the reactions are as follows [13]:

$$Ca(OH)_2 + 2C_6H_{12}O_6 \rightarrow Ca(C_6H_{11}O_6)_2 + 2H_2O \tag{1.11}$$

$$Ca(C_6H_{11}O_6)_2 + \{CO_2\} + H_2O \rightarrow (glucose)_x(CaCO_3)_y(CaO)_z \tag{1.12}$$

$$(glucose)_x(CaCO_3)_y(CaO)_z + \{CO_2\} + H_2O \rightarrow CaCO_3 + C_6H_{12}O_6 \tag{1.13}$$

2. **Strengthening by Silicate Formation**

The calcium hydroxide and silica with water give the hydrated calcium silicate, which provides the strength, and the reaction is as follows [13]:

$$Ca(OH)_2 + SiO_2 + (n-1)H_2O = CaSiO_3 \cdot nH_2O \tag{1.14}$$

3. **Hydraulic Strengthening**

Binder used mainly Portland cement and clinker. This binder with water gives high strength in few days due to calcite formation.

$$2CaO + \{CO_2\} + H_2O = CaCO_3 + Ca(OH)_2 \tag{1.15}$$

1.3.6 Testing of Agglomerates

The characteristics of the burden materials are the important factor for the quality of the charge materials that affect the output and efficiency of a blast furnace operation. The properties of burden materials have bearing on its performance during handling, until it is charged in the furnace, and subsequently on its behaviour inside the furnace. These properties are as follows:

1. Room temperature physical properties,
2. Reducibility,
3. Physical behaviour during reduction at high temperatures.

1.3.6.1 Room Temperature Physical Properties

- Drop test,
- Porosity,
- Shatter test,
- Abrasion test,
- Compressive strength test.

Drop Test

In the drop test, the briquettes or pellets are dropped repeatedly from a height of 45.7 cm on a 10 mm thick steel plate until they break. The number of drops, before breaking of the pellet, is counted and noted down. The final value is taken as the average of four to five such test values. Drop tests are performed for green as well as hardened pellets/briquettes. The higher the drop value, higher is the strength of the sample.

Porosity

This is the most difficult property to measure and is expressed as the volume of pores as a percentage of the total volume of the material tested. There are two types of pores, like open and closed (Fig. 1.22). Open pores are accessible to fluids, whereas the closed pores cannot.

Open pores have one of their ends on the outer surface of the particle. These open-end pores allow the movement of fluids (gas/liquid) to the interior location of the particle permitting chemical reaction.

Closed pores are deep seated and do not open to the surface of the particle. These closed pores do not offer any site for chemical reaction by fluids. However, these closed pores act as a good thermal barrier and increase the heat insulating power of the material.

The porosity of pellets is measured by a simple technique based on Archimedes principle. It is a very important property from the reduction point of view. The reducibility of an ore or pellet increases as the porosity is increased. Porosity is vital in green pellets, which are intended for subsequent heat hardening or reduction. Basically, it is governed by the particle size of the feed stock but may be affected by balling. The porosity of the pellet was calculated as follows [13].

$$\text{Porosity}\ (\%) = \left(\frac{\text{True density} - \text{Apparent density}}{\text{True density}}\right) \times 100 \tag{1.16}$$

Apparent density is the ratio of mass to volume of a single solid particle including closed pores (i.e. volume of solid material + volume of closed pores within the particle), and true density is the ratio of mass to volume of a single solid particle without any pores or cavities.

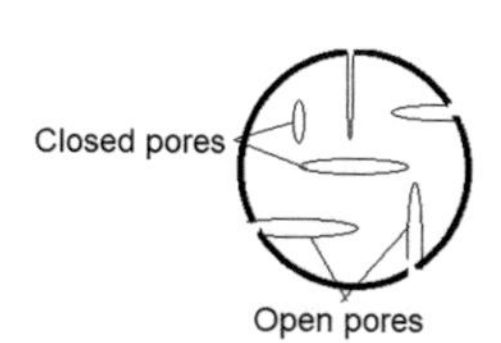

Fig. 1.22 Pores in material

Shatter Test

This test consists of a certain amount of materials (+10 mm size) which are dropped from a standard height (i.e. 2 m) on a 10 mm thick steel plate for four times. The amount of the materials is passed through certain sieves expressed as percentage of the original weight is indicated as the shatter index. It gives physical degradation of agglomerated materials under impact, and it determines the stability of materials.

For sinter:

$$\text{Shatter index}(\%) = \left(\frac{\text{Weight of} - 5\,\text{mm fraction in kg}}{\text{Initial weight of sample in kg}}\right) \times 100. \tag{1.17}$$

For pellets/briquettes:

$$\text{Shatter index }(\%) = \left(\frac{\text{Weight of} - 100\,\text{mesh fraction in kg}}{\text{Initial weight of sample in kg}}\right) \times 100 \tag{1.18}$$

Abrasion Test

This is also called tumbling test. This is to assess the strength of material including its abrasion resistance. Abrasion test essentially consists of tumbling a standard weight of material of certain size (+10 mm) in a standard drum; tumbling is carried out at a standard speed (30 rpm) for a fixed number of revolutions. The percentage material passing through or retained on a certain sieve is the abrasion index.

For sinter:

$$\text{Abrasion index }(\%) = \left(\frac{\text{Weight of} - 5\,\text{mm fraction in kg}}{\text{Initial weight of sample in kg}}\right) \times 100 \tag{1.19}$$

For pellets/briquettes:

$$\text{Abrasion index }(\%) = \left(\frac{\text{Weight of} - 100\,\text{mesh fraction in kg}}{\text{Initial weight of sample in kg}}\right) \times 100 \tag{1.20}$$

When the drum is rotated, the material moves up due to frictional force and tends to slide down by the gravity. This causes wear due to abrasion. The lifters (projected plate) push piece of material and carry it to a height, and still it slides and falls over other pieces of material at the bottom of the drum. During free fall, it causes impact and renders breaking of particle into smaller fragments. The process continues during drum rotation period. After the given time, the size fractions are screening by sieves.

Degradation of materials and consequent generation of fines (−5 mm for sinter and −100 mesh for pellets) is an inherent feature of the behaviour of materials during its handling. The extent of which materials break down during handling has an important influence on the quality of fines arriving at the blast furnace. If these fines find their way in the furnace, it will have adverse effects on furnace operation and if it is screened out it will involve additional cost in handling and its re-agglomeration.

Compressive Strength Test

The direct measurement of compressive strengths of lump ore or sinter is difficult due to irregular shape. Compressive strength of hardened pellets/briquettes (which are more regular shapes) at room temperature can be measured with better accuracy. When a spherical-shaped material (i.e. pellets) is subjected to compressive load, then the stress generated at one point of contract causes its fracture into fragments depending on the nature of material and number of cracks generated. Maximum load withstands by the material before brackets that give the compressive strength of the material.

1.3.6.2 Reducibility

1. **For ordinary pellets**:

The reducibility is a measure of oxygen removal rate from iron ore/oxide by CO or H_2 gas. The reducibility can be measured by degree of reduction (α) with reduction time. If the reduction takes place at constant temperature, then it is called *isothermal reduction*. If the temperature change during heating and reduction takes place, then it is called *non-isothermal reduction*. Weight loss method is generally used to determine the degree of reduction. Schematic diagram for set-up of reduction study is shown in Fig. 1.23.

$$\text{Degree of reduction}\,(\alpha) = \left(\frac{\text{Weight of oxygen removed from the sample}}{\text{Total removable oxygen present in the sample}}\right) \times 100 \qquad (1.21)$$

$$= \left(\frac{\text{Weight loss of the sample}}{\text{Total removable oxygen present in the sample}}\right) \times 100 \qquad (1.22)$$

$$= \left(\frac{W_1 - W_2}{W_O^i}\right) \times 100 = \left(\frac{\Delta W_O}{W_O^i}\right) \times 100 \qquad (1.23)$$

where

W_1 initial weight of the sample before reduction,
W_2 final weight of the sample after reduction,
ΔW_O weight loss due to oxygen loss during reduction,
W_O^i total removable oxygen present in the sample = $W_1 \times f_{ore} \times \rho_{ore} \times f_O$,

where

f_{ore} fraction of iron ore present in the sample,
ρ_{ore} purity of iron oxide (Fe_2O_3) in iron ore,
f_O fraction of oxygen present in the pure iron oxide (Fe_2O_3).

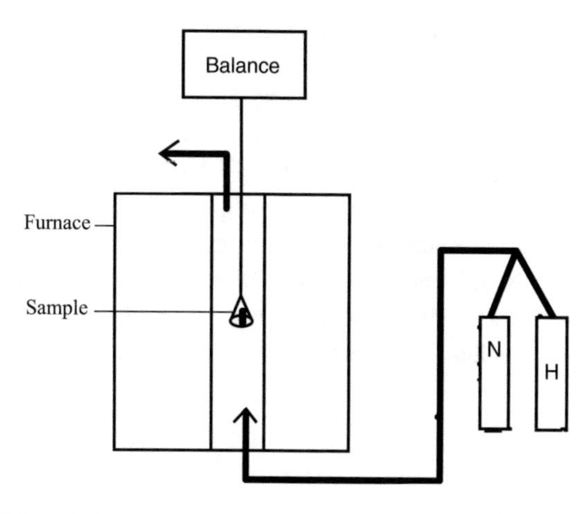

Fig. 1.23 Set-up for reducibility study

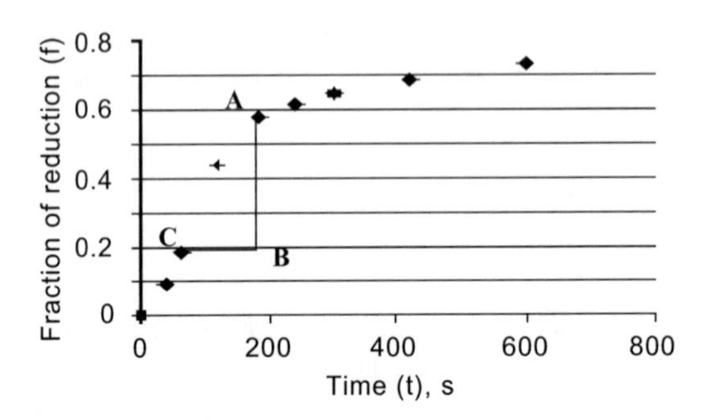

Fig. 1.24 Fraction of reduction (*f*) versus time (*t*)

Fraction of reduction (*f*) can be calculated as follows:

$$\text{Fraction of reduction } (f) = \left(\frac{\text{Weight of oxygen removed from the sample}}{\text{Total removable oxygen present in the sample}}\right) \quad (1.24)$$

$$= \left(\frac{\Delta W_{\text{O}}}{W_{\text{O}}^{i}}\right) \quad (1.25)$$

Reduction rate can be obtained by plotting fraction of reduction (*f*) versus time (*t*) (Fig. 1.24). The slope of the line is taken as index of reducibility of ore, i.e. reduction rate can be expressed as $\left(\frac{df}{dt}\right)$ per second.

From Fig. 1.24, the reduction rate $\left(\frac{df}{dt}\right) = \left(\frac{AB}{BC}\right)$.

2. **For composite pellets**:

Cold-bonded iron ore–coal/char composite pellets consisting of powder mixtures of iron ore, carbonaceous reductants such as coal or char, and some binder, seem to have commercial prospect as feed material for ironmaking.

For reduction of iron oxides by carbon, a major difficulty with experiments is that the extent of the reduction cannot be found out directly from the weight loss of the sample, since this is made up of both oxygen and carbon losses [14]. Furthermore, for ore–coal composite pellets, the weight loss of the sample arises not only from oxygen and carbon losses, but also the loss of volatile matter and residual moisture present in pellet. Since only weight loss of the sample is not sufficient, some additional measurements are required for estimating the degree of reduction (α), which is defined as in Eq. (1.22).

For determining the degree of reduction (α) for carbothermic reduction in powder mixtures of iron ore and reductant, research workers [14–22] have employed various techniques. The product gas consists of only CO and CO_2, if the reductant is carbon in the form of graphite or dead burnt char, and the binder is absent. For such a situation, some investigators assumed composition of product gas. This allowed them to estimate either the oxygen loss of the ore from the total weight loss in mixture or the rate of oxygen loss from volumetric flow rate of product gas at STP.

However, it has been well established that the composition of the CO–CO_2 product gas mixture varies considerably during reaction and actual gas composition cannot be predicted without measurements. Therefore, an improved procedure consisting of measurement of both flow rate as well as composition of product gas as function of time was adopted by some investigators, if CO and CO_2 are the only product gases. Such measurements allow calculations of rate of weight loss of oxygen (W_O^0) and that of carbon (W_C^0) separately. Therefore, total oxygen loss (ΔW_O) can be estimated with the help of Eq. (1.26) as follows:

$$\Delta W_O = \int_0^t W_O^0 dt \tag{1.26}$$

where t is time in second; hence, the degree of reduction (α) can be determined by Eq. (1.23).

Otsuka and Kunii [15] investigated reduction of iron oxide by graphite in flowing argon. They analysed the product gas by gas chromatograph at intervals of time. Abraham and Ghosh [16] studied reduction in iron oxide–graphite powder mixture as well as oxide pellet graphite powder system. They basically followed the approach of Otsuka and Kunii [15], but with different measurement technique. In their investigation, a sensitive capillary flowmeter was employed to measure the rate of evolution of product gases (i.e. CO and CO_2). Gas composition was determined by an oxygen sensor employing a calcia-stabilized zirconia solid electrolyte tube. This allowed determination of X_{CO} and X_{CO_2}, in the gas phase from measured oxygen potential of gas (where X_i is volume fraction of gas, i).

Some other research workers adopted different procedure. They chemically analysed the reduced mass and used this information either alone or with overall weight loss data to estimate the degree of reduction (α). Among them, the procedure employed by Bryk and Lu [17] was the most general and is applicable to all types of carbonaceous reductant with or without binder. They studied reduction of commercial magnetic iron ore concentrates by high-volatile coals. Samples were taken out after various times in the hot zone and chemically analysed. Total iron, metallic iron, FeO and Fe_3O_4 percentages were determined. Degree of reduction (α) was calculated based on these data.

Mookherjee et al. [18] carried out reduction of iron ore fines surrounded by coal or char fines. They measured the overall degree of reduction (α) by using equation of Chernyshev et al. [19] from the total iron analysis in ore and reduced mass.

$$\alpha = K_1 \left(\frac{\%Fe_T^r - \%Fe_T^i}{\%Fe_T^r \times \%Fe_T^i} \right) \times 100 \tag{1.27}$$

where K_1 is the ratio of weight of iron to that of oxygen in initial ore, $\%Fe_T^i$ is the per cent of total iron in initial ore and $\%Fe_T^r$ is the per cent of total iron in reduced mass.

Gonzales and Jeffes [20] have proposed a method of calculation to assess the chemical composition of partially reduced iron oxides. The method is based on the observed relationship between the degree of reduction and the corresponding degree of metallization of samples. The correlation can be used to determine the proportions of Fe^0, Fe^{2+} and Fe^{3+} from total iron analysis of pellets after reduction. The method of determining the composition of reduced samples requires an accurate initial chemical analysis of the material, including gangue and loss on ignition.

However, the above simplified procedures proposed by either Chernyshev et al. [19] or Gonzales and Jeffes [20] are not valid for a powder mixture of iron ore and carbonaceous reductant. Therefore, the only satisfactory general procedure employed so far, to the best of Dutta and Ghosh [14]'s knowledge, consists of chemical analysis of the reduced mixture for total iron, metallic iron and iron present as Fe^{2+} or FeO and Fe_3O_4 separately as employed by Bryk and Lu [17].

Weight loss method is used to determine the degree of reduction in composite pellets. Dutta and Ghosh [14] developed an alternate procedure to determine degree of reduction by hydrogen reduction of partially reduced composite pellet. This method is simpler but seems to be quite satisfactory and general in connection with their investigation on cold-bonded composite pellets consisting of fines of iron ore and coal or char fines. The assumption behind the present method is that only the remaining oxygen present in the partially reduced composite pellet will react with hydrogen, and thus, the weight loss would correspond to residual oxygen present in reduced pellet only.

In Dutta and Ghosh [14]'s procedure, further reduction of reduced composite pellets was carried out in a thermogravimetric set-up by flowing hydrogen gas at 750 °C for 1 h. The pellet should be broken into small pieces. A time of 1 h was enough to complete hydrogen reduction for reduced composite pellet.

The degree of reduction (α) for composite pellet was calculated by the following equation:

$$\alpha = \left(\frac{\Delta W_O}{W_O^i}\right) \times 100 = \left(\frac{W_O^i - \Delta W_O^H}{W_O^i}\right) \times 100 \quad (1.28)$$

where

ΔW_O^H total weight loss, i.e. oxygen loss during hydrogen reduction = $W_2 - W_3$.

W_3 the weight of pellet after hydrogen reduction.

Hence, from total weight loss data of hydrogen reduction of reduced pellet and initial weight of composite pellet before non-isothermal/isothermal reduction, the degree of reduction can be calculated using Eq. (1.28).

However, it is to be noted that there are two possible sources of error [14]:

(i) Loss of carbon due to reaction of residual char with residual oxygen of ore in pellet during hydrogen treatment of partially reduced pellet:

$$C(s) + O_2(ore) = \{CO_2\} \quad (1.29)$$

(ii) Loss of carbon due to the reaction:

$$C(s) + 2\{H_2\} = \{CH_4\} \quad (1.30)$$

For assessment of carbon loss due to reaction of oxygen in ore with carbon, it is to be noted that carbothermic reduction rate is orders to magnitude slower than hydrogen reduction rate at 750 °C. Secondly, there is no much contact between carbon and oxide since most of the oxides have already been reduced to iron by solid reductant. Therefore, error from this source is expected to be insignificant. Even then, to confirm this point, a few samples of partially reduced composite pellets were kept in flowing argon gas at 750 °C. Changes of weight were insignificant in one hour, thus confirming the above expectation.

As already stated, another source of measurement error may be loss of carbon by reaction with hydrogen. Gasification with hydrogen occurs according to reaction (1.30). Equilibrium calculations

Table 1.5 Thermodynamic data for methane formation [14]

Temperature (°C)	ΔG^0 (kJ/mol)	p_{CH_4}, bar (equilibrium)
800	27.42	0.042
750	21.84	0.066
700	16.31	0.105
650	10.83	0.164

based on the standard free energy change of reaction (1.30) are presented in Table 1.5 for a total pressure of 1 bar. It shows that equilibrium p_{CH_4} is significant.

Wang et al. [21] also developed an equation to calculate the degree of reduction (α) of iron ore–coal composite pellets as follows:

$$\alpha = \left[\left(\frac{4}{7W_O^i}\right) \times \{\Delta W_t - (W_1 \times f_v)\}\right] \times 100 \tag{1.31}$$

Now this Eq. (1.31) is further modified by Sah and Dutta [22] with incorporating the following: Total removable oxygen present in the sample can be calculated as:

$$W_O^i = W_1 \times f_{ore} \times \rho_{ore} \times f_O \tag{1.32}$$

and f_v can be calculated as:

$$f_v = f_{coal} \times f_{vm} \tag{1.33}$$

The fractional weight loss (f_{wl}) can be represented by:

$$f_{wl} = \left(\frac{\text{Total weight loss of composite pellet}}{\text{Initial weight of composite pellet}}\right) = \left(\frac{\Delta W_t}{W_1}\right) \tag{1.34}$$

Hence, Eqs. (1.32)–(1.34) are substituted in Wang et al. [21]'s Eq. (1.31) giving:

$$\alpha = \left(\frac{4}{7}\right) \times \left[\frac{f_{wl} - (f_{coal} \times f_{vm})}{(f_{ore} \times \rho_{ore} \times f_O)}\right] \times 100 \tag{1.35}$$

where

- f_{wl} fractional weight loss,
- f_{coal} fraction of coal present in composite pellet,
- f_{vm} fraction of volatile matters present in coal,
- f_{ore} fraction of ore present in composite pellet,
- ρ_{ore} purity of iron oxide (Fe_2O_3) in ore,
- f_O fraction of oxygen present in pure Fe_2O_3.

Equation (1.31) considers total release of volatiles in composite pellet while calculating degree of reduction. However, the release of volatiles from composite pellet is gradual and occurs over a wide range of temperatures (300–927 °C). Therefore, the calculated values of the degree of reduction (α), using Wang et al. [21]'s Eq. (1.31), at temperature below 927 °C turn out to be lower values than the actual degree of reduction of composite pellet. Further, non-isothermal reduction of composite pellet by thermal analysed does not provide continuous recording of separate release of volatiles. Considering the above points, a term f_{vr} is incorporated by Sah and Dutta [22] in Eq. (1.35) as follows:

$$\alpha = \left[\left(\frac{4}{7}\right) \times \left[\frac{f_{wl} - (f_{coal} \times f_{vm} \times f_{vr})}{(f_{ore} \times \rho_{ore} \times f_{O})}\right]\right] \times 100 \quad (1.36)$$

where f_{vr} represents fraction of volatiles released at time, t or at a particular temperature during reduction.

Hence, Eq. (1.36) is the modified equation of Wang et al. [21], which has been used to calculate the degree of reduction (α) by Sah and Dutta [22]. This Eq. (1.35) is directly used to calculate the degree of reduction (α) of composite pellets by weight loss method. This seems to be very simple method.

1.3.6.3 Physical Behaviour During Reduction at High Temperatures

- Volume change,
- Compressive strength at high temperature.

Volume Change

During reduction at high temperature, volume changes of the sample take place. The volumes of the sample are measured before and after reduction.

$$\text{Volume change } \% = \left(\frac{(\text{Initial volume } - \text{Final volume})}{\text{Initial volume}}\right) \times 100 \quad (1.37)$$

Compressive Strength at High Temperature

Compressive strength of hardened pellets/briquettes (which are more regular shapes) at high temperature after reduction can be measured with special furnace attachment to the tensile machine. When a spherical-shaped material (i.e. pellets) is subjected to compressive load at high temperature, then the stress generated at one point of contract causes its fracture into fragments depending on the nature of material and number of cracks generated.

Probable Questions

1. What are the types of iron ores? Name and discuss in brief.
2. Which factors decide the quality of an iron ore for its smelting in BF?
3. What do you mean by metallurgical coal?
4. Although Indian coking coal has high ash content, still it can be used for better environment friendly due to what?
5. What are the main functions of coke in a blast furnace?
6. What are the properties that depend value of coke (as a BF fuel)?
7. 'Reactivity of coke has a significant influence on the reduction process', explain.
8. What are the ideal burden qualities for hot metal production?
9. Discuss the basic principle of coke oven battery with stamp-charging.
10. Write short notes on (i) dry cooling of coke, (ii) blending with low ash coking coals and (iii) formed coke.

11. Why a BF cannot run without a critical amount of coke? Explain.
12. What do you mean by 'insoluble in limestone'? What are the maximum limits?
13. Define the term 'basicity' and explain how is it different from V-ratio.
14. What are the problems with Indian raw materials for BF ironmaking?
15. What are the different processes of agglomeration? Discuss in brief.
16. What are the objects of sintering? What are the process variables for sintering?
17. 'By charging super-flux sinter in the BF, limestone can be eliminated from the feed'. Why?
18. Write the basic principles of palletization process. Compare disc and drum pelletizers.
19. Compare between palletization and sintering processes.
20. What are the functions of a binder? What are the factors for selecting the binder?
21. What is the mechanism of ball formation in palletization process? State the factors that affect size of the pellets produced.
22. What do you mean by bonding mechanism for pellet formation?
23. What do you understand by cold-bonded pellets?
24. What do you mean by composite pellets? What are the merits offered, when they are used as burden material in ironmaking?
25. Explain briefly the different methods of testing of agglomerated products. Also give in brief the test procedures.
26. What are the tests you will recommend for the product of disc pelletizer, before charging to the BF?
27. What is reducibility? How it is evaluated?

Examples

Example 1.1 The initial weight of iron ore pellet is 1.325 g. The pellet is reduced by hydrogen at 750 °C for 30 min. After reduction, weight of pellet is decreased to 1.015 g. Find out the degree of reduction.

Given: Iron ore (Fe_2O_3) contains 65% total Fe.

Solution

Molecular weight of $Fe_2O_3 = 2 \times 56 + 3 \times 16 = 160$

Out of 160 part Fe_2O_3, iron content 112 parts

$$100\,\text{part}\quad \left[\frac{(112 \times 100)}{160}\right] = 70\,\text{parts i.e } 70\%\,\text{Fe}$$

Out of 160 part Fe_2O_3, oxygen content 48 parts

$$100\,\text{part}\quad \left[\frac{(48 \times 100)}{160}\right] = 30\,\text{part i.e } 30\%\,\text{Oxygen}$$

$$\text{Degree of reduction}(\alpha) = \left[\frac{(O_2 \text{ removed from the sample}) \times 100}{(\text{Total removable } O_2 \text{ present in the sample})}\right]$$
$$= \left[\frac{(\text{weight loss}) \times 100}{(\text{Total removable } O_2 \text{ present in the sample})}\right]$$
$$= \left[\frac{\{(W_1 - W_2) \times 100\}}{W_O}\right]$$

where

W_1 initial weight of the sample,
W_2 final weight of the sample,
W_O total removable O_2 present in the sample = $W_1 \times f_{ore} \times \rho_{ore} \times f_O$
f_{ore} fraction of ore presents in sample,
ρ_{ore} fraction of purity of iron oxide in ore,
f_O fraction of oxygen present in iron oxide.

70 part of Fe content in 100 part of iron ore (Fe_2O_3)

$$65\text{ part} \quad \left[\frac{(100 \times 65)}{70}\right] = 92.86\%$$

i.e. purity of iron oxide in ore = 92.86%

$$\text{So } \rho_{ore} = \left(\frac{92.86}{100}\right) = 0.9286$$
$$f_O = \left(\frac{30}{100}\right) = 0.3$$
$$f_{ore} = \left(\frac{100}{100}\right) = 1[\text{since there is no binder in the pellet, so } 100\% \text{ iron ore}]$$

Since $W_O = W_1 \times f_{ore} \times \rho_{ore} \times f_O = 1.325 \times 1 \times 0.9286 \times 0.3 = 0.3691$ g.
Therefore, degree of reduction $\alpha = \left(\frac{(1.325 - 1.015) \times 100}{0.3691}\right) = \mathbf{83.98}\%$.

Example 1.2 The initial weight of iron ore pellet is 1.5 g. The pellet is reduced by hydrogen at 750 °C for 30 min. After reduction, weight of the pellet comes down to 1.15 g. The pellet contains 3% binder, and iron ore contains 93% Fe_2O_3. Find out the degree of reduction?

Solution

Since degree of reduction $(\alpha) = \left[\frac{\{(W_1 - W_2) \times 100\}}{W_O}\right]$

where

W_1 initial weight of the sample,
W_2 final weight of the sample,
W_O total removable O_2 present in the sample = $W_1 \times f_{ore} \times \rho_{ore} \times f_O$

$$f_{ore} = \text{fraction of ore presents in sample} = \left[\frac{(100-3)}{100}\right] = 0.97$$

$$\rho_{ore} = \text{fraction of purity of iron oxide in ore} = \left(\frac{93}{100}\right) = 0.93$$

$$f_O = \text{fraction of oxygen present in iron oxide} = 0.3$$

Now $W_O = W_1 \times f_{ore} \times \rho_{ore} \times f_O = 1.5 \times 0.97 \times 0.93 \times 0.3 = 0.4059$ g.
Therefore, degree of reduction $\alpha = \left[\frac{\{(W_1 - W_2) \times 100\}}{W_O}\right] = \left[\frac{\{(1.50 - 1.15) \times 100\}}{0.4059}\right] = \mathbf{86.23\%}$.

Example 1.3 Iron ore–coke composite pellet contains 12% coke and 3% binder. Initial weight of pellet is 1.55 g. Composite pellet undergoes reduction in nitrogen atmosphere at 1000 °C for 1 h. After reduction weight comes down to 1.27 g. Reduced pellet is then treated again with hydrogen gas at 750 °C for 30 min, after hydrogen treatment weight of reduced pellet further comes down to 1.13 g.

Calculate degree of reduction for composite pellet in nitrogen atmosphere.
Given: Ore contains 92% Fe_2O_3, and coke contains 75% carbon.

Solution
As per Eq. (1.27):

$$\text{Degree of reduction}\,(\alpha) = \left[\frac{\{(W_O^i - W^H) \times 100\}}{W_O^i}\right]$$

where W_O^i = total removable O_2 present in the sample $= W_1 \times f_{ore} \times \rho_{ore} \times f_O$

$$W^H = W_2 - W_3 = 1.27 - 1.13 = 0.14\,\text{g}$$

Now $W_O^i = W_1 \times f_{ore} \times \rho_{ore} \times f_O = 1.55 \times 0.85 \times 0.92 \times 0.3 = 0.3636$ g

$$\left[\text{Since}\, f_{ore} = \left[\frac{\{100 - (12+3)\}}{100}\right] = \left[\frac{\{100 - 15\}}{100}\right] = \left(\frac{85}{100}\right) = 0.85\right]$$

$$\text{Degree of reduction}\,(\alpha) = \left[\frac{\{(W_O^i - W^H) \times 100\}}{W_O^i}\right]$$

$$= \left[\frac{\{(0.3636 - 0.14) \times 100\}}{0.3636}\right] = \mathbf{61.5\%}.$$

Example 1.4 Iron ore–coal composite pellet contains 20% coal and 3% binder. Initial weight of pellet is 1.6758 g. Composite pellet undergoes reduction in nitrogen atmosphere at 950 °C for 1 h. After reduction weight comes down to 1.370 g. During handling, pellet is broken; weight of broken pellet is 1.085 g before hydrogen treatment at 750 °C for 30 min, and after hydrogen treatment, weight of reduced pellet further comes down to 0.9568 g. Calculate degree of reduction for composite pellet in nitrogen atmosphere.

Given: Ore contains 85% Fe_2O_3, and coal contains 60% carbon.

Solution

As per Eq. (1.27):

$$\text{Degree of reduction}\,(\alpha) = \left[\frac{\{(W_O^i - W^H) \times 100\}}{W_O^i}\right]$$

where W_O^i = total removable O_2 in the sample = $W_1 \times f_{ore} \times \rho_{ore} \times f_O$.

Since during handling pellet is broken, a correction factor (ε) will come.

$$\text{Correction factor}\,(\varepsilon) = \left[\frac{(W_3 - W_4)}{W_3}\right]$$

where

W_3 weight of broken pellet = 1.085 g

W_4 weight of reduced pellet after hydrogen treatment = 0.9568 g.

Therefore, correction factor $(\varepsilon) = \left[\frac{(W_3 - W_4)}{W_3}\right] = \left[\frac{(1.085 - 0.9568)}{1.085}\right] = 0.1182$

So $W^H = W_2 \times \varepsilon = 1.370 \times 0.1182 = 0.1619$ g.

Now $W_O^i = W_1 \times f_{ore} \times \rho_{ore} \times f_O = 1.6758 \times 0.77 \times 0.85 \times 0.3 = 0.329$ g.

Therefore, degree of reduction $(\alpha) = \left[\frac{\{(W_O^i - W^H) \times 100\}}{W_O^i}\right] = \left[\frac{\{(0.329 - 0.1619) \times 100\}}{0.329}\right] = \mathbf{50.8}\%$.

Example 1.5 Iron ore (blue dust)–coal composite pellet contains 15% coal and 5% binder. Initial weight of pellet is 1.85 g. Composite pellet undergoes reduction in nitrogen atmosphere at 1000 °C for 1 h. After reduction weight comes down to 1.29 g.

Calculate degree of reduction for composite pellet in nitrogen atmosphere.

Given: Blue dust contains 95.7% Fe_2O_3, 1.6% SiO_2 and 1.8% Al_2O_3. Coal contains 49.7% fixed carbon, 32.5% VM and 11.6% ash.

Solution

As per Eq. (1.34):

$$\alpha = \left(\frac{4}{7}\right) \times \left[\frac{f_{wl} - (f_{coal} \times f_{vm})}{(f_{ore}\, x\, \rho_{ore}\, x f_O)}\right] \times 100$$

Now $W_O^i = W_1 \times f_{ore} \times \rho_{ore} \times f_O = 1.85 \times 0.8 \times 0.957 \times 0.3 = 0.425$ g.

Since f_{ore} = [{100 – (15 + 5)}/100] = 0.8

$$\text{Fractional weight loss} = f_{\text{wl}} = \left(\frac{\Delta W_t}{W_1}\right) = [(W_1 - W_2)/W_1] = [(1.85 - 1.29)/1.85] = 0.303$$

$$\text{Therefore, } \alpha = \left(\tfrac{4}{7}\right) \times \left[\tfrac{f_{\text{wl}} - (f_{\text{coal}} \times f_{\text{vm}})}{(f_{\text{ore}} \times \rho_{\text{ore}} \times f_{\text{O}})}\right] \times 100 = \left(\tfrac{4}{7}\right) \times \left[\tfrac{0.303 - (0.15 \times 0.325)}{(0.8 \times 0.957 \times 0.3)}\right] \times 100$$

$$= \left(\frac{4}{7}\right) \times \left[\frac{0.303 - (0.049)}{(0.8 \times 0.957 \times 0.3)}\right] \times 100 \left(\frac{4}{7}\right) \times \left[\frac{0.254}{(0.23)}\right] \times 100 = \mathbf{63.11}\%.$$

Example 1.6 Iron ore (blue dust)–coal composite pellet contains 25% coal and 5% binder. Initial weight of pellet is 1.82 g. Composite pellet undergoes reduction in nitrogen atmosphere at 1000 °C for 1 h. After reduction weight comes down to 1.15 g.

Calculate degree of reduction for composite pellet in nitrogen atmosphere.

Given: Blue dust contains 95.7% Fe_2O_3, 1.6% SiO_2 and 1.8% Al_2O_3. Coal contains 42% fixed carbon, 24% VM and 28% ash.

Solution

As per Eq. (1.34):

$$\alpha = \left(\frac{4}{7}\right) \times \left[\frac{f_{\text{wl}} - (f_{\text{coal}} \times f_{\text{vm}})}{(f_{\text{ore}} \times \rho_{\text{ore}} \times f_{\text{O}})}\right] \times 100$$

Now $W_{\text{O}}^i = W_1 \times f_{\text{ore}} \times \rho_{\text{ore}} \times f_{\text{O}} = 1.82 \times 0.7 \times 0.957 \times 0.3 = 0.366$ g.
Since $f_{\text{ore}} = [\{100 - (25 + 5)\}/100] = 0.7$.

$$\text{Fractional weight loss} = \text{f}_{\text{wl}} = \left(\frac{\Delta W_t}{W_1}\right) = [(W_1 - W_2)/W_1] = [(1.82 - 1.15)/1.82] = 0.368$$

Therefore,

$$\alpha = \left(\frac{4}{7}\right) \times \left[\frac{f_{\text{wl}} - (f_{\text{coal}} \times f_{\text{vm}})}{(f_{\text{ore}} \times \rho_{\text{ore}} \times f_{\text{O}})}\right] \times 100 = \left(\frac{4}{7}\right) \times \left[\frac{0.368 - (0.25 \times 0.24)}{(0.7 \times 0.957 \times 0.3)}\right] \times 100$$

$$= \left(\frac{4}{7}\right) \times \left[\frac{0.368 - (0.06)}{(0.7 \times 0.957 \times 0.3)}\right] \times 100 = \left(\frac{4}{7}\right) \times \left[\frac{0.308}{(0.2)}\right] \times 100 = \mathbf{88.0}\%.$$

Problems

Problem 1.1 The initial weight of iron ore pellet is 1.52 g. The pellet contains 3% binder. The pellet is reduced by hydrogen at 700 °C for 30 min. After reduction, weight of pellet is decreased to 1.315 g. Find out the degree of reduction.

Given: Ore contains 94% Fe_2O_3.

[Ans: 49.3%]

Problem 1.2 Iron ore pellet contains 2% bentonite. The initial weight of iron ore pellet is 2.435 g. The pellet is reduced by hydrogen at 800 °C for 60 min. After reduction, weight of pellet is decreased to 1.84 g. Find out the degree of reduction.

Given: Iron ore (Fe_2O_3) contains 64% total Fe.

[Ans: 90.9%]

Problem 1.3 Iron ore–char composite pellet contains 15% char and 2% binder. Initial weight of pellet is 1.546 g. Composite pellet undergoes reduction in nitrogen atmosphere at 1000 °C for 1 h. After reduction weight comes down to 1.269 g. Reduced pellet is then treated again with hydrogen gas at 750 °C for 30 min, after hydrogen treatment weight of reduced pellet further comes down to 1.125 g.

Calculate degree of reduction for composite pellet in nitrogen atmosphere.

Given: Ore contains 90% Fe_2O_3, and char contains 80% carbon.

[Ans: 58.44%]

Problem 1.4 Iron ore–char composite pellet contains 20% coal and 5% binder. Initial weight of pellet is 1.75 g. Composite pellet undergoes reduction in nitrogen atmosphere at 1000 °C for 1 h. After reduction weight comes down to 1.42 g. During handling pellet is broken, weight of broken pellet is 1.42 g before hydrogen treatment at 750 °C for 30 min and after hydrogen treatment weight of reduced pellet further comes down to 0.965 g. Calculate degree of reduction for composite pellet in nitrogen atmosphere.

Given: Ore contains 66.5% total Fe, and coal contains 65% carbon.

[Ans: 38.94%]

Problem 1.5 Iron ore (blue dust)–coal composite pellet contains 9.8% coal and 8.2% binder. Initial weight of pellet is 1.8056 g. Composite pellet undergoes reduction in nitrogen atmosphere at 1000 °C for 1 h. After reduction weight comes down to 1.467 g.

Calculate degree of reduction for composite pellet in nitrogen atmosphere.

Given: Blue dust contains 95.7% Fe_2O_3, 1.6% SiO_2 and 1.8% Al_2O_3. Coal contains 49.7% fixed carbon, 32.5% VM and 11.6% ash.

[Ans: **37.86%**]

Problem 1.6 Iron ore (blue dust)–coal composite pellet contains 9.6% coal and 10.4% binder. Initial weight of pellet is 2.048 g. Composite pellet undergoes reduction in nitrogen atmosphere at 1000 °C for 1 h. After reduction weight comes down to 1.687 g.

Calculate degree of reduction for composite pellet in nitrogen atmosphere.

Given: Blue dust contains 95.7% Fe_2O_3, 1.6% SiO_2 and 1.8% Al_2O_3. Coal contains 42% fixed carbon, 24% VM and 28% ash.

[Ans: **38.13%**]

References

1. *Mineral Commodity Summaries* (U.S. Geological Survey, Reston, Virginia, Jan 2018)
2. https://en.wikipedia.org/wiki/Energy_value_of_coal
3. A. Ghosh, A. Chatterjee, *Ironmaking and Steelmaking: Theory and Practice* (PHI learning Pvt Ltd, New Delhi, 2008)

4. J. Madias, M. de Cordova, *Proceedings AISTech 2013*, p. 253. https://www.researchgate.net/publication/263887823
5. http://ispatguru.com/dry-cooling-of-coke/
6. P. Cavaliere (ed.), *Ironmaking and Steelmaking Processes* (Springer International Publishing, Switzerland, 2016), p. 27
7. https://www.bid-on-equipment.com/processing/used-milling-machines/240190drum-pelletizer-.htm
8. T.C. Eisele, S.K. Kawatra, Mineral processing & extractive. Metall. Rev. **24**, 1 (2003)
9. Y. Chokshi, *M.E. Thesis* (Met & Mats Engineering Department, Faculty of Technology and Engineering, M. S. University of Baroda, India, 2012)
10. Centre for Techno-Economic Mineral Policy Options (C-TEMPO) (Ministry of Mines, Government of India), *Development of Iron Ore Pelletization Industry in India* (Jan 2011)
11. S.K. Dutta, A. Ghosh, Trans. Indian Inst. Met. **48**, 1 (1995)
12. A. Ghosh, *Proceedings of International Conference on Alternative Routes of Iron and Steelmaking* (Perth, Australia, 1999), p. 71
13. J. Srb, Z. Reziekova, *Pelletization of Fines* (Elsevier Science Publishers, Amsterdam, 1988)
14. S.K. Dutta, A. Ghosh, ISIJ Int. **33**, 1104 (1993)
15. K. Otsuka, D. Kunii, J. Chem. Eng. Japan **2**, 46 (1969)
16. M.C. Abraham, A. Ghosh, Ironmaking Steelmaking **6**, 14 (1979)
17. C. Bryk, W.K. Lu, Ironmaking Steelmaking **13**, 70 (1986)
18. S. Mookhrjee, H.S. Ray, A. Mukherjee, Ironmaking Steelmaking **13**, 229 (1986)
19. A.M. Chernyshev, N.K. Karnilova, Yu.V. Tarasenko, *Steel USSR* **7**, 133 (1977)
20. O.G. Dam Gonzales, J.H.E. Jeffes, *Ironmaking Steelmaking* **14**, 217 (1987)
21. Q. Wang, Z. Yang, J. Tian, W. Li, J. Sun, Ironmaking Steelmaking **25**, 443 (1998)
22. R. Sah, S.K. Dutta, Trans. Indian Inst. Met. **64**, 583 (2012)
23. http://www.911metallurgist.com/askus/?referrername = https://www.911metallurgist.com/blog/different-types-of-iron-ore
24. https://images.search.yahoo.com/search/images/schematicdiagramofcokeovenbattery
25. http://www.zentralkokerei.de/zks/produktion/koksherstellung/index.shtml.en. Accessed Feb 2013
26. Y. Chokshi, S.K. Dutta, Iron & Steel Review **58**, 19 (2014)
27. Y. Chokshi, S.K. Dutta, Iron Steel Rev. **56**, 214 (2012)
28. R.H. Tupkary, V.R. Tupkary, *An Introduction to Modern Iron Making,* 4th edn. (Khanna Publishers, Delhi, 2012)

Blast Furnace Process

2

The objective of the blast furnace (BF) is to produce hot metal. The blast furnace is a tall, vertical shaft furnace which uses coke to reduce iron ores. The product, a hot metal (which is impure iron), is suitable as feed material for steelmaking. There are different charging systems for BF discussed in detail. Out of that, nowadays, bell-less top (BLT) system is very popular.

2.1 Outline of Blast Furnace Process

The principal objective of the blast furnace (BF) is to produce hot metal at a higher rate. The only critical operating parameter is the temperature of the hot metal and slag which must be greater than 1425 °C for these products to be tapped from the furnace in the molten state. Metal composition is not a critical feature of the blast furnace process because virtually all blast furnace hot metal is subsequently refined to steel. It is, however, controlled at steelmaking plant specifications by appropriate adjustments of slag composition and furnace temperature.

The iron blast furnace is a tall, vertical shaft furnace which uses carbon, mainly in the form of coke to reduce iron ores. The product, a hot metal (which is impure iron), is suitable as feed material for steelmaking to refine it. A schematic view of a typical blast furnace plant is shown in Fig. 2.1.

The raw materials of the blast furnace are (i) solids (ore, coke, flux) which are charged from the top of the furnace and (ii) pre-heated air (i.e. hot blast) which is passed through tuyeres near the bottom of the furnace. Metallurgical coke supplies most of the reducing gas and heat for ore reduction and smelting operation. Air is preheated to between 925 and 1325 °C by hot stoves and in some cases enriched with oxygen to give blast containing up to 25 vol% O_2. The hot blast burns the descent coke in front of the tuyeres at 1525 °C to provide heat for (i) reduction reactions and (ii) heating and melting of the charge materials and products, i.e. hot metal and slag. The products of BF hot metal and slag are in molten conditions. The main product of the blast furnace, hot metal, is tapped from the furnace at regular intervals (or continuously in the case of very large furnaces) through one of several tap holes near the bottom of the hearth.

Pre-heated blast enters the tuyeres through the refractory lined *bustle pipe*, which is like a horizontal circular ring around the furnace. *Iron notch* is the tap hole for hot metal. It is kept sealed by refractory clay. For tapping, the clay seal is opened by a mechanical device. After tapping, the iron notch is again sealed by clay material using the same device. *Slag notch* is the hole for tapping molten slag. It is above the iron notch, since the slag has a lower density compared to molten iron and floats above the hot metal in the hearth.

S. K. Dutta and Y. B. Chokshi, *Basic Concepts of Iron and Steel Making*,
https://doi.org/10.1007/978-981-15-2437-0_2

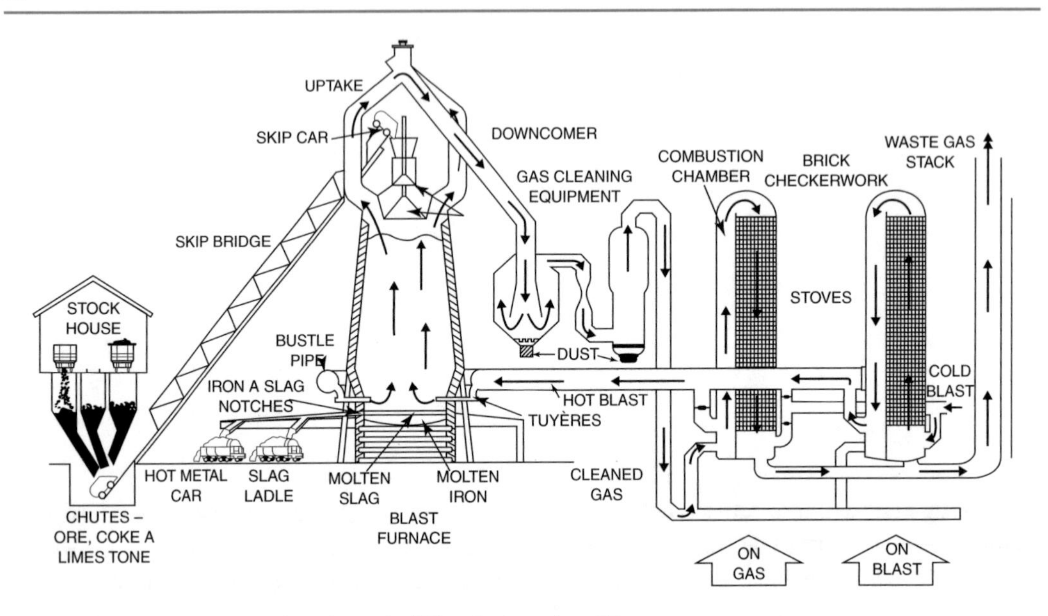

Fig. 2.1 Schematic representation of a typical blast furnace plant [1]

The hot metal is tapped through the *tap hole* several times a day into *ladles*. Hot metal which is produced in BF is the intermediate form through which almost all iron must pass in the production of steel. The hot metal is transferred to the steelmaking shop by *metal mixer* for further refining. The ladles are lined with refractory. Excess hot metal is cast into *pig iron* in a pig iron casting machine for further use as feedstock in foundries or in other steelmaking shops for a wide variety of castings. The molten slag is tapped from time to time through the *slag notch* (i.e. separate tap hole) into the slag ladle and is used as feedstock for the manufacture of slag cement, etc. The top gas is known as *blast furnace gas*. It has a considerable caloric value since it contains carbon monoxide gas. The dust must be removed first in a gas-cleaning unit. The gas is stored and then mostly utilized in the blast furnace shop itself for pre-heating of air in stove and running turbines to drive air blowers.

2.2 Constructional Features of BF

Figure 2.2 shows schematic sketch of a blast furnace indicating different sections, and Fig. 2.3 shows the general constructional features of a blast furnace. It is circular in cross section and around 30–40 m in height. The outer shell of BF is made of steel plates, and refractory lining is at the inside of shell. Nowadays, the steel shell is of welded construction rather than the earlier form of rivetted construction. The tall structure has been made *free standing*, i.e. the only support is provided by the foundation. The furnace interior is broadly divided into different sections:

- Stack: It is the upper portion of the BF, whose wall slopes going outwards as goes downwards. It is the zone in which the burden is completely solid. The charge materials are heated by descending from 200 °C at the stock line level to nearly 1100–1200 °C at the bottom of the stack. To ensure free fall of the charge material, as it expands progressively with the progressive rise in temperature, the cross section of the furnace is uniformly increased to almost double the size from the stock line to the mantle level. Since most of the iron oxide reduction occurs in the stack region.

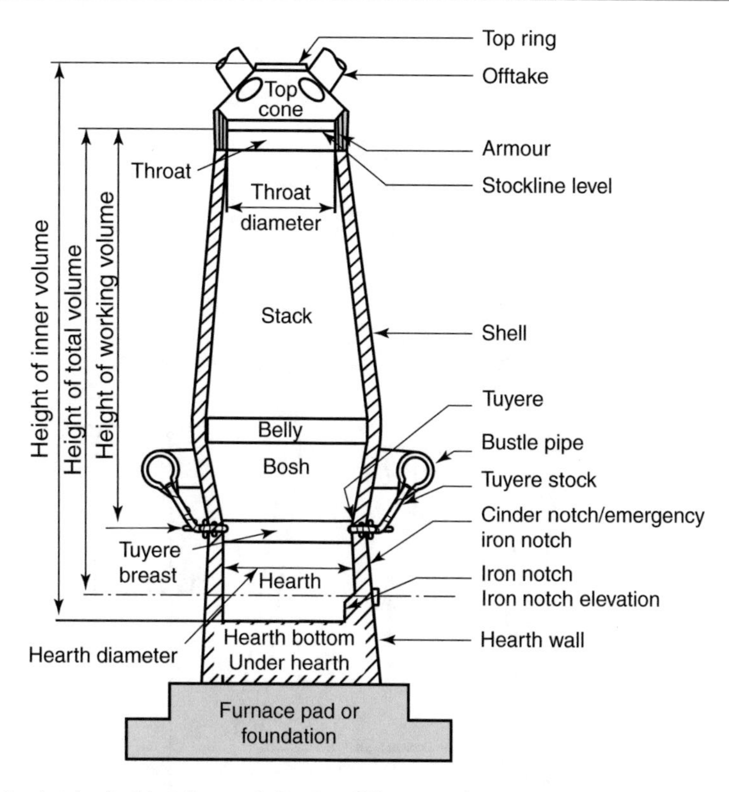

Fig. 2.2 Schematic sketch of a blast furnace indicating different sections

The success of the blast furnace process depends on the efficiency of the counter-current gas–solid reaction in the stack.

- Belly: It is the cylindrical portion below the stack and above the bosh region. The furnace walls are parallel (to some extent) in this region.
- Bosh: It is below the belly and sloping inwards going downwards. The charge materials (except coke) begin to soften and fuse as they come down into the bottom of the stack. The gangue of iron ore, ash of coke and flux combine to form the slag. The furnace walls in this region tapered down to reduce the sectional area by about 20–25% in harmony with the resultant decrease in the apparent volume of the charge. The burden permeability in this region is mainly maintained by the presence of solid coke. Therefore, this dictates that coke should have adequate strength and proper size for efficient operation. Any degradation of coke leads to decrease permeability in the bosh region, and that adversely affects the operation of the blast furnace.
- Tuyeres: Tuyere and combustion zone are located below bosh and above hearth zone. By the time the charge descends into the area near the tuyeres, except the central column of coke (which is solid), the entire charge is molten. The oxygen of the blast burns coke to form reducing gas, CO in front of each tuyere. Thus, there is a *raceway* in front of each tuyere (as shown in Fig. 2.4), which is first horizontal and then smoothly changes its direction to vertical while expanding over the entire cross section of the furnace [2]. Hot blast is blown into the blast furnace via tuyeres. A tuyere is a cooled copper conical pipe numbering up to 12 for smaller furnaces, and up to 42 for bigger furnaces through which pre-heated air (up to more than 1200 °C) is blown into the furnace.

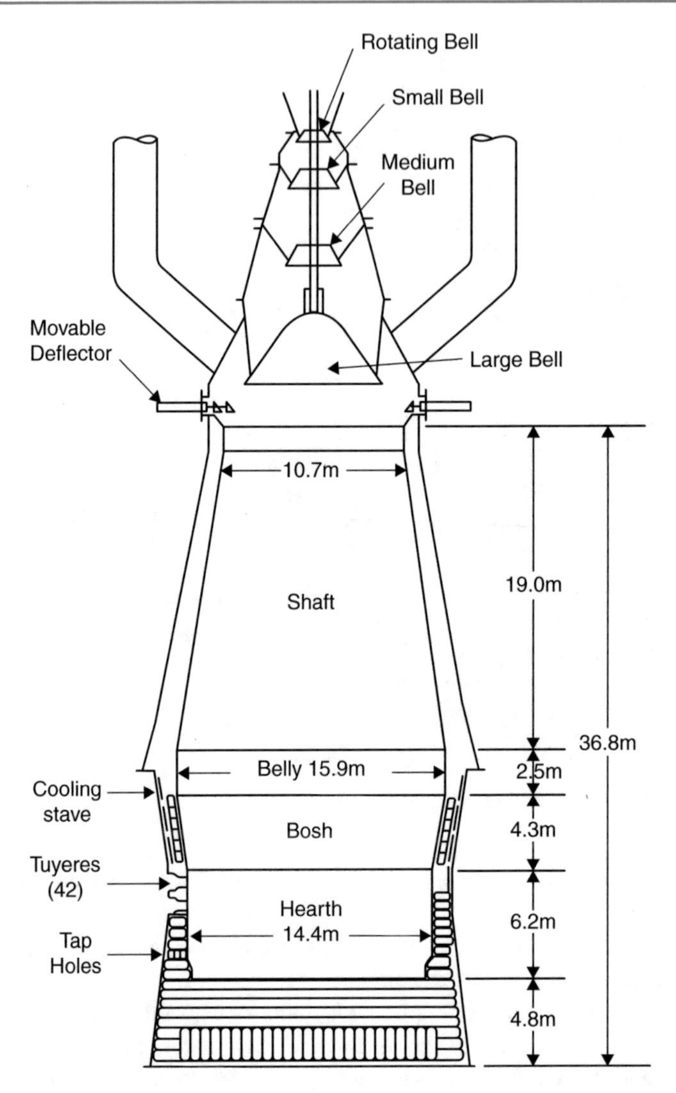

Fig. 2.3 Typical profiles of blast furnace with four-bell system and movable deflectors [1] (working volume and internal volume are 3930 and 4620 m^3, respectively)

- Hearth: It is the bottom cylindrical portion of the BF, below the bosh and tuyere regions. Although most of the coke burns at the tuyere level, a small fraction descends even into the hearth (to form the *dead man zone*, which is undissolved solid coke particles either sits on the hearth or floats just above it). Carbon is dissolved in the metal to its near saturation limit. The entire charge (except *dead man zone* which is in solid form) is molten and tends to stratify into slag and metal layers in the hearth from where these are tapped separately. The cross section of the furnace below the tuyeres decreases since the liquids are dense without pores and voids, thus leading to decrease in volume. The walls of the hearth are parallel, and the hearth is the smallest cross section of the BF.

The general trend of increasing blast furnace size-related operational parameters is shown in Table 2.1. Operating details of large modern blast furnaces in the world are summarized in Table 2.2.

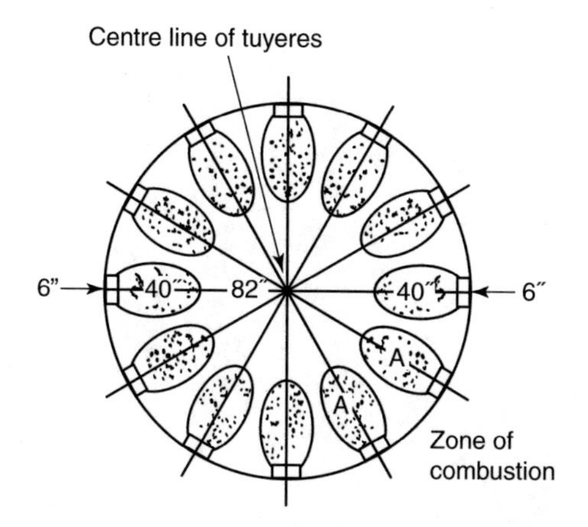

Fig. 2.4 Horizontal cross-sectional view of tuyere zone

Table 2.1 Impact of operational parameter on size of BF

Parameter	BFs		
	Small	Medium	Large
Inner volume (m^3)	1550	3300	5700
Hearth diameter (m)	8.5	12.0	15.5
Number of tuyeres	24	32	42
Production (t/day)	3358	7150	12,350
Blast volume (Nm^3/min)	2565	5500	9434
Blast volume/tuyere (Nm^3/min)	106	172	232
Injection rate (kg/thm)	150	150	150
Injection rate/tuyere (t/h)	0.9	1.4	1.8

Table 2.2 Operational features of some modern BF in the world

Parameter	Steel plants					
	Kimitsu-3 (Japan) [2]	Nippon Steel (Japan) [2]	Posco (Korea) BF-6 [2]	Tata Steel (India) BF-G [2]	Vizag Steel Plant (India) BF-3 [3]	JSW steel limited, Vijayanagar (India) BF-4 [4]
Production (t/day)	10,233	10,051	8600	5150	8000	9800
Inner volume (m^3)	4450	4063	3800	2648	3800	4019
Working volumc (m^3)	3790	NA	3225	2308	3300	3445
Hearth diameter (m)	14.1	13.4	13.2	11.0	12.0	13.2
Productivity (t/m^3/day)	2.7	2.47	2.66	2.2	1.8	2.82
Top pressure (kg/cm^2)	2.25	2.2	2.5	1.3	1.8	2.25

(continued)

Table 2.2 (continued)

Parameter	Steel plants					
	Kimitsu-3 (Japan) [2]	Nippon Steel (Japan) [2]	Posco (Korea) BF-6 [2]	Tata Steel (India) BF-G [2]	Vizag Steel Plant (India) BF-3 [3]	JSW steel limited, Vijayanagar (India) BF-4 [4]
O_2 enrichment (%)	4.0	2.4	2.0	4.6	4.0	7.91
Charge, % sinter (S), ore (O), pellets (P)	50(S) + 50(P)	93(S) + 7(P)	85(S) + 15(O)	70(S) + 30(O)	70(S) +30(O) Or 70(S) + 18–20(O) + 10–12(P)	70(S) + 15(P) + 15(O)
Al_2O_3 in sinter (%)	1.54	1.84	1.85	2.4	2.0–2.5	3.4
Coke rate (kg/thm)	365	392	390	410	530 (without PCI) 430 (with PCI)	390
Coke ash (%)	9.5	10.2	11.0	15.4	13–14	12.5
PCI rate (kg/thm)	125	71 (oil)	100	120	100	170
Slag rate (kg/thm)	236	286	320	300	280–300	395
Al_2O_3 in slag (%)	16.7	15.3	14.0	19.2	18.0	19.0
Blast temperature (°C)	1180	1278	1200	1080	1100	1140

2.3 Temperature Profile of BF

Exothermic combustion of coke by oxygen from air gasifies carbon into reducing gas, CO, and also generates the heat. The highest temperature zone of the furnace (1900–2000 °C) is at the level of tuyeres. As the hot reducing gas travels upwards, it heats up the downward solid charges as well as participates in various reactions at different zones of the furnace. The approximate temperature levels at different heights of the furnace are indicated in Fig. 2.5. It is shown that the softening/melting zone is in an area where temperatures are between 1100 and 1450 °C.

The major reactions of BF are classified as following:

- Removal of moisture from the raw materials,
- Reduction of iron oxides by CO (i.e. indirect reduction),
- Gasification of carbon by CO_2,
- Dissociation of $CaCO_3$,
- Reduction of FeO by carbon (i.e. direct reduction),
- Reduction of some other oxides present as gangue in the ore by carbon,
- Combustion of coke in front of tuyeres.

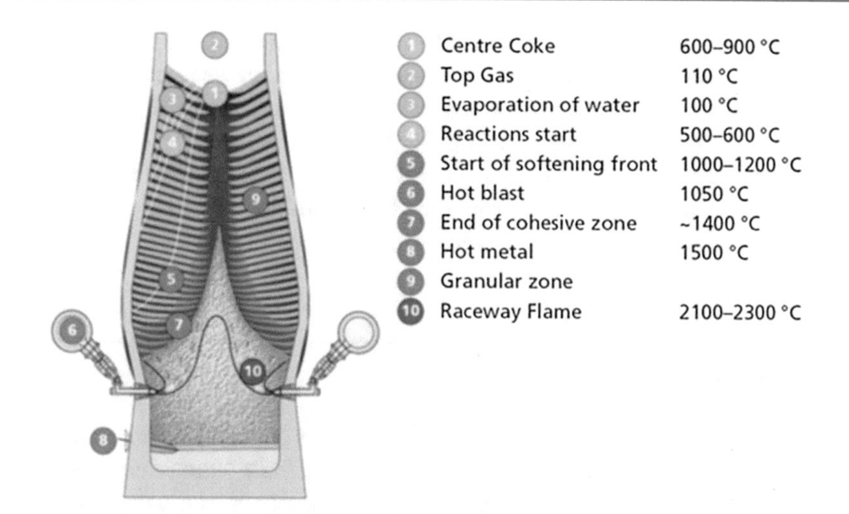

Fig. 2.5 Temperature profile in a blast furnace (iron ore starts melting at 1100–1150 °C and slag become fully liquid at 1350–1400 °C) [5] (reproduce with permission from IOS Press BV, Netherlands)

The outputs from the BF are as follows:

- Molten impure iron (i.e. hot metal),
- Molten slag,
- Gas containing CO, CO_2, N_2, moisture and some dust particles at a temperature of around 200 °C.

2.4 Function of Charged Materials in BF

(i) Iron-bearing materials: The function of the iron-bearing materials is to supply the iron, which represents about 93.5–95% of the hot metal. Major iron-bearing materials are lump ore (contains 52–65% Fe), sinter (contains 52–65% Fe) and pellets (contains 60–67% Fe); minor iron-bearing materials are mill scale and steelmaking slag (which contain 20–25% Fe).

(ii) Coke: Coke has three major functions inside the BF: (i) as a fuel to supply the heat, (ii) as a reducer to produce reducing gas CO and (iii) as a spacer to maintain permeable charge which allows the gases to pass through it smoothly. The function of coke is to produce the heat requirement for smelting and to supply the reducing agents (i.e. solid carbon and CO gas) for reducing iron-bearing materials. In addition to this, it also supplies the carbon that dissolves in the hot metal (about 30–45 kg of carbon for every tonne of hot metal).

The shape of the coke and the size distribution of the particles are the decisive factors for the permeability of the coke bed, for ascending gas as well as for the descending burden. The lowest flow resistance is obtained when larger coke is being used of high uniformity. Fines have a strong decreasing effect on the harmonic mean size and so increase the bulk resistance of the gas flow. Although excellent blast furnace operations are reported with screening at 24 mm^2, there are also plants where screening even at 40 mm is preferred.

(iii) Fluxes: The function of the limestone and/or dolomite is threefold: (a) to supply calcined lime, (b) to form a fluid slag with the coke ash, gangue of iron ore and any other impurities of the charged materials and (c) to form a slag of such chemical composition that will provide a degree of control of sulphur content in the hot metal.

(iv) Air: The air blast helps to burn part of the fuel to produce heat for the chemical reactions involved and for smelting the iron, while the balance of the fuel and part of the gas from the combustion remove the oxygen combined with the metal.

2.5 Charging System of BF

In the blast furnace process, iron-bearing materials (e.g. lumps iron ore, sinter/pellets, mill scale and steelmaking slag), coke (fuel as well as reducer) and flux (limestone and/or dolomite) are charged by the skip car or conveyor belt from the top of the furnace. The materials are charged in the sequence OCSOCC (i.e. ore-coke-stone-ore-coke-coke). Hot blast is blown from the tuyeres, along with hydro-carbon gas, and oil or powder coal is injected to the tuyeres.

The blast furnace is working on the counter-current principle. Descending solid charge meets an opposite current of ascending gases; due to that progressively heating and reduction of iron ore take place. The production rate of a BF depends on:

(i) The rate of reduction of iron ore and
(ii) The rate of heating of the burden.

These two factors are not independent but are related to (a) the quantity of blast and (b) permeability of burden. The rate of reduction and heating of the burden depends upon the degree and time of contact of gases with the burden. So, the burden inside the furnace should have uniform and good bulk permeability. A proper charging mechanism is required for more uniform permeable burden distribution.

Earlier double-bell charging system had long been a familiar feature of the BF. Nowadays, large BFs (>3500 m^3) are used. Large BFs have significant advantages in terms of productivity, raw material transportation, performance and maintenance requirements and environmental friendliness. Furnaces have working volume of up to 4500 m^3 and employ the rather different charging mechanism. The reasons for dissatisfaction with the traditional two-bell charging method for large furnace are that (i) uniform burden distribution is not easily achieved with two bells and (ii) accelerated wear and consequent gas leakage occur when a furnace is operated with high top pressure. Bell-less top (BLT) charging systems are the new development of charging devices.

2.5.1 Two-Bell Charging System

The two-bell charging system consists of a revolving material distribution, a small bell and a large bell as shown in Fig. 2.6. The diameter of large bell is usually smaller than the stock line diameter. The lower edge of the upper face of the bell forms a seal against the bottom edge of the large bell hopper. The bells are connected by a rod and move in the vertical direction by means of air cylinders.

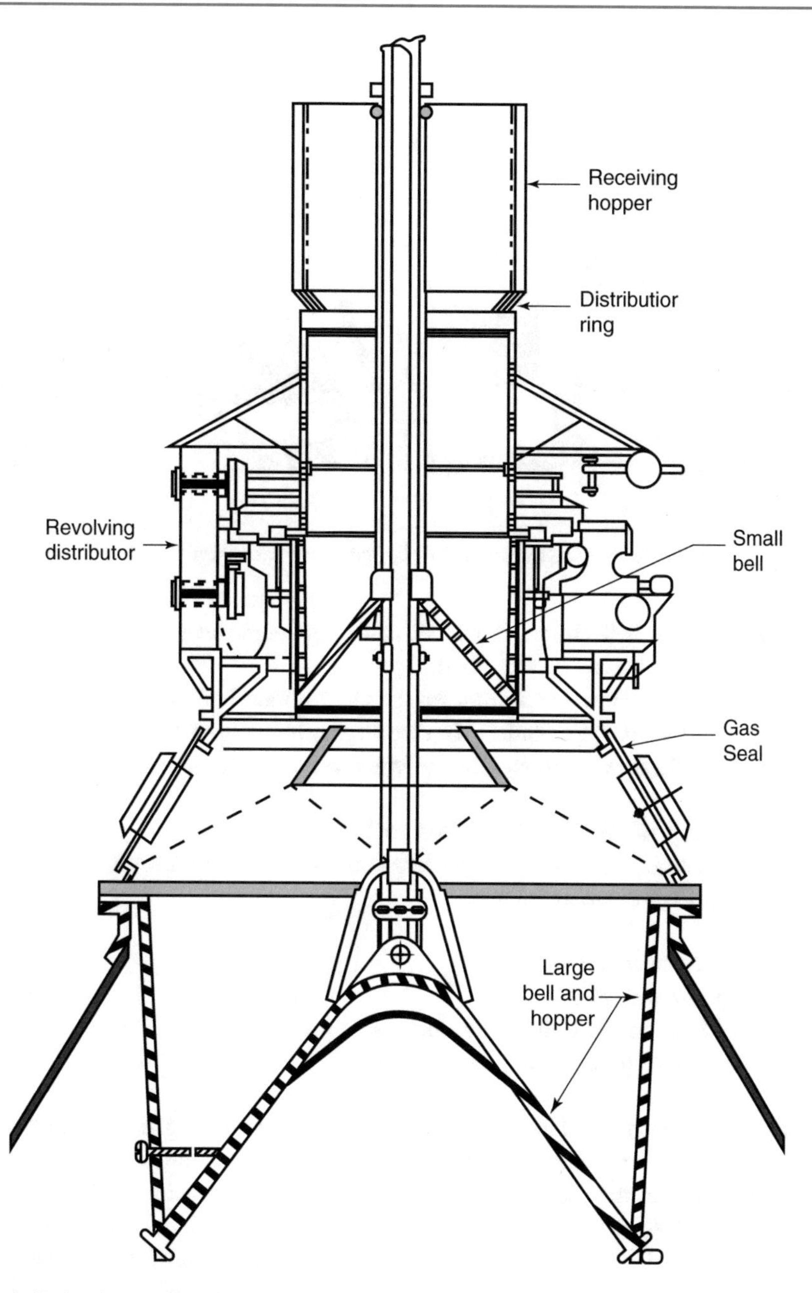

Fig. 2.6 Two-bell charging system [6] (reproduce with permission from Ispatguru)

The furnace charging is done in four steps by two-bell charging system [6] (as shown in Fig. 2.7).

Step 1 The charge materials are taken to the furnace top either by a skip car or by a conveyor belt, and the materials are delivered to a receiving hopper. Small bell and large bells both are in closed condition. The charge materials from skip car or conveyor belt are dumped in hopper above the small bell. Gas flowing from top of furnace through uptakes located in the dome (top cone).

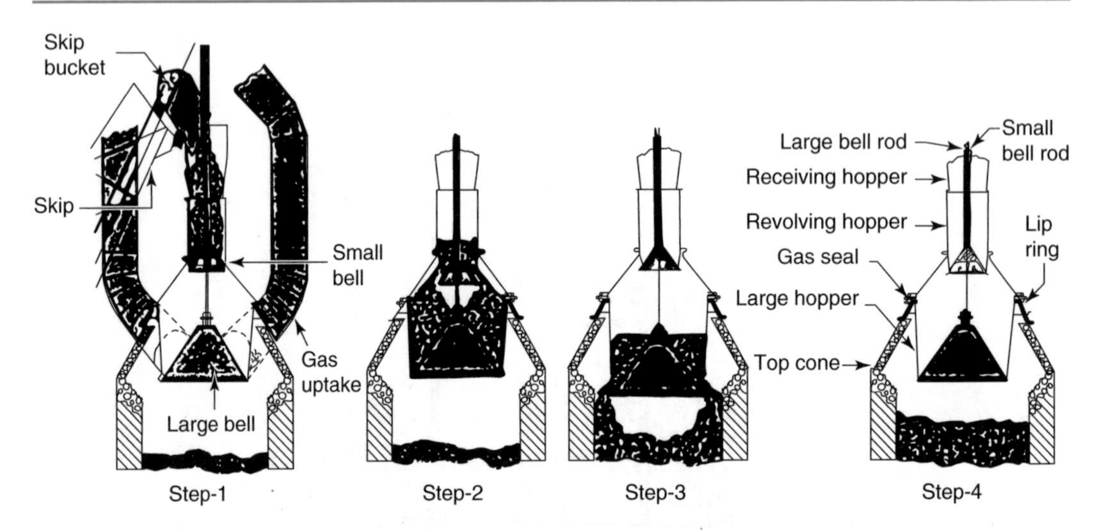

Fig. 2.7 Four steps of two-bell charging system in BF [6] (reproduce with permission from Ispatguru)

Step 2 With the large bell closed, the small bell is lowered down, and the charge material is dropped to the hopper of large bell.

Step 3 The small bell is closed to prevent gas escape to atmosphere. Now the large bell is lowered down, and the charge materials are discharged into the blast furnace.

Step 4 Both the bells are closed, and the system is ready for further charging.

2.5.1.1 Distribution of Charge Materials to BF

Better burden distribution is necessary to get maximum productivity from the BF. Various factors affecting burden distribution are as follows:

(i) Selection of raw materials,
(ii) High driving rate (i.e. blast),
(iii) Low thermal load on the walls of the furnace,
(iv) Smooth charging of burden from the top of the BF.

Burden distribution can be used to control the blast furnace gas flow. The conceptual framework of the use of burden distribution is rather complex since the burden distribution is the consequence of the interaction of properties of the burden materials with the charging mechanism.

Figure 2.8 shows the angles of repose of the various materials used in a blast furnace. Coke has the steepest angle of repose, pellets have the lowest angle of repose, and sinter is in between. Hence, in a pellet charging furnace, the pellets have the tendency to slide to the centre.

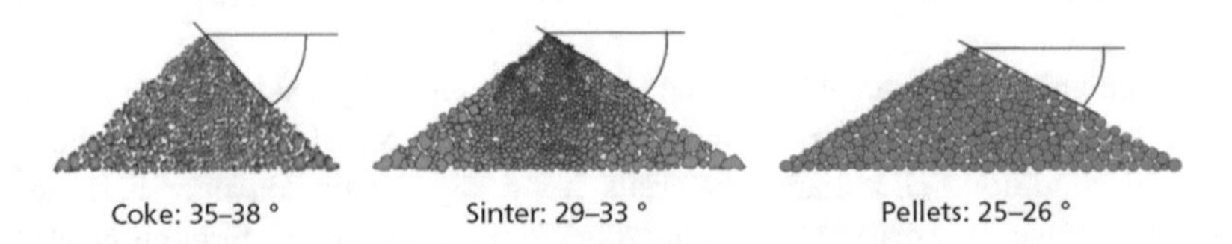

Fig. 2.8 Angles of repose of the various raw materials [5] (reproduce with permission from IOS Press BV, Netherlands)

Fines concentrate at the point of impact, and the coarse particles flow down-hill, while the fine particles remain below the point of impact. This mechanism, known as segregation, is also illustrated in Fig. 2.8.

If the charge particles are of similar size and shape, the burden would be automatically a uniformly permeable burden irrespective of the way the materials are charged in the BF. This is ideal situation. But actually, BF charge consists of different sizes of coke and iron ore, also as sinter, pellets and limestone, with different physical properties, and it is difficult to distribute them at the top of stack zone, so that the entire vertical and horizontal cross sections of the BF would offer equal resistance to the gas flow.

The areas containing the coarser materials offer less resistance to the passage of the furnace gases resulting in higher temperature and better reduction in these areas which tend to descent faster than areas containing finer particles inside the BF. In general, the courser particles tend to segregate in the centre of the furnace, and the fines segregate at or near the wall depending upon the clearance between the bell and the furnace wall.

The size of the large bell should be viewed in relation to the throat diameter of the furnace. The way the distribution is affected by the clearance between the bell and the BF wall. The more is the clearance the fines' pile-up is located further away from the wall (i.e. M type, in Fig. 2.9), the lesser is the gap between the bell corner and the throat wall tends to segregate the sizes (i.e. V type) and prevents normal distribution from being achieved. Figure 2.9 shows two types of stock line profiles. It may be noted that there is some segregation of small and large particles. This is due to their different trajectories when they fall into the furnace by opening of the large bell.

By reducing the range of lump sizes, gas/solid contact is enhanced, but of course it is impossible to eliminate size variations completely, with respect to layer thickness and particle segregation. These variations can be minimized by alterations to the charging sequence, but non-uniform burden distribution remains one of the drawbacks of the two-bell system.

As far as possible, iron ore should be present in the areas of maximum gas flow for efficient reduction. The system of charging, the level of the charging and the size of the charge are to be

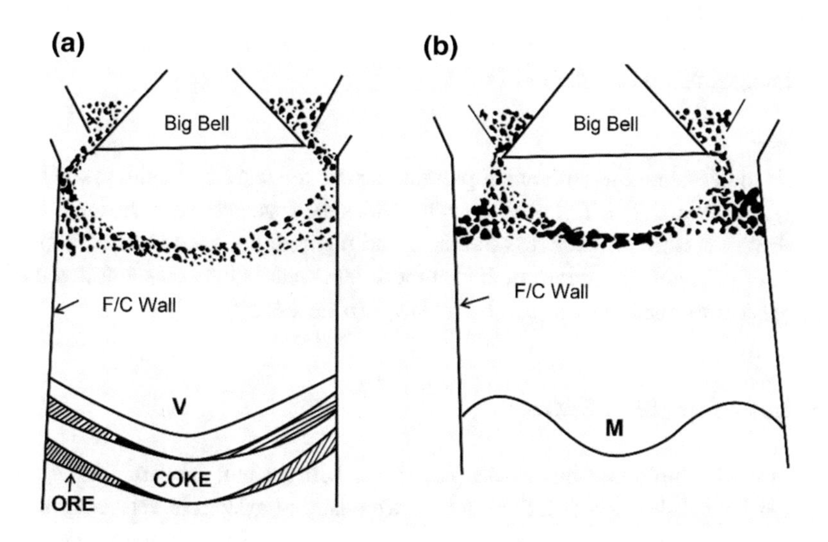

Fig. 2.9 Burden profiles in BF showing ore and coke layers: **a** smaller clearance between the bell and the BF wall and **b** larger clearance between the bell and the BF wall

selected to have maximum utilization of gas, subject to ensuring smooth performance of the BF. Good size ranges are as follows:

- Ore: 10–35 mm,
- Sinter: 5–50 mm,
- Pellets: 9–16 mm,
- Coke: 10–50 mm,
- Limestone: 10–50 mm.

Accelerated larger bell wear is a consequence of the use of high top pressure to improve furnace productivity. The iron production rate is directly related to the rate of coke combustion, since it is gasification of coke at the tuyeres that allows the burden to descent. The rate of coke combustion is dependent on the supply of oxygen via the air blast. In normal furnace operation, the gas pressure is around 2.5 atmosphere at tuyere level and 1.1 atmosphere at the throat, whereas with high top pressure, the corresponding values can be range up to 3.75 and 3.0 atmosphere, respectively. Although the dust content of the furnace top gas is highly reduced, the combination of a positive internal pressure with a temperature of around 300 °C and with some dust remaining in the gas causes serious erosion of the large bell, necessitating redesign of the charging equipment to be used with large modern furnaces, such as (i) four-bell system and (ii) bell-less top (BLT) system.

2.5.1.2 Modification of Bell-Type Charging System

There have been subsequent modifications for improvement of charge distribution of the bell-type charging, as follows:

- Two-bell type with movable throat armour.
- Four-bell charging device.

Two-Bell Type with Throat Armour Systems

With the advent of the movable throat armour systems in which a movable deflector is introduced into the stream of material falling from the large bell, a much wider cross section of the stock line can be controlled. Figure 2.10 shows the mechanism of armour movement. The position of armour can be changed by turning it around a fulcrum, O.

Four-Bell System

One approach is to eliminate the differential pressure across the large bell using more than two bells, e.g. four-bell system (Fig. 2.11). The medium bell divides the space between the large and small bells into two chambers. Pressure drops across the large and medium bells are reduced almost to zero by the introduction of compressed air into both chambers, and wear is thus concentrated upon the small bell, which is easiest to maintain and can be replaced (if necessary).

2.5.2 Bell-Less Top (BLT) System

Old furnaces are still employing buckets/skips for two-bell system; modern furnaces employ belt conveyor system for bell-less top (BLT) system. Skip charging system is very common, hopper for small bell. Actual distribution of the raw materials on the large bell, prior to their being dropped inside, affects the distribution in the furnace. Several devices have been introduced, revolving the hopper for better distribution.

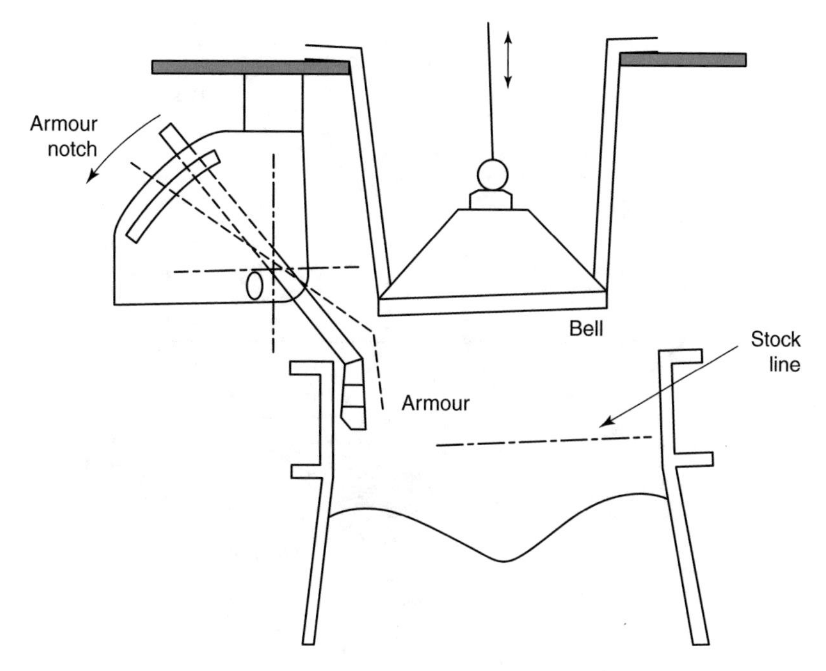

Fig. 2.10 Mechanism of armour movement

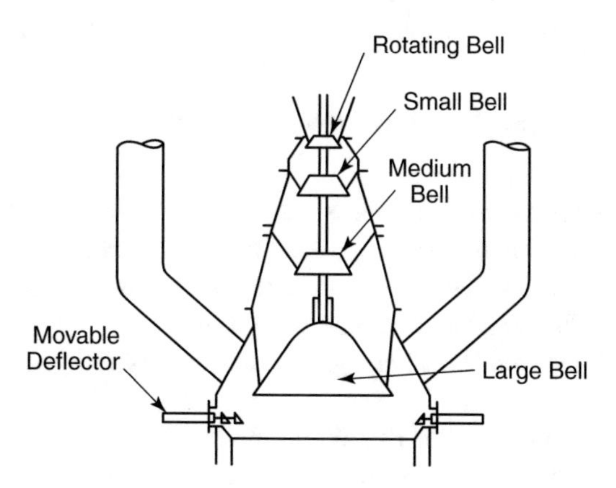

Fig. 2.11 Four-bell system

Combine control over burden distribution with the containment of gas at pressures above atmospheric is well maintained. The bell-less top system is shown in Fig. 2.12. This is unique design in which the large bell is replaced by a rotating distributor chute which has no sealing function whatsoever. The angle of inclination of rotating distributor chute can be adjusted. The problems of distribution associated with large bell are eliminated. A rotating chute is provided inside the furnace top cone. All the materials are charged via holding hoppers with seals at its top and bottom, which are charged and discharged alternately, while the third is acting as a spare. Regulating gates in each hopper are provided to control the rate of charging to facilitate uniform distribution on the stock line. The hoppers are sealed at top and bottom by valves, and the internal pressure can be raised or lowered as necessary to avoid a pressure differential across them when they are opened.

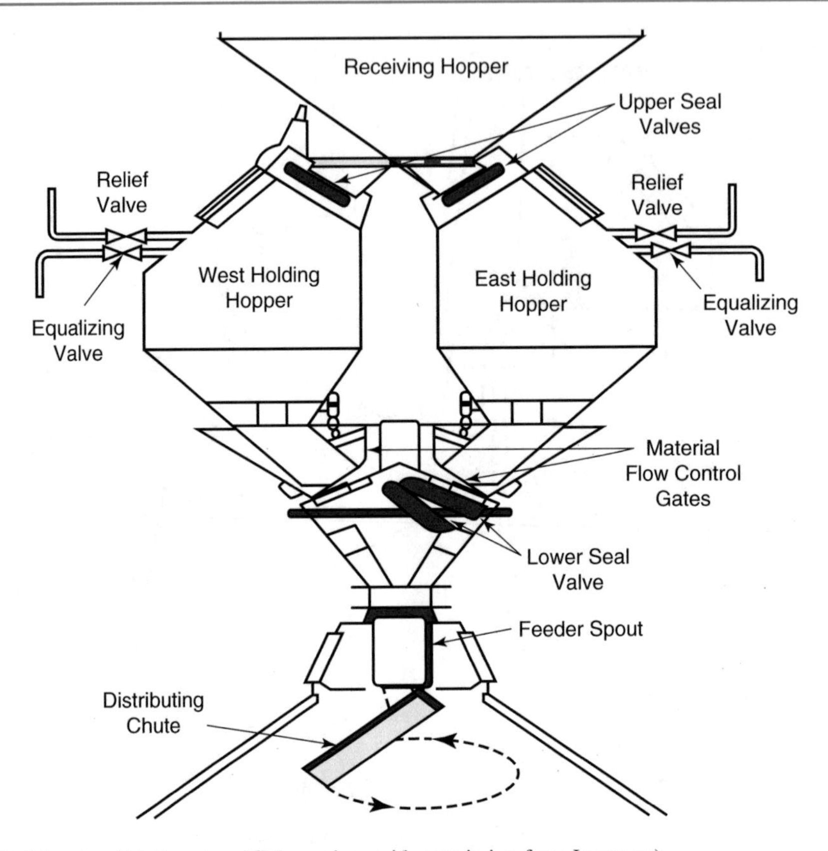

Fig. 2.12 Bell-less top (BLT) system [6] (reproduce with permission from Ispatguru)

The rate of material flow from each hopper is controlled by an adjustable gate and charged to the stack level through a rotating chute whose angle of inclination can be adjusted. This arrangement gives good control on burden distribution since successive portions of the charge can be placed on different annular areas within the furnace and reduced heat loss.

The advantages of bell-less top (BLT) charging system are as follows:

- It allows continuous charging of the BF. While the rotating chute is distributing, the materials from one lock hopper bin and the other hopper bin can be filled.
- The problem of gas sealing under a high-pressure operation is solved.
- Charge distribution flexibility is more with a small amount of mechanical equipment.
- Improved BF operational stability and efficiency, feeding for better hot metal chemistry control.
- Increases in the BF productivity.
- Reduces BF coke rate and helps to achieve higher injection rates of pulverized coal.
- Contributes to higher campaign life due to reduction of heat loads on BF wall.
- Access to any part of the system is easier, and hence, one or two parts can be changed even during normal shutdown of the furnace.
- Wearing parts are rather few and inexpensive, and hence, these can be regularly changed during routine maintenance.
- Gives more operational safety and easy control over various charging patterns.

- Largely reduces the maintenance time and frequency of maintenance of top equipment. The chute can be replaced within a short period of time.
- The top equipment is light and simple construction compared to other high-pressure top charging system.

However, the height of the BLT top equipment is more than the two-bell-type charging system. BLT charging system can be integrated with skip car or conveyor belt; the conveyor belt system is mostly preferred. During the operation of the blast furnaces equipped with BLT charging system, the skip car or conveyor belt brings the charging materials to the receiving hopper. The material is then filled in the lock hopper bin which is then sealed and pressurized to the furnace top operating pressure. The lock hoppers are used alternately, that is, one is being filled while other is being emptied. By design, the seal valves are always out of the path of material flow to prevent material abrasion. This reduces the probability of sealing problem. To control the rate of discharge, the flow control gate opens to predetermined positions for the different types of charging materials [6]. Lock hopper bins are lined with replaceable wear plates. The lower seal valves and material flow gates are in a common gas tight housing with the material flow chute, which directs the material through a central discharge spout located in the main gear housing. The charging materials through a BLT charging system are shown in a Fig. 2.13.

2.5.2.1 Compare BLT with Two-Bell System

The top of the blast furnace is closed, as modern blast furnaces tend to operate with high top pressure. There are two different systems (both types are shown in Fig. 2.14):

- The two-bell system is often equipped with a movable throat armour. A movable throat armour is used when a bell top charging system is utilized, to gain some control over the burden distribution of the charged materials.
- The bell-less top allows easier burden distribution with the rotating chute.

Compare charging systems of bell-less top type, and two bells with armour type are shown in Fig. 2.15. With respect to charging conditions, two systems are differed as follows:

- Discharging behaviour from hopper,
- Falling trajectory in respect to charging plane,
- Number of burdens charged one after another,
- Burden charging rate.

These factors cause differences in burden profile, particle size segregation, percolation and mixed layer formation.

In a bell-less top, the possibility exists to distribute the fines in the burden over various points of impact by moving the chute to different vertical positions. Coke can be brought to the centre by programming of the charging cycle. With a two-bell charging system, there is less possibility to vary the points of impact and fines will be concentrated in narrower rings. Modern blast furnaces with a double-bell charging system are mostly equipped with movable armour, which give certain flexibility with respect to distribution of fines and the ore-to-coke ratio over the diameter, especially at the wall. However, its flexibility is inferior to the more versatile bell-less top system.

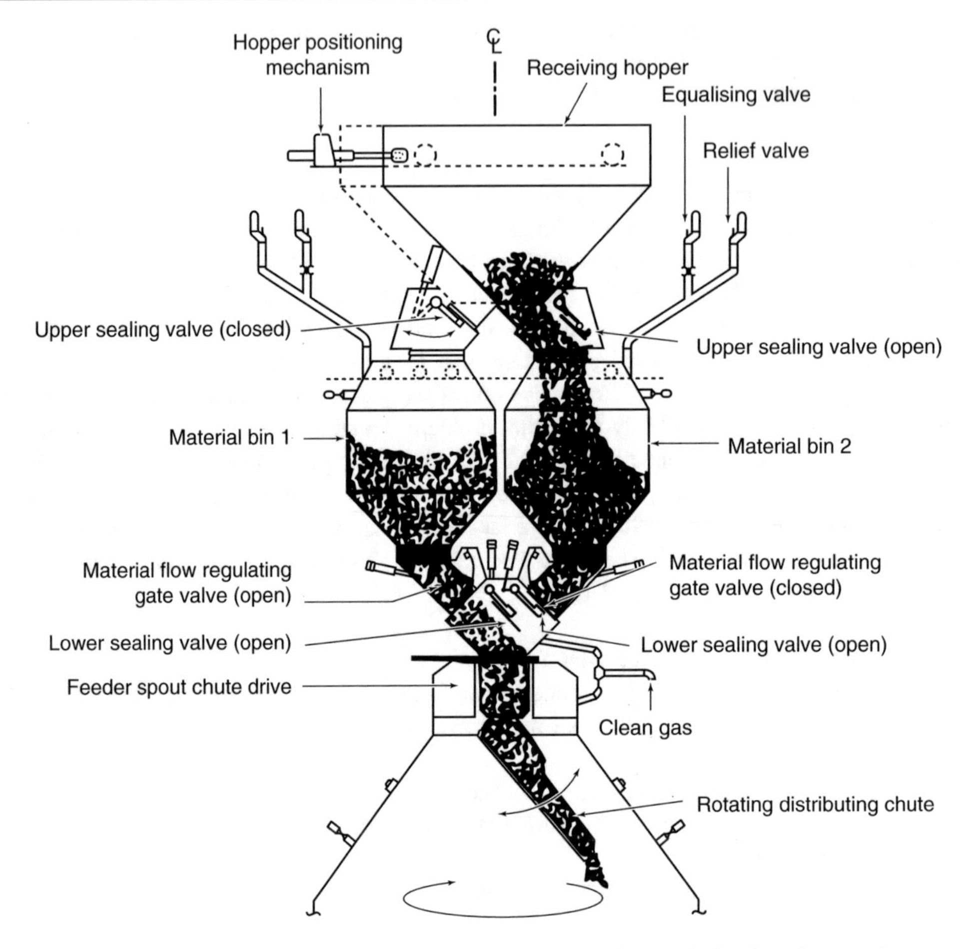

Fig. 2.13 Bell-less top charging system with materials [6] (reproduce with permission from Ispatguru)

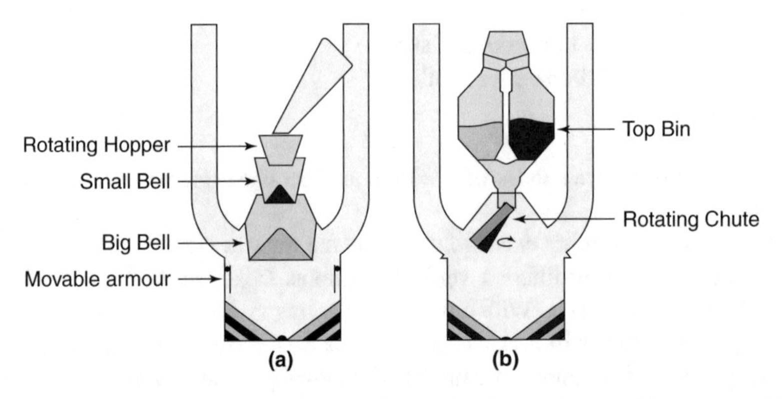

Fig. 2.14 Modern blast furnace top charging systems [5], **a** two-bell top with movable armour and **b** bell-less top (reproduce with permission from IOS Press BV, Netherlands)

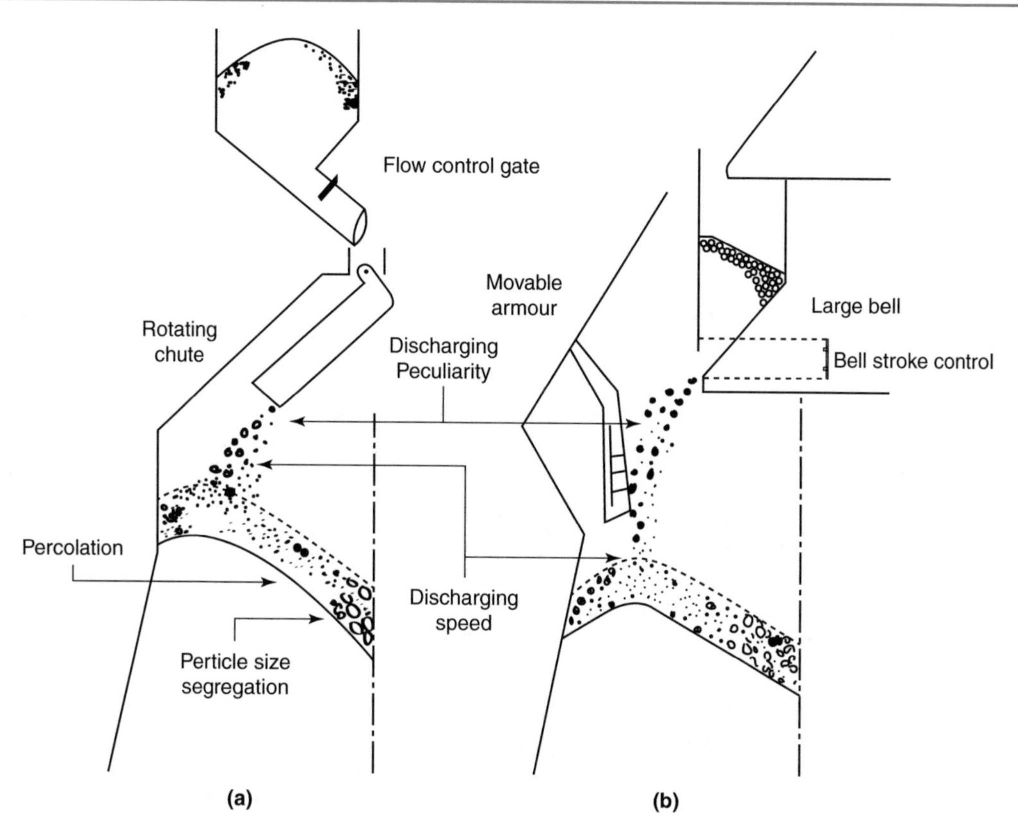

Fig. 2.15 Comparing cross-sectional charging system of **a** bell-less top type and **b** two bells with armour type

2.6 Size of Charge Particles

The packing density or bulk density increases, and correspondingly, the permeability decreases with increase in the range of sizes of particle in a bed. The permeability of the bed will be decreased with increase in the range of sizes of raw materials in general and in particular increasing the fines in the charge. The lower permeability all through the cross section results in decreased gas flow through the burden, i.e. lower driving rate at a given pressure of blowing. The size ranges of raw materials for BF are shown in Sect. 2.5.1.1.

Initially, clean ore and coke layers are taken as a starting point. However, since the average diameter of coke 40–50 mm is much larger than that of pellets and sinter (typically under 15 mm and 25 mm, respectively), burden components dumped on a coke layer will tend to form a mixed layer. This mixed layer will have permeability comparable with the ore layer.

The counter-current interaction between the reducing gas and the ore burden is considered. The reducing gas removes more oxygen from the iron ore, the more efficient is the blast furnace process. Consequently, intimate contact between the reducing gas, and the ore burden is very important. To optimize this, contacting the permeability of the ore burden must be as high as possible. The ratio of the gas flowing through the ore burden and the amount of oxygen to be removed from the burden must also be in balance.

Experience has shown that many problems in the blast furnace are the consequence of low permeability ore layers. Therefore, the permeability of the ore layers across the diameter of the furnace is a major issue. The permeability of an ore layer is largely determined by the amount of fines (under 5 mm) in the layer. Generally, most of the fines are generated from sinter if it is present in the charged burden or from lump ores. The problem with fines in the furnace is that they tend to concentrate in rings in the furnace. As fines are charged to the furnace, they concentrate at the point of impact where the burden is charged. They are also generated by low temperature reduction–disintegration. Thus, it is important to screen the burden materials well, normally with 5 or 6 mm screens in the stock house, and to control the low temperature reduction–disintegration characteristics of the burden.

Probable Questions

1. Draw schematic diagram of BF with different zones.
2. Discuss the tap holes and tuyere assembly of a BF.
3. What are the function of charged materials in BF?
4. Discuss the modern charging mechanism of BF.
5. Discuss the bell-less top charging mechanism of BF.
6. What are the size ranges of charged materials in BF?
7. What are the modification of bell-type charging system? Discuss.
8. Compare bell-less top with two-bell charging system.
9. What is high top pressure operation in a BF? Discuss the merits and demerits of high top pressure operation in a BF.
10. How high top pressure can be control by bell-less top charging system?

References

1. J.G. Peacey, W.G. Davenport, in *The Iron Blast Furnace*, 1st edn. (Pergamon Press Ltd., Oxford, England, 1979)
2. A. Ghosh, A. Chatterjee, in *Ironmaking and Steelmaking: Theory and Practice* (PHI learning Pvt Ltd., New Delhi, 2008)
3. A.K. Mishra, in *Vizag Steel Plant* (*India*). Personal communication
4. R. Sah, in *JSW Steel*. Personal communication (Vijayanagar, Bellary)
5. M. Geerdes, H. Toxopeus, C. Vliet, in *Modern Blast Furnace Ironmaking*, 2nd edn. (IOS Press BV, Amsterdam, Netherlands, 2009)
6. http://ispatguru.com/blast-furnace-top-charging-systems/

3 Blast Furnace Reactions

Blast furnace reactions play an important role in BF production. Heat and reducing gas are generated in tuyere zone. Reducing gas goes upwards, solid charge materials come down, and counter-current interaction reactions take place at the upper zones. Slag-metal reactions take place at the hearth.

3.1 Blast Furnace Reactions

Blast furnace reactions can be divided according to different sections of the furnace (Fig. 3.1):

1. Stack reactions,
2. Bosh reactions,
3. Tuyere reactions or combustion zone reactions,
4. Hearth reactions.

Combustion zone reactions are taken first due to heat as well as reducing gas generation in this zone. Reducing gas goes upwards, solid charge materials come down, and counter-current interaction reactions take place at the upper zones.

3.1.1 Tuyere Reactions or Combustion Zone Reactions

The blast furnace must produce a definite quantity of heat for each tonne of iron production. This heat comes from the combustion of the coke, from exothermic chemical reactions taking place within the furnace and from the sensible heat in the pre-heated blast. Thus, the combustion of the coke is one of the most important reactions in the BF process.

The air (contains 23 wt% oxygen and 77 wt% nitrogen) enters the furnace through tuyeres situated just above the hearth. Air is pre-heated to a temperature 800–1200 °C. The combustion of coke takes place in front of tuyeres in the presence of air. In this process, the oxygen in the blast and carbon in the coke are transformed into gaseous carbon monoxide. The resulting gas has a high flame temperature between 2100 and 2300 °C. Coke in front of the tuyeres is consumed, thus creating voidage; this is known as *tuyere raceway zone*. Figure 3.2 shows combustion taking place as a balloon-like space in front of each tuyere. There are two zones at the tuyere raceway zone: (i) zone A, the gas phase, consists primarily of O_2, CO_2 and N_2, and (ii) zone B, the gas phase, consists of CO_2, CO and N_2.

S. K. Dutta and Y. B. Chokshi, *Basic Concepts of Iron and Steel Making*,
https://doi.org/10.1007/978-981-15-2437-0_3

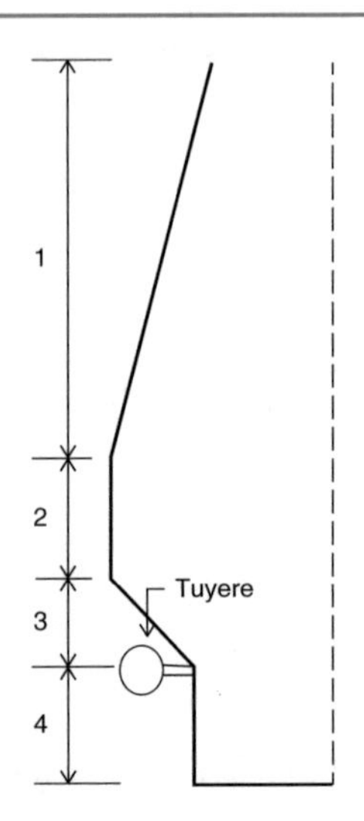

Fig. 3.1 Different sections of BF (1: stack, 2: bosh, 3: tuyere and 4: hearth)

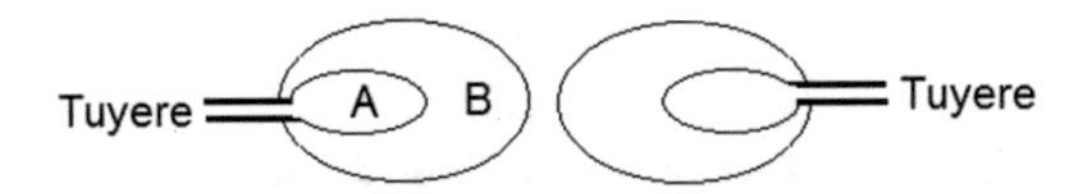

Fig. 3.2 Tuyere raceway zone

In zone A, carbon in coke reacts with oxygen in air to form CO_2:

$$C(s) + \{O_2\} = \{CO_2\}, \quad \Delta H^0{}_{298} = -394.13\,\text{kJ/mol of C} \tag{3.1}$$

This reaction (3.1) is strongly exothermic and releases a large amount of heat to the process, and temperature increases above 1650 °C. As the gases enter the zone B, the oxygen is depleted, and CO_2 reacts with carbon because (i) there are present large amount of carbon particles and (ii) CO_2 is not stable at above 1000 °C. For this reason, if any CO_2 forms it reacts immediately with carbon to form CO gas.

$$C(s) + \{CO_2\} = 2\{CO\}, \quad \Delta H^0{}_{298} = 170.7\,\text{kJ/mol of C} \tag{3.2}$$

This reaction (3.2) is endothermic and has great importance for producing a lot of reducing gas, although this reaction (3.2) absorbs a large proportion of the heat generated by the reaction (3.1). The reaction (3.2) is called the *solution loss reaction* because carbon, which would otherwise be burned at the tuyeres to produce heat, is lost by reacting with CO_2 gas. This means increase in coke consumption to overcome the loss of carbon as well as loss of large amount of heat. This reaction is also known as *Boudouard reaction or gasification reaction.*

The net result of the combustion process is the sum of the reactions (3.1) and (3.2):

$$2C(s) + \{O_2\} = 2\{CO\}, \quad \Delta H^0{}_{298} = -111.715\,\text{kJ/mol of C} \tag{3.3}$$

The reaction (3.3) is exothermic and may be the overall combustion reaction taking place in the blast furnace.

Any moisture present in the blast will also react with carbon in coke:

$$C(s) + \{H_2O\} = \{CO\} + \{H_2\}, \quad \Delta H^0{}_{298} = 135.14\,\text{kJ/mol of C} \tag{3.4}$$

The hydrogen, which is formed by the reaction (3.4), and any unreacted water vapour join the gas stream and pass upwards through the furnace. The reaction (3.4) is endothermic, and it may be thought that the presence of water vapour in the blast would be detrimental to the process.

According to the reaction (3.3), one volume of oxygen produces two volumes of CO gas; oxygen is supplied by air containing 21% O_2 and 79% N_2 (by volume). Hence, the ascending gases at the tuyeres level should contain 34.7% CO and 65.3% N_2 (by volume). The hydrogen, formed by decomposition of the water vapour in the blast, must be added to the above figure; the hydrogen is generally approximating about 1%. Therefore, the theoretical analysis of the gases at the combustion zone should be 35% CO, 1% H_2 and 64% N_2, which is very near the practical figure.

The burden descends in the blast furnace from top to bottom. For the burden to descend, voidage must be created in the following location of the furnace:

1. Coke is gasified in front of the tuyeres, thus creating voidage at the tuyere region.
2. The hot gas ascends the furnace and melts the burden material. So, the volume of burden material is drastically decreased at the melting zone.
3. The dripping hot metal consumes carbon. It is used for carburization of the iron as well as for the direct reduction reactions, so below the melting zone coke is consumed.

When the very hot gas ascends through the furnace, it carries out a few important functions as follows:

- Heats up the coke in the bosh/belly region,
- Melts the iron ore in the burden, creating voidage,
- Heats up the material in the shaft zone of the furnace,
- Removes oxygen of the ore burden by chemical reactions,
- Upon melting, the iron ore produces hot metal and slag, which drips down through the coke slits to the hearth.

In the dripping zone, the hot metal and slag consume coke, creating voidage. Additional coke is consumed for final reduction of iron oxide, and carbon dissolves in the hot metal, which is called carburization.

In the blast furnace process, iron ore and reducing agents (i.e. coke) are transformed into hot metal and slag, which is formed from the gangue of the ore and the ash of coke. Hot metal and liquid slag do not mix due to their density difference and remain separate from each other with the slag floating on top of the heavier liquid iron.

By the time charge comes down into this zone, except coke, the entire charge is in molten condition. Highest temperature (1900–2000 °C) of the furnace is at this tuyere region.

3.1.2 Reactions in the Stack

In the stack region, ores are heated and started reduction due to direct contact with ascending hot gases (which contain 55–60% N_2, CO, CO_2, H_2, H_2O and traces of hydrocarbon like CH_4 etc.). These ascending gases react with solid charge on the stack region.

(i) Gas–solid reduction reactions:

The velocity of the descending raw materials is much slower than that of the ascending hot gases. As these solids descend, they are gradually pre-heated. At the top of the furnace, where the temperature may vary from 200 to 300 °C, the moisture and any combined water are removed from raw materials.

The reduction of the iron ore by CO gas commences at about 200 °C, according to the nature of the ore, is accelerated as the temperature increases and removes oxygen from the iron ore by chemical reactions.

The reduction of Fe_2O_3 took place in stages (above 570 °C), viz:

$$Fe_2O_3 \rightarrow Fe_3O_4 \rightarrow FeO \rightarrow Fe.$$

$$3Fe_2O_3(s) + \{CO\} = 2Fe_3O_4(s) + \{CO_2\}, \quad \Delta H^0{}_{298} = -52.43\,\text{kJ/mol of CO} \tag{3.5}$$

$$xFe_3O_4(s) + x\{CO\} \leftrightarrows 3Fe_xO(s) + x\{CO_2\}, \quad \Delta H^0{}_{298} = 40.46\,\text{kJ/mol of CO} \tag{3.6}$$

$$Fe_xO(s) + \{CO\} \leftrightarrows xFe(s) + \{CO_2\}, \quad \Delta H^0{}_{298} = -18.54\,\text{kJ/mol of CO} \tag{3.7}$$

where $0.92 < x < 0.98$. The reactions (3.6 and 3.7) are reversible and a function of temperature.

The reduction of Fe_2O_3 took place in stages (below 570 °C), viz:

$$Fe_2O_3 \rightarrow Fe_3O_4 \rightarrow Fe.$$

At temperature below 570 °C, then instead of reactions (3.6 and 3.7) the following reaction (3.8) takes place:

$$Fe_3O_4(s) + 4\{CO\} = 3Fe(s) + 4\{CO_2\}, \quad \Delta H^0{}_{298} = -3.79\,\text{kJ/mol of CO} \tag{3.8}$$

Although the reduction of iron ore takes place in three stages, it is often convenient for the reduction process and for calculations on the thermal and chemical balance of the blast furnace to consider only the overall reduction of ferric oxide to metallic iron. This is accomplished by combining reactions (3.5)–(3.7) to give reaction (3.9).

$$Fe_2O_3(s) + 3\{CO\} = 2Fe(s) + 3\{CO_2\}, \quad \Delta H^0{}_{298} = -9.19\,\text{kJ/mol of CO} \tag{3.9}$$

(ii) Dissociation reactions:

As the descending raw materials are going on, the temperature of materials is increased. At about 700–800 °C, the following reactions take place:

$$CaCO_3(s) = CaO(s) + \{CO_2\}, \quad \Delta H^0{}_{298} = 161.3\,\text{kJ/mol} \tag{3.10}$$

This reaction (3.10), which is endothermic, is accelerated by increasing temperature. At the lower part of the stack, the temperature is about 900 °C, and reaction (3.10) is practically completed at 900 °C. Calcium carbonate should completely decompose at this level. But this does not happen because of kinetic limitations, and significant decomposition occurs only at 1000–1100 °C. Decomposition of limestone is endothermic and requires heat to be supplied. During the progress of decomposition, a layer of porous CaO forms on the outer layer of limestone. This layer has very poor thermal conductivity and, consequently, slows down heat transfer into the interior of the limestone lumps. This factor also affects the rate of decomposition.

$$(\mathrm{Ca,Mg})\mathrm{CO_3(s)} = (\mathrm{Ca,Mg})\mathrm{O(s)} + \{\mathrm{CO_2}\} \quad (3.11)$$

$$\mathrm{FeCO_3(s)} = \mathrm{FeO(s)} + \{\mathrm{CO_2}\} \quad (3.12)$$

$$\mathrm{Fe(OH)_2(s)} = \mathrm{FeO(s)} + \{\mathrm{H_2O}\} \quad (3.13)$$

(iii) Carbon deposition reaction:

Since below 1000 °C, CO_2 gas is more stable than CO gas, at low temperature (200–300 °C), CO gas is transformed into CO_2 gas and carbon. At the upper stack region (200–300 °C), the dissociation of CO takes place to form CO_2 and carbon (shoot formation) in the presence of Fe and FeO.

$$2\{\mathrm{CO}\} \rightarrow \{\mathrm{CO_2}\} + \mathrm{C(s)} \quad (3.14)$$

The temperature, at the bottom of stack, is 1100–1200 °C. Charge begins to soften and fuse at the bottom of stack region.

3.1.3 Bosh Reactions

In the bosh, the reduction of iron ore is completed, and the oxides and metal are partially melted down. At the upper bosh region, ore consists of partially reduced oxides with variable degree of reduction. But all these particles have a common single feature; that is, outer layer of most of the particles contains metallic iron of a different thickness.

Since the direct reduction of iron oxide is highly endothermic, it requires high temperature.

$$\mathrm{FeO(s)} + \mathrm{C(s)} = \mathrm{Fe(s)} + \{\mathrm{CO}\}, \quad \Delta H^0{}_{298} = 153.89\,\mathrm{kJ/mol\ of\ C} \quad (3.15)$$

Besides iron oxide, other oxides entered in the BF bosh region, such as (i) P_2O_5, MnO and SiO_2 and (ii) Al_2O_3, MgO and CaO.

(i) These oxides are easily reducible by BF gas (CO) at 1300–1400 °C (as shown in Fig. 3.3) and also by direct carbon of the coke at the higher temperature.

$$\mathrm{MnO} + \{\mathrm{CO}\} = \mathrm{Mn} + \{\mathrm{CO_2}\}, \quad \Delta H^0{}_{298} = 102.31\,\mathrm{kJ/mol\ of\ CO} \quad (3.16)$$

$$\mathrm{SiO_2} + 2\{\mathrm{CO}\} = \mathrm{Si} + 2\{\mathrm{CO_2}\}, \quad \Delta H^0{}_{298} = 153.35\,\mathrm{kJ/mol\ of\ CO} \quad (3.17)$$

$$\mathrm{P_2O_5} + 5\{\mathrm{CO}\} = 2\mathrm{P} + 5\{\mathrm{CO_2}\}, \quad \Delta H^0{}_{298} = 34.73\,\mathrm{kJ/mol\ of\ CO} \quad (3.18)$$

$$\mathrm{MnO} + \mathrm{C(s)} = \mathrm{Mn} + \{\mathrm{CO}\}, \quad \Delta H^0{}_{298} = 273.52\,\mathrm{kJ/mol\ of\ C} \quad (3.19)$$

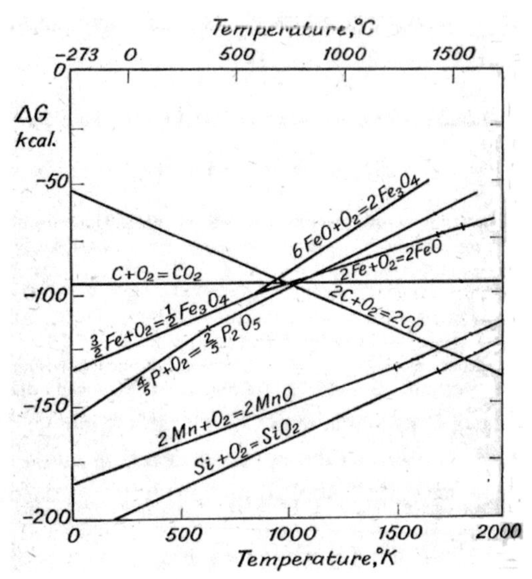

Fig. 3.3 Free energy of formation of some oxides

$$SiO_2 + 2C(s) = Si + 2\{CO\}, \quad \Delta H^0{}_{298} = 324.05\,\text{kJ/mol of C} \tag{3.20}$$

$$P_2O_5 + 5C(s) = 2P + 5\{CO\}, \quad \Delta H^0{}_{298} = 205.43\,\text{kJ/mol of C} \tag{3.21}$$

Since all above reactions are endothermic, these reactions are favoured by a high temperature and proceed in the hotter zones of the furnace. At bosh region, melting of charge, except coke, takes place; gangue and calcined flux combine to form basic slag.

The iron and phosphorous lines, on the Ellingham diagram (Fig. 3.3), are so close to each other that the entire phosphorus in the charge materials gets reduced along with iron in the blast furnace. In the blast furnace, the conditions are made sufficiently reducing to recover 99.5% of the iron charged, and almost complete reduction of phosphorus pentoxide takes placed.

Above 1200 °C, phosphorus is in vapour state and outer layer of iron absorbs this P as well as Mn and Si. P_2O_5 and MnO are nearly completely reduced, and SiO_2 is only partly reduced. Excess carbon also dissolves in it causing lowering the melting point of iron over a considerable range. At the upper bosh region, where temperature nearly 1200 °C, dissolution of the elements take place in iron (which is in the semi-molten condition).

(ii) These oxides (Al_2O_3, MgO and CaO) are not easily reducible. So at 1300–1400 °C, Al_2O_3, MgO, CaO, SiO_2 (partly) and MnO (very small amount) are the remaining oxides present. SiO_2, MnO and FeO are dissolved in each other because these are producing a highly acidic slag which is very rich in silica at the upper bosh region, so the viscosity of the slag is also very high. Gradually when the slag travels to the lower bosh region, it dissolves the higher melting oxides like CaO, MgO and Al_2O_3 making it more and more basic and more fluid. The dissolution of CaO in the slag is a slow process and strongly temperature dependent and is never complete until and unless the slag has been heated to a high temperature, which is only possible at the lower region at the BF. Upon smelting, the iron ore produces hot metal and slag, which come down through the coke slits to the hearth region.

3.1.4 Hearth Reactions

Molten metal and slag are collected in the hearth. Since the specific gravities of slag and molten metal are 2.5 and 7.0, respectively, slag is floated on the top of the molten metal layer. Metal contains 3.5–4.0% C, 0.5–2.5% Si, 0.4–2.0% Mn, 0.05–2.5% P and 0.04–0.1% S. Slag contains CaO, MgO, Al_2O_3, SiO_2 and MnO and traces of P_2O_5 and FeO. When a droplet of metal comes into metal layer through slag layer, it comes in contact with slag, and then slag-metal reactions take place.

(i) Carbon dissolution:

Although most of the carbon in coke burns at the tuyere region, a certain fraction is descent into hearth and deposited in *dead man zone*. Since the metallic iron in the BF hearth remains in contact with carbon in dead man zone for a sufficiently long time before the particular metallic elements under consideration are tapped out of the BF, it is very natural to expect carbon dissolves in metallic iron in equilibrium proportion as detected by the phase diagram of Fe–C system. BF hot metal is always in equilibrium with carbon. Hot metal contains 3.5–4.0% C; this is less than maximum carbon (6.67%) present in Fe–C phase diagram; why?

There is no shortage of carbon in BF, but due to the presence of other elements (e.g. Si, P, S, etc.) in hot metal, solubility of carbon in iron is decreased, and the activity of C is increased by Si, P, S which form stronger bonds with iron than with carbon (as shown in Fig. 3.4). Carbon content in hot metal depends upon, in general, Si, P, S, etc., and in particular Si content (0.2–1.2%) in hot metal. That is why Indian hot metal contains only 3.0–3.5% C due to the presence of 0.75–1.25% Si. Entire charge is molten in hearth, and separation of slag and metal takes place.

(ii) Mn reaction:

$$(\text{MnO}) + [\text{C}] = [\text{Mn}] + \{\text{CO}\} \tag{3.19}$$

As the basicity of slag increases, there is sharp increase in activity of MnO, so more Mn goes to the metal phase; i.e. forward reaction proceeds. Normally, basicity of BF is 1.1–1.2. It is possible to obtain the value of 0.85 for the ratio of [Mn]/(MnO) at a basicity of 1.2 and temperature of 1500 °C,

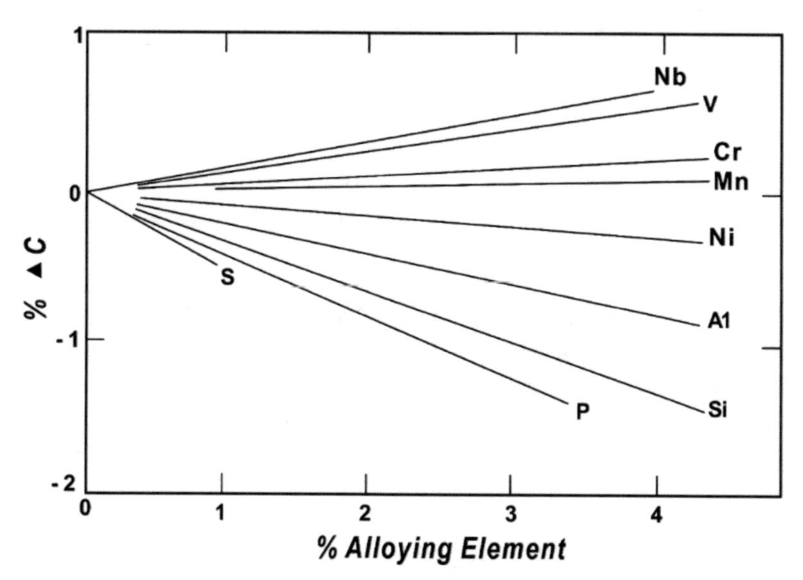

Fig. 3.4 Effect of alloying elements on the solubility of carbon in liquid iron [1]

i.e. high per cent (85%) of recovery of Mn in metal. But in actual BF, at 1500 °C and basicity of 1.2, the ratio is nearly exceeding 60% and the immediate conclusion is that equilibrium has not attained, so far Mn is concerned.

(iii) Si Reaction:

$$(SiO_2) + [C] = [Si] + \{CO\} \tag{3.20}$$

(iv) S Reaction:

$$[FeS] \leftrightarrows (FeS) \tag{3.22}$$

$$(FeS) + (CaO) \leftrightarrows (FeO) + (CaS) \tag{3.23}$$

$$(FeO) + [C] = [Fe] + \{CO\} \tag{3.24}$$

$$[FeS] + (CaO) + [C] = [Fe\} + (CaS) + \{CO\}, \quad \Delta H^0{}_{298} = 136.51\,\text{kJ/mol} \tag{3.25}$$

Conditions for de-sulphurization:

(a) High temperature: Since reaction (3.25) is endothermic in nature, high temperature is required.
(b) Basic slag: Since sulphur in metal is removed as (CaS), high basic slag is required.
(c) Reducing atmosphere: Reducing atmosphere is required for reduction of (FeO) to maintain the forward direction of the reaction (3.24); otherwise, concentration of FeO increased in slag, so backward reaction (3.23) would be occurred; i.e. de-sulphurization would not take place. This reduction reaction (3.24) is done by dissolved carbon or carbon in solution.
(d) Good fluidity of slag: The rate of diffusion of the reactants and products depends upon the viscosity of the slag, and again viscosity of slag depends upon its composition and temperature. Hence, increase in diffusion rate (i.e. rate of de-sulphurizaion) can be achieved by: (i) increase in temperature and (ii) decrease in viscosity of slag, i.e. increase in fluidity of slag.
(e) High Si content in metal: By increasing temperature, Si content in metal is also increased; this [Si] reduced the (FeO) and helped de-sulphurization; i.e. by increasing Si content in metal, S in metal is decreased by decreasing backward reaction (3.23):

$$\{2(FeO) + [Si] = 2[Fe] + (SiO_2)\}.$$

3.2 Slag

The combined molten oxides, which are the product of smelting and refining processes, are known as *slag*. In other words, oxide melt containing at least two oxides of opposite chemical nature should be called *slag*. The temperature and composition of the slag determine physical properties of slag like viscosity, thermal conductivity and surface tension. These properties of the slag affect the intensity of erosion of the refractory lining of the furnace and dissolution of the lining materials in the slag.

The slag-forming components may be divided into three categories according to the properties of materials:

1. Acid oxides: SiO_2, P_2O_5, TiO_2, V_2O_5, B_2O_3, etc.,
2. Basic oxides: CaO, MgO, FeO, MnO, Na_2O, K_2O, etc.,
3. Amphoteric oxides: Al_2O_3, Fe_2O_3, Cr_2O_3, V_2O_3, etc.

The ability of a slag to retain oxides is generally expressed as the ratio of basic oxides to acid oxides and is variously represented as:

$$V\text{ratio} = \left(\frac{\%\text{CaO}}{\%\text{SiO}_2}\right) \tag{3.26}$$

$$\text{Modified}V\text{ratio} = \left(\frac{\%\text{CaO}}{\{\sum\%(\text{SiO}_2 + \text{Al}_2\text{O}_2 + \text{P}_2\text{O}_5)\}}\right) \tag{3.27}$$

$$\text{Basicity}, B = \left(\frac{\{\sum\%(\text{all basic oxides})\}}{\{\sum\%(\text{all acid oxides})\}}\right) \tag{3.28}$$

$$\text{Generally}, B = \left(\frac{\{\sum\%(\text{CaO} + \text{MgO})\}}{\{\sum\%(\text{SiO}_2 + \text{Al}_2\text{O}_2 + \text{P}_2\text{O}_5)\}}\right) \tag{3.29}$$

Acid slag: V or $B < 1$; basic oxide, CaO, may or may not be present at all.
Basic slag: V or $B > 1$.

3.2.1 BF Slag

In blast furnace, metallic iron starts absorbing carbon from the coke in the lower part of stack and it becomes liquid at a low temperature (1300 °C). In the hearth, molten iron contains about 4% C with a liquidus temperature lower than 1200 °C. Major constituents of BF slag are CaO, Al_2O_3, SiO_2 and MgO. Minor constituents of BF slag are MnO, TiO_2, FeO and alkali oxides.

Al_2O_3 and SiO_2 in slag come from gangue of iron ore and coke ash; CaO and MgO come from fluxes (e.g. limestone and dolomite). Liquid slag is much more viscous than liquid iron. Viscosity varies on the composition and temperature: overall variation from 2.5 to 2000 poises (i.e. 0.25–200 kg m^{-1} s^{-1}). Liquid slag should have as low viscosity as possible, lower than 2 poises (0.2 kg m^{-1} s^{-1}).

There are difficulties in viscous slag as follows:

1. Pose difficulties during slag-metal separation in hearth.
2. Slow down the rates of slag-metal reactions.
3. Do not flow down properly from the bosh to the hearth.
4. Hinder smooth upward flow of gas through the burden.

Slag basicity is another important parameter. Transfer of S and Si from metal to slag is easy if the slag is basic. Highly basic slag has higher viscosity as well. These two contradictory requirements are met in the BF slag by maintaining a *V* ratio (wt% CaO/wt% SiO_2) of the hearth slag between 1.0 and 1.25. BF slag contains 4–10% MgO; MgO lowers down both liquidus temperature and viscosity of the slag. Since Indian iron ore contains high Al_2O_3/SiO_2 ratio, so high Al_2O_3 goes to the BF slag. Hence a large amount of MgO is required to add to the slag for lower down the viscosity. The slag contains about 30–32% CaO, 30% SiO_2, 25–30% Al_2O_3 and 7–10% MgO.

At any slag composition, its viscosity decreases with increase in its temperature. To obtain a fluid, free running slag, a minimum hearth temperature is required; this is known as *critical hearth temperature*. It should be as low as possible to cut down the heat requirement as well as to increase the hearth lining life. CaO–SiO_2–Al_2O_3–MgO slag is enough fluid at 1400–1450 °C.

Sinter is charged as burden material to the BF and contains large amount of iron oxide. Sinter may be acid, self-fluxed or super-fluxed having CaO/SiO_2 ratio less than one, equal to one and above two, respectively. First slag in BF forms in the bosh region at around 1200–1300 °C, and may or may not contain any CaO. Slag at the bosh region is known as *bosh slag*. Melting in such a low temperature is only possible because of the formation of FeO–SiO_2 compound in large amounts; it has low liquidus temperature; this is known as *primary slag*. As the primary slag trickles down through the bed of solids, its temperature rises; FeO gets reduced, and slag dissolves more CaO. At tuyere, the coke burns thereby releasing coke ash containing SiO_2 and Al_2O_3; this makes the slag at the tuyere level high in SiO_2 (i.e. acid slag). Final composition adjustments occur during the passage of bosh slag into the hearth, during which FeO content becomes very low.

3.3 Modern Concept of BF Process

The BF is essentially a counter-current reactor for heat and mass exchanges between the gaseous and solid phases. Heat is transferred from the gas phase to the charging materials, and oxygen is transferred from the iron ore to the gas phase. The purpose of BF operation is to heat up and to reduce iron ore to produce hot metal at minimum cost. Hot reducing gases move up (i.e. ascends), and the solid charge moves down (i.e. descend) through the furnace. Due to this intimate gas–solid charge contact, heat transfer from gas to solid charge is accompanied by oxygen and carbon transfer from solid charge to gas. Besides, the distribution of gas is determined by the permeability of charge in the furnace. The counter-current nature of the movements makes the overall process extremely efficient.

The BF has three main functions:

1. *Chemical reactor* for the reduction of iron ore,
2. *Melter* to melt the metal and slag,
3. *Combustion chamber* to produce large amount of heat and reducing gases.

Apart from these, BF also serves three active works:

(i) Reduction of iron oxides,
(ii) Heat transfer from hot gas to solids, and
(iii) Proper gas utilization.

Compared with a shaft furnace such as Midrex, the stack portion of a blast furnace is much less productive. This is because the blast furnace has the following disadvantages:

- More than half of the gas is nitrogen.
- Only about 6% of the gas is hydrogen.
- About half of the space of the stack is filled with coke.

Professor Lu [2] proposed that a BF may be divided into five zones as per the movement of materials from solid to liquid phase (as shown in Fig. 3.5).

1. Lumpy or granular zone,
2. Softening and melting zone,
3. Dripping (or dropping) zone,
4. Raceways zone, and
5. Hearth zone.

3.3.1 Lumpy or Granular Zone

(a) Below 600 °C → pre-heating and pre-reduction take place.
(b) 600 < T < 950 °C →

(i) Indirect reduction of iron oxide by CO and H_2:

$$3Fe_2O_3(s) + \{CO\} = 2Fe_3O_4(s) + \{CO_2\} \tag{3.5}$$

$$xFe_3O_4(s) + x\{CO\} = 3Fe_xO(s) + x\{CO_2\} \tag{3.6}$$

$$Fe_xO(s) + \{CO\} \leftrightarrows xFe(s) + \{CO_2\} \tag{3.7}$$

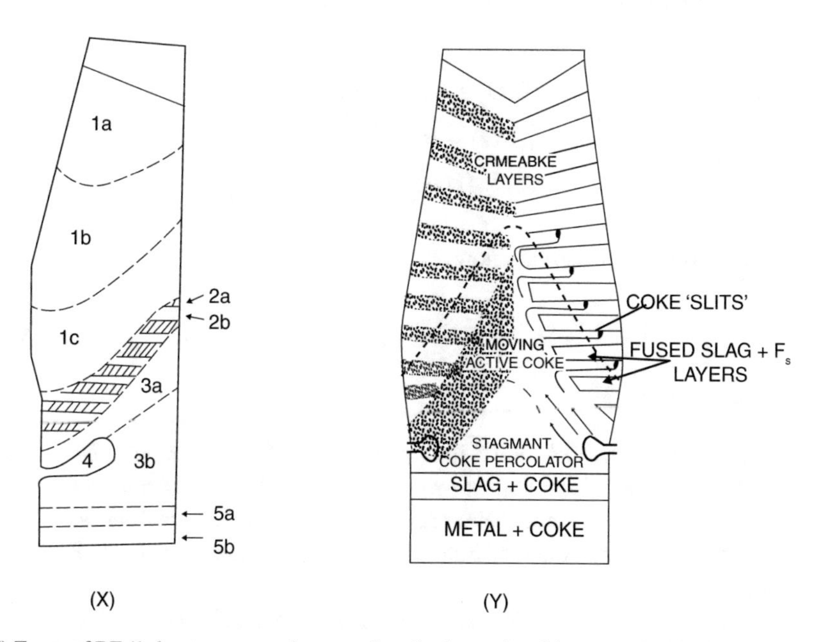

Fig. 3.5 (X) Zones of BF (1: lumpy or granular zone, 2: softening and melting zone, 3: dripping (or dropping) zone, 4: raceway zone and 5: hearth zone); (Y) vertical cross section of zones within BF

$$3Fe_2O_3(s) + \{H_2\} = 2Fe_3O_4(s) + \{H_2O\} \tag{3.30}$$

$$xFe_3O_4(s) + x\{H_2\} = 3Fe_xO(s) + x\{H_2O\} \tag{3.31}$$

$$Fe_xO(s) + \{H_2\} = xFe(s) + \{H_2O\} \tag{3.32}$$

(ii) Calcination of limestone:

$$CaCO_3(s) = CaO(s) + \{CO_2\} \tag{3.10}$$

(c) $950 < T < T_{Softening} \rightarrow$

(i) Direct reduction:

$$Fe_xO(s) + C(s) = xFe(s) + \{CO\} \tag{3.15}$$

(ii) Gasification of carbon (solution loss reactions) by CO_2 and H_2O:

$$C(s) + \{CO_2\} = 2\{CO\} \tag{3.2}$$

$$C(s) + \{H_2O\} = \{H_2\} + \{CO\} \tag{3.4}$$

3.3.2 Softening and Melting Zone

(a) Area where ore starts to soften and melt is known as the softening and melting zone. The formation of cohesive layers of partially reduced and partially molten iron ore takes place. Metal oxides, in primary slag, are being reduced, and the nearby carbon is gasified by CO_2 and H_2O gases.

$$(FeO) + C(s) = [Fe] + \{CO\} \tag{3.33}$$

$$C(s) + \{CO_2\} = 2\{CO\} \tag{3.2}$$

$$C(s) + \{H_2O\} = \{H_2\} + \{CO\} \tag{3.4}$$

(b) Coke slits are a passage for gaseous flow.

3.3.3 Dripping (or Dropping) Zone

(a) Area where there is present coke, liquid iron and slag is called the *active coke* or *dripping zone*. Semi-fluidized region into which liquids (iron and slag), coke and fragments of cohesive layers drip. In the dripping zone, the hot metal and slag consume coke, creating voidage. Additional coke is consumed for final reduction of iron oxide, and carbon dissolves in the hot metal, which is called carburization.
(b) The *dead man zone*, which is a stable pile of solid coke particles in the hearth of the furnace. The coke in dead man zone does not take part in any action or reaction; that is why this zone is known

as dead man zone. Dead man zone through which liquids drop comes down to the hearth. Reduction of metalloids and de-sulphurization is taking place. It is the final stage of iron oxide reduction.

3.3.4 Raceway Zone

Blast, injectants and coke are converted to hot reducing gases and delivered to the central part of the furnace. The coke is gasified in front of tuyeres and creates void.

$$C(s) + \{O_2\} = \{CO_2\}, \quad \Delta H^0{}_{298} = -394.13\,\text{kJ/mol of C} \tag{3.1}$$

$$C(s) + 1/2\{O_2\} = \{CO\}, \quad \Delta H^0{}_{298} = -111.72\,\text{kJ/mol of C} \tag{3.3}$$

3.3.5 Hearth Zone

(a) The container for liquids and coke where slag/metal/coke/gas reactions take place. Metal droplets pass through slag/coke layer.
(b) Liquid metal/coke layer in which little chemical reaction takes place.

3.4 Direct and Indirect Reduction

The influence of direct and indirect reductions on coke rate is well known. There is an optimum degree of direct reduction which leads to minimum coke rate and maximum degree of utilization of reducing gases.

In actual practice, the degree of direct reduction is more than the optimum value, any decrease of the degree of direct reduction leads to an improvement of coke rate. At temperature above 950 °C, the reduction of wustite by CO is accompanied by the carbon gasification reaction.

$$FeO(s) + \{CO\} = Fe(s) + \{CO_2\}, \quad \Delta H^0{}_{298} = -18.54\,\text{kJ/mol of CO} \tag{3.7}$$

$$C(s) + \{CO_2\} = 2\{CO\}, \quad \Delta H^0{}_{298} = 85.35\,\text{kJ/mol of CO} \tag{3.2}$$

The net effect of these two reactions (3.7 and 3.2) is equivalent to the direct reduction:

$$FeO(s) + C(s) = Fe(s) + \{CO\}, \quad \Delta H^0{}_{298} = 151.25\,\text{kJ/mol of C} \tag{3.15}$$

Only when solid wustite encounters with carbon, direct reduction could take place according to reaction (3.15).

When direct reduction takes place through gaseous intermediates, i.e. by reactions (3.7 and 3.2), the rate of direct reduction is same as the solution loss reaction (3.2).

Above 1100 °C, any CO_2 gas formed by reaction (3.7) is immediately converted to CO by reaction (3.2). Theoretical estimations show that in BF, rate of CO_2 formation by reaction (3.7) is more than the rate of its consumption by reaction (3.2) up to about 1100 °C. Thereby, an increase in the rate of wustite reduction or decrease in the rate of solution loss reaction will lead to a decrease in the degree of direct reduction. Since at a temperature below 1100 °C, the rate of reaction (3.2) is directly proportional to the volume of coke, any process modification which leads to lower coke rate also leads to lower degree of direct reduction and higher degree of utilization reducing gases.

3.5 Tuyere Flame Temperature (TFT)

Free running slag is required for efficient BF operation, and for this, a minimum hearth temperature must be maintained. This minimum temperature is known as the critical hearth temperature and is about 1400–1450 °C.

The main source of heat is the raceway zone, i.e. the combustion of carbon in front of the tuyeres. To ensure rapid heat transfer, the combustion zone temperature must be maintained at a much higher temperature normally in the range of 1800–2000 °C. This temperature is known as the *tuyere flame temperature* (TFT). While the requirement of free running fluid slag sets the lower limit of TFT, the upper limit is dictated by other requirements such as the following:

1. Extremely high TFT would result in the formation of FeO-rich slag at a somewhat upper level of the furnace, before the entire FeO gets reduced. As the charge descends, the FeO in the slag gets reduced, and the slag becomes richer in CaO; thus, it becomes more viscous.
2. SiO_2 is reduced, and SiO gas is formed in the raceway. In steelmaking, lower Si in hot metal is desirable, i.e. TFT be kept low.
3. Higher TFT encourages the vaporization of alkalis and thereby increases the problem of alkali vapour recirculation.

Therefore, TFT should be maintained within a range which is decided by the experience gathered by operating a specific furnace.

3.6 Raceway Adiabatic Flame Temperature (RAFT)

The heat content (i.e. sensible heat) of the flame and its temperature can be calculated. This is known as *raceway adiabatic flame temperature* (RAFT). The RAFT is the theoretically estimated temperature in raceway zone in front of the tuyeres if there is no heat loss. Flame temperature is normally in the range of 2000–2300 °C and is influenced by the raceway conditions. The RAFT is controlled by blast parameters like:

(1) Temperature,
(2) Oxygen content,
(3) Humidity content, and
(4) Fuel injectant rate.

First two parameters increase the value of RAFT, and last two parameters decrease the value of RAFT. The flame temperature in the raceway is the temperature that the raceway gas reaches as soon as all carbon, oxygen and water have been converted to CO and H_2. The flame temperature is a theoretical concept, since not all reactions are completed in the raceway. From a theoretical point of view, it should be calculated from a heat balance calculation over the raceway. For practical purposes, linear formulas have been derived as follows (as per AIST):

$$\begin{aligned}\text{RAFT} = {} & 1489 + 0.82 \times \text{BT} - 5.705 \times \text{BM} + 52.778 \times (\text{OE}) \\ & - 18.1 \times \left(\frac{\text{Coal}}{\text{WC}}\right) \times 100 - 43.01 \times \left(\frac{\text{Oil}}{\text{WC}}\right) \times 100 \\ & - 27.9 \times \left(\frac{\text{Tar}}{\text{WC}}\right) \times 100 - 50.66 \times \left(\frac{\text{NG}}{\text{WC}}\right) \times 100 \end{aligned} \tag{3.34}$$

where

BT blast temperature in °C,
BM blast moisture in g/m^3 STP dry blast,
OE oxygen enrichment (% O_2),
Oil dry oil injection rate in kg/thm,
Tar dry tar injection rate in kg/thm,
Coal dry coal injection rate in kg/thm,
NG natural gas injection rate in kg/thm,
WC wind consumption in m^3/thm.

Table 3.1 gives some basic rules with respect to flame temperature effects.

The top gas temperature is governed by the amount of gas needed in the process; the less gas is used, the lower the top gas temperature and vice versa. Less gas per tonne hot metal results in less gas for heating and drying the burden.

$$\begin{aligned}\text{Heat content of flame} = {} & \text{Mass of gas in the flame} \\ & - \text{specific heat of gas} \times (\text{RAFT} - 298).\end{aligned} \tag{3.35}$$

The calculation of RAFT is a very useful tool for furnace control. It is known that all inputs through the tuyeres influence the RAFT. The moisture of the air blast causes reaction with carbon:

$$C(s) + \{H_2O\} = \{H_2\} + \{CO\}, \quad \Delta H^0{}_{298} = 135.14\,\text{kJ/mol of C} \tag{3.4}$$

which is endothermic and thus lower RAFT. On the other hand, oxygen enrichment of the air blast increases RAFT, since the volume of gas in the flame is less because of less nitrogen. RAFT should be maintained in between 1900 and 2100 °C. Table 3.2 shows the effect of changes in furnace conditions on RAFT.

Table 3.1 Effects of flame temperature

	Unit	Change	RAFT (°C)	Top gas temperature (°C)
Blast temperature	°C	+100	+65	−15
Coal	kg/t	+10	−30	+9
Oxygen	%	+1	+45	−15
Moisture	g/m^3 (STP)	+10	−50	+9

Table 3.2 Effects of changes in furnace conditions on RAFT [3]

Change in operating variable	Change of RAFT (°C)
1. Blast temperature raised by 100 °C	+82
2. Blast oxygen raised by 1%	+53
3. Blast moisture raised by 5 g/N m^3	−28

In modern blast furnace practice, simple blast alone is not blown through the tuyeres but blast is accompanied by other injectants like steam, oxygen, auxiliary fuels, fluxes and iron ore/Mn ore fines. In order to know the effect of these variables on coke rate and productivity, the term RAFT (raceway adiabatic flame temperature) should be understood.

RAFT has two specific temperature zones: the average temperature typical for coke gases and coke in circulation in front of the tuyere and the maximum temperature that is a stationary zone at approximately 0.3–0.5 m from the front of the tuyere. The average RAFT can be measured with an infrared thermal camera mounted in the sight of the air nozzle and seeing the spotlight from the wind. The camera must be calibrated correctly because the measured temperature is the effect of the radiation of gases and incandescent coke. This may be between 1450 and 1700 °C. Its continuous measurement can provide information about the thermal status of the blast furnace and signal the occurrence of some process malfunctions by decreasing this temperature.

The maximum RAFT in the oxidizing zone can be measured with a cooled lance inserted through the blowpipe elbow and through the tuyere. Some data: water velocity in the probe >10 m/s, thermocouple W–Re reinforced in a Mo pipe through which a small inert gas flow (Ar) the thermocouple distance from the cooled lance is 50–70 mm, the diameter of the probe so that its surface is less than 15% of the minimum section of the free cross section of the tuyere, the lance body is made of the Cu pipe and, if possible, a ceramic coating in the plasma jet as the thermal protection is very useful, the water circulation system in the lance is pipe in pipe system, or a longitudinal diaphragm in which the water comes to the top of the probe on the opposite side and returns to the other half of the section, can also be fitted a gas sampling socket with 4 mm copper pipe with walls of 1 mm. Maximum measurement time is 7–15 min.

The lance must have connections to: a temperature reading device in the thermocouple signal, a $CO_2/CO/H_2$ gas analyser, a sample filter for the gasses to be passed to the analyser, a rotameter to indicate and maintain a very low flow rate of argon as the thermocouple protection gas.

Indian blast furnaces work with comparatively lower RAFT values in the range of 1900–2180 °C, with respect to Japanese furnaces having 2150–2350 °C. Indian blast furnaces have lower values mainly due to (a) poor quality of burden, (b) low percentage of agglomerate in the burden and (c) poor slag characteristics in bosh region.

Higher production rate was attained at 2100 °C of RAFT value. The RAFT should be between 2050 and 2150 °C for improved production and smooth furnace operation. Higher RAFT (>2200 °C) reduced the production level.

3.7 Modern Trends of BF Practice

The choice and development of technology are largely dictated by the availability of raw materials, their quality, energy for the process and its techno-economic feasibility. The ironmaking technology of blast furnace has dominated the world scenario as most economic and widespread resource of molten iron (i.e. hot metal) used for steelmaking. However, this technology has been facing a problem of shortage of metallurgical coke throughout the world. The economics of BF ironmaking mainly depends on the coke rate and productivity. Raw materials account for over 60% of the cost of hot metal production in BF, and the principal cost component is cost of the coke.

Hence, developments have been made in BF to design and practice with objective like: (i) reduction of coke rate and (ii) increase of productivity.

Main developments are as follows:

1. Large capacity of furnaces,
2. Better prepared burden (including better quality coke),
3. Better distribution of charge materials in the furnace,
4. Higher blast rate and temperature,
5. Oxygen enrichment of blast,
6. Humidification of blast,
7. Auxiliary fuel injection through the tuyeres,
8. Pulverized coal injection (PCI),
9. Lime dust injection through the tuyeres, and
10. High top pressure.

3.7.1 Large Capacity of Furnaces

The operational costs of a large capacity of BFs are lower than smaller capacities of BFs. Hence, the size of the BFs has been progressively increasing and BFs of smaller capacities are no longer economically attractive. The same is true with the auxiliary units like coke ovens, sintering plants, stoves, etc. Size of a BF can be compared either in terms of either its hearth diameter or the total useful volume. Earlier diameter of hearth and useful volume were 5.6–8.9 m and 1000 m^3, respectively; now that were 12–15 m and >5000 m^3, respectively. Daily production was also increased from 1000 to 12,000 t/day. The specific production capacity increased from 0.8–1.0 to >2.5 t/m^3/day.

3.7.2 Burden Preparation

It is generally required that better quality burden preparation should be adopted at every iron and steel plant. Large capacity of BF requires better quality of coke. The quality of coke can be improved by the following:

- Better preparation and more extensive blending of coals,
- The use of low volatile additives,
- Pre-heating of coal blend (30–40% with other low ash coal) before carbonization,
- Faster rates of carbonization at higher temperature,
- Partial briquetting of charges,
- Manufacture of formed coke (i.e. synthetically prepared coke),
- Mechanical stabilization of oven coke,
- Compacted coke oven charge, i.e. stamps charging, and
- Dry quenching of coke in nitrogen or CO_2 atmosphere.

3.7.3 Better Distribution of Burden

The blast furnace is working on the counter-current principle. The descending solid charge meets a current of ascending reducing gases, and the reduction of iron ore along with its progressive heating must take place during this passage. All the reactions are taking place inside the BF, and the reduction of iron ore is the most important and most difficult one also. The production rate of a BF is directly determined by: (i) rate of reduction of iron oxide and (ii) rate of heating of the materials.

Both factors are related to the quality of blast introduced in the furnace, i.e. its driving rate. The efficiency with which both requirements are met in the furnace stack alternately determines the furnace production rate. Stack is essentially a diffusion-controlled counter-current gas–solid reactor in which the ascending hot reducing gases reduce the iron oxide and heat up the descending burden. The rate of reduction and heating of the burden depends upon the degree and time of contact of gases with the burden. Burden should have good permeability. Uniformity of distribution of reducing gases throughout the cross section of the furnace is an essential feature for efficient reduction and heating of the burden. Since the gases are passing through the furnace at a high speed, by obtaining uniform resistance of charge to the ascending gases over the entire cross-section of the burden, this is the only way of achieving uniform gas-solid contact. This can be achieved only by obtaining a uniform permeable burden inside the furnace. Uniform permeable burden can be obtained by better burden distribution. Hence, distribution of the charge materials in the throat of the furnace acts a very dominant role for the performance of a furnace. After charging, materials are rolled due to (i) size and (ii) density of the material. The course size materials are more rolling tendency than small size. Densities of ore, coke and limestone are 5–6 g/cm^3, 1.5 g/cm^3 and 3.0–3.5 g/cm^3, respectively. Hence, rolling tendency of coke particles is more due to low density, and ore has less tendency of rolling due to high density. Since densities of materials cannot be altered but sizes can be changed. Hence, sizes of materials may be chosen in such a way that rolling tendencies can be offset to some extent.

3.7.4 Blast Temperature

Since part of the fuel energy in the outgoing gases is pumped back into the furnace in the form of pre-heated blast, the thermal efficiency of the process could be improved by increasing the rate of blowing of hot blast as well as its temperature.

The direct effect of blast temperature is to save coke and thereby improve productivity. India operates at blast temperature in the range of 600–1100 °C compared to more than 1100 °C (i.e. 1200–1300 °C) in other countries. This low level of blast temperature can normally be attributed to two reasons, viz. (i) lower acceptable RAFT and (ii) inability of the existing stoves to provide higher temperature.

The decrease in coke rate is achieved in BF in other countries with increase in blast temperature. By changing stove design, dome temperature can be increased to 1500–1600 °C (on gas) and dropped down to 1250 °C at the end of blast (on blast). Higher hot blast temperature reduces coke consumption in the BF. A higher hot blast temperature is often used in addition with humidified blast, so that flame temperature in the combustion zone of the BF is still within proper limits.

3.7.5 Oxygen Enrichment of Blast

For every unit weight of coke burnt at the tuyere by the blast, nearly 4/5 part weight of nitrogen of the blast are also heated to nearly 2000 °C. The presence of 79% nitrogen by volume in the blast restricts the temperature generated in the combustion zone. This temperature can be increased by decreasing the nitrogen content of the blast, i.e. by oxygen enrichment of the blast. An enrichment of only 2% (by weight) oxygen reduces the nitrogen burden by about 4 units per unit weight of coke, and a higher temperature would be possible. For every percent increase in oxygen content of blast, there is increase in production rate of about 3–4% and a marginal saving in coke rate.

The blast is enriched with oxygen primarily to increase the productivity by bringing down the tuyere gas volume/thm. This, however, increases the RAFT steeply. The ability of oxygen enrichment to increase the RAFT can be utilized for injection of coal as well as natural gas through the tuyeres to reduce the coke rate. Combined use of oxygen enrichment and humidification of blast offers a unique method of BF process control.

3.7.6 Humidification of Blast

A uniform and steady RAFT is on the basic prerequisite for smooth BF operation. RAFT is sensitive to moisture content of the blast, and moisture varies from season to season. Pre-heating of blast not only adds heat energy in the furnace, but also increases the RAFT. There is a limit to which RAFT can be allowed to increase, depending upon the burden characteristics and furnace profile, and beyond which it is detrimental to furnace operation.

Blast temperature can be increased still further without increase in RAFT if equivalent coolants are added along with the blast. Steam is one such additive because of its endothermic reaction with carbon:

$$C(s) + \{H_2O\} = \{CO\} + \{H_2\}, \quad \Delta H^0{}_{298} = 135.14\,\text{kJ/mol of C} \tag{3.4}$$

The presence of moisture in the blast generates double the volume of reducing gas per mol of carbon burnt. As per reaction (3.4), for every carbon burnt one mol of CO and an additional mol of H_2 will be available. More moisture the more will be this additional hydrogen available. Kinetically, hydrogen reduction of iron ore is faster than by CO. The presence of moisture helps to burn coke at a faster rate. Some of the endothermic heat of moisture disintegration is compensated by exothermic reduction of iron oxide by hydrogen. For an increase of 20 g/N m^3 moisture in blast, endothermic heat can be compensated by a rise of 200 °C in the blast pre-heat.

The endothermic nature of reaction between carbon and moisture is utilized with the following objectives:

(i) To decrease the erratic movement of the furnace when the flame temperature increases beyond a limit,
(ii) To use higher blast temperature while maintaining the RAFT constant,
(iii) To maintain uniformity in level of moisture in the blast which fluctuates seasonally, thus disrupting thermal regime of the furnace,
(iv) With increase in blast humidity, hydrogen in the bosh gas increases which in turn decreases the coke rate.

Most of the world's highly productive BFs operate with blast humidity ranging between 15 and 20 g/N m^3 mainly due to superior raw materials which are capable of accepting higher RAFT (2100–2300 °C).

Although increase in blast humidity decreases the RAFT, the simultaneous decrease in coke rate, Si, S content in hot metal could be attributed to the better efficiency of gas utilization. Japanese have also gone in for higher steam injection to produce low Si hot metal.

3.7.7 Auxiliary Fuel Injection

The necessity to adopt fuel injection in a BF arises from the fact that coke is not only costlier but also scarce and hence it should be replaced by other cheaper and readily available fuels. The heat-producing function of coke is partially replaced by injection auxiliary fuels in the tuyeres. They are readily available for combustion in front of the tuyeres. The auxiliary fuels are solid, liquid or gaseous. Solid fuel is pulverized coal, and liquid fuels are light oils, naphtha or heavy oil; gaseous fuels are natural gas, coke oven gas or BF gas that are injected through the tuyeres. The choice of the type of fuel for injection almost entirely depends on its availability, economy and feasibility of injection. The function of coke as a fuel is minimized by injection auxiliary fuels in the tuyeres. Fuel is burnt in front of the tuyeres to generate the heat and to compensate that lost in decreasing the coke rate.

To optimize cost and improve productivity, the heat-producing function of coke is partially replaced by injecting auxiliary fuels in the tuyeres, i.e. to improve blast furnace economy by reducing the coke rate. In blast furnace ironmaking, 50% of the hot metal cost can be attributed to the cost of coke. Injection of auxiliary fuels in the blast furnace aimed, worldwide, to achieve a reduction of coke rate or energy cost and thus the cost of hot metal.

The primary purpose of using injectants with the blast is profitability which depends upon the relative price of coke and injectants, and the amount of coke that can be saved per unit of the later, i.e. upon the replacement ratio.

$$\text{Replacement ratio} = \frac{\text{Coke saved, kg/thm}}{\text{Injectant, kg or N m}^3/\text{thm}} \tag{3.36}$$

In addition, such injections generally cause a smoother furnace movement especially when coolant additives are used and give rise to a higher productivity especially with oxygen-containing additives, e.g. humidified or oxygenated blast or oxygen plus fuel addition. Oxygen enrichment of the blast along with fuel injection is required to achieve full advantages of fuel injection. The additives used as part of replacement of coke are those which cannot be charged from the top. They may be: (i) gaseous, such as oxygen, moisture, natural gas, coke oven gas and reformed gas; (ii) liquid fuels, such as light oil, heavy oil, tar and naphtha; and (iii) solid fuels, such as pulverized coal and anthracite.

The amount of fuel injected in a BF varies from place to place, on same furnace as high as 70–80 kg of oil or 80–200 N m^3 of natural gas/thm. Coke saved depends upon the increase of blast temperature. For constant blast temperature, replacement ratio is about 2.0 kg of coke/kg of oil. The productivity increases by (a) about 2% up to 50 kg of oil/thm, and (b) 1.2–2.0% up to 80 kg of coal dust/thm. The replacement ratio calculated for H blast furnace, Tata Steel (volume: 3814 m^3, capacity 2.5 Mtpa), ranges from 0.9 to 1.1, with coal injection rates of about 200 kg/thm decreasing coke rates about 350 kg/thm [4].

$$\text{Replacement ratio} = \left(\frac{0.75\,\text{to}\,1.0\,\text{kg of coke}}{\text{N m}^3\ \text{of natural gas}}\right) \tag{3.37}$$

$$= \left(\frac{0.8\,\text{kg of coke}}{\text{N m}^3\ \text{of coke oven gas}}\right) \tag{3.38}$$

$$= \left(\frac{0.68\,\text{kg of coke}}{\text{kg of coal dust}}\right) \tag{3.39}$$

$$= \left(\frac{0.9\,\text{to}\,1.1\,\text{kg of coke}}{\text{kg of oil}}\right) \tag{3.40}$$

These injections affect the following furnace parameters:

1. Flame temperature,
2. Bosh gas composition and volume,
3. Top gas temperature and CO/CO_2 ratio and hence the efficiency,
4. Operational efficiency, i.e. the proportion of direct and indirect reduction,
5. Slag volume and its basicity, and
6. Overall production efficiency.

3.7.8 Pulverized Coal Injection (PCI)

Injection of pulverized coal into the BF is one of the most promising options to reduce the coke consumption. The efficient utilization of in-plant generated gases and fines as a source of heat and reducing agent can greatly enhance the overall efficiency of steel industry. PCI is one of essential methods to enhance the BF profitability. Due to the ease of use, oil and natural gas were popular injectants in the 1960s but due to the oil crises in the 1970s many companies stopped the oil injection and turned to coal injection since the 1980s. Nowadays, the vast majority of BFs all over the world apply PCI due to the relatively lower cost of coal compared to other fuels besides the beneficial effect on the BF efficiency [5].

Injection of coal dust through tuyeres of BF is for partial substitution of coke by non-coking coal [6]. Higher PCI rate helps in reducing operational costs and at the same time increases productivity. Injection of coal in Indian BF is considered quite relevant in view of the country's less reserves of coking coal and comfortable reserves of non-coking coal. Coals are injected via lances into the tuyeres, and gasified and ignited in the raceway (as shown in Fig. 3.6). Coal at a level of 175 kg/thm has been injected steadily into many BFs of the world.

The raceway can be classified into three main zones: (i) PC devolatilization zone, (ii) oxidation or combustion zone, and (iii) solution loss reaction zone.

The concentration of oxygen sharply decreases in the oxidation zone due to its reaction with carbon of coke and coal to produce CO/CO_2. Figure 3.7 shows a schematic representation of the PC injection and the subsequent reactions in the raceway [5].

The amount of coke is replaced by unit weight of coal powder, this ratio being defined as the replacement ratio. The higher the replacement ratio, the better will be the results of coal injection. The replacement ratio depends on the varieties and qualities of the coals using the fineness of the coal powder and on the operational techniques of the furnace and injection system. To obtain a high

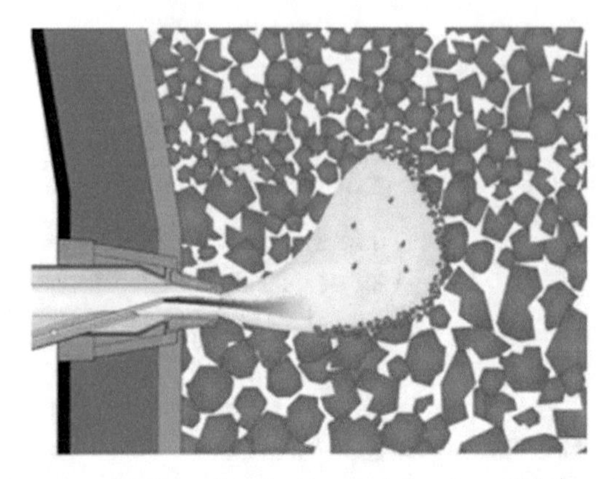

Fig. 3.6 Coal injection in the tuyeres [6] (reproduced with permission from IOS Press BV, Netherlands)

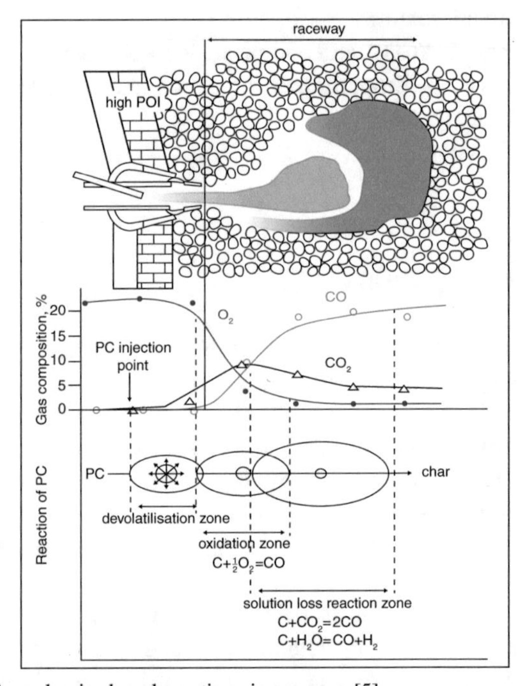

Fig. 3.7 Schematic diagram for pulverized coal reactions in raceway [5]

replacement ratio, a high quality of coal is required preferably with a low ash content, which also reduces the coke rate and a fine particle size to increase the specific surface area of coal and hence accelerate the combustion.

The most important arguments for the pulverized coal injection (PCI) in a blast furnace are as follows:

- Lower consumption of expensive coking coals. Cost savings by lowering the coke rates. Cost of coke is substantially higher than that of coal; moreover, the use of an injectant allows higher blast temperatures to be used, which also leads to a lower coke rate.
- Increased productivity from using oxygen-enriched blast.
- Greater flexibility in BF operation; PCI allows the flame temperature to be adjusted, and the thermal condition in the furnace can be changed much faster.

- Improved consistency in the quality of the hot metal.
- Decrease of the CO_2 footprint, i.e. the amount of CO_2 produced per tonne of steel.

After the enrichment of blast with oxygen, attempts were made to operate the furnace with high injection rates ranging between 120 and 135 kg/thm. With oxygen enrichment (up to 3–3.5%) and stable raw material characteristics, up to 130 kg/thm of coal can be injected successfully with a replacement ratio of 0.8–0.9. Further increase in injection rates would result in incomplete combustion of coal in the raceway resulting in excessive loss of carbon particles through top gas, poor stack permeability resulting in poor blast acceptance and hence productivity and unstable furnace operations. The largest operation blast furnace in RSP, India (in 2013) BF # 5, has a useful volume of 4060 m^3, and hot metal production is 9500 tonnes per day with a pulverized coal injection (PCI) system designed for 200 kg/thm [7]. JSPL Raigarh, India BF # 2, stabilized coal injection rates to more than 200 kg/thm with high oxygen enrichment of the blast, and they are planning to increase the pulverized coal injection up to 250 kg/thm.

Oxygen injection in the tuyeres accelerates combustion of PCI which helps increase the temperature of reducing gas and as well increasing the reduction rate. The increased PCI injection helps in reduction of coke and also cost of overall fuel. With an increase in coal injection rate, BF permeability tends to decrease. As a result, wind volume reduction is observed. To compensate for resultant loss in production, oxygen enrichment in the blast is increased. Increase in PCI vis-à-vis decrease in coke rate leads to increase in ore/coke ratio and lowers the RAFT temperature, and to maintain the constant RAFT temperature, steam injection needs to be lowered. At JSW Steel, the PCI in BF # 3 and # 4 is 150 kg/thm [8].

Most of the plants use blends of coal for injection. Blending allows for (financial) optimization of coal purchases. For example, a company with a grinding mill for hard coals can use a considerable percentage of softer coals by blending it into hard coals. In doing so, an optimized value can be obtained. Blending dilutes the disadvantages of coal types. Every material has disadvantages like high moisture content, sulphur and phosphorous level, a relatively poor replacement ratio and so on.

The permanent efforts aimed at reducing the costs of ironmaking have led to an increasing portion of substitute materials for coke, which has mainly been coal injected through the tuyeres. Nowadays, blast furnaces with total coal injection rates more than 200 kg/thm are operated with coke consumptions of less than 400 kg/thm. At these high coal injection rates, coke is subjected to more rigorous conditions in the blast furnace. Dissection of furnaces taken out of operation and probing and sampling through the tuyeres of furnaces in operation have allowed the assessment of the extent of coke degradation in the furnace. Coke degradation is controlled by the properties of feed coke, i.e. mechanical stabilization, resistance to chemical attack (solution loss, alkalis and graphitization), and by the blast furnace operating conditions. At high coal injection rates, the amount of coke present in the furnace decreases and the remaining coke is subjected to more vigorous mechanical and chemical conditions: increased mechanical load as the ore/coke ratio becomes higher; increased residence time at high temperatures; increased solution loss reaction (CO_2 formation); and alkali attack. More severe coke degradation during its descent from the furnace stock line into the hearth can therefore be expected at high coal rates.

However, high coal injection rates can also affect the direct reduction reactions as follows:

1. Coal injection increases hydrogen content, and at elevated temperatures (800–1100 °C), hydrogen is a very effective reducing agent in gas reduction of iron oxides.
2. The unburnt soot remaining after the raceway is more reactive than coke and used for direct reduction in preference to coke.
3. The alkali cycle is reduced because of the elimination of alkali through the hot furnace centre.

Therefore, at high coal injection rates, the attack of coke by direct reduction reactions may also decrease. This is beneficial for coke integrity in the lower part of the furnace.

Coals are injected through the tuyeres, gasified and ignited in the raceway. The coal is in the raceway area only for a very short time (in milliseconds), and so the characteristics of the gasification reaction are very important for the effectiveness of a pulverized coal injection (PCI) system. Coal gasification consists of several steps: (i) the coal is heated, and the moisture evaporates; (ii) gasification of the volatile components then occurs after further heating; (iii) the volatile components are gasified and ignited, which causes an increase in the temperature. These steps occur sequentially with some overlap. The speed of gasification increases as: (a) the volatility of the coals increases, (b) the size of the coal particles decreases, and (c) the better the mixing of blast and coal.

At high PCI operation, about 40% of the reductant is injected via the tuyeres. Therefore, it is important to control the amount of coal per tonne of hot metal as accurate as the coke rate is controlled.

The heat requirement of the lower furnace is a special topic when using PCI. Not only coal is used for producing the reduction of gases but use of coal influences the heat balance in the lower part of furnace. The heat of the bosh gas must be sufficient to melt the burden: define the *melting heat* as the heat needed to melt the burden. The heat requirement of the burden is determined by the *pre-reduction degree* or how much oxygen has still to be removed from the burden when melting. The removal of this oxygen requires a lot of energy. The *melting capacity* of the gas is defined as the heat available with the bosh gas at a temperature over 1500 °C. The melting capacity of the gas depends on: (i) the quantity of tuyere gas available per tonne of hot metal; especially when using high volatile coal, there is a high amount of hydrogen gas in the bosh gas; and (ii) the flame temperature in the raceway.

The flame temperature is also determined by coal rate, coal type, blast temperature, blast moisture and oxygen enrichment. The percentage of oxygen in the blast can be used to balance the heat requirements of the upper and lower part of the furnace. The balance is dependent on the local situation. It depends on burden and quality of coke, and type of coal used. For the balance, there are some technical and technological limitations. For higher injection rates, more oxygen is required. The higher the oxygen injection, the higher is the productivity of the furnace. Minimum oxygen requirement for sufficient PCI combustion and smooth operation of blast furnace, excess air ratio should not be less than 0.9.

$$\text{Where excess air ratio} = \left(\frac{\text{oxygen supplied}}{\text{stoichiometric oxygen required}}\right).$$

The limitations are given by:

- Too low top gas temperature: If top gas temperature becomes too low, it takes too long for the burden to dry and the effective height of the blast furnace shortens.
- Too high flame temperature: If flame temperature becomes too high, burden descent can become erratic.
- Too low flame temperature: Low flame temperature will hamper coal gasification and melting of the ore burden.
- Technical limitations to the allowed or available oxygen enrichment.

High PCI rate leads to:

- Decrease in RAFT,
- Increase in ore/coke ratio,
- Decrease in stoichiometric oxygen ratio,
- Rise in pressure drop at tuyere and in furnace,
- High slag rate.

3.7.9 Lime Injection

Lime injection can control the quality of bosh slag, which produced in the bosh region has direct effect on the productivity of the furnace. If the slag is more viscous, then it affects the aerodynamics adversely, and as a result the productivity falls. Increase of bosh slag viscosity by even 1 poise may decrease the productivity by about 0.5%.

Flux is added in the BF charge to flux the gangue of the iron ore/sinter and the ash of the coke. The entire flux requirement is met by adding limestone as a separate constituent in the charge or via sinter in part or full. Where coke ash is very high like Indian coke, the bulk of the slag is formed from the ash of the coke. Major portion of this ash is released only after the combustion of coke in the tuyere region. The lime proportion of the slag-formed bosh region is far too large than required to maintain the desired basicity. In other words, the basicity of the slag that is formed before bosh region is very high and consequently its viscosity will be very high. Such highly viscous slag poses several problems: (i) it comes down slowly through the coke bed, (ii) it interferes with the aerodynamics adversely and decreases the rate of reduction of ore and rate of heating of the charge in the stack, and (iii) it decreases the productivity.

In fact, the lime required to flux the ash of coke is ideally required at the tuyere level. If it can be provided at that portion, then the slag formed up in the stack and in bosh regions will be almost the same type of slag as formed at the end. This will not be more viscous than necessary and hence will readily come down through the coke bed.

All such slag-related problems can be solved to a great extent if part of the total lime requirement can be injected through the tuyeres in a way similar to fuel injection. Lime will be available where it is needed most, i.e. flux to ash of coke. The flux is added: (i) via sinter and (ii) rest by injection through the tuyeres. In that case, no free limestone may be required in the charge mix. The powder lime, not limestone powder, is injected through the tuyeres. That means the heat requirement for calcination is done outside of the BF. Thus, coke rate in BF is decreased accordingly.

3.7.10 High Top Pressure

The gaseous reduction in a blast furnace would increase if the static pressure in the furnace is increased by throttling back of the discharge gas pressure. This increase in the furnace top gas pressure is known as *high top pressure*. The adoption of high top pressure in a furnace provided with standard two-bell top leads to a differential pressure across the big bell. The pressure below the bell is high and that above the bell is practically atmospheric pressure. As a result, the bell would get pressed against its hopper seat and would not lower down. The big bell can be lowered only after admitting gas under pressure in the chamber above the big bell (i.e. between the two bells) to develop zero pressure differential across the big bell. Similarly, the small bell could be opened only after withdrawing the gas pressure from below the small bell, thereby bringing it down to the normal atmospheric pressure. The adoption of high top pressure therefore requires a system of pressuring and de-pressuring the big bell hopper often to maintain the high top pressure while working the furnace in a normal way.

The bell-less top charging system solves the problem of gas sealing under a high pressure operation. The hoppers are sealed at top and bottom by valves, and the internal pressure can be raised or lowered as necessary to avoid a pressure differential across them when they are opened.

High top pressure of the order of 1 kg/cm^2 gauge is most commonly adopted even on the existing old furnaces. The problems and difficulties multiply with increasing top pressure. Several modified top charging systems have therefore been designed to obtain reliable sealing while charging the furnace. The efficiency of sealing and the duration of its life, besides other less important factors, dictate the value to which the top pressure could be raised. It has been estimated that the limit of effective top pressure is about 3 kg/cm^2 gauges for obtaining increased production rate and decreased coke consumption.

Advantages for high top pressure are as follows:

1. Increasing production rate due to increase time of contact of reducing gas and solid, because of decrease velocity of gases in the furnace.
2. Increasing the reduction rate of iron oxide.
3. Improving the efficiency of utilization of reducing gas results in saving of coke.
4. More uniform operation with lower and more consistent hot metal silicon content has been obtained by high top pressure.
5. High top pressure may be applied at constant blowing rate to obtain fuel saving with lesser increase in output.
6. High top pressure markedly decreases channelling and dust losses in the effluent gases.
7. At high top pressure, the Boudouard equilibrium moves to left due to decreasing the coke consumption.

BF designers should provide equipment to operate effectively at top pressure up to 3 kg/cm^2 gauge with adequate life. Average of gas density is directly proportional to the average pressure in the furnace. By increasing top pressure, (i) gas pressure at the tuyeres is also increased, (ii) loss of carbon (due to solution loss reaction, $CO_2 + C = 2CO$) is decreased and (iii) silicon in hot metal is decreased.

Productivity can be increased by increasing gas flow rate in the furnace, and fuel consumption can be decreased.

3.8 Transfer of Silicon and Sulphur

Silicon and sulphur contents of iron droplets begin to increase rapidly from upper or middle level of bosh region, reach their maximum value at the tuyere level and then decrease to their final value during passage of droplets through the slag. Manganese content of droplets continues to increase; simultaneously silicon has reached its maximum and falls to its final value after passing through the slag.

It was observed that silicon content of hot metal increases with the increase in ash content of coke and coke-to-ore ratio, so ash of coke plays an important role in silicon transfer. Silicon transfer to metal is via the formation of silicon monoxide (SiO) gas from the coke ash in the high-temperature region of the tuyere zone. Silicon picked up by the metal is oxidized by the iron oxide and manganese oxide in slag.

$$[\mathrm{Si}] + 2(\mathrm{FeO}) = (\mathrm{SiO_2}) + 2[\mathrm{Fe}] \tag{3.41}$$

$$[\mathrm{Si}] + 2(\mathrm{MnO}) = (\mathrm{SiO_2}) + 2[\mathrm{Mn}] \tag{3.42}$$

Turkdogan and co-workers found that at 1600 °C, SiO_2 in coke ash forms SiC which is stable phase above 1506 °C at 1 atm p_{CO} and a_{SiO_2} is one. Other research workers also observed that all free SiO_2 present in coke ash can form SiC under BF operating conditions.

$$\mathrm{SiO_2(s)} + 3\mathrm{C(s)} = \mathrm{SiC(s)} + 2\{\mathrm{CO}\} \tag{3.43}$$

At high temperature, SiC reacts with CO:

$$\mathrm{SiC(s)} + \{\mathrm{CO}\} = \{\mathrm{SiO}\} + 2\mathrm{C(s)} \tag{3.44}$$

It is observed that this reaction can approach equilibrium even under BF conditions.

Overall ash and carbon reaction: addition of reaction (3.43) and reaction (3.44):

$$\mathrm{SiO_2(s)} + \mathrm{C(s)} = \{\mathrm{SiO}\} + \{\mathrm{CO}\} \tag{3.45}$$

Sulphur is present in lime as CaS. So, sulphur evaporates as SiS:

$$\mathrm{CaS(s)} + \{\mathrm{SiO}\} = \mathrm{CaO(s)} + \{\mathrm{SiS}\} \tag{3.46}$$

SiO and SiS vapours thus formed react with metal and slag resulting in a transfer of the respective species.

Sulphur is also present in coke as FeS, and it can also form SiS:

$$\mathrm{FeS(s)} + \mathrm{C(s)} + \{\mathrm{SiO}\} = \{\mathrm{SiS}\} + \{\mathrm{CO}\} + \mathrm{Fe(s)} \tag{3.47}$$

It can be shown that the decomposition of FeS is thermodynamically favoured in the high-temperature zone of BF and evaporation of S is more favourable on kinetic consideration.

$$\mathrm{FeS(s)} = \mathrm{Fe(s)} + 1/2\{\mathrm{S_2}\} \tag{3.48}$$

At least, 50% SiO_2 and all S present in coke can evaporate (as SiO and SiS) by the time coke comes to tuyere zone.

The mechanism of silicon transfer clearly shows that silicon content of metal could be reduced by reducing the activity of SiO_2 in coke ash and by increasing iron oxide content in slag. Silicon pickup by metal is less when basicity of coke (i.e. CaO content increased in ash) is increased.

The oxidation of silicon in metal by FeO during the passage of droplets through the slag layer is a highly exothermic reaction. One percentage drop of silicon in metal can lead to an increase of metal and slag temperature by about 80 °C.

3.9 Aerodynamics

The productivity of a BF is largely dictated by its aerodynamics characteristics. The BF can be conveniently divided into two parts from the point of view of aerodynamics the stack or dry zone and the bosh or the wet zone.

The gas distribution in the BF has great influence on the major functions of the furnace heat exchange and chemical reactions. Inside the furnace, there are essentially main three zones: a lumpy zone, a cohesive zone and a dripping zone. The cohesive zone is mostly either in the inverted 'V' shape or 'W' shape. This separates the low-temperature lumpy zone from high-temperature dripping zone.

As coke occupies about 60–70% of the furnace volume and being larger of the two constituents, ore and coke, the overall permeability of stack region is governed by ore burden, the lower permeability constituent of the charge. The nature of burden means particle size distribution and the shape of the particles play an important role in determining the permeability of stack of the BF.

In the lumpy zone, the gas can flow through the entire cross section of the furnace and radial gas distribution can be roughly estimated from Ergun's equation assuming that there is no radial pressure gradient.

$$\frac{\mathrm{d}p}{\mathrm{d}z} = 150\frac{(1-\varepsilon)^2}{\varepsilon^3}\mu\frac{RT}{pM}\frac{G}{(Q\mathrm{dp})^2} + 1.75\frac{(1-\varepsilon)}{\varepsilon^3}\frac{RT}{pM}\frac{G^2}{(Q\mathrm{dp})} \tag{3.49}$$

where

p is the pressure, ε is the porosity of burden, μ is the viscosity of gas,
M is the molecular weight and G is the mass flow rate of gas/unit area,
Q is sphericity factor, R is gas constant, T is temperature, dp is particle diameter.

Ergun's equation suggests that the radial gas distribution is strongly related to porosity and particle size and distribution in the radial direction. Thereby, the gas distribution in the lumpy zone is primarily governed by the particle size distribution in the charge and burden distribution at the stock level. Since ore and coke are having widely different permeabilities, the distribution of ore/coke strongly affects the gas distribution.

Non-uniform gas distribution leads to non-uniform temperature and utilization of reducing gases across the cross section of the furnace. The regions having higher coke-to-ore ratio offer lower resistance to gas flow, and in these regions temperature isotherms move up.

The cohesive layer formed by the softened ferrous burden has only a small residual permeability which depends on raw material quality and degree of reduction of materials. Therefore, gas flows around these layers through alternate coke layers or coke slits. The shape and size of cohesive zone play an important role in the gas distribution in the furnace. The shape and size of cohesive zone depend on the burden distribution. The thickness of cohesive zone depends on the softening–melting characteristics of burden under BF conditions.

In the dripping zone, gas flows primarily through the annulus formed by the *dead man* and inner boundary of cohesive zone and finally it goes out through the coke slits.

3.10 BF Productivity

Last 50–60 years, the expansion in technology has been so rapid that the furnace diameters have almost doubled and production increased about fivefold. Since the increase in height of furnace is limited for the mechanical strength of the burden. Hence, the modern high-capacity furnaces explained outwards rather than upwards to increase the inner volume.

Table 3.3 Production of pig iron in top five countries in world (Mt) [9]

Country	2013	2014	2015	2016	2017	2018
China	748.1	716.5	695.9	698.2	748.3	771.1
Japan	83.8	83.9	81.0	80.2	78.3	77.3
India	51.4	55.2	58.4	63.7	66.8	71.5
Russia	50.1	51.5	52.6	51.9	52.0	51.2
South Korea	41.0	46.9	47.6	46.3	47.1	47.1
World total	1206.7	1188.0	1162.4	1166.7	1218.2	1246.6

Productivity depends upon the amount of carbon burned in unit time at the tuyeres and the tuyere carbon (coke) consumed for producing a unit of iron. The amount of coke burned is usually called the throughput or coke burning rate or driving rate per day. The coke consumed for a unit is termed as coke consumption rate or simply coke rate in kg per thm. World's pig iron production is shown in Table 3.3.

The production rate of BF, also known as smelting rate or output, is depended on coke burning efficiency and denoted as:

$$P = \frac{Q}{K} \tag{3.50}$$

where

P is the productivity, thm/day
Q is the coke burned, tonnes/day
K is coke consumed, tonnes/thm.

From Eq. (3.50), P is increased as increasing Q and or decreasing K, i.e. to achieve a high production rate:

(I) By increasing coke burning rate, which can be increased by:

(a) Increased oxygen supply:

(i) Increase blast volume: An increased blast volume rate means an increased oxygen input in unit time which results in an increased coke burning rate. The coke burning rate is proportional to oxygen input.
A large blast volume will be accepted by the furnace if the permeability of the stock column is increased. The influences of various factors on the blast volume rate are as follows: by using sinter and pellets, and by maintaining proper coke size and strength.

(ii) Oxygen-enriched blast: Incorporating of oxygen in blast increases the percentage of oxygen in the blast and therefore increases the coke burning rate per unit volume of blast, i.e. per unit time. Oxygen enrichment lowers the nitrogen content of the blast and decreases the blast volume per unit of carbon burned as well as the bosh gas volume per tonne of hot metal. Oxygen enrichment blast would give increased coke burning rate, and high increase in productivity is observed.

(iii) Humidified blast: Incorporating of steam increases the percentage of oxygen in the blast, therefore increases the coke burning rate and produces a bosh gas richer in CO and H_2; according to the reaction:

$$C(s) + \{H_2O\} = \{CO\} + \{H_2\}, \quad \Delta H^0{}_{298} = 135.14\,\text{kJ/mol of C} \tag{3.4}$$

Moisture reduces the flame temperature, due to endothermic reaction, while oxygen increases that. Since moisture lowers the flame temperature, it is necessary to increase the blast temperature (8–10 °C per 1 g H_2O/N m^3 blast).

(iv) High top pressure: The purpose of high top pressure is to introduce more oxygen to burn more carbon by blowing more air and at the same time maintaining the linear gas velocity (and pressure drop).

(b) By decreasing volume, viscosity and increased density, surface tension of slag.

(II) By decreasing coke rate: (i) preparation and concentration of charge materials, (ii) increase in blast temperatures, (iii) blast modifications, (iv) super-(metalized) burden, (v) stack gas injection (not used) and (vi) fuel additives.

Probable Questions

1. Classify the blast furnace reactions on the basis of zones and discuss.
2. What do you understand by Tuyere's raceway zone? Discuss the reactions at that zone.
3. C(s) + $\{CO_2\}$ = 2{CO}, this reaction is called the *solution loss reaction. Why?* What is the other name of the reaction?
4. What is the role of coke in a BF? Why a blast furnace cannot run without a critical amount of coke? Explain.
5. What are the different reactions taking place at the stack and bosh regions of the BF? Discuss.
6. What are the conditions for de-sulphurization of hot metal in BF?
7. Discuss BF slag. What is the role of Al_2O_3 in BF slag?
8. How many zones are divided the BF according to Professor W. K. Lu? Draw and discuss.
9. What are 'dead man zone' and 'coke slits'? What are their significances?
10. What do you mean by direct and indirect reductions during BF ironmaking?
11. What do you understand by (a) TFT and (b) RAFT? How these factors affect the BF performance in ironmaking?
12. Discuss modern trends of BF practice.
13. Discuss the principles of (a) oxygen enrichment of blast and (b) humidification of blast.
14. Is it desirable to adopt oxygen enrichment along with humidification of blast? Why?
15. Discuss pulverized coal injection through tuyeres and its benefits in BF.
16. Discuss mechanism of transfer of silicon in BF.
17. How productivity of a BF is depending on its aerodynamics characteristics?

Examples

Example 3.1 Dissociation of limestone is given by reaction:

$$CaCO_3 = CaO + CO_2, \quad \Delta G^0 = 168{,}406 - 143.93T\,\text{J}$$

Find the temperature if external pressure is (i) 1 atm, (ii) 1.5 atm, (iii) 2 atm and (iv) 0.5 atm.

Solution

(i) For 1 atm, i.e. $P_{CO_2} = 1$
$k = P_{CO_2} = 1$
$\Delta G^0 = -RT \ln k$
Therefore, $168{,}406 - 143.93T = -8.314 \times T \times \ln 1 = 0$
$T = 1170\,\text{K} = \mathbf{897}\,°C$

(ii) For 1.5 atm, i.e. $P_{CO_2} = 1.5$
$k = P_{CO_2} = 1.5$
$\Delta G^0 = -RT \ln k$
Therefore, $168{,}406 - 143.93T = -8.314 \times T \times \ln 1.5 = -3.37T$
$T = 1198.1\,\text{K} = \mathbf{925.1}\,°C$

(iii) For 2 atm, i.e. $P_{CO_2} = 2$
$k = P_{CO_2} = 2$
$\Delta G^0 = -RT \ln k$
Therefore, $168{,}406 - 143.93T = -8.314 \times T \times \ln 2 = -5.76T$
$T = 1218.86\,\text{K} = \mathbf{945.86}\,°C$

(iv) For 0.5 atm, i.e. $P_{CO_2} = 0.5$
$k = P_{CO_2} = 0.5$
$\Delta G^0 = -RT \ln k$
Therefore, $168{,}406 - 143.93T = -8.314 \times T \times \ln 0.5 = 5.76T$
$T = 1125\,\text{K} = \mathbf{852}\,°C$.

Example 3.2 An ironmaking blast furnace produces hot metal of the following composition: 94% Fe, 2.2% Si, 3.8% C. The composition of the iron ore is 84% Fe_2O_3, 9% SiO_2, 3% Al_2O_3, 4% moisture. Limestone is charged 500 kg/thm production. Limestone contents are 95% $CaCO_3$ and 5% SiO_2, Coke is charged 900 kg/thm. Composition of coke is 84% C, 10% SiO_2, 3% Al_2O_3, 3% moisture. Calculate (i) weight of ore used to produce one tonne of hot metal and (ii) weight of slag produced per tonne of HM (Assume that there is no iron loss in the slag phase.)

Solution

Basis of calculation: One tonne hot metal production.
Assumptions: There is no Fe loss in slag.

Iron Balance: Fe input = Fe output
Fe from iron ore = Fe in hot metal [let W_0 = weight of iron ore]
$0.84 \times W_0 \times \left(\frac{112}{160}\right) = 0.94 \times 1000$ [since mol wt of $Fe_2O_3 = 2 \times 56 + 3 \times 16 = 160$]
Therefore, $\mathbf{W_0 = 1598.64\ kg}$

Silica Balance: SiO_2 input = SiO_2 output
SiO_2 from iron ore + SiO_2 from limestone + SiO_2 from coke = Si in HM + SiO_2 in slag
$0.09 \times 1598.64 + 0.05 \times 500 + 0.1 \times 900 = 0.022 \times 1000 \times \left(\frac{60}{28}\right) + W_{SiO_2}$ [since mol wt of $SiO_2 = 28 + 2 \times 16 = 60$]
Therefore, $\mathbf{W_{SiO_2} = 211.73\,kg}$

Alumina Balance: Al_2O_3 input = Al_2O_3 output
Al_2O_3 from iron ore + Al_2O_3 from coke = Al_2O_3 in slag
$0.03 \times 1598.64 + 0.03 \times 900 = W_{Al_2O_3}$
Therefore, $\mathbf{W_{Al_2O_3} = 74.96\,kg}$

Lime Balance: CaO input = CaO output
CaO from limestone = CaO in slag
$0.95 \times 500 \times \left(\frac{56}{100}\right) = W_{CaO}$ [since mol wt of $CaCO_3 = 40 + 12 + 3 \times 16 = 100$]
Therefore, $\mathbf{W_{CaO} = 266.0\,kg}$
Hence weight of slag produce $= W_{SiO_2} + W_{Al_2O_3} + W_{CaO}$
$= 211.73 + 74.96 + 266.0 = \mathbf{552.69\,kg}$

Example 3.3 An ironmaking blast furnace produces hot metal of the composition: 93.6% Fe, 2.1% Si, 3.6% C, 0.7% Mn. The composition of the iron ore is 78% Fe_2O_3, 9% SiO_2, 5% Al_2O_3, 1% MnO, 7% moisture. Coke rate is 850 kg/thm. Composition of coke is 80% C, 10% SiO_2, 5% Al_2O_3, 5% CaO. Enough limestone is charged to make 45% CaO in slag, and composition of limestone is 100% $CaCO_3$. Calculate (i) weight of ore used per tonne of hot metal production, (ii) weight of slag produced per tonne of HM and (iii) weight of limestone used to produce given slag composition.

Solution

Basis of calculation: One tonne hot metal production.
Assumptions: There is no Fe loss in slag.

Iron Balance: Fe input = Fe output
Fe from iron ore = Fe in hot metal
$0.78 \times W_0 \times \left(\frac{112}{160}\right) = 0.936 \times 1000$
Therefore, $\mathbf{W_0 = 1714.29\,kg}$

Silica Balance: SiO_2 input = SiO_2 output
SiO_2 from iron ore + SiO_2 from coke = Si in HM + SiO_2 in slag
$0.09 \times 1714.29 + 0.1 \times 850 = 0.021 \times 1000 \times \left(\frac{60}{28}\right) + W_{SiO_2}$
Therefore, $\mathbf{W_{SiO_2} = 194.29\,kg}$

Alumina Balance: Al_2O_3 input = Al_2O_3 output
Al_2O_3 from iron ore + Al_2O_3 from coke = Al_2O_3 in slag
$0.05 \times 1714.29 + 0.05 \times 850 = W_{Al_2O_3}$
Therefore, $\mathbf{W_{Al_2O_3} = 128.21\,kg}$

MnO Balance: MnO input = MnO output
MnO from iron ore = Mn in HM + MnO in slag
$0.01 \times 1714.29 = \left(\frac{0.7}{100}\right) \times 1000 \times \left(\frac{71}{55}\right) + W_{MnO}$ [since mol wt of MnO = 55 + 16 = 71]
Therefore, $\mathbf{W_{MnO}}$ **= 8.11 kg**

Now, slag content 45% CaO and 55% (SiO_2 + Al_2O_3 + MnO)
$W_{SiO_2} + W_{Al_2O_3} + W_{MnO} = 194.29 + 128.21 + 8.11 = 330.61\,\text{kg}$
Fifty-five parts (SiO_2 + Al_2O_3 + MnO) contain 100 parts of slag
Therefore, 330.61 kg $\left(\frac{100 \times 330.61}{55}\right) = 601.11\,\text{kg}$
Hence, weight of slag = **601.11 kg**

$$\text{Therefore CaO in slag} = \text{weight of slag} - \text{weight of } (SiO_2 + Al_2O_3 + MnO)$$
$$= 601.11 - 330.61 = \mathbf{270.5\,kg}$$

CaO Balance: CaO input = CaO output
CaO from limestone + CaO from coke = CaO in slag
$W_{CaCO_3} \times \left(\frac{56}{100}\right) + 0.05 \times 850 = 270.5$
Therefore, weight of limestone, $\mathbf{W_{CaCO_3} = 407.14\,kg}$.

Example 3.4 BF produced hot metal (3.5% C, 2.0% Si, 0.75% Mn). The composition of iron ore is 80% Fe_2O_3, 9% SiO_2, 6% Al_2O_3, 1% MnO, 4% moisture. Coke rate is 750 kg/t. Composition of coke is 85% C, 10% SiO_2, 3% Al_2O_3, 2% CaO; 550 kg of limestone is used. Limestone contains 92% $CaCO_3$, 5% SiO_2 and 3% Al_2O_3. Find out (i) weight of ore used to produce one tonne of HM and (ii) weight of slag produced and composition of slag.
Assume that there is 5% iron loss in the slag phase w.r.t. HM production.

Solution
Basis of calculation: One tonne hot metal production.

Iron Balance: Fe input = Fe output
Fe from iron ore = Fe in hot metal + Fe loss in slag
$0.8 \times W_0 \times \left(\frac{112}{160}\right) = 0.9375 \times 1000 + 0.05 \times 1000$
(Since Fe in HM = 100 – (3.5 + 2 + 0.75) = 93.75%)
Therefore, $\mathbf{W_0}$ **= 1763.39 kg**
Fe + 1/2O = FeO
56 72
Fifty-six parts of Fe form 72 parts of FeO
50 $\left(\frac{72 \times 50}{56}\right) = \mathbf{64.29\,kg\,of\,FeO}$

Silica Balance: SiO_2 input = SiO_2 output
SiO_2 from iron ore + SiO_2 from limestone + SiO_2 from coke = Si in hot metal + SiO_2 in slag
$0.09 \times 1763.39 + 0.05 \times 550 + 0.1 \times 750 = 0.02 \times 1000 \times \left(\frac{60}{28}\right) + W_{SiO_2}$
Therefore, $\mathbf{W_{SiO_2} = 218.35\,kg}$

Alumina Balance: Al_2O_3 input = Al_2O_3 output
Al_2O_3 from iron ore + Al_2O_3 from coke + Al_2O_3 from limestone = Al_2O_3 in slag
$0.06 \times 1763.39 + 0.03 \times 750 + 0.03 \times 550 = W_{Al_2O_3}$
Therefore, $\mathbf{W_{Al_2O_3} = 144.80\,kg}$

MnO Balance: MnO input = MnO output
MnO from iron ore = Mn in hot metal + MnO in slag
$0.01 \times 1763.39 = \left(\frac{0.75}{100}\right) \times 1000 \times \left(\frac{71}{55}\right) + W_{MnO}$
Therefore, $\mathbf{W_{MnO} = 7.95\ kg}$

Lime Balance: CaO input = CaO output
CaO from limestone + CaO from coke = CaO in slag
$0.92 \times 550 \times \left(\frac{56}{100}\right) + 0.02 \times 750 = W_{CaO}$
Therefore, $\mathbf{W_{CaO} = 268.36\ kg}$

$$\begin{aligned}\text{Hence weight of slag produce} &= W_{FeO} + W_{SiO_2} + W_{Al_2O_3} + W_{MnO} + W_{CaO} \\ &= 64.29 + 218.35 + 144.80 + 7.95 + 268.36 = \mathbf{703.75\,kg}\end{aligned}$$

Composition of slag = 9.14%FeO, 31.03%SiO_2, 20.57%Al_2O_3, 1.13%MnO and 38.13%CaO.

Example 3.5 A blast furnace produces hot metal containing 3.5% C, 1.5% Si and 95% Fe. Iron ore contains 80% Fe_2O_3 and 20% gangue (SiO_2 + Al_2O_3). The coke rate is 850 kg/thm, and coke contains 80% fixed carbon and 20% ash. Flux rate is 450 kg/thm and contains 90% $CaCO_3$, and rest is SiO_2. BF gas contains $\left(\frac{CO}{CO_2}\right)$ ratio = $\left(\frac{24}{16}\right)$.
Calculate (i) weight of iron ore used, (ii) weight of slag produced and (iii) volume of BF gas.

Solution

Basis of calculation: One tonne hot metal production.
Assume that there is no iron loss in the slag phase.

(i) **Iron Balance**: Fe input = Fe output
Fe from iron ore = Fe in pig iron
$W \times 0.8 \times 0.7 = 0.95 \times 1000 = 950\,\text{kg}$
Therefore, weight of iron ore, **W = 1696.4 kg**

(ii)
$$\begin{aligned}\text{Weight of slag} &= (\text{Gangue in iron ore} + \text{ash in coke} \\ &\quad + SiO_2 \text{ in flux} + \text{CaO in flux}) \\ &\quad - (SiO_2 \text{ equilivalent Si in hot metal}) \\ &= [(1696.4 \times 0.2) + (850 \times 0.20) + (450 \times 0.1) \\ &\quad + \left\{450 \times 0.9 \times \left(\frac{56}{100}\right)\right\}] - \left\{1000 \times 0.015 \times \left(\frac{60}{28}\right)\right\} \\ &= 781.08 - 32.14 = \mathbf{748.94\,kg}\end{aligned}$$

(iii) Carbon in BF gas = carbon input from coke − carbon in HM

$$= [(850 \times 0.8) - (1000 \times 0.035)] = 680 - 35 = 645\,\text{kg}$$

$$= \left(\frac{645}{12}\right) = 53.75\,\text{kg mol.}$$

Mol fraction of CO = $\left(\frac{24}{40}\right) = 0.6$
Mol fraction of CO_2 = $\left(\frac{16}{40}\right) = 0.4$
Therefore, CO gas formation = 53.75 × 0.6 = **32.25 kg mol**
and CO_2 gas formation = 53.75 × 0.4 = **21.50 kg mol**

Oxygen required from air = [(oxygen required to produce CO and CO_2) − oxygen from ore]

$$= \left[\{(32.25 \times 1/2) + 21.5\} - \left\{1696.4 \times 0.8 \times 0.3 \times \left(\frac{1}{32}\right)\right\}\right]$$

$$= 37.63 - 12.72 = \mathbf{24.91\,kg\,mol}$$

Nitrogen in air = $24.91 \times \left(\frac{79}{21}\right) = \mathbf{93.68\,kg\,mol}$

Therefore total BF gas = $CO + CO_2 + N_2 = 32.25 + 21.50 + 93.68 = 147.43\,\text{kg mol}$

$$= 147.43 \times 22.4 = \mathbf{3302.41\,Nm^3}.$$

Example 3.6 An ironmaking blast furnace produces hot metal of the composition: 94% Fe, 1.5% Si, 4% C, 0.15% Mn, 0.1% P and 0.05% S. The charge contains two types of haematite ores: (i) ore A has a high silica content, and (ii) ore B has high limestone content. The coke rate is 800 kg per tonne of hot metal produced. The slag basicity $\left(\frac{CaO}{SiO_2}\right)$ must be 1.4 in order to achieve low sulphur content in the hot metal, and the $\left(\frac{CO}{CO_2}\right)$ ratio in the exit gases is 2. Calculate per tonne of hot metal produced: (a) the weight of each ore required, (b) the weight of slag produced and sulphur content of the slag (wt %) and (c) the volumes of blast and exit gases.

Composition	Ore A	Ore B	Coke
Fe	55.3	45.5	0.7
C	–	–	86.0
Mn	0.10	0.07	–
Si	–	–	–
P	0.04	0.08	–
S	0.15	0.05	0.5
SiO_2	13.0	4.0	3.0
$CaCO_3$	7.0	30.0	5.6 (CaO)

There is no loss of sulphur or fine ore in the exit gases

Solution

Basis of calculation: One tonne hot metal production.
Assume that there is no iron loss in the slag phase.

(i) **Iron Balance**: Fe input = Fe output
Fe from iron ore A + Fe from iron ore B + Fe from coke = Fe in hot metal

$$W_1 \times 0.553 + W_2 \times 0.455 + 800 \times 0.007 = 1000 \times 0.94$$
$$\text{Or } 0.553W_1 + 0.455W_2 + 5.6 = 940.0 \quad (1)$$
$$\text{Or} W_1 + 0.823W_2 = 1689.69$$

(ii) **CaO Balance**: CaO input = CaO output
CaO from iron ore A + CaO from iron ore B + CaO from coke = CaO in slag
$[W_1 \times 0.07 + W_2 \times 0.3] \times \left(\frac{56}{100}\right) + 800 \times 0.056 = \text{CaO in slag}$
[Since $\left(\frac{CaO}{CaCO_3}\right) = \left(\frac{56}{100}\right)$]

$$\text{Therefore, CaO in slag} = 0.039W_1 + 0.168W_2 + 44.8 \quad (2)$$

(iii) **SiO_2 Balance**: SiO_2 input = SiO_2 output
SiO_2 from iron ore A + SiO_2 from iron ore B + SiO_2 from coke = Si in hot metal + SiO_2 in slag
$W_1 \times 0.13 + W_2 \times 0.04 + 800 \times 0.03 = [1000 \times 0.015] \times \left(\frac{60}{28}\right) + SiO_2 \text{ in slag}$

$$\text{Therefore, } SiO_2 \text{ in slag} = 0.13W_1 + 0.04W_2 - 8.14 \quad (3)$$

Since slag basicity $\left(\frac{CaO}{SiO_2}\right) = 1.4$

From Eqs. (2) and (3), we get: $\left(\frac{(0.039W_1 + 0.168W_2 + 44.8)}{(0.13W_1 + 0.04W_2 - 8.14)}\right) = 1.4$

$$\text{By simplification: } W_1 - 0.784W_2 = 393.53 \quad (4)$$

By subtracting Eq. (4) from Eq. (1), we get: $W_2 = \mathbf{806.57\,kg}$
Putting value of W_2 in Eq. (1): $W_1 = \mathbf{1025.88\,kg}$
From Eq. (2): CaO in slag $= 0.039W_1 + 0.168W_2 + 44.8$
$= 0.039 \times 1025.88 + 0.168 \times 806.57 + 44.8 = 220.52\,\text{kg}$

From Eq. (3): SiO_2 in slag $= 0.13W_1 + 0.04W_2 - 8.14$
$= 0.13 \times 1025.88 + 0.04 \times 806.57 - 8.14 = 157.49\,\text{kg}$

(iv) **S Balance**: S input = S output
S from iron ore A + S from iron ore B + S from coke = S in hot metal + S in slag
$W_1 \times 0.0015 + W_2 \times 0.0005 + 800 \times 0.005 = 1000 \times 0.0005 + \text{S in slag}$
$\text{S in slag} = 1025.88 \times 0.0015 + 806.57 \times 0.0005 + 4.0 - 0.5 = 5.44\,\text{kg}$

Therefore, weight of slag $= \text{CaO in slag} + \text{SiO}_2 \text{ in slag} + \text{S in slag}$

$= 220.52 + 157.49 + 5.44 = \mathbf{383.45\,kg}$

$\%\text{S in slag} = \left(\frac{5.44}{383.45}\right) \times 100 = \mathbf{1.42\%}$

(v) **C Balance**: C input = C output

C in coke + C in $CaCO_3$ in ores = C in hot metal + C in gases

$800 \times 0.86 + [(1025.88 \times 0.07) + (806.57 \times 0.30)] \times \left(\frac{12}{100}\right) = 1000 \times 0.04 + \text{C in gases}$

Therefore, C in gases = **685.66 kg** = 685.66/12 = **57.138 k mol.**

Since $\left(\frac{CO}{CO_2}\right)$ ratio in the exit gases is 2,

$\mathbf{X_{CO}}$ **= 38.092 k mol** and $\mathbf{X_{CO_2}}$ **= 19.046 k mol.**

Since volume of 1 k mol of gas = 22.4 m^3.

Therefore, volume of CO and CO_2 gases are 853.26 and 426.63 m^3 respectively.

Oxygen Balance: O_2 from air + O_2 from minerals = O_2 in exit gas

$$O_2 \text{ from minerals} = O_2 \text{ from ores} + O_2 \text{ from CaCO}_3 + O_2 \text{ from SiO}_2 \quad (5)$$

$O_2 \text{ from ores} = [(1025.88 \times 0.553) + (806.57 \times 0.455)] \times \left(\frac{48}{112}\right) = \mathbf{400.41\,kg}$

$$O_2 \text{ from CaCO}_3 = [(1025.88 \times 0.07) + ((806.57 \times 0.30)] \times \left(\frac{48}{100}\right) + \left[(800 \times 0.056) \times \left(\frac{16}{56}\right)\right]$$

$$= 150.62 + 12.8 = 163.42\,\text{kg}$$

Since CaO in slag = 220.52 kg,

O_2 from CaO in slag $= 220.52 \times \left(\frac{16}{56}\right) = 63.01\,\text{kg}$

Therefore, actual O_2 from $CaCO_3$ = 163.42 − 63.01 = **100.41 kg**

$$O_2 \text{ from SiO}_2 = [(1025.88 \times 0.13) + ((806.57 \times 0.04) + (800 \times 0.03)] \times \left(\frac{32}{60}\right)$$

$$= (133.36 + 32.26 + 24.0) \times \left(\frac{32}{60}\right) = 101.13\,\text{kg}$$

Since SiO_2 in slag = 157.49 kg,

O_2 from SiO_2 in slag = $157.49 \times \left(\frac{32}{60}\right) = 83.99\,\text{kg}$

Therefore, actual O_2 from SiO_2 = 101.13 − 83.99 = **17.13 kg**.

From Eq. (5): $O_2 \text{ from minerals} = O_2 \text{ from ores} + O_2 \text{ from CaCO}_3 + O_2 \text{ from SiO}_2$

$= 400.41 + 100.41 + 17.13 = 517.95\,\text{kg} \approx \mathbf{16{,}186\,mol}$

O_2 in exit gas $= O_2$ in CO gas $+ O_2$ in CO_2 gas $= \left(\frac{38{,}092}{2}\right) + 19{,}046 = \mathbf{38{,}092\,mol}$
Since O_2 from air $= O_2$ in exit gas $- O_2$ from minerals

$$= 38{,}092 - 16{,}186 = \mathbf{21{,}906\,mol} = \mathbf{21.906\,k\,mol} = \mathbf{490.69\,m^3}$$

Volume of air $= \left(\frac{490.69}{0.21}\right) = \mathbf{2336.64\,m^3}$
Volume of nitrogen $= 2336.64 - 490.69 = \mathbf{1846\,m^3}$

$$\text{Volume of BF gas: CO} = 853.26\,\text{m}^3\,(27.30\%)$$
$$CO_2 = 426.63\,\text{m}^3(13.65\%)$$
$$N_2 = 1846\,\text{m}^3(59.05\%)$$

Therefore, total volume of BF gas = **3125.89 m^3**.
Ratio of exit gas volume to air = $\left(\frac{3125.89}{2336.64}\right) = \mathbf{1.34}$.

Example 3.7 Hot metal, which contains 4.3% C, 1.5% Si, 0.6% Mn and 0.31% P, is produced with a consumption of 900 kg coke and 225 kg of limestone per tonne of HM. The coke contains 75% carbon. The carbon ratio $\left(\frac{CO}{CO_2}\right)$ is 1.7. The loss of carbon in flue dust is 15 kg per tonne of HM. What is the percentage of carbon burnt at the tuyeres? Also, find out the composition of top gas.

Solution

Carbon balance: C input = C output
C from coke + C from limestone = C in HM + C in gas + C in flue dust
$(0.75 \times 900) + (0.12 \times 225) = (0.043 \times 1000) + \text{C in gas} + 15$
Therefore, C in gas = 675 + 27 – (43 + 15) = 644 kg

$$CaCO_3 = CaO + CO_2$$
$$100 \qquad\qquad 22.4\,\text{N m}^3$$
$$100\,\text{kg limestone produce } 22.4\,\text{N m}^3\,CO_2$$
$$225\,\text{kg} \qquad \left(\frac{(22.4 \times 225)}{100}\right) = 50.4\,\text{N m}^3\,CO_2$$

Since $\left(\frac{CO}{CO_2}\right)$ is 1.7, let % $CO_2 = x$, % CO = $1.7x$

Therefore, $x + 1.7x = 100$ or $x = 37.04\%$ = % CO_2 and % CO = 62.96%.
Since C in gas = 644 kg, out of that C coming from limestone 27 kg.

Hence, C from coke to form gases = 644 − 27 = **617 kg**.
Now, out of that 37.04% form CO_2 and 62.96% form CO.

So, C forms CO_2 from coke = 617 × 0.3704 = 228.54 kg.
and C forms CO from coke = 617 × 0.6296 = 388.46 kg.

$$2C + O_2 = 2CO$$

$$2 \times 12 \quad 22.4 \quad 2 \times 22.4$$

24 kg of C form 44.8 N m^3 CO gas

$$388.46\,\text{kg} \quad \left(\frac{(44.8 \times 388.46)}{24}\right) = \mathbf{725.13\,Nm^3\,CO\,gas}$$

To burnt 24 kg C, oxygen require 22.4 N m^3

$$388.46\,\text{kg} \quad \left(\frac{(22.4 \times 388.46)}{24}\right) = \mathbf{362.56\,Nm^3}$$

Again

$$C + O_2 = CO_2$$

$$12 \quad 22.4 \quad 22.4$$

12 kg C form 22.4 N m^3 CO_2 gas

$$228.54\,\text{kg} \quad \left(\frac{(22.4 \times 228.54)}{12}\right) = \mathbf{426.61\,Nm^3\,CO_2\,gas}$$

To burnt 12 kg C, oxygen require 22.4 N m^3

$$228.54 \quad \left(\frac{(22.4 \times 228.54)}{12}\right) = \mathbf{426.61\,Nm^3}$$

Total oxygen require = 362.56 + 426.61 = **789.17 Nm^3**

21 N m^3 oxygen content in 100 N m^3 of air

$$789.17 \quad \left(\frac{100}{21}\right) \times 789.17 = \mathbf{3757.94\,Nm^3\,of\,air}$$

Hence nitrogen content = 3757.94 − 789.17 = **2968.77 Nm^3**

Total top gas contain = $CO + CO_2 + N_2$

= 725.13 N m^3 CO gas + 426.61 N m^3 CO_2 gas from coke

+ 50.4 N m^3 CO_2 from limestone

+ 2968.77 N m^3 N_2 = **4170.91 Nm^3**

Top gas contains 17.39% CO, 12.29% CO_2 and 71.17% N_2.
Hence, the percentage of carbon burnt at the tuyeres = $\left(\frac{617}{675}\right) \times 100 = \mathbf{91.41\%}$.

Example 3.8 The volume of the blast per tonne of hot metal was found to be 2868 m^3. The coke used per tonne of hot metal was 900 kg (coke contained 88% carbon). Calculate the per cent of carbon of the coke that was burned at the tuyeres of B.F.

Solution

$$2C + O_2 = 2CO$$
$$2 \times 12 \quad 22.4$$

2868 m^3 of air contains 0.21 × 2868 m^3 = 602.28 m^3 of oxygen.
Again 22.4 m^3 oxygen requires to consume 24 kg of carbon

$602.28\,m^3 \qquad \left(\frac{(24\times602.28)}{22.4}\right) = 645.3\,\text{kg of C}$

Carbon contain in coke $= 0.88 \times 900 = 792\,\text{kg}$
Therefore, per cent of carbon burned at the tuyeres of B.F = $\left(\frac{(645.3\times100)}{792}\right) = \mathbf{81.48\%}$.

Example 3.9 In the representative equation: $Fe_2O_3 + xCO = 2Fe + 3CO_2 + (x - 3)CO$, the ratio of $\left(\frac{CO}{CO_2}\right)$ is $\left(\frac{(x-3)}{3}\right)$, must be high and will usually be between 1.5 and 2.0.

The blast furnace operator desires the ratio of $\left(\frac{CO}{CO_2}\right)$ to be 1.8:1, ignoring the additional CO_2 that will be supplied by decomposition of limestone.

Calculate (i) the value of x in the above equation, (ii) the weight of carbon requires per tonne of iron reduced, and (iii) the volume of air requires per tonne of iron reduced.

Solution

Since the ratio of $\left(\frac{CO}{CO_2}\right)$ is 1.8:1.
Therefore, $\left(\frac{(x-3)}{3}\right) = \left(\frac{1.8}{1}\right)$ or $x - 3 = 5.4$ or $\mathbf{x = 8.4}$.

$$\text{Hence equation become: } Fe_2O_3 + 8.4CO = 2Fe + 3CO_2 + 5.4CO \qquad (1)$$

$$\text{Again } 2C + O_2 = 2CO \text{ or } 8.4C + 4.2O_2 = 8.4CO \qquad (2)$$

From the above, reactions to reduce Fe_2O_3 to form 2Fe require 8.4C.
That is, (2 × 56) kg of Fe production requires (8.4 × 12) kg of carbon

$1000\,\text{kg} \qquad \left(\frac{(8.4\times12)}{(2\times56)} \times 1000\right) = \mathbf{900\,kg}$

From Eq. 2: (2 × 12) kg carbon requires 22.4 N m^3 of oxygen

$900\,\text{kg} \qquad \left(\frac{(22.4\times900)}{24}\right) = 840\,\text{N m}^3 \text{ of oxygen}$

Again, air contains 21% oxygen.
So, 21 N m^3 of oxygen remains in 100 N m^3 of air

$840\,\text{N m}^3 \text{ of oxygen} \qquad \left(\frac{(100\times840)}{21}\right) = \mathbf{4000\,Nm^3\,of\,air}$.

Example 3.10 The hot metal produced in the blast furnace has the composition 3.6% C, 1.4% Si and rest Fe. The input materials consist of iron ore, coke and pure $CaCO_3$ (as flux). The iron ore analyses 84% Fe_2O_3, 10% SiO_2 and 6% Al_2O_3, and the coke contains 10% SiO_2, 5% Al_2O_3 and 85% carbon; 800 and 400 kg of coke and flux are used per tonne of iron production, respectively. Calculate the weight of the ore charged and the weight of slag produced per tonne of iron produced.

Solution

Basis of calculation: One tonne hot metal production.
Assume that there is no iron loss in the slag phase.

Iron Balance: Fe input = Fe output
Fe from iron ore = Fe in hot metal
$W \times 0.84 \times 0.7 = 0.95 \times 1000 = 950\,\text{kg}$
Therefore, weight of iron ore, W = **1615.65 kg**

Silica Balance: SiO_2 from ore + SiO_2 from coke = SiO_2 equivalent Si in hot metal + SiO_2 in slag
$(0.1 \times 1615.65) + (0.1 \times 800) = (0.014 \times 1000) \times \left(\frac{60}{28}\right) + W_{SiO_2}$
$161.57 + 80 = 30 + W_{SiO_2}$
Therefore, weight of SiO_2 in the slag, $W_{SiO_2} = 211.57\,\text{kg}$

Alumina Balance: Al_2O_3 input = Al_2O_3 output
Al_2O_3 from iron ore + Al_2O_3 from coke = Al_2O_3 in slag
$(0.06 \times 1615.65) + (0.05 \times 800) = W_{Al_2O_3}$
Therefore, weight of Al_2O_3 in the slag, $W_{Al_2O_3} = 136.94\,\text{kg}$

CaO Balance: CaO input = CaO output
CaO from $CaCO_3$ = CaO in slag
$\left(\frac{52}{100}\right) \times 400 = W_{CaO}$
Therefore, weight of CaO in the slag, W_{CaO} = 208 kg

Hence, the weight of slag produced per tonne of iron produce
$= W_{SiO_2} + W_{Al_2O_3} + W_{CaO} = 211.57 + 136.94 + 208 = \mathbf{556.51}\,\text{kg}.$

Example 3.11 Calculate the oxygen required for PCI injection into blast furnace. The ultimate analysis of coal is: 82% C, 4.64% H, 0.67% S and 3.85% O.

Solution

Major combustion reaction taking place is as follows:

$$\begin{array}{ccccc} C & + & O_2 & \rightarrow & CO_2 \\ 12 & & 32 & & 44 \end{array}$$

12 kg carbon requires 32 kg oxygen and produces 44 kg CO_2.
So, 1 kg C requires $\left(\frac{32}{12}\right)$ kg oxygen and produces $\left(\frac{44}{12}\right)$ kg CO_2.
1 kg C requires 2.67 kg oxygen and produces 3.67 kg CO_2.

$$\begin{array}{ccccc} H_2 & + & 1/2O_2 & \rightarrow & H_2O \\ 2 & & 16 & & 18 \end{array}$$

1 kg of H_2 needs 8 kg of O_2, and it produces 9 kg of H_2O.

$$\begin{array}{ccccc} S & + & O_2 & \rightarrow & SO_2 \\ 32 & & 32 & & 64 \end{array}$$

32 kg of S needs 32 kg of O_2, and it produces 64 kg of SO_2.
1 kg of S needs 1 kg of O_2, and it produces 2 kg of SO_2.
The oxygen is required stoichiometrically as follows:

82% C $\rightarrow$ 0.82 $\times$ 2.67 = 2.19 kg of oxygen,
4.64% H $\rightarrow$ 0.0464 $\times$ 8 = 0.37 kg of oxygen,
0.67% S $\rightarrow$ 0.0067 $\times$ 1 = 0.0067 kg of oxygen,
3.85% O $\rightarrow$ 0.0385 kg

$$\begin{aligned} \text{Stoichiometric oxygen required} &= 2.19 + 0.37 + 0.0067 - 0.0385 \\ &= \mathbf{2.53\,kg} \text{ of oxygen required.} \end{aligned}$$

Problems

Problem 3.1 A blast furnace produced hot metal of composition: 4% C, 0.9% Mn, 2% Si and 93% Fe. The amount of limestone used ¼th of the weight of the ore and is composed of 96% $CaCO_3$, 2% $MgCO_3$ and 2% SiO_2; 800 kg of coke is used, and coke contained 88% C, 1% Al_2O_3, 8% SiO_2 and 3% moisture. Ore contained 56% Fe, 8% SiO_2, 4% MnO, 6% Al_2O_4, 2% moisture.

Calculate: (i) weight of iron ore used per tonne of hot metal production and (ii) weight and composition of slag produced.

[Ans: Wt of iron ore = **1660.71 kg** and wt slag = **551.9 kg**]

Problem 3.2 A blast furnace produces hot metal containing 3.6% C, 1.2% Si and 95.2% Fe. The iron ore contains 90% haematite and 10% silica. The coke rate is 480 kg/thm, and the coke contains 95% carbon and 5% SiO_2. Calculate the weight of iron ore required per thm. Find out the weight of SiO_2 in the slag/thm.

[Ans: Wt of iron ore = **1511.11 kg** and wt of SiO_2 in the slag = **149.4 kg**]

Problem 3.3 If an iron ore (Fe_3O_4) has 50% Fe content then what amount of coke (contain 80% fixed carbon) is required for reducing one tonne of iron ore to iron by the following reaction:

$$Fe_3O_4 + 4CO = 3Fe + 4CO_2$$

[Ans: Coke required = **178.57 kg**]

Problem 3.4 The volume of the blast per tonne of hot metal was found to be 2750 m^3. The coke used per tonne of hot metal was 800 kg. The coke contained 85% carbon. Calculate the per cent of the carbon of the coke that is burned at the tuyeres. [Ans: **90.99%**].

Problem 3.5 A blast furnace operator wishes to increase his hearth temperature by enriching the blast to 25 vol% oxygen. What weight and volume (N m^3) of pure oxygen must he inject per 1000 N m^3 of dry air blast? [Ans: 53.3 N m^3 or 76.1 kg].

References

1. C. Bodsworth, *Physical Chemistry of Iron and Steel Manufacture* (CBS Publishers & Distributors, Delhi, 1988)
2. W.K. Lu, in *Proceedings of International Symposium on BF Ironmaking*, Jamshedpur, India, Nov 1985
3. R.D. Walker, *Modern Ironmaking Methods* (The Institute of Metals, London, 1986)
4. S. Kundu et al., IIM Metal News **19**(5), 18 (2016)
5. T. Kamijou, M. Shimizu, PC combustion in blast furnace, in *Advanced Pulverized Coal Injection Technology and Blast Furnace Operation*, vol. 1 (Pergamon, Amsterdam, The Netherlands, 2000), p. 63
6. M. Geerdes, H. Toxopeus, C. Vliet, *Modern Blast Furnace Ironmaking*, 2nd edn. (IOS Press BV, Amsterdam, Netherlands, 2009)
7. Cover Story, Iron Steel Rev. **57**(11), 29 (2014)
8. R. Sah et al., Steel Tech. **12**(2), 27 (2018)
9. *World Steel in Figures 2019* (World Steel Association)
10. J.G. Peacey, W.G. Davenport, *The Iron Blast Furnace*, 1st edn. (Pergamon Press Ltd., Oxford, England, 1979)
11. R.H. Tupkary, V.R. Tupkary, *An Introduction to Modern Iron Making*, 4th edn. (Khanna Publishers, Delhi, 2012)
12. A. Ghosh, A. Chatterjee, *Ironmaking and Steelmaking: Theory and Practice* (Prentice-Hall of India, New Delhi, 2008)

4 Furnace Auxiliaries

The blast furnace (BF) generates gas which contains many fine particles as well as substantial calorific value. To use the raw BF gas, it is necessary to clean it by using certain processes. Air pre-heated to temperatures between 1000 and 1250 °C is produced in the hot blast stoves. The control of BF depends on the variables like heat, control of slag and choice of raw materials.

4.1 Cleaning of BF Gas

The blast furnace (BF) generates gas at the furnace top which is an important by-product of the BF process, and this is known as raw BF gas. This BF gas is at the temperature and pressure existing at the BF top and usually contaminated with many fine particles, i.e. dust and water vapour. This BF gas has substantial calorific value. Before further use of the raw BF gas, it is necessary to clean and cooled to reduce gas volumes and moisture content by using certain processes which reduces the solid particles. The BF gas contains mainly carbon monoxide (CO) and nitrogen. It is used as fuel gas in the hot blast stoves for heating blast as well as acts as supplemental fuel in the steel plant.

The analysis of the blast furnace gas with pulverized coal injection (PCI) in BF is shown in Table 4.1. The process systems for the gas cleaning are either wet-cleaning or dry-cleaning system. Wet-cleaning system is the more commonly used system for the BF gas cleaning [1].

The primary function of the blast furnace gas-cleaning system is to remove solid particles from the gas. In addition, the system also cools the gas to reduce its moisture content, thus increasing its calorific value [1]. The recovered sludge/dust contains relatively high amount of iron oxide and carbon; thus, this can be recycled through the sinter/pelletizer plants. BF gas leaves the furnace via uptakes and a downcomer. The top gas is passed through a dust catcher and wet-/dry-cleaning system to remove dust.

4.1.1 Dust Catcher

The object of the dust catcher is, as implied by its name, to remove as much as possible of the flue dust blown over from the furnace, with which the gas is heavily loaded. Dust catchers are situated between the furnace and the stoves. The general principles of a dust catcher are shown in Fig. 4.1. The single downcomer from the top of the furnace carries the gas to the dust catcher, where it enters through the top by a vertical pipe that is centrally located inside the dust catcher and extends nearly to

S. K. Dutta and Y. B. Chokshi, *Basic Concepts of Iron and Steel Making*,
https://doi.org/10.1007/978-981-15-2437-0_4

Table 4.1 Typical analysis of BF gas with PCI [1]

Constituent	Unit	Value
CO	vol%	20–24
CO_2	vol%	18–23
N_2	vol%	52–57
H_2	vol%	1.5–4.5
SO_2	mg/m^3	10–30
NH_3	mg/m^3	5–21
Chloride	mg/m^3	50–200
Oxides of N_2	mg/m^3	3–12

the bottom. This pipe flares at its lower extremity like an inverted funnel. As the gas passes downwards through this pipe, its velocity (therefore its ability to carry dust) is decreased, and most of the coarser dust drops out of the gas stream and is deposited in the cone at the bottom of the dust catcher. Since the bottom of the dust catcher is closed, and the gas outlet is near the top, the direction of the travel of the gas must reverse 180°. This sudden reversal in the direction of flow causes more of the fine dust to get settle down [1]. Precipitation of the dust depends on two factors: (i) change of direction of the gas flow and (ii) sudden increase of volume due to pressure drop. This causes the velocity of the gas to be reduced, whereby the dust is precipitated to some degree. Efficiency of the dust catcher is approximately 60–75%, depending upon the type of ores used and the blast volume.

After dust catcher, the gas is sent to secondary gas-cleaning stage. Next BF gas is cleaned either by wet-type gas-cleaning system or by dry-type gas-cleaning system. In the wet-type gas-cleaning plant, BF gas washed of dust in scrubbers in several stages, while in dry-type gas-cleaning plant, filter bags are used for removal of fine particles of dust from the gas.

4.1.2 Primary Cleaning or Wet-Cleaning

Aim is to wet the dust particles and wash them out of the gas with water. Incidentally, the gas is cooled to about the temperature of the wash water, and any moisture more than saturation at this temperature is precipitated. Wet cleaners include: (1) venture washer, (2) stationary spray tower, (3) revolving spray tower and (4) centrifugal machines.

4.1.2.1 Venture Washer

The venture washer is a vertical-type unit. Gas from the main enters the top of the unit and passes downwards. As the gas passes through the narrow throat of the unit, it is sprayed with water. There are two sets of water sprays, one operating at low pressure and entering the unit at right angles to the gas flow and the other operating at high pressure and directed upwards at an angle of 110° to the gas flow.

4.1.2.2 Stationary Spray Tower

The stationary spray tower consists of a supported steel cylinder with conical bottom and conical top with the gas entering at a point near the bottom of the cylinder and leaving through a centrally located outlet on top of the top cone (as shown in Fig. 4.2). Inside the cylinder are three or four banks of ceramic tile which split up the rising gas and tend to prevent channelling. Above each bank are water sprays which uniformly cover the cross section of the washer with a falling rain.

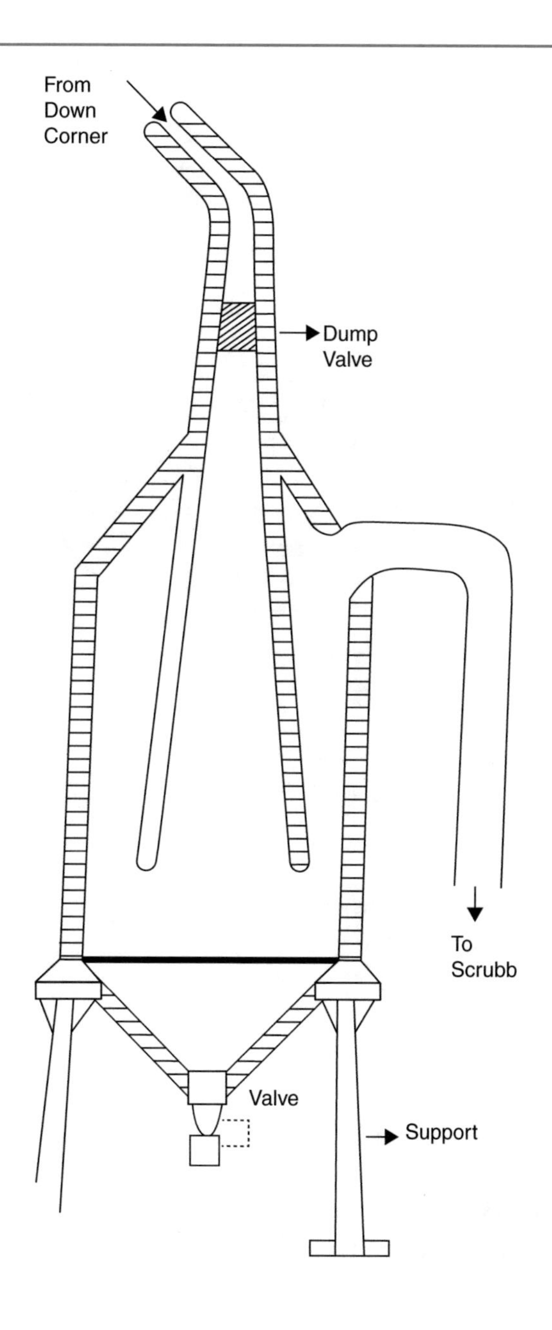

Fig. 4.1 Dust catcher

4.1.2.3 Revolving Spray Tower

The revolving spray tower consists of a stationary horizontal cylinder, through the centre of which is slowly revolving shaft carrying perforated dices which are half-submerged in water. The gas flows above the water and is formed through the perforations, which are charged with water, whereby freshwater is constantly brought into contact with the gas.

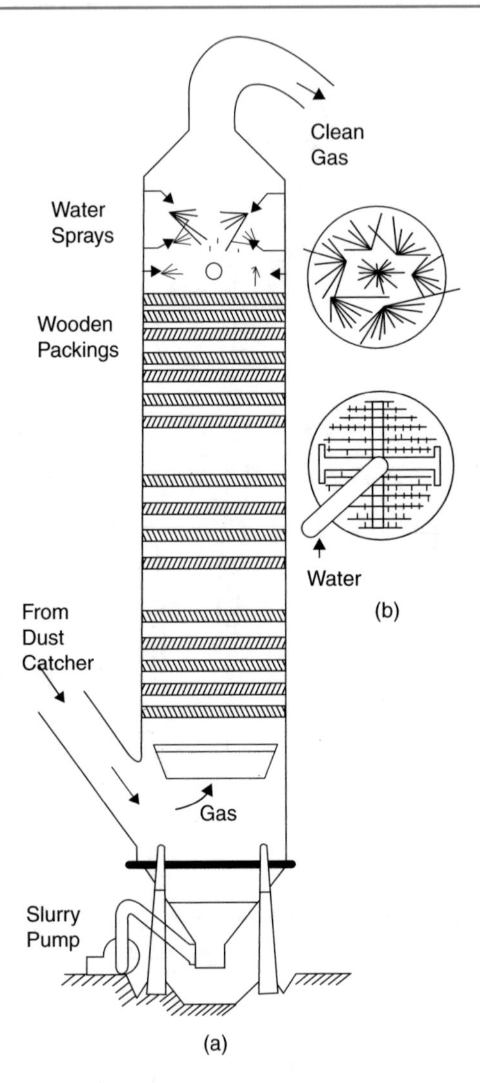

Fig. 4.2 Wet scrubber for cleaning of BF gas. **a** Details of baffles and **b** details of water spray arrangement

4.1.2.4 Centrifugal Machines

The centrifugal machines consist of rapidly rotating cylinders; gas enters at one end of the cylinder and water at the periphery.

4.1.3 Secondary Cleaning

It is carried out in both wet and dry conditions. Wet-type cleaners are electrostatic precipitators, high-speed disintegrators or Theisen disintegrators.

4.1.3.1 Electrostatic Precipitator (ESP)

It is based on the principle that under an action of high applied static voltage, the dust particles in a gas phase acquire electrostatic charge and are attracted to an electrode of opposite polarity where they are collected and washed out. The unit consists of either tubes or plates as one electrode and wires,

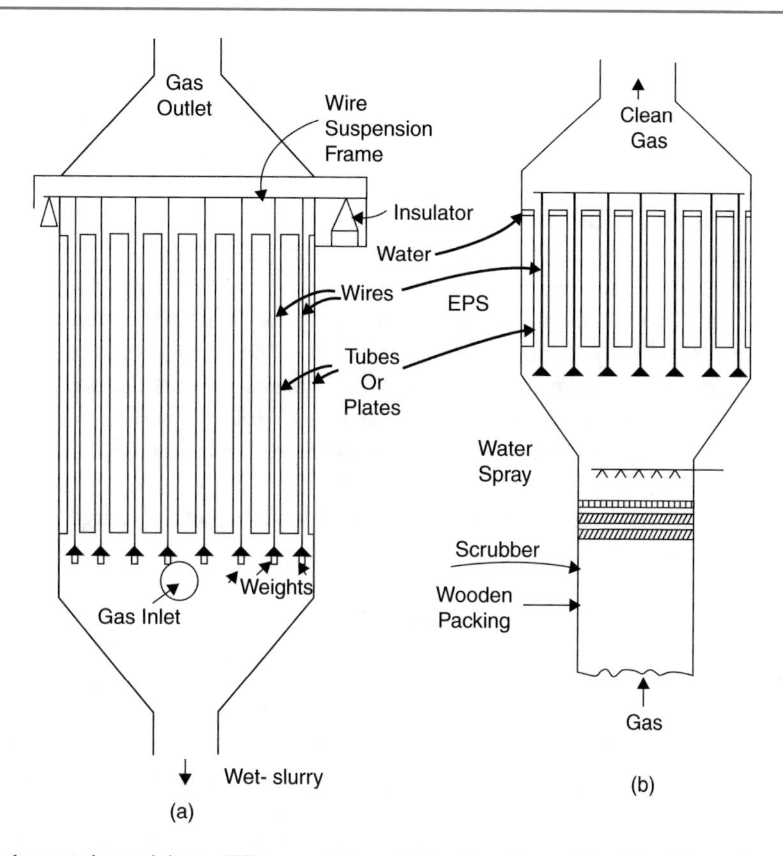

Fig. 4.3 Wet electrostatic precipitator (ESP), **a** as independent unit and **b** combination with scrubber

centrally located in each tube or parallel with the plates, as the other electrode as shown in Fig. 4.3. Rectified high voltage is applied across these two electrodes. The gas enters from the bottom and rises upwards through a system of each cell, while a thin film of water flows over the surface of the tube or plates from inside downwards. The dust particles on ionization get attracted to these wet surfaces and are washed down. The dust particles are precipitated on the electrode, hence the name. The slurry is led to a thickener. The cleaning efficiency is however better in the case of electrostatic precipitator, and hence it is practically universally adopted.

4.1.3.2 High-Speed Disintegrators or Theisen Disintegrators

High-speed disintegrators essentially mix water and gas by a high-speed rotor. Blades are provided on both rotor and the stator bodies from inside to achieve this objective. Water is atomized to almost moisty condition. The centrifugal action of the rotor forces the wet dust to stator walls where it forms part of the slurry and is removed. The cleaned gas from the space goes out into a water separator. The dust content of the cleaned gas is 0.015–0.02 g/m^3.

Theisen disintegrator is essentially a fan that brings water and gas in intimate contact and separates them in opposite direction. It is now out of date and not in use.

4.1.4 Dry-Cleaning

In dry-cleaning, one of the aims is to remove the dust without cooling, thus conserving the sensible heat. Dry cleaners include: (1) electrical precipitator and (2) metallic wool pad filter bags.

4.1.4.1 Electrical Precipitator (ESP)

The ESP is also used widely for cleaning the dry effluent gases before they are let off into the atmosphere as per the pollution control. The cleaning of BF gas by electrical means depends on the principle that when a gas containing foreign particles is passed through a tube between two electrodes, one charged to a high potential and the other earthed, the foreign particles in the gas become charged, because of the ionization of the gas, and are impelled towards the earthed electrode. The direct current used in the precipitators is of the order of 130–160 mA and 30–45 kv. The degree to which the gas can be cleaned depends on the (i) electrodes, (ii) gas velocity and (iii) voltage employed.

Two essential factors for electrical precipitation are: (i) imparting of the charge to the particles and (ii) provision of a suitable field.

4.1.4.2 Metallic Wool Pad Filter Bags [1]

The filter bag is normally having 8–16 sections, which are arranged in two rows, where one or two sections are standby, one more section is in cyclic purge nitrogen-cleaning mode and other sections are in gas-cleaning mode. For the filtering material, a fabric with high refractory, wear and tear resistance is used. Besides, after chemical treatment the fabric is easy to clean from dust during purge cleaning since it has improved water and oil repellent properties. The fabric also prevents occurrence of electrostatic current. Maximum filtering rate is 1.0–1.5 m^3/min, and the gas temperature is 50–280 °C (the system can also operate stably, i.e. reliably for 2 h at gas temperature of 300 °C). The normal dust content of blast furnace gas after cleaning is usually not more than 3 mg/m^3. Service life of fabric bags is generally up to two years. Dry-cleaning of blast furnace gas by bag filters has found the wide application in China.

4.1.5 Comparing of Dry- and Wet-Cleaning

Dry gas cleaning has the following differences over the wet gas cleaning using venturi scrubbers [1]:

- Absence of water in the cleaning process eliminates or reduces several blast furnace waterworks facilities, e.g. sludge pumping station of gas-cleaning system, sludge settling tanks and flocculators, pump station for slurry transfer, circulating pump station of turnaround cycle of gas cleaning, etc.
- Temperature of gas after cleaning is in the range of 100–120 °C, which is 50–70 °C higher than the temperature of the gas after wet gas cleaning. Also, the humidity of gas is reduced by 50–60 g/m^3, which together is equal to an increase in caloric value of blast furnace gas by 210–250 kJ/m^3.
- Reduction in the dust content in blast furnace gas to 2–3 mg/m^3 that improves the service life of hot blast stoves and enhances the performance of top pressure recovery turbine.
- Environmental improvement in the steel plant due to a better dust removal from the gas and elimination of sludge handling facilities.

4.2 Hot Blast Stoves

Air pre-heated to temperatures between 1000 and 1250 °C is produced in the hot blast stoves and is delivered to the furnace via a hot blast main, bustle pipe, tuyere stocks and finally through the tuyeres. The hot blast reacts with coke and injectants. The high gas speed forms the area known as the raceway in front of the tuyeres.

Blast furnace stoves, usually three per furnace (two are working, and one is under repair), are brick-lined regenerators of heat enclosed in a circular steel shell with a flat bottom and a dome-shaped top. Function of stove is to pre-heat the blast before its entries into the furnace through the tuyeres. The result of pre-heating the blast is to intensify and speed up the burning of the coke at the tuyeres with a consequent reduction in the coke required for the smelting operation; this reduction in coke consumption is more than would correspond merely to the additional sensible heat carried in by the heated blast. Use of hot blast has become universal, due to that remarkable increases in BF production and lower fuel consumption.

The blast is heated (800–1200 °C) at the chamber which is known as *hot blast stove*. The interior of the stove is of firebrick construction and is divided into two sections: (i) combustion chamber and (ii) checker work. Figure 4.4 shows the hot blast stove.

The stove is a tall cylindrical (height: 20–36 m and diameter: 6–8 m) steel shell, lined with insulating bricks. The combustion chamber is an open space extending the whole height of the stove.

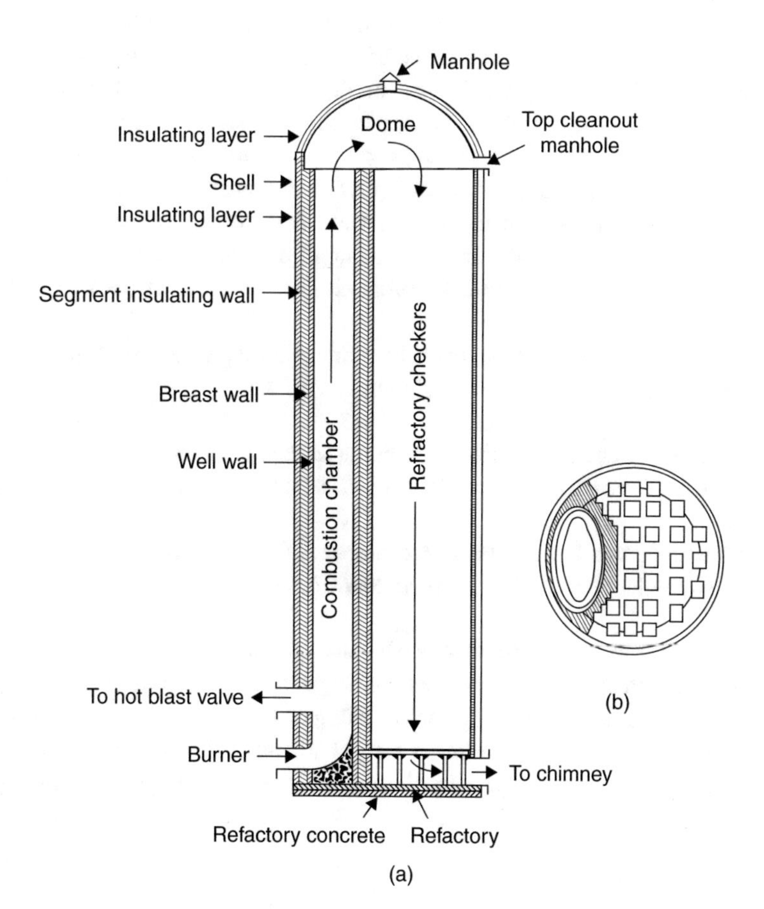

Fig. 4.4 Side combustion stoves: **a** vertical section and **b** horizontal section showing details of checker work

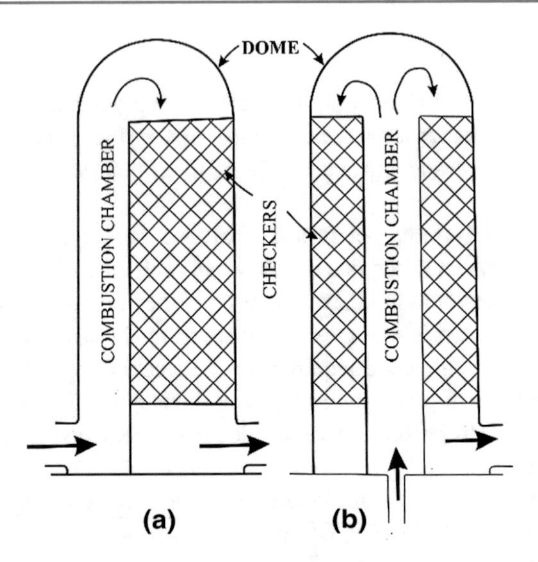

Fig. 4.5 BF stoves, **a** side combustion type and **b** central combustion type

A supply of air and gas is admitted at the bottom of this chamber in which combustion takes place, i.e. *on gas*. The hot products of combustion gases then pass through the checker work, to which they transfer their heat to the checker work; heat is therefore stored in that brickwork. When it is desired to heat the blast, the gas and air supply are cut off, and the cold blast is passed through the stove in the opposite direction, i.e. *on blast*. During its passage through the stove, the blast absorbs heat from the brickwork. Therefore, the stoves are alternatively *on gas* and *on blast*; at least two stoves must be provided for each furnace. One stove is heating up, while the other is pre-heating the blast. Such an arrangement, however, is not too satisfactory, as wide fluctuations in result of blast temperature. Three stoves, one pre-heating the blast and two heating up, give a more regular blast temperature. Sometimes, prefer to provide four stoves per furnace. The provision of the fourth stove simplifies repair periods, which are inevitable.

There are mainly two types of stoves: (a) side combustion type and (b) central combustion type. The location of the combustion chambers is shown in Fig. 4.5.

(a) Side combustion type: The combustion chamber is located at the one side of the stove and oval in cross section. This is known as *side combustion type*. It is American designs. It is mostly used and more popular.
(b) Central combustion type: The combustion chamber is in the centre, and the checker work is all around it in an annular shape. This is called *central combustion type*.

The checker work is usually built up of specially shaped bricks. These bricks are designed to give the maximum heating surface and, at the same time, to provide an opening sufficiently large to allow the free passage of the gases. When using unclean BF gas, the size of these openings is of supreme importance, owing to the tendency for dust in the gas to cause choking. The use of small checker openings, however, necessitates the use of thoroughly cleaned BF gas if choking up is to be prevented.

There are three main methods of stove operation:

(1) A two-stove system, with one stove *on gas* and other *on blast*, but this system is very unsatisfactory since any trouble with one stove immediately affects the furnace operation.
(2) A three-stove system, with two *on gas* and one *on blast*, which is the more normal practice at present.
(3) A four-stove system, with two *on gas* and two *on blast*, which is receiving serious consideration in many localities. If trouble is experienced with any stove, it is possible to revert to a normal three-stove operation.

The stove height will be considerably less in a four-stove than in a three-stove system, but a much more complex system of main valves and controls is required. The valves used on stoves are of considerable importance. Defective valves not only lead to heat losses but may poison the atmosphere and cause explosions.

4.3 Blast Furnace Control

4.3.1 Control of Temperature

Controlling the blast furnace with the aim of two objects:

1. The smooth operation of the furnace,
2. The correct grade and quality of the products.

The control of BF depends on the variables like heat, control of slag and choice of raw materials. The hearth temperature depends on two factors: (i) the amount of fuel in the burden and (ii) the temperature of the blast.

This blast temperature can be understood by the following methods:

1. Appearance through the tuyeres: If the furnace is working cold, a reddish colour will be seen and the coke can be readily discerned as it plays in the blast. If the hearth is very hot, a white dazzling light will be seen, which makes the coke not discernible.
2. Appearance of the slag: The relative temperature of the slag can be estimated as it flows from the furnace, by judging its fluidity and colour.
3. Appearance of the metal: Metal flows from a furnace with a clear, rapid flow, free from scum, and does not easily skull the runner. Iron from a cold furnace runs sluggishly, tending to skull the runner. It is generally accompanied by a considerable amount of scum.

If the furnace is working in cold condition, the temperature may be adjusted by:

(i) Increasing the blast temperature, such as putting on a fresh stove,
(ii) Reducing the volume of blast, if insufficient fuel is charged, the quantity of blast may be excessive for coke reaching the tuyeres zone,
(iii) Increasing the coke in the burden and charging a coke blank (at extreme cases).

A cold furnace is an indicative of insufficient fuel in the burden. A hot furnace can be corrected by:

(i) Increasing the volume of the blast, thereby melting more iron and increasing the output,
(ii) Increasing the burden, whereby the fuel cost per tonne is reduced.

4.3.2 Control of Composition

The composition of the blast furnace's product can be controlled by the following means:

1. Silicon: Silicon is largely a function of temperature, although it is to some extent affected by the basicity of the slag. Naturally, a highly basic slag will tend to hold the silica in combination as a silicate. Temperature, however, has a much more important effect. A high temperature favours the reduction of silicon and produces a high silicon–iron.
   Silica is reduced by solid carbon and is not affected by CO gas. Therefore, the silicon content of the iron is normally controlled by varying the proportion of fuel in the burden. A high hearth temperature will produce a high silicon and low-sulphur hot metal, whereas a basic slag will give hot metal of low silicon and sulphur.
2. Phosphorus: Phosphorus can only be controlled by selection of the raw materials, since at least 90% of the phosphorus passes into the metal. It is probably safer to assume that all the phosphorus is absorbed by the hot metal.
3. Sulphur: Sulphur is controlled by temperature and the basicity of the slag, a high temperature and a basic slag favouring low sulphur. The production of low sulphur–iron, however, is best achieved, when possible, by selecting low-sulphur raw materials.
4. Manganese: The amount of manganese which enters the hot metal depends on the hearth temperature and the composition of the slag. The reduction of MnO is endothermic; therefore, a high hearth temperature will favour a high percentage of manganese passing into the hot metal.
5. Carbon: The presence of silicon and phosphorus tends to reduce the carbon content in hot metal, whereas manganese tends to increase it (as shown in Fig. 3.4).

$$\text{Total carbon}\,(\%) = 4.5 - 0.25 \times \%\text{Si} - 0.3 \times \%\text{P} + 0.03 \times \%\text{Mn}. \tag{4.1}$$

4.4 BF Cooling Arrangements

Provision of good-quality refractory lining is by itself no guaranty that the furnace life would be more. For longer life and normal functioning of the BF, cooling of lining particularly in the hearth and bosh regions is essential. It keeps the shell temperature within limits and prevents expansion of lining in normal working conditions.

Advantages:

(i) Maintained proper temperature gradient across the lining thickness,
(ii) Reduction of effective lining thickness to get more useful volume for the same size of the furnaces.

Cooler designs are depending on:

(i) With size of the furnace and
(ii) Location on the furnace where it is to be installed.

Types of cooler:

1. Box cooler: That is to be inserted in the lining,
2. Spray cooler: That is used externally on the shell,
3. Water jacket: That is used on the shell,
4. Cooling pipes: That are for air or water circulation for cooling the bottom.

(1) Box cooler: Box coolers are three types: (i) cantilever type, (ii) L-type and (iii) plate type.

Designs of box cooler for stack, bosh and hearth regions are different. Cast iron boxes, with steel tubes embedded inside, are used for stack cooling; nowadays, copper coolers are used.

Copper boxes with copper tubes embedded inside are used for belly, bosh and hearth since cooling must be more effective at these areas than in the stack.

For smaller furnaces, cantilever-type coolers are used. There are 12 rows with 24 coolers of each row which are staggered in the checker fashion. Coolers are inserted in the lining.

For larger furnaces, L-type coolers are used. There are 7 rows with 40 coolers of each row. Plate-type flat copper coolers are used to cool the lining in the bosh and hearth walls of the furnace.

(2) Water sprays: Recent trend is to use external water sprays to cool the shell. This is more effective if carbon lining is used. Shells are cooled by water sprays halfway down the stack to the bottom. Upper half of the stack is cooled by the cantilever-type coolers.

In addition to the inserted coolers, external water cooling of almost the entire shell is adopted, since carbon is used for lining the hearth or hearth and bosh regions. Thermal conductivity of graphite is nearly 18 times superior to ordinary carbon blocks. Hence for effective cooling of the hearth, graphite blocks are preferred than carbon blocks.

(3) Water jacket: External water spray cooling of the hearth is rather difficult because of the oblique walls. Use of water jacket for cooling the bosh and belly is recommended. Cooling jackets are welded onto the shell jacket through which water passes in an upward direction. Lining thickness is reduced to 600–800 mm.
(4) Cooling pipes: Air or water circulation for cooling the bottom is done with the help of piping.

Probable Questions

1. What is the scheme for BF gas cleaning? Discuss in brief.
2. Explain the set-up used for semi-fine cleaning of BF gas.
3. Discuss the working and construction of the set-up used for effective reutilization of a clean gas for ironmaking.

4. Discuss the primary cleaning and secondary cleaning of BF gas.
5. Compare dry and wet-cleaning processes of BF gas.
6. What are types of blast furnace stoves?
7. Draw a neat diagram of blast furnace stove and explain its working principle.
8. Discuss (i) control of temperature and (ii) control of composition in a BF.

References

1. http://ispatguru.com/cleaning-of-blast-furnace-gas/
2. R.H. Tupkary, V.R. in Tupkary *An Introduction to Modern Iron Making*, 4th edn. (Khanna Publishers, Delhi, 2012)
3. A. Ghosh, A. Chatterjee, in *Ironmaking and Steelmaking: Theory and Practice* (PHI learning Pvt Ltd, New Delhi, 2008)

5 Operation of Blast Furnace

Some operation of the furnace is done before starting of the furnace, or before shutting down for breakdown of the furnace. If demand of hot metal is decreased to the extent, production is no longer required, or for major repairs; the decision may be made to blow out a furnace. There are some operational problems, faced by BF operators, described. And BF refractory is also discussed.

5.1 Operation of the Furnace

5.1.1 Blowing In

After making or major repairing of blast furnace, the process of starting the furnace is known as *blowing in*. It is carried out in four steps, viz. (a) drying, (b) filling, (c) lighting and (d) operating until routine production is established.

(a) Drying: Newly constructed furnace and stoves are carefully dried before they are put into service. The preliminary drying of the lining is most important. The reason is to avoid as much as possible extreme thermal shock to the structure and to drive off the vast amount of water absorbed by the brick during construction and contained in the slurry used in brick laying.

(i) Where natural gas is available, the gas pipe is placed in the lower combustion chamber, starts with a small flame and increases gas input for several days until a small quantity of BF gas can be used, keeping the natural gas as a pilot light. This is usual method for stoves. It is desirable to slowly increase the heat for at least 10 days to 2 weeks in a new stove before starting to bring the unit up to operating temperatures.

(ii) Another method of drying or heating, where natural gas is not available, is a wood fire built in the bottom of the combustion chamber. The wood fire requires consistent attention until wall temperatures are enough to insure proper combustion of BF gas.

(iii) The best method for drying of a BF is the using of hot blast. It is simple, and drying is done under control ways. In this method, the conventional hot blast system is used, similarly to an operating furnace except that initial blast temperature is held about 205 °C and wind volume at slow blast level. Temperature is slowly increased for a couple of days to 427 °C and maintained for a few additional days. The entire operation can be done for a week. In single furnace plants, the above method cannot be entirely ruled out because of unavailable BF gas.

S. K. Dutta and Y. B. Chokshi, *Basic Concepts of Iron and Steel Making*,
https://doi.org/10.1007/978-981-15-2437-0_5

(iv) Another method for lack of a better name it is called hearth fire method. It simply consists of a wood, coke or coal fire in the hearth of the furnace and controlled similarly to the other methods with tuyere shutter and bleeders.

(b) Filling: At the end of the drying operation, furnace's bells and bleeders are opened; blow pipes taken down, some tuyeres and coolers are removed; remaining of drying process is cleared away. In a relatively short time, the interior of the furnace will have to cool sufficiently to permit person entering the hearth to prepare for filling. In the event, if hearth fire is used for drying, all ashes and remaining residues are removed.

(i) Some activities are taking place, the final chore before filling is started. Coke is then charged up to the level of the coolers by the regular filling equipment like skips, small and big bells. Tuyeres and sometimes blow pipes may be installed at this time, but probably the blow stock caps will not be closed until time for the wind to be put on and the furnace lighted.
(ii) One practice was the use of cord wood. Four-foot lengths were very carefully arranged in the bottom, log-cabin style; each layer 90° from the next, or layers were set up on end or combinations of the two were used until the hearth and bosh were filled to the mantle. Care was exercised to provide soft kindling wood around the tuyeres and frequently charcoal, oil-soaked waste or shavings were preferred to start the fire. A heavy scaffold was built in the hearth to hold the cord wood, coke and burden. When the scaffold burned out and collapsed, stock movement was assured, at least for the moment.
(iii) But in returning to current practice with the initial coke at tuyere label, a second coke blank without flux (limestone) is charged in quantity to fill the bosh to about an equal amount is charged with limestone and siliceous material, possibly some BF slag.

(c) Lighting: Any one of several methods may be used to light a blast furnace; however, hot blast lighting consists of blowing hot blast into the furnace; low blast volume is applied, and within a matter of minutes, all tuyeres are become bright. BF coke is easily ignited and blast temperature at around 370 °C.
(d) Operating until routine production is established: The moment blast is turned into the tuyeres, and the furnace is lighted; official operation is underway. Except for the first four or five days when higher than normal temperatures are maintained, things begin to settle down to routine. It may well be assumed that except for an infrequent shutdown for a few minutes, like the replacement of copper cooling nozzles and possibly one major repair outage lasting for 4 or 5 days to replace the bells, otherwise continuous operation may be expected for the next 5–7 years.

5.1.2 Banking

Plans for an extended shutdown or interruption to furnace operation take place by a breakdown. Scheduled repair or because business conditions indicate a pause in production is desirable may influence management to bank blast furnace. The temporary shutdown of blast furnace operation is known as *banking*. The word banking is applied because of similarity to the operation of banking a fire. The origin is lost in antiquity; however, generally it means covering a fire either with ashes or fresh fuel to restrict air, reduce the combustion rate and the preserve the heat for future use.

Occasionally it is necessary; because of labour troubles, shortage of raw materials supplies, extensive repairs, etc.; to suspend the operation of the furnace temporarily. This is done by banking the furnace, which consists of taking off the blast, closing all air inlets and smoothing the stock with fine ore. During the period that the furnace is banked, some combustion will take place even through all efforts to exclude air are taken. This is overcome by charging a large blank of coke (65%) before taking off the blast. As soon as the coke blank has reached the top of the bosh, the blast is taken off and the furnace is tapped clear of hot metal and slag, leaving nothing to solidify during the stoppage.

5.1.3 Blowing Out

If demand of hot metal is decreased to the extent, production is no longer required, or for major repairs, the decision may be made to blow out the furnace. Blowing out is accomplished in two general ways: (i) charging is stopped and the stock allowed to descend until a minimum remains and (ii) iron bearing components are removed from the burden and silica gravel or coke is substituted until all iron is reduced. The latter method is preferred since the blow out can be accomplished faster, and the longer time required to take out is partly compensated for by the longer blow out time required in the former method.

5.2 Operational Problems of BF

There are some operational problems faced by BF operators, such as:

1. Hot spots,
2. Scaffolding,
3. Slipping,
4. Breakouts.

5.2.1 Hot Spots

Hot spots may develop in the stack, due to local failure of the lining. The hot gas ascending from the tuyeres must contact the descending charge to transfer the heat to it and react with it chemically. If all the fines are segregated into region, the permeability of that region will be reduced to such an extent that most of the gas will bypass it, and the burden materials will not be heated properly. If too much coarse materials are placed close to the wall, most of the gas will flow up along the walls and will rapidly wear away the brick lining. In some instances, the heat may become so intense that the steel shell may reach a red heat; such a condition is known as a *hot spot*.

When this occurs, the furnace can sometimes be kept in blast by the liberal use of water on the shell until the stack cooling plate can be inserted in the location of the hot spot to solidify a scab in that area.

5.2.2 Scaffolding

Scaffolding may occur near the top of the bosh. This condition is often due to irregularities in the working of the furnace, such as low fuel ratio, increased refractoriness of the slag, increased moisture in the blast, or a change of blast temperature.

If the zone of fusion is suddenly lowered, the pasty mass at its top tends to adhere to the encircling wall, with the result that an incrustation is formed that projects towards the centre of the furnace. This mass offers obstruction to both the gases and the descent of the stock.

5.2.3 Slipping

Slipping occurs due to an initial wedging or bridging of the stack in the furnace. When this occurs, the material underneath continues to move downward, and a void is created. The void tends to increase in size until the bridge collapses, causing a sudden downward movement of the stock above. In severe cases, this causes a sudden increase in gas pressure and an effect like an explosion. There is reason to believe that hanging or wedging or bridging may be due to at least four separate causes.

(i) Fine carbon deposited by the reaction, $2\{CO\} \rightarrow \{CO_2\} + C(s)$, may fill the voids between large particles of ore and thus impede counter-current flow of gases and materials.
(ii) Previously fused slag may solidify, causing a large impervious mass that interrupts the smooth movement of the charging materials.
(iii) Where the coke contains considerable fines, these may be wedged between the larger lumps by the flow of gases and cause both arching of the solid material and resistance to flow of the gases.
(iv) Alkali vapours may condense in the upper part of the stock and cement with the solid material into large impervious masses.

5.2.4 Breakouts

Breakouts are due to failure of the walls of the hearth, as a result hot metal and/or slag may flow out of the furnace in uncontrolled manner. Slag breakout may be chilled by streams of water, and the slag hole is closed by laying brick, ramming a plastic cement or asbestos rope. Metal breakout may be controlled by closing tap hole by refractory material.

5.3 BF Refractory

High-duty fireclay bricks have the life less than a few years. Thickness of lining depends on the furnace size and location in the furnace. The chief causes of failure of the lining are as follows:

1. Attack of CO gas,
2. Action of alkali vapours,
3. Action of basic and alkaline slag,
4. Abrasion by solids, liquids and gases,
5. Temperature,

6. Action of molten metal,
7. Operational condition and depend on design,
8. Blowing in procedure.

These factors may not be operative at all zones in the furnace, e.g. lining must withstand abrasion by solid materials and attack by CO gas at stack region. Lining must stand high temperature, erosion by ascending gas and attack of molten basic and alkaline slag at bosh region. Lining must stand action of molten slag and metal at hearth zone.

Stack Lining: Lining should have a very good abrasion resistance and resistance to CO gas attack; since temperature is low, i.e. refractoriness is relatively less important, but good dense refractory is ideal. High-fired, super-duty (35–40% Al_2O_3) fireclay bricks are used. 60% Al_2O_3 fireclay bricks are used for lower part of stack zone. Fireclay bricks are made by machine moulding under high pressures and de-aired conditions; these lead to a high bulk density of the refractory.

Bosh Lining: Due to high temperature and chemical reactions in this region, lining should be good refractoriness; it should be resistance to action of basic and alkali slag. High-fired, super-duty (45–65% Al_2O_3) fireclay bricks are used. Carbon blocks for lining possess better properties of high thermal conductivity than high-duty fireclay bricks. Carbon-lined walls can be cooled by spray coolers or water jackets.

Hearth Lining: Lining should prevent breakouts. Earlier fireclay bricks suffered from frequent breakouts due to Fe oxidation by gases and that penetrated fireclay bricks to lower down the melting point of refractory; ultimately failure occurs.

$$FeO + SiO_2 = FeSiO_3 \tag{5.1}$$

Carbon and tar mixture have excellent resistance to breakouts. Carbon has high refractoriness, high thermal conductivity, high abrasion resistance and high bulk density with low porosity. Carbon hearths are air cooled or water cooled at the bottom. Graphite–silicon carbide bricks are given excellent performance both for bosh and hearth.

Advantages of carbon lining:

1. Increase in overall life of the furnace,
2. Minimum breakouts,
3. Cooling arrangement becomes simpler,
4. More uniform wear of the lining,
5. Stack cooling may not be necessary,
6. Clean surface in contact with slag and metal,
7. Relatively thinner lining.

All these advantages compensate for more than the additional cost of lining. Gases and steam can damage the carbon lining:

$$\{CO_2\} + C(s) = 2\{CO\} \tag{5.2}$$

$$\{H_2O\} + C(s) = \{H_2\} + \{CO\} \tag{5.3}$$

Leaking from water coolers can lead to the formation of steam, due to contact of water with the red-hot lining. Carbon and tar mixture are used at slag and metal tap hole regions.

Probable Questions

1. What is meant by 'banking' of a blast furnace? How it is achieved?
2. Discuss (i) blowing in, (ii) banking and (iii) blowing out operations of BF.
3. Discuss operational problems of BF.
4. Discuss the following operational troubles in a BF: (i) pillaring, (ii) breakout, (iii) flooding and (iv) channelling. Also suggest the suitable remedies in each case.
5. Discuss BF refractory.
6. Why carbon blocks are used at the hearth of BF?

References

1. J.S. Kirkaldy, R.G. Ward, in *Aspects of Modern Ferrous Metallurgy* (University of Toronto Press, London, 1964)
2. R.H. Tupkary, V.R. Tupkary, in *An Introduction to Modern Iron Making*, 4th edn. (Khanna Publishers, Delhi, 2012)

Part II
Alternate Methods of Ironmaking

Introduction

Smelting of iron ores in modern blast furnaces (BFs) essentially depends on the availability of abrasion-resistant coke of satisfactory physical and chemical characteristics. The metallurgical coal is therefore the most valuable commodity. Coking coal is not only costly, and the prices are upwards. But it is also a scarce commodity since its deposits are fast getting exhausted. This has promoted to induce new technology for ironmaking and innovate the operation to reduce the coke consumption in BF.

Therefore, alternative routes are developed to reduce/smelt iron ores using other alternative sources of energies like non-coking coal, charcoal, natural gas, or any other type of fuel, and electricity. Iron ore smelted to form molten iron i.e. hot metal (impure metal) which essentially to refine and finally produce steel. Any alternative route of iron production should be such that the production of steel should finally be technologically feasible and economically viable.

The driving force behind the interest in alternative ironmaking processes is as follows [1]:

- Exploitation of market opportunities,
- Improvement in the productivity of the existing iron- and steel-producing units,
- Increase in operational flexibility in terms of the raw materials used, minimum capacity for efficient operation, etc.,
- Efforts to utilize less expensive and easily available raw materials, such as non-coking coals and non-agglomerated ores,
- Providing opportunities for utilizing waste (which contains iron oxide) generated by the existing steel production units,
- Upgradation in the metallic charge relative to what is in current usage,
- Improvement in steel quality,
- Reduction in the capital and operating costs,
- Increasing attention towards the deleterious impact of emissions from coke ovens and sinter plants.

Direct reduction (DR) processes are mostly gas-based, which are globally dominated by Midrex technology (65% of world DR production). Ore (fines, pellets or lumps) is reduced in solid state and solid sponge iron produced, unlike BF process where hot metal is produced. Sponge iron or direct reduced iron (DRI) offers an alternative steel production route, DR-EAF to BF-BOF and scrap-EAF routes. DR process offers an attractive route due to its low capital investment and its suitability to local raw material situations. Despite the advantages, the DR route suffers from drawbacks such as the small scale of operations, which acts as a barrier for energy efficiency investments. Also, sponge iron is prone to re-oxidation unless passivated or briquetted. Sponge iron or DRI production is common in the Middle East, South America, India, Mexico, etc.

Blast furnaces have played a major role in producing hot metal because of their high efficiency, high productivity and high degree of gas utilization. Blast furnace ironmaking is still the dominant process to produce liquid iron from ore throughout the world. But BF process depends on scarce metallurgical coal and several process steps like coke making, sintering, etc. To minimize capital investment and reduce harmful impact on environment, *smelting reduction* (SR) technologies are coming up to offer such alternatives to blast furnace ironmaking.

Smelting reduction (SR) has been receiving considerable attention. SR processes have the potential to complement BFs for producing hot metal. In fact, for iron production in limited tonnages where the cost is often prohibitive if small-capacity BFs are utilized, SR processes may be the preferable option. The most attractive feature of SR is that it is totally independent of coke. The long-standing need of cokeless hot metal production, thereby decreasing the deleterious impact of excessive CO_2 generation, has been effectively addressed by the advent of SR [2].

The energy requirement of the DR-EAF route compared with the BF-BOF route is shown in Table II.1. The amount of electricity required for melting DRI in EAF also makes the process less efficient in terms of energy use (18 GJ/t of crude steel compared with 16 GJ/t in the BF-BOF route [3], as shown in Table II.1). The clear potential of this proven technology, however, is the removal of the need for coke ovens and reducing CO_2 emissions.

Table II.1 World's best practice of primary energy intensity values for iron and steel [3]

Production step	Process	BF-BOF	SR-BOF	DRI-EAF	Scrap-EAF
Material preparation	Sintering	2.2		2.2	
	Pelletizing		0.8	0.8	
	Coking	1.1			
Ironmaking	Blast furnace smelting reduction	12.4			
	Direct reduced iron		17.9	9.2	
Steelmaking	Basic oxygen furnace	−0.3	−0.3		
	Electric arc furnace			5.9	5.5
	Refining	0.4	0.4		
Total (GJ/t steel)*		15.8	18.8	18.1	5.5

*(Values are GJ /t steel, and the primary energy includes electricity generation, transmission and distribution losses of 67%.)

These alternative routes are broadly classified as follows [2]:

1. Production of sponge iron processes, solid-state reduction of iron ore using cheaper solid fuels and gaseous fuels,

2. Smelting reduction processes, which operate in two stages, i.e. first pre-reduction and then reduction melting in the second stage; energy requirements are met by the use of coal-burning with oxygen gas,
3. Low shaft furnace, i.e. mini-BF, in which the strength of ore or coke is not so important; non-metallurgical coal can be used as a fuel,
4. Charcoal BF, using charcoal as a fuel,
5. Submerged electric arc furnace, using poor quality of coke as reducing agent and thermal requirement of the process to be met by electrical energy.

There has been a great effort to develop alternate ironmaking technologies such as ***direct reduction*** and ***smelting reduction processes***. These have triggered the development of alternative routes for ironmaking which are not dependent on the use of metallurgical coal. Investigations into alternative routes of ironmaking were initiated with the objective of utilizing the locally available raw materials (Table II.2). As of now, there are basically two types of alternative ironmaking processes which are in commercial operation:

1. *Direct reduction (DR) processes,*
2. *Smelting reduction (SR) processes.*

Table II.2 Comparison of ironmaking and steelmaking processes [4]

Ironmaking				Steelmaking	
Process	Feedstock	Reductant	Product	Reactor	Location
Blast Furnace, MBF	Lump ore, sinter and pellets	Metallurgical coke	Hot metal	BOF, EAF	Coastal, near mine side
Rotary kiln	Lump ore	Non-coking coal	DRI	EAF/IF, BOF	Mine side, central
Midrex, HyL	Lump ore, pellets	Natural gas	DRI/HBI	EAF/IF, BOF	Gas-rich area
Fastmet	Iron ore concentrate, BF/BOF/EAF dust → pellets	Non-coking coal	DRI	EAF, BOF	Steel mill or central
Fastmelt	Iron ore concentrate, BF/BOF/EAF dust → pellets	Non-coking coal	Hot metal	EAF, BOF	Steel mill or central
ITmk3	Iron ore concentrate → pellets	Non-coking coal	Iron nuggets	EAF, BOF	Mine side, central
Corex	Lump ore, pellets/sinter	Non-coking coal	Hot metal	EAF, BOF	Steel mill or central
Finex	Fine iron ore	Non-coking coal	Hot metal	EAF, BOF	Steel mill or central

References

1. S.K. Dutta, *JPC Bulletin on Iron & Steel*, VII(11), Nov 2007, p. 23
2. A. Chatterjee, Proc. Inter. Seminar on *Alternative Routes for Ironmaking in India, Steel Tech*, (Kolkata, 2009), p. 5

3. H.Y. Sohn and Y. Mohassab, *Ironmaking and Steelmaking Processes: Greenhouse Emissions, Control, and Reduction:* Pasquale Cavaliere (Ed), Chapter: 25: Greenhouse Gas Emissions and Energy Consumption of Ironmaking Processes, (Springer International Publishing, New York, 2016), p. 427
4. S.K. Dutta and R. Sah, *Alternate Methods of Ironmaking (Direct Reduction and Smelting Reduction Processes).* (S. Chand & Co Ltd, New Delhi, April 2012)

Raw Materials for DR Processes 6

6.1 Introduction

Direct reduction (DR) processes are very sensitive to chemical and physical characteristics of raw materials used in the process. Iron ore or pellets, reductant (i.e. non-coking coal or natural gas) and limestone/dolomite are the main raw materials for DR technology. For the successful operations, the process of DR technology has specified the characteristics of the raw materials to be used in the process.

The principal raw materials used in DR/sponge iron processes are iron ore (lump, pellets or fine), non-coking coal or natural gas, and dolomite or limestone.

6.2 Iron Ore

6.2.1 Characteristics of Iron Ore

Lumps or pellets should have high iron content, low gangue content, good mechanical strength, readily reducible and non-decrepitating variety. The characteristics of the iron ore feed to the reactors which are as following:

- Chemical composition,
- Reduction properties and
- Physical characteristics.

6.2.1.1 Chemical Composition

The key input materials required for sponge iron production are the iron ore. Since sponge iron-making is a solid reduction process, the gangue in the ore is retained in the product. Due to oxygen removal, there is a reduction in weight by about 30%, and the percentage of gangue material goes up by about 1.4 times. Hence, it is very important to choose a higher grade of iron ore input.

In all sponge iron processes, the only noteworthy chemical change which takes place is the removal of oxygen from the iron oxide in iron ore charged. Since no melting or refining occurs, all impurities in the ore feed get concentrated in the reduced product. As a result, in any iron ore chosen

S. K. Dutta and Y. B. Chokshi, *Basic Concepts of Iron and Steel Making*,
https://doi.org/10.1007/978-981-15-2437-0_6

for DR process, the total iron content should be as high as possible, and the gangue content should not only be a minimum, but the gangue should have a composition which is acceptable to the user of the final product. In general, an ore with iron content higher than 65% is preferred. The gangue content of sponge iron, i.e. its silica plus alumina level, should be as low as possible. Unfortunately, very low silica contents in the ore often result in excessive swelling and decrepitation together with an increased tendency towards sticking within the reduction reactor, which directly affects the furnace space availability, i.e. productivity. Further, if the silica content of the sponge iron made is inadequate, it could be more prone to degradation during handling. The alumina content is not so critical from the point of view of reduction within the furnace but being an impurity, which is acidic in nature, it would require extra fluxes during steelmaking. Thus, normally the total amount of silica plus alumina in the ore feed should not be allowed to exceed 4%; in some cases, it may be too high. Experience with many selected hard lump haematite ores (containing over 67% Fe) has shown that such ores are often not easily reducible and have poor decrepitation characteristics, whereas medium hard ores of relatively poorer quality (containing over 64–66% Fe) are more likely to be suitable for sponge iron production. Thus, a compromise may have to be made between the gangue content of the ore and its reducibility or decrepitation characteristics while choosing the ore feed for sponge iron processes.

The phosphorus content of the ore is also extremely important and should be as low as possible (preferably below 0.03%) because there is no removal of phosphorus during the reduction processes. Some ores, e.g. many Indian ores, generally content between 0.04 and 0.08% phosphorus, and if they are used, the phosphorus content of sponge iron would be at least 0.055% and even as high as 0.10% which is substantially higher than phosphorus in scrap normally used in steelmaking. Like phosphorus, the sulphur content of the ore should be low (below 0.02%) since during reduction, there could be some pickup of sulphur. The sulphur content of many iron ores available in the world is generally quite low (0.01–0.02%) and consequent low sulphur in sponge iron is a positive feature, particularly if the phosphorus levels are high.

The specific requirements of lime (CaO) and magnesia (MgO) in the ore feed, if charged as pellets, are generally determined by the slag characteristics required for steelmaking, up to 2.5% CaO and 1.0% MgO are usually acceptable. Higher percentages of these constituents (used as a binder in some pellets) may have an adverse effect on the reducibility of the pellets. Similarly, though titania (TiO_2) is a slag forming constituent, in excess amounts, it may have a deleterious effect on reducibility thereby imposing limits on the maximum degree of metallization which can be achieved in a given reactor. In general, the titania content should not exceed 0.15% in the ore feed. Steelmaking grade sponge iron requires iron ore having a minimum of 65% total Fe, maximum of 4.0–5.0% gangue (SiO_2 and Al_2O_3), and low sulphur and phosphorous content (less than 0.05% each).

6.2.1.2 Reduction Properties

The iron ore should have adequate reducibility and favourable decrepitation characteristics if any sponge iron process must be techno-economically feasible. The retention time required to reduce iron oxide to sponge iron of a specific quality depends, to a large extent, on the reducibility of the ore. There exists a direct correlation between reducibility and retention time, and hence, the productivity of any given reactor. The higher is the reducibility, the lower is the retention time required and consequently, the higher is the productivity.

The actual breakup of the ore is the combined effect of thermal fragmentation as well as reduction degradation. It has been observed that most lump ores are susceptible to some amount of thermal (the result of a stress release phenomenon) which takes place when the ore is heated from room temperature to a temperature of around 400 °C. With a high thermal decrepitation index, the extent of generation of fines even with gradual heating can be as high as 8–10% leading to a low overall yield of sponge iron and limited reactor's space availability. Reduction degradation is a chemical phenomenon which takes place during the initial stages of reduction when haematite is converted to magnetite and expansion occurs. Depending on the type of the ore, fines generation owing to reduction decrepitation may be as low as 3–4% or as high as 10–15%.

Another important characteristic of the ore feed, if it is in the form of pellets, is its swelling index since it can influence the descent of the bed and the bed permeability. This index is a measure of the magnitude of expansion that takes place during the process of reduction and is expressed as the percentage change in the original pellet volume. For smooth furnace operation, it is desirable that the swelling index of pellets should be reasonably low (below 18%).

6.2.1.3 Physical Characteristics

The important physical features of the ore feed are its size range, abrasion index and compression strength (particularly in the case of pellets). The optimum size of the iron ore is determined essentially by its reducibility characteristics; with higher reducibility, the top size can be increased. In the case of highly decrepitating type ores, a larger size may be necessary so that even after degradation, the generation of fines (below 3 mm) is minimal. In the case of lump ores, a size range of 8–20 mm is normally preferred with a mean size of around 12 mm whereas, in case of pellets, the same is 6–22 mm with a majority of 9–16 mm sized pellets. Fines in any form are undesirable. In rotary kilns, they promote ring formation inside the kiln, thereby affecting the space of kiln availability and lead to an increased tendency of the ore particles to segregate from the rest of the bed. Careful screening is essential to guarantee removal of the adherent fines which is particularly difficult in the case of clay bearing lump ores. Abrasive of the ore can also contribute to generation of fines because of physical disintegration arising out of the tumbling action. To minimize the generation of fines in the kiln, the abrasion index of the ore should obviously be as high as possible. The fine concentrates cannot be processed in sponge iron reactors such as Midrex, HyL, Rotary hearth and Rotary kiln.

Pellets having a more regular shape and superior reduction characteristics are superior to lump ores as the iron oxide feedstock, especially because of a more uniform size, better chemical consistency as well as the guaranteed higher furnace availability. Extra cost of pellets is compensated by smoother operation, higher furnace availability, higher degree of metallization and higher net yield of sponge iron because of lower generation of fines, depends on the relative costs of ore and pellets and would, therefore, vary from case to case.

Tables 6.1, 6.2 and 6.3 show the typical chemical, physical and reduction characteristics of the ore to be used in sponge iron processes.

Table 6.1 Characteristics of iron ores for use in coal-based sponge iron process [1]

Chemical constituent	Percentage (%)	
Fe	62.0 min	
$SiO_2 + Al_2O_3$	7.0 max	
CaO + MgO	2.0 max	
S	0.03	
P	0.08 max	
$\sum$ Pb, Zn, Cu, Sn, Cr and As	0.02 max	
Physical		
A. Size range (mm)	**Percentage (%)**	
−5	5 max	
+25	10 max	
B. Tumbler index	**Tolerable (%)**	**Preferable (%)**
+5 mm	80	92
−28 mesh	10	5
C. Reducibility (40%)		
% per min	0.5	0.6
D. Decrepitation		
−5 mm	20	15
−0.5 mm	5	3

Table 6.2 Typical chemical composition of ore used in gas-based sponge iron process [1]

Weight (%)	Pellets	Lump ore
Fe (total)	67–69	66–68
SiO_2	0.9–1.0	0.5–0.7
Al_2O_3	0.2–0.3	0.3–1.3
CaO	1.1–1.2	0.06 max
MgO	0.80	0.05
P	0.015	0.03–0.06
S	0.006	0.005–0.008
Moisture	1.5 max	4.0 max

Table 6.3 Typical physical characteristics of ore used in gas-based sponge iron process [1]

Screen analysis (weight %) mm	Pellets	Lump ore
50–75		5 max
30–63		93 max
+15	10 max	
8–15	85 max	
−8	5 max	
−6.3		7 max
Bulk density (t/m^3)	2.0–2.1	2.0–2.6
Compression strength (kg/pellet)	270 min	
ISO tumbler test (% by wt) (mm)		
+6.3	95 min	
−0.5	4 max	

6.3 Coal

The non-coking coals should have high fixed carbon and volatiles content, but should have low ash, sulphur and moisture. The fusion point of coal ash should be high. The coal should be highly reactive but should have low coking and swelling indices.

Non-coking coals are used for alternative ironmaking processes like direct reduction (DR) processes and smelting reduction (SR) processes. Non-coking coals are needed for sponge iron production in rotary kiln process. Around 900–1000 kg of coal is consumed to produce one tonne of sponge iron [2]. Coal is required as a reductant as well as source of heat. Cost of coal is around 50% of the production cost.

Coals for sponge iron production should not have agglomeration tendency (caking properties), as it will result in accretion problems in rotary kilns. Coals of higher reactivity are preferred as they permit operation of the furnace at lower temperature with higher kiln output. Lower rank coals in the range of bituminous high volatile yield char of higher reactivity. Generally, coals with higher fixed carbons are preferred since they provide more carbon for reduction and have higher calorific value. (*Calorific value* is the energy content which is measured as the heat released on complete combustion of fuel in air or oxygen, expressed as the amount of heat per unit weight of fuel, i.e. kJ/kg.) But higher fixed carbon means lower volatile matter (VM), thereby lowering the char reactivity. Ash content should be as low as possible, as it affects the kiln throughput and heat requirement adversely. High moisture also reduces kiln productivity and increases the energy requirement. Since sulphur of coals is picked by sponge iron, it should be as low as possible. Further, melting characteristics of ash need to be considered while selecting coals for sponge iron production. Table 6.4 gives the criteria for selection of coals for sponge iron production in rotary kilns. Table 6.5 shows characteristics of non-coking coals suitable for coal-based sponge iron process.

The following characteristics are of importance in selecting the coal for sponge iron production:

1. Reactivity,
2. Ash content,
3. Ash fusion temperature,
4. Volatile matter and sulphur content and
5. Coking and swelling indices.

Table 6.4 Typical coals for sponge iron process [2]

Criteria	Unit	IS standard	Manufactures' experience
Fixed carbon	%	100 − (VM + Ash)	40 (min)
Volatile matter (VM)	%	25–35	32
Ash	%	24 (max)	22 ± 2
Moisture	%	4 (inherent)	8 (total)
Reactivity	cc/g s	1.75	2.2
Initial deformation temperature of ash	°C	>1150	>1200
Calorific value	kJ/kg	23,012	21,757
Coking index		>1	>1

Table 6.5 Characteristics of non-coking coals suitable for coal-based sponge iron process [1]

Chemical analysis	%
Ash	15 ± 2
Volatile matter (VM)	32 ± 2
Fixed carbon	By difference: [100 − (VM + Ash)]
Sulphur	1.0 max
Inherent moisture	6.0 max
Other properties	
Initial deformation temperature of ash (°C)	>1050
Coking index	>5
Swelling index	>3

6.3.1 Reactivity

The reactivity of the reductant is a measure of its ability to react with carbon dioxide to form carbon monoxide according to *Boudouard reaction*:

$$C(s) + \{CO_2\} = 2\{CO\}, \Delta H^0_{298} = 170.7\,\text{kJ/mol of C} \tag{6.1}$$

i.e. for conversion of CO_2 to CO. The reactivity has a significant influence on the reduction process, particularly the operating temperature of the reactor. The reactions of ore reduction as well as coal gasification take place simultaneously within a rotary kiln. Reactivity of the coal is the main factor affecting the kinetics of the process. Coals with high reactivity values are preferred. This type of coal makes it possible to operate the kiln at lower temperature and to achieve higher kiln productivity. Therefore, the decisive factor in rotary kiln is the ability of any coal to react with carbon dioxide to form carbon monoxide, a parameter quantified by the coal reactivity. In general, coals with high reactivity are preferred as these allow kiln operations at relatively lower temperatures. Lower kiln operating temperatures automatically decrease the tendency towards formation of accretions or ring formation inside the kiln because the difference between the solid and gas temperature inside a kiln decreases and the chances of fine materials sticking to the kiln refractory diminish.

6.3.2 Ash Content

In any reduction process, the ash of the reductant is undesirable burden. With high ash content in coal, a greater proportion of the furnace volume is occupied by waste/inert materials, thereby reducing the volume available for the oxide charge. This affects the productivity adversely. Further, high ash content in coal also enhances the tendency towards ring formation in rotary kilns.

High ash content in coal means higher ratio of coal and iron ore, which ultimately reduces the kiln capacity and hence decrease productivity to a certain extent. Hence, the ash content of the coal should be as low as possible. From the metallurgical standpoint, ash contents up to 35% are tolerable. In general, the ash content should preferably be below 20% and more than 24% ash bearing coals should not even be considered if the process is to be competitive.

6.3.2.1 Ash Fusion Temperature

Ash fusion temperatures give an indication of the softening and melting behaviour of coal ash when the coal is exposed to high temperatures. Fusion temperatures are typically measured by heating the coal sample in the form of a cone at four defined points, under both reducing and oxidizing conditions. These four temperatures are as follows [4]:

(i) *Initial deformation temperature* (IDT) is the temperature at which the point of the cone begins to get rounded.
(ii) *Softening temperature* (ST), it is also called the spherical temperature, is the temperature at which the base of the cone is equal to its height.
(iii) *Hemispherical temperature* (HT) is the temperature at which the base of the cone is twice its height.
(iv) *Fluid temperature* (FT) is the temperature at which the cone has spread as a fused mass.

Another important characteristic of any non-coking coal is the initial deformation temperature (IDT) of the ash, especially under mild reducing conditions. Normally, the IDT of coals should be at least 100^0 higher than the maximum furnace operating temperature. Since under mild reducing conditions, the IDT value has decreased by 50–80°, all ash fusion temperatures for assessing the suitability of coals for DR process should be determined under reducing conditions.

The ash fusion temperature of the coal is an important parameter. Coal having ash softening temperature above 1300 °C is preferred. High ash and low ash softening temperatures for Indian non-coking coal have restricted its usage in many sponge iron processes. Melting characteristics of the ash are also important for the sponge iron production.

6.3.3 Volatile Matter and Sulphur Content

The value of fuel ratio, which is the ratio of fixed carbon to volatile matter in naturally occurring coal, should not exceed 1.8 and should preferably be within 1.5 for selecting adequately reactive coal. It is desirable to have a coal with less than 15% ash, about 25–35% volatile matter and sulphur content less than 1.0%.

Sulphur in coal can occur as iron pyrite (FeS), calcium sulphate ($CaSO_4$) and organic sulphur compounds with carbon and hydrogen. During heating, approximately half of the pyritic sulphur is converted into H_2S and volatilized at 600 °C, while $CaSO_4$ is reduced to CaS and forms a constituent of COS and H_2S, and thus gets volatilized. The organic sulphur, however, is not affected by charring up to 1000 °C and is responsible for sulphur pickup by sponge iron. This sulphur can normally be scavenged by dolomite/limestone added in very fine form to the coal feed into any rotary kiln. The sulphur content in any coal selected should normally be below 1.0% and many non-coking coals which are available easily fulfil this requirement.

To ensure sufficiently low sulphur content in the sponge iron, the sulphur content of the coal should not exceed 1.5%. Coals with sulphur content less than 1.0% are preferred, which ensure sulphur content in sponge iron below 0.02%. However, sulphur pick up in sponge iron can be effectively controlled by addition of dolomite to the charge materials.

6.3.4 Coking and Swelling Indices

These values should be as low as possible to avoid problems of formation of large masses with low density within the reactor. Coals with a free swelling index above 3.0 should be rejected; ideally, the free swelling index of coals used in DR process should not exceed 1.0. Similarly, coals with a coking index of 5.0 (maximum) and preferably below 3.0 should only be used as they are strictly non-coking coals.

6.4 Natural Gas

In all gas-based DR processes, the reductant is natural gas. Natural gas consists primarily of carbon and hydrogen compounds in the form of hydrocarbons. Natural gas has 92–96% CH_4. Table 6.6 shows the typical composition of natural gas. If natural gas was to be directly used for gas-based DR processes, the problems would be encountered, like (i) very slow reduction as compared to H_2 and CO, and (ii) due to high-temperature carbon soot formation and consequent process problems, like choking, and unfavourable thermal balance in the reactor owing to endothermic effects. Hence, natural gas (i.e. CH_4) must be reformed into CO and H_2. Reforming reactions are endothermic nature; hence, temperature must be high and require a catalyst. The de-sulphurization of natural gas is necessary before reforming, because the nickel reformer catalysts are very sensitive to even low concentrations of sulphur. Sulphur is poisoning the Ni (catalyst), so sulphur in natural gas must be as low as possible.

6.4.1 Reforming Reaction

Methane (CH_4) is not used directly for reduction of iron ore because:

- Rate of reaction is very slow,
- Due to high temperature, decomposition of methane takes place and carbon soot is formed:

$$\{CH_4\} = C(s) + 2\{H_2\}, \Delta H^0_{298} = 73.79\,\text{kJ/mol of } CH_4 \tag{6.2}$$

Table 6.6 Typical composition of natural gas [3]

Component	Volume (%)
Methane	93.310
Ethane	1.650
Propane	0.240
Butane	0.030
Pentane	0.003
Hexane	Traces
CO_2	0.600
O_2	0.060
N_2	3.570
Gross calorific value	33,472–43,514 kJ/Nm^3
Net calorific value	30,125–39,330 kJ/Nm^3

Therefore, natural gas cannot be used in its virgin form for sponge iron process. It must be converted into a mixture, predominantly of H_2 and CO, to increase the calorific value and to increase the proportion of reducing gas in relation to the oxidizing gases (like CO_2 and H_2O). This step of conversion into H_2 and CO is known as reforming, that are taken part in reduction of iron oxide. Reforming takes care of the problem of thermal balance during reduction of iron ore since reduction of iron ore with a 75% H_2 and 25% CO mixture has almost zero heat effect on the system (exothermic CO reduction is balanced by endothermic reduction by H_2 at this proportion of the two gases).

Hence, methane must be reformed into CO and H_2 in presence of catalyst (Ni or Al_2O_3) at 850–950 °C. There are four types of reformations as follows:

- Steam reformation:

$$\{CH_4\} + \{H_2O\} = \{CO\} + 3\{H_2\}, \Delta H^0_{298} = 206.3\,\text{kJ/mol of } CH_4 \tag{6.3}$$

- Oxygen reformation:

$$\{CH_4\} + 1/2\{O_2\} = \{CO\} + 2\{H_2\}, \Delta H^0_{298} = -35.7\,\text{kJ/mol of } CH_4 \tag{6.4}$$

- Carbon dioxide reformation:

$$\{CH_4\} + \{CO_2\} = 2\{CO\} + 2\{H_2\}, \Delta H^0_{298} = 247.4\,\text{kJ/mol of } CH_4 \tag{6.5}$$

- Combination of steam and oxygen reformation:

$$2\{CH_4\} + \{H_2O\} + 1/2\{O_2\} = 2\{CO\} + 5\{H_2\}, \Delta H^0_{298} = 85.3\,\text{kJ/mol of } CH_4 \tag{6.6}$$

- Combination of carbon dioxide (30–40%, from Midrex shaft exit gas) and steam (60%) reformation:

$$2\{CH_4\} + \{CO_2\} + \{H_2O\} = 3\{CO\} + 5\{H_2\}, \Delta H^0_{298} = 226.7\,\text{kJ/mol of } CH_4 \tag{6.7}$$

Reformer chamber is heated by burning process gas (e.g. 50% H_2, 25% CO, 20% CO_2, 3% CH_4 and 2% N_2) at the burners (1100 °C). Reformer chamber is full of 280–300 INCONEL tubes which are packed with nickel (Fig. 6.1). Each tube has 0.25 m internal diameter and 10 m long. Natural gas is pre-heated to 370 °C in a coil in the reformer convection section and is then passed through a de-sulphurization system to reduce the sulphur content to permissible levels for the reformer catalyst operation. The de-sulphurization is necessary, because the nickel reformer catalysts are very sensitive to even low concentrations of sulphur. After reforming of natural gas, the composition of reformed gas is 60% H_2, 35% CO, 2.5% CO_2, 1.5% CH_4 and 1% N_2.

Reformation tube material is a stringent factor, if it is made of ceramic that will be cracked due to high temperature (850–950 °C). Because ceramic material has poor thermal shock resistance power and since the temperature is high (>900 °C), so chance of cracking is more. The ceramic tubes are also porous in nature; there is possibility to leakage of poison gas (i.e. CO). To avoid these problems, tubes are made of high-temperature alloys, e.g. INCONEL (80% Ni, 13% Cr and 7% Fe). For this reason, reforming process is a costlier process. The life of nickel catalyst is 7 years only due to:

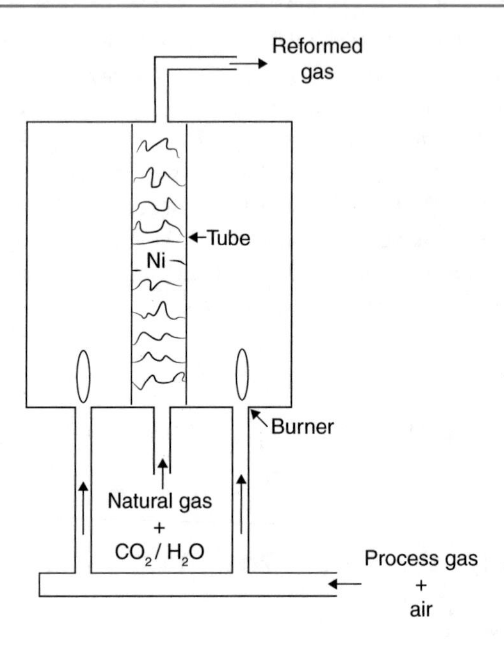

Fig. 6.1 Reforming chamber for natural gas

(i) The carbon (i.e. soot) deposits on the Ni catalyst that makes Ni inactive:

$$\text{Below}\,600\,^{\circ}\text{C} \rightarrow \{\text{CO}\} + \{\text{H}_2\} = \text{C(s)} + \{\text{H}_2\text{O}\}, \Delta H^0_{298} = -135.1\,\text{kJ/mol of CO} \quad (6.8)$$

$$\{\text{CH}_4\} = \text{C(s)} + 2\{\text{H}_2\}, \Delta H^0_{298} = 73.79\,\text{kJ/mol of CH}_4 \quad (6.9)$$

(ii) Sulphur poisoning of Ni:

$$\text{Ni(s)} + \{\text{S}\} = \text{NiS(s)} \quad (6.10)$$

The recommended analysis of natural gas to be used in gas-based sponge iron process is given in Table 6.7. For efficient gas reforming, the extent of higher hydrocarbons is limited to: C_2H_6 below 25%, C_3H_8 below 4% and C_4H_{10} below 2%. Higher amounts of these constituents increase the possibility of carbon deposition on the catalyst used for reforming. To prevent carbon deposition, the

Table 6.7 Typical composition of natural gas for gas-based sponge iron process [4]

Chemical analysis	Volume (%)
CH_4	75 min
C_2H_6	0–25
C_3H_8	0–4
C_4H_{10}	0–2
$+C_4$ hydrocarbons	0–0.5
CO_2	Max 20
N_2	Max 20
S	Max 20 ppm by wt.

amount of water vapour in the gas mixture exposed to the catalyst must be increased. Excessive amounts of nitrogen and carbon dioxide do not harm the reforming process but more than 20% carbon dioxide results in excess top gas, which only must be unnecessarily heated and cooled. Every 10% increase in nitrogen results in approximately 2% higher fuel consumption. Sulphur in the natural gas, either as H_2S (g) or as COS (g), can poison the nickel catalyst owing to the formation of NiS (s). Therefore, natural gas rich in sulphur must be de-sulphurized prior to reforming.

The natural gas required for sponge iron production should be very low in sulphur, low in higher hydrocarbons and unsaturated hydrocarbons. Natural gas shows the advantage over the coal based on CO_2 emissions. Natural gas generates only 49 t/TJ CO_2 emissions whereas coals generate about 90–94 t/TJ CO_2 emissions [5]. Hence, natural gas is an ideal energy source for ironmaking and as well as electricity generation.

Main advantages of gaseous fuels:

- It is clean source of fuel and no ash content (i.e. no residue after combustion),
- Easy to handle and
- Good combustion.

6.5 Other Raw Materials

Limestone or dolomite is used in rotary kilns as the de-sulphurizing agent. Since most iron ores contain very low sulphur (0.01–0.02%), the sulphur pickup in sponge iron is almost fully from the coal charged; even the best non-coking coals generally contain 0.5–0.6% sulphur. It has been reported that dolomite is a better fluxing agent compared with either limestone alone or a mixture of dolomite and limestone of the same size. However, both dolomite and limestone are acceptable provided the size is in the range of 1–4 mm. For minimizing accretion formation inside the kiln without sacrificing efficient de-sulphurization, it is essential that most of the de-sulphurizer (at least 60%) is 1–2 mm in size with very little of −1 mm particle. Limestone should contain minimum 46% CaO and maximum 6% SiO_2, and dolomite should contain minimum 28% CaO, 20% MgO and maximum 6% SiO_2. Dolomite is mainly used as a de-sulphurizing agent to prevent the absorption of sulphur by the hot sponge iron. The sulphur is released by the burning of coal inside the furnace.

6.6 Sizes of Raw Materials

The lump ore can be charged with a particle size of 5–25 mm. Average size of ore is around 12 mm. Size of pellets for iron ore is preferred 6–22 mm, with majority in the range of 9–16 mm. In the earlier days, the iron ore size was kept at 5–20 mm and was washed in a scrubber, but presently it has become a standard norm to use 5–18 mm ore as feed for a large kiln without scrubbing and/or washing. This has resulted in reducing the cost of iron ore fed to the kiln. The consumption of iron ore has also decreased from about 1600 kg per tonne of sponge iron to 1500 kg levels mainly due to a better understanding of the process, improvements of the equipment and increased use of automation.

The optimum size of non-coking coal is approximately 5–15 mm. Average sizes of limestone and dolomite are around 1–4 mm. The initial specifications for dolomite were 1–4 mm, later it was found that 4–8 mm dolomite was far more suitable by which the consumption can be reduced by 50%. This was mainly because lot of dolomite fines were being lost to waste gases and with 4–8 mm fraction this loss was minimized.

6.7 Composite Pellets

The two major raw materials required for sponge iron production are iron ore and coal/natural gas. India has adequate reserves of iron ore. But the country has inadequate infrastructure for catering to the iron ore demands of all the sponge iron/steel plants. The excessive fines, generated from the iron ore crushing units or during mining, are mostly going to waste. Indian iron ore deposits are partly soft and friable in nature. So, they contain a good amount of super fines (−200 mesh) rich in iron content (65% and above). These are known as *blue dust*. Estimated reserves of blue dust in India are around 550 Mt. Apart from this around 4 Mt of fine slimes containing 55 to 60% Fe are generated every year by ore washing plants in India [6]. Large amount of mill scale is also generated during hot rolling and heat treatment of steels. India also has good reserves of non-coking coal, but limited reserves for natural gas and coking coal. Fines of coal and coke are generated during mining and coking, respectively. Utilization of these fines (i.e. iron ore, coal/coke fines) for extracting metal is of vital concern for resource conservation and pollution control.

Therefore, if ***composite pellets*** are made by cold bonding technique from blue dust/iron ore fine/mill scale and coal/coke/char fines, then utilization of fines is taking place. The composite pellets should have sufficient strength to withstand high temperature and stresses in reduction furnaces [6]. The main advantage of composite pellets is utilization of cheap resources (i.e. fines) without the need for high temperatures such as sintering, coke making and conventional pellet making by heat hardening. Reduction is much faster with composite pellets than with ordinary iron ore pellets or lumps. This lowers the residence time in rotary kiln by a factor of 6–8, thus improving productivity.

The composite pellets are used as feed material in rotary hearth furnace, where reduction takes place within 10 min to produce iron nuggets. The process is known as ironmaking technology mark 3 (***ITmk3***), jointly developed by Kobe Steel, Japan and Midrex Direct Reduction Corporation, USA. Midrex Technologies in partnership with its parent company, Kobe Steel Ltd. have developed the FASTMET process as a coal-based reduction technology, applicable for processing iron ore as well as iron oxide-containing materials like steel mill wastes.

The advantages of using composite pellets in reduction processes are as follows [6]:

- Very fast reduction due to intimate contact of iron ore fines and coal/coke/char fines,
- Higher rate of production,
- Reduction in energy consumption,
- Utilization of fines, which are cheap and are being generated in large quantities, awaiting suitable disposal and utilization and
- Promising prospect for small-scale ironmaking with higher production rate.

Probable Questions

1. Discuss characteristics of iron ore for sponge iron production.
2. Discuss characteristics in selecting the coal for sponge iron production.
3. Why ash fusion temperatures of coals are important for coal-based sponge iron production?
4. Why natural gas cannot be used directly for sponge iron production?
5. What are the types of reformations for natural gas?
6. "Natural gas is required to be reformed before using for sponge iron production". Why?
7. Why ceramic material cannot used for reformation tube? Which material is used for that?
8. Discuss the principle of natural gas reforming.

9. Why sulphur content in natural gas should be max 20 ppm for sponge iron production?
10. What advantages are getting by using composite pellets as feed material for sponge iron production instead of lump/pellets of iron ore?

References

1. G.R. Singh, B.N. Mukhopadhyay, R.C. Khowala, S. Bhattacharyya, Steel Scenario **13**(3), 8 (2004)
2. S.K. Gupta, C. Bohm, Iron Steel Rev. **39**, 31 (1995)
3. B.B. Dutta, Inst. Eng. India J. CH **73**, 69 (1993)
4. S.K. Dutta, R. Sah, in *Alternate Methods of Ironmaking* (*Direct Reduction and Smelting Reduction Processes*) (S. Chand & Co Ltd., New Delhi, Apr 2012)
5. J. Kopfle, in *Green Steel Summit* (Midrex Technology, May 2010)
6. S.K. Dutta, A. Ghosh, Trans. Indian Inst. Met. **48**(1), 1 (1995)

7 Sponge Iron

Processes that produce iron by reduction of iron ore in solid state are generally classified as direct reduction (DR) processes and the products also referred to as direct reduced iron (DRI) or sponge iron. Different processes of coal-based and gas-based are discussed in detail. Forms of sponge iron, characteristics of sponge iron and re-oxidation of sponge iron are described. Uses of sponge iron are also discussed.

7.1 Introduction of Sponge Iron

The industrial development programmed of any country, by and large, is based on its natural resources. Depleting resources of coking coal, the world over and especially in India, is posing a threat to the conventional (BF-BOF/LD) route of ironmaking and steelmaking. During the Last four decades, a new route of ironmaking has been rapidly developed for direct reduction of iron ore to metallic iron by using non-coking coal/natural gas. This product is known as ***sponge iron***, which can be further used for cast ironmaking/steelmaking. Processes that produce iron by reduction of iron ore in solid state are generally classified as ***direct reduction (DR) processes*** and the products also referred to as ***direct reduced iron (DRI) or sponge iron***.

7.2 Definition of Sponge Iron

Sponge iron means porous iron produced by ***direct reduction process***. ***Direct reduction (DR) process*** is a solid-state reaction process (i.e. solid–solid or solid–gas reaction) by which removable oxygen is removed from the iron ore, using coal or reformed natural gas as reductants, below the melting and fusion point of the lump ore or agglomerates of fine ore [1]. The external shape of the ore remains unchanged. Duc to removal of oxygen, there is about 27–30% reduction in weight, a honeycombed microstructure remains which have suggested the name ***sponge iron*** (means solid porous iron, lumps/pellets, with many voids filled with air).

Sponge iron is obtained when iron ore (generally lump haematite iron ore, pellets) is reduced to metallic form in solid state. Since there is no melting, external shape of raw material is retained. Colour changes from red to black. The true density changes from about 3.5 to 4.4 g/cm^3. The true density of pure iron is 7.8 g/cm^3. Thus, there is about 45–56% reduction in true volume, and this is manifested in the formation of pores throughout the interior of sponge iron pieces.

S. K. Dutta and Y. B. Chokshi, *Basic Concepts of Iron and Steel Making*,
https://doi.org/10.1007/978-981-15-2437-0_7

The direct reduction processes have some basic advantages as follows [2]:

1. They are easily adaptable to small as well as moderately large size steelmaking units.
2. Their capital cost requirement is low.
3. They can work with off-grade raw materials.
4. They can utilize small sources of iron ore near the steel plants, thus reducing transportation cost.
5. Their products are uniform in size, as well as they contain low levels of tramp metallic elements (0.02%) compared to scrap (0.13–0.73%).

Why are sponge irons required to produce?
Because:

- Sponge iron substitute for steel scrap → due to shortage of scrap,
- Shortage of coking coal → non-coking coal or natural gas can be used for sponge iron production,
- Low investment cost → DR-EAF route 30–35% cheaper than BF-BOF route,
- Production cost of steel → DR-EAF route 10–15% cheaper than BF-BOF route.

7.3 Sponge Iron Processes

Based on the *types of reductant* used, the *sponge iron processes* can be broadly classified into two groups:

1. Using solid reductant, i.e. *coal-based direct reduction (DR) process*, and
2. Using gaseous reductant, i.e. *gas-based direct reduction (DR) process*.

Processes of sponge iron are summarized as shown in Table 7.1.

7.3.1 Coal-Based Processes

In coal-based DR processes, non-coking coal is used as reducing agent. In the solid reduction processes, iron oxides together with solid reductant (non-coking coal) are charged into the reactor. The generation of reducing gas (mainly CO) takes place in the reduction reactor, and the product must be separated from excess reductant, ash and/or sulphur-absorbing materials (lime and dolomite) by magnetic separation after discharge at low temperature, which makes product handling more complicated. Because of the presence of these substances in sponge iron, hot briquetting and hot feeding

Table 7.1 Sponge iron processes [3]

Process	Type of reactor	Type of ore use	Type of reductant	Rank[a]
Midrex	Shaft	Lump/pellet	Gaseous	1
HyL	Retort	-Do-	-Do-	3
SL/RN, ACCAR, CODIR, etc.	Rotary kiln	-Do-	Solid	2
HIB	Fluidized bed	Fine	Gaseous	4

[a]Rank is in terms of popularity and production in the world

to the steelmaking are not possible for coal-based process. Magnetic separator also does not work at high temperatures to separate the hot sponge iron.

Direct reduction (DR) processes like *SL/RN* (1970), *ACCAR* (1973), *Krupp* (1973), *CODIR*, *TDR* (1975), etc., using rotary kiln as a reactor and coal as a reductant, have been developed worldwide. India, due to its large reserve of non-coking coal, has shown keen interest in pursuing these technologies using such coal as a cheap energy source for highest sponge iron production. These processes are highly sensitive on types of raw materials used.

Out of the coal-based technologies, SL/RN technology of Lurgi GmbH, West Germany, has been the most successful one. Little success has been achieved by other coal-based technologies, namely CODIR technology of Krupp Industrietechnik, now Mannesmann Demag, Germany; DRC technology of Davy McKee, USA; ACCAR technology of Allis-Chalmers, now Boliden-Allis, USA; and the TDR, the only indigenously developed technology of Tata Steel, India.

Main advantages of coal-based processes are [3]:

(i) They do not require high-grade coking coal which is scarcely available.
(ii) They can use non-coking coal.
(iii) They can be installed at lower capacity.
(iv) They can be easily installed at centre where small reserves of coal and iron ore are available.
(v) Modules of small-scale operation are available.

Main disadvantages of coal-based processes are [3]:

(i) Coal-based processes have lower economy of scale.
(ii) High energy consumption (16.0–21.0 GJ/t).
(iii) Low carbon content in the product (<1.0%).
(iv) Lower productivity (0.5–0.9 t/m^3/day).
(v) Hot feeding to the steelmaking furnace is not possible, due to the presence of residual char and ash with the product.

Types of coal-based processes are as follows:

1. Rotary kiln-based processes:
   SL/RN, CODIR, ACCAR, DRC, TDR, SIIL, OSIL, Jindal.
2. Shaft furnace-based processes:
   Kinglor, Metor, NML, vertical retort.
3. Rotary hearth furnace-based processes:
   FASTMET, ITmk3, INMETCO, Comet.

The basic equipments required for the coal-based processes are as follows:

1. Raw materials handling and feeding system,
2. Rotary kiln with air injection and coal blowing systems,
3. Rotary cooler,
4. Product separation and handling systems,
5. Waste gas-cleaning system,
6. Briquetting system,
7. Electrical system,
8. Instrumentation and control system.

7.3.2 Gas-Based Processes

Reformed natural gas is used as reducing agent. Iron ore limps or pellets are reduced in the solid state, and oxygen from iron oxide is removed by gaseous reducing agent (i.e. CO and H_2). Natural gas is reformed at 950 °C, in the presence of catalysts (Ni or Al_2O_3), to produce reducing gases CO and H_2, which take part in reduction of iron oxide. The reducing gas, H_2, CO or mixtures of H_2 and CO are introduced into the reactor at elevated temperature (up to 1000 °C) and pressure (up to 5 bars). If CH_4 is present in the reducing gas which results in carburization of the reduced product.

The processes based on gaseous reduction are confined to their areas where natural gas is available in large quantity at reasonable price. Commercialized processes, which used reformed natural gas as reducer, are ***Midrex*** (1969) and ***HyL*** (II—1957 and III—1980). The most widely used technologies for DRI production are Midrex and HyL, both using natural gas. The most successful technologies were gas-based moving bed technologies of Midrex Corporation of USA, the gas-based static bed HyL I technology of Hojalata Y Lamina of South Africa (HyL) and HyL III process of HyL.

Types of gas-based processes are as follows:

1. Retort processes:
   HyL I, Hoganas.
2. Shaft furnace processes:
   Midrex, HyL III, Plasmared, Armco, Purofer, NSC, HyL IV.
3. Fluidized bed processes:
   FINMET, FIOR, Circored.

The above processes have their own merits and demerits; gas-based processes have the advantages as follows:

- Higher thermal efficiency,
- Much larger unit size and higher productivity,
- Greater benefit of economies of scale,
- Lower energy consumption (10.5–14.5 GJ/t), and
- Higher carbon content in the product (>1.0%).

Gas-based processes account 82.4% of the world's sponge iron/DRI production, out of that Midrex alone contributes 64.8% production in 2017 (as shown in Table 7.2). This dominance of gas-based processes is not without reason. The gas-based processes offer distinct advantages over the coal-based processes as follows [3]:

Table 7.2 World sponge iron/DRI production by process (Mt/year) [4]

Process	2015		2016		2017		2018	
	Total	%	Total	%	Total	%	Total	%
Midrex	45.77	63.1	47.14	64.8	56.44	64.8	63.82	63.5
HyL	11.62	16.0	12.66	17.4	14.71	16.9	15.58	15.5
Other gas-based	0.51	0.70	0.22	0.30	0.26	0.7	0.70	0.7
Coal-based	4.74	20.2	12.74	17.5	15.33	17.6	20.30	20.2
Total	**72.64**	**100.0**	**72.76**	**100.0**	**87.1**	**100.0**	**100.49**	**99.9**

1. **Less capital cost**: The capital cost per tonne of installed capacity, in case of coal-based plants, is 1.8–2 times as high as in the gas-based plants.
2. **High productivity**: The throughput rates for gas-based sponge iron plants are much higher than in coal-based plants. The productivity for gas-based plants can be as high as 11 t/m^3/day as against merely 0.5 to 0.9 t/m^3/day in case of coal-based plants.
3. **Better quality**: The quality of the sponge iron produced in terms of metallization and carbon content is also higher in case of gas-based plants.
4. **Energy efficiency**: The gas-based processes have over the years improved considerably and are highly energy efficient.
5. **Better plant availability**: The gas-based processes are by now highly standardized. Gas is a clean source of fuel, and there are no problems of ash content, as in case of coal which leads to ring formation at the kiln and other attendant problems. The gas-based plants do not suffer from the maintenance problems often encountered in case of rotary kilns, and therefore the plant availability is better.
6. **Environmental pollution**: Since gas is a clean source of fuel, it is advantageous from the environmental pollution point of view.

It must be mentioned that the gas-based processes use comparatively a high-cost resource like natural gas, and on the other hand coal-based processes are gaining grounds where abundant coal is available. Coal-based process is the only answer in places where natural gas is not available.

The basic equipments required for the gas-based plants are as follows:

1. Natural gas reformer,
2. Reduction reactor/shaft,
3. Reducing gas heater,
4. CO_2 removal unit,
5. Reducing and cooling gas compressors,
6. Cooling water circuits for equipment and for process gas,
7. Boiler feed water system,
8. Inert gas generation system,
9. Raw materials and product handling systems,
10. Instrumentation and control system.

Technological developments by Midrex in the gas reforming system have enabled them to eliminate certain equipments like steam generation system, auxiliary boilers, boiler feed water treatment system, reformed gas quenching system, CO_2 removal system and reducing gas heaters.

7.4 Reactions

7.4.1 Coal-Based

7.4.1.1 Reduction Reaction

The reduction of iron ore occurs at 900–1000 °C in the presence of coal. Although it is solid–solid reaction, it takes place via gas–solid reaction:

$$3Fe_2O_3(s) + \{CO\} = 2Fe_3O_4(s) + \{CO_2\}, \Delta H^\circ_{298} = -52.43\,\text{kJ/mol of CO} \quad (7.1)$$

$$Fe_3O_4(s) + \{CO\} = 3FeO(s) + \{CO_2\}, \Delta H^\circ_{298} = 40.46\,\text{kJ/mol of CO} \quad (7.2)$$

$$FeO(s) + \{CO\} = Fe(s) + \{CO_2\}, \Delta H^\circ_{298} = -18.54\,\text{kJ/mol of CO} \quad (7.3)$$

$$\{CO_2\} + C(s) = 2\{CO\}, \Delta H^\circ_{298} = 85.35\,\text{kJ/mol of CO} \quad (7.4)$$

Hence, overall reaction is:

$$Fe_2O_3(s) + 3C(s) = 2Fe(s) + 3\{CO\}, \Delta H^\circ_{298} = 162.06\,\text{kJ/mol of CO} \quad (7.5)$$

7.4.1.2 Carburization

$$3Fe(s) + 2\{CO\} = Fe_3C(s) + \{CO_2\}, \Delta H^\circ_{298} = -85.33\,\text{kJ/mol of CO} \quad (7.6)$$

$$3Fe(s) + C(s) = Fe_3C(s), \Delta H^\circ_{298} = 0.034\,\text{kJ/mol of C} \quad (7.7)$$

Reduction is carried out by the reducing gas CO at below 1000 °C. The sponge iron produced is porous in nature and has nearly the same size and shape as the original iron ore or pellets. The metallic iron absorbs carbon to form iron carbide (Fe_3C) according to reactions (7.6) and (7.7).

7.4.2 Gas-Based

Reformed natural gas is used as reducing agent in gas-based processes. Iron ore limps or pellets are reduced in the solid state, and oxygen from iron oxide is removed by gaseous reducing agents (CO and H_2) which take part in reduction of iron oxide.

7.4.2.1 Reduction Reaction

Apart from reactions (7.1–7.3), there are some reductions also occurred by H_2 gas:

$$3Fe_2O_3(s) + \{H_2\} = 2Fe_3O_4(s) + \{H_2O\}, \Delta H^\circ_{298} = -55.31\,\text{kJ/mol of } H_2 \quad (7.8)$$

$$Fe_3O_4(s) + \{H_2\} = 3FeO(s) + \{H_2O\}, \Delta H^\circ_{298} = 37.57\,\text{kJ/mol of } H_2 \quad (7.9)$$

$$FeO(s) + \{H_2\} = Fe(s) + \{H_2O\}, \Delta H^\circ_{298} = -21.42\,\text{kJ/mol of } H_2 \quad (7.10)$$

7.4.2.2 Carburization

$$3Fe(s) + \{CO\} + \{H_2\} = Fe_3C(s) + \{H_2O\}, \Delta H^\circ_{298} = -135.11\,\text{kJ/mol of CO} \quad (7.11)$$

$$3Fe(s) + \{CH_4\} = Fe_3C(s) + 2\{H_2\}, \Delta H^\circ_{298} = 73.82\,\text{kJ/mol of } CH_4 \quad (7.12)$$

Most of the above reactions are endothermic, and hence heat must be provided for these reactions to occur. Hence, heat transfer plays a very dominant role in the sponge iron production processes.

7.5 Coal-Based Processes

Coal-based processes are broadly classified into two categories:

1. Rotary kiln process,
2. Rotary hearth process.

7.5.1 Rotary Kiln Process

Rotary kiln processes are SL/RN, ACCAR, Krupp, CODIR, TDR, etc.; out of these, SL/RN process is most common. The full name of the SL/RN process is the Stelco, Lurgi, Republic Steel, National Lead process, which was conceived about 1960 to produce high-grade direct reduced iron (DRI). It can be considered as a modification of the Krupp-Renn process, which was also a rotary kiln process for the beneficiation of low-grade ores. The Krupp-Renn process made it possible to use low-grade iron ores and a variety of carbonaceous fuels, such as coal, coke and char, to produce a highly (98%) metallized product. From the experience of both the Waelz (a rotary kiln process to recover zinc from low-grade ores) and the Krupp-Renn processes, the Krupp DR process which is now known as CODIR process (CODIR stands for Coal-Ore-Direct-Iron-Reduction) emerged. There are so many similar types of processes developed. Basically, all processes are same in principle, slightly different from process to process. Hence, here discussion is only the basic rotary kiln process in general.

7.5.1.1 Equipment

General size of kiln is 4–6-m diameter (D) and 60–125 m long (L), and L/D ratio is about 15–20. Size of biggest kiln is 6.1-m diameter and 125 m long. Inclination of the kiln is about 1–4% of the length. Rotation is about 0.5–4.0 rpm. Figure 7.1 shows flow sheet of SL/RN process. Sized lump iron ore

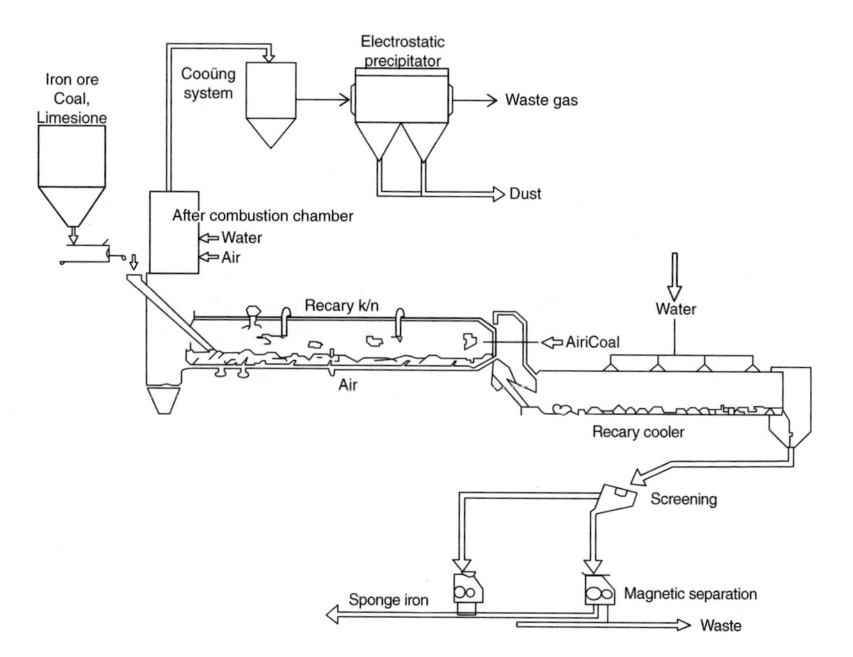

Fig. 7.1 Flow sheet of SL/RN process [3] (reproduced with permission from Authors)

(or pellets) and a relative course fraction of non-coking coal are charged into the kiln from the feeding side. Coal not only acts as a reducing agent, but also acts as fuel to supply the heat required for maintaining the temperature of the charge within the kiln. A finer fraction of coal is introduced from the discharge end of the kiln through burner along with the air to supply the heat and complete the reduction. Since by the time the charge travels to around 70% of the length of the kiln, very little course coal is available to complete the last stages of reduction. Feed of raw materials is done from one end which is at a higher level. They travel from one end to other end under gravity due to the rotating motion. They pass through several heating zones, and the reduced iron ore comes out from the other end of the kiln. The 50–60% of the volume of kiln is empty, where volatile matter of coal and other gases is burnt in the presence of oxygen from air and produces heat. The 40% return coke is charged back in the kiln to make the process economical. A flux (limestone/dolomite) is added along with coal to control the sulphur level in sponge iron. CaO in flux forms CaS by reaction with sulphur. Control air is provided throughout the length of the kiln to get uniform temperature.

Air is blown into the kiln, with shell air fans through burners, nozzles and central burner towards the feeding end. The shell air fans suck the air and are introduced into kiln along the centre line of the kiln through stainless steel burner tubes. Shell air fans in the kiln are at an angle of 90° to the central axis of the kiln. The gases from burner tubes move opposite to the discharging material. The blowing air is controlled by closing or opening the dampers of the shell air fan. At the discharge end, fine coal is injected through a coal injector tube compressive through rotary airlock feeder and injection coal throw pipe.

Most of reactions are endothermic, so heat should be provided in the kiln. It is done as the following:

- Burning of gas, oil or coal powder in a burner though the discharge end to produce heat,
- By burning volatile matter of coal within the kiln,
- Excess carbon monoxide emerging from the charge reacts with oxygen within the kiln and generates a lot of heat,
- Exit gas from kiln is rich in CO and volatile matter of coal. In the presence of oxygen, partial combustion of them takes place to produce electricity.

The heat is also produced by burning volatile matter of coal and excess carbon monoxide emerging from the charge. This is done by introducing controlled amount of air in the kiln free board in the pre-heat and reduction zones of the kiln. Air is provided throughout the whole length of the kiln through the tuyeres to get uniform temperature. Tuyeres are operated alternately (like 1, 3, 5, 7, etc. or 2, 4, 6, 8, etc.). Tuyeres are designed (as shown in Fig. 7.2) in such a way that when they are under the charge materials; the dust particles do not penetrate in the tuyeres. Otherwise the particles enter the tuyeres that chocked the tuyeres. Part of coal is introduced from the discharge end of the kiln to supply heat and maintaining reducing atmosphere. At the discharge end, reducing atmosphere prevents re-oxidation of porous and hot sponge iron, as well as to control degree of metallization and carbon content of sponge iron. The separation of the product is more or less similar in all the coal-based processes and involves screening and magnetic separation for removal of non-magnetic ash, char and used flux for de-sulphurizer.

Fine coal, which contains high volatile hydrocarbon, is injected by compressed air. Thus, the volatile hydrocarbon in the coal is released and cracked $[2\{C_nH_m\} \rightarrow 2nC(s) + m\{H_2\}]$ which acts as an additional and very effective reducing agent (i.e. H_2). Temperature of the charge is about 1000–1050 °C, depending on reactivity of coal. The kiln is lined with high alumina refractory (70% Al_2O_3) having a thickness of 0.2 m, and it can withstand 1600–1700 °C and have high abrasion resistance.

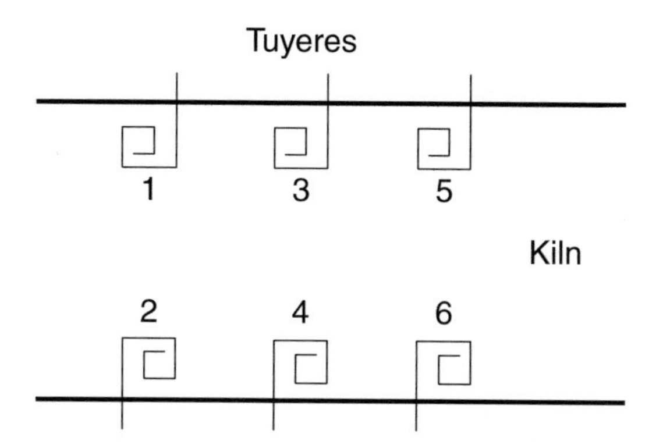

Fig. 7.2 Design of tuyeres in rotary kiln

7.5.1.2 Principle

All rotary kiln-based DR processes operate on the counter-current principle; i.e. the gases move in a direction opposite to the flow of solids. Various operations and processes like transport, mixing, charge separation by size, heating, gas generation and reduction occur both in parallel and in series. Segregation of the charge materials (ore, coal, flux) is taking place because size and density differences, as well as due to the slope and rotation of the kiln, must be prevented to the extent possibly, by adopting the appropriate kiln design and through proper operating measures. Another area of critical importance in rotary kiln operation is the prevention/minimization of localized areas of high temperature. Unless this is successfully carried out, partial fusion of coal ash takes place, leading to the formation of accretions (build-ups) on the kiln lining as well as clusters of fused materials within the bed. The difference between the gas phase temperature and the charge bed temperature (normally about 100–150 °C) has a major influence on both these phenomena. To prevent such occurrences, the temperature of the charge bed inside the kiln must be confined to a maximum of around 950–1050 °C so that the ash in coal does not fuse and the entire charge remains strictly in the solid state. High reactivity of coal encourages reaction of CO_2 with solid carbon in the bed resulting in the formation of reducing gas CO. Since this is a highly endothermic reaction, high reactivity automatically ensures that the gas temperature does not exceed the solid bed temperature by more than 1050 °C.

The charge moves through the kiln in counter-current to the hot gases and during the process gets heated to the reduction temperature. About 45% of coal is charged with iron ore from inlet end and rest 55% from discharge end of the kiln. Depending on the properties of ore and coal, the temperature in isothermal zone is maintained between 1000 and 1050 °C. After retention time of 8–10 h in the kiln, the sponge iron is discharged via a transfer chute into a cooler. Here, the material is cooled below 100 °C by indirect cooling in rotary cooler by water. Since no water comes in direct contact with the product in the cooler, there is very little chance of re-oxidation of sponge iron. Since sponge iron is magnetic in nature, it can be easily separated from the non-magnetic materials (consisting of coal ash, char and excess flux) by using magnetic separators. With the rapid and frequent increase in the cost of the main energy source, i.e. coal, control of specific consumption of coal has become a major task for economic operation of the plant. Based on C/Fe ratio of 0.42–0.44, the coal input in the process varies between 1.10 and 1.20 tonne per tonne of sponge iron, depending on the ash content of the coal. Typical consumption of raw materials is shown in Table 7.3.

The length of kiln is divided into two zones: (i) pre-heated zone and (ii) reduction zone (as shown in Fig. 7.3). The hot gas heated up the solid, and the gap between solid charge temperature and gas temperature is low at the end of the pre-heating zone. This temperature gap is depending on reactivity of coal. In the pre-heating zone, some reduction will occur. $Fe_2O_3 \rightarrow Fe_3O_4$ and $Fe_3O_4 \rightarrow FeO$ steps

Table 7.3 Typical consumption of raw materials per tonne of sponge iron production [5]

Iron ore	1.8 t
Coal	1.3 t
Dolomite	0.06 t
Air	1950 Nm^3
Electrical power	1.55 kWh
Heat	16.7–23.0 GJ

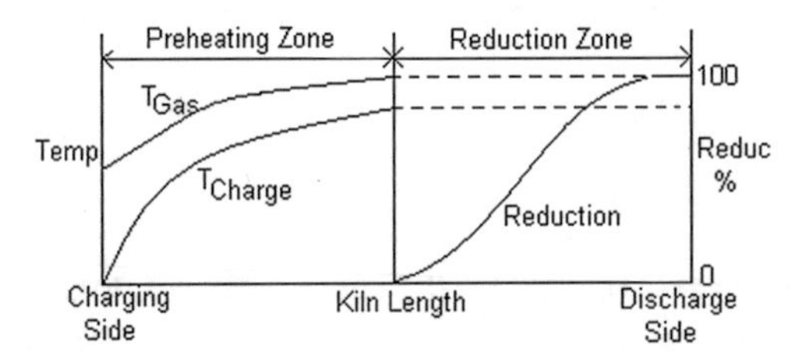

Fig. 7.3 Temperature and degree of reduction as a function of kiln length [3]

of reduction are completed in the pre-heating zone. In this zone, devolatilization and charring of coal, and decomposition of limestone also take place. In the reduction zone, FeO → Fe step of reduction occurs.

After the charge has stayed in the kiln for about 8–10 h, in this period the charge has been exposed to the temperature above 900 °C for about 5–7 h. The product is cooled up to 120 °C by way of direct or indirect cooling. Discharge product of the kiln should be cooled in inert or reducing atmosphere, since sponge iron is porous and hot at that time; otherwise, sponge iron is oxidized. Reduced and cooled product, which consists of sponge iron, coal ash and unreacted coal char, is screened, and the oversize is subjected to magnetic separation to obtain clean sized sponge iron while the non-magnetic portion of materials, that is unreacted coal char, is recirculated and coal ash is discarded. When hot sponge iron is coming out from kiln, it contains coal ash and unreacted coal char; that is why coal-based sponge iron cannot be made hot briquetted iron (HBI), or hot charging to the steelmaking furnace. If at high temperature magnet is available, then sponge iron can be separated from coal ash and unreacted coal char at hot condition; on that time, HBI can be produced or hot charging is possible.

For coal-based sponge iron plant, iron ore requirement to produce sponge iron with different rates is shown in Fig. 7.4. As the production rate increases, the iron ore requirements are increased. Iron ore requirement also varies with total iron content in the ore (Fig. 7.5). When the total iron content in the ore increases, the iron ore requirements are decreased; i.e. high-purity iron ore is less than the low-purity iron ore to produce sponge iron. Figure 7.6 shows the coal requirement varies with fixed carbon content in coal. As the fixed carbon content in coal increases, rate of coal consumption also decreases; i.e. requirement of good-quality coal will be less than the poor quality coal.

Rotary kiln stands to gain in productivity by about 20% if pellets are charged. Rotary kiln can run on 100% lump ore without any problem; 100,000 tonne per annum (tpa) rotary kiln [300–330 tonne per day (tpd)], which is a common size, can generate around 80–85 km^3 of gas at 950–1000 °C, from which 7.5 MW of power can be generated. For a 500 tpd, 150 ktpa plant, which is the largest single

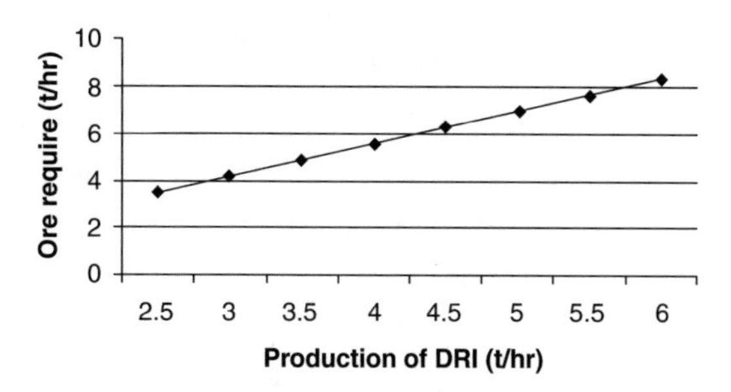

Fig. 7.4 Iron ore requirement for sponge iron production [3] (*Basis* Total Fe present in ore and sponge iron is 65.1 and 90%.)

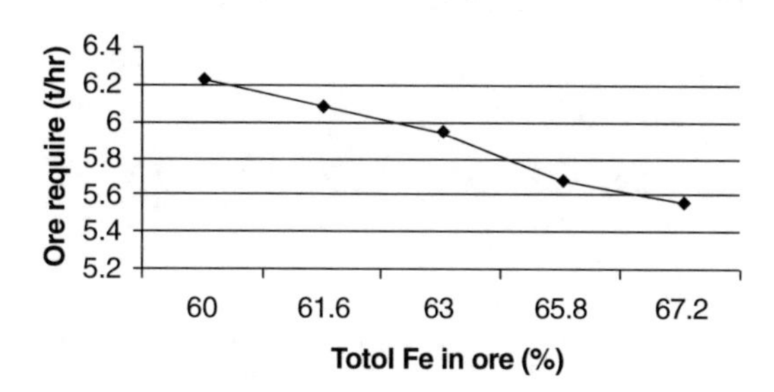

Fig. 7.5 Iron ore requirement versus total Fe in ore [3] (*Basis* Sponge iron production at the rate of 4.16 t/hr)

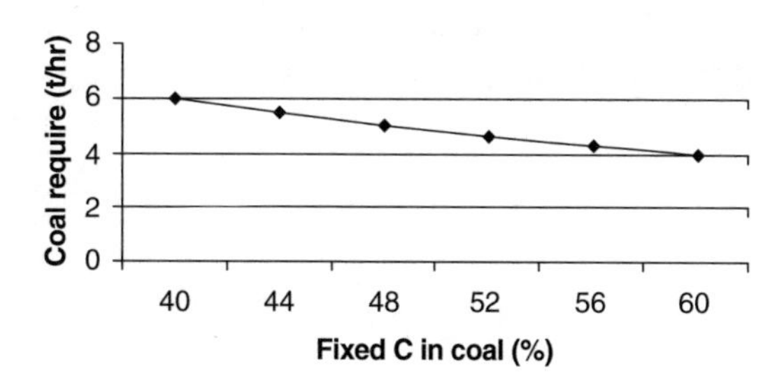

Fig. 7.6 Coal requirement versus fixed carbon in coal [3] (*Basis* Sponge iron production at the rate of 4.16 t/hr)

kiln that is available at present, the corresponding figures are 110 km^3 and 10 MW of power. More than 70% of the power generated becomes surplus after meeting the requirements of the entire DR plant.

7.5.1.3 Rate of Reduction

Within the charge, two reactions take place simultaneously:

1. Gasification of coal and
2. Reduction of iron ore.

Gasification reaction controlled overall reaction.

Gasification reaction:

$$C(s) + \{CO_2\} = 2\{CO\} \tag{7.13}$$

Therefore,

$$r_c = k_c(C^{\circ}_{CO_2} - C^{eq,C}_{CO_2}) \tag{7.14}$$

where

r_c the rate at which carbon is going to convert into CO gas,
k_c the rate constant,
$C^{\circ}_{CO_2}$ the concentration of CO_2 gas in gas mixture,
$C^{eq,C}_{CO_2}$ the concentration of CO_2 gas in equilibrium with carbon.

Reduction reaction:

$$Fe_xO(s) + \{CO\} = xFe(s) + \{CO_2\} \tag{7.15}$$

Therefore,

$$r_o = k_o(C^{eq,Fe}_{CO_2} - C^{\circ}_{CO_2}) \tag{7.16}$$

where

r_o the rate at which oxygen from ore is converted to gas,
$C^{eq,Fe}_{CO_2}$ the concentration of CO_2 gas in equilibrium with iron.

Assuming: $r_c = r_o$.
Therefore,

$$k_c(C^{\circ}_{CO_2} - C^{eq,C}_{CO_2}) = k_o(C^{eq,Fe}_{CO_2} - C^{\circ}_{CO_2}) \tag{7.17}$$

Hence,

$$\begin{aligned} C^{\circ}_{CO_2} &= \left[\frac{k_o C^{eq,Fe}_{CO_2} + k_c C^{eq,C}_{CO_2}}{k_c + k_o}\right] \\ &= \left[\frac{C^{eq,Fe}_{CO_2} + \left(\frac{k_c}{k_o}\right) C^{eq,C}_{CO_2}}{1 + \left(\frac{k_c}{k_o}\right)}\right] \end{aligned} \tag{7.18}$$

From Eqs. (7.16) and (7.18):
Therefore,

$$r_o = k_o . C^{eq,Fe}_{CO_2}\left[1 - \frac{\left\{1 + \left(\frac{k_c}{k_o}\right)\left(\frac{C^{eq,C}_{CO_2}}{C^{eq,Fe}_{CO_2}}\right)\right\}}{\left\{1 + \left(\frac{k_c}{k_o}\right)\right\}}\right] \tag{7.19}$$

where k_o and k_c are the measured reducibility and reactivity of coal, respectively.

Case I: $k_c \ll k_o$, then $\left(\frac{k_c}{k_o}\right)$ is a very small value which tends to zero.

Hence, $r_o \approx 0$; therefore, no reduction will be occurred.

Case II: $k_c \gg k_o$, then $\left(\frac{k_c}{k_o}\right) \geq 1$.

Therefore,

$$r_o = k_o(C_{CO_2}^{eq,Fe} - C_{CO_2}^{eq,C}) = r_o(\text{maximum}) \tag{7.20}$$

i.e. to get high productivity in the kiln.

Since $k_c \gg k_o$, i.e. the reactivity of coal should be higher than reducibility of ore.

7.5.1.4 Problems of Rotary Kiln Process

There are some problems in rotary kiln process as follows:

1 Low productivity,
2 Building up of accretions or ring formation,
3 Erratic operation,
4 Sponge iron contains low carbon,
5 Heat loss through exit gas.

(1) **Low productivity**: Rotary kiln has low productivity (0.5 to 0.7 t/m^3/day) than modern BF (2.0 to 2.5 t/m^3/day). That is due to: (i) furnace is only partially filled (40%), i.e. a lot of space is unutilized, (ii) low rate of reduction, (iii) lack of mixing of ore and reductant and (iv) movement of the bed is gentle, so the mixing is not proper.
Productivity can be improved:

- High reactivity coal can be used.
- Pickup of ore with high reducibility.
- Gasification rate depends on temperature, so higher the temperature of kiln it will be improved. By properly controlling the combustion, maximum possible temperature can be kept within the kiln as long a length as possible.
- Try to induce some gaseous reductant (H_2), especially towards the later stages of reduction. At later stage by introducing some H_2 in the kiln, it will be effective to reduce the inner part of ore. Diffusivity of H_2 is higher (5–6 times) than by CO.
- Only solution of remarkable increase in productivity is by using iron ore-coal composite pellets. Since residence time in the reduction zone is 40–60 min for iron ore-coal composite pellet than 4–5 h for ordinary case; due to higher rate of reduction of iron ore-coal composite pellets than ordinary pellets/lump ore.

(2) **Building up of accretions or ring formation**: Accretion or ring formation means the process of growing together (i.e. separate particles) into one, i.e. deposit on refractory surface. The presence of fines accelerates the formation of these deposits. These are occurred due to fusion mass, and some of coal ash has low fusion temperature. It is not that entire ash fuses. Due to fusion of ash and some other reaction, this build-up occurs. If accretion occurs, then the kiln must be stopped. This problem has overcome by adopting proper selection of coal so that ash should not fuse at that

low temperature. Coal, having ash's softening temperature above 1300 °C, is preferred. If operation temperature is more, then also accretion will be more that is why very high temperature in kiln cannot be adopted to improve the gasification reaction.

(3) **Erratic operation**: Erratic means irregular operation having no certain course. This is due to lack of proper mixing of raw materials.

(4) **Low carbon content of sponge iron**: Carbon content is less than 1% for rotary kiln product. This can be improved by injection of hydrocarbon gas under the solid bed. Due to dissociation of hydrocarbon gas ($2C_nH_m = 2nC + mH_2$), the nascent carbon, which is generated, is absorbed by porous sponge iron; hence, carbon in sponge iron may go up to 1%; generally, carbon in sponge iron is 0.3–0.5% without injection of hydrocarbon gas.

(5) **Heat loss through exit gas**: Exit gas left the kiln at very high temperature (heat content is about 6.28 GJ/t), and it can be used to pre-heat the materials or to generate the electric energy.

7.5.2 Rotary Hearth Process

The rotary hearth furnace (RHF) processes are not a new technology. For decades, they have been successfully used in a variety of industrial applications including heat treatment, calcination of petroleum coke, waste treatment, high-temperature non-ferrous metal recovery, etc. The problem with the use of RHF technology for the direct reduction of iron-bearing materials is not with the RHF itself; it is with the way it is being applied. If the RHF is correctly integrated into the process and the coal-based direct reduction technology is applied correctly, then the quality alternative iron can be produced economically. The process will be energy efficient and environmentally friendly.

There are developments of some processes based on rotary hearth furnace (RHF), viz.

(1) FASTMET process (1990), (2) SIDCOMET process (1990s) and (3) ITmk3 (2004). Among, all these processes, FASTMET process has been a commercial success. ITmk3 and FASTMET processes are noteworthy direct reduction processes which involve simple operation, lower unit consumption, lower production cost and superior environmental compatibility. The typical features of the RHF-based processes are as follows:

- Basically, the commercial scale is 0.5 Mt per annum on product basis.
- The main raw materials are iron ore concentrate or iron-bearing waste/dust of iron and steel industries and non-coking coal fine. Use of fines is as green pellets or briquettes. Though it depends on the quality, approximately 1.5 and 0.5 t of iron ore and non-coking coal, respectively, are used.
- Direct use of coal as reducing agent in the pellets. The heat requirement is met through firing gas to heat the furnace.
- These processes are developed to supply iron nuggets or hot metal for those countries which endowed with non-coking coal.
- These processes produce iron nuggets or hot metal which is the premium quality feedstock for EAF and/or BOF.
- Plant can be located next to the area of EAF or BOF in steelmaking facilities, for the efficient charge of hot metal.

7.5.2.1 FASTMET Process

The success of rotary hearth furnace (RHF)-based DRI technology has led to the development of FASTMET process. Midrex Technologies in partnership with its parent company, Kobe Steel Ltd., has developed the FASTMET process as a coal-based reduction technology; applicable for processing iron ore as well as iron oxide containing materials such as steel mill wastes. From both economical and environmental points of view, FASTMET process is very attractive as a proven technology for dust recycling. FASTMET process shows excellent promise for application to Indian *blue dust* [3].

FASTMET is a coal-based iron ore reduction process and a descendant of the Heat Fast process, originally developed in 1965 as a joint venture between Midland-Ross Corporation (the predecessor of Midrex), National Steel Corporation and Hanna Mining Company. In the original Heat Fast process, RHF was used to process dried, carbon-containing magnetite pellets. The process was successfully demonstrated at the Cooley pilot plant facility in Minnesota, where 50–70% metallized sponge iron was produced at the rate of 1.7 t/hr. In the early 1990s, Midrex has shown its interest on solid carbon (i.e. coal)-based reduction process using RHF. Studies confirmed that the economics for RHF-based process were attractive. In the early 1992, Midrex decided to construct and operate a 160 kg/hr, FASTMET pilot plant at Midrex Technical Center in Pineville, NC, to simulate the reduction process [6]. Over 100 trials were run at the pilot plant from 1992 to 1994. Based on the success of the pilot plant operation, in 1995 Midrex and Kobe Steel Limited (KSL) constructed an 8.5-m diameter RHF of 2.5 tonne per hour (tph); FASTMET plant at Kobe Steel's Kakagawa works in Japan [7].

The rotary hearth furnace (RHF) consists of a flat, refractory hearth rotating inside a high-temperature, circular tunnel kiln. The feed to the RHF consists of composite pellets made from a mixture of iron-bearing oxides and a carbon source such as coal, coke fines, charcoal or other carbon-bearing solid. The feed pellets are placed on the hearth, one or two layers thick. Burners located above the hearth provide heat required to raise the pellets to reduce temperature and start the process. The burners are fuelled with natural gas, fuel oil, waste oil or pulverized coal. Most of the heat required for maintaining the process is supplied by combustion of volatiles, which are liberated from the coal, and combustion of carbon monoxide, which is produced by the reaction of carbon with metallic oxides. The pellets are fed and discharged continuously and stay at the hearth for only one revolution, typically 6–12 min, depending on the reactivity of the mixture and target product quality. The product can be cold DRI, hot DRI, HBI or hot metal depending on end use requirements.

FASTMET uses a rotary hearth furnace to convert steel mill wastes and iron oxide fines to highly metallized sponge iron. Carbon contained in the wastes or added as coal, charcoal or coke is used as the reductant. Combustion of volatiles from the reductant and carbon monoxide from the reduction reaction supplies the primary energy to the RHF. The FASTMET process is extremely energy efficient, as all fuel energy is consumed within the FASTMET RHF (i.e. 100% post-combustion).

Figure 7.7 shows the flow sheet of FASTMET process. The FASTMET process is a solid reductant-based rotary hearth furnace (RHF) process. Iron ore concentrate fines along with reductant fines (such as charcoal or other carbon-bearing solid) are pelletized, to form iron oxide–carbon composite pellets, dried and then charged directly to the RHF. If the mill scale or other wastes of steel plant are used as feedstock of iron oxide, then that are briquetted along with solid reductant before charging to the RHF. As the hearth rotates, the pellets/briquettes are heated to 1250–1400 °C by combustion of natural gas, oil or pulverized coal. The pellet layers on rotary hearth are one to three pellets depth, and burners and post-combustion of CO provide the heat to raise the pellets to reduction temperature. CO is generated from the carbon present in the charcoal or other carbon-bearing solid in the composite pellets as well as by the combustion of liquid or gaseous fuels in burners installed above the rotating hearth. Table 7.4 shows consumption of raw materials for FASTMET process. The agglomerates containing the solid reductant get reduced to metallic iron. The reduction is accomplished by intimate contact between the carbon and iron oxide particles within the pellets/briquettes at relatively high temperature. Since the rate of reduction is very fast, the residence time of the charge in the hearth is typically as less as 6–12 min, during which 90–95% of the iron oxide is reduced.

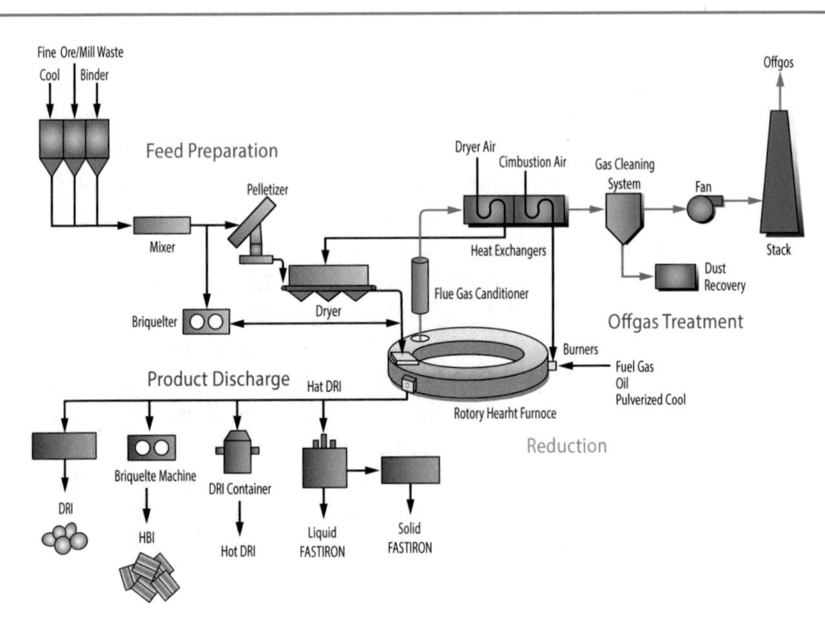

Fig. 7.7 Flow sheet of FASTMET process [3] (reproduced with permission from Authors)

Table 7.4 Consumption of raw materials for FASTMET [3]

Raw materials	Unit consumption
Iron ore (67% Fe), t	1.2–1.35
Coal, t	0.3–0.4
Auxiliary fuel, GJ	2.93
Electrical power kWh	100
Water, m^3	2

The key features of the FASTMET process are:

- FASTMET sponge iron even with a high gangue content is continuously fed to the electric furnace (i.e. melter).
- The reduction time in the RHF at the FASTMET stage is as short as 12 min.
- Energy is supplied to the melter by an electric arc, and molten iron is tapped at 1450–1550 °C for use in EAF or BOF.
- Steel plant's wastes like BF/BOF dust, mill scale, filter cake, etc., can be processed and can recycle them, for producing a cost-effective iron product.

The sponge iron produced is continuously discharged at around 1000 °C from the furnace, either into refractory lined container for hot charging to the melt shop or into briquetting machines to produce HBI, or directly cooled in inert atmosphere to get sponge iron pellets. The carbon content in the product can be controlled between 1.0 and as high as 6.0%, if required. Additional heat from exit gas of the process can be recovered by producing electrical power.

7.5.2.2 ITmk3 Process

The ***ironmaking technology mark 3*** (**ITmk3**) process is the third generation of ironmaking technology—the first two generations comprising blast furnace and direct reduction processes. ITmk3 represents the next generation of modern ironmaking technology, processing iron ore fines into almost pure pig iron nuggets within ten minutes. The result is a conveniently sized, slag-free material ideally suited for further processing by conventional technologies into high-quality steel products and foundry iron castings [3]. The Mesabi Nugget demonstration plant, commissioned in July 2004, achieved continuous, reliable production of pig iron nuggets under commercial operation conditions. Ten thousand metric tonnes of quality pig iron nuggets were produced during the four-test campaign. The produced nuggets were consumed in the steelmaking operations at various North American locations. Figure 7.8 shows the ITmk3 process.

The ITmk3 process is an ideal process for iron ore mining companies to supply pig iron grade nuggets directly to the EAF steelmaking industry. ITmk3 nuggets are a metallurgically clean, dust-free source of alternative iron for high-quality steelmaking in EAF. The ITmk3 process emits 20% less CO_2 than blast furnace operations. ITmk3 nuggets are not prone to re-oxidation and do not require special handling during shipment. Because of their convenient form, they can be continuously fed to the steelmaking furnace for higher productivity and lower liquid steel cost. Kwik Steel represents an extension and expansion of modern ironmaking technology that combines the best of natural gas-based Midrex direct reduction with ITmk3-type technology.

In **ITmk3**, the iron ore fines and non-coking coal are formed into green iron ore–coal composite pellets. ITmk3 uses the same type of mixing and agglomeration steps and rotary hearth furnace (RHF) as FASTMET. The composite pellets are fed to a rotary hearth furnace and heated to 1300–1450 °C; at this temperature range, the pellets are reduced to form iron nuggets. The temperature of RHF is raised, thereby melting the reduced iron and enabling it to easily separate from the gangue. This ironmaking process takes only 10 min against 10 h in BF process and 8 h in rotary kiln process. Iron and slag get separated, and product is called *nuggets*. Iron nuggets can be fed directly into BOF or EAF as pure iron source and substitute of scrap; by substituting scrap, it can dilute tramp elements like Cu, Pb, Sn, Cr, etc.

In the process, all the iron oxide is completely reduced and no iron oxide (i.e. FeO) remains in the nugget. The silicon, manganese and phosphorus contents in the product depend on raw materials. The product sulphur level also depends on the sulphur contained in the reductant; however, it is often possible to reduce the sulphur level remaining in the nugget an acceptable range (typically <0.05%). Quality of iron nuggets is as follows:

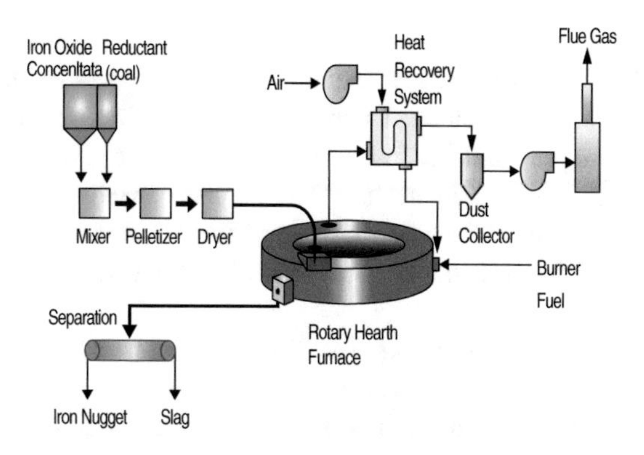

Fig. 7.8 Flow sheet of ITmk3 process [3] (reproduced with permission from Authors)

Quality of iron nuggets

Metallic iron, %	Carbon, %	Sulphur, %	Phosphorus, %	Size, mm
96–97	2–3	0.05–0.07	0.01–0.02	5–25

Some of the key features of ITmk3 are as follows [3]:

- The process produces iron nuggets within a very short reduction time (10–12 min) in RHF.
- Impurities are removed to some extent in the form of slag.
- The iron nuggets contain high iron (96–97%), high carbon (2.0–3.0%) and low (0.05%) sulphur, but in solid form.
- Coal consumption is reported to be 500 kg/t iron nuggets, which is fairly low.
- The process is flexible as far as the types of raw materials are concerned.
- ITmk3 reactors emit at least 20% less carbon dioxide than blast furnace. The overall NO_x, SO_x and particulate matter emissions are also lower.
- The iron nuggets can be fed directly to the EAF/IF to make steel along with steel scrap or sponge iron.
- The iron nuggets are easy to transport and handle. High in density, they do not re-oxidize or generate fines.

7.6 Gas-Based Processes

Gas-based processes are broadly classified as follows:

1. Midrex process,
2. HyL process,
3. Purofer process,
4. FINMET process,
5. HIB process.

7.6.1 Midrex Process

The Midrex process (***Mid***land ***R***oss ***Ex***perimental) was developed by the Surface Combustion Division of Midland-Ross Corporation in the mid-1960s. The first commercial Midrex plant was installed near Portland, Oregon, and started production in 1969. The plant included two shaft reduction furnaces of 3.4 m inside diameter and had a total capacity of 300,000 tpa. The average energy consumption of this early plant was approximately 15 GJ/t of sponge iron [3]. The Midrex Division became a subsidiary of Korf Industries in 1974. Midrex was subsequently acquired by Kobe Steel Ltd. in 1983.

The flow sheet for Midrex is shown in Fig. 7.9. The main components of the process are the shaft furnace, the gas reformer and the cooling gas system. The temperature and composition of the gas to the shaft furnace are controlled to maintain optimum bed temperature for reduction, degree of metallization, carbonization level and efficient utilization of the reducing gas.

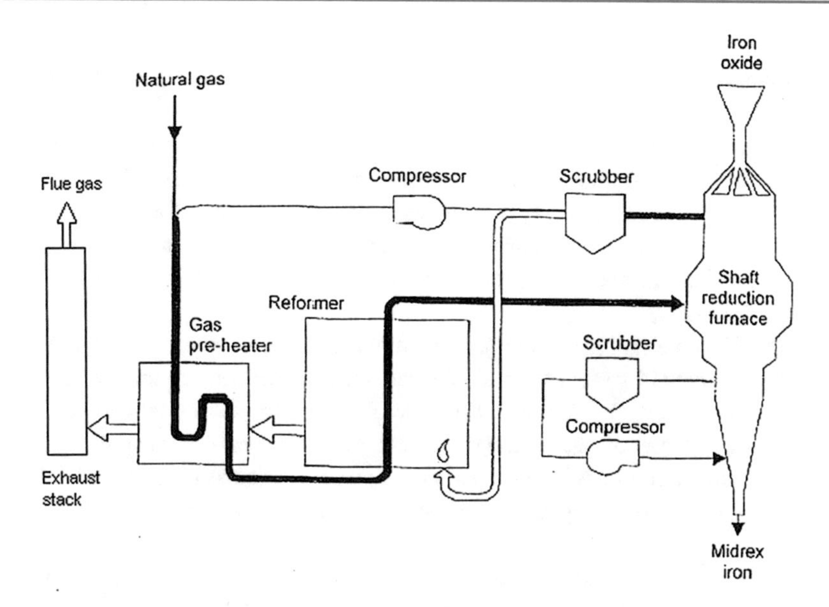

Fig. 7.9 Flow sheet of Midrex process with gas reformer [3] (reproduced with permission from Authors)

7.6.1.1 Equipment

The charge of Midrex shaft consists of around 60% pellets and 40% lump ore of a particular type. Pellets are the preferred feedstock owing to their superior physico-chemical characteristics compared with lump ores. As a result, in most cases, a minimum amount of pellet usage becomes mandatory. The process essentially consists of reducing the iron ore pellets and lump by reformed natural gas in a vertical shaft furnace. It is a continuous counter-current gas–solid reactor. Iron ore lump and pellets are fed in the shaft from the top, and the product sponge iron is removed from the bottom of shaft furnace.

Reduction of iron ore by reducing gases, in a shaft furnace, is according to counter-current principle. The main characteristic is given by the production of reducing gas by recirculation of the top gas from the shaft and using part of it as a reforming agent for natural gas. The process can be broken down into three main steps which are interdependent and influence each other:

1. Production of reducing gas,
2. Reduction of iron ore,
3. Cooling of end product.

7.6.1.2 Production of Reducing Gas

(I) Reforming: Reducing gas is produced by continuous catalytic reforming of natural gas with CO_2 and H_2O contents in the recirculated shaft furnace top gas, thus allowing the reforming process to work without additional input of external reforming media like CO_2, air or steam. The reforming reactions are:

$$\text{Major:}\{CH_4\} + \{CO_2\} = 2\{CO\} + 2\{H_2\}, \Delta H^\circ_{298} = 247.4\,\text{kJ/mol of } CH_4 \quad (7.21)$$

$$\text{Minor:}\{CH_4\} + \{H_2O\} = \{CO\} + 3\{H_2\}, \Delta H^\circ_{298} = 206.3\,\text{kJ/mol of } CH_4 \quad (7.22)$$

The gas volume is increased as per reactions (7.21) and (7.22). The reducing reactions in the shaft do not affect gas volume, so that the surplus gas together with additional fuels can be used for heating the reformer. Heating of the reformer is necessary for endothermic reforming reactions, temperature losses of reformer and supplying sensible heat to the reducing gas for heat requirements in shaft furnace. The product is a reducing gas with about 85–90% ($CO + H_2$) and 10–15% ($CO_2 + H_2O$).

(II) Gasification: Since natural gas is the primary fuel source for gas-based sponge iron production, countries, where natural gas is not available, can use coal gasifier to generate the reducing gases for sponge iron production. The typical gasifier fuels are low-cost coal and petroleum refining by-products (heavy fuel oil and petroleum coke), which are easily transported and available at economically attractive prices in most parts of the world. The basic objective of the gasification process is to generate reducing gases for the Midrex plant.

In general, gasification can be defined as a partial oxidation process in which a carbonaceous fuel (gas, liquid or solid) reacts at high temperature and usually at high pressure with oxygen and possibly steam to produce a synthesis gas. The synthesis gas contains mostly hydrogen and carbon monoxide but also some water vapour, carbon dioxide, methane, nitrogen and sulphur compound (H_2S, COS). The synthesis gas exiting the gasifier is cooled and scrubbed. Most of the water vapour is removed by cooling, and the remaining entrained particulates are removed by scrubbing. Gasification provides several environmental benefits. Midrex plants have been designed for H_2/CO ratios ranging from 0.5 to 4.0. Typically, synthesis gas produced by gasification [7] has a H_2/CO ratio in the range of 0.5–1.1.

7.6.1.3 Reduction of Iron Ore

Iron ore reduction is performed in the vertical, cylindrical shaft furnace by the H_2 and CO components of the reducing gas, while the iron ore is continuously charged from the top by skip. The reduction gas enters the shaft furnace at the bottom of the reduction zone with a temperature of 800–900 °C. Higher temperature could cause formation of clusters and cracking in the descending ore burden. The charge takes about six hours to pass through the reduction zone, where it is pre-heated and reduced by counter-current contact with the reducing gas. The residence time, for obtaining 95% metallized sponge iron which varies depending upon gas composition, flow rate, type of iron ore, etc., is between 5 and 6 h. The carbon present in the sponge iron depends on ($CO + H_2$)/ ($CO_2 + H_2O$) ratio. The higher the ratio, the more the carbon is deposited from CO and CH_4 decomposition.

$$2\{CO\} = C(s) + \{CO_2\} \quad (7.23)$$

$$\{CH_4\} = C(s) + 2\{H_2\} \quad (7.24)$$

7.6.1.4 Cooling of Product

As the metallized material passes down through the lower portion (i.e. conical section) of the shaft, it is cooled by inert gas. Because the sponge iron is leaving the reducing zone in hot condition, controlled amount of natural gas mixed with processed off-gas is added to the cooling gas. The reforming reaction of natural gas, which is endothermic, helps to cool the hot sponge iron and provides additional hot reducing gas, some of which enters the reduction zone where its sensible heat and reducing power are recovered. In this manner, the amount of reducing gas required from the reformer can be decreased by 8–10%. The sponge iron is cooled to 50 °C before it is discharged. The gas is withdrawn at the top of the shaft. This gas then passes through a scrubber and cooler, after which it contains about 70% hydrogen and carbon monoxide. It is then used for firing the reformers and for mixing with natural gas to make the process gas mixture that is fed to the reformers.

The top gas, at 350–400 °C, leaves the reduction furnace containing about 30% (CO_2 + H_2O) and enters a five-stage venture cooler scrubber where it is cooled to about 80 °C, water vapour is condensed, and the dust is removed to less than 0.02 grains of solid/m^3. Part of this gas (60–70%) goes to the compressor and then to a mixing chamber where it is mixed with fresh natural gas. The mixture passes through a reforming furnace where it is converted to reducing gas containing 95% (CO + H_2). Low-carbon sponge iron (<1.5% C) is directly cooled using a circulating stream of cooled exhaust gas introduced in the conical section, before cold sponge iron is discharged. When higher carbon sponge iron (up to 4.0% C) must be produced, natural gas is introduced along with cooling gas into the conical section. Natural gas readily decomposes in the presence of highly reactive metallic sponge iron, thereby generating nascent carbon, which gets absorbed by the porous product before it is discharged. In both cases, the final product is sponge iron with 93–94% metallization, with the desired carbon content (1.0–1.5%). Midrex technology is the commercially most successful and popular DR technology today.

7.6.2 HyL Process

In the HyL process developed by Hojalata y Lamina S.A. (Hylsa) of Monterrey, Mexico, lump ore and fired pellets are reduced in fixed bed retorts by reformed natural gas in 1953. This development came up in 1957 to a four (fixed bed)-reactor prototype unit, using steam-reforming of natural gas for generation of H_2 and CO gases for reduction. The first commercial HyL plant was installed at Monterrey and started production late in 1957. This plant has a capacity of 200 tpd of sponge iron, and the reactors are approximately 2.5 m in diameter and hold approximately 15 t of ore in a 1.5-m-deep bed. The reactors in the more recent plants are 5.4 m in diameter and 15 m high. Design capacity is approximately 1900 tpd of sponge iron having an average reduction of approximately 90%. The energy consumption in the most recent plants is 14.9 GJ/t of 90% reduced sponge iron [3]. Two stage reduction of lump ore or pellets is carried out in retorts by using a mixture of H_2 and CO gases which are obtained by catalytic reforming of natural gas by steam. Figure 7.10 shows flow sheet of HyL I/II process.

7.6.2.1 Process Technique

Each HyL plant consists of gas reformer and four reactors; each reactor is identical. In 12 h cycle, each reactor goes through 4 steps of about 3 h each. In step 1, reactor loading follows reactor discharging. In step 2 (secondary reduction), valves are opened to pass partially depleted reducing gas through the gas fixed pre-heater. From the pre-heater, the reducing gas enters the burner section where pre-heated air is injected to raise the reducing gas temperature to 1040 °C. This hot gas flows down through the charge, reducing ores/pellets, and exits through quencher (about 40% of combined oxygen is removed). The off-gas from this step is not recycled but is used as fuel gas throughout the plant. In step 3 (primary reduction), valves are remotely operated to switch from partially depleted reducing gas to fresh reformer gas coming from the reactor being cooled. This gas is also pre-heated, and this strong reducing gas reduces almost all the FeO present in the charge to metallic Fe. The quenched off-gas from this step is piped to step 2 above. In step 4 (cooling and carburizing), cold reducing gas directly from the reformer is passed through the reactor without pre-heating. The passage of the cold and fresh reducing gas through the product helps in cooling it and introducing

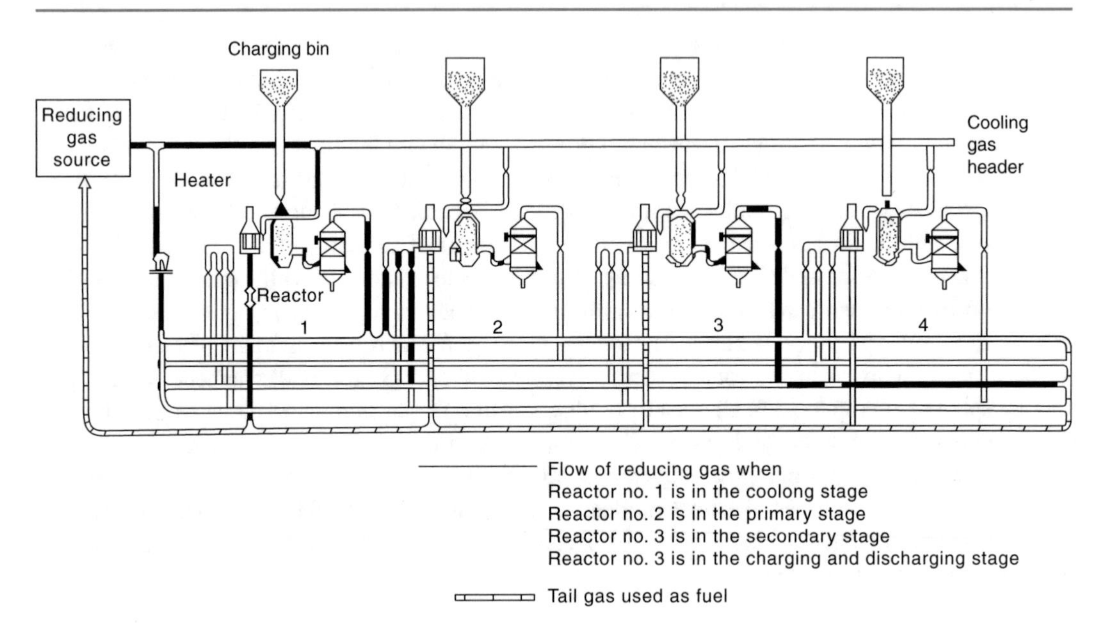

Fig. 7.10 Flow sheet of HyL I/II process [3]

carbon into it. The off-gas from this cooling step is quenched and is piped to a reactor in step 3 above. Mostly, cooling of the gas to condense the product water and then heating before entering the reactor are followed.

7.6.2.2 Production of Reducing Gas

The reducing gas is produced in a steam–natural gas reformer. The high-pressure natural gas is first de-sulphurized with activated carbon to prevent poisoning of the catalyst in the reformer. The gas is then mixed with superheated steam. This mixture is pre-heated and then passed through heated Cr–Ni alloy tubes containing Ni catalyst. The tubes are heated to provide the endothermic heat required in the reaction:

$$\{CH_4\} + \{H_2O\} = \{CO\} + 3\{H_2\}, \Delta H^\circ_{298} = 206.3\,kJ/mol\ CH_4 \tag{7.22}$$

7.6.2.3 Reducing Gas Temperature and Pressure

The reducing gases for both the primary and secondary steps are first pre-heated to about 800 °C in tubular heaters. After that, the gas is elevated to a temperature of 980–1230 °C by injecting controlled quantity of heated air into the reducing gas steam (for partial combustion).

The pressure zones are as follows:

First retort—4.5 kg/cm^2 abs,
Second retort—3.5 kg/cm^2 abs,
Third retort—2.5 kg/cm^2 abs.

7.6.2.4 Gas Requirements

Gas requirements for reducing iron oxide pellets are as follows:

Degree of metallization of iron oxide pellets	Average reducing gas requirement (Nm^3/t of Fe total)
86.0	1050
92.4	1250

Every HyL reduction plant has a fixed volume input of reducing gas per hour [80–90% of ($CO + H_2$)].

7.6.2.5 HyL III Process

*H*ojalata *Y L*amina of South Africa (HyL) developed moving bed HyL III technology and converted a few of the plants based on earlier fixed bed or HyL I technology. After many years of research, HyL announced a new process called HyL III (III signifying the third generation of HyL reactors, though II was not a separate process, only an improvement of HyL I). HyL III process is shown schematically in Fig. 7.11.

Main distinguishing features of HyL III process are:

1. High-pressure operation (5.5 kg/cm^2) for better process control and higher throughput,
2. Independent operation of the reforming and reduction areas and independent recycling circuits in the reduction and cooling sections of the reactor; thus, reformer and reformer catalyst are unaffected by sulphur and dust.

The principal change introduced (over HyL I) was the replacement of the four fixed bed reactors by a single moving bed reactor. The HyL III process can operate with 100% pellets, 100% lump ore or mixtures thereof; normally, 70% pellets and 30% lump ore are charged. In some cases, it is extremely important to add 5% non-sticking ore in the feed, check the sticking tendency of the pellets and thereby improve the furnace performance in terms of uniform descent of the burden.

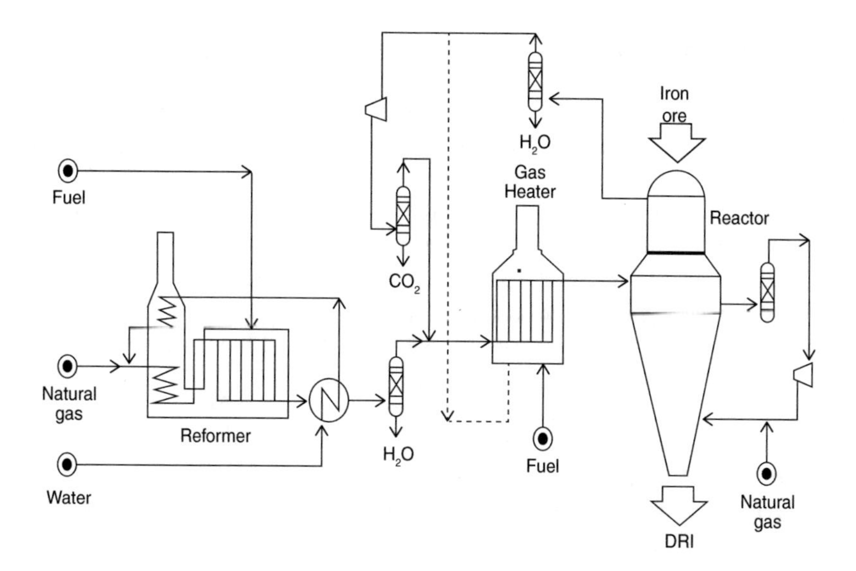

Fig. 7.11 Flow sheet of HyL III process [3] (reproduced with permission from Authors)

Reducing gas generation: Natural gas is pre-heated to 370 °C in a coil in the reformer convection section and is then passed through a de-sulphurization system to reduce the sulphur content to permissible levels for the reformer catalyst operation. The de-sulphurization is necessary, because the nickel reformer catalysts are very sensitive to even low concentrations of sulphur. Now, this gas is mixed with a predetermined amount of superheated steam. The steam–carbon ratio depends on the catalyst used, the composition of gas and the operating temperature. Typical values of this ratio range between 2.1 and 2.4. This ratio has been chosen to prevent carbon deposit on the catalyst.

The steam–natural gas mixture is pre-heated to about 620 °C in the reformer convection section and then enters the reformer radiant section. The reformer consists of a set of stainless steel pipes packed with a nickel-based catalyst. The reforming reactions are carried out at 830 °C.

$$\{C_nH_{(2n+2)}\} + n\{H_2O\} \leftrightarrows n\{CO\} + (2n+1)\{H_2\} \tag{7.25}$$

At the same time, water gas shift equilibrium is established:

$$\{CO\} + \{H_2O\} \leftrightarrows \{CO_2\} + \{H_2\} \tag{7.26}$$

The overall reaction proceeds to equilibrium which depends on temperature, pressure and steam to carbon ratio. The reformed gas leaves the reformer via the effluent chamber at 830 °C and 7.8 bar pressure and enters the reformer quench boiler where steam is generated by recovering sensible heat from hot reformed gas. Here, the temperature of the reformed gas is lower to 300 °C. Quenching of the hot reformed gas is necessary to get a gas with a low content of vapour (condensations of vapour in the reformed gas). The reformer product gas has typically the following composition (by volume): 73% H_2, 17% CO, 6% CO_2, 1% H_2O and 3% CH_4.

Reduction: Before the fresh produced cold reformed gas enters the reduction gas heater, it is mixed with the stream of recycled gas and is heated up to the desired gas temperature of 900–950 °C. This gas is fed directly to the reactor at a pressure of 5.5 bars. While the process operates at high pressure, the feeding of the iron ore to the reactor is achieved by means of the charging scaling mechanism, which consists of 1 bin and a set of special valves designed for the mass flow of lump ore or pellets. The iron ore is fed through the top of the reactor and flows downwards by the effect of gravity. The rate of flow is controlled by a rotary valve located at the bottom of the reactor. The hot reducing gas is injected at the middle of the reactor and flows upwardly counter-current to the solid flow.

From the reduction zone, the solids flow downwards to an intermediate zone in the reactor known as isobaric zone. This zone is designed to control and minimize the mixing of the reducing gas with the cooling gas. The appropriate control of this zone of prime importance is for effective control of product quality. The product sponge iron is cooled by counter-current cooling gas in cooling section from where the cooled product is removed continuously.

The reducing gas is leaving the reactor as top gas at approximately 450 °C and 5.0 bar. From the reduction part of the reactor, it is then scrubbed in the reduction quench tower: (1) for condensation of the vapour in the top gas and (2) cleaning the gas from dust. Then, it is sent to the recycle gas compressor.

Some of the advantages of the HyL III process are given below [3]:

1. Natural gas reforming is carried out using only steam in the presence of a catalyst.
2. To maintain the highest possible temperature for reduction, a combustion chamber is incorporated in the reducing gas inlet to the reactor. Partial combustion of natural gas also improves the

carburizing efficiency of the gases and makes it possible to achieve reduction temperatures more than 900 °C. This also enables HyL III units to produce high carbon-containing sponge iron.

3. The reducing zone in the upper part of the shaft is separated from the cooling/carburization zone in the lower part by an isobaric zone, which prevents the gases from mixing in the cooling and reducing zones. This helps in independent control of metallization and carbon content of sponge iron. Accordingly, it is possible to produce sponge iron (or HBI) with different carbon levels to meet the specific demands of steelmakers.
4. High-pressure operation allows the equipment size in the gas handling plant to be reduced. It also lowers the energy requirements.
5. The reforming section is independent of the reduction section, thus allowing the reformer to operate stably and maintaining the appropriate operating conditions reliably over long periods.
6. The process gas is not recycled through the reformer resulting in longer life of the catalyst. As a result, high sulphur inputs can be tolerated, without any deleterious effect on the equipment or on the quality of sponge iron, without having to utilize a bypass route.
7. HyL III plants are available in capacities between 0.25 and 2.0 Mtpa.
8. The energy consumption varies between 9.0 and 10.0 GJ/t of sponge iron.

7.6.2.6 HyL IV M Process

It is a shaft furnace process with self-reforming, in which reforming of natural gas occurs within the reduction reactor itself and the metallic iron in sponge iron acting as the catalyst. This allows in situ reforming of natural gas to proceed in parallel with the reduction of iron oxide and carburization of sponge iron. No separate gas reformer is required. Figure 7.12 shows a schematic representation of HyL IV M process. HyL IV M process employs a moving bed shaft furnace (similar to HyL III) to reduce iron ore pellets and lump ore at normal reduction temperature and at intermediate pressure. The process can produce cold/hot sponge iron as well as HBI. The first HyL IV M plant started production in 1998 in Monterrey, Mexico; hence, the suffix M stands for *M*exico.

The process starts with the injection of natural gas, which together with recycled gas enters a humidifier, where the required amount of steam is supplied. This gas mixture then enters a heater where its temperature is increased to above 900 °C. The hot gas mixture along with oxygen is

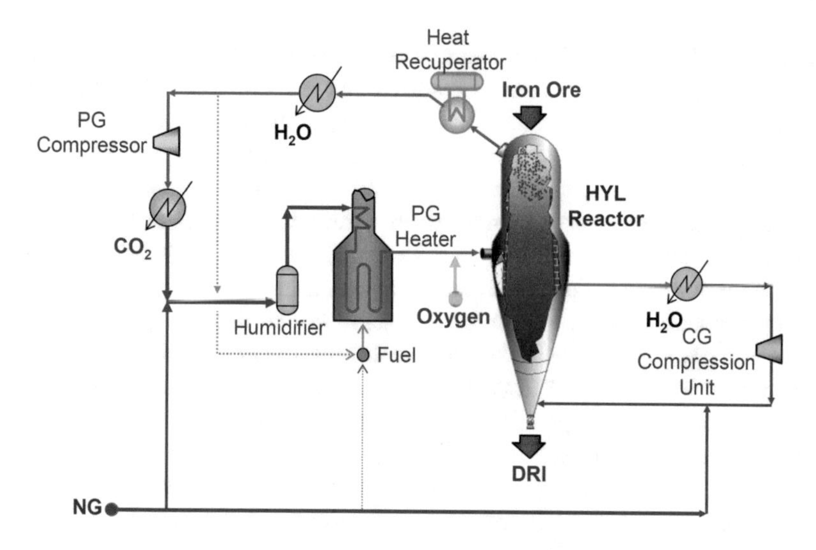

Fig. 7.12 HyL IV M process [3] (reproduced with permission from Authors)

introduced at the bottom of the shaft furnace. Oxygen helps in partial combustion of the reducing gas to increase the temperature above 1020 °C. In the lower part of the reduction zone, in situ reforming reactions occur, when the hot gas comes in contact with metallic sponge iron. The reacted exit gas leaves from the top of the reactor. Steam is condensed and removed before the gas is compressed and sent for CO_2 removal, after which it is mixed with natural gas for further use.

7.6.3 Purofer Process

In a shaft furnace, counter current reduction of ores is carried out with reducing gases. Such gases are obtained by oxidizing reforming of hydrocarbon gases. Catalytic reforming of natural gas with air and the temperatures required is reached by the stored regenerative heat of the reforming chamber and the catalyst as well by the reaction heat of the partial oxidation of methane. Reforming with air has the advantage that N_2 contained in the reducing gas acts as the heat carrier for supplying heat to the reduction shaft.

Reforming of natural gas with recycled top gas: It gives better overall utilization of the reducing gas. The carbon monoxide and hydrogen, components of top gas, are heated in the reformer and recycled to the shaft; moreover, carbon dioxide replaces oxygen of air as the reforming medium.

Reforming reactions with natural gas and oxygen (air), steam or recycled top gas as the reforming medium are:

$$\text{Air}: \quad \{CH_4\} + 1/2\{O_2\} = \{CO\} + 2\{H_2\} \tag{7.27}$$

$$\text{Top gas}: \quad \{CH_4\} + \{CO_2\} = 2\{CO\} + 2\{H_2\} \tag{7.28}$$

$$\{CH_4\} + \{H_2O\} = \{CO\} + 3\{H_2\} \tag{7.26}$$

The two reformers operate alternately. One reformer is heated by burning part of the top gas to a temperature of about 1400 °C. The heat is stored in the mass of the catalyst.

7.6.4 FINMET Process

Fluidized Iron Ore Reduction (FIOR) process was developed for the continuous reduction of iron ore fines by reformed natural gas in a chain of fluidized bed reactors. The original FIOR process was substantially improved which as a result of it's to develop FINMET process. The FINMET process uses a train of four fluidized bed reactors in which the gas and solids moving in counter-current directions come in contact throughout the entire reactor train. The feed concentrates (−12 mm iron ore fines) are charged to the reactor train at the topmost reactor via a pressurized lock hopper system. The ore fines are fed continuously to the reactors out of the lower lock hopper, which is always maintained at the reactor pressure of 11–13 bar. In the topmost reactor, the feed is pre-heated to 550–570 °C by the reducing gas coming from lower reactor. Pre-heating, dehydration and reduction of different stages take place as the feed is transferred downwards to the subsequent reactors. The gas required for reduction is a mixture of recycled top gas and fresh reformer make-up gas provided by a standard steam reformer. The composition of the recycle gas entering the last reactor is adjusted in accordance with the desired carbon content of the product. The temperature in the last reactor is maintained at around 780–800 °C so that the final reduction to 93% metallization takes place. Hot fine sponge iron at around 650 °C is transported to briquetting machine by a sealed system for producing ***hot briquetted iron*** (HBI).

7.6.5 HIB Process

The ***H***igh ***I***ron ***B***riquette (HIB) fluidized bed direct reduction process is a modified version of the Nu-iron process that was developed by the United States Steel Corporation in 1953. The HIB plant was started at Venezuela in 1968, and in 1971 the plant was ready for operation. In the HIB process, fine iron ore is reduced in two stages fluidized bed reactor and then product in hot briquetted form. Steam–natural gas reformers are used to generate the reducing gas. These are followed by heat exchangers to cool the gas and remove the unreacted water vapour. The gas is then heated to the desired temperature in a direct fired tubular heater. The off-gas from the reduction process is cleaned and used as a fuel for the reformer and the reducing gas heater.

The reduction reactor is divided into two stages. Reduction of Fe_2O_3 to FeO is accomplished in the first stage by contact with spent gas from the second stage. Reduction of FeO to Fe is accomplished in the second stage by contact with fresh reducing gas. The fine reduced product from reactor is carried by an inert gas pneumatic lift to a sealed surge bin above the briquetting rolls. Screw feeders extract the fines from this bin, pre-compact the fines and feed them into the briquetting rolls. Undersized material is screened out and recycled to the surge bin, and the briquettes are transported to a shaft cooler where they are cooled with inert gas.

7.7 Forms of Sponge Iron

Sponge iron has been fast gaining ground throughout the world since the 1980s, mainly because of the shortage of coking coal for blast furnace and steel scrap for steelmaking. Sponge iron is produced by direct reduction of iron ore by using non-coking coal/natural gas. World production of sponge iron is shown in Table 7.5. World sponge iron/DRI production increased by 13.4 Mt over 2017, exceeding 100 Mt in 2018, according to data collected by Midrex Technologies, Inc. and audited by World Steel Dynamics [4]. From 2016 to 2018, World DRI production had increased by 38%, which was the largest increased in any two year period since 1985. This significant increase is credited to the impact of new plants in Iran, Russia and USA, as well as improved conditions in Egypt and India. The greatest increases in sponge iron/DRI production were seen in India and Iran.

The major part of sponge iron production is used as a substitute for scrap in steelmaking. Sponge iron is consumed in three primary product forms, namely lump, pellets and hot briquettes (as shown in Fig. 7.13). The other secondary product form is cold briquettes made from sponge iron fines. Hot briquette form is popularly known as hot briquetted iron (HBI). HBI is a combined solid form of sponge iron lump and pellets, which are hot pressed at 700–800 °C, immediately after its production and before discharge. The most important characteristics of HBI are its high density and lower specific area which improves the resistance to re-oxidation and makes it easier to handle. Due to high density, the charging of HBI in furnace and penetration from the slag layer are much easier and

Table 7.5 Sponge iron production for top five countries of the world (Mt) [4]

Country	2014	2015	2016	2017	2018
India	17.31	17.68	18.47	22.34	28.11
Iran	14.55	14.55	16.01	20.55	25.75
Saudi Arabia	6.46	5.80	5.89	5.74	6.00
Russia	5.35	5.44	5.70	6.99	7.90
Mexico	5.98	5.50	5.31	6.01	5.97
World (total)	74.59	72.64	72.71	87.10	100.49

Fig. 7.13 Different forms of direct reduced iron: **a** Lump, **b** Pellets, and **c** HBI [3] (reproduced with permission from Authors)

Table 7.6 Typical chemical and physical characteristics of sponge iron [2]

Chemical properties	Coal-based	Gas-based		
	Lump	Pellets	HDRI	HBI
Fe(m) %	86.7	83–88	83–88	85–88
Fe(t) %	93.2	91–93	91–93	91–93
Metallization %	88–93	92–95	92–95	92–95
C %	0.15–0.20	1.0–2.5	1.0–2.5	0.9–1.5
S %	0.01	0.005–0.015	0.005–0.015	0.002–0.006
P %	0.05	0.02–0.4	0.02–0.4	0.02
Total gangue materials %	4.7	1.8–4.0	1.8–4.0	3.5–4.0
Bulk density (t/m^3)	1.6–1.9	1.6–1.9	1.6–1.9	2.4–2.8
Apparent density (kg/m^3) × 10^3	3.4–3.6	3.4–3.6	3.4–3.6	5.0–5.5
Product temperature (°C)	50–80	50–80	600–700	50–80
Typical size (mm)	4–20	4–20	4–20	30 × 50 × 110

Table 7.7 Physical properties of sponge iron [3]

Product forms	Carbon (%)	Metallization (%)	Apparent density (kg/m^3) × 10^3	Bulk density (t/m^3)	Shape and size (mm)	Weight (g)	Relative fine generation
Lump	0.2 Max	93 max	3.0	1.7	Irregular, 3–25	3–5	More
Pellets	1.0 Min	92 min	3.5	1.6–2.0	Spherical, 4–20	3–4	Less
HBI	1.0–1.5	93.5	5.0	2.6–2.7	Pillow-like, 35 × 50 × 110	450–750	Minimum

melting is faster. Tables 7.6 and 7.7 compare the chemical composition and physical properties of different forms of sponge iron. Being uniform in size, sponge iron is easy to transport, handle and permit continuous charging in electric furnace steelmaking. The direct reduction process consists of removing the larger part of the oxygen contained in the iron oxide and some sulphur, but otherwise no other refining and melting occur in the process. Consequently, all the original impurities of the ore, i.e. all the gangue, remain in the product, which stresses the importance of having high-quality feed materials.

In DR processes, iron oxide is reduced below the melting point of iron, involving no molten slag phase. Therefore, all gangue elements of the iron ore remain in the DRI and need to be separated in the form of slag at the EAF. This increases the electrical energy and electrode consumption of the EAF compared to all scrap melting (secondary steelmaking). This energy consumption can be reduced by immediate transfer of hot sponge iron/DRI to the EAF melt shop. By doing this, the heat content in hot sponge iron/DRI lowers the cost of melting the sponge iron/DRI in the EAF, considerably cutting the energy costs and electrode consumption. The energy consumption in natural gas-based sponge iron/DRI production is well known and established to be 10–11.4 GJ/t of sponge iron/DRI. Natural gas-based sponge iron/DRI production also leads to lower CO_2 emissions, with emissions ranging from 0.77 to 0.92 tonne of CO_2 per tonne of steel (compared with $\sim$1.9 tonne of CO_2 per tonne of steel for the BF-LD/BOF route), depending on the type of electricity used.

Generally, there are three forms of sponge iron/DRI [2] used:

1. Cold sponge iron/DRI (CDRI),
2. Hot sponge iron/DRI (HDRI),
3. Hot briquetted iron (HBI).

7.7.1 Cold Sponge Iron/DRI (CDRI)

Most direct reduction plants are built to produce cold sponge iron/CDRI. After reduction, the sponge iron/DRI is cooled in the lower part of the shaft furnace (for gas-based) to about 50 °C or cooled in other kiln (for coal-based) to about 80 °C. This material is typically used in a nearby steel plant and must be kept dry to prevent re-oxidation and loss of metallization. CDRI is ideal for continuous charging to the electric arc furnace (EAF).

7.7.2 Hot Sponge Iron/DRI (HDRI)

Hot sponge iron/HDRI can be transported to an adjacent EAF (within plant) at about 650 °C to take advantage of the sensible heat content of hot sponge iron/HDRI, which allows the steelmaker to increase productivity and reduce consumption of electricity and production cost. There are three methods for hot sponge iron/HDRI transport: hot transport conveyor, hot transport vessels and HOTLINK. By charging hot sponge iron/DRI [8], electricity savings are 20 kWh/t liquid steel per 100 °C and up to 120–140 kWh/t; productivity of EAF increases up to 20%.

7.7.3 Hot Briquetted Iron (HBI)

Hot briquetted iron (HBI) is traditionally known as a compacted form of sponge iron/DRI at 700–800 °C, consistent chemical and physical characteristics. HBI is 50% denser (density $\geq$ 5 g/cm^3) than sponge iron pellets and lump and having less tendency for re-oxidation. HBI product form was developed for the need of a direct reduction product to transport through sea for distant consumers. This enables HBI to be stored and transported without special precautions under the International Maritime Organization (IMO) code for shipping solid bulk cargoes. It can be used in the BF, BOF and EAF. Hot briquetting of sponge iron has been practised on an industrial scale for more than three decades and is now the preferred method of preparing sponge iron for storage and transport internationally [9].

Now, dedicated merchant HBI plants are in operation in Venezuela, Russia, India, Malaysia and Libya. Qatar Steel Company operates a hot discharge plant that includes briquetting capability and is currently exporting HBI. HBI is manufactured to be shipped over great distances and melted in a variety of iron and steel processes. It is available throughout the year unlike scrap, which tends to have a collection season. The chemical composition of HBI is certified by the producer, and ISO quality standards are strictly followed. Physical characteristics are the real reasons that HBI was created. Its heavy mass easily allows rapid penetration of the furnace slag layer. HBI is 100 times more resistant to re-oxidation than conventional sponge iron/DRI and will pick up 75% less water. HBI generates less fines, which provides greater value to users and reduces safety concerns during handling and shipping. Size and shape of HBI are compatible with standard materials handling equipment, and HBI can be batch charged or continuously fed to a melting furnace. Because HBI is produced from natural iron ores and pellets, it is a source of clean, highly metallized iron units.

Advantages of HBI are as follows [9]:

1. Attractive cost structure when compared with price of imported scrap in countries that have abundant and inexpensive natural gas resources,
2. Availability of annual supply contracts avoids price spikes of spot market,
3. Year-round production (no collection season),
4. Well-defined, consistent chemistry with guaranteed specifications,
5. Low residual content (Cu, Ni, Cr, Mo, Sn, Pb and V),
6. Dilutes impurities in lower-quality scrap,
7. Blends with other metallic for best total charge economics,
8. Applicable to full range of products, from rebar to sheet steel,
9. Continuous feeding maximizes power-on time and increases bath weight,
10. Promotes foamy slag and reduces EAF nitrogen level,
11. Shields refractory to reduce damage,
12. Excellent for AC or DC furnaces, long or short arc operation,
13. Density greater than 5.0 g/cm^3, which allows for rapid penetration to the slag layer in the furnace,
14. Higher thermal and electrical conductivity for faster melting,
15. Less fine generation for added value to the customer,
16. Less reactive to water for safer river/sea and road transport,
17. Easy to handle, store and transfer in all types of weather and with standard material handling equipment.

7.8 Characteristics of Sponge Iron

The main chemical characteristics of sponge iron are [3]:

1. Metallization,
2. Carbon content,
3. Gangue content and
4. Impurities (S, P, etc.) and tramp elements.

7.8.1 Metallization

The degree of reduction of sponge iron is usually expressed as the per cent metallization of the product. It is the ratio of the metallic iron present in sponge iron/DRI divided by the total iron present in sponge iron. The degree of metallization depends greatly on the type of reduction process being used. The degree of metallization varies from 85 to 95% depending on the process adopted for sponge iron production. Low degree of metallization leads to economic disruption such as higher energy consumption, higher slag volume, more heat time and lower per cent yield during steelmaking.

$$\text{Degree of metallization, }\% = \left(\frac{\text{Free metallic Fe present in sponge iron}}{\text{Total Fe present in sponge iron}}\right) \times 100 \qquad (7.29)$$

7.8.2 Carbon Content

The control of the carbon content of sponge iron depends on the DR process being used. Those based on solid reductants usually give low carbon content (<0.5%), and this is inherent to the process itself. However, with the gaseous reduction processes the carbon level can be adjusted, within limits, to any desired value. The carbon content of gas-based sponge iron is varied between about 1.0 and 3.0%, corresponding to a degree of metallization of 85–95%, respectively. The carbon as Fe_3C is more desirable than loose carbon fines or soot which may not be useful to the process. Oxygen present in the sponge iron is in the form of FeO, which reacts vigorously with carbon in the molten bath to make boil that improves heat transfer, slag-metal contact and homogeneity of the molten bath. Thus, higher percentage of carbon is required in sponge iron and hence steelmakers prefer gas-based sponge iron (content 1.0–2.5% C) instead of coal-based sponge iron (0.2% C). Any unreduced iron oxide present in sponge iron, during steelmaking, enters the slag.

It is possible to produce sponge iron (or HBI) with different carbon levels to meet the specific demands of steelmakers. This has given rise to the concept of *equivalent metallization* since the carbon content of sponge iron has a strong influence on its melting behaviour during steelmaking.

$$\begin{aligned}\text{Equivalent metallization} &= \text{actual metallization} \\ &\quad + \text{five times the carbon content in sponge iron}\end{aligned} \qquad (7.30)$$

7.8.3 Gangue Content

The gangue in sponge iron that substitutes for normal slag-producing agents does not penalize the steelmaking operation and is usually between 2 and 4%. However, the actual amount depends on the proportion of silica (SiO_2) in the gangue which must be fluxed and on the percentage of iron units in the charge that is derived from sponge iron. All the gangue content of the iron ore fed to the DR plant is finally landed to the sponge iron. Since sponge iron becomes competitive with steel scrap, the gangue content of the sponge iron had to be reduced appreciably. The iron ore feeds for DR processes usually contain less than 2% SiO_2 and 1% Al_2O_3. Moreover, certain additions of either limestone or dolomite are made to improve the behaviour of the iron ore pellets during reduction. Therefore, some lime and magnesia are found in sponge iron.

7.8.4 Impurities and Residual Elements

The residual elements usually found in steel scrap (Cu, Zn, Pb, Sn, As, Cr, Ni, Mo) are not often present in any noticeable quantities in iron ore deposits, and therefore, these elements are found only as traces (<0.02%) in sponge iron. The impurities that can be found in sponge iron are the alkalis (Na_2O and K_2O), titanium oxide, sulphur and phosphorus. The first three are usually present in quantities, which are small enough not to have any bearing on the steelmaking practice [e.g. ($Na_2O + K_2O$) < 0.1% and TiO_2 < 0.1%]. The sulphur content of sponge iron is also relatively low, depending on the amount of sulphur in the fuels and reductants used. Sulphur contents of sponge iron vary from less than 0.005% in the direct reduction processes employing sulphur-free gas to approximately 0.02% in direct reduction processes employing sulphur-bearing coal and limestone together with iron ore in the charge mix. Phosphorus is not eliminated during direct reduction process, and therefore, the quantities found in sponge iron will be directly a function of those contained in the iron ore. Phosphorus content normally found in sponge iron ranges between 0.01 and 0.04%.

7.9 Quality of Sponge Iron

The portions of iron oxide and gangue in sponge iron above certain minimums increase the power requirements in the EAF/IMF compared to an equivalent quantity of scrap. A portion of the iron oxide reacts with carbon in the EAF/IMF to produce metallic iron and carbon monoxide according to the reaction:

$$\mathrm{FeO(s) + C(s) = Fe(s) + \{CO\}}, \Delta \mathrm{H}^0_{298} = 151.26\,\mathrm{kJ/mol\ of\ C} \tag{7.31}$$

This is an endothermic reaction. Furthermore, the gangue and the associated flux require energy for melting. However, a metallization that is too high decreases both the fuel efficiency and the productivity of the direct reduction process. The metallization of sponge iron normally ranges between 85 and 95% depending on the process and on the reducibility of the original iron oxide. Based on the level of metallization, the carbon content of the sponge iron is controlled in some direct reduction processes to be between 1.0 and 2.5% to facilitate reduction of FeO by carbon during melting. This increases iron recovery, and the CO generated promotes foamy slag practice in the EAF/IMF. Another impact on the operation of the EAF pertains to fines in the sponge iron which affect iron recovery, increase the dust loading and contaminate electrical parts. The allowable fines are usually approximately 5% of −5 mm. A major advantage of sponge iron in EAF operations is the absence of contaminating residuals such as copper and tin. Phosphorus, manganese and vanadium contained in the ore remain in the gangue in the gas-based processes and are usually not a factor in electric steelmaking operations.

7.10 Re-oxidation of Sponge Iron

Sponge iron is produced, in the solid state, by gaseous or solid reduction of iron ore. Due to removal of oxygen, a honeycombed structure remains which has a very large surface area per unit weight. Hence, sponge iron has an inherent tendency to re-oxidize back to its native stage. The main reasons for such behaviour are (i) extremely high surface area-to-volume ratio and (ii) poor thermal conductivity due to porosity and gangue content of sponge iron (approximate 2.092 kJ/mhK). Freshly prepared sponge iron is highly prone to oxidation whenever it encounters air. The heat generated in the oxidation reaction increases the susceptibility of oxidation, thereby starting a sort of chain reaction and ultimately leading to the fire of sponge iron. Hence, storage and handling of sponge iron are major concerns of the material.

The re-oxidation of sponge iron, caused by the influence of water and oxygen, is an exothermic reaction. Thus, an oxygen pickup of 0.1% would mean a rise of about 35 °C under adiabatic conditions. This makes sponge iron highly reactive in oxygen-containing atmosphere. It is even more prone to re-oxidation in the presence of moisture. However, these reactions of sponge iron in ambient atmosphere are very slow and therefore may result in very little overheating. Because the initial reaction is slow, it promotes passivation that hinders further reaction. The passivated material is stable under ambient conditions. Generally, freshly produced sponge iron exposed to air reacts quickly within a few minutes and the reaction decays very rapidly almost to zero, resulting in slight loss of degree of metallization (>1%); this tends to passivate or stabilize the sponge iron for subsequent handling. The passivated oxide layers form a barrier between the iron and air, thus retarding further oxidation. Heat is formed during this natural ageing, or passivation is insufficient to cause any problem during further handling. This scenario is true only if the ambient air is dry and less than 65 °C. Due to poor thermal conductivity of sponge iron, the heat generated (as shown in Table 7.8) during re-oxidation reaction cannot be dissipated away, ultimately leading to the development of hot spots or even auto-ignition of sponge iron piles.

Sponge iron is chemically reactive, and this makes it dangerous to store or shipment in bulk. In the case of some stockpiles that have wet sponge iron buried deeply, high-temperature re-oxidation reaction has occurred. The heat generated by the exothermic re-oxidation reaction (Table 7.8) within the pile does not have an opportunity to dissipate. As a result, increase in temperature causes a corresponding increase in reaction rates leading to auto-ignition. Overseas transportation of sponge iron has therefore often resulted in corroding the cargo, even sometimes to the extent of setting ship ablaze, because initial reaction between sponge irons with sea water may yield hydrogen and heat.

Chemical reactions involved in re-oxidation of sponge iron are as follows:

(i) Oxidation in dry air:

$$4Fe(s) + 3\{O_2\} = 2Fe_2O_3(s) \tag{7.32}$$

Table 7.8 Heat effects during re-oxidation of sponge iron [3]

Reaction	Heat of formation at 25 °C (298 K) and 1.0 atm (kJ/mol)	Amount of heat produced per kg sponge iron (kJ/kg)
I Exothermic:		
$0.95\,Fe + 1/2\,O_2 = Fe_{0.95}O$	−264.43	−4970.59
$3Fe + 2\,O_2 = Fe_3O_4$	−1116.71	−6648.38
$2Fe + 3/2\,O_2 = Fe_2O_3$	−821.32	−7334.55
$Fe + 1/2\,H_2O + 3/4\,O_2 = FeOOH$	−558.98	−9983.02
$Fe + H_2O + 1/2\,O_2 = Fe(OH)_2$	−568.19	−10,146.20
$Fe + 3/2\,H_2O + 3/4\,O_2 = Fe(OH)_3$	−825.08	−14,731.86
II Endothermic:		
$3\,Fe + 4\,H_2O = Fe_3O_4 + 4\,H_2$	+27.61	+164.43
$Fe + 2H_2O = Fe(OH)_2 + H_2$	+3.35	+59.83

(ii) Oxidation in the presence of moisture:

$$2Fe(s) + 3\{H_2O\} = Fe_2O_3(s) + 3\{H_2\} \quad (7.33)$$

(iii) Oxidation in the presence of dissolved oxygen:

$$2Fe(s) + 2\{H_2O\} + \{O_2\} = 2Fe(OH)_2(s) \quad (7.34)$$

The following recommendations were made to prevent sponge iron burning:

- Sponge iron should be kept dry always. Sponge iron should be naturally aged for at least 72 h before loading to the ship. The natural passivation is caused by the formation of an oxide layer on the surface of sponge iron that stops further oxidation.
- Sponge iron should not be loaded if the temperature is greater than 65 °C.
- Naked flame should be removed from the storage areas.
- Nitrogen purging should be done in the ship hutch.
- Hydrogen and oxygen concentration should be measured continuously.
- Higher degree of metallization for sponge iron should be preferred.

Precautions should be taken during loading:

(a) Loading, steering and unloading during rain must be avoided; this is extremely important.
(b) Holds must be dry and clean, and shall be of all seal and watertight.
(c) Holds must not contain steam lines or other heat sources.
(d) Holds must have watertight hutches to avoid ingress of water during voyage; sea water is particularly very aggressive.

7.10.1 Preventive Measures

There are two types of prevention for re-oxidation of sponge iron:

- Natural protection and
- Artificial protection.

7.10.1.1 Natural Protection

There are two types of natural protection as follows:

(i) Weathering, and
(ii) Air passivation.

(i) **Weathering**: Storing sponge iron prior to shipping allows time for a protective skin to develop which significantly improves sponge iron stability. It would be normally resistant to dry re-oxidation up to 50–65 °C. Large quantities of freshly produced sponge iron should be stored in silos purged with inert gas or allowed to complete its ageing time in covered state. Some heat is produced during ageing, which takes places rather rapidly, and the height of a layer of fresh material should be limited to 1.5 m to allow for quick dissipation of heat. Aged sponge iron can be stored safely in bulk if the product is kept dry either under roof or covered by a two-layered canvas tarpaulin. Such tarpaulin is preferred because it will breathe; polyethylene or the other plastic is not desirable because it does not breath, and moisture will condense on the underside. Long exposure to natural weathering (about 3 months) has been found to effectively ward off re-oxidation by growth and development of a protective envelope of oxide layer.

(ii) **Air Passivation**: It involves raising the temperature of the sponge iron to about 100 °C by reacting air with a bed of sponge iron under controlled conditions. A thin oxide film is formed which makes it resistant to dry oxidation up to passivation temperature.

7.10.1.2 Artificial Protection

There are two types of artificial protection as follows:

(i) **Chemair process,**
(ii) **Coating processes,**
(iii) **Densification of sponge iron.**

(i) **Chemair Process**: It is a combination of chemical treatment followed by air passivation at about 80 °C. That is, it consists of chemical solution treatment, air passivation and cooling of the sponge iron in a vertical shaft dryer/cooler.

(ii) **Coating Processes**:

(a) **Wax Coated**: Wax wetting is done by soaking sponge iron in molten paraffin wax at 110–120 °C in a stainless steel or aluminium tank for about 3–5 s to develop protective coating. The amount of wax consumed is only 5–6% by weight, depending on porosity of sponge iron and the temperature involved. Wax burning takes place with rise in temperature up to 400 °C. Such burning of wax results in deposition of carbon in the pores and all over the surface of sponge iron providing protection. Wax coating has the ability to prevent re-oxidation of sponge iron up to about 400 °C.

(b) **Lime Coated**: It involves dry rolling of sponge iron in a mixture of 95% burnt lime powder and 5% iron oxide powder followed by controlled moisture. Thus, a coating like mortar is formed on the surface of sponge iron.

(c) **Sodium Silicate Coated**: Sponge iron is immersed in 4% sodium silicate solution for an hour and dried at 120 °C for an hour. The loss in metallization in a coated product is about 1.5 times less than that on an uncoated sponge iron. It is found that sodium silicate coating is superior to lime coating in terms of passivation achieved. However, unlike lime, sodium silicate addition is not acceptable to steelmakers.

(iii) **Densification of Sponge Iron**: By hot pressing of sponge iron (lumps and pellets) at 700–800 °C, hot briquetted iron (HBI) is formed. This results in substantial improved physical properties, namely higher densities (apparent and bulk, as shown in Table 7.7) and decrease in its surface area and porosity. Subsequently, HBI is quenched in a corrosion inhibiting chemical before storage, thus producing a material resistant to both re-oxidation and corrosion.

7.11 Use of Sponge Iron

Due to shortage of scrap in the 1980s, sponge iron is used as feed materials in steelmaking. Sponge iron acts as burden enrichment in BF, as a coolant in LD/BOF and as a charge material for EAF/IMF. The quality steel contains total summation of carbon, sulphur, nitrogen and hydrogen 100–150 ppm (i.e. 0.01–0.015%), by using sponge iron as feed materials in EAF that can be achieved.

7.11.1 Use of Sponge Iron/DRI in BF

Due to its physical properties, HBI is the preferred form of sponge iron/DRI for blast furnace use. It can increase hot metal production and lower coke consumption and can be used as up to 30% of the BF charge without significant equipment or process changes.

This is highly desirable when:

- Hot metal availability is insufficient to meet demand.
- Hot metal requirements are not matched to blast furnace output.
- One blast furnace is offline for maintenance.

Using HBI to increase BF burden metallization has an environmental benefit as well. A 10% increase of the burden metallization results in (i) 8% production increase and (ii) 7% decrease of the coke rate, which in turn reduces CO_2 emissions [8] (as shown in Fig. 7.14). HBI has been part of normal operation of a commercial blast furnace in the USA for nearly 30 years and has been used in other blast furnaces around the world.

Use of sponge iron/DRI in the blast furnace is done by AK Steel of Middletown, Ohio, USA [10]. For nearly three decades, AK Steel has been adding hot briquetted iron (HBI) to the charge mix of their blast furnace. This is quite similar to the practice of adding prepared scrap or other metallic sources of iron (used grinding balls, for instance) to a blast furnace. It greatly enhances the

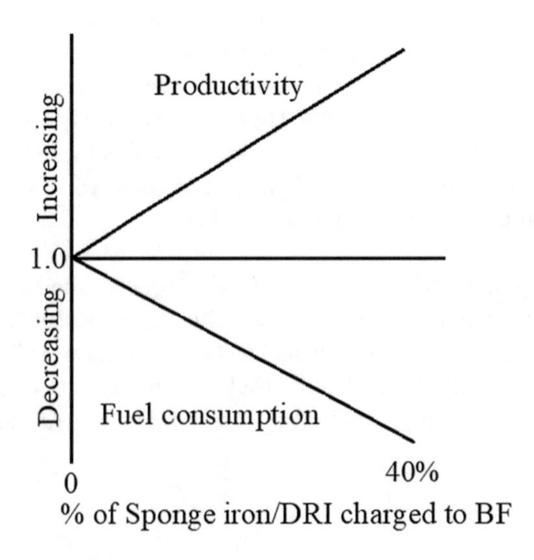

Fig. 7.14 Addition of sponge iron/DRI effect in BF productivity and fuel consumption

productivity of the furnace while simultaneously achieving vast savings of fuel (on per tonne of hot metal basis).

For the past few years, AK Steel, USA, has typically charged 30% of the burden as metal, primarily HBI from Venezuela, together with some B-scrap. The furnace productivity has averaged over four tonnes per day per cubic metre of working volume, the best in the world. Similarly, total fuel consumption is remarkably lowered, to about 440 kg per tonne of hot metal. The extraordinary advantage this gives AK Steel is primarily the increase in productivity.

Since AK Steel began using HBI on a regular basis, nearly every integrated work in the USA has also employed it as blast furnace charge. However, the others typically only use it when they need a production boost or when one blast furnace is down in a work which has multiple furnaces. Some mills in Western Europe have also tested the concept. In each of the cases in North America and in Europe, the focus was on improving production rates. In Japan, at least two of the major integrated steelmakers have conducted extensive tests (hundreds of thousands of tonnes of HBI each) with the focus being CO_2 savings; in China, where over half of the world's iron is produced, a major feasibility study regarding CO_2 saving by the steel industry targeted blast furnace usage of HBI [2]. Benefits of HBI use in BF are: (i) increased productivity, (ii) decrease in coke and other reductants/fuels and (iii) decrease in CO_2 emissions.

7.11.2 Use of Sponge Iron/DRI in LD/BOF

HBI is the form of sponge iron/DRI best suited for use in the LD/BOF because of its bulk density and physical strength. It is a preferred alternative to scrap in the initial cold furnace charge because [11]:

- Residual levels are lower.
- Bulk density is higher.
- Mass and heat balances are more accurate.
- Steel chemistry is easier to control.

Advantages of using HBI in BOF as are follows:

1. The cooling effect of HBI is approximately 25–35% greater than the cooling effect of scrap steel, due to higher heat of fusion of HBI (394 kJ/kg) than scrap (298 kJ/kg).
2. There is no increase in slopping relative to using all scrap as the cold charge.
3. There is no sculling on the lance.
4. HBI can be used for low sulphur steel production.
5. HBI has lower levels of tramp elements than steel scrap.

HBI may be charged to the LD/BOF from either: (i) an overhead bin or (ii) the charging box as up to one-third of the cold charge.

Charging of HBI was done in the same way as scrap charging in LD/BOF. The following general observations were made during trials at Rourkela Steel Plant (RSP), India [12]:

1. Charging time of HBI was same as scrap.
2. The size of HBI has been found to be more convenient for charging compared to scrap, and impact over lining was less.
3. Smooth operation; no abnormalities have been observed.
4. No significant slopping was observed.

5. Increased of hot metal and flux consumption.
6. Relative to scrap, the cooling effect of HBI is higher.
7. FeO in slag increased.
8. Decreased in turn-down temperature and steel yield.

7.11.3 Use of Sponge Iron/DRI in EAF

In DR processes, iron oxide is reduced below the melting point of iron, i.e. solid-state reduction, no molten metal and slag formation. Therefore, all gangue elements of the iron ore remain in the sponge iron/DRI and need to be separated through the slag in the EAF. This increases the electrical energy and electrode consumption of the EAF compared to steel scrap melting (secondary steelmaking). This energy consumption can be reduced by immediate transfer of hot sponge iron/DRI to the EAF melt shop. By doing this, the heat from the direct reduction process lowers the cost of melting the DRI in the EAF, considerably cutting the energy costs and electrode consumption.

EAF steelmaking and sponge iron/DRI have been closely associated for half a century. They have grown up together from producing low-cost, carbon steel long products to making high-quality flat products that meet the most exacting specifications. Today, sponge iron/DRI is still used primarily to make long products, such as reinforcing bars and structural steel where scrap supplies are limited and costly to import; however, the chemical purity of sponge iron/DRI makes it far more valuable. EAF steelmakers have experienced DRI's true value as a scrap supplement [11] that dilutes undesirable contaminants in the charge when the EAF is called upon to make high-quality flat products and low nitrogen steels.

Sponge iron/DRI is ideally suited for use as either a replacement or a supplement for scrap in the EAF. Sponge iron/DRI provides EAF operators the flexibility to tailor their furnace charges to achieve the desired product quality at the lowest cost per tonne of liquid steel. The benefits of sponge iron/DRI go beyond their chemical properties. They can be either charged in scrap buckets or continuously fed to an EAF. When batch charged, the density of CDRI and HBI can eliminate the need for a second or third scrap charge, which results in lower energy losses, fewer interruptions in power-on time, improved tap-to-tap time and increased productivity. Typically, cold sponge iron (CDRI) or HBI can be used as up to 30% of the batch charge. Continuous charging systems are used when CDRI, HBI or hot sponge iron (HDRI) makes up higher percentages of the total charge. Most commonly, a bucket of scrap is charged initially and then the sponge iron/DRI is continuously fed. This allows the furnace roof to remain closed and the power to stay on during most of the heat. Maintaining a consistent temperature improves bath heat transfer and speeds up metallurgical reactions and reduced charge to tap time.

There are two primary benefits of charging hot sponge iron (HDRI) into an EAF: lower specific electricity consumption and increased productivity (15–20% or higher). The energy savings occur because less energy input is required to heat the DRI to melting temperature, which results in a shorter overall melting cycle. The rule of thumb is that electricity consumption can be reduced about 20 kWh/t liquid steel for each 100 °C increase in DRI charging temperature. Thus, the savings when charging at over 600 °C can be 120–140 kWh/t liquid steel [2]. An additional benefit of a shorter overall melting cycle is a reduction in electrode consumption (at least 0.5–0.6 kg/t liquid steel) and refractory consumption (at least 1.8–2.0 kg/t liquid steel). The increased productivity from HDRI charging is significant. Use of HDRI reduces the tap-to-tap time, resulting in a productivity increase of up to 20% versus charging DRI at ambient temperature.

Melting of sponge iron/DRI in EAF is greatly influenced by factors like carbon content and degree of metallization of DRI. Carbon in sponge iron/DRI reacts with unreduced iron oxide (present in sponge iron) giving CO evolution from liquid bath (i.e. carbon boil) which is helpful for subsequent removal of hydrogen and nitrogen gases. Therefore, higher carbon content in sponge iron/DRI is always desired by steelmakers. If carbon is less than required amount, it must be externally added to the melt [13]. In certain cases, more carbon may be desirable to speed up tap-to-tap times; however, high levels can often be counterproductive and increase the duration of the tap-to-tap cycles. This can happen when the residual carbon is more than what is needed to optimize the steel cycle time. Any excess carbon (defined as carbon beyond that which is necessary to meet the specific steel requirement of the steel product being produced) must be burned off (de-carburized), thus causing a decrease in EAF productivity.

Advantages of higher carbon contained in sponge iron are:

- Easy to produce foamy slag,
- Reduction in carburizer consumption,
- Better heat transfer, lower power, electrode and refractory consumption,
- Higher carbon boil to reduce nitrogen and hydrogen in steel,
- Low tap-to tap time, increases productivity.

Metallization of sponge iron/DRI has an important factor in steelmaking; increasing metallization in DRI decreased the energy consumption for melting of sponge iron/DRI in EAF. Sponge iron/DRI having lower metallization value has relatively higher unreduced iron oxide content. Generation of higher slag volume consumes extra heat in melting. As a thumb rule, loss of one per cent of metallization of sponge iron/DRI consumes extra energy [1] of 15 kWh/tonne of liquid steel.

7.11.4 Use of Sponge Iron/DRI in IMF

Emphasis is given for quality steel production in induction melting furnace (IMF) [8]:

- Good weldability,
- Proper balancing of charge mixes for controlling tramp elements,
- Correct carbon equivalent,
- Correct addition of alloying elements to achieve the desired mechanical properties, as per the standard specification,
- Control on sulphur and phosphorus.

Role of gas-based sponge iron/DRI/HBI in IMF steelmaking:

- Limited scope for refining of steel in IMF to achieve low levels of sulphur and phosphorus,
- Only way is to prepare a balanced charge mix for control of sulphur and phosphorus,
- 100% scrap melting for making quality steel not practical in this route, gas-based sponge iron/DRI/HBI can replace judicious amount of scrap,
- Coal-based sponge iron/DRI also has limitations in use due to its high sulphur and phosphorus content, and low carbon and metallization.

7.12 Environmental Benefits of Sponge Iron/DRI

Blast furnace ironmaking uses a carbon source such as coal or coke to generate metallic iron by removing the oxygen, thus producing CO_2 as a by-product. Natural gas-fuelled direct reduction plants make metallic iron by using hydrogen molecules to remove 2/3 of the oxygen atoms and carbon to remove the remainder. Accordingly, the generation of CO_2 is cut by 2/3 relative to standard blast furnace ironmaking. Since blast furnaces produce about 93% of the world's iron, there clearly is a very large potential for lowering CO_2 emissions by replacing the carbon intensive ironmaking of blast furnaces with the much lower carbon footprint of natural gas-based direct reduction ironmaking. Coupled with an EAF, a sponge iron/DRI plant can make any variety of steel normally associated with integrated BF-LD/BOF steelmaking. Many organizations have advocated this to help the steel industry reduce its CO_2 emissions [11]; however, it is not feasible that DR-EAF will completely replace BF-LD/BOF steelmaking.

Another means of lowering the carbon footprint of ironmaking in BF-LD/BOF operations is to pre-reduce a portion of the iron in a sponge iron/DRI furnace prior to charging the iron to a BF. Removing the oxygen (reduction) is the big energy step and is accomplished using natural gas and then the lesser energy step, and melting is done in the BF. This has been demonstrated at many locations over the past decades with as much as 30% of the BF charge being sponge iron/DRI in the form of HBI. HBI is best for use in BFs since the overburden pressure of a large BF may crush non-briquetted DRI. By using HBI, and thus greatly lowering the carbon footprint of a blast furnace, the direct reduction industry can help increase the sustainability of the BF industry.

Another way sponge iron/DRI can help lower the carbon footprint of ironmaking is to maximize the coke rate or the amount of iron produced from a given amount of coal. When coke is made for use by a blast furnace, volatiles are driven off the coal and the remaining carbon and ash become the coke. Some of these volatile gases are used to heat the coke ovens. The remaining volatile gases may be used for a variety of purposes including making DRI. CO_2 emissions for different steelmaking processes are shown in Fig. 7.15. CO_2 emission is directly reduced from 1959 kg/t liquid steel to 760.3 kg/t liquid steel for BF-LD/BOF route to EAF 30% HBI process (i.e. about 60% less CO_2 emissions/t of steel in DR-EAF route than BF-BOF route).

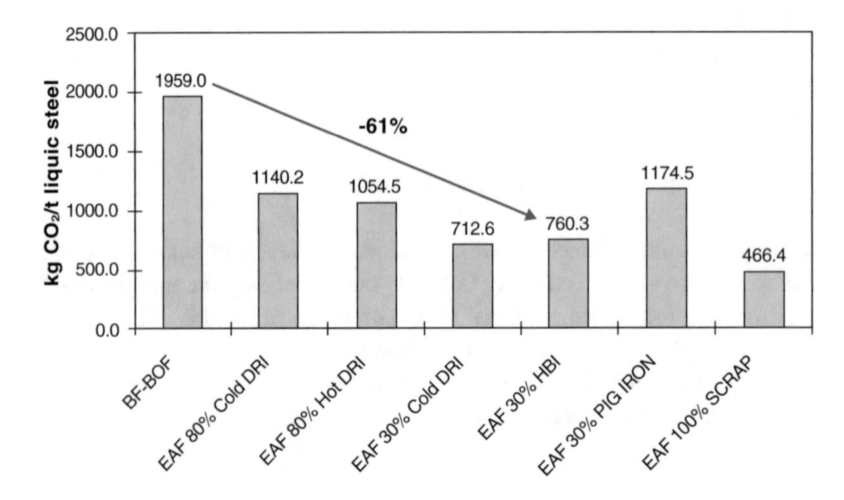

Fig. 7.15 CO_2 emissions for different steelmaking processes [13]

7.13 Iron Carbide

Iron carbide is developed through an environmentally friendly process. Joint venture between Nucor and U.S. Steel utilized iron carbide through converter-based steelmaking. Based on fluidized bed technology using iron ore and natural gas, the reduction of iron ore fines is done with natural gas at a temperature of about 600 °C under a pressure of 1.8 atmospheres; it produces iron carbide (Fe_3C). The iron carbide is a metallic iron (93%) product with 7% carbon.

The conversion of iron ore fines (size: −1 to + 0.1 mm) to iron carbide takes place in a circulating fluid bed reactor, where the pre-heated ore reacts with the reducing constituents of the natural gas (i.e. hydrogen, carbon monoxide and methane) as per the following reaction:

$$3Fe_2O_3(s) + 5\{H_2\} + 2\{CH_4\} = 2Fe_3C(s) + 9\{H_2O\} \quad (7.35)$$

The natural gas is processed through a gas shift reactor for generating the desired reducing constituents. The product is cooled to room temperature in a separate cooler. The cooled iron carbide product is exposed to dry magnetic separation before being sent to product storage. Typical product quality indicates 93–97.5% Fe_3C, 4% Fe_2O_3 and 2–3% gangue. The processed gas is scrubbed, heated and recycled to the reactor (as shown in Fig. 7.16). Energy input to the process is reported to be 12.6 GJ/t-product.

Iron carbide has been found to possess the following specific characteristics [14]:

1. It is inherently stable, hard, non-friable and non-pyrophoric and does not require any passivation like in sponge iron through hot briquetting, to avoid re-oxidation,
2. Ore fines can be used directly for producing carbide,
3. No sticking tendency or build-up in reactor due to fusion of metallic iron,
4. Magnetic separation of gangue feasible,
5. 100% utilization of reagents with closed loop process,

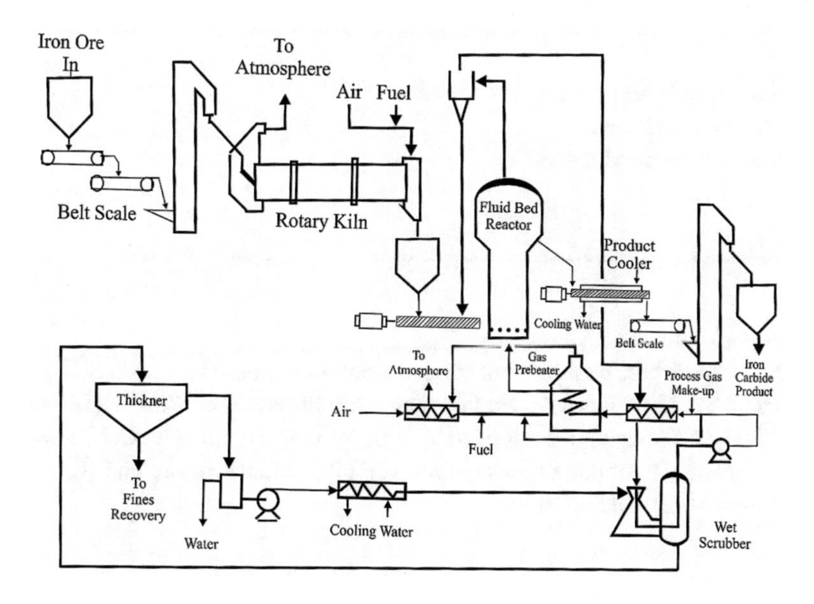

Fig. 7.16 Flow sheet for iron carbide production [14] (reproduced with permission from *IIM Metal News*)

6. It has a high melting point (1837 °C) and dissolves in liquid steel instead of melting. Iron carbide grains dissolve in less than 1 s in molten steel,
7. Being in fine form, it is amenable to inject into EAF,
8. Iron carbide contains high carbon (6–7%),
9. Iron carbide is completely free of sulphur and residual metals, such as copper, zinc, tin, chromium, elements that trouble many steelmakers. It helps better steel quality.

Probable Questions

1. What do you mean by sponge iron? How DR processes are classified? What do you mean by "solid-state reduction"?
2. Discuss the principle and process for production of sponge iron by coal-based process.
3. (i) Discuss the principle of rotary kiln process for sponge iron production. (ii) What are the problems of rotary kiln process and how to overcome them?
4. Discuss the salient features of rotary kiln process.
5. Why coal-based sponge iron cannot form HBI?
6. Discuss the principle of Midrex process for production of HBI.
7. 'Sponge iron produced by gas-based process contains more carbon than by coal-based process'. Why?
8. Discuss the advantages of gas-based process over the coal-based process.
9. Discuss the HyL III process.
10. What do you understand by HBI? How that are produced? Discuss the advantages of HBI.
11. Discuss the characteristics of sponge iron.
12. 'Sponge iron has an inherent tendency to re-oxidize back to its native stage'. Why?
13. What are prevention methods for re-oxidation of sponge iron?
14. Discuss the applications of sponge iron in iron and steel industries.
15. 'Steelmakers are preferred gas-based sponge iron than coal-based sponge iron'. Why?
16. Discuss environmental benefits of sponge iron which acts as feed material in steelmaking furnaces.
17. Discuss the rotary hearth furnace processes.
18. Discuss the ITmk3 process.
19. Discuss iron carbide production.

Examples

Example 7.1 An iron ore has the following composition: 58.1% $FeCO_3$, 11.5% SiO_2, 11.7% $CaCO_3$, 10.5% Al_2O_3 and 8.2% H_2O. It is calcined in a rotary kiln furnace fired with coke containing 84% C, 13% ash and 3% H_2O. Enough air is supplied to burn the coke completely. Coke requires 80 kg per tonne of ore. Calculate (i) volume of air required at STP per tonne of ore and (ii) volume of gases generated per tonne of ore calcinated.

Solution

Since coke contains 84% C,

80 kg coke contains → 0.80 × 84 = 67.2 kg of carbon

$$\begin{array}{ccccc} C & + & O_2 & = & CO_2 \\ 12 & & 22.4\,m^3 & & 22.4\,m^3 \end{array}$$

12 kg of C requires 22.4 m^3 of O_2

$$67.2 \quad \left(\frac{22.4 \times 67.2}{12}\right) = 125.44\,m^3 \text{ of } O_2 \text{ or } CO_2 \text{ form}$$

21 m^3 of O_2 content in 100 m^3 of air

$$125.44\,m^3 \text{ of } O_2 \quad \left(\frac{100 \times 125.44}{21}\right) = \mathbf{597.33\,m^3\ of\ air}$$

Volume of air required at STP per tonne of ore = **597.33 m^3**

Amount of N_2 = amount of air − amount of O_2 = 597.33 − 125.44 = **471.89 m^3**
Amount of CO_2 form by burning of coke = 125.44 m^3

Iron ore contains 58.1% $FeCO_3$ and 11.7% $CaCO_3$.
That is, one tonne ore contains 581 kg $FeCO_3$ and 117 kg $CaCO_3$.

$$\begin{array}{ccccc} FeCO_3 & = & FeO & + & CO_2 \\ 116 & & & & 22.4\,m^3 \end{array}$$

116 kg of $FeCO_3$ forms 22.4 m^3 of CO_2

$$581\,kg \quad \left(\frac{22.4 \times 581}{116}\right) = 112.19\,m^3 \text{ of } CO_2$$

$$\begin{array}{ccccc} CaCO_3 & = & CaO & + & CO_2 \\ 100 & & & & 22.4m^3 \end{array}$$

100 kg of $CaCO_3$ forms 22.4 m^3 of CO_2

$$117\,kg \quad \left(\frac{22.4 \times 117}{100}\right) = 26.21\,m^3 \text{ of } CO_2$$

$$\text{Total } CO_2 \text{ form} = CO_2 \text{ form coke} + CO_2 \text{ form } FeCO_3 + CO_2 \text{ form } CaCO_3$$
$$= 125.44 + 112.19 + 26.21 = \mathbf{263.84\,m^3}$$

18 kg of H_2O = 22.4 m^3 of H_2O

$$(0.03 \times 80)\text{kg of } H_2O \text{ from coke} = \left(\frac{22.4 \times 2.4}{18}\right) m^3 \text{ of } H_2O = 2.99\, m^3 \text{ of } H_2O \text{ form coke}$$

18 kg of H_2O = 22.4 m^3 of H_2O

$$82\,\text{kg} \quad \left(\frac{22.4 \times 82}{18}\right) m^3 \text{ of } H_2O = 102.04\, m^3 \text{ of } H_2O \text{ form ore.}$$

Total moisture = 2.99 + 102.04 = **105.03 m^3**
Therefore,

$$\text{total amount of gases } (N_2 + CO_2 + H_2O) = 471.89 + 263.84 + 105.03$$
$$= \mathbf{840.76\,m^3}$$

Example 7.2 Total iron content in ore and DRI are 65.1 and 90%, respectively. If the production rate of a rotary kiln furnace is 4.16 t/hr, what is the rate of iron ore charged to the rotary kiln furnace?

Solution
Fe Balance:
Fe from ore = Fe in DRI

$$(T_{Fe,ore}/100) \times W_{ore} = (T_{Fe,DRI}/100) \times W_{DRI} \qquad (1)$$

Since $W_{DRI} = 4.16\,t/hr$, $T_{Fe,DRI} = 90\%$ and $T_{Fe,ore} = 65.1\%$
So from Eq. (1): (65.1/100) × W_{ore} = (90/100) × 4.16 = 3.744
Therefore, $W_{ore} = (3.744/0.651) = 5.75\,t/hr$.
The rate of iron ore charged to the rotary kiln furnace is 5.75 t/hr.

Example 7.3 Find out the change of free energy $(\Delta G^\circ_{r,298})$ and minimum temperature for the following reactions:

(a) $3Fe_2O_3 + CO = 2Fe_3O_4 + CO_2$
(b) $Fe_3O_4 + CO = 3FeO + CO_2$
Given:

$$3Fe_2O_3 = 2Fe_3O_4 + 1/2O_2, \Delta G^\circ_r = 249.45 - 0.14\,T\,kJ$$
$$Fe_3O_4 = 3FeO + 1/2O_2, \Delta G^\circ_r = 312.21 - 0.125\,T\,kJ$$
$$C + 1/2O_2 = CO, \Delta G^\circ_r = -111.71 - 0.088\,T\,kJ$$
$$C + O_2 = CO_2, \Delta G^\circ_r = -394.13 - 8.4 \times 10^{-4}\,T\,kJ$$

Solution

$$\text{(a)}\quad \begin{aligned} 3Fe_2O_3 &= 2Fe_3O_4 + 1/2O_2, \Delta G^\circ_{Fe_2O_3} \\ CO &= C + 1/2O_2, -\Delta G^\circ_{CO} \\ C + O_2 &= CO_2, \quad \Delta G^\circ_{CO_2} \end{aligned}$$

$$3Fe_2O_3 + CO = 2Fe_3O_4 + CO_2, \Delta G^\circ_f = (\Delta G^\circ_{Fe_2O_3} - \Delta G^\circ_{CO} + \Delta G^\circ_{CO_2})$$

Therefore,

$$\begin{aligned} \Delta G^\circ_f &= \left[(249.45 - 0.14\,T) - (-111.71 - 0.088\,T) + (-394.13 - 8.4 \times 10^{-4} T)\right] \\ &= \mathbf{-32.97 - 0.052\,T\,kJ} \end{aligned}$$

At equilibrium, $\Delta G^\circ_f = 0$
Hence, $-32.97 - 0.052\ T = 0$.
Therefore, **T = −634 K = −361 °C**.

$$\text{(b)}\quad \begin{aligned} Fe_3O_4 &= 3FeO + 1/2O_2, \quad \Delta G^\circ_{Fe_3O_4} \\ CO &= C + 1/2O_2, \quad -\Delta G^\circ_{CO} \\ C + O_2 &= CO_2, \quad \Delta G^\circ_{CO_2} \end{aligned}$$

$$Fe_3O_4 + CO = 3FeO + CO_2, \Delta G^\circ_f = (\Delta G^\circ_{Fe_3O_4} - \Delta G^\circ_{CO} + \Delta G^\circ_{CO_2})$$

Therefore,

$$\begin{aligned} \Delta G^0_f &= \left[(312.21 - 0.125\,T) - (-111.71 - 0.088\,T) + (-394.13 - 8.4 \times 10^{-4} T)\right] \\ &= \mathbf{29.79 - 0.037\,T\,kJ} \end{aligned}$$

At equilibrium, $\Delta G^\circ_f = 0$
Hence, $29.79 - 0.037\ T = 0$.
Therefore, **T = 805 K = 532 °C**.

Example 7.4 Calculate the minimum per cent of carbon monoxide required to reduce FeO at 727 °C and (i) one atmospheric pressure and (ii) 1.5 atm. The reduction is taking place as follows:

$$FeO + CO = Fe + CO_2 \quad \Delta G^\circ = -22802.8 + 28.45\,T\,J/mol$$

Solution

$T = 727 + 273 = 1000$ K
$\Delta G^\circ = -22802.8 + 28.45 \times 1000 = 5647.2\,J/mol$

Since $\Delta G^\circ = -RT \ln k$
Therefore, $5647.2 = -8.314 \times 1000 \times \ln k$
$\ln k = -0.6792$
Therefore, $k = \mathbf{0.507}$.
Again, $k = \left(\frac{p_{CO_2}}{p_{CO}}\right)$

Case I

$p_{CO_2} + p_{CO} = 1\,\text{atm}$

Let p_{CO} = x atm, p_{CO_2} = 1 − x

Therefore, $\left(\frac{p_{CO_2}}{p_{CO}}\right) = \left(\frac{1-x}{x}\right) = k = 0.507$

Therefore, x = 0.6635 atm.

Per cent of CO = **66.35%**

Case II

$p_{CO_2} + p_{CO} = 1.5\,\text{atm}$

Let p_{CO} = x atm, p_{CO_2} = 1.5 − x

Therefore, $\left(\frac{p_{CO_2}}{p_{CO}}\right) = \left(\frac{1.5-x}{x}\right) = K = 0.507$

Therefore, x = 0.9953 atm.

Per cent of CO = **99.53%**

Example 7.5 Find the change in the percentage of hydrogen in a mixture of 75% H_2 and 25% CH_4 in equilibrium with carbon if the total pressure changes from 1 to 0.1 atm. What is the temperature of this equilibrium at 1 atm pressure?

Given: $CH_4 = C + 2\,H_2$, $\Delta G^\circ = 90165.2 - 109.45\,T\,J$

Solution

Equilibrium constant, $k = \left(\frac{p^2_{H_2}}{p_{CH_4}}\right)$

Total pressure = 1 atm = p_{H2} + p_{CH4}

Since 75% H_2 and 25% CH_4, i.e. p_{H2} = 0.75, and p_{CH4} = 0.25

Hence, $k = \left(\frac{p^2_{H_2}}{p_{CH_4}}\right) = \left(\frac{0.75^2}{0.25}\right) = 2.25$, and ln k = 0.8109.

Since $\Delta G^\circ = -RT \ln k = -8.314 \times T \times \ln k$

Therefore, $90165.2 - 109.45\,T = -8.314 \times T \times 0.8109 = -6.74\,T$

Hence, **T = 877.88 K = 604.88 °C.**

Now, total pressure changes from 1 atm to 0.1 atm.

So, $p_{H_2} + p_{CH_4} = 0.1\,\text{atm}$

Let $p_{H_2} = x$, $p_{CH_4} = (0.1 - x)$

Since ΔG^o = −RT ln k

Therefore, $90165.2 - 109.45\,T = -8.314 \times T \times \ln k$

Or $90165.2 - 109.45 \times 878 = -8.314 \times 878 \times \ln k$ (since T = 878 K)

Or ln k = 0.813 or k = 2.25

Since $k = \left(\frac{p^2_{H_2}}{p_{CH_4}}\right) = \left(\frac{x^2}{0.1-x}\right) = 2.25$

Or $x^2 + 2.25x - 0.225 = 0$

$$\text{Roots} = \left(\frac{\{-b \pm \sqrt{(b^2 - 4ac)}\}}{2a}\right)$$

Therefore, $x = \left(\frac{\{-2.25 \pm \sqrt{(2.25^2 - 4.1.(-0.225))}\}}{2X1}\right)$

Hence, $x = 0.096$, so $\mathbf{H_2 = 9.6\%}$.

Therefore, the change in the percentage of hydrogen in a mixture of 75% H_2 and 25% CH_4 in equilibrium with carbon, if the total pressure changes from 1 atm to 0.1 atm, is 9.6%.

Example 7.6 Calculate the equilibrium CO: CO_2 ratio and %CO at 900 °C according to reaction:

$$FeO + CO = Fe + CO_2$$

Given:

$$C + O_2 = CO_2 \quad \Delta G^\circ = -394,100 - 0.84\,T\ J$$
$$C + 1/2O_2 = CO \quad \Delta G^\circ = -111,000 - 87.65\,T\ J$$
$$Fe + 1/2O_2 = FeO \quad \Delta G^\circ = -259,600 + 62.55\,T\ J$$

Solution

$$FeO = Fe + 1/2O_2 \quad \Delta G^\circ = 259,600 - 62.55\,T$$
$$CO = C + 1/2O_2 \quad \Delta G^\circ = 1111,000 + 87.65\,T$$
$$C + O_2 = CO_2 \quad \Delta G^\circ = -394,100 - 0.84\,T$$

$$FeO + CO = Fe + CO_2 \quad \Delta G^\circ = -23500 + 24.26\,T = -23500 + 24.26 \times 1173 = 4956.98\,J$$

Since $\Delta G^\circ = -RT \ln k$

Therefore, $4956.98 = -8.314 \times 1173 \times \ln k$.

$\ln k = -508$

Therefore, $k = 0.6015$.

Again, $k = \left(\frac{p_{CO_2}}{p_{CO}}\right)$

So, $\left(\frac{p_{CO_2}}{p_{CO}}\right) = \left(\frac{1}{k}\right) = \left(\frac{1}{0.6015}\right) = \mathbf{1.66}$

Therefore, the equilibrium CO: CO_2 ratio is **1.66**.

Let $p_{CO} = x$, since $p_{CO} + p_{CO_2} = 1$.

Therefore, $p_{CO_2} = 1 - x$.

Hence, $k = \left(\frac{p_{CO_2}}{p_{CO}}\right) = \left(\frac{1-x}{x}\right) = 0.6015$
Therefore, $x = \left(\frac{1}{1.6015}\right) = 0.624$, i.e. CO = **62.4%**.

Example 7.7 Find out whether FeO can be reduced by H_2/H_2O mixture containing 60% H_2 and 40% H_2O at 727 °C.

$$Fe + 1/2O_2 = FeO \quad \Delta G^\circ = -259,600 + 62.55\,T\ J$$
$$H_2 + 1/2O_2 = H_2O \quad \Delta G^\circ = -246,000 + 54.8\,T\ J$$

Solution

$$FeO = Fe + 1/2O_2 \quad \Delta G^\circ = 259,600 - 62.55\,T$$
$$H_2 + 1/2O_2 = H_2O \quad \Delta G^\circ = -246,000 + 54.8\,T$$
$$FeO + H_2 = Fe + H_2O \quad \Delta G^\circ = 13600–7.75\,T = 13600–7.75 \times 1000 = 850\,J$$

$\Delta G^\circ = -RT \ln k$
Therefore, $5850 = -8.314 \times 1000 \times \ln k$.
$\ln k = -0.7036$
$k = 0.4948 = \left(\frac{p_{H_2O}}{p_{H_2}}\right), p_{H_2O} + p_{H_2} = 1\,\text{atm}, \text{Let } p_{H_2} = x$

Therefore, $0.4948 = \left(\frac{1-x}{x}\right)$
$x = 0.6689$ atm
Pct of hydrogen = 66.89%

Since 60% H_2 is present in mixture, equilibrium H_2 requires minimum 66.89%; hence, FeO cannot be reduced by H_2 and H_2O mixture containing 60% H_2 and 40% H_2O at 727 °C.

Another thing should be noted that ΔG^0 value is positive. Hence, from that value also it can be concluded that reduction is not feasible at all at 727 °C.

Example 7.8 Iron sample is heated at 1000 °C in an atm of hydrogen which contains some moisture. Find out the maximum permissible water vapour in hydrogen to avoid oxidation of iron sample.

Given: $FeO + H_2 = Fe + H_2O \quad \Delta G^\circ = 12761.2–7.03T\ J/mol$

Solution

$T = 1000\ °C = 1273\ K$
$\Delta G^\circ = 12761.2–7.03\,T = 12761.2–7.03 \times 1273 = 3812.01\,J/mol$

Since $\Delta G^\circ = -RT \ln k$
Therefore, $3812.01 = -8.314 \times 1273 \times \ln k$
$\ln k = -0.3602$

$k = 0.6975 = \left(\frac{p_{H_2O}}{p_{H_2}}\right) = \left(\frac{1-x}{x}\right)$ [let $p_{H_2} = x$]

Therefore, $x = 0.5891$, i.e. $H_2 = 58.91\%$.

So, $H_2O = 100 - 58.91 = 41.09\%$

Hence, the maximum permissible water vapour in hydrogen to avoid oxidation of iron sample is **41%**.

Problems

Problem 7.1 Total iron content in ore and DRI are 66% and 92%, respectively. If the production rate is 5 t/hr, what is the rate of iron ore charged to the rotary kiln furnace?

[Ans: 6.97 t/hr.]

Problem 7.2 Hydrogen is passed through a furnace at 800 °C, containing FeO (solid), which is undergoing reduction to form Fe. Find out the percentage of utilization of H_2 gas, assuming that equilibrium prevails.

Given: $\Delta G_r^\circ = 8368\,\text{J/mol}$ at 800°C.

[Ans: 71.87%]

Problem 7.3 Find out whether FeO can be reduced by H_2 and H_2O mixture containing 60% H_2 and 40% H_2O at 600 °C.

$$\text{Fe} + 1/2\text{O}_2 = \text{FeO} \quad \Delta G^\circ = -259{,}600 + 62.55\,\text{T J}$$
$$\text{H}_2 + 1/2\text{O}_2 = \text{H}_2\text{O} \quad \Delta G^\circ = -246{,}000 + 54.8\,\text{T J}$$

[Ans: Since 60% H_2 is present in mixture, equilibrium H_2 requires minimum 71.94%; hence FeO cannot be reduced by H_2 and H_2O mixture containing 60% H_2 and 40% H_2O at 600 °C.]

References

1. S.K. Dutta, P.J. Roy Choudhury, Inst of Engg (India). J. MM **66**, 91 (March 1986)
2. S.K. Dutta, JPC Bull Iron & Steel **XV** (3), 13 (March 2015)
3. S.K. Dutta, R. Sah, Alternate Methods of Ironmaking (Direct Reduction and Smelting Reduction Processes). S. Chand & Co Ltd, New Delhi (April 2012)
4. Midrex: 2018 World Direct Reduction Statistics (May 2019)
5. A. Chatterjee, Proc Inter Seminar on Alternative Routes for Ironmaking in India. Steel Tech, September 2009, Kolkata, p. 5
6. J. McClelland, JPC Bull Iron & Steel **IV** (1), 109 (January 2004)
7. R.B. Cheeley, Direct from Midrex, 3rd Quarter (1999)
8. J. Kopfle, Green Steel Summit. Midrex Technology (May 2010)
9. S.K. Dutta, A.B. Lele, Iron & Steel Review **55**(3), 162 (2011)
10. A. Hassan, F. Griscom, Direct from Midrex, 3rd Quarter, p. 6 (2008)
11. DRI Products & Applications, Midrex (January 2015)
12. A.K. Sarkar et al., Steel India **17**(2), 45 (1994)

13. P. Tatia, Paper presented at National Metallurgist Day-Annual Technical Meeting of Indian Institute of Metals, November 2014, Puna, India (Privately Collected)
14. A. Ganguly, R.H.G. Rau, V.S. Shah, IIM Metal News **19**(2), 13 (1997)
15. M. Paswan, C. Mukherjee, IIM Metal News **12**(1), 6 (2009)

8 Smelting Reduction Processes

Blast furnace is still very popular in producing hot metal because of their (i) high energy efficiency, (ii) high productivity, (iii) high degree of gas utilization and (iv) long service life. Main weaknesses of blast furnace are (i) use of good-quality raw materials, (ii) requirement of coke making and sinter making processes and (iii) generation of air pollutant such as SO_x, NO_x, dioxin and dust. Blast furnace ironmaking is still the dominant process to produce liquid iron from ore. But its dependence on metallurgical coal and prepared burden has driven steelmakers to look for alternative processes of ironmaking throughout the world. Smelting reduction (SR) technologies are coming up to offer such alternatives to blast furnace ironmaking.

8.1 Need of Smelting Reduction

Iron and steel industries produce large quantities of fine waste material as by-product. This fine waste can be disposed off as landfill or sintered/palletized. These materials include blast furnace dust, blast furnace sludge, basic oxygen furnace precipitator dust, EAF dust, mill scale, fired pellet screenings, fine metallic scrap, coke breeze and coal/coke dust. Nowadays, most steelmakers are searching for appropriate ways to recycle steelmaking wastes. Accumulations of steel plants dust and its disposal have become a worldwide problem. There are two aspects of BF/BOF dust recycling: one is to recycle valuable resources (i.e. higher metallization of iron oxides which would reduce the burden on the melting process and high removal ratio of zinc which would reduce the concentration of zinc within the recycling loop), and another is to reduce environmental pollution.

While several proven natural gas-based ironmaking processes are readily available and new technologies are under development, these processes are generally only competitive where natural gas prices are low and will remain low over the lifetime of the plant. Such cheap gas is only available in few locations. Coal-based ironmaking technologies have no territorial limitation, but they are very limited in number, and by virtue of the more difficult processing requirements generally exhibit limitations in size of module and raw material specification. They also demand a high degree of operating skills and process control and have lower availability and higher maintenance requirements than natural gas-based processes.

Blast furnace has played a major role in producing hot metal because of their (i) high energy efficiency, (ii) high productivity, (iii) high degree of gas utilization and (iv) long service life. Main weaknesses of blast furnace are (i) use of high-grade raw materials, (ii) requirement of coke making and sinter making processes and (iii) generation of air pollutant such as SO_x, NO_x, dioxin and dust.

S. K. Dutta and Y. B. Chokshi, *Basic Concepts of Iron and Steel Making*,
https://doi.org/10.1007/978-981-15-2437-0_8

Blast furnace ironmaking is still the dominant process to produce hot metal from ore. But its dependence on metallurgical coal and the prepared burden has driven worldwide steelmakers to look for alternative processes for ironmaking to avoid dependence on scarce metallurgical coal and reduce process steps like cokemaking in coke oven battery, sintering, etc., to minimize capital investment involved and reduce harmful impact on environment. Smelting reduction (SR) technologies are coming up to offer such alternatives to blast furnace ironmaking.

Smelting reduction (SR) processes are the latest development in pig iron production. The SR process emerged during the 1990s. In SR processes, iron ore and coal are added directly to a metal-slag phase where the ore is reduced. The SR process combines the gasification of coal with the smelting reduction of iron ore. Energy consumption of smelting reduction is lower than that of blast furnace, as coking step is avoided and the need for ore preparation is reduced. Examples of this process are Finex and Hismelt. Although SR process reduces *greenhouse gas* (GHG) emissions, the process demands higher energy for ironmaking step (18.7 GJ/t of crude steel) [1] that exceeds the total energy required for the entire steelmaking process through the BF-BOF route (15.8 GJ/t).

All smelting reduction processes aim to produce hot metal, without using either coke or high-grade iron ore as the feedstock. However, it needs to be stressed that only a few SR processes that have been conceptualized till date have reached commercial scale. This has happened even though these new processes are one generation ahead of classical blast furnaces in terms of their intrinsic process versatility as well as superior environmental friendliness. Hence, the smelting reduction processes deserve special attention since they use non-coking coal for the production of hot metal. However, it is pertinent to mention that only a few of the SR processes have reached commercial scale. Reduction of iron ore into metallic iron is the first step in the production process of steel. It is also the most energy-intensive step. At a high temperature, the iron ore is reduced to molten iron. The molten iron is subsequently converted into molten steel, cast into semi-finished products and rolled and shaped into finished products.

There are two major conventional routes (routes I and II in Fig. 8.1) for producing finished steel. Both routes play an important role in the development of smelting reduction (SR) technology. First route is an integrated steel plant where iron ore is reduced in a blast furnace (BF) and subsequently refined to steel. A schematic diagram illustrating various process routes to steelmaking is shown in Fig. 8.1.

Second route is known as a mini steel plant. In most mini steel plants, recycled/purchased steel scrap and/or sponge iron/direct reduced iron (DRI) is melted in electric furnaces and further processed into final products. The mini steel plant route is considerably less energy-intensive than the traditional integrated steel plants. In the mid-1960s, direct reduction (DR) technology was projected as the best possible alternative to the dominant BF technology route. Several processes based on natural gas and non-coking coal using iron oxide lump/pellets of fines as feedstock were developed, a few of which matured to the stage of commercial exploitation. Advantages of sponge iron, as feed material, are well known. Even though DR technology could eliminate some of the inherent drawbacks of the blast furnace, it had its own limitations: (i) its solid product, (ii) small module size, (iii) product pyrophoricity, (iv) low productivity, etc. Location of the sponge iron or HBI plant is limited to regions where coal/natural gas is available at a reasonable price. Coal-based DR processes have also been adopted for commercial production, but they have their own limitations, the important ones are lower levels of productivity, low carbon content in the product, inability to use fines, ring formation in the kiln and its solid state. Coal-based sponge iron/DRI is not suitable for high-capacity operation due to restriction in module capacity.

The need for an alternate ironmaking technology also arises out of the demand to conserve the depleting reserves of good-quality coking coal, to make use of non-coking coal reserves dispersed across the country and to take advantage out of large price differences between imported metallurgical

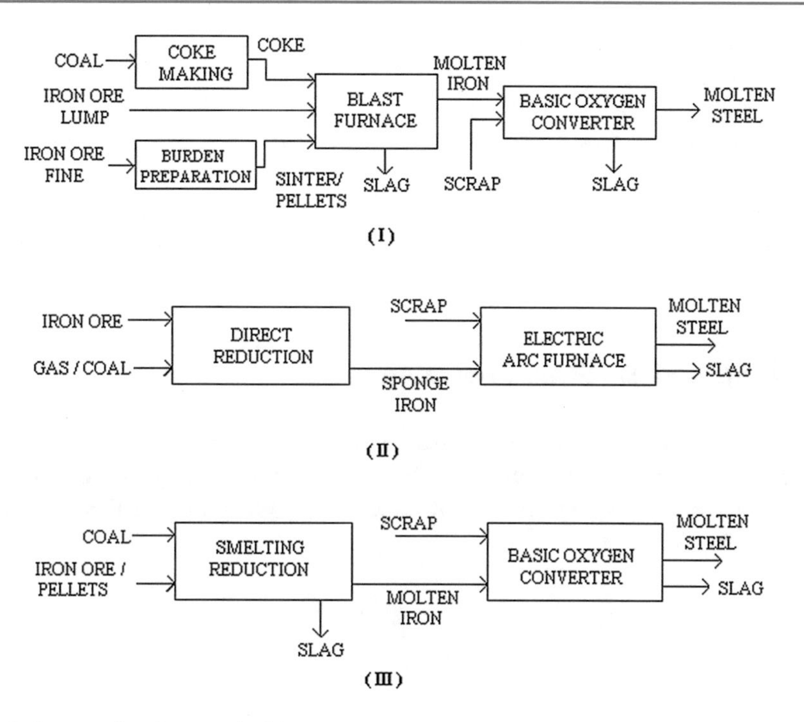

Fig. 8.1 Block diagram of various steelmaking process routes

coal and non-coking steam coal. The scenario forced innovators to look for alternative coal-based ironmaking routes.

To overcome the above, scientist and technologists have been working for quite some time specially in the second half of this century to develop suitable process of ironmaking, called *smelting reduction* (SR) process, using non-coking coals directly. Iron ore can be pellets, lump ore or fines depending on the process route. Smelting reduction processes of liquid ironmaking drawing considerable attention due to SR processes are currently being developed as an alternative route to BF ironmaking with the following objectives [2]:

1. To utilize low-grade solid fuels and ore fines,
2. To produce hot metal at lower cost and by low capital investment,
3. To reduce pollutants emission,
4. To achieve flexibility regarding input of raw materials and selection of operating parameters and
5. To install small hot metal production units to meet fluctuating market demand and increase production in small steps of capital investment.

8.1.1 Why Are Smelting Reduction Processes Required?

In 1992, Chatterjee [3] wrote a book *Beyond the Blast Furnace* on direct reduction and smelting reduction processes. Chatterjee wrote that the blast furnace ironmaking has achieved near perfect maturity through intensive developments that have taken place around the world. There are some threats to the blast furnace route as follows:

- Very stick raw material requirements,
- High capital requirement,
- Lack of flexibility and
- Strick environmental policies.

Chatterjee concluded that coke is the biggest threat for blast furnace because of the need for high coking coal and the environmental issues of cokemaking. After 25 years, still, the situation is not different.

8.2 Significance of Smelting Reduction

There is a revolution in the iron and steel industry with the development of new smelting technology for the high production of liquid iron directly from iron ore fines and agglomerates. These processes have the potential to completely replace the conventional blast furnace production of hot metal as they offer advantages as follows [5]:

1. Use of non-coking coal rather than metallurgical coke,
2. Use of iron ore fines rather than lump ores or sinter,
3. Economic operation at small scale (250,000–500,000 tpy),
4. Lower capital costs,
5. Flexibility in operation and materials use and
6. Greater environmental control.

8.3 Principle of SR Processes

The basic principle of a SR process is to melt the pre-reduced iron ore/pellets with non-coking coal and oxygen in a reactor. SR technology consists of a pre-reduction unit and a smelting unit. In pre-reduction unit, ore/pellets/fines are partially reduced and pre-heated using the gas generated in the smelting unit. The partially reduced ore is injected or fed into smelting vessel containing iron–carbon melt and slag [4].

The overall reaction is:

$$(\mathrm{FeO}) + \mathrm{C(s)} = [\mathrm{Fe}] + \{\mathrm{CO}\} \tag{8.1}$$

There is a general agreement that the above reaction takes place in two stages.

First stage → slag–gas reaction:

$$(\mathrm{FeO}) + \{\mathrm{CO}\} = [\mathrm{Fe}] + \{\mathrm{CO_2}\} \tag{8.2}$$

Second stage → carbon gasification reaction:

$$\mathrm{C(s)} + \{\mathrm{CO_2}\} = 2\{\mathrm{CO}\} \tag{8.3}$$

The following four possible rate-limiting steps may be visualized for this reaction (8.1):

- Mass transfer of FeO in the form of Fe^{2+} [i.e. (FeO) $\rightarrow \left(Fe^{2+}\right) + \left(O^{2-}\right)$] in the slag phase,
- Gas–slag reaction at the interface,
- Gas diffusion in the gas layer separating slag and solid carbon, and
- Gasification reaction at the solid–gas interface.

8.4 Classification of SR Processes

Based on the type of fuel used, the smelting reduction processes can be broadly classified into two categories as follows [5]:

(i) Processes utilizing coal and electricity and
(ii) Processes utilizing oxygen and coal.

8.4.1 Processes Utilizing Coal and Electricity

The smelting reduction processes utilize electricity and coal as the source of energy and reductant. Based on the number of stages involved, these processes can be divided mainly into two types, namely (i) single-stage process and (ii) two stages process.

8.4.1.1 Single-Stage Process

The reactor vessel is fed with cold iron ore without any pre-reduction and coal as reductant. In single-stage process, both reduction and smelting take place in the electric smelter [5]. The off-gases from the electric smelter are used for internal generation of electricity as shown in Fig. 8.2. The process offers greater flexibility compared with any other oxy-coal process.

8.4.1.2 Two Stages Process

The direct reduced iron (DRI) produced in a reduction unit is smelted in the electric smelter [5]. The off-gases from the reduction unit are used for internal generation of electricity as shown in Fig. 8.3. This process offers greater flexibility compared with other oxy-coal processes.

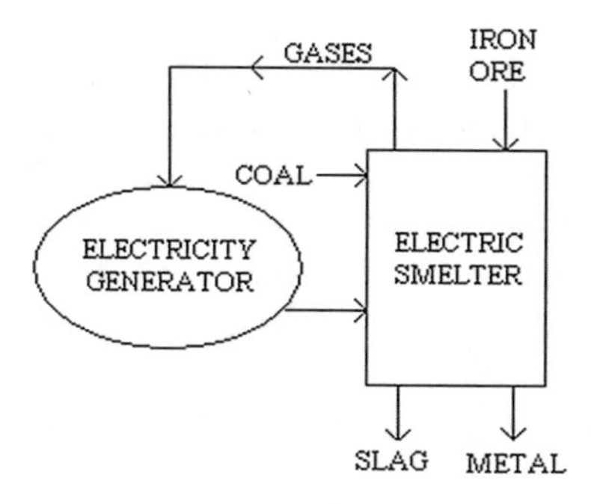

Fig. 8.2 Single-stage smelting reduction process [5]

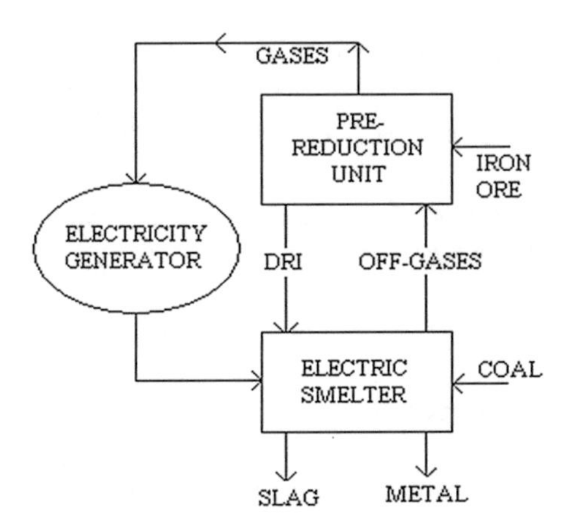

Fig. 8.3 Two stages smelting reduction process [5]

8.4.2 Processes Utilizing Oxygen and Coal

The smelting reduction processes utilizing oxy-coal combustion as the source of energy cum reductant. Based on the number of stages involved (i.e. according to their thermo-chemical configuration), these processes can be divided into three types as follows [5]:

(i) Single stage,
(ii) Two stages and
(iii) Three stages processes.

8.4.2.1 Single-Stage Process

The single-stage process is the simplest of all the iron bath processes and theoretically represents the ideal smelting reduction configuration where hot metal is produced in a single reactor. In the single-stage process, entire metallurgical reactions are carried out in a single reactor where iron ore, coal and oxygen are fed [5]. The gases evolved from the molten bath are post-combusted to a very high degree (around 70%) and the major portion of heat is transferred back to the molten bath. Refractory erosion is likely to take place and foamy slag is to be controlled during operation. The oxygen and coal requirements are higher compared with two stages and three stages processes. Romelt and Ausmelt processes belong to this category. Figure 8.4 shows the schematic diagram of the single-stage process.

8.4.2.2 Two Stages Process

In two stages processes, pre-reduced iron ore is smelted in the separate reactor and most of the gas generated during the operation is utilized to pre-reduce the ore as well as to supply heat energy to the entire system through post-combustion, e.g. Corex, HIsmelt, DIOS, AISI-DOE, etc. Here, hot pre-reduced iron oxide generally having low degree of reduction is charged into smelting reactor along with coal and oxygen/pre-heated air. The gases evolving from the molten bath is post-combusted by oxygen/pre-heated air to a moderate degree (around 50%) inside the smelter [5]. The heat of the post-combustion is efficiently transferred from the gas phase to the molten bath. HIsmelt process belongs to this category. Figure 8.5 shows the schematic diagram of the two stages process.

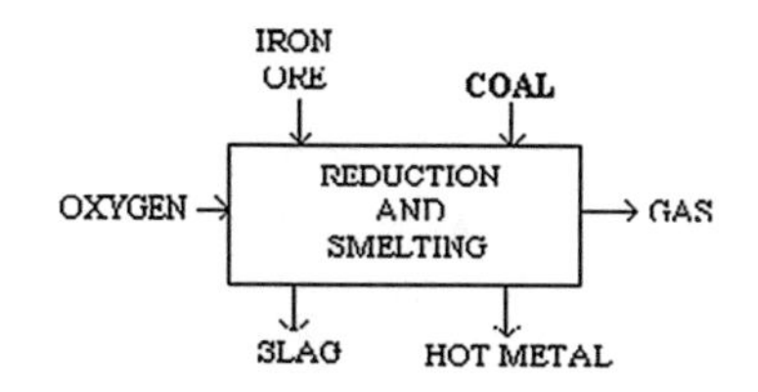

(a) Single stage

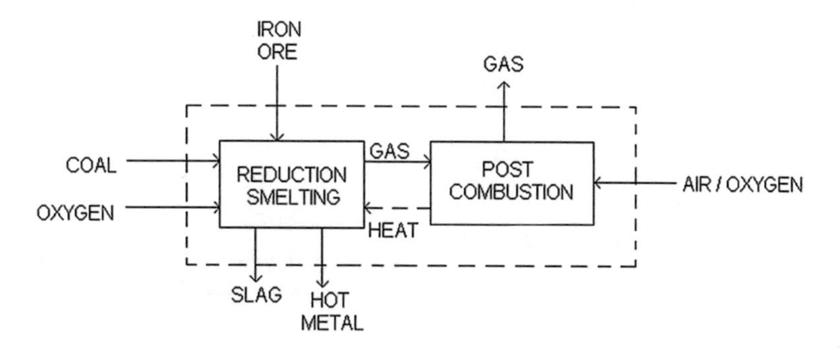

(b) Single stage with post combustion.

Fig. 8.4 Single-stage smelting reduction process [5]

8.4.2.3 Three Stages Process

In three stages process, a separate gasifier is coupled with the smelting unit and the pre-reduction unit, to gasify coal producing CO, H_2, CH_4, etc. The process involves an additional step of gasification/gas reformation between smelting reactor and pre-reduction reactor [5]. The presence of carbon in gasification zone helps in reducing the temperature of the melter off-gases without the loss of energy. Figure 8.6 shows the schematic diagram of the three stages process.

8.5 Advantages of Smelting Reduction Processes

Smelting reduction processes have many advantages with respect to raw materials, energy cost, investment cost, economy of scale and environmental compatibility. The SR processes are expected to have the following advantages [5]:

(i) Lower capital cost due to the lower economic scale of operation,
(ii) Lower operating cost primarily owing to the use of non-coking coal,
(iii) Highly energy efficient,
(iv) Higher smelting intensity with higher productivity (due to faster reaction kinetics, increased transport rate due to convection, and an increase in the convection rate due to enlargement of specific phase contact areas in dispersed phases),
(v) Ability of some processes to direct utilize of iron ore fines,
(vi) No necessity of expensive coking coal,

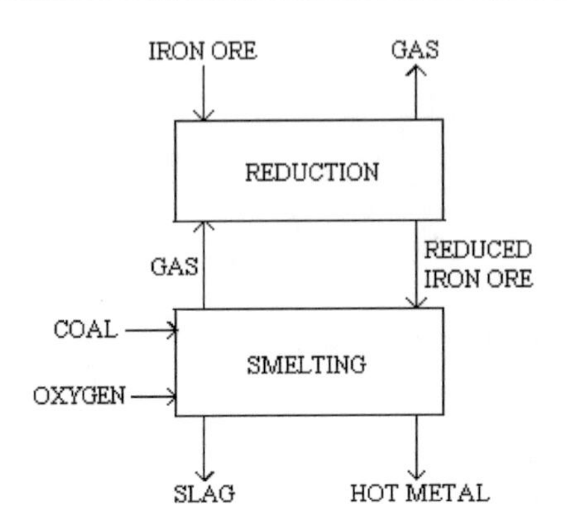

(a) Two stages

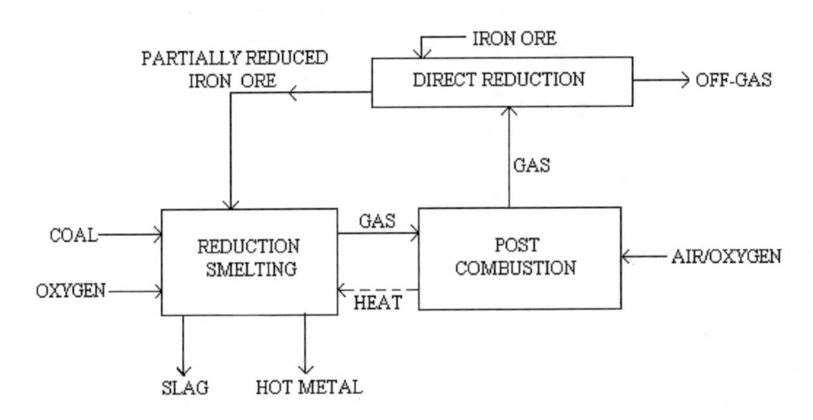

(b) Two stages with post combustion

Fig. 8.5 Two stages smelting reduction process [5]

(vii) Possible elimination of sinter/pellet and coke oven plants,
(viii) Decrease in manpower requirement and reduced operational cost,
(ix) Environmental friendliness (due to lower emissions) since no coke ovens and sinter plants,
(x) Maintain the same level of hot metal quality as obtained in the blast furnace,
(xi) Better control over process parameters,
(xii) Flexibility in selection of thermo-chemical configuration,
(xiii) Faster reduction,
(xiv) Recycling of existing stockpiles, in-plant dusts, sludge and other reverts wherever and to the extent possible and
(xv) Better energy economy, i.e. electricity generation from the off-gases (some SR processes yield rich off-gases, which can be used for the generation of electricity).

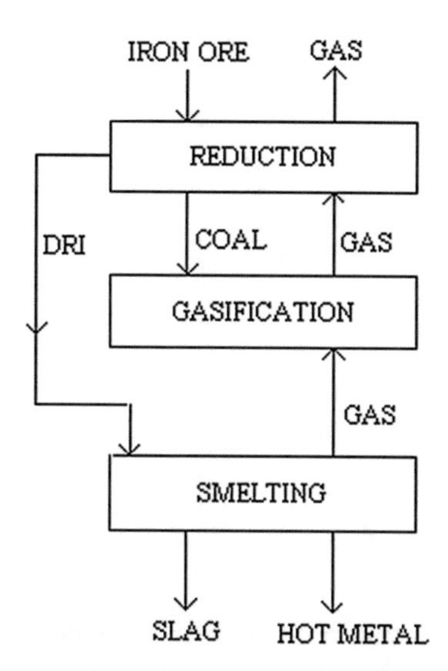

Fig. 8.6 Three-stage smelting reduction process [5]

8.6 Limitations of Smelting Reduction Processes

Followings are the limitations of the smelting reduction processes [5]:

(i) Consumption of a large amount of oxygen, the generation of which requires a high amount of power consumption (0.6 KWh/Nm^3 of oxygen),
(ii) The process generates a large amount of high calorific value export gas and the economic viability of the process depends on its efficient utilization,
(iii) The maximum module size of a SR unit may be limited to 1 Mt per annum and may not be the best alternative if large iron making capacities are to be generated,
(iv) Pre-reduction of ore fines is mandatory and
(v) Needs highly efficient post-combustion necessary to meet the energy requirement of the smelting processes.

8.7 Major *Smelting Reduction* (SR) Processes

A list of major *Smelting Reduction* (SR) processes is given below. Except Corex, nonc of them have gone beyond the laboratory or pilot plant scale. So far, Corex is the only SR process which is commercialized and is in operation at several places.

1. Corex process,
2. Romelt process,
3. DIOS process,
4. HIsmelt process,
5. Ausmelt (AusIron) process,

6. AISI—DOE process,
7. Redsmelt process,
8. Iron dynamics process,
9. Kawasaki SR (Star) process,
10. FASTMELT process,
11. FINEX process,
12. Tecnored process,
13. Inred process,
14. Plasma-smelt process and
15. Cyclone converter

8.7.1 Corex Process

Corex is an industrially and commercially proven smelting reduction process developed by Siemens-VAI (VOEST-ALPINE GmbH, Austria) for the cost-efficient and environmentally friendly production of hot metal directly from iron ore and non-coking coal. The Corex process was developed in the late 1970s and its feasibility was confirmed during 1980s. Corex developed from an initial pilot plant in Brazil (Korf Stahl, late 1970s) to a development plant during 1981–1989 at Kehl in Germany (60,000 tpa). 1986–1998 it progressed to a demonstration plant in South Africa (C1000, rated at 0.3 Mtpa). Following the first industrial application of a Corex C-1000 plant (production of 1,000 thm/d) at ISCOR, Pretoria Sorks, South Africa, four C-2000 plants (production of 2,000 thm/d) were subsequently put into operation at POSCO, Pohang Works, South Korea; Mittal Steel, South Africa; and at Jindal Vijayanagar Steel Ltd. (now JSW Limited), India [5]. Two Corex units, each having rated capacity of 0.8 Mtpa hot metal, are in operation at JSW Steel's Vijayanagar Works. The first true commercial C-2000 plant (Pohang, Korea) was commissioned in 1995. Corex took around 30 years to get from concept trials to full commercial C-2000 operation. ESSAR Steel is constructing two Corex units at Hazira, Gujarat, India, which are expected to produce approximately 1.74 Mtpa hot metal. The Corex process is the most successful smelting reduction development. Several industrial Corex plants are operational today, the largest with a capacity of 1.5 M thm/year.

The Corex process differs from the conventional blast furnace route in that non-coking coal can be directly used for iron ore reduction and smelting process, eliminating the need for coke oven plants. The use of lump ore or pellets also eliminates the need for sintering units. The process combines pre-reduction of ore in a shaft furnace, to a level of about 80% with final reduction and melting in a melter/gasifier.

Viewing the process from the coal-route perspective, non-metallurgical coal is directly charged into the melter-gasifier via a lock-hopper system. Due to the high temperatures predominating in the dome of the melter-gasifier (more than 1000 °C), a portion of the hydrocarbons released from the coal during devolatilization are immediately dissociated to carbon monoxide and hydrogen. Undesirable by-products such as tars and phenols are destroyed and therefore cannot be released to the atmosphere. Combustion with oxygen injected into the melter-gasifier results in the generation of a highly efficient reducing gas [5]. The reducing gas exiting from the melter-gasifier consists mainly of CO and H_2 gases with fine coal, ash and iron dust. This dust is largely removed from the gas stream in a hot-gas cyclone and is then recycled to the process. Through the addition of cooling gas, the reducing gas temperature is adjusted to its optimum working range. After leaving the hot-gas cyclone, the reducing gas is then blown into the reduction shaft via a bustle, reducing the iron ores in counter-flow.

The process is not completely independent of coking coals. The exact coke requirement is unclear. It may range from 50 to 200 kg/thm. However, the coke addition improves productivity.

8.7.1.1 Description

In the Corex process (Fig. 8.7), all metallurgical work is carried out in two separate process reactors—the reduction shaft and the melter-gasifier. The reduction shaft is located above the melter-gasifier and is used to metallize incoming ore. Lump ore, pellets or a mixture thereof and additives (limestone and dolomite) are first charged into a reduction shaft via a lock-hopper system where they are reduced to sponge iron by hot reduction gases (850 °C) at a pressure of 3 bar, generated by the melter-gasifier, moving in the counter-flow direction. About 5–6% of coke is also added to the shaft to avoid clustering of the burden inside the shaft due to sticking of ore/pellets and to maintain adequate bed permeability. The iron-bearing material is reduced to sponge iron over 70–90% metallization in the shaft [5]. The hot sponge iron at about 700–800 °C along with partially calcined limestone and dolomite is continuously fed into the melter-gasifier through downpipes for smelting. Non-coking coal (>59% fixed carbon, 25–27% volatile matter, <11% ash and <0.6% sulphur; calorific value >29,000 kJ/kg) is fed directly into the melter-gasifier. Mean particle size is 20–25 mm. Pure oxygen, predominantly blown through the tuyeres, burns the fixed carbon/char of the coal and generates heat and reducing gases. The process operates at low post-combustion (<10%) and high pre-reduction (85–95%). The hot metal and slag formed in the process are collected in the hearth and tapped periodically. The quality of the hot metal is equivalent to that produced in a blast furnace. The hot gas generated in the melter-gasifier, containing about ~70% CO and ~25% H_2, is used as reducing gas in the shaft. The top gas from the shaft and residual reducing gas is subsequently made available as a highly valuable export gas (with a net calorific value of approximately 7,500–8000 kJ/m^3 at STP) suitable for wide range of applications. Coal consumption rate of around 1000 kg/thm is significantly higher than that of a blast furnace (typically 600–700 kg/thm under Indian condition). The melter-gasifier operates in a strongly reducing mode (like a blast furnace), resulting in most of the phosphorus entering to the metal. The reduction shaft and melter-gasifier

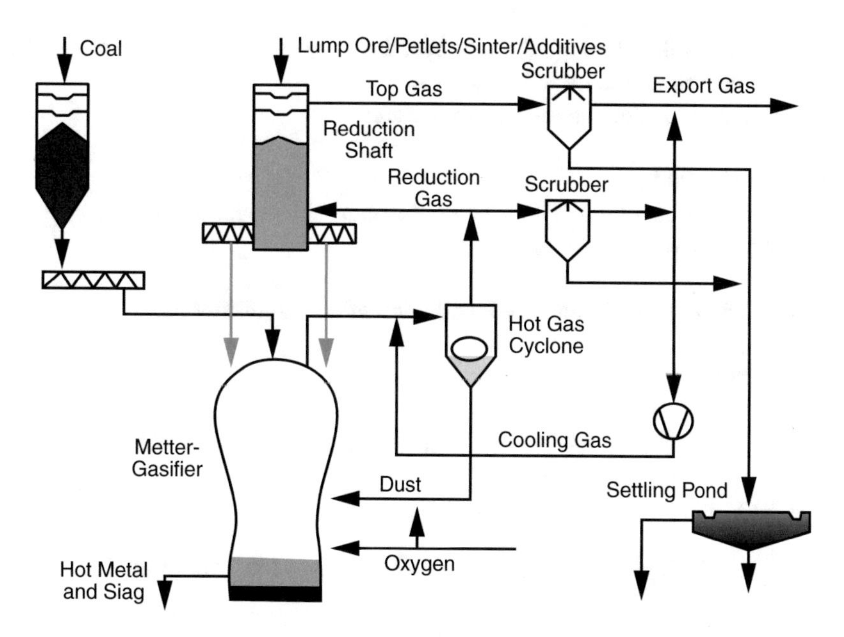

Fig. 8.7 Corex process flow sheet [5] (reproduced with permission from Authors)

work at an elevated pressure of approximately 5 bars. Corex is heavily reliant on export gas credits to make the economics work. For C-2000 unit, the volumes of the reduction shaft and melter-gasifier are ~600 and 2200 m^3, respectively.

8.7.1.2 Reactions

Reduction shaft: The primary reactions taking place inside the shaft are as follows:

(i) Reduction of iron ore by CO and H_2 and transforming the iron oxides to metallic iron:

$$Fe_2O_3 \rightarrow Fe_3O_4 \rightarrow FeO \rightarrow Fe \quad (8.4)$$

(ii) Calcination of limestone and dolomite:

$$CaCO_3(s) \rightarrow CaO(s) + \{CO_2\}, \Delta H^{\circ}_{298} = 179 kJ/mol \quad (8.5)$$

$$CaMg(CO_3)_2(s) \rightarrow CaO \cdot MgO(s) + 2\{CO_2\} \text{ (endothermic reaction)} \quad (8.6)$$

(iii) Carbon deposition reaction and formation of Fe_3C:

$$2\{CO\} \rightarrow \{CO_2\} + C(s), \Delta H^{\circ}_{298} = -85.35 kJ/mol \text{ of } CO \quad (8.7)$$

$$3Fe(s) + 2\{CO\} \rightarrow Fe_3C(s) + \{CO_2\}, \Delta H^{\circ}_{298} = -85.33 kJ/mol \text{ of } CO \quad (8.8)$$

Melter-gasifier: The melter-gasifier can largely be divided into three reaction zones as follows [5]:

(i) Gaseous free board zone (upper part or dome),
(ii) Char bed (middle part above oxygen tuyeres) and
(iii) Hearth zone (lower part below oxygen tuyeres).

Due to continuous gas flow through the char bed, there also exists a fluidized bed in the transition area between the char bed and the free board zone.

Reactions taking place inside the melter-gasifier are:

(i) Drying of coal (100 °C),
(ii) Devolatilization of coal (200–950 °C) and liberation of methane and higher hydrocarbons and
(iii) Decomposition of volatile matter.

Other reactions taking place in the free board zone of melter-gasifier are:

(1) Boudouard reaction:

$$\{CO_2\} + C(s) = 2\{CO\}, \Delta H^{\circ}_{298} = 170.7 kJ/mol \text{ of } C \quad (8.9)$$

(2) Water–gas reaction:

$$\{H_2O\} + C(s) = \{CO\} + \{H_2\}, \Delta H^{\circ}_{298} = 135.15 kJ/mol \text{ of } C \quad (8.10)$$

(3) Shift reaction:

$$\{CO\} + \{H_2O\} = \{CO_2\} + \{H_2\}, \Delta H^{\circ}_{298} = -35.56 kJ/mol\ of\ CO \quad (8.11)$$

(4) Decomposition of undecomposed limestone and dolomite,
(5) Residual reduction of iron oxide,
(6) Direct reduction of FeO into the DRI takes place by carbon in the char bed,
(7) Combustion of charcoal by oxygen and
(8) Melting and formation of hot metal and slag.

8.7.1.3 Efficiency and Advantages [6]

The efficiency of the Corex process primarily depends on the following:

(i) Size and chemical analysis of the raw materials, especially the coal,
(ii) Low CO_2 in the reduction gas to ensure higher metallization of the DRI,
(iii) Optimum distribution of oxygen between the tuyeres and the dust burners,
(iv) Permeability of the char bed,
(v) High system pressure and
(vi) Higher melting rate operation.

The main advantages of the Corex process are:

1. Cost-efficient and environmentally friendly process, since SO_x, NO_x and dust emissions are much lower than that from blast furnace route,
2. Use of non-treated, non-coking coal,
3. Direct use of iron ore fines up to 15% of total iron-bearing charge,
4. Flexibility in operation, easy shutdown and restart of the plant within one hour, (capable of operating at 50–115% of normal capacity),
5. Specific melting capacity is higher than that of BFs, productivities of the order of 3–3.5 $t/m^3/d$ can be achieved,
6. High level of automation and
7. Considerably lower investment and production costs.

Weaknesses of Corex process are:

(i) It has limitations in distributing the coal and DRI in optimized manner in melter-gasifier, and this results in more peripheral flow of hot gases,
(ii) It has low refractory lining life in high-temperature zone of melter-gasifier, and this affects the campaign life of the furnace,
(iii) It is provided with very sophisticated gas-cleaning facilities,
(iv) The system is maintenance oriented, including cooling gas compressor for recycling part of Corex gas for cooling the hot gases from melter-gasifier,
(v) Transfer of hot DRI and hot dust recycling are hazardous especially during their maintenance period and
(vi) The process is sensitive to the quality of input materials and coal containing low ash, and medium volatile matter is suitable for this process.

The following are some of the unique features of Corex operation at JSW Steel Ltd. [5] (formerly Jindal Vijayanagar Steel Limited), Torangallu, Bellary, Karnatka, India:

(1) It uses non-coking coal predominantly,
(2) The gas generated from the process contains high calorific value $\sim$8368 kJ/Nm^3 which is used in power generation and as a fuel gas,
(3) Excellent quality of hot metal,
(4) Easy shutdown and restart of the plant (within an hour),
(5) Direct use of iron ore fines up to 15% of the total iron-bearing charge and
(6) Cost-effective and environmental-friendly process.

8.7.1.4 Process Economy

Since coke oven and sinter plants are not required for the Corex process, substantial cost savings are achieved in the production of hot metal. The extent of such savings depends, of course, on the local site conditions, however, can be in the range of up to 20%.

8.7.1.5 Environmental Aspects

Corex plant emissions contain only insignificant amounts of NO_x, SO_2, dust, phenols and sulphides. Emission values already exceed by far future European standards. Also, wastewater emissions from the Corex process are far lower than those in the conventional blast furnace route. These environmental features are additional key reasons for the attractiveness of the Corex process.

8.7.1.6 Use of Export Gas

After the cleaning and cooling of the top gas which exits the reduction shaft, it is then available for use in a number of industrial applications. These include electrical power generation, production of sponge iron and for heating purposes throughout the iron and steel works. In the chemical industry, the Corex export gas can also serve as a feedstock for many other applications.

8.7.1.7 Relevance Under Indian Condition

In Indian context, Corex process has special relevance. Firstly, India is a power-starved country. Secondly, the thermal power plants, the main source of power supply, produce huge stockpiles of fly ash which is going to pose a serious environmental problem. Corex process, on the other hand, produces gaseous fuel for power plants and slag as waste. The technology of producing slag cement is well established. Hence, an Ironmaking-Power Plant-Slag cement complex can be considered as independent units and the overall economics of complex should be considered instead of a single Corex unit. The performance of Corex technology indicates a bright future for the smelting reduction process. The present pace of developments would pave the way towards the lower-cost steel production.

8.7.2 Romelt Process

Romelt is the ferrous off-shoot of a Russian non-ferrous smelting technology known as Vanyukov. The process is developed by Professor V. A. Romenets of Moscow Institute of Steel and Alloys for ***melting*** of iron (hence the name) and 0.3 Mtpa pilot plant was set up in Novolipetsk Iron and Steel Works, Lipetsk, Russia, in 1984 [5]. This process concept is rather unique in so far as it involves smelting and reduction together in a single rectangular-shaped reactor (like a small open-hearth steelmaking furnace) shown in Fig. 8.8.

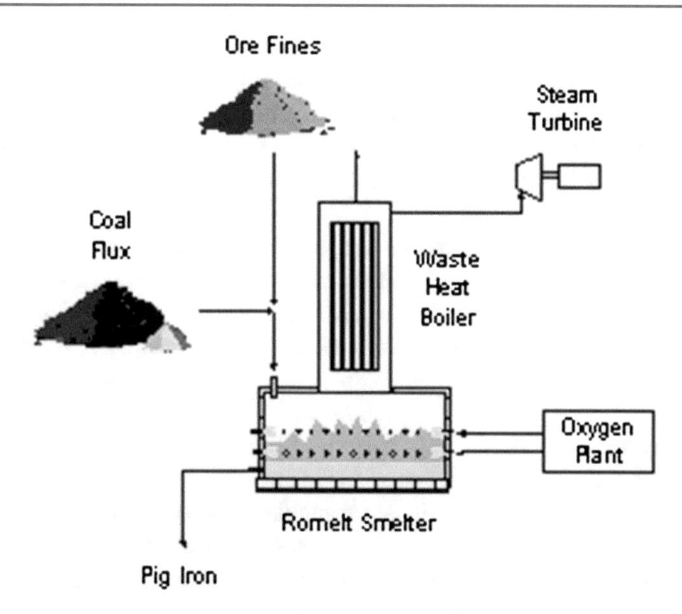

Fig. 8.8 Romelt process flow sheet [5] (reproduced with permission from Authors)

Romelt has travelled a long and bumpy road from initial concept development in the late 1970s. This includes a series of AISI-funded trials at Novolipetsk in 1994 and an unsuccessful marketing campaign (1994–1998) by Kaiser Engineering (now Hatch) to establish a commercial plant in North America. NSC Engineering also funded trials at Novolipetsk in 1995 and attempted to establish a Romelt plant in Japan. Although Romelt achieves many of the objectives of direct smelting, it fails to satisfy the wishlist in terms of coal consumption. Its large waste-heat output has Corex-like issues in terms of utilization [5].

Some of the important points for Romelt are as follows [7]:

- It can accept iron ore in sizes 0–20 mm without any pre-treatment,
- Non-coking coal (0–20 mm) with moisture less than 10% is acceptable, volatile matter less than 20% is preferred, but higher volatile matter coal can also be used,
- The coal and oxygen consumptions are rather high, typically, 1500–2000 kg coal (22% ash)/thm and 1100–1200 Nm^3 oxygen/thm,
- All the reactions take place in a single vessel, the operation of which is relatively simple with 70% post-combustion of the exit gas taking place within the process itself,
- Only 60% of the input phosphorus reports to the hot metal; 30% phosphorus goes to slag and 10% forms a part of the exit gas,
- The refractory life is limited owing to the intense temperature and reducing condition existing within the furnace and
- The capacity of the plant may be varied from 1000–3000 tpd.

It is a single-stage process operating at subatmospheric pressure where liquid iron is produced in a molten slag bath reactor directly using iron-bearing materials and non-coking coal. It requires minimum preparation of raw materials and there is no limitation of size range and moisture content. It requires oxygen as the primary gasification medium. In this process, reduction of iron oxide takes place in foaming slag phase. It has simple design requiring fewer numbers of unit operations and unit processes compared to other processes under development. It is environmentally more acceptable compared to the conventional routes. Higher post-combustion (i.e. >70%) with high heat transfer efficiency is the principal driving force of the process. The productivity of Romelt furnace basically depends upon achieving the post-combustion level of the order of 70%. In the Indian context, the process has significance in that it does not depend upon coking coal and is able to accept coal with high ash.

There is a little restriction on raw material usage with respect to size and type of raw material input. The suitable charge material for the Romelt process is lumps ore including fines (preferred size range <20 mm), mill scale, sludge, pre-reduced pellets and coal as reductant. Coal contains 10% max. moisture and 20% max. volatile matter, <20% ash. Either lime (<5 mm size) or limestone can be used as flux.

8.7.2.1 Description

It is essentially a single-stage process in which reduction of iron oxide is carried out in a single molten bath reactor. The process is carried out in a rectangular reaction chamber (with refractory lined bottom and sides up to metal level) having water-cooled panels for sides and roof. The high degree of post-combustion generates additional heat to sustain the reduction reaction in an auto-thermic manner. Large surface area generates due to high agitation of bath captures majority of post-combustion heat. Slag and metal droplets ejected in the free board also help in the heat transfer process. Beneath the agitated slag layer (somewhat below the lower row of tuyeres), there is relatively calm layer of slag where metal droplets coalesce, and slag metal separation occurs.

The process, shown in Fig. 8.8, operates either at ambient pressure or under slightly negative pressure and does not require special sealing equipment. Although the process itself is simple, it consumes large amounts of coal and oxygen, while giving off hot exit gases which contain a large amount of waste heat that to be recovered by the formation of steam in a boiler system. A waste-heat boiler for recovery of heat of outgoing gases in the Romelt unit is an additional element. In the waste-heat boiler, complete post-combustion of gases to carbon dioxide and water is accompanied with emission of respective amount of energy and recovery of sensible heat of waste gases, their temperature dropping from 1600–1700 °C to 250–300 °C. Therefore, the Romelt plant is power-cum-technological unit that may incorporate a power block for generation of electrical power.

8.7.2.2 Important Features for Romelt Process

Important features for Romelt process are as follows [7]:

(1) It is single-stage SR process directly using iron-bearing materials and non-coking coal for production of liquid iron,
(2) It required minimum preparation of raw material, no limitation of size range and moisture content,
(3) It has flexibility in the use of wide range of iron-bearing materials including lumps ore and ore-fines pellets,
(4) It eliminates coke ovens, sintering plants,
(5) It required oxygen as the primary gasification medium,
(6) It has simple design required fewer number of unit operations compared to other SR processes under development,
(7) It is environmentally more acceptable compared to the conventional routes, and since the furnace operates at slightly negative pressure, there is hardly any pollution safety risk,
(8) Unlike other SR processes practicing high post-combustion, Romelt process does not depend upon the existence of thick slag foam for the successful operation of the process,

(9) Proportion of oxygen blown through the lower level of tuyeres is also small, and this factor adds to the relative safer operation of the process,
(10) The amount of foam being maintained on the slag surface is small compared to other SR processes practicing high post-combustion. In fact, the foam is like windsurf produced on seashores, and therefore, there is practically no change of the foaming phenomenon going out of control.

8.7.3 DIOS Process

Japan Iron and Steel Federation promotes the ***Direct Iron Ore Smelting*** (DIOS) process in collaboration with Center for Coal Utilization and a consortium of eight Japanese Steelmakers. Elemental studies were conducted for the first three years, and a large integrated pilot plant was operated from 1993 to 1996. A 500 tpd, i.e. 150,000 tpa capacity pilot plant has been successfully in operation at NKK, Keihin steel plant since 1993. The pilot plant, located at NKK's Keihin works, was essentially a modified 250 t BOF. Based on the pilot plane data, performance of a 6000 tpd commercial plant has been worked out and following benefits are envisaged with respect to similar capacity blast furnace [8]:

1. Reduction in construction cost by 35%,
2. Reduction in production cost by 19%,
3. Reduction in CO_2 emission by 4–5% and
4. Reduction in net energy consumption by 3–4%.

The process is a combination of smelting reduction furnace (SRF) and two-stage pre-reduction furnace (PRF). Iron ore fines of up to 8 mm size and coal up to 25 mm size are used. Iron ore fines are pre-heated at approximately 600 °C in a fluidized bed preheater, then preliminarily reduced to 27% at 780 °C in a fluidized bed pre-reduction furnace. The DIOS process has the following characteristics [5]:

(1) Iron ore and coal fines can be directly used,
(2) Coke is not used, so there are no restrictions on the kind of coal,
(3) Process can be halted and restarted easily, affording excellent flexibility and
(4) The intensity of bath stirring may be optimized.

The DIOS process operates at a pressure of 1–2 bar and uses top feed of coal and ore as shown in Fig. 8.9a. Unlike Romelt, it uses a top lance for oxygen injection and has a fluidized bed system for iron ore pre-reduction. In the smelter, most of the reduction occurs in a foamy slag layer. Theoretically, this combination can achieve the coal rates necessary to compete with a modern blast furnace. The calorific value of export gas is low (4602 kJ/Nm^3) compared to Corex process and can be used for power generation. Figure 8.9b shows the physical images of SRF (DIOS).

8.7.4 HIsmelt Process

The ***high-intensity smelting*** (HIsmelt) process has its origins in the early 1980s when Rio Tinto Limited identified the potential to adapt the high scrap rate steelmaking and iron bath coal gasification

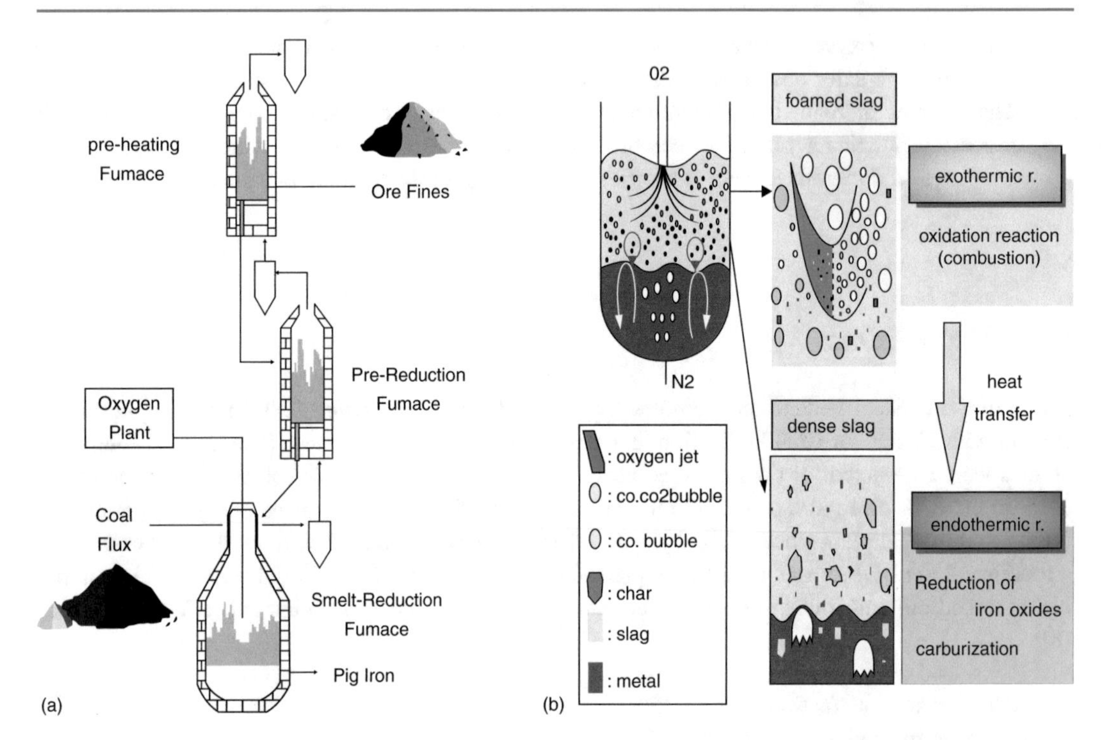

Fig. 8.9 **a** DIOS process flow sheet [5], **b** physical images of SRF (DIOS) [5] (reproduced with permission from Authors)

processes developed by Klockner Werke to the direct smelting of iron ore. This process was developed by CRA Ltd. (Australia) and Midrex Corporation (USA) in cooperation with CSIRO (Australia), Klockner (Germany) and Davy McKee (UK) but presently it has been taken over by Rio Tinto Group. The original version of HIsmelt started out as a modification of bottom-blown K-OBM (steelmaking) process. Initial trials were conducted in the early 1980s on a 60 t K-OBM converter at Maxhutte in Germany [9]. Thereafter, a 10 t horizontal rotating pilot plant was built at the same site. This unit was operated through the second half of the 1980s and closed in 1990. At this point, work began on a larger pilot plant (100,000 tpa) in Perth, Australia. The larger pilot plant was run for around five years (early to mid-1990s), during which it became clear that the original concept was flawed. Iron ore pre-reduction based on an Ahlstrom Flux flow CFB system failed to operate, and refractory wear in the upper regions of the smelter was unacceptably high. First commercial 0.8 Mt/a HIsmelt plant is in Kwinana, Western Australia [10]. It was hot commissioned between April and October 2005 and has come a long way since then.

The HIsmelt process is based on direct injection of iron ore and coal into a metal-slag bath, with smelting in the lower region and post-combustion (using oxygen-enriched hot blast) in the upper region. At the heart of the process is splash-driven heat transfer between the upper (high oxygen potential) combustion zone and the lower (low oxygen potential) smelting zone. The HIsmelt process combines pre-reduction and pre-heating of fine ore in a fluidized bed with final reduction in a smelter. The smelter operates with enriched hot blast, generated in blast furnace type stoves. A unique feature of the process is the submerged injection of coal and pre-reduced ore using the so-called solid injection lances (SIL). Submerged solids injection creates a highly stirred slag phase and even throws large amounts of slag into the top space, so called slag fountain. This promotes the transfer of heat

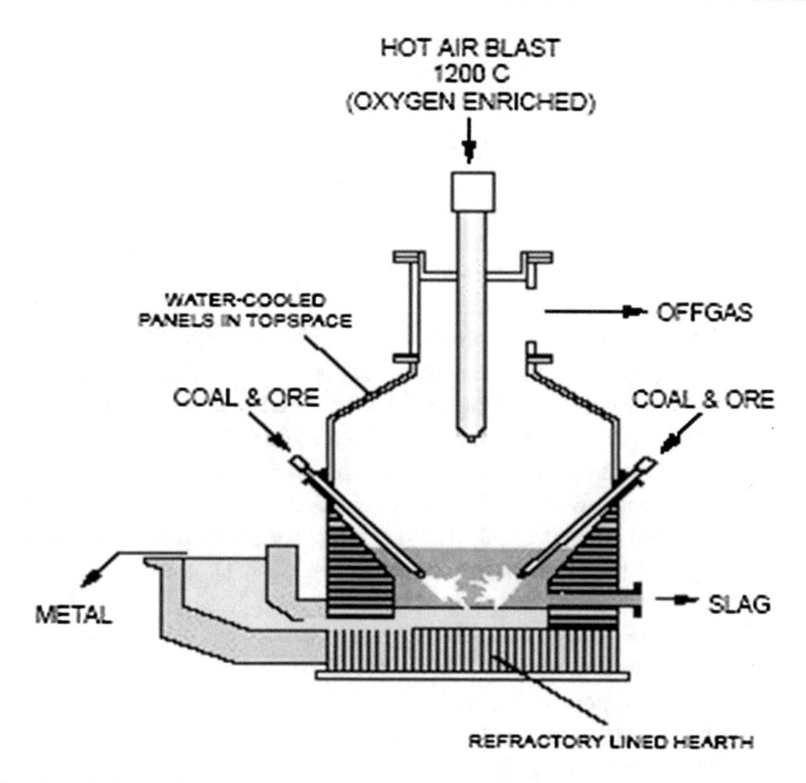

Fig. 8.10 HIsmelt vertical vessel [5] (reproduced with permission from Authors)

from the post-combustion zone to the bath. With respect to heat transfer efficiency, the process is superior to other smelting reduction processes.

A rejuvenated version of HIsmelt based on top injection is shown in Fig. 8.10. This involved a fixed vertical vessel with water-cooled panel in the regions where refractory wear had previously caused major issues and simplified engineering that was capable of scale up. A thick slag layer is situated above the metal bath. Iron ore fines, coal and fluxes are injected directly into the melt in the smelt reduction vessel. The process uses iron ore fines directly with non-coking coal (50–70% fixed carbon, 10–38% volatile matter and 5–12% ash; less than 10% ash is preferred). A thick slag layer is situated above the metal bath. Upon contact with the iron bath, dissolution of the carbon in the coal occurs, which react with the oxides in the iron-bearing feeds, forming carbon monoxide. Rapid heating of the coal also results in cracking of the coal volatiles releasing hydrogen. The CO and H_2 is post-combusted with oxygen [5]. The sensible and chemical energy in the off-gas can be utilized for pre-heating, pre-reduction and/or calcination of the metallic feed and fluxes. One of its most important achievements was demonstration of plant availability of around 99%. The phosphorous removal capacity of the process has been readily demonstrated with an average of 85–95% of the input reporting to the slag phase.

The HIsmelt process configuration has several unique features [10]:

- The method of solids injection, using high-velocity lances, means that capture efficiency in the melt is very high and even ultra-fines can be used directly,

- The FeO level in the slag (5–6%), in conjunction with carbon in metal at around 4%, creates conditions for strong migration of phosphorus from metal to slag. Typically, 80–90% of the phosphorus is transferred to slag,
- Coal performance has virtually no dependence on particle morphology since the coal is grounded to fine for injection and
- The production rate of 75–80% of capacity, with coal rates (as injected basis) of around 800 kg/thm.

The major difference with respect to other SR processes is that this process uses oxygen-enriched air and the vertical SR vessel operates at pressure slightly below atmospheric pressure, thus there is no chance of leakage of gases. The hot metal does not contain any silicon, as silicon reduction does not take place during the process and, therefore, the hot metal is more suited for steelmaking. The dioxin problem in exhaust gases has also not been observed during the pilot plant operation because of the more efficient gas-cleaning systems. Based on the pilot plant data, two modules of commercialization have been designed (6 m and 8 m hearth diameter vessels for 0.5 and 1.5 Mtpa capacity, respectively). It is economic at smaller scale (0.5–1.0 Mtpa) and meets future environmental standard.

The main advantages of the HIsmelt process are as follows [11]:

(i) Economic at smaller scale (0.5–1.0 Mtpa),
(ii) Saleable to single units producing 2 Mtpa and higher,
(iii) Uses a wide range of non-coking coal,
(iv) Direct use of low cost, readily available raw materials,
(v) Simple, flexibility and controllable,
(vi) High energy efficiency and balanced energy flow sheet,
(vii) High level of process intensity,
(viii) High-quality hot metal product,
(ix) Meets future environmental standards,
(x) Air-based process and therefore no need for large oxygen production facility and
(xi) Plant availability of greater than 99%.

Compared to hot metal from blast furnace, hot metal of HIsmelt has the following potential advantages [12]:

(i) Reduced flux consumption,
(ii) Increased liquid steel yield,
(iii) Increased productivity due to a reduced blowing time,
(iv) Production of high-quality (low phosphorous) grades,
(v) Reduced refractory consumption,
(vi) Decreased consumption of ferro-alloys and
(vii) Production of ultraclean steel.

HIsmelt may be differentiated from so-called *deep slag* smelters such as DIOS, AISI and Romelt in that it injects feed solids significantly deeper into the melt. This leads to stronger mixing in the vessel, with hardly any temperature gradients in the liquid [5]. As a result, melt temperature and hence phosphorus removal can be better controlled. The new smelter demonstrated (repeatedly) that it could remove 90–95% of the phosphorus to slag. These results provided a strong confidence boost leading into the commercialization phase. After 20 years of development phase involving multiple pilot plants, HIsmelt has recently emerged as the only ferrous direct (bath) smelting process thus far to proceed to the commercial status. Rio Tinto, together with Nucor Steel, Mitsubishi and Shougang Steel as joint venture partner, is now in the process of building a 0.8 Mtpa plant in Perth, Western Australia.

8.7.5 AusIron Process

The process is developed by Ausmelt Ltd., Victoria, Australia, and a 2 tonnes/h capacity pilot plant had been built at Whyalla in South Australia. It is a single-stage process operated under subatmospheric pressure. The developers have long experience in non-ferrous metallurgy using vertical injection lances.

Process description: The AusIron process is a bath smelting process which directly processes ferrous-bearing feed materials to produce molten iron. The process, as shown in Fig. 8.11, is based on top submerged lance technology where coal, ferrous feed, flux and oxygen are injected directly into the smelter slag bath. Ferrous materials rapidly dissolve in the slag bath, while the reductant coal reacts with iron oxides contained in the slag. The turbulence generated by submerged combustion provides efficient mixing, resulting in high smelting rates [5]. The carbon monoxide produced by smelting reactions, together with reductant coal volatiles and residual fuel from lance tip combustion, is post-combusted immediately above the slag bath using oxygen-enriched air delivered through a shroud around each lance. The evolution of gases from the lance tips generates a cascade of slag droplets above the bath providing a large surface area for efficient recovery of post-combustion energy. Complete combustion of the fuel gases is achieved within the furnace maximizing energy recovery and avoiding the production of difficult low calorific value fuel gases.

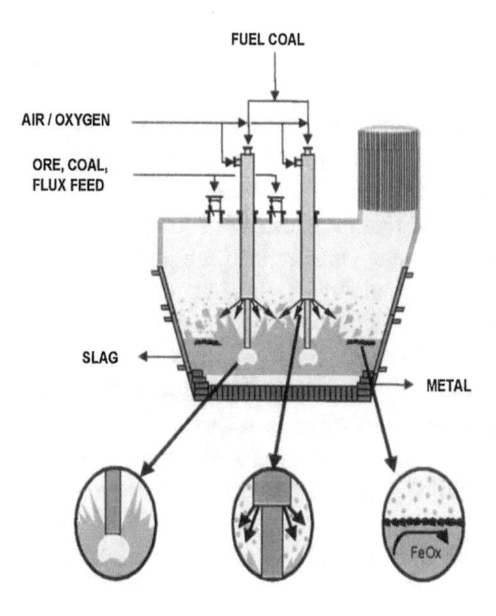

Fig. 8.11 AusIron smelter [5] (reproduced with permission from Authors)

The key attributes of the technology are as follows [13]:

1. Direct use of non-coking coals,
2. Low-grade lump ore/fines and any grade of coal can be used and, therefore, no pelletizing or sintering is required,
3. It can use ferrous residues from mineral processing or steel plant waste products, as low-cost feed to produce hot metal,
4. Typical hot metal chemistry is: 2.5–4.2% C, 0.02–0.04% Si, 0.02–0.08% S and <0.08% P. High-quality metal product is suitable for both electric and oxygen steelmaking,
5. Unique feature of the process is production of low phosphorous and low sulphur hot metal from high phosphorous iron ore,
6. An efficient, single-stage process not reliant on pre-heating or pre-reduction of ore feeds,
7. The furnace operates at subatmospheric pressure allowing simple feeding and tapping arrangements,
8. Dust carry-over is low, typically less than 1% of the feed rate,
9. The absence of coke ovens and sinter/pellet plants reduces environmental issues typically associated with ironmaking,
10. The use of a single-stage, highly efficient smelter reduces capital requirements for new iron-making capacity and permits effective use at smaller scales of operation than conventional processes,
11. The lance-based operations offer the ability to start, stop and idle the process easily and
12. A waste-heat boiler system, directly coupled, can generate enough electrical power to supply the ironmaking facility and associated oxygen plant, with surplus power available for sale.

8.7.6 FINEX Process

In Corex process, use of agglomerated iron-bearing feed is seen as a key shortcoming. FINEX is an attempt to address this by replacing the reduction shaft with a fine ore-based fluidized bed system. In the FINEX process, the reduction shaft of the Corex is replaced by a series of fluidized beds. This enables the FINEX process to use fine ores instead of lump ore or pellets. As a result, the process requires neither coke making nor ore agglomeration. In a joint research and development project [14] POSCO, Korea, and Primetals Technologies, Austria, have developed FINEX process with the objective of utilizing the fines generated during processing of the feed required for Corex unit. In Corex process, a large amounts of iron ore fines (−8 mm) and coal fines (−6 mm) remain unutilized, causing difficulties. Thus, FINEX process is a fine ore-based smelting reduction process. It is a technology combining a gas-based reduction in a series of fluidized bed reactors and a reduction smelting in a melter gasifier. It is rated as the most advanced commercial ironmaking technology.

The key features of the FINEX process are as follows:

1. Direct use of fine ore (less than 8 mm) without prior agglomeration, more than 60% of ore, worldwide, in this size,
2. Direct use of non-coking coal, higher availability and lower costs than coking coal; 10% coke is often used to ensure hearth permeability,
3. Minimum coal consumption and high gas utilization,
4. After fluidized bed reduction, fine DRI is hot briquetted before melting in a melter-gasifier,
5. Non-coking coal fines are also briquetted before use in the melter-gasifier and

6. Saving of 15–20% of hot metal production cost in comparison to blast furnace route, same hot metal quality as from blast furnace and
7. Lowering environmental pollutions and to increase the flexibility in terms of operation and the choice of raw materials.

The FINEX process is an innovative, next-generation iron making technology. Molten iron is produced directly using iron ore fines and non-coking coal rather than processing through sintering and cokemaking. Because the preliminary processing of raw materials is eliminated, the construction of the FINEX plant will cost 8% less to build than a blast furnace facility of the same scale. Furthermore, a 17% reduction in production costs is expected since lower-priced raw materials can be used in the FINEX process. The commercialization of FINEX will be a major turning point for global steel production technology.

In FINEX process, the fine ore is reduced to DRI in four stages fluidized bed system that precedes the melter-gasifier. The reactors in first and second stages are primarily used to pre-heat the ore fines to the reduction temperature. In third reactor (in third stage), fine ore is pre-reduced to a degree of reduction of about 30%. The final reduction to DRI (degree of reduction of 85–90%) takes place in the fourth reactor (i.e. fourth stage). Melting and carburization of the hot DRI to liquid hot metal and slag is performed in the melter-gasifier. POSCO has actively promoted commercialization of the technology.

POSCO completed construction of the 2000 tpd demonstration plant that utilizes a next-generation environmentally friendly ironmaking technology and entered into operation on May 2003. First, 1.5 Mtpa commercial plant had demonstrated in April 2007. Another 2.0 Mtpa, third-generation FINEX (3G) plant was installed at POSCO, Pohang Works [15] and has been operating satisfactorily since January 2014.

In the FINEX process (Fig. 8.12), the hot metal is produced by two steps: (i) a series of fluidized bed reactors where iron ore fines are reduced to direct reduced iron (DRI), compacted and (ii) then transported to a melter gasifier where melting of DRI take place by gasification of coal and coal briquettes, which also supply the necessary heat energy.

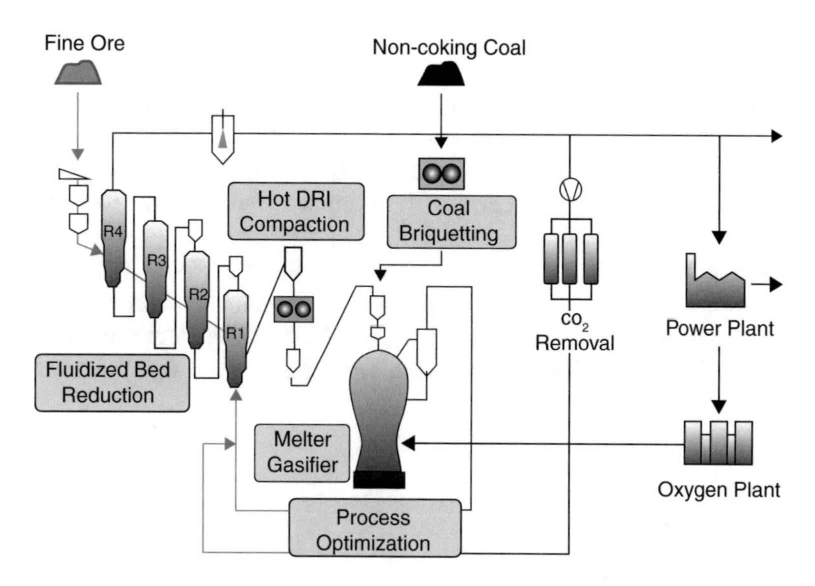

Fig. 8.12 Flow diagram of FINEX process [16]

Fine iron ore and additives (limestone and dolomite) are dried and charged to three stages fluidized bed system where the iron ores are progressively reduced in counter-current flow of reducing gas to produce fine DRI with the degree of reduction in the range of 60–85%. Subsequently, the fine DRI is compacted in hot condition and charged along with coal briquettes continuously to the melter gasifier. By means of gravity, the hot compacted DRI falls through the dome zone onto the char bed, where that is heated, finally reduced and melting take place in addition to all other metallurgical reactions. The liquid hot metal and slag are collected at the melter-gasifier and tapped time-to-time like blast furnace.

Coal and coal briquettes are charged from the top of the melter gasifier, dried and degassed in the upper char bed area and finally gasified with oxygen blown in at the tuyere zone. The gasification reaction supplies the energy required for the metallurgical reactions and for the melting of compacted DRI and coal ash to produce hot metal and slag.

The exit gas consists mainly of CO and H_2 at a temperature of 1000 °C, the gas is subsequently cooled and de-dusted. The conditioned reducing gas is then supplied to the fluidized bed system for indirect reduction of the iron ore. The used reducing gas leaving the fluidized bed reactors can be partly recycled.

The hot metal composition obtained was: 4.3–4.5% C, 0.8–0.9% Si and 0.017–0.020% S. The FINEX process is more advanced and the new plant produces molten iron directly using cheaper and abundant iron ore fine and bituminous coal (non-coking coal). The iron ore fine used for FINEX is bountiful and cheaper. This results in the process being 15% cheaper in production cost than the blast furnace production method. In addition, FINEX is being praised as an eco-friendly process because sulphur and nitrogen oxide emissions are reduced to 10% of the amount generated from the conventional blast furnace method.

Emission of pollutants will also be drastically reduced. The process is reported to have 4% less CO_2 reduction, as compared to BF route. In addition to arsenic dust, levels of SO_x and NO_x will only be 8% and 4%, respectively, of the emissions occurring with the blast furnace process. FINEX has been recognized as an environmentally friendly process, which will increase POSCO's future competitiveness as raw materials are decreasing and environmental regulations are tightening. POSCO will position FINEX as its core technology, enabling the company to stand out against others in the global steel industry and increasing its competitiveness. The company will focus on successfully commercializing FINEX and reinforcing its security system for protecting technology. It is the objective of the POSCO to replace several of their smaller blast furnaces at the Pohang Works site with FINEX technology. Considerable savings in the consumption of energy and resources as well as in the reduction of emissions can be achieved since no coking plant, no sintering plant and no pelletizing plant are necessary for hot metal production. Coal consumption of the process is less than 700 kg/thm. An additional energy reduction of 1.3 GJ/thm is reported by utilizing off-gases after CO_2 removal.

Therefore, depending on the location, the operating cost of the FINEX is up to 20% lower in comparison to a similar sized blast furnace route. Typical consumption rates of FINEX process are shown in Table 8.1. As a second valuable product, FINEX provides an export gas with a medium calorific value of 5,500–6,250 kJ/m^3 (STP) which can be used as a fuel for energy production, heating purposes, etc.

Table 8.1 Typical consumption of FINEX process [15]

Materials	Amount
Ore, kg/thm	1,600–1,650
Coal (dry), kg/thm	750–850
Additives, kg/thm	230–250
Oxygen, m^3/thm (STP)	480–530
Nitrogen, m^3/thm (STP)	350–400
Make up water, m^3/thm	1.5 approx.
Electric energy, kWh/thm	200–230
Export gas credit, GJ/thm	9.0

8.7.7 FASTMELT Process

In the FASTMELT process, FASTMET DRI is melted to produce hot metal, hence the name *FASTMELT*. FASTMET DRI (HBI) is melted by electric energy to produce hot metal, hence another name of the process *FASTMELT*.

The key features of the FASTMELT process are as follows:

- FASTMET DRI even with a high gangue content is continuously fed to the electric furnace (melter),
- The reduction time in the rotary hearth furnace (RHF) at the FASTMET stage is as short as 12 min,
- Energy is supplied to the melter by an electric arc, and molten iron is tapped at 1450–1550 °C for charging in EAF or LD/BOF,
- Steel plant's wastes like BF/BOF dust, mill scale, filter cake, etc., can be processed and recycled them, thus greatly reducing the volume to be disposed off and producing a cost-effective iron product.

FASTMELT also uses a rotary hearth furnace (RHF) but adds an electric iron melting furnace to take the FASTMET process one more step. In the FASTMELT process, hot sponge iron/DRI produced via the FASTMET process is fed to a specially designed melter, the electric ironmaking furnace (EIF), for production of a high-quality hot metal known as *Fastiron*. The EIF used in the FASTMELT process provides many advantages over conventional submerged arc furnace (SAF). Typical SAF process cannot provide the most effective method of producing hot metal from FASTMET DRI. The FASTMELT melter, EIF, was therefore developed based on the properties of FASTMET DRI, a relatively high specific melting rate and the target analysis for the molten product. Table 8.2 provides a comparison of the EIF and SAF. In the EIF, a major portion of sulphur is removed during preparation of the heat. Silica contained in the gangue and coal ash can also be reduced, providing silicon-level control in the Fastiron. The carbon content can be varied by adjusting the carbon addition in the feed materials to the EIF. This provides for the desired carbon content in the liquid Fastiron without the requirement to add carbon in the EIF. Alloys such as ferrosilicon or ferromanganese can be added to the EIF if desired. Typical chemistry of Fastiron is given as [17]: 4.54% C, 0.47% Si, 0.1% Mn, 0.013% S and 0.036% P.

The FASTMELT process is an attractive option for many applications. Highly metallized and high-temperature FASTMET DRI is fed directly in a proprietary melter (EIF) to produce blast furnace grade hot metal. This Fastiron can be charged into BOF or EAF or can be cast into pig iron. Off-gas

Table 8.2 Comparison of EIF and SAF [18]

	EIF	SAF
Source of heat	Electric arc	Slag resistance
Charge materials	FASTMET DRI	Low metallization DRI, lime, iron oxide and carburizers
Material feeding	Rapid/direct from RHF	Choke feed
Metallization of DRI	85–95%	60–80%
Energy input	550 kWh/t	800 kWh/t
Hot metal composition	4.2–4.8% C, 0.3–0.7% Si, 0.03% S	3.5–3.8% C, 0.5–1.5% Si, 0.05% S

from the EIF, primarily CO, is recycled to the FASTMET process where it is used as fuel. In the EIF, a major portion of the sulphur is removed during preparation of the heat. Silicon contained in the gangue and coal ash can also be reduced, providing silicon-level control in the Fastiron. The carbon content can be varied by adjusting the carbon addition in the feed materials to the RHF. The EIF was designed to satisfy the following objectives: (i) effective melting of FASTMET DRI, (ii) removal of gangue, (iii) reduction of residual FeO to Fe, (iv) de-sulphurization and (v) continuous operation.

Steelmakers face problems in operating, permitting and repairing blast furnaces, coke ovens and sinter plants. FASTMELTcan enable integrated plants to produce enough hot metal while shutting down some or all these facilities. Because the RHF and EIF operating units are designed for high efficiency and minimum export heating value, the process operation costs do not require any off-gas energy credits to be competitive, which also minimizes the overall capital expenditure.

FASTMELT can be used to economically convert low-grade iron ores and wastes into high-quality hot metal without extensive beneficiation. The FASTMELT process produces hot metal with the lowest energy consumption and least green gases of any coal-based ironmaking process. FASTMELT can replace blast furnace ironmaking with lower operating cost and greater flexibility in feed selection. This Fastiron is refined in EAF or BOF for making steel and developed two processes like FASTEEL process (combining FASTMELT and EAF) and FASTOX process [19] (combining FASTMELT and BOF).

Energy consumption is most important in today's era of high energy costs and environmental concern. Table 8.3 compares FASTMELT process with other ironmaking processes on an equivalent GJ per tonne of hot metal basis.

Table 8.3 Energy consumption for different processes [20]

Energy (GJ per tonne of hot metal)	FASTMELT	HIsmelt	Corex	Redsmelt	Blast furnace
Coal	12.25	19.47	14.85	–	–
Natural gas	2.68	2.20	0.50	2.17	–
Electricity	6.11	3.38	0.83	6.65	–
Subtotal	21.03	25.05	32.10	23.67	23.36
Off-gas energy credit	4.56	3.38	13.20	5.10	5.35
Total energy	16.47	21.67	18.90	18.57	18.01

Probable Questions

1. What do you mean by smelting reduction processes. State the basic principle of SR processes. Discuss the advantages of SR processes.
2. How SR processes are classified? Discuss.
3. What are the alternate methods of hot metal production?
4. Discuss the basic principle of Corex process.
5. What are the difference in COREX and mini blast furnace processes?
6. Discuss the basic principle of HIsmelt process.
7. Discuss the basic principle of FINEX process.
8. Discuss the basic principle of FASTMELT process.

References

1. H.Y. Sohn, Y. Mohassab, Ironmaking and Steelmaking Processes: Greenhouse Emissions, Control, and Reduction, in *Greenhouse Gas Emissions and Energy Consumption of Ironmaking Processes (p 427)*, ed. by P. Cavaliere (Springer, New York, 2016)
2. S.K. Dutta, *JPC Bulletin on Iron and Steel* **VII**(11) 23 (2007)
3. A. Chatterjee, *Beyond the Blast Furnace* (CRC Press, Boca Raton, 1992)
4. A. Ghosh, in *Proceedings of International Conference on Alternative Routes to Iron and Steel* (Jamshedpur, India, Jan 1996), p. A17
5. S.K. Dutta, R. Sah, *Alternate Methods of Ironmaking (Direct Reduction and Smelting Reduction Processes*) (S. Chand & Co Ltd, New Delhi, April 2012)
6. G.R. Singh, B.N. Mukhopadhyay, R.C. Khowala, S. Bhattacharyya *Steel Scenario* **13**(3) 8 (Jan–Mar 2004)
7. A. Ghosh, in *Proceedings of International Conference on Alternative Routes to Iron and Steel* (Jamshedpur, India, Jan 1996), p. A45
8. H.S. Ray, A.K. Jouhari, R.K. Galgali, P. Datta, B. Bhoi, V.N. Misra, in *Proceedings of International Conference on Advances in Materials and Processing* (IIT, Kharagpur, India, Feb 2002), p. 505
9. P.D. Burke, S. Gul, in *Proceedings of International Conference on Smelting Reduction for Ironmaking* (Bhubaneswar, India, Dec 2002), p. 61
10. N. Goodman, in *Proceedings of International Seminar on Alternative Routes for Ironmaking in India* (Kolkata, India, Sep 2009), p. 15
11. D. Macauley, in *Proceedings of International Conference on Alternative Routes to Iron and Steel* (Jamshedpur, India, Jan 1996), p. A67
12. P. Bhattacharya, S.S. Chatterjee, B.N. Singh, S. Prasad, in *Proceedings of International Conference on Alternate Route of Iron and Steelmaking* (Perth, Australia, Sep 1999), p. 151
13. *Information booklet on AusIron Process,* Ausmelt Limited, 40th NMD and 56th ATM of IIM (Vadodara, India, Nov 2002)
14. E. Ottenschlaeger, D. Siuka, J. Wurm, H. Freydorfer, in *Proceedings of International Conference on Direct Reduction and Direct Smelting* (Jamshedpur, India, Oct 2001), p. 105
15. N. Rein et al., Steel Tech **12**(2), 19 (2018)
16. R.S. Brahma, S.K. Bhattacherjee, A.K. Agrawal, Steel. Tech. **12**(2), 13 (2018)
17. S. Chakraborty, A.K. Ray, S.K. Ray (Steel Scenario **13**(3), 30 (2004)
18. J.C. Simmons, K.L. Shoop, J.M. McClelland, *Direct from Midrex* (3rd Quarter, 2003), p. 3
19. G.E. Hoffman, *Direct from Midrex* (4th Quarter, 2000), p. 5
20. R.M. Klawonn, G.E. Hoffman, *Direct from Midrex* (2nd Quarter, 2005), p. 4

Alternate Ironmaking 9

Blast furnace process continues to be the principal producer of hot metal for steel plant even today due to its size, volume of production, techno-economics, energy utilization efficiency and reasonable quality of hot metal compared to other alternative process of hot metal production. The extremely high thermal efficiency of a conventional BF is essentially due to the tall shaft where processing of the ore and pre-heating as well as reduction takes place before it descends into the melting zone. The shaft height can be reduced without impairing the thermal efficiency if the heating and smelting zones in the furnace are compressed by accelerating the heat exchange and chemical reactions, i.e. sharper thermal and reducing potential gradients are obtained. Mini-blast furnace, charcoal blast furnace, electrothermal process and ELRED process are the alternate ironmaking processes.

9.1 Low Shaft Furnace

Blast furnace process continues to be the principal producer of hot metal for steel plant even today due to its size, volume of production, techno-economics, energy utilization efficiency and reasonable quality of hot metal compared to other alternative process of hot metal production.

Although sponge iron/DRI is popular on commercial level, production of liquid iron has its own merit and economical advantages. The extremely high thermal efficiency of a conventional BF is essentially due to the tall shaft where processing of the ore and pre-heating as well as reduction takes place before it descends into the melting zone. The effective height of the shaft required to carry out the processing of the ore depends on the reducing power and the temperature of the furnace gas generated at the tuyere level. In a normal BF, the gas generated at the tuyere level contains nearly 60% nitrogen which is inert and acts only as a heat transfer agent. Now question arises, whether it is feasible in reducing the shaft height without impairing the thermal efficiency of the process. It means that the gases from the low shaft furnace should go out of the furnace as a temperature and CO concentration which are comparable. If this is feasible, then poorer-quality coke or even non-metallurgical coal could be used as a fuel to run such furnace. The shaft height can be reduced without impairing the thermal efficiency if the heating and smelting zones in the furnace are compressed by accelerating the heat exchange and chemical reactions, i.e. sharper thermal and reducing potential gradients are obtained. Use of oxygen-enriched blast would produce higher temperature and furnace gas containing lower nitrogen content and higher reducing potential [1]. Consequently, the necessary heat transfer and reduction would be achieved over a shorter shaft. Figure 9.1 shows the low shaft furnace.

S. K. Dutta and Y. B. Chokshi, *Basic Concepts of Iron and Steel Making*,
https://doi.org/10.1007/978-981-15-2437-0_9

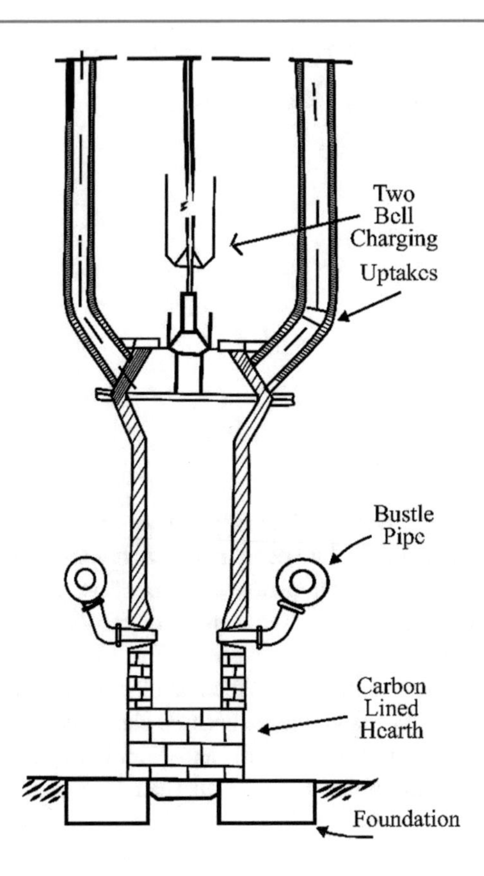

Fig. 9.1 Low shaft furnace

Based on the above reasons, the idea of low shaft furnace was developed by a group of seven countries, viz. France, Italy, Belgium, Luxemburg, Holland, Austria and Greece, who jointly set up the first even experimental low shaft furnace at Ougree, near Liege, in Belgium in 1950s. The raw materials that were tried included low-grade lump ores, concentrates obtained by upgrading the lean ores, rich ore fines, poor-quality coke and lignite. The difficulties were anticipated like high thermal losses, low indirect reduction, insufficient conditioning of the charge before reaching the tuyere zone, incapability of high top pressure application, etc. Because of these trials, a few commercial low shaft furnace plants were set up in these countries in the 1950s, cross-sectional furnaces are in operation; rectangular cross section is very popular. The hearth area is generally around 8–10 m^2. Nearly 500–700 Nm^3 of blast is blown to generate 8000–11,000 Nm^3 of top gas. Average consumption of lignite is more than 2000 kg/thm which is higher compared to coke consumption of BF. Low shaft furnace can be charged lumps or briquettes. Briquettes are made from suitable mixture of fines of ore, limestone and lignite with tar as binder. When briquettes are used, it prevents condensation of tar at the top. This is evolved when briquettes get pre-heated at the furnace top. Therefore, top gas temperature is kept rather high, i.e. 350–400 °C. The use of briquettes not only assures uniform distribution of the charge but also generates sufficient permeability for efficient heating and reduction.

The low shaft furnace plants at Coble, Germany, had ten furnaces each producing 80–100 tpd and producing 250,000 t of foundry grade iron per annum. The furnace had a 11 m^2 rectangular cross section at the tuyere level and an effective height of 4.8 m. The blast was pre-heated to 780–800 °C. It had a total eight tuyeres, three each on the two long sides and one each on the two short sides. The furnace is operated on a bedded charge of small lumps of iron ore, limestone and lignite coke briquettes produced by low temperature carbonization.

Advantages of low shaft furnace [1]:

1. It can take fine, friable, low grade and such inferior type of ores in the charge.
2. Fuels of inferior grade like lignite can be successfully used.
3. The furnace top gas, a by-product, is a rich fuel (i.e. it contains high caloric value) which can be economically used elsewhere in the plant or partly for pre-heating the blast.
4. The blast pressure necessary to run the low shaft furnace is of the order of 2 kg/cm^2 which is much lower than that for normal BF, and hence, costly high-capacity turbo blowers are not necessary.
5. Pre-heating of blast can be done in a recuperator, and hence, costly stoves are not necessary.
6. Control of smelting operation is easy and is exercised by control of blast temperature, pressure and volume.
7. The furnace can be readily started and stopped without any expansive and prolonged procedure involved.
8. Production rate is around 2.5 t/m^3 of useful volume of the furnace, which is much more than that of BF.
9. The retention time of the charge is only 1–2 h as compared to 8–10 h in a BF.

Disadvantages:

1. Briquettes may be charged for efficient production that adds to the cost of raw material preparation.
2. Daily production of an individual unit is low (100–200 tpd).

9.2 Mini-blast Furnace

Although several DR processes are developed, their products are solid sponge iron or HBI. Liquid iron, as the product, has its own economic advantages in steelmaking. Hence, mini-blast furnace (MBF) is the best alternative to a conventional BF. It is a small shaft furnace in which inferior quality of coke can be used. Lower-size fraction coke, which is screened off as the nut size fraction of regular coke, is used as a fuel for such small shaft furnace, popularly known as mini-blast furnace or MBF. These are widely used in Germany, Bulgaria, Vietnam, Brazil, China and India.

Mini-blast furnaces have the advantages of lower investment requirement and shorter implementation time and can accept coke with less stringent quality stipulations as compared to bigger blast furnaces. The hot metal produced is also of good quality. Consequently, such furnaces are playing a unique role in the development of mini-integrated steel plants.

Mini-blast furnace (MBF) available has working volume in the range of 75–750 m^3 and has capacity to produce 100–750 tonnes of hot metal per day. However, quite often, the working volume of MBF is in the range of 100–370 m^3, and corresponding production capacities are in the range of 40,000–200,000 tpa. MBF generally runs on 100% lump ore with 750–850 °C as hot blast temperature. MBF suffers from low heat transfer and low indirect reduction efficiency because of the short stack height. The off-gas is richer in CO gas to the extent of 28–30% with a CO/CO_2 ratio of 3.0 against 1.5 obtained in regular BF. The lining of furnace is heavy-duty medium alumina bricks. It is not water cooled through cooling plates (like regular BF). Still the life of lining is 3–5 years.

Table 9.1 Differences of the MBF and regular blast furnaces

Parameter	MBF	Regular BF
Size, m^3	75–750	2000–5000
Working height, m	12–16	18–30
Blast pressure, kg/cm^2	1.3	>2.0 (1 atm = 1 kg/cm^2)
Blast pre-heating temperature, °C	750–850	1000–1200
Lining	Medium alumina	Special fireclay
Cooling	External sparely	Internal (using cooling plates)
Raw material	Lumps or lump + sinter	Lumps or lump + sinter or lump + pellets
Coke size, mm	15–50	30–80
Quality of burden	Medium	High
Productivity, $t/m^3/d$	1.5–2.0	1.25–2.0
Coke rate, kg/thm	650–900	500–650

The trial was carried out by NML-Tata steel [2], India, with 15% replacement of iron ore lump with the iron ore–coal composite briquette. The hot metal and slag analysis was the same as in the regular operation. The slag rate increased about 12.9%. Since iron ore–coal composite pellets have higher reduction rate, using of composite pellets in MBF will improve the efficiency of reduction reaction. The iron ore–coal composite pellets were used in Japanese blast furnace [3] on trial bases; it was found that the reduction of carbon consumption at the blast furnace per 1 kg C/thm of carbon derived from carbon of composite is 0.23 kg C/thm.

Nowadays, MBFs are becoming popular due to following reasons [4]:

1. Increasing demand of pig iron,
2. Decreasing availability of scrap for steelmaking,
3. Shortage and increasing cost of electrical power, use of hot metal as feed into EAF,
4. Comparatively simple and economical production in MBF and
5. Installing any location.

Differences between the MBF and regular blast furnaces are shown in Table 9.1.

9.3 Charcoal Blast Furnace

Ironmaking technology started off with the use of wood charcoal as the only fuel and reductant. Charcoal was used as a fuel as well as a reductant in blast furnace. Although inferior in quality, it became a universal fuel for smelting of iron ores until the eighteenth century. The use of charcoal had to be abandoned due to restriction of cutting forest trees and burning the wood for producing charcoal. Since the forests were getting wiped out at a faster rate than their natural growth. Coke was introduced as fuel in 1619, but the use of charcoal continued with decreasing rate. Charcoal blast furnaces are continued for operation in Brazil, Sweden, China and some countries. The high-grade iron is produced in charcoal blast furnaces [5], combined with its lower operation cost and capital investment, in contrast with that of coke blast furnaces.

Most of these furnaces were less than 100 m^3 useful volumes. Height and size of the blast furnace could not be increased due to low strength of charcoal. Very low bulk density of charcoal resulted in its occupying substantial useful volume of the furnace. As a result, the productivity of charcoal blast furnace was relatively low. The slag volume, however, was much less, i.e. in the range between 100 and 160 kg/thm because of very low ash (3–5%) content of charcoal. But the calorific value (dry basis) of charcoal falls between 28.5 and 30.1 kJ/kg; bulk density (dry basis) is 220–280 kg/m^3. Since sulphur is non-existent in charcoal and iron ore is of high grade, sulphur must not be removed and hence slag may be acidic with CaO and MgO-to-SiO_2 ratio between 0.8 and 0.9, as against 1.3–1.4 in the coke blast furnace.

Since charcoal is much more reactive than coke:

$$C(s) + \{CO_2\} = 2\{CO\} \tag{9.1}$$

this reaction begins around 750–800 °C in the charcoal blast furnace, as against 900–950 °C in the case of coke BF. Also due to higher reactivity of charcoal, the degree of indirect reduction is higher than in the coke BF. Thus, the lower stack height is possible.

The low bulk density of charcoal (250 kg/m^3 as against 550 kg/m^3 for coke) makes that about 75% of the furnace volume is occupied by the reductant. So, the permeability of the burden is widely determined by the charcoal, which must be carefully screened (at least 95 mm), charged in separate layers and should be uniform in size. Differences between the charcoal and coke blast furnaces are shown in Table 9.2.

The outstanding quality of charcoal pig iron is well known, as a raw material for both the foundry and the melt shop. It has features like having low P and S, low content of residuals, absence of oxides and other impurities, high density and the high value of this metallic raw material as compared to other. The chemical energy content of charcoal pig iron is in the form of about 4% C and 0.5% Si. In an electric arc furnace, properly prepared to make use of oxygen and with a scrap pre-heater, this energy will be fully used and saving of electric energy.

Table 9.2 Differences of the charcoal and coke blast furnaces [5]

Parameter	Charcoal BF	Coke BF
Temperature in the reserve zone	800 °C	950 °C
Residence time of ore in the reserve zone	Half of coke BF	
Apparent density	250 kg/m^3 for charcoal	550 kg/m^3 for coke
Volume taken by the ore	15%	30%
Blast temperature	800 °C	1100–1200 °C
Slag	100 to 160 kg/thm	300 to 500 kg/thm
Basicity of slag (CaO + MgO)/SiO_2	0.8–0.9, since no S present in charcoal	1.3–1.4
Refractory	Cheaper quality (50%, in very special case 70% Al_2O_3) due to lower blast and flame temperatures	High quality (>80% Al_2O_3) due to higher blast and flame temperatures
Top gas temperature	90–120 °C	150–200 °C
Productivity	400–800 tpd	1500–2500 tpd
Investment	Low investment (half of coke BF)	High investment

9.4 Electrothermal Process

The idea of using an electric furnace to produce hot metal is not new. Electric furnace has been producing hot metal and ferro-alloys since the early twentieth century. Submerged arc furnaces (SAFs) are known to be capable of smelting reduction of iron oxides by carbon and producing hot metal typically with 3.5–4.0% C. SAFs are fed with layers of oxides and carbon floating as a charge burden on top of the slag layer [1]. Heat for the process is generated by passing electric current through the slag from electrode to electrode.

Electric arc supplies the heat, instead of burning of coke, in electrothermal smelting processes; that is, thermal requirement is met by electric arc. But coke or other forms of carbon is still necessary because the reduction of iron oxide takes place predominantly by the classical endothermic reaction.

$$Fe_xO_y(s) + yC(s) = xFe(s) + y\{CO\} \tag{9.2}$$

Electric smelting of pig iron is developed in the beginning of the twentieth century in Italy and Sweden. Twelve furnaces of 10 MVA capacity were in operation in Italy and Sweden in the early nineteen twenties. Visvasvaraya Iron and Steel Ltd (formerly Mysore Iron and Steel Co), Bhadrawati, Karnataka, was pioneer to install electrothermal process in India on 1952 with two furnaces of 100 tpd each. 100 tpd capacity furnace was installed at Sandur Manganese and Iron Ores Ltd, Sandur, Karnataka, India. Maharashtra Electrosmelt, Chandrapur, India, also has two furnaces of 33 MVA.

The electrothermal production of hot metal is carried out in three electrodes submerged arc furnace of the fixed or non-tilting type as shown in Fig. 9.2. Transformer capacity is an indication of the size of the furnace, primary voltage of the order of 33 kV or more, and low secondary adjustable voltage with high current like 150 kA is used.

The furnace is a steel shell inside lined with basic refractories like magnesite or chrome-magnesite. At the lower portion of the furnace is lined with carbon bricks. The furnace is closed from the top by roof lined with fireclay bricks, since arcing is surrounded by charge materials and slag. Roof has five holes, three for electrodes, one for charging and another for exit gas (modern anti-pollution restrictions necessitate gas collection and cleaning, before discharge into the atmosphere).

Modern furnaces are provided with self-baking electrode, since these are cheaper, and they can be of any desired dimension to carry the desired current. Electrode is produced by putting electrode paste inside of steel cylinders, and baking is done during operation of the furnace. The electrode paste is made from coke or calcined anthracite and bituminous binder. As the electrode is consumed and can slip in, fresh steel cylinder is fixed to it from top and electrode paste is filled into that. The electrode paste gets slowly backed in situ to form a hard electrode before it is consumed.

Most submerged arc furnaces are fed with premixed charges which must be proportioned accurately be weighting each raw material individually. Charge consists of iron ore, coke or cheaper fuel (almost half that of the normal BF) and limestone. Chemistry of the charges is adjusted to finally produce a thin, fluid slag at the operating temperature of the furnace. Flux requirement is therefore reduced in proportion to the decrease in the actual ash reduction in the charge. Higher basicity up to 1.7–1.8 can be used depending upon the sulphur content in the charge. The fines do affect the operation adversely, and hence, that should be eliminated. Iron ore–coal composite pellets can be used as charge material.

The reactivity of the reductant is important since reduction of iron oxide must take place directly by carbon in relatively shorter duration. The electrical conductivity of the reductant is also equally important since during smelting, a bed of reductant is formed under the electrode and the resistance to passage of current generates most of the heat produced in the furnace. Size and proportion of

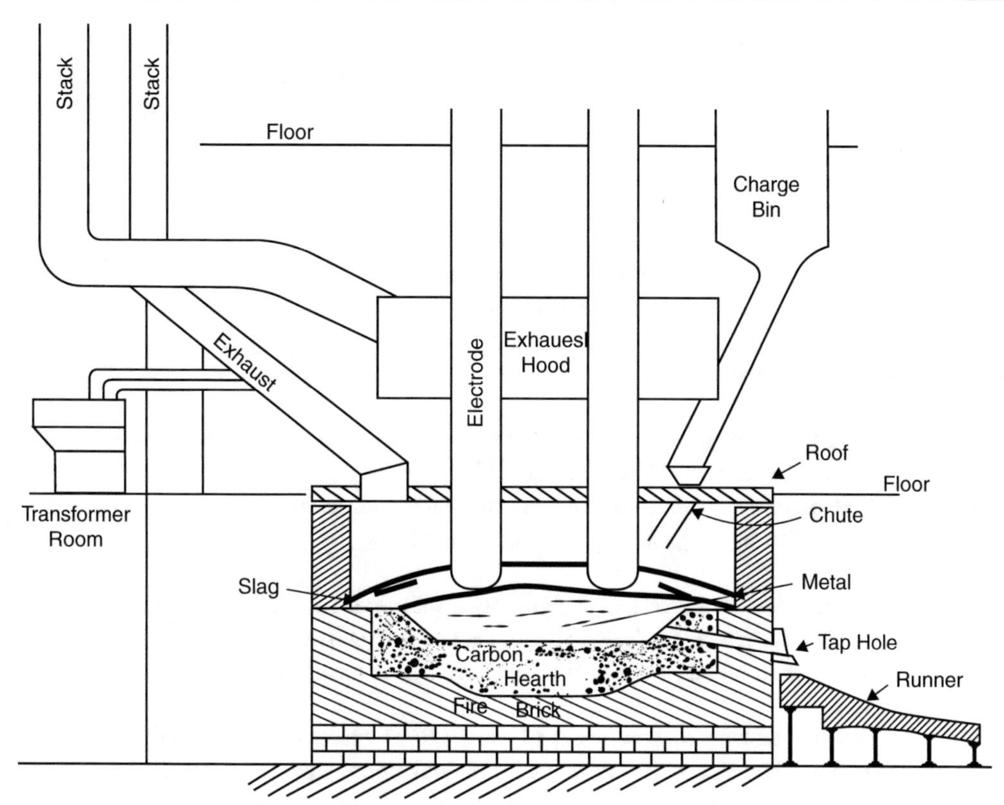

Fig. 9.2 Submerged electric arc furnace

reductant influence the smelting temperature, and these are adjusted to obtain the desired bath temperature.

Reactivity and electrical conductivity of the reductant are important parameters, and choice of the reductant must be made in the light of the above properties. Most of the heat is developed within one metre from the electrode tip. Size of reductant bed must be under the electrode, and its electrical conductivity thus decides the temperature of the furnace. Proportion, quality and size of reductant mixed with the charge must therefore be closely controlled to obtain the desired optimum thermal profile in the furnace which in turn will decide the existent of furnace reaction and finally the composition of hot metal produced. One or two tap holes may be provided, depending upon the size of the furnace, to tap hot metal and slag. Hot metal and slag flow out from the same hole and are separated afterwards outside the furnace through slag dam. The slag dam is detained and diverted the slag from the runner into slag line, while the hot metal is flowing through under the dam.

Electric power consumption is generally 2200–2500 kWh/thm, which is very high. Use of self-fluxing sinter dccreases power consumption by 10–15%. Pre-heating the charge by using furnace gas has also reducing the power consumption by up to 25%.

The indirect reduction can contribute up to 20% reduction, which is reflected in the furnace gas composition (72–75% CO, 13–15% CO_2, 7–10% H_2, 1–3% N_2 and 2–3% hydrocarbons). The furnace gas has a calorific value of around 10,460 kJ/Nm^3, and nearly 600–700 Nm^3 of gas is generated per tonne of hot metal. If all iron oxide is reduced by carbon, i.e. direct reduction, the CO content of furnace gas can be above 90–95%. But as the proportion of indirect reduction increases, more CO gets converted to CO_2. If highly reducible self-fluxing type sinter is used, the CO proportion can fall to 50–52% and CO_2 rising to 40%. Slag is very much depending on the burden chemistry, i.e. composition of ore, limestone and ash of the coke. It is varied from practice to practice and plant to plant. General composition of slag is 40–45% CaO, 25–30% SiO_2, 2–10% MgO and 5–25% Al_2O_3.

9.5 ELRED Process

The ELRED process was developed in Sweden by Stora Kopparberg (STORA) in 1971; STORA reached a cooperation agreement with ASEA, covering theoretical and experimental investigation of the process. The ELRED process is a method for producing liquid iron by a two-stage reduction of iron ore concentrate with coal [6].

First stage: Fine-grained concentrates are pre-reduced (up to 60–70%) in a fluidized bed with gas generated from coal powder and air in the same bed.
Second stage: Fine-grained product from the first stage undergoes final reduction and smelting to iron under the plasma created by electrode in a DC arc furnace.

An important feature of the process is that both the concentrates and the coal can be used without previous agglomeration in a sintering/palletization plant or coking in a coke oven plant. In absence of these, it is estimated that liquid iron can be produced at a cost approximately 20% less that of modern BF.

Figure 9.3 shows a schematic flow sheet of the ELRED process, whose basic concept is the reduction of iron ore in two stages: (i) first stage for pre-reduction in the solid state in the presence of excess of carbon to a partly metallized product containing carbon and (ii) final reduction stage for

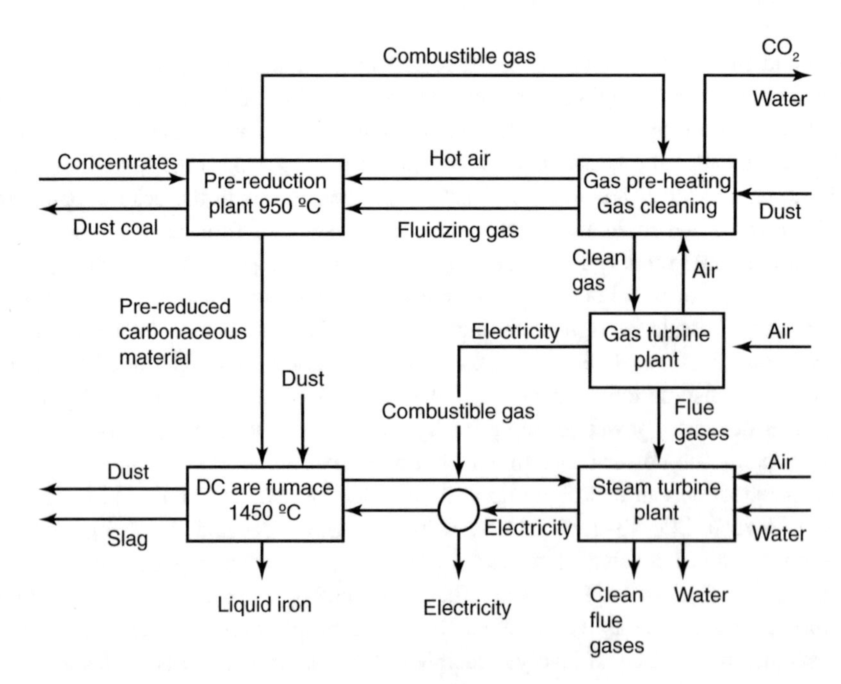

Fig. 9.3 Schematic flow diagram of the ELRED process [6] (reproduced with permission from *IIM Metal News*)

smelting of this product using electrical energy in a DC arc furnace. The flue gases from both stages are used to generate electrical energy.

Ore concentrate fine grained (<0.1 mm) having iron content >65% Fe. Iron ore which contains high P can be used. Dust coal (anthracite, lignite) is preferable (size 0.2–0.3 mm) that should be dried before injected into the fluidized bed.

I Pre-reduction Stage:

The pre-reduction takes place under pressure in a circulating fluidized bed (as shown in Fig. 9.4), which differs from the classic fluidized bed primarily through the substantially higher gas velocity. The excess of carbon in the bed and the high gas velocity prevent sticking. Coal powder and air are injected direct into the bed and generate the temperature 950–1000 °C. CO and H_2 mixture as well as excess C, in the form of fine-grained char, are act as reducing agents. The following reactions are occurred:

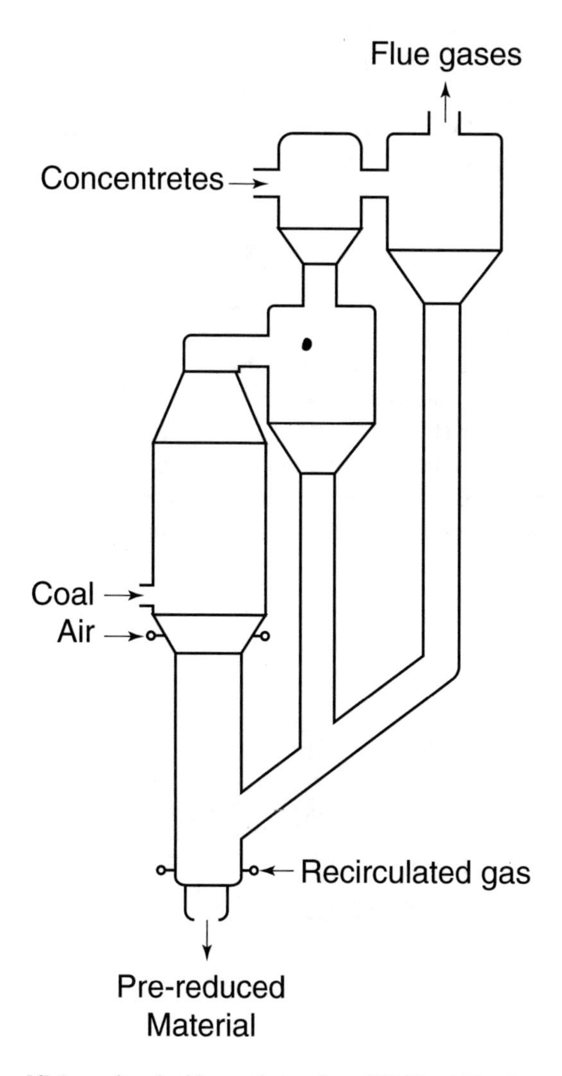

Fig. 9.4 Pre-reduction stage [6] (reproduced with permission from *IIM Metal News*)

$$C(s) + \{O_2\} = \{CO_2\} \tag{9.3}$$

$$C(s) + \{CO_2\} = 2\{CO\} \tag{9.4}$$

$$2\{C_nH_m\} = 2nC(s) + m\{H_2\} \tag{9.5}$$

The main component of the pre-reduction plant is a reactor vessel with internal refractory lining. It has an inside diameter of 3–10 m and a height of about 25 m. The heat content of the exit gas from the reactor is utilized for different purpose, such as pre-heating of air and concentrate of ore. After dust, moisture and CO_2 have been removed, 30–50% of the gas is utilized for the fluidizing in the lower part of the reactor. The remaining of gas is used to generate electricity. Degree of metallization of the product is controlled to 60–70% by adjusting the residence time and the temperature in the reactor.

II Final Smelting Stage:

Final smelting takes place in a DC arc furnace. A hollow graphite electrode, located in the centre of the furnace roof, acts as cathode and liquid melt as well as hearth acts as anode (as shown in Fig. 9.5). The arc, which is submerged in the foaming slag, extends vertically down towards the bath. Pre-reduced charge material at 600–700 °C passes through the whole of the electrode and falls on the hot plasma. Melting, carburizing and final reduction take place very quickly at high plasma temperature.

$$[FeO] + [C] = [Fe] + \{CO\} \tag{9.6}$$

$$3[Fe] + [C] = [Fe_3C] \tag{9.7}$$

Basicity of slag is maintained about 1.2. Liquid iron is taped time to time, 30–50% is left in the furnace for helping the smelting operation. Liquid iron contains 3–4% C, 0.05% Si and Mn. Half of the S and most of the P present in charge materials are remained in the liquid iron. The energy content in the flue gases from the pre-reduction and final reduction stage is used for power generation by a gas turbine and a steam turbine. The gas turbine is driven by cleaned gas from the pre-reduction stage. The gas from the final reduction stage is also utilized to generate steam.

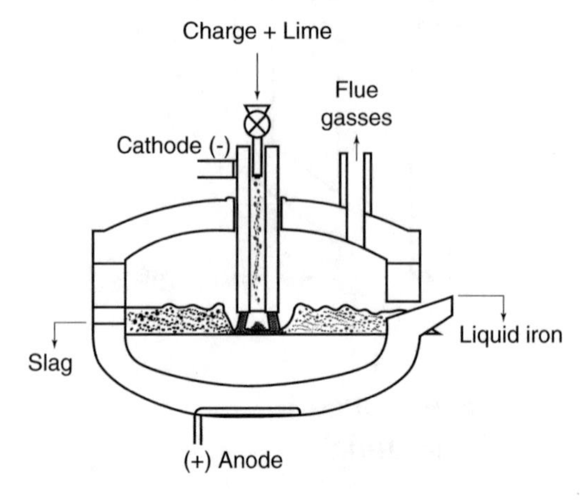

Fig. 9.5 DC—arc furnace [6] (reproduced with permission from *IIM Metal News*)

9.6 KR Process

KR process was developed by KORF Engineering GmbH, Germany, and Austrian Voest-Alpine AG in 1978. It is eliminated coke making step in BF and uses directly coal as fuel as well as reducing agent. Reduction technology to produce hot metal on coal base introduced under the designation KR (i.e. coal reduction) process under the new trademark *Corex*. First commercial plant is established in 1988 at South Africa.

KR plant represents a BF divided up into a (i) reduction shaft furnace and (ii) melter gasifier. Hot reducing gas (consists of 95% CO + H_2) generated in the melter gasifier is directly conveyed to the reduction shaft to reduce lump ore or pellets and discharge hot sponge iron. In the melter gasifier, hot sponge iron is melted to hot metal by the excess heat generated during the partial oxidation of coal [7]. Coal (up to 12 mm) is charged into the melter gasifier and is gasified by using oxygen.

Probable Questions

1. Discuss the basic principle of low shaft furnace. Discuss the advantages of low shaft furnace.
2. Discuss the basic principle of charcoal blast furnace.
3. Discuss the differences between the MBF and regular blast furnaces.
4. Compared the charcoal blast furnace and coke blast furnace.
5. Discuss the basic principle of electrothermal smelting process.
6. Discuss the basic principle of submerged arc furnace.
7. Discuss the basic principle of ELRED process.

References

1. R.H. Tupkary, V.R. Tupkary, *An Introduction to Modern Iron Making*, 4th edn. (Khanna Publishers, Delhi, 2012)
2. Y. Rajshekar, T. Venugopalan, IIM Metal News **19**(5), 26 (2016)
3. H. Yokoyama et al., ISIJ Int. **52**(11), 2000 (2012)
4. S. Das Gupta, Proc of Inter. *Seminar on Alternative Routes for Ironmaking in India* (Steel Tech, Sep 2009, Kolkata, India), p. 1
5. R. Weber, H.C. Pfeifer, D. Nose, *Proceedings of International Sym*posium *on Blast Furnace Ironmaking*, Nov 1985, Jamshedpur, India), p. 121
6. P. Collin, H. Stickler, IIM Metal News **3**(1), 8 (1981)
7. *Steel Times International*, **10**(1), 63 (1986)

Part III
Physical Chemistry of Ironmaking

Thermodynamics of Reduction 10

Success and efficiency of a process are depending on the thermodynamic calculation of the reactions. How much reducing gas (CO) required to reduce iron oxides at different stages and temperatures that should be known before the actual process, i.e. how much reductant is required for the process.

10.1 Reduction of Metal Oxide

Reduction of metal oxide by CO gas is based on a reversible reaction (general form):

$$MO\,(s) + \{CO\} \leftrightarrows M(s) + \{CO_2\}, \Delta G_{10.1} \tag{10.1}$$

Reaction (10.1) can take place according to the combination of two reversible reactions:

$$2MO(s) \leftrightarrows 2M(s) + \{O_2\}, \Delta G_{10.2} \tag{10.2}$$

$$2\{CO_2\} \leftrightarrows 2\{CO\} + \{O_2\}, \Delta G_{10.3} \tag{10.3}$$

The equilibrium constants for reactions (10.1)–(10.3) are:

$$k_{10.1} = \frac{a_M\ p_{CO_2}}{a_{MO}\ p_{CO}} \tag{10.4}$$

Consider that M and MO are pure solid at STP, so $a_M = a_{MO} = 1$,
So, Eq. (10.4) can be modified as:

$$k_{10.1} = \frac{p_{CO_2}}{p_{CO}} \tag{10.5}$$

Similarly

$$k_{10.2} = \frac{{a_M}^2 P_{O_2}}{{a_{MO}}^2} = p_{O_2} \tag{10.6}$$

S. K. Dutta and Y. B. Chokshi, *Basic Concepts of Iron and Steel Making*,
https://doi.org/10.1007/978-981-15-2437-0_10

$$k_{10.3} = \frac{p_{CO}{}^2 P_{O_2}}{P_{CO_2}{}^2} \tag{10.7}$$

Addition of reaction (10.2) and reverse reaction (10.3):

$$\begin{aligned}
&2MO(s) \leftrightarrows 2M(s) + O_2(g), \Delta G_{10.2} \\
&2CO(g) + O_2(g) \leftrightarrows 2CO_2(g), -\Delta G_{10.3} \\
\hline
&2MO(s) + 2CO(g) = 2M(s) + 2CO_2(g), \Delta G_{10.2} - \Delta G_{10.3} \\
&\text{Or} \quad MO(s) + CO(g) \leftrightarrows M(s) + CO_2(g), \Delta G_{10.1}
\end{aligned} \tag{10.1}$$

Therefore,

$$\Delta G_{10.1} = 1/2(\Delta G_{10.2} - \Delta G_{10.3}) \tag{10.8}$$

Thus, Eq. (10.8) is controlled by the overall reaction (10.1). If the value of $\Delta G_{10.1}$ is (−) ve, then the direction of the reaction goes towards the left to right, i.e. forward direction.

Since

$$\Delta G = \Delta G^0 + \text{RT} \ln k \tag{10.9}$$

Since the value of ΔG of a reaction depends on the initial and the final (i.e. equilibrium) state if conditions are not standard.

Therefore,

$$\Delta G = \Delta G^0 + \text{RT}(\ln k' - \ln k) \tag{10.10}$$

where k' and k are the initial and the final (i.e. equilibrium) state.

Hence,

$$\Delta G_{10.2} = \Delta G^0_{10.2} + \text{RT}\left(\ln p'_{O_2} - \ln p_{O_2}\right) \tag{10.11}$$

$$\Delta G_{10.3} = \Delta G^0_{10.3} + \text{RT}\left[\ln\left(\frac{p'_{CO}{}^2}{p'_{CO_2}{}^2} p'_{O_2}\right) - \ln\left(\frac{p_{CO}{}^2 P_{O_2}}{P_{CO_2}{}^2}\right)\right] \tag{10.12}$$

where $p'_{O_2}, p'_{CO}, p'_{CO_2}$ are the partial pressures of the gaseous components of the initial state while P_{O_2}, p_{CO}, P_{CO_2} are the partial pressures of the gaseous components of the final (i.e. equilibrium) state.

Now, the values of $\Delta G_{10.2}$ and $\Delta G_{10.3}$ from Eqs. (10.11) to (10.12) are substituting in Eq. (10.8):

$$\begin{aligned}
\Delta G_{10.1} &= 1/2(\Delta G_{10.2} - \Delta G_{10.3}) \\
&= 1/2(\Delta G^0_{10.2} - \Delta G^0_{10.3}) + \text{RT}\left[\ln\left(\frac{p_{CO}}{p_{CO_2}}\right) - \ln\left(\frac{p'_{CO}}{p'_{CO_2}}\right)\right] \\
&= \Delta G^0_{10.1} + \text{RT}\left[\ln\left(\frac{p_{CO}}{p_{CO_2}}\right) - \ln\left(\frac{p'_{CO}}{p'_{CO_2}}\right)\right]
\end{aligned} \tag{10.13}$$

Thus, Eq. (10.13) is the general equation expressing the thermodynamics of the reduction of an oxide or oxidation of a metal by a gaseous reactant.

The following condition of the reaction of the system under consideration:

(I) Reduction of the oxide:

$$\Delta G_{10.1} < 0, \text{ i.e. } (-)\text{ ve ; i.e. } \Delta G_{10.3} > \Delta G_{10.2} \text{ [in Eq. (10.8)]}$$

$$\text{and} \left(\frac{p'_{CO}}{p'_{CO_2}}\right)_{g.ph} > \left(\frac{p_{CO}}{p_{CO_2}}\right)_{eq.} \quad \text{[in Eq. (10.13)]} \tag{10.14}$$

(II) Oxidation of metal:

$$\Delta G_{10.1} > 0, \text{ i.e.}(+)\text{ve; i.e. } \Delta G_{10.3} < \Delta G_{10.2} \text{ [in Eq. (10.8)]}$$

$$\text{and} \left(\frac{p'_{CO}}{p'_{CO_2}}\right)_{g.ph} < \left(\frac{p_{CO}}{p_{CO_2}}\right)_{eq.} \quad \text{[in Eq. (10.13)]} \tag{10.15}$$

(III) Equilibrium of the process:

$$\Delta G_{10.1} = 0, \text{ i.e. } \Delta G_{10.3} = \Delta G_{10.2} \text{ [in Eq. (10.8)]}$$

$$\text{and} \left(\frac{p'_{CO}}{p'_{CO_2}}\right)_{g.ph} = \left(\frac{p_{CO}}{p_{CO_2}}\right)_{eq.} \quad \text{[in Eq. (10.13)]} \tag{10.16}$$

10.2 Phase Stability Diagrams

10.2.1 Fe–C–O System

Figure 10.1 shows the phase stability diagram or Fe–C–O system. This is also known as *fish-tail diagram* (due to looking like the tail of a fish). The reduction of iron oxide by carbon monoxide gas takes place in three stages at temperature above 567 °C (at point *X*):

$$\overset{\text{I}}{Fe_2O_3 \rightarrow} \overset{\text{II}}{Fe_3O_4 \rightarrow} \overset{\text{III}}{FeO \rightarrow} Fe$$

$$3Fe_2O_3(s) + \{CO\} = 2Fe_3O_4(s) + \{CO_2\} \tag{10.17}$$

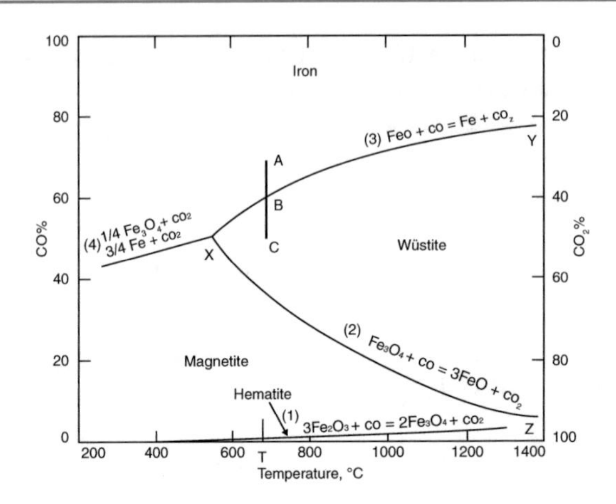

Fig. 10.1 Fe–C–O system [2] (reproduced with permission from Authors)

$$\mathrm{Fe_3O_4(s)} + \{\mathrm{CO}\} = 3\mathrm{FeO(s)} + \{\mathrm{CO_2}\} \tag{10.18}$$

$$\mathrm{FeO(s)} + \{\mathrm{CO}\} = \mathrm{Fe(s)} + \{\mathrm{CO_2}\} \tag{10.19}$$

The reduction of iron oxide by carbon monoxide gas takes place in two stages at temperature below 567 °C (at point *X*):

$$\mathrm{Fe_2O_3} \xrightarrow{\text{I}} \mathrm{Fe_3O_4} \xrightarrow{\text{II}} \mathrm{Fe}$$

$$3\mathrm{Fe_2O_3(s)} + \{\mathrm{CO}\} = 2\mathrm{Fe_3O_4(s)} + \{\mathrm{CO_2}\} \tag{10.17}$$

$$\mathrm{Fe_3O_4(s)} + 4\{\mathrm{CO}\} = 3\mathrm{Fe(s)} + 4\{\mathrm{CO_2}\} \tag{10.20}$$

For all reactions (10.17)–(10.20), the equilibrium constant:

$$k_{\mathrm{eq}} = \frac{p_{\mathrm{CO_2}}}{p_{\mathrm{CO}}} \tag{10.21}$$

Therefore,

$$\left(\frac{p_{\mathrm{CO}}}{p_{\mathrm{CO}} + P_{\mathrm{CO_2}}}\right) = \left(\frac{1}{1 + k_{\mathrm{eq}}}\right) \quad \text{as} \quad f(T) \tag{10.22}$$

The equilibrium composition of the gas phase for the reactions of reduction of iron oxide is given in Fig. 10.1 as function of temperature. If the composition of the gas mixture at a given temperature is different from an equilibrium one, the system becomes unstable and its components react in the forward or backward direction depending on the composition of the gas mixture with respect to the equilibrium composition. The condition governing the reduction of an oxide is dependent upon

$\left(\frac{p_{CO}}{p_{CO_2}}\right)_{g.ph}$ (i.e. in the initial state, point A (in Fig. 10.1) for FeO reduction) and $\left(\frac{p_{CO}}{p_{CO_2}}\right)_{eq}$ (i.e. in the final state, point B for FeO reduction in equilibrium with Fe at temperature, *T*).

Hence,

(i) At point A for temperature, *T*:

$$\left(\frac{p_{CO}}{p_{CO_2}}\right)_{g.ph} > \left(\frac{p_{CO}}{p_{CO_2}}\right)_{eq}, \quad \text{so FeO is reduced to Fe.} \tag{10.23}$$

(ii) At point C:

$$\left(\frac{p_{CO}}{p_{CO_2}}\right)_{g.ph} < \left(\frac{p_{CO}}{p_{CO_2}}\right)_{eq}, \tag{10.24}$$

i.e. oxidation takes place, metal will be oxidized; so, FeO does not reduce at all, instead of that if there is any reduced metal (Fe) present that will be oxidized.

(iii) At point B:

$$\left(\frac{p_{CO}}{p_{CO_2}}\right)_{g.ph} = \left(\frac{p_{CO}}{p_{CO_2}}\right)_{eq}, \tag{10.25}$$

so FeO–Fe equilibrium is attained, no further reaction takes place.

Line *XY* corresponds with the equilibrium coexistence of gas with FeO–Fe, line *XZ* corresponds with the equilibrium coexistence of gas with Fe_3O_4–FeO line, *OX* corresponds with the equilibrium coexistence of gas with Fe_3O_4–Fe and line very close to *x*-axis corresponds with the equilibrium coexistence of gas with Fe_2O_3–Fe_3O_4.

In Fig. 10.1, according to the curve (*XY*), large amount of CO gas is required in contact with FeO–Fe at 900 °C. FeO will be reduced until the gas mixture contains 64% of CO and 36% of CO_2. At the same temperature (i.e. 900 °C), according to the curve (*XZ*), the equilibrium mixture over Fe_3O_4–FeO is 18% CO and 82% CO_2, whereas the equilibrium mixture over Fe_2O_3–Fe_3O_4 is below 1% CO. Therefore, the greatest carbon monoxide requirement in the reduction of Fe_2O_3 to Fe is the reduction of the FeO.

Overall reaction:

$$1/2\ Fe_2O_3(s) + 3/2\ \{CO\} = Fe(s) + 3/2\ \{CO_2\} \tag{10.26}$$

10.2.2 Fe–H–O System

Figure 10.2 shows the phase stability diagram for hydrogen reduction or Fe–H–O system. The reduction of iron oxide by hydrogen gas also takes place in three stages at temperature above 567 °C (at point K):

$$Fe_2O_3 \xrightarrow{\text{I}} Fe_3O_4 \xrightarrow{\text{II}} FeO \xrightarrow{\text{III}} Fe$$

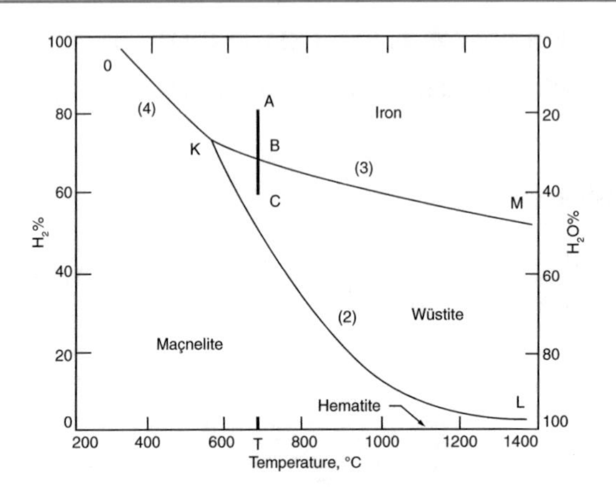

Fig. 10.2 Fe–H–O system [2] (reproduced with permission from Authors)

$$(1)\quad 3Fe_2O_3(s) + \{H_2\} = 2Fe_3O_4(s) + \{H_2O\} \tag{10.27}$$

$$(2)\quad Fe_3O_4(s) + \{H_2\} = 3FeO(s) + \{H_2O\} \tag{10.28}$$

$$(3)\quad FeO(s) + \{H_2\} = Fe(s) + \{H_2O\} \tag{10.29}$$

The reduction of iron oxide by hydrogen gas takes place in two stages at temperature below 567 °C (at point K):

$$\begin{array}{ccccc} & \text{I} & & \text{II} & \\ Fe_2O_3 & \rightarrow & Fe_3O_4 & \rightarrow & Fe \end{array}$$

$$(1)\quad 3Fe_2O_3(s) + \{H_2\} = 2Fe_3O_4(s) + \{H_2O\} \tag{10.27}$$

$$(4)\quad Fe_3O_4(s) + 4\{H_2\} = 3Fe(s) + 4\{H_2O\} \tag{10.30}$$

For all reaction, the equilibrium constant,

$$k_{eq} = \left(\frac{p_{H_2O}}{p_{H_2}}\right)_{eq} \tag{10.31}$$

Therefore,

$$\frac{p_{H_2}}{p_{H_2} + p_{H_2O}} = \left(\frac{1}{1 + k_{eq}}\right) \text{as} f(T) \tag{10.32}$$

The equilibrium composition of the gas phase for the reactions of reduction of iron oxide is given in Fig. 10.2 as function of temperature. The condition governing the reduction of an oxide is dependent upon $\left(\frac{p_{H_2}}{p_{H_2O}}\right)_{g.ph}$ (i.e. in the initial state, point A (in Fig. 10.2) for FeO reduction) and $\left(\frac{p_{H_2}}{p_{H_2O}}\right)_{eq}$ (i.e. in the final state, point B for FeO reduction in equilibrium with Fe at temperature, T).

Hence,

(i) At point A for temperature, T:

$$\left(\left(\frac{p_{H_2}}{p_{H_2O}}\right)_{g.ph} > \left(\frac{p_{H_2}}{p_{H_2O}}\right)_{eq} \rightarrow \text{i.e. FeO reduction take place, to form Fe.} \quad (10.33)$$

(ii) At point C:

$$\left(\frac{p_{H_2}}{p_{H_2O}}\right)_{g.ph} < \left(\frac{p_{H_2}}{p_{H_2O}}\right)_{eq} \quad (10.34)$$

→ i.e. oxidation takes place, metal will be oxidized; so, FeO does not reduce at all, instead of that if there is any reduced metal (Fe) present that will be oxidized.

(iii) At point B:

$$\left(\frac{p_{H_2}}{p_{H_2O}}\right)_{g.ph} = \left(\frac{p_{H_2}}{p_{H_2O}}\right)_{eq} \quad (10.35)$$

→ so equilibrium is attained, no further reaction takes place.

Line KL corresponds with the equilibrium coexistence of gas with Fe_3O_4–FeO, line KM corresponds with the equilibrium coexistence of gas with FeO–Fe, line OK corresponds with the equilibrium coexistence of gas with Fe_3O_4–Fe and line very close to x-axis corresponds with the equilibrium coexistence of gas with Fe_2O_3–Fe_3O_4.

In phase diagram, gas phases are not considered, but in phase stability diagram, gas phases are considered. With increase in temperature, hydrogen will be more active as a reductant according to the diagram.

Overall reaction:

$$1/2\,Fe_2O_3(s) + 3/2\,\{H_2\} = Fe(s) + 3/2\,\{H_2O\} \quad (10.36)$$

10.3 Reduction of Iron Oxides

Steps of reduction: $Fe_2O_3 \rightarrow Fe_3O_4 \rightarrow Fe_xO \rightarrow Fe$

where $0.83 < x < 0.955$, since FexO is not a stoichiometrically balanced compound.

The reduction of iron oxide by carbon monoxide or hydrogen takes place in three steps at above 567 °C:

$$\begin{array}{lccccccc} & & \text{I} & & \text{II} & & \text{III} & \\ & Fe_2O_3 & \rightarrow & Fe_3O_4 & \rightarrow & FeO & \rightarrow & Fe \\ \text{Oxygen content}: & 30.1\% & & 27.7\% & & 22.3\% & & 0\% \end{array}$$

The important point is be noted that even after reduction of Fe_2O_3–FeO stage, 22.3% oxygen (starting oxygen 30.1%) still remains, which is removed only at the final stage reduction of wustite (FeO) to metallic iron.

The most common reducing agents used in iron oxide reduction process are carbon monoxide, hydrogen and mixtures of these two gases. The thermodynamics of iron oxide reduction deals primarily with the equilibrium between its oxides and these reducing gases. Thermodynamics calculation shows the feasibility of the reaction, whether reaction moves to forward direction or not. Feasibility of reaction is indicated by a decrease in free energy. An Ellingham diagram is drowned (Appendix: Figure A-1) the standard free energy changes of reactions and their variation with temperature. The term standard refers to a state of reference, generally 1 atmosphere pressure and a temperature of 25 °C (298 K). Figure 10.3 is a representation of Ellingham diagram of a few key oxides which are of major interest in iron oxide reduction. The lines of oxides lying higher sides represent a less stable oxide as compared to the lines of lower sides. It can be seen that above 727 °C, the line of CO_2 formation from CO to O_2 lies slightly above the line of FeO formation from Fe to O_2. This means that under standard state, the following reaction cannot proceed forward direction:

$$FeO(s) + \{CO\} \leftrightarrows Fe(s) + \{CO_2\} \tag{10.19}$$

This reaction (10.19) is the key reaction in sponge ironmaking and the reaction is made to move in the forward direction by keeping a large excess of CO gas over CO_2 gas, CO/CO_2 ratio required to be over 2.52 at 1000 °C for reaction to proceed [1], i.e. by decreasing the activity of CO_2.

Compared with blast furnaces, the reduction potential of direct reduction is considerably lower, and the reduction takes place at a much lower temperature. The net result is that while in blast furnace

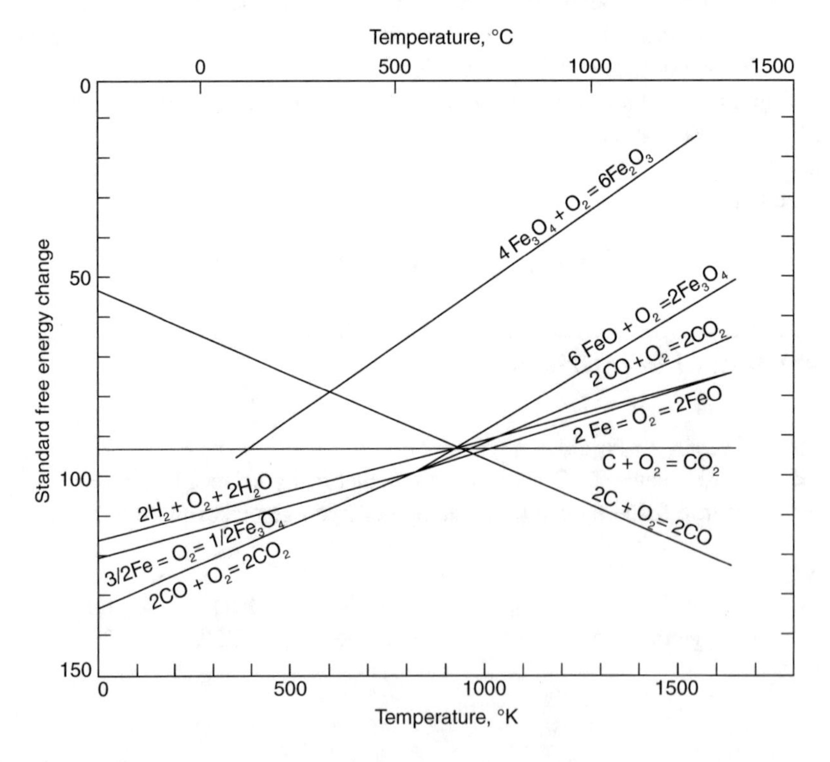

Fig. 10.3 Free energy versus temperature diagram for main oxides

ironmaking, FeO is completely reduced to iron (there is no FeO even in the slag phase); the DR processes are incapable of reducing FeO completely and some FeO is always present in sponge iron [2]. Gas-based sponge iron contains less FeO than coal-based sponge iron because of the presence of hydrogen gas in reformed natural gas.

10.3.1 Reduction by Carbon Monoxide

Three reduction reactions by carbon monoxide gas with their heats of reaction at 25 °C (298 K) are as follows:

$$3Fe_2O_3(s) + \{CO\} = 2Fe_3O_4(s) + \{CO_2\},\ \Delta H^0_{298} = -52.43\,\text{kJ/mol of CO} \tag{10.17}$$

$$Fe_3O_4(s) + \{CO\} = 3FeO(s) + \{CO_2\},\ \Delta H^0_{298} = 40.46\,\text{kJ/mol of CO} \tag{10.18}$$

$$FeO(s) + \{CO\} = Fe(s) + \{CO_2\}, \Delta H^0_{298} = -18.54\ \text{kJ/mol of CO} \tag{10.19}$$

The heat of reactions at 25 °C (298 K) was calculated with the data (Appendix 3) taken from the standard reference book [3]. For reactions (10.17) and (10.19), the heat of reactions is negative, which means that the reactions are exothermic nature, i.e. heats are generated, whereas for reaction (10.18), heat of reaction is positive, i.e. the intermediate stage of reduction of magnetite to wustite is endothermic nature that requires considerable amount of heat.

Since wustite (FeO) is metastable below 567 °C, iron oxide (Fe_2O_3) reduction takes place in two stages at below 567 °C:

$$Fe_2O_3 \xrightarrow{\text{I}} Fe_3O_4 \xrightarrow{\text{II}} Fe$$

$$3Fe_2O_3(s) + \{CO\} = 2Fe_3O_4(s) + \{CO_2\}, \Delta H^0_{298} = -52.43\,\text{kJ/mol of CO} \tag{10.17}$$

$$1/4Fe_3O_4(s) + \{CO\} = 3/4Fe(s) + \{CO_2\},\ \Delta H^0_{298} = -3.79\,\text{kJ/mol of CO} \tag{10.20}$$

This reaction (10.20) is also exothermic nature. However, most DR processes operate at temperature above 600 °C, so the reduction reaction (10.20) is only of minor interest.

The free energy change for reaction (10.19) (i.e. $FeO(s) + CO(g) = Fe(s) + CO_2(g)$) at 727 °C (1000 K) can be calculated as follows:

Since ΔG^0_{1000} for FeO $= -199.535$ kJ, ΔG^0_{1000} for CO = -199.368 kJ,
and ΔG^0_{1000} for $CO_2 = -394.97$ kJ (From Appendix 2)
Since,

$$\begin{aligned}\Delta G^0_{10.19,1000} &= \Sigma\Delta G^0_{1000,\text{Product}} - \Sigma\Delta G^0_{1000,\text{Reactant}} \\ &= \left[\left(\Delta G^0_{1000,CO_2}\right) - \left(\Delta G^0_{1000,FeO} + \Delta G^0_{1000,CO}\right)\right] \\ &= [(-394.970) - \{(-199.535) + (-199.368)\}] \\ &= [(-394.970) - (-398.903)] = 3.933\ \text{kJ/mol of Fe}\end{aligned} \tag{10.37}$$

Similarly, the values of $\Delta G^0_{10.19,1273}$ and $\Delta G^0_{10.19,1473}$ for reaction (10.19), can be calculated, are 9.793 kJ and 14.085 kJ, respectively. By increasing the temperatures, the values of the free energy changes (for reaction 10.19) are also increased. Feasibility of reaction is indicated by a decrease in free energy. That means reaction (10.19) cannot proceed forward direction according to thermodynamics calculation. But the reaction is made to move in the forward direction by keeping a large excess of CO gas over CO_2 gas, i.e. by decreasing the activity of CO_2. Hence, this reaction (10.19) is kinetically controlled rather than thermodynamic control.

The equilibrium constant and the gas composition at equilibrium for reaction (10.19) can be calculated at 727 °C (1000 K) as follows:

Since,

$$\Delta G^0_{10.19,1000} = -\mathrm{RT}\ \ln k_{10.19} \tag{10.38}$$

$$\ln k_{10.19} = \left[\frac{-\Delta \mathbf{G}^0_{\mathbf{r},\,1000}}{\mathbf{RT}}\right] = \left[\frac{\mathbf{-3933}}{(\mathbf{8.314 \times 1000})}\right] = -0.473$$

Therefore, $k_{10.19} = 0.6231$

Again

$$k_{10.19} = \left[\frac{(\boldsymbol{a}_{Fe} \times \boldsymbol{a}_{CO_2})}{(\boldsymbol{a}_{FeO} \times \boldsymbol{a}_{CO})}\right] = 0.6231$$

Assuming that the iron and wustite are pure solids, their activities ($\mathbf{a_{Fe}}$ and $\mathbf{a_{FeO}}$) are equal to unity, and the activities of the two gases to be equal to their partial pressures, then

$$k_{10.19} = \left(\frac{\boldsymbol{p}_{CO_2}}{\boldsymbol{p}_{CO}}\right) = 0.6231$$

If it is further assumed that the sum of the partial pressures of the two gases is equal to the total pressure, i.e. one atmosphere, then: $p_{CO_2} + p_{CO} = 1$ atm.

It is now possible to calculate these partial pressures and the gas composition:

$$p_{CO} = 0.6161\,\text{atm and } p_{CO_2} = 0.3839\,\text{atm.}$$

The equilibrium gas composition at 727 °C (1000 K) is 61.61% CO and 38.39% CO_2 by volume. Hence for the reduction of FeO [reaction (10.19)] at 727 °C (1000 K), the gas composition should be consisting of more than 61.61% CO by volume (as shown in Fig. 10.4). If the gas phase contains another gas such as nitrogen, the ratio of CO to CO_2 will remain the same (i.e. 1.6), but their percentage composition will decrease by an amount equal to the percentage of the nitrogen gas.

To reduce wustite, the temperature and gas composition point must be lie in the area where iron is the stable phase (in Fig. 10.4), for example, at 800 °C, gas composition must be consisting of 80% CO and 20% CO_2. Changes in temperature will also affect equilibria and the direction in which the reactions will proceed.

Figure 10.4 also contains the curve showing the CO–CO_2 composition and temperature equilibrium for the reaction:

$$\{CO_2\} + C(s) = 2\{CO\}, \quad \Delta H^0_{298} = 170.7\ \text{kJ/mol of C} \tag{10.39}$$

This is commonly known as the *Boudouard reaction* or *gasification reaction*. This reaction has very important consequences in iron oxide reduction when carbon is used as the reductant, because it regenerated the reducing gas by converting CO_2 to CO. It is strongly endothermic, and since the

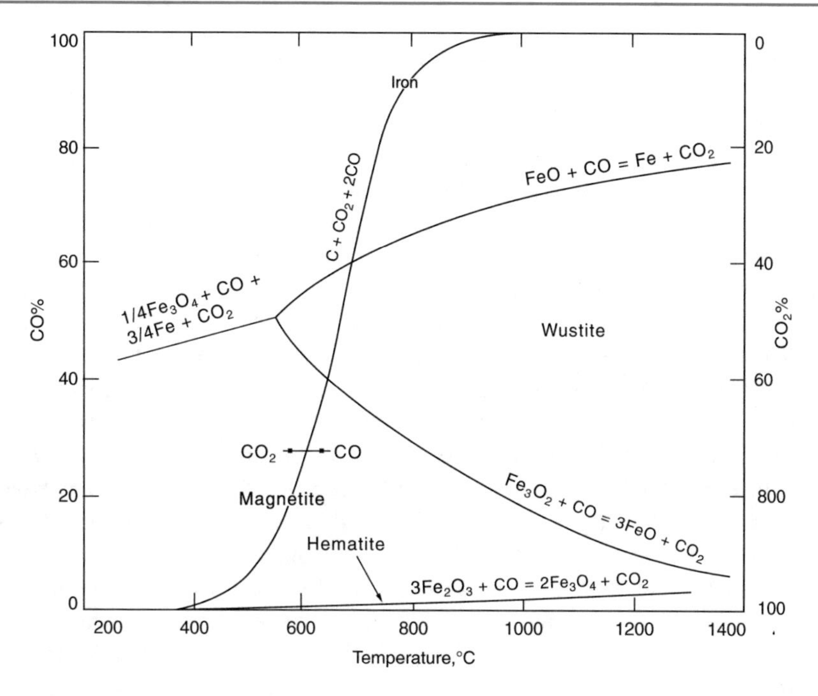

Fig. 10.4 Fe–O–C system and S-curve

activation energy of an endothermic reaction must be larger than the heat of reaction, the rate of reaction of Eq. (10.39) is sensitive to crystalline structure and impurities of carbon and temperature. To the left of this curve in Fig. 10.4, which is at low temperature, carbon monoxide is an unstable gas. Carbon monoxide decomposes into carbon dioxide and deposits carbon in the form of soot [for reverse reaction of Eq. (10.39)]. This soot accounts for some of the carbon found in sponge iron. At high temperatures (above 1000 °C), carbon dioxide is an unstable gas, so it reacts with carbon to produce carbon monoxide. Again, this is an endothermic reaction and is favoured by high temperature.

Note that at temperature above 1000 °C, the reaction is essentially completed, i.e. 100% CO is formed. This means that thermodynamically, carbon dioxide cannot stable at temperature above 1000 °C in presence of carbon. Therefore, this suggests that CO cannot reduce FeO [as reaction (10.19)] at temperature above 1000 °C. However, reduction does take place, and this appears to be on account of reactions (10.19) and (10.39) taking place separately. The two reactions can be combined as follows to indicate the overall reaction:

$$\mathrm{FeO(s)} + \{\mathrm{CO}\} = \mathrm{Fe(s)} + \{\mathrm{CO_2}\}, \quad \Delta H^0_{298} = -18.54\ \mathrm{kJ/mol\ of\ CO} \tag{10.19}$$

$$\{\mathrm{CO_2}\} + \mathrm{C(s)} = 2\{\mathrm{CO}\}, \quad \Delta H^0_{298} = 85.35\ \mathrm{kJ/mol\ of\ CO} \tag{10.39}$$

$$\mathrm{FeO(s)} + \mathrm{C(s)} = \mathrm{Fe(s)} + \{\mathrm{CO}\}, \quad \Delta H^0_{298} = 66.81\ \mathrm{kJ/mol\ of\ CO} \tag{10.40}$$

Reaction (10.40) is often referred to as *direct reduction of iron oxide by carbon,* whereas reaction (10.19) and reaction (10.39) are known as *indirect reduction of iron oxide by carbon* and *gasification reaction of carbon,* respectively.

Note that the Boudouard equilibrium curve in Fig. 10.4 crosses the wustite–iron line at 700 °C and the magnetite–wustite line at 650 °C. This means that thermodynamically [4] wustite cannot be reduced at temperature below 700 °C and magnetite cannot be reduced at below 650 °C because the carbon monoxide decomposes into carbon dioxide and carbon.

In the case of reaction (10.17) (i.e. $3Fe_2O_3(s) + \{CO\} = 2Fe_3O_4(s) + \{CO_2\}$), the equilibrium gas compositions are 0.0029%, 0.0068% and 0.01% CO for temperature 727 °C, 1000 °C and 1200 °C, respectively. These variations are too small so that the curve for reaction (10.17) appears to very close to the temperature axis. In the areas between the curves, one of the solid phases like magnetite, wustite or iron is stable. As for example, if the temperature is 900 °C and gas phase contains 10% CO and 90% CO_2, area magnetite is the stable phase. This means that gas composition reaction (10.17) will proceed to the forward direction at 900 °C, i.e. haematite will have reduced to magnetite, but reaction (10.18) (i.e. $Fe_3O_4(s) + \{CO\} = 3FeO(s) + \{CO_2\}$) will proceed to the backward direction, i.e. wustite will be oxidized to magnetite. Similarly, at 900 °C and gas composition is 40% CO and 60% CO_2, this point lies in the area where wustite is the stable phase. This means that gas composition reaction (10.18) will proceed to the forward direction at 900 °C, i.e. magnetite will be reduced to wustite, but reaction (10.19) [i.e. $FeO\,(s) + \{CO\} = Fe(s) + \{CO_2\}$] will proceed to the backward direction, i.e. iron, if present, will be oxidized to wustite. For wustite reduction, the temperature and gas composition must be lie in the area where iron is the stable phase, i.e. at 900 °C and gas composition is 80% CO and 20% CO_2.

10.3.2 Reduction by CO and H_2 Mixtures

Reducing gases (CO and H_2) produce by reforming natural gas is widely used for the direct reduction process. Reforming reactions are shown in Sect. 6.4.1. Considering the reactions proceed simultaneously in the same system at 927 °C as follows:

$$FeO(s) + \{CO\} = Fe(s) + \{CO_2\}, \quad \Delta H^0_{298} = -18.54\ \text{kJ/mol of CO} \tag{10.19}$$

$$FeO(s) + \{H_2\} = Fe(s) + \{H_2O\}, \quad \Delta H^0_{298} = -21.42\ \text{kJ/mol of } H_2 \tag{10.29}$$

In the reduction of Fe_3O_4 to FeO and further to Fe with solid carbon, a minimum reaction rate has been observed before the nucleation of iron. The time required for the nucleation of metallic iron from FeO is largely dependent on the amount of initial oxygen present in the oxide. The quantity of oxygen removed in the nucleation period increases with increasing initial oxygen content. Iron nucleated on FeO grows linearly with time for reduction with both hydrogen and carbon monoxide. However, the growth rate of iron on FeO, when hydrogen is used, is approximately forty times higher than that at the same partial pressure of carbon monoxide at 800 °C. This is because iron forms a thin and dense layer during hydrogen reduction, while in the case of carbon monoxide, the layer is thick and porous [2]. This is the advantage in the case of gas-based DR processes where both carbon monoxide and hydrogen gases are present.

The retarding effect on nucleation occurs only at the FeO–Fe step and not during the formation stage of Fe_3O_4 and FeO from Fe_2O_3 and Fe_3O_4. Because both Fe_3O_4 and FeO have a cubic structure and are crystallographically similar. Consequently, the transformation in this stage occurs without any incubation period. However, during the reduction of Fe_2O_3 to Fe_3O_4, the lattice transforms from rhombohedral to cubic; therefore, the reduction behaviour between Fe_2O_3 and Fe_3O_4 is different. The nucleation and growth processes of iron oxide are very important because they influence the structure of the reduced phase, which in turn, affects the subsequent reduction rate.

10.4 Reaction in BF

10.4.1 Gas Concentration Within Stack of BF

Air blast enters the furnace at pre-heat temperature of 500–800 °C and the oxygen potential is atmospheric ($p_{O_2} = 0.21$ atmosphere). These conditions set the starting point A (in Fig. 10.5) of the composition changes which are marked on the diagram by the heavy line.

Immediately in front of tuyeres, the oxygen of the blast reacts with the coke; the temperature rises rapidly to about 1900 °C and the oxygen potential is lowered towards the C–CO line, which is the lowest attainable potential under the prescribed conditions. Before this limit is reached, the gas will probably come into contact and react with unreduced iron oxides; this reaction raises the oxygen potential of the gas towards the appropriate iron–oxygen line. Hence, the upper limit is fixed by the Fe–FeO line.

Alternate contact with ore and coke particles causes the gas composition to cycle between these limits as the gas rises the furnace and its temperature falls by thermal energy transfer to the burden. Point B is where gas discharged from the furnace at about 250 °C.

Limestone decomposes in the stack region of the furnace:

$$\begin{aligned} CaCO_3(s) &= CaO(s) + \{CO_2\}, \quad \Delta H^0_{298} = 179 \text{ kJ/mol} \\ \Delta G^0_{10.41} &= 168{,}500 - 144T \text{ J/mol} \end{aligned} \tag{10.41}$$

At temperature 850 °C at lower part of stack, p_{CO_2} in the gas in the range of between 0.3 and 0.4 atmospheres.

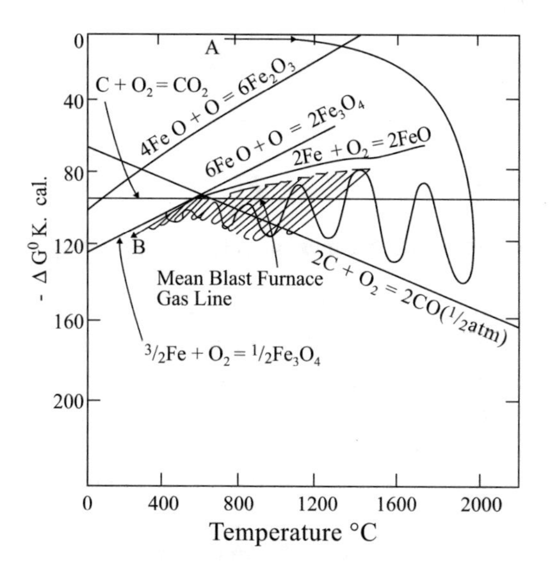

Fig. 10.5 Changes in the composition of gas during its ascent through the blast furnace [13]

The equilibrium constant,

$$k_{10.41} = \left(\frac{a_{CaO}\,p_{CO_2}}{a_{CaCO_3}}\right) = p_{CO_2} \tag{10.42}$$

Since a_{CaO} and a_{CaCO_3} are one for solid and pure state.
At temperature 850 °C i.e.1123 K:

$$\Delta G^0_{10.41} = 168,500 - 144 \times 1123 = -\text{RT}\ \ln k_{10.41}$$

So $\ln k_{10.41} = -0.727$ and $k_{10.41} = 0.48 = p_{CO_2}$ in equilibrium.

Since p_{CO_2} (actual) < p_{CO_2} (equilibrium), limestone should completely decompose at this level. But this does not happen because of kinetic limitation.

Decomposition of limestone is endothermic reaction, so it requires high heat supply, so significant decomposition occurs only at 1000–1100 °C. During decomposition, a layer of porous CaO forms on the outer layer of limestone. This layer has very poor thermal conductivity and consequently slows down heat transfer into the interior of the limestone lump and affects the rate of decomposition.

10.4.2 Raceway Zone

Coke is the only material that comes down into this zone in the form of solid. It gets pre-heated to 1500 °C by the time it reaches the tuyere zone. This highly pre-heated coke burns, in front of tuyere, with the help of pre-heated blast at 800–1250 °C; temperature increases to 1900–2000 °C. This is the main heat source and a source of reducing gas.

The main reaction in the tuyere zone is the combustion of coke:

$$\text{C(s)} + \{\text{O}_2\} = \{\text{CO}_2\}, \quad \Delta H^0_r = -393\ \text{kJ/mol of C} \tag{10.43}$$

Presence of excess of coke reaction:

$$\{\text{CO}_2\} + \text{C(s)} = 2\{\text{CO}\}, \quad \Delta H^0_r = 172.4\ \text{KJ/mol of C} \tag{10.39}$$

Overall reaction:

$$2\text{C(s)} + \{\text{O}_2\} = 2\{\text{CO}\}, \quad \Delta H^0_r = -110.3\ \text{kJ/mol of C} \tag{10.44}$$

A certain proportion of the coke is consumed in the stack, i.e. lumpy zone in solution loss reaction (10.39).

20–40% heat requirement of the process is met by the sensible heat in the blast. It may be noted that the combustion of C to CO_2 [reaction (10.43)] releases much more heat than the conversion to CO [reaction (10.44)]. Hence from the point of view of thermal efficiency, the formation of CO_2 to the maximum possible extent is preferable.

From reaction (10.43):

$$\begin{aligned}\Delta G^0_{10.43} &= -\text{RT}\ \ln k_{10.43} = -\text{RT}\ \ln\left(\frac{p_{CO_2}}{p_{O_2}}\right)\\ &= -394{,}100 - 0.84T\ \text{J/mol}\end{aligned} \tag{10.45}$$

At tuyere temperature, 1900 °C (i.e. 2173 K):

$$\ln k_{10.43} = \{(394,100 + 0.84 \times 2173)/(8.314 \times 2173)\} = 21.915$$

Therefore, $k_{10.43} = 3.29 \times 10^9$
Similarly, for reaction (10.44) at 1900 °C (i.e. 2173 K):

$$\begin{aligned}\Delta G^0_{10.44} &= -\mathrm{RT}\ \ln k_{10.44} = -\mathrm{RT}\ \ln\left(\frac{p_{\mathrm{CO}}}{p_{\mathrm{O_2}}^{\frac{1}{2}}}\right) \\ &= -111{,}700 - 87.65T\ \mathrm{J/mol}\end{aligned} \quad (10.46)$$

Therefore, $\ln k_{10.44} = 16.725$ and $k_{10.44} = 1.83 \times 10^7$.

The values of $k_{10.43}$ and $k_{10.44}$ are very large; therefore, at equilibrium, from Eqs. (10.45) and (10.46), p_{O_2} is negligible. In other words, the oxygen is almost completely converted into CO and CO_2 at the tuyere level.

Reaction (10.39) is a famous *Boudouard reaction*. In BF ironmaking, the forward reaction (10.39) is known as the solution loss reaction. In general, it is also referred to as the *gasification reaction*. As the temperature increases, CO becomes more and more stable compared with CO_2. At tuyere, where temperature is very high, CO is the only stable oxide of carbon.

The composition of gas as it enters the bottom of stack can be calculated assuming that air contains 21% O_2 and 79% N_2; the entire amount of O_2 gets converted into CO. For 100 mols of air, the number of mols of N_2 and CO would be 79 and 42 mols; after the reaction [2C (s) + $\{O_2\}$ = 2{CO}].

$$\text{Therefore, \% volume of CO} = \left(\frac{42}{121}\right) \times 100 = 34.7\%$$

$$\text{\% volume of N}_2 = \left(\frac{79}{121}\right) \times 100 = 65.3\%$$

There is always some moisture in the blast; it is more in rainy season and less in summer. Moisture reacts with hot coke in the tuyere area as:

$$\{\mathrm{H_2O}\} + \mathrm{C(s)} = \{\mathrm{CO}\} + \{\mathrm{H_2}\}, \quad \Delta H_r^0 = 176.5\,\mathrm{kJ/mol\ of\ H_2O} \quad (10.47)$$

It [reaction (10.47)] is an endothermic reaction but generates two reducing gases.

In tuyere, (i) injection of stream to control the RAFT, (ii) injection of hydrocarbon, in form of natural gas or oil, and (iii) injection of pulverized coal, coal decomposes hydrocarbon. All these are the sources of hydrogen in BF.

10.4.3 Bosh and Hearth

Composition of hot metal production in BF: 3.5–4.0% C, 0.2–1.25% Si, 0.2–1.0% Mn, 0.02–0.05% S and 0.1–0.4% P. Composition of hot metal differs from country to country and even from region to region in the same country. As far as control of hot metal composition is concerned, the following need to be noted:

1. Some amount of Mn is desirable in HM,
2. Control of C, P is not possible,
3. S and Si should be controlled. Si should be maintained below 0.6%, if possible. As far as S is concerned, it should be as low as possible. When productivity increases, S content can increase due to lack of time for proper de-sulphurization. Hence, external de-sulphurization is preferred.

The formation of a slag of desirable properties is of considerable important to:

- Control HM composition,
- Obtain sufficiently fluid slag at as low a temperature as possible,
- Make the slag suitable for use in cementmaking.

10.4.3.1 Silicon Reaction

Coke burns in the BF raceway in front of tuyeres, thus releasing coke ash which contains SiO_2. Some molten slag containing high SiO_2 and FeO also drips through the raceway. The temperature of raceway zone is about 1900–2000 °C. In the presence of carbon and depending on the activity of Si, the following reaction occur:

$$SiO_2(\text{coke ash}) + C(s) = \{SiO\} + \{CO\} \tag{10.48}$$

SiO_2 initially forms SiC, which then forms SiO gas:

$$SiO_2(s) + 3C(s) = SiC(s) + 2\{CO\} \tag{10.49}$$

$$SiC(s) + \{CO\} = \{SiO\} + 2C(s) \tag{10.50}$$

At 1500 °C, p_{SiO} at equilibrium with reaction (10.48) is approximately 10^{-4} atmosphere. This makes SiO vapour unstable when it rises upwards in the bosh region. Therefore, in this region, Si gets transferred into liquid iron by the reaction of SiO with dissolved C in liquid iron as follows:

$$\{SiO\} + [C] = [Si] + \{CO\} \tag{10.51}$$

$$\text{Or by decomposition as:}\quad 2\{SiO\} = (SiO_2) + [Si] \tag{10.52}$$

$$\text{Other reaction:}\quad (FeO) + \{SiO\} = [Fe] + (SiO_2) \tag{10.53}$$

The product (SiO_2) joins the slag phase.

Here, the parentheses denote the liquid slag phase, the square brackets denote the liquid metal phase and curly brackets denote the gas phase.

Molten metal droplets react with the slag in hearth while passing through the slag layer.

For Si, the slag-metal reaction at hearth:

$$(SiO_2) + 2[C] = 2[Si] + 2\{CO\} \tag{10.54}$$

For which, equilibrium constant,

$$k_{10.54} = \left(\frac{[a_{Si}^2]p_{CO}^2}{(a_{SiO_2})[a_C^2]}\right) = \left(\frac{[wt\%\ Si^2]p_{CO}^2}{(a_{SiO_2})}\right) \tag{10.55}$$

Since

$$a_C = 1, \text{ and } \log k_{10.54} = -\left(\frac{30{,}935}{T}\right) + 20.455 \tag{10.56}$$

Si–MnO reaction, which is also a slag-metal reaction at hearth:

$$2(MnO) + [Si] = 2[Mn] + (SiO_2) \tag{10.57}$$

Now, question arises whether reaction of Si–SiO_2 with C (Eq. 10.54) or with Mn (Eq. 10.57) attains equilibrium in the BF hearth. By experiment: (i) it is found that equilibrium is not attained; (ii) actual Si content of hot metal is a few times higher than that predicted by the slag-metal equilibrium.

Hot metal with a low Si content is desirable for efficient steelmaking. This can be achieving the same on the mechanism of Si reaction, steps are as follows:

1. Decreasing the extent of SiO formation by:
   - Lowering ash content in coke, as well as decreasing coke rate,
   - Lowering the value of RAFT,
   - Lowering the activity of SiO_2 in coke ash by lime injection through tuyeres.
2. Decreasing Si absorption by liquid iron in the bosh by enhancing the absorption of SiO_2 by the bosh slag. This can be achieved by:
   - Increasing bosh slag basicity,
   - Lowering the bosh slag viscosity by operating at lower basicity.
3. Removal of Si from metal by slag-metal reaction at the hearth by:
   - Lowering the hearth temperature,
   - Producing a slag of optimum basicity and fluidity.

10.4.3.2 Sulphur Reaction

80% of sulphur enters the BF through coke as CaS and FeS in coke ash. Remaining comes through the other burden materials. In BF: (i) 80–85% S input leaves BF with the slag, (ii) 10–15% S leaves BF to flue dust and top gas, and (iii) 2–5% leaves BF as dissolved in hot metal.

Earlier in steel grade, hot metal contains 0.04% S. Due to Concast, upper limit comes down to 0.025% in hot metal. Earlier all S removal had to be completed within the BF. Current practices are (i) external de-sulphurization of hot metal and (ii) de-sulphurization during secondary steelmaking.

Figure 10.6 shows the variation of weight percent of metal sulphur and slag sulphur at various heights from the tuyeres. The behaviour of sulphur in qualitatively similar to the Si, i.e. the highest S in metal is at the tuyere level.

Sulphur in coke ash undergoes the following reactions in the raceway:

$$CaS\ (\text{in coke ash}) + \{SiO\} = \{SiS\} + (CaO) \tag{10.58}$$

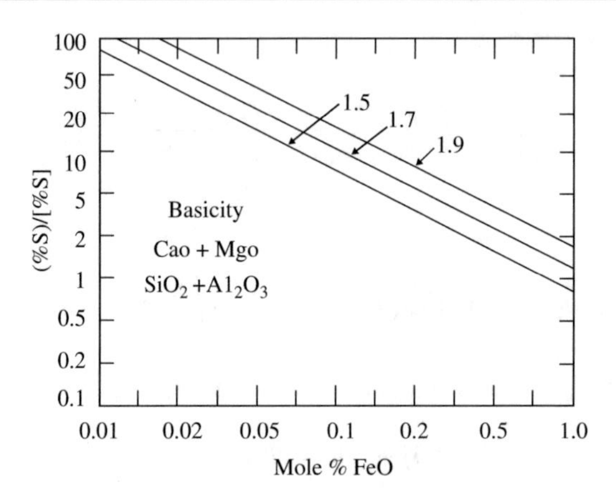

Fig. 10.6 Influence of basicity and iron oxide content of slag on a partition of sulphur [12]

$$\text{FeS (in coke ash)} + \{\text{SiO}\} + \text{C(s)} = \{\text{SiS}\} + \{\text{CO}\} + [\text{Fe}] \tag{10.59}$$

In the bosh and belly regions, $\{SIS\}$ decomposes as:

$$\{\text{SiS}\} = [\text{Si}] + [\text{S}] \tag{10.60}$$

Small amount of S is also absorbed by the bosh slag.

Sulphur also forms other volatile compounds, e.g. COS, CS which are also carried up into the gas stream.

In the hearth, the slag-metal reaction may be written in ionic form:

$$\left(\text{O}^{2-}\right) - 2\text{e} = [\text{O}] \tag{10.61}$$

$$[\text{S}] + 2\text{e} = \left(\text{S}^{2-}\right) \tag{10.62}$$

$$\text{Overall reaction: } [\text{S}] + \left(\text{O}^{2-}\right) = \left(\text{S}^{2-}\right) + [\text{O}] \tag{10.63}$$

Assuming Henrian behaviour for all species except (O^{2-}):

$$k_{10.63} = \left(\frac{\{(\%\text{S})[\%\text{O}]\}}{\{[\%\text{S}](a_{\text{O}^{2-}})\}}\right) \tag{10.64}$$

Equilibrium partition coefficient for sulphur:

$$(L_\text{S})_\text{eq} = \left(\frac{(\%\text{S})}{[\%\text{S}]}\right)_\text{eq} = k_{10.63}\left(\frac{(a_{\text{O}^{2-}})}{[\%\text{O}]}\right) \tag{10.65}$$

$$\text{Or} \quad \left(\frac{(\%S)}{[\%S]}\right) \alpha \left(\frac{(a_{O^{2-}})}{[\%O]}\right) \tag{10.66}$$

Now,

$$[Fe] + [O] = (FeO) \tag{10.67}$$

Since $a_{Fe} = 1$, therefore, $k_{10.67} = \left(\frac{(\%FeO)}{[\%O]}\right)$

$$\text{or} \quad [\%O] = \left(\frac{(\%FeO)}{k_{10.67}}\right) \tag{10.68}$$

By combining Eqs. (10.65) and (10.68):

$$(Ls)_{eq} = k_{10.63} \cdot k_{10.67} \cdot \left(\frac{(a_{O^{2-}})}{(\%FeO)}\right) \tag{10.69}$$

Therefore,

$$(Ls)_{eq} \; \alpha \; (a_{O^{2-}}) \tag{10.70}$$

i.e. equilibrium partition coefficient for S is directly proportional to basicity of slag.
and,

$$(Ls)_{eq} \; \alpha \; \left(\frac{1}{(\%FeO)}\right) \tag{10.71}$$

i.e. equilibrium partition coefficient for sulphur is inversely proportional to (%FeO).

$(a_{O^{2-}})$ increases with increasing free O^{2-} ion concentration in the slag, which again increases with slag basicity. Therefore, equilibrium partition coefficient for sulphur will increase with increasing basicity and will decrease with increasing FeO content of slag (as shown in Fig. 10.6).

CaO is a much more powerful base and de-sulphurizer than MgO (about 100 times stronger). For adequate de-sulphurization, the oxygen content of the metal should be very low. This is achieved in the BF hearth by the reaction of dissolved oxygen in hot metal with strong oxide formers: C, Si, Mn dissolved in liquid metal.

$$(CaO) + [S] + [C] = (CaS) + \{CO\} \tag{10.72}$$

$$(CaO) + [S] + [Mn]/1/2[Si] = (CaS) + (MnO)/1/2(SiO_2) \tag{10.73}$$

Based on reaction (10.72), an empirical correlation has been proposed [5]:

$$\text{Log}(L_S)_{eq} = 1.35\left(\frac{1.79(\%CaO) + 1.24(\%MgO)}{1.66(\%SiO_2) + 0.33(Al_2O_3)}\right) - \log p_{CO} - \left(\frac{8130}{T}\right) + 4.15 \tag{10.74}$$

It appears that reaction (10.73) also attains equilibrium in the blast furnace hearth. Based on the above, the strategy for making low sulphur hot metal is the same as that for silicon, except that higher temperatures promote better de-sulphurization.

10.4.3.3 Manganese Reaction

Whereas the lowest contents of silicon and sulphur in hot metal are desired, it is necessary to maximum manganese recovery in hot metal. Manganese is an input into BF through iron ore and sometimes through the deliberate addition of manganese ore (MnO_2).

$MnO_2 \rightarrow Mn_3O_4 \rightarrow MnO$ in BF stack by indirect reduction. MnO is more stable than FeO, but less stable than SiO_2. MnO is reduced by C in coke at about 1400 °C:

$$(MnO) + C(s) = [Mn] + \{CO\} \tag{10.75}$$

This reaction is possible only above 1400 °C; therefore, MnO reduction by C occurs in the hearth region when liquid slag containing MnO flows down through the coke bed or during reaction in the hearth between metal and slag:

$$2(MnO) + [Si] = 2[Mn] + (SiO_2) \tag{10.57}$$

The equilibrium constant,

$$k_{10.57} = \left(\frac{[a^2_{Mn}](a_{SiO_2})}{(a^2_{MnO})[a_{Si}]}\right) \tag{10.76}$$

Manganese content in hot metal is 0.2–1.5%; recovery of manganese in the hearth can be increased by increasing (i) a_{MnO} in the slag by using higher basicity, (ii) metal temperature and (iii) Si content in hot metal.

For basic slag [$\left(\frac{CaO}{SiO_2}\right) > 1$]: recovery of Mn is 60–70% and for acid slag [$\left(\frac{CaO}{SiO_2}\right) < 1$]: recovery of Mn is 50–60%.

10.5 Carbon Deposition on Sponge Iron

As stated earlier that carbon is formed at low temperatures by the backward reaction of reaction (10.39) [i.e. 2{CO} = C (s) + {CO_2}]. Although thermodynamics calculation shows that carbon deposition is highly favoured by low temperatures and high carbon monoxides concentrations, it is unlikely that this reaction ever goes completely to a state of equilibrium because reaction rates are very slow at low temperatures and because the deposition of carbon requires the nucleation of carbon on some catalytic site which may not always be available. Undoubtedly, some of the carbon found in sponge iron is soot formed by this reaction. In addition, iron carbide (Fe_3C) can be formed by the reaction [4]:

$$3Fe(s) + 2\{CO\} = Fe_3C(s) + \{CO_2\} \tag{10.77}$$

Like carbon deposition, reaction (10.77) is also favoured by low temperatures and high carbon monoxides concentrations.

The carbon present in the gas-based sponge iron depends on ($CO + H_2$)/($CO_2 + H_2O$) ratio. The higher the ratio, the more the carbon is deposited from CO and CH_4 decomposition within the pores of sponge iron.

$$2\{CO\} = C(s) + \{CO_2\} \tag{10.78}$$

$$\{CH_4\} = C(s) + 2\{H_2\} \tag{10.79}$$

10.6 Mechanism of Smelting Reduction of Iron Oxides

When iron ore–coal/char/coke composite pellets are used as charging material in smelting reduction process. Then kinetically reduction of oxides in composite pellet is significantly by carbon in a mixture of iron oxide and carbon. If char or coke is used as reductant in composite pellet, then this is wholly carbothermic reduction. If, on the other hand, coal is used, then reduction will be partly by gases evolved (due to pyrolysis of coal) especially hydrogen and carbon monoxide [6, 7]. Now, it has been universally accepted that carbothermic reduction takes place via the reactions (10.17)–(10.19), i.e. through gaseous intermediates. The reduction kinetics is expected to be enhanced due to the presence of reductant in situ which causes shortening of diffusion distances of reductant and availability of a large number of reactions sites [8]. Diffusion of reducing gases into the lump, pellet depends on the porosity. Reduction reaction in composite pellets is controlled by gasification reaction and mass transfer, which involve both chemical reaction and pore diffusion. The reducing gases diffuse through the micropores or cracks of the product layer to reach reaction interface. The overall rate is sometimes affected by the rate of heat transfer as well since it is endothermic in nature [9].

Beyond solid-state reduction of haematite/magnetite to wustite, smelting reduction essentially involves reduction of molten FeO by CO. The partially reduced ore is injected or fed into the smelting reactor containing iron–carbon melt and slag [10].

The overall reaction is:

$$(FeO) + C(s) \text{ or } [C] \rightarrow [Fe] + \{CO\} \tag{10.80}$$

There is a general agreement that the above reaction takes place in two stages:

First stage; slag–gas reaction:

$$(FeO) + \{CO\} \rightarrow [Fe] + \{CO_2\} \tag{10.19}$$

Second stage; carbon–gasification reaction:

$$C(s) + \{CO_2\} \rightarrow 2\{CO\} \tag{10.39}$$

The following four possible rate-limiting steps may be visualized for this reaction (10.80) between FeO in molten slag and solid carbon [11]:

(i) Mass transfer of FeO (i.e. Fe^{2+} and O^{2-} ions) from slag phase to slag–gas interface,
(ii) Chemical reaction at slag–gas interface,
(iii) Gas diffusion in the gas layer separating slag and solid carbon, and
(iv) CO_2—carbon reaction (chemical reaction at carbon–gas or metal–gas interface).

The liquid phase gives rise to far higher transport rates owing to convection, and a remarkable increase in the conversion rate because of enlargement in the specific phase contact area. The latter is a direct consequence of the dispersed nature of the phases. These two major advantages of SR processes accrue because of the formation of liquid phases.

From studies on the oxidation and reduction behaviour of pure molten iron oxide by CO/CO_2 mixtures at 1500 °C, it has been concluded that the rate controlling step for both oxidation and reduction is the interdiffusion of iron and oxygen atoms within the melt. It has also been found that mass transport plays an important role in the reduction of kinetics. The overall reaction rate is proportional to the square root of the gas flow rate. Therefore, all our efforts are made to increase the amount of gas that is available for reduction.

To generate a sufficient amount of reducing gas, all SR processes consume large amount of reductant (normally coal). Having generated the large volume of gas that is required, its effective utilization becomes extremely important. However, this gives rise to one of the inherent deficiencies of SR, i.e. with most SR reactor configurations given the productivity requirements, the entire gas cannot be fully utilized in the process, and rich gas at a high temperature leaves the SR reactor. For this reason, the use of the export gas in any SR process has a marked influence on the cost of the hot metal produced. Without enough credit for the off-gas, the cost of hot metal produced by smelting reduction can be as much as 40–50% higher than that of blast furnace hot metal, despite starting with less expensive and inferior grade raw materials. The reverse is the case if adequate credit can be obtained from the utilization of the off-gases. The most convenient way to utilize the exit gas is to generate electrical power [12]. Alternatively, the exit gas can be fed to a shaft furnace direct reduction unit, located adjacent to the smelting reduction reactor.

Heat loss in a two-stage SR process occurs because the smelter off-gas must be cooled from around 1600 °C to 800 °C, representing an energy loss of 1.8 GJ per tonne of hot metal, before it can be used for reduction. However, if carbon is present in the gas reaction zone, it can help in reducing the temperature of the smelter off-gases, without loss of energy because of the endothermic reactions (10.39) and (10.47). The thermodynamics of the carbon–oxygen system is such that below 1000 °C, the equilibrium gas mixture in contact with carbon contains increasing quantities of CO_2 with decreasing temperature [12]; it contains 5% and 15% CO_2 at 900 °C and 800 °C, respectively.

Probable Questions

1. Discuss the thermodynamics of metal oxide reduction. What is the condition of the reduction reaction of the oxide with respect to CO/CO_2 ratio?
2. State and discuss the probable mechanism by which an iron ore is reduced in a $CO–CO_2$ mixture.
3. Draw the diagram showing equilibrium CO/CO_2 ratio in contact with carbon and iron oxides at various temperatures. What is the diagram known as? Below 570 °C, FeO is not formed during reduction of Fe_3O_4. Why? Discuss the significant of the diagram for reduction of iron ore.
4. With the help of Fe–H–O system, find out the conditions of reduction of oxide and oxidation of metal? Discuss the importance of this system.
5. Discuss the changes in the composition of gas in stack region of the BF.
6. Discuss the silicon equilibrium in BF. Explain why at high blast temperature silicon content of metal increases.
7. What are the thermodynamics conditions required for de-sulphurization of iron in BF? Discuss.
8. What are the conditions required for maximum recovery of manganese in metal of a BF?

Examples

Example 10.1 Find out the heat of reaction ($\Delta H^0_{r,298}$) of the following reactions:

(i) $$3Fe_2O_3 + CO = 2Fe_3O_4 + CO_2$$

(ii) $$Fe_3O_4 + CO = 3FeO + CO_2$$

(iii) $$FeO + CO = Fe + CO_2$$

Given: $\Delta H^0_{f,298}$ for Fe_2O_3, Fe_3O_4, FeO, CO and CO_2 are—821.32, −1116.71, −264.43, −110.54 and −393.51 kJ/mol, respectively.

Solution

(i) $$3Fe_2O_3 + CO = 2Fe_3O_4 + CO_2$$

Therefore,

$$\begin{aligned}\Delta H^0_{r,298} &= \left(2\Delta H^0_{Fe_3O_4} + \Delta H^0_{CO_2}\right) - \left(3\Delta H^0_{Fe_2O_3} + \Delta H^0_{CO}\right)\\ &= [2(-1116.71) + (-393.51)] - [3(-821.32) + (-110.54)]\\ &= [-2233.42 - 393.51] - [-2463.96 - 110.54]\\ &= [-2626.93 - (-2574.5)] = \mathbf{-52.43\ kJ/mol\ of\ CO/CO_2}\\ &= \mathbf{-17.48\ kJ/mol\ of\ Fe_2O_3}\\ &= \mathbf{-26.22\ kJ/mol\ of\ Fe_3O_4}\end{aligned}$$

(ii) $$Fe_3O_4 + CO = 3FeO + CO_2$$

Therefore,

$$\begin{aligned}\Delta H^0_{r,298} &= \left(3\Delta H^0_{FeO} + \Delta H^0_{CO2}\right) - \left(\Delta H^0_{Fe3O_4} + \Delta H^0_{CO}\right)\\ &= [3(-264.43) + (-393.51)] - [-1116.71 + (-110.54)]\\ &= [-7932.9 - 393.51] - [-1116.71 - 110.54)]\\ &= [-1186.8 - (-1227.25)] = \mathbf{40.45\ kJ/mol\ of\ Fe_3O_4/CO/CO_2}\\ &= \mathbf{13.48\ kJ/mol\ of\ FeO}\end{aligned}$$

(iii) $$FeO + CO = Fe + CO_2$$

Therefore,

$$\begin{aligned}\Delta H^0_{r,298} &= \left(\Delta H^0_{Fe} + \Delta H^0_{CO_2}\right) - \left(\Delta H^0_{FeO} + \Delta H^0_{CO}\right)\\ &= [0 + (-393.51)] - [-264.43 + (-110.54)]\\ &= -393.51 - [-264.43 - 110.54] = [-393.51 - (-374.97))\\ &= \mathbf{-18.54\ kJ/mol\ of\ FeO/Fe/CO/CO_2}\end{aligned}$$

Example 10.2 Reduction of iron oxide in blast furnace takes place in three stages:

$$Fe_2O_3 \rightarrow Fe_3O_4 \rightarrow FeO \rightarrow Fe.$$

The reactions are as follows:

(i) $3Fe_2O_3 + CO = 2Fe_3O_4 + CO_2, \Delta G^0_1 = -32969.92 - 53.85T$ J
(ii) $Fe_3O_4 + CO = 3FeO + CO_2, \Delta G^0_2 = 29790.08 - 38.28T$ J
(iii) $FeO + CO = Fe + CO_2, \Delta G^0_3 = -22802.8 + 24.27T$ J
(iv) $Fe_3O_4 + 4CO = 3Fe + 4CO_2, \Delta G^0_4 = -38618.32 + 34.52T$ J

Find out the percentage of CO in CO–CO_2 mixture in equilibrium with

(a) $Fe_2O_3-Fe_3O_4$, (b) Fe_3O_4-FeO, (c) $FeO-Fe$ at 1000 K.
(d) Fe_3O_4-Fe at 800 K.
(e) Find out temperature at which Fe, FeO and Fe_3O_4 exist in equilibrium.

Solution

(a) ***Equilibrium constant***, $k_1 = \left[\frac{a^2_{Fe_3O_4} a_{CO_2}}{a^3_{Fe_2O_3} a_{CO}}\right] = \left(\frac{p_{CO_2}}{p_{CO}}\right)$

At Std state, $a_{Fe_3O_4} = 1$ and $a_{Fe_2O_3} = 1$
Since $\Delta G^0_1 = -32969.92 - 53.85T\ J = -32969.92 - 53.85 \times 1000 = -86819.92\ J$
Again $\Delta G^0_1 = -RT \ln k_1$
Therefore, $-86819.92 = -8.314 \times 1000 \times \ln k_1$
$\ln k_1 = 10.4426$
$k_1 = 34290.28$
Let $p_{CO} = x$, since $p_{CO} + p_{CO_2} = 1$
Therefore, $p_{CO_2} = 1 - x$
Since $k_1 = \left(\frac{p_{CO_2}}{p_{CO}}\right) = 34290.28 = \left(\frac{1-x}{x}\right)$
Therefore, $x = 2.916 \times 10^{-5}$
Hence, the percentage of CO in CO–CO_2 mixture in equilibrium with $Fe_2O_3 - Fe_3O_4 = \mathbf{2.916 \times 10^{-3}\%}$

Similarly, we can calculate the following:

(b) Equilibrium constant, $k_2 = \left(\frac{p_{CO_2}}{p_{CO}}\right)$

Since $\Delta G_2^0 = 29790.08 - 38.28\,T = 29790.08 - 38.28 \times 1000 = -8490$ J
Again $\Delta G_2^0 = -\text{RT ln } k_2$
Therefore, $-8490 = -8.314 \times 1000 \times \ln k_2$
$\ln k_2 = 1.021$
$k_2 = 2.776$
Let $p_{CO} = x$, since $p_{CO} + p_{CO_2} = 1$
Therefore, $p_{CO_2} = 1 - x$
Since $k_2 = \left(\frac{p_{CO_2}}{p_{CO}}\right) = 2.776 = \left(\frac{1-x}{x}\right)$
Therefore, $x = 0.2648$
Hence, the percentage of CO in CO–CO_2 mixture in equilibrium with Fe_3O_4–FeO = **26.48%**

(c) Equilibrium constant, $k_3 = \left(\frac{p_{CO_2}}{p_{CO}}\right)$

Since $\Delta G_3^0 = -22802.8 + 24.27T = -22802.8 + 24.27 \times 1000 = 1468$ J
Again $\Delta G_3^0 = -\text{RT} \ln k_3$
Therefore, $1468 = -8.314 \times 1000 \times \ln k_3$
$\ln k_3 = -0.1766$
$k_3 = 0.838$
Let $p_{CO} = x$, since $p_{CO} + p_{CO_2} = 1$
Therefore, $p_{CO_2} = 1 - x$
Since $k_3 = \left(\frac{p_{CO_2}}{p_{CO}}\right) = 0.838 = \left(\frac{1-x}{x}\right)$
Therefore, $x = 0.544$
Hence, the percentage of CO in CO–CO_2 mixture in equilibrium with FeO–Fe = **54.4%**

(d) Equilibrium constant, $k_4 = \left(\frac{p_{CO_2}}{p_{CO}}\right)^4$

Since $\Delta G_4^0 = -38618.32 + 34.52T = -38618.32 + 34.52 \times 800 = -11002$ J
Again $\Delta G_4^0 = -\text{RT} \ln k_4$
Therefore, $-11002 = -8.314 \times 800 \times \ln k_3$
$\ln \text{k}_4 = 1.654$
$k_4 = 5.228 = \left(\frac{p_{CO_2}}{p_{CO}}\right)^4$, so $\left(\frac{p_{CO_2}}{p_{CO}}\right) = (5.228)^{1/4} = 1.512$
Let $p_{CO} = x$, since $p_{CO} + p_{CO_2} = 1$
Therefore, $p_{CO_2} = 1 - x$
Since $\left(\frac{p_{CO_2}}{p_{CO}}\right) = 1.512 = \left(\frac{1-x}{x}\right)$
Therefore, $x = 0.3981$
Hence, the percentage of CO in CO–CO_2 mixture in equilibrium with Fe_3O_4–Fe = **39.81%**

(e) $Fe_3O_4 + CO = 3FeO + CO_2, \Delta G_2^0 = 29790.08 - 38.28T$ J

$Fe + CO_2 = FeO + CO, -\Delta G_3^0 = -22802.8 + 24.27T$ J

$Fe_3O_4 + Fe = 4FeO, \Delta G^0 = (\Delta G_2^0 - \Delta G_3^0) = 52592.88 - 62.55T$

At equilibrium $\Delta G^0 = 0$
Therefore, $52592.88 - 62.55T = 0$
Hence, **T = 840.81 K ≈ 567.81 °C**
Therefore, at **567.81 °C** Fe, FeO and Fe_3O_4 will exist in equilibrium.

Example 10.3 Find the oxygen pressure of 75% CO and 25% CO_2 mixture at 1 atmospheric pressure and 1500 K. Will this gas be able to reduce Fe_2O_3 and FeO at 1500 K?

Given: $2CO_2 = 2CO + O_2 \quad \Delta G^0 = 565,258.4 - 173.64T$ J
Oxygen pressure for Fe_2O_3 at 1500 K = 10^{-2} atm,
Oxygen pressure for FeO at 1500 K = 10^{-12} atm.

Solution

$$\Delta G^0 = 565,258.4 - 173.64T = 565,258 - 173.64 \times 1500 = 304,798 \text{ J}$$
$$\Delta G^0 = -\text{RT} \ln k$$

Therefore, $304,798 = -8.314 \times 1500 \times \ln k$
$\ln k = -24.44$
$k = 2.43 \times 10^{-11}$
$2CO_2 = 2CO + O_2$

Equilibrium constant, $k = \left[\frac{p_{CO}^2 p_{O_2}}{p_{CO_2}^2}\right] = \left[\frac{(0.75)^2 p_{O_2}}{(0.25)^2}\right]$

Therefore $p_{O_2} = k.\left(\frac{p_{CO_2}^2}{p_{CO}^2}\right) = 2.43 \times 10^{-11} \times \left[\frac{(0.25)^2}{(0.75)^2}\right] = 2.7 \times 10^{-12}$ atm.

Since oxygen pressure for Fe_2O_3 at 1500 K is 10^{-2} atm which is greater than equilibrium oxygen pressure, 2.7×10^{-12} atm, mixture of gases will reduce the Fe_2O_3. But oxygen pressure for FeO at 1500 K is 10^{-12} atm which is lower than equilibrium oxygen pressure, 2.7×10^{-12} atm, hence mixture of gases will not reduce FeO; instate of that FeO will be oxidized.

Example 10.4 For the reaction: $C + CO_2 = 2CO, \Delta G^0 = 170.71 - 0.17T$ kJ

Find the temperature of which carbon would be in equilibrium with a mixture of CO and CO_2 (at one atm. pressure) containing 70% CO.

Solution

The equilibrium constant of the reaction: $k = \left[\frac{p_{CO}^2}{a_C p_{CO_2}}\right] = \left[\frac{p_{CO}^2}{p_{CO_2}}\right]$

Since a_C = 1, 70% CO, i.e. $p_{CO} = 0.7$ and $p_{CO_2} = 0.3$; Since total pressure = 1 atm.

Therefore, $k = \left(\frac{(0.7)^2}{0.3}\right) = 1.633$

Since

$$\Delta G^0 = -\mathrm{RT}\ \ln k = -8.314T \times \ln(1.633) = -(8.314 \times 0.49)T = -4.08T$$
$$(170.71 - 0.17) \times 10^3 T = -4.08T$$
$$(170 - 4.08)T = 170{,}710$$

Therefore, $T = 1028.9\ K = 755.9 = \mathbf{756\ ^\circ C}$

Example 10.5

(i) $Fe(1) + 1/2\ O_2(g) = FeO(1)\quad \Delta \mathbf{G_i^0} = -232714.08 + 45.31T$ J
ii) $C(s) + 1/2\ O_2(g) = CO(g)\quad \Delta \mathbf{G_{ii}^o} = -111712.8 - 87.65T$ J
iii) $C(s) + O_2(g) = CO_2(g)\quad \Delta \mathbf{G_{iii}^\circ} = -394132.8 - 0.84T$ J

From the above reactions, find out the ΔG_T^0 of the reaction: $FeO(s) + CO(g) = Fe(s) + CO_2(g)$ at 1000 °C.

Solution

Taking reverse equation (i): $FeO(1) = Fe(1) + 1/2O_2(g) - \Delta \mathbf{G_i^0}$
Taking reverse equation (ii): $CO(g) = C(s) + 1/2O_2(g) - \Delta \mathbf{G_{ii}^o}$
Taking equation (iii): $C(s) + O_2(g) = CO_2(g)\quad \Delta \mathbf{G_{ii}^0}$

By adding: $FeO(l) + CO(g) = Fe(1) + CO_2(g)$

$$\begin{aligned}\Delta \mathbf{G_T^o} &= -\Delta \mathbf{G_i^o} - \Delta \mathbf{G_{ii}^o} + \Delta \mathbf{G_{iii}^o} = -(-232714.08 + 45.31T) - (-111712.8 - 87.65T)\\ &\quad + (-394132.8 - 0.84T)\\ &= \{(232714.08 + 111712.8) - 394132.8\} + \{-45.31 - 0.84 + 87.65\}T\\ &= -49705.92 + 41.5T = -49705.92 + 41.5 \times 1273\\ &= 3123.58\,\mathrm{J} = 3.12\,\mathrm{kJ}\end{aligned}$$

Since value of $\Delta \mathbf{G_T^o}$ **is** (+) ve, reaction will not be possible in forward direction, reaction will take place in backward direction.

Problems

Problem 10.1 Find out the heat of reaction ($\Delta H_{r,298}^0$) of the following reactions:

(i) $3Fe_2O_3 + H_2 = 2Fe_3O_4 + H_2O$
(ii) $Fe_3O_4 + H_2 = 3FeO + H_2O$
(iii) $FeO + H_2 = Fe + H_2O$

Given: $\Delta H_{f,298}^0$ for Fe_2O_3, Fe_3O_4, FeO, and H_2O are −821.32, −1116.71, −264.43, and −241.81 kJ/mol respectively.

[Ans: (i) −3.48 kJ/mol of Fe_2O_3, (ii) 81.61 kJ/mol of Fe_3O_4, (iii) 22.61 kJ/mol of FeO.]

Problem 10.2 Find out the heat of reaction ($\Delta H^0_{r,298}$) of the following reactions:

(a) $C + CO_2 = 2CO$

(b) $C + H_2O = H_2 + CO$

Given: $\Delta H^0_{f,298}$ for CO_2, CO, and H_2O are −393.51, −110.54 and −241.81 kJ/mol respectively.
[Ans: (a) 172.42 kJ/mol of C, (b) 131.27 kJ/mol of C.]

Problem 10.3 For the reaction: $CO_2 + C = 2CO$, $\Delta G^0 = 170707.2 - 174.47T$ J

(a) Find the temperature at which graphite will be in equilibrium with CO and CO_2 at 1 atm pressure containing (i) 30% CO, (ii) 60% CO.
(b) Find similar values for a total pressure of 0.1 atm.
(c) Find composition of CO and CO_2 in equilibrium with graphite at 1000 K.
[Ans: (a) (i) 618.31 °C, (ii) 700.55 °C; (b) (i) 537.31 °C, (ii) 656.95 °C; (c) 69.38% CO and 30.62% CO_2].

Problem 10.4 For the reaction: C (s) + CO_2 (g) = 2CO (g) at 1100 K and 1 atm, the equilibrium mixture contains 91.6% CO and 8.4% CO_2 by volume. (i) Calculate equilibrium constant for the reaction and (ii) the partial pressure of CO_2 gas if partial pressure of CO is changed to 10^{-4} atm.
[Ans: (i) 9.989, (ii) 1.0×10^{-9}].

Problem 10.5 The reactions are as follows:

(i) $3Fe_2O_3 + CO = 2Fe_3O_4 + CO_2, \Delta G^0_1 = -32969.92 - 53.85T$ J
(ii) $Fe_3O_4 + CO = 3FeO + CO_2, \Delta G^0_2 = 29790.08 - 38.28T$ J
(iii) $FeO + CO = Fe + CO_2, \Delta G^0_3 = -22802.8 + 24.27T$ J

Find out the percentage of CO in CO–CO_2 mixture in equilibrium with

(a) Fe_2O_3–Fe_3O_4, (b) Fe_3O_4–FeO, (c) FeO–Fe at 1200 K.

[Ans: (a) $\mathbf{5.647 \times 10^{-3\%}}$, (b) 16.55% and (c) 65.33%].

Problem 10.6 Calculate ΔG^0_T and the equilibrium CO/CO_2 ratio for the reduction of wustite at 900 °C according to the reaction: FeO(s) + CO (g) = Fe (s) + CO_2 (g).
Given:

(i) $Fe(I) + 1/2O_2(g) = FeO(1) \quad \mathbf{\Delta G^o_i} = -232714.08 + 45.31T$ J
(ii) $C(s) + 1/2O_2(g) = CO(g) \quad \mathbf{\Delta G^o_{ii}} = -111712.8 - 87.65T$ J

(iii) $C(s) + O_2(g) = CO_2(g) \quad \mathbf{\Delta G^o_{iii}} = -394132.8 - 0.84T$ J

[Ans: −1026.42 J and 0.9].

References

1. K.K. Prasad, H.S. Ray, *Advances in Rotary Kiln Sponge Iron Plant* (New age International Publishers, New Delhi, 2009)
2. S.K. Dutta and R. Sah, *Alternate Methods of Ironmaking* (*Direct Reduction and Smelting Reduction Processes*) (S. Chand & Co Ltd, New Delhi, April 2012)
3. O. Kubaschewski, C.B. Alcock, *Metallurgical Thermochemistry*, 5th edn (Pergamon press, Oxford, 1979)
4. R.L. Stephenson, R.M. Smailer, in *Direct Reduced Iron: Technology and Economics of Production and use* (The Iron & Steel Society of AIME, USA, 1980)
5. S.S. Gupta, A. Chatterjee, in *Blast Furnace Ironmaking* (Tata Steel, India, 1991)
6. J.K. Wright, I.F. Taylor, D.K. Philip, Mineral Engineering **4**, 983 (1991)
7. Rene Cypres and Claire Soudan-Moinet: *Fuel,* **60**, 33 (Jan 1981)
8. A. Ghosh, in *Proceedings of the International Conference on Alternative Routes of Iron and Steelmaking*, Sep 1999, Perth, Australia, p 71
9. A. Ghosh, S.K. Ajmani, in *Proceedings of the Symposium on Kinetics of Metallurgical Processes,* 1987, IIT, Kharagpur, India, p 1
10. A. Ghosh, in *Proceedings of the Workshop on Production of Liquid Iron Using Coal,* Aug 1994, Bhubaneswar, India, p 139
11. R.K. Paramguru, R.K. Galgali, H.S. Ray, Metall. Mat. Trans. B **28B**, 805 (1997)
12. A. Ghosh, A. Chatterjee, *Ironmaking and Steelmaking: Theory and Practice* (PHI learning Pvt Ltd, New Delhi, 2008)
13. C. Bodsworth, *Physical Chemistry of Iron and Steel Manufacture* (CBS Publishers & Distributors, Delhi, 1988)
14. L.V. Bogdandy, and H.J. Engell, *The Reduction of Iron Ores* (Springer, Berlin, 1971)

Kinetics

11

Productivity of a process is depending on the kinetic calculation of the reactions. What are the rates of reduction for iron oxides at different stages and temperatures that should be known before the actual process. The reduction mechanism should be known to understand the process.

11.1 Reduction by Gases

Reduction takes place, whether it is gaseous reduction or solid state by gas/gases. Most of the materials found in nature are sedimentary, metamorphic in nature, and hence all the natural mineral/ores are mostly porous in nature.

$$\begin{aligned} \text{Porosity} &= \text{volume fraction of voids} \\ &= \frac{\text{Total volume of voids in lump of ore/pellet}}{\text{Total volume of the lump of ore/pellet}} \end{aligned} \tag{11.1}$$

All pores are not equal in size. Inside the grain, there are micropores; between the grains there are macropores.

According to reduction mechanism, different models are proposed as follows:

1. Topochemical reduction: It occurs at layer by layer. It occurs at the surface; reaction zone moves from outer surface to towards the core.
2. Internal reduction: Reduction occurs throughout the ore.
3. Grain model: Individual grains act as reacting side.

11.1.1 Interfacial Reaction Control

Most of the fundamental kinetic information came from studies of reduction of single spherical pellet. The rates were mostly followed by noting the weight loss of the iron oxide at intervals of time in thermo-gravimetric set up. Spherical iron oxide pellet is surrounded by the reducing gas and hydrogen (as shown in Fig. 11.1a).

S. K. Dutta and Y. B. Chokshi, *Basic Concepts of Iron and Steel Making*,
https://doi.org/10.1007/978-981-15-2437-0_11

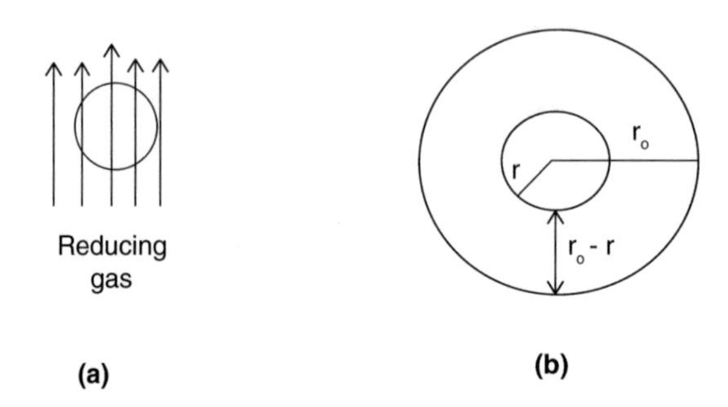

Fig. 11.1 Reduction of oxide by reducing gas

The kinetic steps are as follows:

1. Transport of hydrogen gas through boundary layer from gas phase,
2. Diffusion of hydrogen gas through pores of iron layer,
3. Chemical reaction at the interface of porous iron and unreacted surface of oxide,
4. Diffusion of water vapour through pores of iron layer,
5. Transport of water vapour through boundary layer.

Steps (1) and (5) are the faster transfer process; it can be verified by high gas flow rate experiment in the laboratory. Steps (2) and (4) can be combined; this combined term is pore diffusion. Pore diffusion rate is much lower than the chemical reaction rate.

Edstrom [1] first put forward a simple workable postulate of reaction mechanism applicable to dense samples of iron oxide in 1953. According to that if a partially reduced haematite sample is examined, it would be found to consist of layers. The outermost layer would consist of metallic iron followed by wustite, magnetite and a core of haematite. This was termed as *topochemical pattern of reduction.* The outermost iron layer would be porous.

Mckewan [2] carried out first time in 1965 the rate measurement extensively under controlled conditions. Mckewan assumed the following points to develop a mathematical model:

1. Topochemical pattern of reduction,
2. Negligible thickness of iron and wustite layers,
3. Reaction is chemical control,
4. Reaction is first-order reversible.

Considering the shape of metal oxide is spherical and that is surrounded by reducing hydrogen gas. Then, reaction takes place on the surface of the oxide pellet with the hydrogen gas (Fig. 11.1b).

$$MO(s) + \{H_2\} = M(s) + \{H_2O\} \tag{11.2}$$

If the above reaction is controlling the kinetic, then it is called chemical reaction control. Iron oxide reduction has been shown to be a surface-controlled reaction. To measure the reaction rate, the surface area of the sample must be known; but the way, the surface area changes with the reduction proceeds that must be known. If the sample is dense and of regular shape, the reaction surface area changes as a function of weight loss can be determined geometrically. Since weight loss easily measured, the reaction rate can also be easily measured for any regular shape particle such as a sphere, slab or cylinder.

Let the rate of reaction is controlled by the reaction interface, i.e. rate of a reaction depends on interfacial area; a rate equation can be derived that will fit any shape of particle. Consider the initial radius of the sphere pellet is r_o, oxygen density is d_o and total weight of the pellet is W_o.

Assume that the rate of formation of a uniform reaction product layer is proportional to the surface area (A) of the remaining oxide at time, t. If W is the weight of the material reacted, then

$$\left(\frac{\mathrm{d}W}{\mathrm{d}t}\right) \alpha A \quad \text{or} \quad \left(\frac{\mathrm{d}W}{\mathrm{d}t}\right) = k_M A \tag{11.3}$$

where k_M is the proportionality constant or rate constant ($\mathrm{mL}^{-2}\,\mathrm{t}^{-1}$), which is a function of temperature, pressure and gas composition.

If r is radius of the unreacted oxide core, then the fractional thickness of the reaction product layer (f) is defined as $\left(\frac{\mathbf{r_o} - r}{\mathbf{r_o}}\right)$ or

$$\mathrm{f}r_o = r_o - r \quad \text{or} \quad r = r_o(1-f) \tag{11.4}$$

So surface area of remaining oxide,

$$A = 4\pi r^2 = 4\pi\{r_o(1-f)\}^2 \tag{11.5}$$

Since material reacted, i.e. weight of oxygen removed from sample = Initial weight –Final weight So

$$\begin{aligned} W &= \left(\frac{4}{3}\right)\pi r_o^3 d_o - \left(\frac{4}{3}\right)\pi r^3 d_o \\ &= \left(\frac{4}{3}\right)\pi r_o^3 d_o - \left(\frac{4}{3}\right)\pi\{r_o(1-f)\}^3 d_o \end{aligned} \tag{11.6}$$

Therefore, differentiating Eq. (11.6) with respect to time (t):

$$\begin{aligned} \left(\frac{\mathrm{d}W}{\mathrm{d}t}\right) &= 0 + 4\pi r_o d_o \{r_o(1-f)\}^2 \left(\frac{\mathrm{d}f}{\mathrm{d}t}\right) \\ &= 4\pi r_o^3 d_o (1-f)^2 \left(\frac{\mathrm{d}f}{\mathrm{d}t}\right) \end{aligned} \tag{11.7}$$

since r_o and d_o are constant, they are not varying with time.

Substituting values of $\left(\frac{\mathrm{d}w}{\mathrm{d}t}\right)$ and A [from Eqs. (11.7) and (11.5)] in Eq. (11.3):

$$r_o d_o \left(\frac{\mathrm{d}f}{\mathrm{d}t}\right) = k_M \tag{11.8}$$

By integrating:

$$r_o d_o \int \mathrm{d}f = k_M \int \mathrm{d}t \quad \text{or} \quad \mathbf{r_o d_o f = k_M t} \tag{11.9}$$

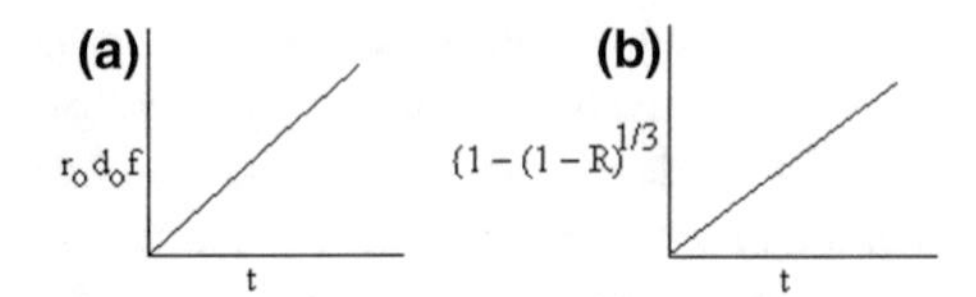

Fig. 11.2 Kinetic plots

Equation (11.9) means that the reaction product layer grows linearly with time (as shown in Fig. 11.2a), i.e. this reduction reaction is *interfacial reaction control.* To use Eq. (11.9), it is necessary to correlate weight change data to the thickness of the reaction product layer.

The *fractional reduction* (R) is defined as the oxygen removed from the oxide divided by the total removable oxygen originally present in oxide sample.

$$R = \left(\frac{W}{\mathbf{W_o}}\right) = \left(\frac{\frac{4}{3}\pi r_o^3 d_o - \frac{4}{3}\pi r^3 d_o}{\frac{4}{3}\pi r_o^3 d_o}\right) = \left\{1-(1-f)^3\right\} \tag{11.10}$$

Therefore,

$$f = \left\{1-(1-R)^{1/3}\right\} \tag{11.11}$$

Again, from Eq. (11.9):

$$r_o d_o f = k_M t \quad \text{or} \quad r_o d_o \left\{1-(1-R)^{1/3}\right\} = k_M t$$

$$\text{or} \quad (\mathbf{1}-(\mathbf{1}-\mathbf{R})^{\mathbf{1/3}}\} = \left(\frac{\boldsymbol{k_M}}{\boldsymbol{r_o d_o}}\right)\mathbf{t} \tag{11.12}$$

Equation (11.12) is known as *Mckewan Equation* or *Mckewan Model.*

From the experimental data, $\{1-(1-R)^{1/3}\}$ versus t plot (Fig. 11.2b) should be straight line that means reduction reaction obeys Mckewan's equation. Mckewan concluded that reduction of oxide was controlled by the interfacial chemical reaction.

11.1.2 Kinetics of Solid–Solid Reaction

Reduction of iron ore by solid carbon, there are two possible ways of reduction are as follows [3]:

1. Direct reduction:

$$Fe_2O_3(s) + 3C(s) = 2Fe(s) + 3\{CO\} \tag{11.13}$$

2. Indirect reduction, i.e. via gas phase by CO gas:

$$Fe_2O_3(s) + 3\{CO\} = 2Fe(s) + 3\{CO_2\} \tag{11.14}$$

It is desirable because here surface area is more compared to reduction with solid carbon.

Gasification reaction or *Boudouard's reaction*:

$$3\{CO_2\} + 3C(s) = 6\{CO\} \tag{11.15}$$

Combination of Eqs. (11.14) and (11.15) become Eq. (11.13).
Generally solid–solid reaction can be represented by as follows:

$$A(s) + B(s) = C(s) + \{D\} \tag{11.16}$$

The kinetic equation first derived by Jander [4] in 1927 for the reaction between $BaCO_3$ and SiO_2:

$$BaCO_3(s) + SiO_2(s) = BaSiO_3(s) + \{CO_2\} \tag{11.17}$$

Other kinetic models appear to be based on one or more of the following assumptions:

1. The surface of the reducing component particle (*A*) is surrounded with particles of the reductant (*B*).

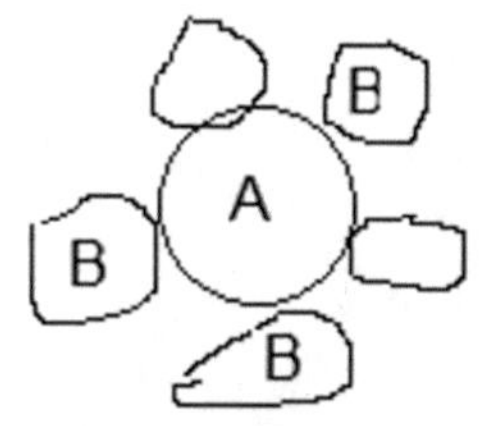

2. The reaction starts only at the points of contact between the two solids (i.e. *A* and *B*) and the rate of reaction is proportional to the number of contact points; this is time dependent.
3. The reductant component (*B*) and the product gas (*D*) can diffuse through the layer of reaction product (*C*) (Figure 11.3 shows the mechanism of solid–solid reaction.).

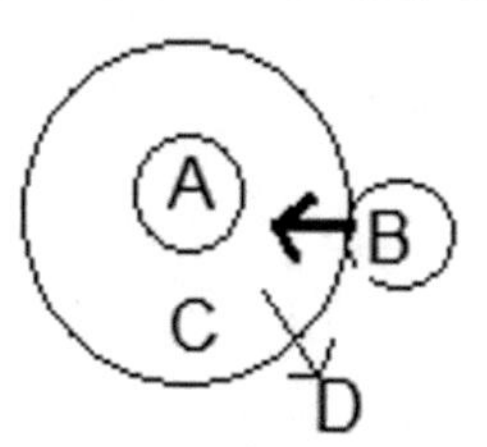

4. The reaction mixture consists of particles of similar size.

If the reaction product grows spherically from each contact points as shown in Fig. 11.4. At any time, *t*; let *x* be the thickness of the product (*C*) layer. According to Fick's first law of diffusion in a solid, the rate of growth of the product (*C*) layer, (dx/dt), is indirectly proportional to the thickness of product layer, *x*; due to diffusion of *B*.

i.e.

$$\left(\frac{dx}{dt}\right) \alpha \left(\frac{1}{x}\right) \tag{11.18}$$

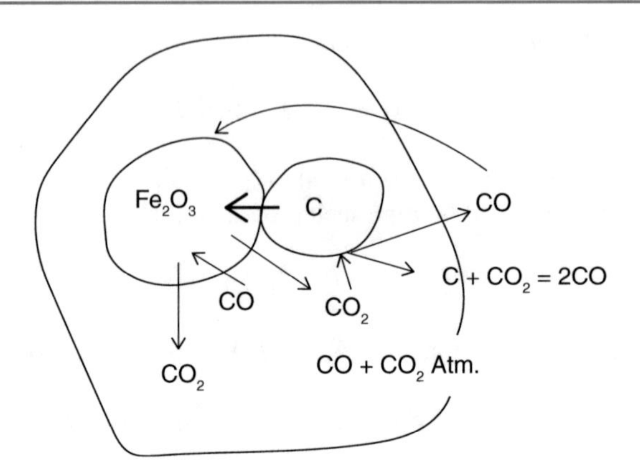

Fig. 11.3 Mechanism of solid–solid reaction

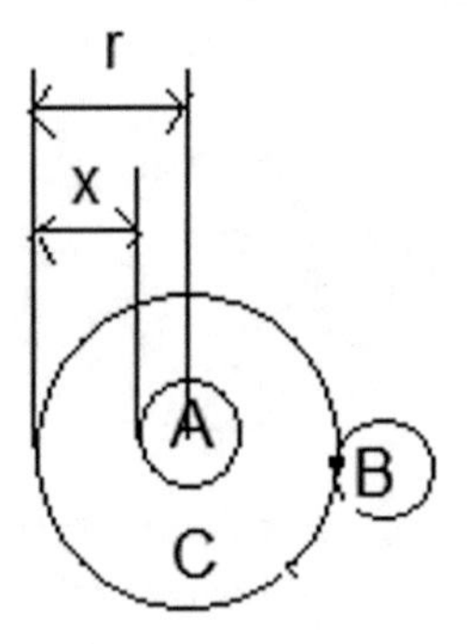

Fig. 11.4 Solid–solid reaction

Therefore, the rate of reaction is given by

$$\left(\frac{dx}{dt}\right) = D\left(\frac{1}{x}\right) \tag{11.19}$$

where D is the diffusion coefficient, or $x\mathrm{d}x = D\mathrm{d}t$.

By integration: $\int x\mathrm{d}x = D\int \mathrm{d}t$

or

$$\left(\frac{x^2}{2}\right) = Dt$$

Or

$$x^2 = 2Dt \tag{11.20}$$

Volume of unreacted core at time, t is given by

$$\left\{\left(\frac{4}{3}\right)\pi(r-x)^3\right\} \quad (11.21)$$

where r is the radius of the original particle.

$$\text{Volume of the original particle is } \left\{\left(\frac{4}{3}\right)\pi r^3\right\}.$$

If f is the fraction of reduction at time, t.

$$\text{Then unreacted volume is } \left\{\left(\frac{4}{3}\right)\pi r^3\right\} - \left\{\left(\frac{4}{3}\right)\pi r^3 \cdot f\right\} = \left\{\left(\frac{4}{3}\right)\pi r^3(1-f)\right\} \quad (11.22)$$

Equating Eqs. (11.21) and (11.22):

$$\left\{\left(\frac{4}{3}\right)\pi(r-x)^3\right\} = \left\{\left(\frac{4}{3}\right)\pi r^3(1-f)\right\}$$

Therefore,

$$(r-x)^3 = r^3(1-f)$$

or

$$(r-x) = r(1-f)^{1/3}$$

Therefore,

$$x = r - r(1-f)^{1/3} = r\left[1-(1-f)^{1/3}\right] \quad (11.23)$$

$$\text{or} \quad x^2 = r^2\left[1-(1-f)^{1/3}\right]^2$$

or $2Dt = r^2\left[1-(1-f)^{1/3}\right]$ [from Eq. (11.20)]

Or,

$$\left[1-(1-f)^{1/3}\right]^2 = \left(\frac{2Dt}{r^2}\right) \quad (11.24)$$

For a given mixture, the constant $\left(\frac{2D}{r^2}\right)$ may be replaced by a single constant, k. Therefore,

$$\left[1-(1-f)^{1/3}\right]^2 = kt \quad (11.25)$$

Equation (11.25) is known as Jander's equation. The left hand side of Eq. (11.25), i.e. $[1-(1-f)^{1/3}]^2$ is plotted against time, t in Fig. 11.5 for iron oxide (Fe_2O_3) reduction by solid carbon. Jander's equation is valid if the curve is straight line. However, the curves show variable slopes, leading to the conclusion that the Jander's equation is unsatisfactory for this case, i.e. Jander's equation is not applicable of solid iron oxide (Fe_2O_3) reduction by solid carbon. Curve-1: when −325 mesh (i.e. fine)

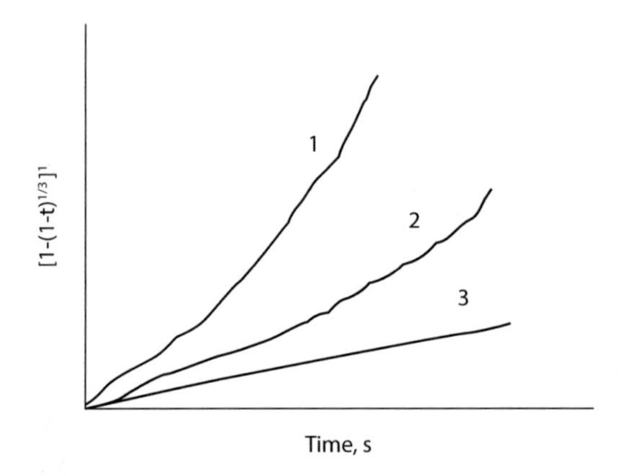

Fig. 11.5 Reduction of iron oxide by solid carbon

carbon is used, curve-2: when −100 mesh (i.e. medium) carbon is used, and curve-3: when −48 mesh (i.e. coarse) carbon is used to reduce Fe_2O_3 at 1310 K.

Hence, the experimental evidence failed to support the theory of solid–solid reaction, a different reaction mechanism involving gas–solid reaction is proposed, i.e. via gaseous reduction through gasification of carbon.

11.1.3 Reduction of Iron Oxides by CO and H_2

The general features of kinetics and mechanism of iron oxide reduction by CO and H_2 are similar. The major difference is that reduction rate by H_2 is 5 to 10 times faster than that by CO. Lump ores and pellets contain iron oxide mostly as Fe_2O_3. The gangue minerals contain primarily SiO_2 and Al_2O_3 besides other minor compounds. The fundamental measure of the extent of reduction is the degree of reduction (α) defined as:

$$\alpha = \left(\frac{\text{weight loss due to removal of oxygen}}{\text{weight of total removable oxygen present in ore}}\right) \times 100 \tag{11.26}$$

Reduction of iron oxide by hydrogen/carbon monoxide gas occurs as a combination of the gas–solid reactions as follows [5]:

$$3Fe_2O_3(s) + \{H_2\}/\{CO\} = 2Fe_3O_4(s) + \{H_2O\}/\{CO_2\} \tag{11.27}$$

$$Fe_3O_4(s) + \{H_2\}/\{CO\} = 3FeO(s) + \{H_2O\}/\{CO_2\} \tag{11.28}$$

$$FeO(s) + \{H_2\}/\{CO\} = Fe(s) + \{H_2O\}/\{CO_2\} \tag{11.29}$$

The reduction reactions occur according to the reactions (11.27)–(11.29). As a result, when haematite is reduced to magnetite, additional porosity develops, enhancing the rate of further reduction of haematite. Hence, haematite is more reducible than magnetite. The higher the reducibility of an iron oxide bearing solid, the faster is the rate of reduction. Hence, the rate of reduction ($d\alpha/dt$) may be considered as a measure of reducibility.

Salient kinetic features that need to be noted are as follows [6]:

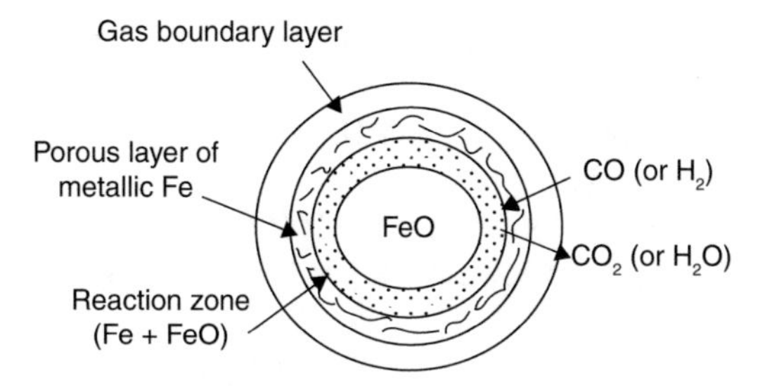

Fig. 11.6 Partially reduced sphere of FeO showing layers and kinetic steps

- Lump ore and pellets are porous solids.
- Reduction is characterized by the formation of a porous product layer.
- Fe_2O_3 is reduced in stages, e.g. $Fe_2O_3 \rightarrow Fe_3O_4 \rightarrow FeO \rightarrow Fe$.
- Additional porosity develops during reduction owing to density differences of the product solids. The relative volumes per unit mass of Fe are:

$$Fe{:}FeO{:}Fe_3O_4{:}Fe_2O_3 = 1{:}1.79{:}2.08{:}2.14$$

The kinetic steps involved as shown in Fig. 11.6 are as follows:

- Transfer of reactant gas (H_2 or CO) to the solid surface across the gas boundary layer around the solid piece,
- Inward diffusion of reduce gas through the pores of the solid,
- Chemical reactions,
- Outward diffusion of the product gas (H_2O or CO_2) through the pores,
- Transfer of the product gas from the solid surface into the bulk gas across the boundary layer.

The overall rate would depend on temperature, gas composition, size of the particle, and nature of the solid in terms of its structure and composition. It is interesting to note that at a fixed temperature and gas composition, and for any type of ore/sinter/pellet burden of a given size range, the reducibility values do not differ by more than a small factor.

11.2 Gasification of Carbon

Reduction of iron ore by solid carbon takes place (via the gaseous phase) in two stages as follows:

(i) Gasification reaction or Boudouard's reaction:

$$C(s) + \{CO_2\} = 2\{CO\} \tag{11.30}$$

(ii) Reduction of iron oxide:

$$\mathrm{Fe}_x\mathrm{O}_y(\mathrm{s}) + \{\mathrm{CO}\} = 2\mathrm{Fe}_x\mathrm{O}_{y-1}(\mathrm{s}) + \{\mathrm{CO}_2\} \quad (11.31)$$

where Fe_xO_y denotes Fe_2O_3, Fe_3O_4, Fe_xO; and Fe_xO_{y-1} denotes Fe_3O_4, Fe_xO, Fe.
Mechanism of gasification reaction:
A two steps mechanism proposed by Reif [7] in 1952 (it is also known as Reif's mechanism):
First step: reversible oxygen exchange between gas phase and carbon surface:

$$\mathrm{C}^1(\mathrm{s}) + \{\mathrm{CO}_2\} \underset{k_\mathrm{b}^1}{\overset{k_\mathrm{f}^1}{\leftrightarrow}} \leftrightarrow C^{1^o} + \{\mathrm{CO}\} \quad (11.32)$$

Second step: carbon gasification stage:

$$C^{1^o} \overset{K_\mathrm{f}^2}{\rightarrow} \{\mathrm{CO}\} \quad (11.33)$$

where C^1 represents a free carbon site on the surface and C^{1^o} denotes a site where oxygen atom is chemisorbed. k_f, k_b are forward and backward reaction rate constants.

Step-1 is reversible, so it is fast. Step-2 proceeds relatively slow, so it is rate-controlling step.

Based on Reif's mechanism, Ergun [8] in 1956 derived the rate expression for gasification of carbon by CO_2 gas, which is known as Ergun's rate equation.

$$\text{Overall rate}(r) = \text{rate of step-2} \simeq k_\mathrm{f}^2\mathrm{C_O} \quad (11.34)$$

where C_O is the concentration of C^{1^o} and r is rate of reaction per unit surface area.

The value of C_O can be determined by applying steady-state approximation as follows:

At steady-state, i.e. concentration of intermediate, e.g., CO and C^{1^o} are not changed with time, i.e. C_O is constant.

Therefore, rate of reaction (r) = rate of step-1 = rate of step-2.

Hence,

$$r = k_\mathrm{f}^2 C_o = k_\mathrm{f}^1 \quad p_{\mathrm{CO}_2} \cdot \mathrm{C_f} - k_\mathrm{b}^1 \quad p_{\mathrm{CO}}\mathrm{C_O} \quad (11.35)$$

where C_f is the concentration of free carbon site on the surface.

Now,

$$\mathrm{C_o} + \mathrm{C_f} = \mathrm{C_T} = \text{constant}$$

Or

$$\mathrm{C_f} = \mathrm{C_T} - \mathrm{C_O} \quad (11.36)$$

where C_T is the total number of carbon sites on the surface.

Combining Eqs. (11.35) and (11.36):

$$k_\mathrm{f}^2\mathrm{Co} = k_\mathrm{f}^1\, p_{\mathrm{CO}_2} \cdot (\mathrm{C_T} - \mathrm{C_O}) - k_\mathrm{b}^1 p_{\mathrm{CO}}\mathrm{C_O}$$

Or,

$$(k_f^2 + k_f^1 p_{CO_2} + k_b^1 p_{CO})\ C_O = k_f^1 p_{CO_2} \cdot C_T$$

Therefore,

$$C_O = \left(\frac{k_f^1 C_T p_{CO_2}}{(k_f^2 + k_f^1 p_{CO_2} + k_b^1 p_{CO})} \right) \quad (11.37)$$

Again combining Eqs. (11.34) and (11.37):

$$r = k_f^2 C_O = k_f^2 \left(\frac{k_f^1 C_T p_{CO_2}}{(k_f^2 + k_f^1 p_{CO_2} + k_b^1 p_{CO})} \right) \quad (11.38)$$

or

$$r = \frac{k_f^1 . C_T . p_{CO_2}}{\left[1 + \left(\frac{k_f^1}{k_f^2}\right) p_{CO_2} + \left(\frac{k_b^1}{k_f^2}\right) p_{CO}\right]} = \frac{k_1 . p_{CO_2}}{1 + k_2 . p_{CO_2} + k_3 . p_{CO}} \quad (11.39)$$

where $k_1 = k_f^1 .\ C_T$ = constant

$$k_2 = \left(\frac{k_f^1}{k_f^2}\right) = \text{constant}$$

$$k_3 = \left(\frac{k_b^1}{k_f^2}\right) = \text{constant}$$

Now, k_2 and k_3 do not dependent on nature of carbon, but k_1 is only dependent on nature of carbon.

The reaction (11.30) has been found to be controlled by the interfacial chemical reaction step below 1100 °C if the size is not too large. Such a conclusion has been drowning from evidences such as (i) high activation energy (300–350 kJ/mol) and (ii) strong influence of even trace amounts of solid and gaseous impurities on reaction rate.

The gasification reaction (11.30) is also of significant interest in carbothermic reduction. The term reactivity is commonly used to denote the speed of the gasification reaction. The higher the reactivity, the faster is the gasification. The reactivity (r_C) can be defined as:

$$r_C = \left(\frac{dF_C}{dt}\right) \quad (11.40)$$

where F_C is the degree of gasification of carbon and t is the time of reaction.

The kinetic steps involved in the gasification of a carbon are as follows:

- Transfer of CO_2 across the gas boundary layer to the surface of the particle,
- Inward diffusion of CO_2 through the pores,
- Chemical reaction on the pore surface,
- Outward diffusion of CO through the pores,
- Transfer of CO into the bulk gas by mass transfer across the boundary layer.

Laboratory experiments have demonstrated that the chemical reaction is much slower compared with the other steps; hence, it constitutes the principal rate-controlling step. Such a conclusion has been arrived at based on the large activation energy involved, strong retarding influence of CO on the rate.

Probable Questions

1. What are the assumptions made by Mckewan to develop a mathematical model? Derive Mckewan equation.
2. Derive the equation for solid–solid reaction.
3. What do you mean by Reif's mechanism for gasification reaction? Derive Ergun's rate equation for gasification reaction.
4. Discuss the factors that affect kinetics of solid-state iron oxide reduction, in the stack of a BF.

Examples

Example 11.1 The initial weight of pure iron oxide (Fe_2O_3) powder is 295.6 mg. Iron oxide powder is reduced by hydrogen at 800 °C. In reduction experiment with time weight loss are noted as follows:

Time (s)	36	60	120	180	240	300	420	600
Wt. loss (mg)	6.69	14.05	31.94	48.14	55.33	59.34	64.94	70.24

Calculate (i) fraction of reduction (f) w.r.t time (t), (ii) draw f versus t graph, (iii) calculate rate of reduction $\left(\frac{df}{dt}\right)$ at initial stage of reduction from the graph.

Solution

$$\text{The fraction of reduction}\,(f) = \left[\frac{(\text{O}_2\text{ removed from the sample})}{(\text{Total removable O}_2\text{ present in the sample})}\right]$$
$$= \left[\frac{(\text{weight loss})}{(\text{Total removable O}_2\text{ present in the sample})}\right]$$
$$= \left[\frac{\{(W_1 - W_2)\}}{W_O}\right]$$

$$\text{where total removable oxygen present in iron oxide} = W_O = W_1 \times f_{ore} \times \rho_{ore} \times f_O$$
$$= 295.6 \times 1 \times 1 \times 0.3 = 88.68\,\text{mg}$$

(Since iron oxide is pure, so f_{ore} and ρ_{ore} are one.)

Time (s)	36	60	120	180	240	300	420	600
Wt. loss (mg)	6.69	14.05	31.94	48.14	55.33	59.34	64.94	70.24
f	0.075	0.158	0.36	0.543	0.624	0.669	0.732	0.792

Plot *f* versus *t*, from initial slop we get rate of reduction:

$$\left(\frac{df}{dt}\right) = \left(\frac{AB}{BC}\right) = \left(\frac{(55.33 - 14.05)}{(240 - 60)}\right) = \mathbf{0.229\,s^{-1}}$$

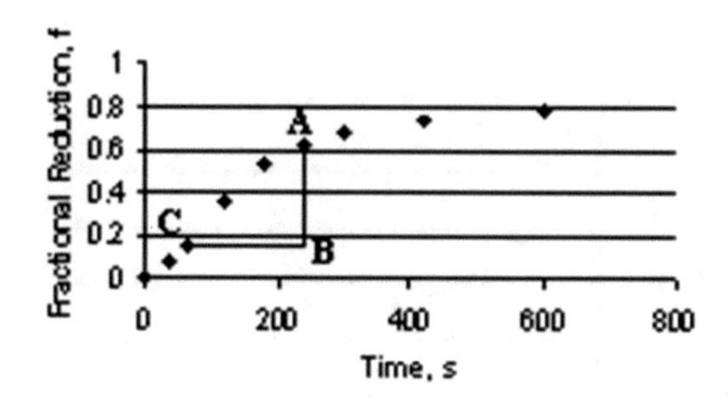

Example 11.2 The following data are obtained from an experiment of reduction of iron ore fine by the hydrogen gas at 973 K. Find out (a) initial rate of reduction from fraction of reduction (*f*) w.r.t time (*t*) graph, (b) order of reaction and rate constant.

Time (s)	36	60	120	180	240	300	420	600
Wt. loss (mg)	4.86	10.06	24.52	32.40	34.75	36.32	38.74	41.30

Given: Total removable oxygen present in iron ore is 56.18 mg.

Solution

$$\text{The fraction of reduction}\,(f) = \left[\frac{\{(W_1 - W_2)\}}{W_O}\right] = \left[\frac{\{(W_1 - W_2)\}}{56.18}\right] \quad (1)$$

Fraction of reductions is calculated by Eq. (1) as follow:

Wt. loss (mg)	*F*	(1 − *f*)	Ln (1 − *f*)	[1/(1 − *f*)]
4.86	0.0865	0.9135	−0.0905	1.095
10.06	0.179	0.821	−0.197	1.218
24.52	0.436	0.564	−0.573	1.773
32.4	0.577	0.423	−0.86	2.364
34.75	0.6185	0.3815	−0.964	2.621
36.32	0.646	0.354	−1.038	2.825
38.74	0.6895	0.3105	−1.17	3.221
41.3	0.735	0.265	−1.328	3.774

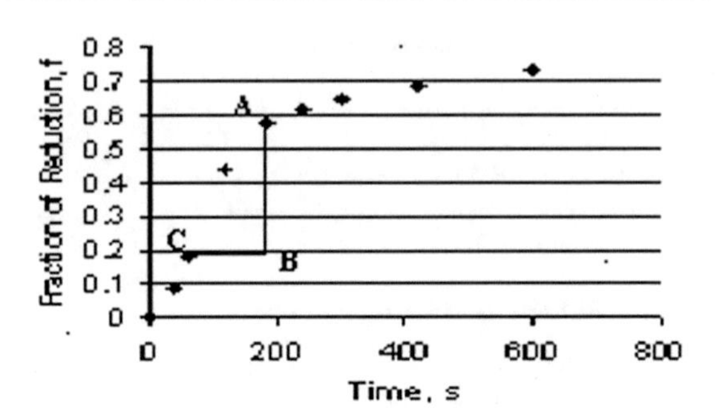

Fig. 11.7 f versus t

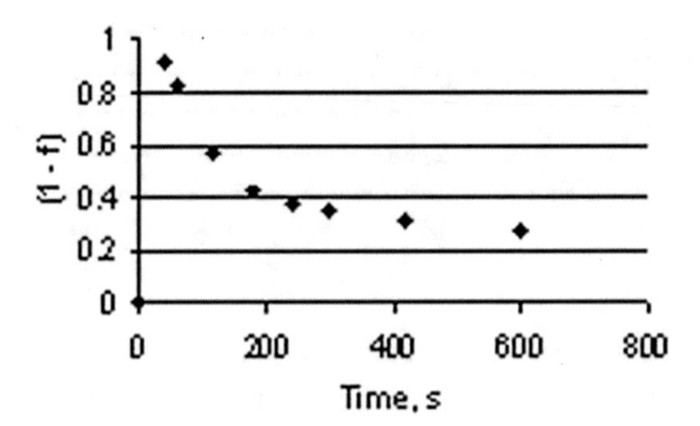

Fig. 11.8 $(1 - f)$ versus t

Now, plot f versus t graph (Fig. 11.7), and from the initial plot we get initial rate of reduction.

$$\text{The initial rate of reduction}\left(\frac{df}{dt}\right) = \left(\frac{\boldsymbol{AB}}{\boldsymbol{BC}}\right) = \left(\frac{(\mathbf{0.6-0.4})}{(\mathbf{180-50})}\right) = \mathbf{1.54 \times 10^{-3}\,s^{-1}}$$

For zero order reaction the equation is:

$$[A] = -kt + [A_o] \tag{2}$$

where $[A] = (1 - f)$, so plot $(1 - f)$ versus t (Fig. 11.8).

In Fig. 11.8 shows the line is not straight line, so reaction is not zero order.

Now, we go for first-order reaction:

$$\ln[A] = \ln[A_0] - kt. \tag{3}$$

So plot $\ln(1 - f)$ versus t (Fig. 11.9).

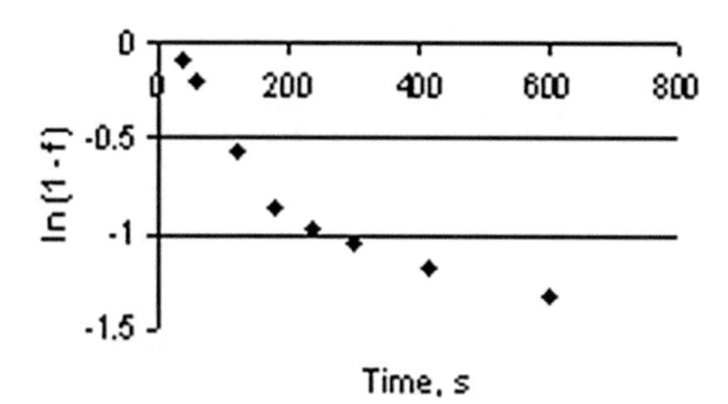

Fig. 11.9 ln $(1 - f)$ *vd t*

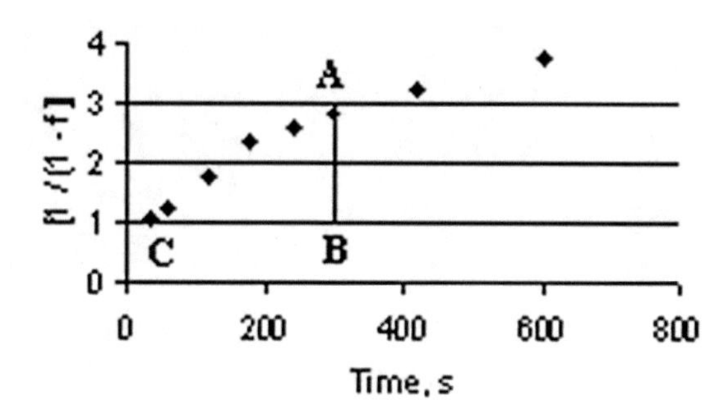

Fig. 11.10 $[1/(1 - f)]$ versus t

Here, also the line is not straight line, so reaction is not first order.
From general equation:

$$\left(\frac{a-x}{n-1}\right)^{1-n} = kt + C \tag{4}$$

Now, we go for second-order reaction, $n = 2$, $x = f$ and $a = 1$;
So Eq. (4) become:

$$\left(\frac{1}{1-f}\right) = kt + C \tag{5}$$

Now, plot $\left(\frac{1}{1-\mathbf{f}}\right)$ versus t (Fig. 11.10). Now, the line is almost straight.

From the slope, we get rate constant = $\left(\frac{AB}{BC}\right) = \left(\frac{(2.8-1.0)}{(290-20)}\right) = \mathbf{6.67 \times 10^{-3}\ s^{-1}}$

Hence, the reduction of iron ore fine by the hydrogen gas is second-order reaction and rate constant is $6.67 \times 10^{-3}\ s^{-1}$.

Example 11.3 Initial weight of pure iron oxide (Fe_2O_3) fines is 296.5 mg. Oxide is reduced by hydrogen gas at 800 °C. Find out: (i) fraction of reduction with respect to time, and (ii) whether these data fit to the Mckewan equation?

The following data are obtained:

Time (s)	36	60	120	180	240	300	420	600
Wt. loss (mg)	6.69	14.05	31.95	48.14	55.14	59.34	64.94	70.24

Solution

Total removable oxygen present in iron oxide = 296.5 × 0.3 = 88.95 mg.

Time (s)	Weight loss (mg)	Fraction of reduction (R)	$(1 - R)^{1/3}$	$[1 - (1 - R)^{1/3}]$
36	6.69	0.075	0.974	0.026
60	14.05	0.158	0.944	0.056
120	31.95	0.359	0.862	0.138
180	48.14	0.541	0.771	0.229
240	55.14	0.620	0.724	0.276
300	59.34	0.667	0.693	0.307
420	64.94	0.730	0.646	0.354
600	70.24	0.790	0.594	0.406

From Eq. (11.12), by plotting $[1 - (1 - R)^{1/3}]$ versus time, line should be straight for Mckewan equation, below figure shows that line is not straight; hence, the given data for reduction of pure iron oxide (Fe_2O_3) fine does not obey the Mckewan equation.

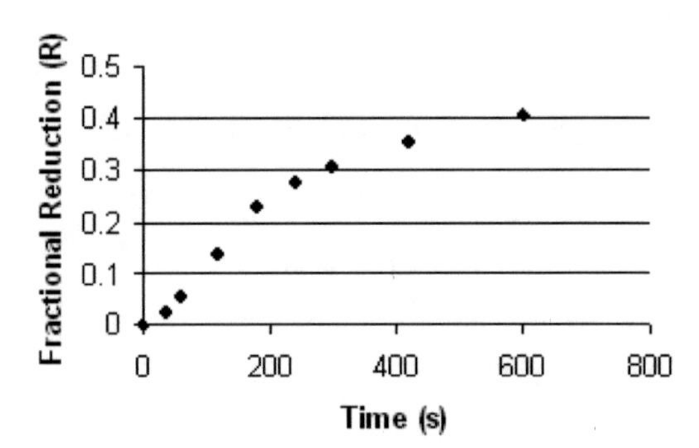

Example 11.4 For gaseous reduction of iron ore fines, the following data are obtained for H_2 reductions. Find the activation energy for H_2 reduction of iron ore.

Temp (°C)	H_2 reduction rate (s^{-1})
898	5.67×10^{-3}
973	5.96×10^{-3}
1048	6.66×10^{-3}
1123	8.80×10^{-3}

Solution

From Arrhenius equation: Rate, $k = A \cdot e^{-(E/RT)}$

Therefore, $\ln k = \ln A - (E/RT)$

Now, plot graph of ln k versus ($1/T$) and we get slope = − (E/R)

So, we calculate as follows:

Temp. (°C)	Temp. (T) (K)	$1/T$	$1/T \times 10^4$	H_2 reduction rate k (s^{-1})	ln k
898	1171	0.00085	8.54	5.67×10^{-3}	−5.17
973	1246	0.00080	8.03	5.96×10^{-3}	−5.12
1048	1321	0.00076	7.57	6.66×10^{-3}	−5.01
1123	1396	0.00072	7.16	8.80×10^{-3}	−4.73

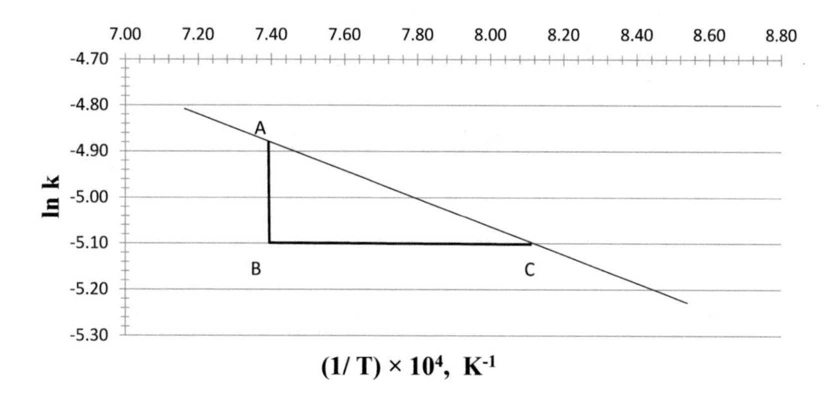

From graph, Slope $= -\left(\frac{E}{R}\right) = -\frac{AB}{BC} = -\left[\frac{(-4.88-(-5.10))}{(7.40-8.12)\times10^{-4}}\right] = -3.06 \times 10^3$

Therefore, $\left(\frac{E}{R}\right) = -3.06 \times 10^3$

Hence, $E = (-3.06 \times 10^3) \times R = (-3.06 \times 10^3) \times 8.314 = -25.404$ kJ

Therefore, the activation energy for H_2 reduction of iron ore is −25.404 kJ.

Example 11.5 For gasification of carbon by carbon dioxide, the following data are given for $CO–CO_2$ gas mixture at 950 °C and 1 atm pressure. Find out the value of k_1, k_2, k_3.

Gas composition	$r \times 10^4$ (s^{-1})
Pure CO_2	1.16
45% CO	0.16
30% CO	0.20

Solution

Since total pressure = 1 atm.

For pure CO_2, $p_{CO_2} = 1$ atm and $p_{CO} = 0$

Putting above value in Eq. (11.39): $r = \frac{k_1 \cdot p_{CO_2}}{1 + k_2 \cdot p_{CO_2} + k_3 \cdot p_{CO}}$

Therefore,

$$1.16 \times 10^{-4} = \left(\frac{k_1}{1 + k_2}\right) \quad \text{(a)}$$

For 45% CO, p_{CO} = 0.45 atm and p_{CO_2} = 0.55 atm
Hence, from Eq. (11.35):

$$0.16 \times 10^{-4} = \left(\frac{0.55k_1}{1 + 0.55k_2 + 0.45k_3}\right) \quad \text{(b)}$$

Again for 30% CO, p_{CO} = 0.3 atm and p_{CO_2} = 0.7 atm
Hence, from Eq. (11.35):

$$0.2 \times 10^{-4} = \left(\frac{0.7k_1}{1 + 0.7k_2 + 0.3k_3}\right) \quad \text{(c)}$$

Equation (a) is divided by Eq. (b): $(1.16/0.16) = \{\left(\frac{k_1}{1+k_2}\right) / \left(\frac{0.55k_1}{1+0.55k_2+0.45k_3}\right)\}$
Or

$$0.45k_3 - 3.44k_2 = 2.99 \quad \text{(d)}$$

Again Eq. (a) is divided by Eq. (c): $(1.16/0.2) = \{\left(\frac{k_1}{1+k_2}\right) / \left(\frac{0.7k_1}{1+0.7k_2+0.3k_3}\right)\}$

$$\text{Or } 3.06 = 0.3k_3 - 3.36k_2 \quad \text{(e)}$$

Eq. (d) is divided by

$$0.45{:}6.64 = k_3 - 7.64k_2 \quad \text{(f)}$$

Equation (e) is divided by

$$0.3{:}1.02 = k_3 - 11.2k_2 \quad \text{(g)}$$

Solving (f) and (g) we get: $\mathbf{k_2 = 1.58}$
Putting k_2 value in Eq. (f) we get: $\mathbf{k_3 = 18.71}$
Putting k_2 and k_3 values in Eq. (a) we get: $\mathbf{k_1 = 2.99 \times 10^{-4}\ s^{-1}}$.

Example 11.6 From an experiment on sulphur removal from molten iron to slag at 1773 K, the following data is obtained:

Time, min	0	4	15	33	63	91	151
[S] %	0.16	0.14	0.132	0.091	0.072	0.033	0.012

Calculate first-order rate constant for de-sulphurization.

Solution

From first-order reaction: ln $[A]$ = ln $[A_0]$ − kt; we can calculate of rate constant (k) by plotting a graph of ln $[A]$ versus t (correlate with: $y = mx + c$)

So we calculate as per following:

Time, min	[S] %	ln ([S] %)
0	0.16	−1.83258
4	0.14	−1.96611
15	0.132	−2.02495
33	0.091	−2.3969
63	0.072	−2.63109
91	0.033	−3.41125
151	0.012	−4.42285

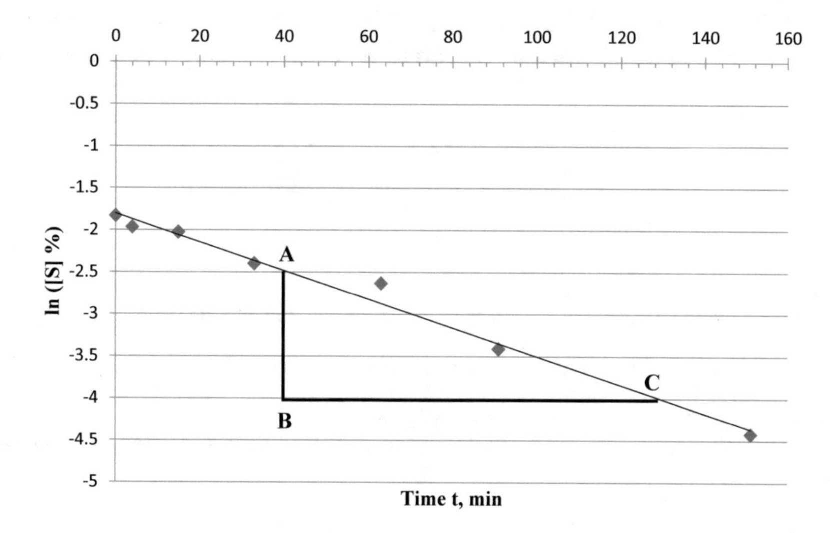

By plotting graph, the slope of line is rate constant (–*k*),

$$\text{Slope}(-k) = -\frac{AB}{BC} = -\left[\frac{(-2.5-(-4.0))}{(40-128)}\right] = -0.01705\ \text{min}^{-1}$$

So, first-order rate constant is 0.01705 min^{-1} for sulphur removal from molten iron to slag.

Problems

Problem 11.1 The initial weight of pure iron oxide (Fe_2O_3) powder is 300 mg. Iron oxide powder is reduced by hydrogen at 800 °C. In reduction experiment with time weight loss arc noted as follows:

Time (s)	35	60	120	180	240	300	420	600
Wt. loss (mg)	6.65	14.0	31.95	48.15	55.35	59.35	64.94	70.24

Calculate (i) fraction of reduction (*f*) w.r.t time (*t*), (ii) draw *f* versus *t* graph, (iii) calculate rate of reduction (d*f*/d*t*) at initial stage of reduction from the graph. [Ans: $3 \times 10^{-3}\ \text{s}^{-1}$].

Problem 11.2 For gaseous reduction of iron ore fines following data is obtained for reduction by CO gas. Find the activation energies for such reductions.

Temp (°C)	Rate (s^{-1})
800	1.04×10^{-3}
900	1.20×10^{-3}
1000	1.72×10^{-3}
1100	2.52×10^{-3}

[Ans: **−43.76 kJ]**

Problem 11.3 Find the rate of the reaction at 300 K, if activation energy of the reaction is 167.36 kJ/mol. By raising the temperature by 100 K, how much reaction rate will increase?

[Ans: $\mathbf{1.927 \times 10^7}$]

References

1. J.O. Edstrom, J. Iron Steel Inst. **175**, 289 (1953)
2. W.M. Mckewan, Chipman Conference on *Steelmaking*, ed. by J.F. Elliott (MIT Press, Cambridge, MA, 1965), p. 141
3. L.V. Bogdandy, H.J. Engell, *The Reduction of Iron Ores* (Springer, Berlin, 1971)
4. W. Jander, Z. Anorg, Allg Chem. **163**, 1 (1927)
5. S.K. Dutta, A. Ghosh, Metall. Mat. Trans. B **25B**, 15 (1994)
6. A. Ghosh, A. Chatterjee, *Ironmaking and Steelmaking (Theory and Practice)* 1st edn. (PHI Learning Pvt. Ltd., New Delhi, 2008)
7. A.E. Reif, J. Phys. Chem. **56**, 773 (1952)
8. S. Ergun, J. Phys. Chem. **60**, 480 (1956)

Part IV
Steelmaking

Historical Steelmaking 12

Steel cannot be produced directly from the iron ore. Steel is a product obtained by refining or purification of impure iron (i.e. hot metal). So, steelmaking is an oxidation process by which refining of hot metal take place. *Steelmaking* is a refining process in which impurity elements are oxidized from hot metal and steel scrap which are charged as raw materials to the furnace. Before modern steelmaking processes are developed, there are three earlier processes such as wrought ironmaking, cementation process, and crucible process are established.

12.1 Introduction of Steelmaking

General flow diagram for steelmaking from iron ore is shown in Fig. 12.1. Iron ore is reduced to molten impure iron (i.e. *hot metal*) in the blast furnace, using carbon of coke as the reducing agent and fuel in blast furnace. In the reduction process, the iron absorbs about 3–4% C and other elements (like Si, Mn, P, S etc.) to form impure iron (i.e. hot metal). By solidification of hot metal into small casting that is known as *pig iron*. Hot metal is directly used as raw material for steelmaking and pig iron is used as raw material for foundry. But most of the steels contain less than 1.0% C, hence the excess carbon and other impurities must be removed from the hot metal to convert into steel. The excess carbon and other impurities are removed by controlled oxidation of hot metal in steelmaking furnace by oxygen to produce carbon steel of the desired carbon content. Various alloying elements (e.g. Cr, Mn, Ni, Mo etc.) may be added alone or combinations to the molten bath during or after the carbon removal to produce alloy steels. Steel cannot be produced directly from an iron ore. Steel is a product obtained by refining or purification of impure iron (i.e. hot metal). So, steelmaking is an oxidation process by which refining of hot metal take place. *Steelmaking* is a refining process in which impurity elements are oxidized from hot metal and steel scrap which are charged as raw materials to the furnace.

Therefore, steelmaking is two stages processes:

- Reduction stage i.e. ironmaking stage (iron ore to hot metal),
- Oxidation stage i.e. steelmaking stage (hot metal to steel).

Energy requirement for steelmaking, with respect to other metals production, is the lowest as shown in Table 12.1.

S. K. Dutta and Y. B. Chokshi, *Basic Concepts of Iron and Steel Making*,
https://doi.org/10.1007/978-981-15-2437-0_12

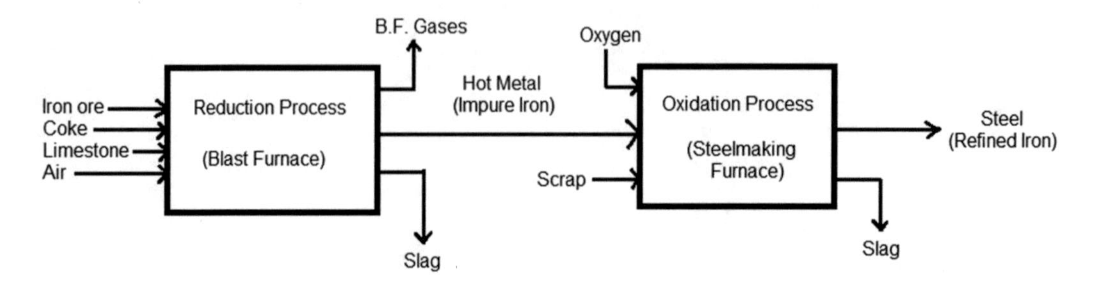

Fig. 12.1 Flow diagram for steelmaking

Table 12.1 Energy requirement for different metals production [1]

Metal	Energy require, GJ/tonne	
	Primary production (from ore)	Secondary production (from scrap)
Mg	372	10
Al	353	13
Cu	116	19
Zn	68	19
Steel	33	14

12.2 Earlier Processes

Pig iron as such is hardly a useful material for practical purpose, because of contains high impurity (particularly carbon), it is brittle in nature. By re-melting and minor composition adjustment it can be use as raw material for cast iron production in foundries. Cast iron is not malleable i.e. ductile nature. The malleable form of iron can be produced by removing most of the impurities present in pig iron and adding some alloying elements.

There are three earlier processes:

1. Wrought ironmaking,
2. Cementation process, and
3. Crucible process.

12.2.1 Wrought Ironmaking

Pure iron melts at 1539 °C, the presence of impurities decreases the melting point of iron, so hot metal is usually tapped from the blast furnace at 1300–1350 °C in the form of molten stage. Before 1850 AD, the maximum attainable furnace temperature was of the order of 1450 °C which was obtained by combustion of fuels with cold air blast. As a result, pig iron could be readily re-melted in that furnace. However, during refining of this pig iron, as the impurity content decreased, the melting point of iron increased. With the progress of refining it become more and more viscous and nearly pasty mass when most of the impurities were removed. In a semi-liquid state clean slag and metal separation was almost impossible. The maximum amount of slag was squeezed out from the pasty mass by mechanical working, finally producing a nearly impurity free iron containing a small amount of mechanically entrapped slag particles. This was known as *wrought iron*, a name related to its fibrous structure. It is quite malleable and was the purest form of iron produced commercially on a large scale on that time.

12.2.2 Cementation Process

The origin of this process is not known, but it had been used in Sheffield, UK for many years. Since wrought iron lack-of hardness required for making cutting tools. It was carburized in solid state by heating it in contact with carbonaceous materials for sufficiently long time to allow carbon to diffuse into the surface of wrought iron. The iron-carbon alloy thus formed at the surface was hard enough for making cutting tools. This was known as cementation process.

When iron bars (75–80 mm wide and 10–15 mm thick) were heated (about 800–900 °C) in a pot (1.22 × 1.22 × 3.66 m) covered with carbon, the iron absorbed carbon. The amount of carbon absorbed depends on the temperature and the length of time for contact of two materials. The product of the cementation process was known as *blister steel*. The blister, which appeared on the surface of the bar and were usually hollow, were formed by liberation of CO gas. Since the wrought iron bars contained a certain amount of slag; the iron oxide (FeO) of the slag reacted with the carbon present to form CO gas.

$$(\mathrm{FeO}) + \mathrm{C(s)} = \mathrm{Fe(s)} + \{\mathrm{CO}\} \tag{12.1}$$

12.2.3 Crucible Process

Cementation process lacked very much in compositional homogeneity. Tools made from such an inhomogeneous material were not uniform in properties. Although steel is used for more than 5000 years, but until 1740 AD it was not produced in the liquid form. Liquid steel gives very good compositional homogeneity.

Despite the modern process in the art of steelmaking, crucible process is still being used very much in the same manner, as it was used many years ago. Its products are usually high grade alloy and special steels. The crucible process is the first method of producing liquid steel and it was introduced by a clockmaker called Huntsman at Sheffield, UK in 1740. Main problems of crucible process, on that time, were as follows:

(i) Obtaining a crucible which had sufficiently refractory to withstand the high temperature and service conditions,
(ii) The provision of a sufficiently high temperature and the production of a sound ingot.

The crucible is made either of fireclay or graphite. Coke fired furnace (i.e. pit furnace) is used for crucible process. When the crucible (which was placed on the furnace previously) and furnace was brought up gradually to a full red heat. Then the feed was charged into the crucible by means of a charging funnel. The lid was then placed on the crucible. Melting took 3–4 h. When melting was completed, then the killing fire was done. The killing process was done by addition of fresh coke and brings the metal up to a high temperature. During that period, carbon react with the fireclay resulting reduction of silica.

$$SiO_2(s) + 2C(s) = Si + 2\{CO\} \quad (12.2)$$

This silicon reduces the oxide of iron.

$$2FeO + Si = 2Fe + SiO_2 \quad (12.3)$$

Now a day, the killing process has been replaced by additions of Fe–Si and Fe–Mn into the melt.

12.3 Modern Steelmaking

(a) In 1855, the first of the new techniques of steelmaking, by blowing air into the hot metal from the bottom of the vessel, was developed by Henry Bessemer. That process of steelmaking is known as *Pneumatic process* (i.e. self-generating heat for the process) or *Acid Bessemer process*. Modern steelmaking processes was developed with the invention of the air blown into the high silicon hot metal by Henry Bessemer.
In 1878, *Thomas process*, a modified Bessemer process was developed to permit processing of the high phosphorous hot metal. It is also known as *Basic Bessemer process*.

(b) Prior to 1850's the maximum furnace temperature was attended of 1400–1450 °C, due to burning of cold fuel with cold air. Pig iron could be readily melted in such furnace, but after refining the product was nearly pure form of iron (melting point 1539 °C), could not however be kept molten condition. Clear slag and metal separation from the pasty mass was impossible. By developing the principle of regeneration of heat from exhaust gases, a part of the heat in the exhaust gases was stored and returned to the furnace as sensible heat in a reversible cycle. Siemens brothers in Germany could utilize the concepts of heat energy and its intensity, modify the concept of heat regeneration principle and applied to existing reverberatory furnace with additional fittings to develop temperature around 1600 °C for the first time in 1861. Product of refining of pig iron could be kept in molten condition. Martin brothers in France design the furnace on the principle of regeneration of heat from exhaust gases. Steelmaking in this furnace was known as *Siemens-Martin process* (in 1862), commonly known as the *Open-Hearth Process*. The process utilizes regeneration of heat transfer to pre-heat air used in a burner and can generate sufficient heat to melt and refine solid pig iron and steel scrap in a reverberatory furnace.

(c) The availability of high tonnage oxygen at reasonable costs after World War II led to the development of various oxygen steelmaking processes including the top blown oxygen process (LD/BOF), LD-AC, Kaldo, Rotor, bottom blown process (OBM) etc.
In 1950, the steel industries in most countries has just recovered from the World War II and only 189 million tonnes (Mt) of crude steel were produced during that year throughout the World [2]. At that time main steelmaking process was the open-hearth process in which 77.5% of the steel were produced. The electric arc furnace represented not more than 6.33% and only important pneumatic process was the basic Bessemer process which produced 10.5% of the total production. It was also being noted that in 1950, 86.5% of the total crude steel were produced by the USA, USSR and the nine countries of the European Common Market.
Between 1950 and 1960, the quantity of steel produced throughout the world increase by 158 Mt, but the still predominant the open-hearth process (68%) and basic Bessemer process (11.2%). In 1960, oxygen steelmaking processes represented only 13.5 Mt i.e. 4% of the total steel production.
The 60's decade were characterized by very important changes in world steel production as follows:

- The rapid increase of the world's crude steel production which reached 595 Mt in 1970.
- The important progress of the crude steel production in Japan from 22 Mt (i.e. 6.34%) in 1960 to 93 Mt (i.e. 15.63%) in 1970.
- The development of the oxygen blowing processes from 13.5 Mt (i.e. 3.89%) in 1960 to 237 Mt (i.e. 39.83%) in 1970.
- Increasing production of carbon steels in electric furnace from 37.5 Mt (i.e. 10.81%) in 1960 to 85 Mt (i.e. 14.29%) in 1970.

The 70's decade were characterized by very important changes in world steel production as follows:

- Increasing in production of the industrialized countries has slowed down sharply between 1970 and 1980.
- World steel production increased by 122 Mt.
- A large part of these additional tonnes were produced in developing countries which have taken an increasing share of the world steel production (from only 7% in 1960 to 16% in 1979).
- The electric furnace has continued to increase its production from 85 Mt in 1970 to 150 Mt in 1979.
- Rapid decrease of the open-hearth process from 230 Mt (i.e. 38.66%) in 1970 to 164 Mt (i.e. 21.93%) in 1978.
- Disappearance of the Bessemer process from 23 Mt (i.e. 3.87%) in 1970 to only 1.5 Mt (i.e. 0.2%) in 1978.
- In the beginning of the 70's decade, the top blown oxygen processes were suddenly challenged by a new comer i.e. the bottom blown process (i.e. OBM/Q-BOP).

(d) After 1970, combined blowing processes are developed. Preheating of scrap with the bottom blowing tuyeres acting as burners during preheating with combustion of liquid or gaseous hydrocarbons. Oxygen is blown from top as well as from bottom; e.g. KMS, LBE, LD/OB, LD-KG etc.

Oxygen steelmaking processes are very popular throughout the world, steel produced by these processes are more than 70% (as shown in Table 12.2). Table 12.3 shows the top five steel producing countries in world.

12.4 Various Steelmaking Routes

There are two major conventional routes for producing steel. Both routes play an important role in the development of smelting reduction (SR) technology. First route: BF → LD/BOF is used by an integrated steel plant where iron ore is reduced in a blast furnace (BF) and subsequently refined in

Table 12.2 World steel production by processes [2]

Process	2013	2014	2015	2016	2017	2018
Oxygen %	73.3	73.9	74.3	73.8	71.5	70.8
Electrical %	26.0	25.6	25.2	25.7	28.0	28.8
Other[a] %	0.7	0.5	0.5	0.5	0.5	0.4
Total (Mt)	1,648.0	1,663.2	1,618.8	1,626.9	1,688.2	1,807.1

[a]Mainly Open hearth

Table 12.3 Five major crude steel producing countries in world (Mt) [2]

Country	2013 ([a])	2014	2015	2016	2017	2018
China	822.0(1)	822.3(1)	803.8(1)	807.6(1)	870.9(1)	928.3(1)
Japan	110.6(2)	110.7(2)	105.1(2)	104.8(2)	104.7(2)	104.3(3)
India	81.3(4)	87.3(4)	89.0(3)	95.5(3)	101.5(3)	106.5(2)
United States	86.9(3)	88.2(3)	78.8(4)	78.5(4)	81.6(4)	86.6(4)
Russia	69.0(5)	71.5(5)	70.9(5)	70.5(5)	71.5(5)	71.7(6)[b]
World total	*1,650.4*	*1,669.5*	*1,620.0*	*1,627.0*	*1,729.8*	*1,808.4*

[a]Rank due to production [b]South Korea 72.5(5)

converter (LD/BOF) to produce steel. Second route: DR → EAF, where iron ore is reduced in a solid state to form sponge iron/DRI/hot briquetted iron (HBI) which is melted in electric arc furnaces (EAF) along with recycled/purchased steel scrap and further processed into final products. Third route: SR → BOF/EAF is used by smelting reduction (SR) process to produce hot metal and subsequently refined in converter process (LD/BOF) or in electric arc furnaces (EAF) for making steels. All these three routes are shown in Fig. 12.2.

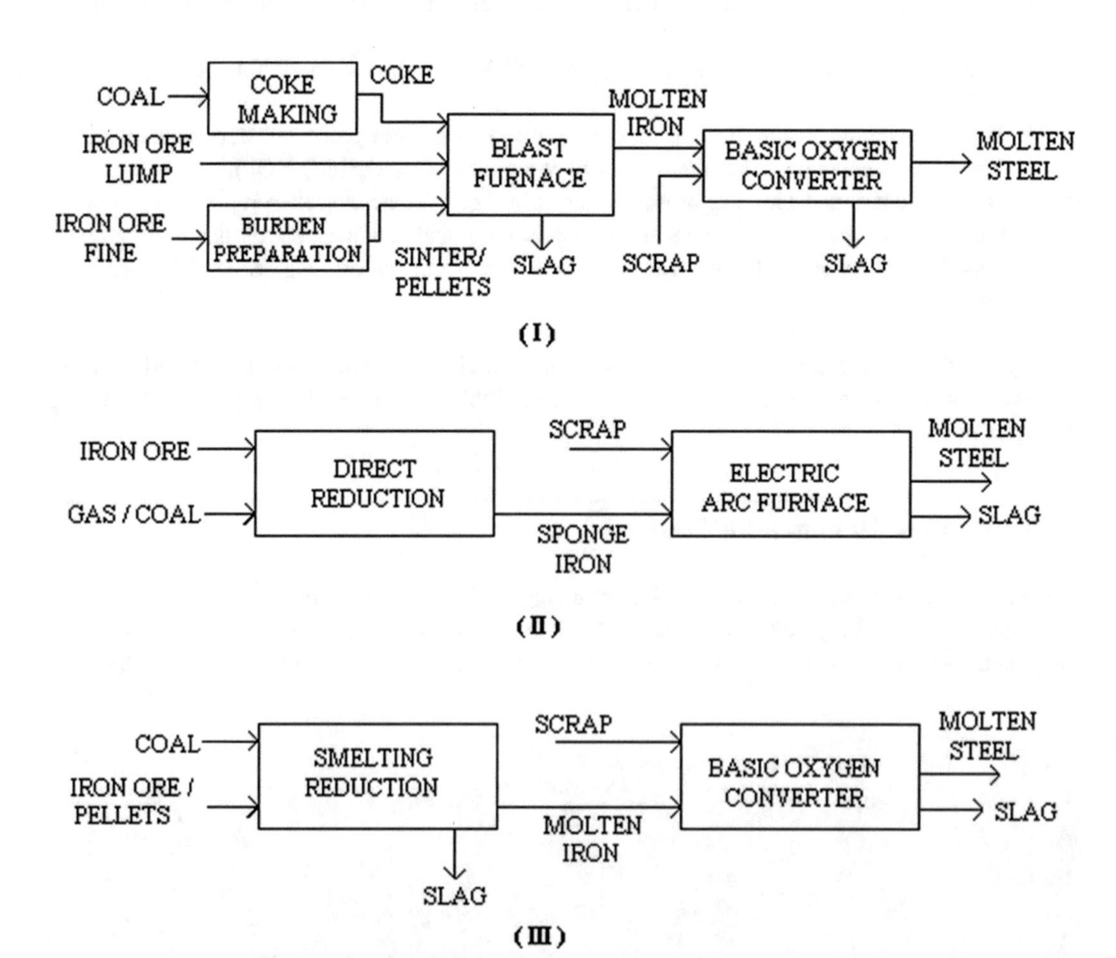

Fig. 12.2 Various steelmaking process routes [3] (reproduce with permission from Authors)

12.5 Sources of Heat in Steelmaking

The temperature of hot metal, which is tapped from blast furnace, is about 1300–1350 °C. Now to produce 0.1% C steel, whose melting point is around 1500 °C, more heat is required. At melting point the metal is more viscous, at high temperature (100 °C more than melting point) the viscosity of metal is less, and fluidity is higher; that means the metal will flow freely. So, the tapping temperature of liquid metal must be 100 °C above the melting point of 0.1% C steel, hence the tapping temperature is required about 1600 °C. But the temperature of hot metal is about 1300 °C.

Therefore, difference in temperature is 300 °C (1600–1300), to increase 300 °C temperature, extra heat must be supplied to the system. This extra heat comes from the heat of oxidation of the reactions (like $Si \rightarrow SiO_2$, $C \rightarrow CO$, $Mn \rightarrow MnO$, $P \rightarrow P_2O_5$ etc., all reactions are exothermic in nature, i.e. heats are generated).

When steel scrap is used, the scrap is to be heated from room temperature to high temperature; for that heat to be supply for: (i) heating the scrap from room temperature to melting point, (ii) latent heat of fusion for melting, (iii) super heat (i.e. melting point to high temperature). If charge 50% hot metal and 50% scrap; so, require a lot of heat. Here the heat of oxidations is not enough to provide all required heat, hence a suitable heat source is required.

Classification of steelmaking processes is shown in Fig. 12.3. Depending upon the heat requirement, there are two types of steelmaking processes are developed:

(1) The heat of oxidations is enough for the process i.e. extra heat from external source is not required at all. This process is known as *Autogenous* or *Pneumatic process.*
(2) The heat of oxidations is not enough for the process i.e. extra heat, from external source, is required by burning some liquid or gaseous fuel but not solid fuel i.e. coke; or using electrical energy. These type processes are known as *Non-autogenous Process.*

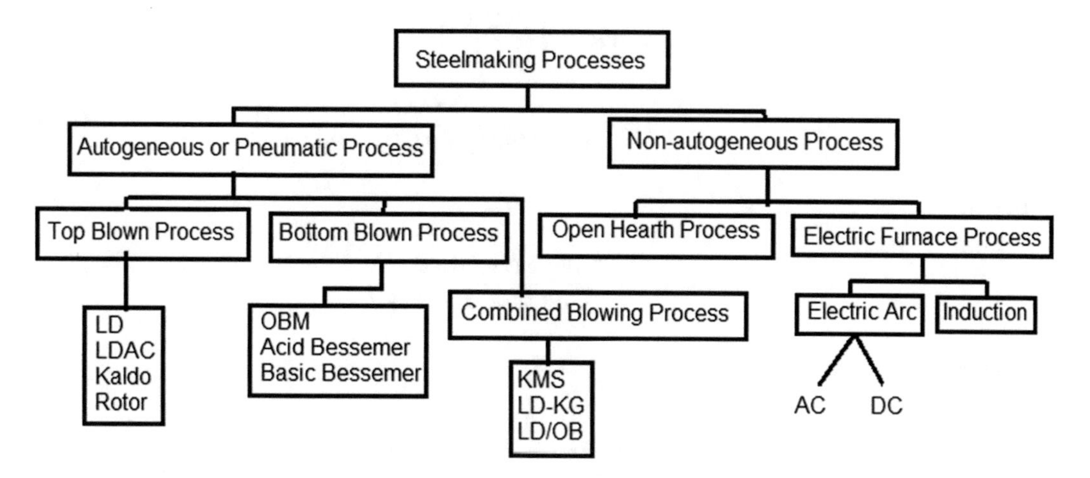

Fig. 12.3 Families of steelmaking processes

12.6 Slag

Slag is an important component in iron and steelmaking processes, due to its direct contact with the metal baths, refractories and the environment. The combined molten oxides (usually a mixture of $CaO–MgO–SiO_2–FeO–Al_2O_3$), which are the by-product of smelting and refining processes, are known as *slag* that can be found floated above the metal bath. In other word, oxide melt containing at least two oxides of opposite chemical nature should be called *slag*. The temperature and composition of the slag determine physical properties of slag like viscosity, thermal conductivity and surface tension. These properties of the slag affect the intensity of erosion of the refractory lining of the furnace and dissolution of the lining materials in the slag.

The role of slag in the steelmaking processes is extremely important. The chemical reactions between the slag and metal are directed towards attaining equilibrium, because of which the composition of the metal at the end of a melt is largely decided by the composition of the slag.

The slag forming components may be divided into three categories according to the properties of materials:

1. Acid oxides: SiO_2, P_2O_5, TiO_2, V_2O_5, B_2O_3 etc.
2. Basic oxides: CaO, MgO, FeO, MnO, Na_2O, K_2O etc.
3. Amphoteric oxides: Al_2O_3, Fe_2O_3, Cr_2O_3, V_2O_3 etc.

The most decisive influence on slag properties are exerted by SiO_2 and CaO. The content of iron oxide, in particular of FeO, in a free state (i.e. not chemical combined in slag phase) determines the activity of FeO, i.e. *oxidizing power of slag*.

The ability of a slag to retain oxides is generally expressed as the ratio of basic oxides to acid oxides; and is variously represented as:

$$\text{V ratio} = \left(\frac{\%\ \text{CaO}}{\%\ \text{SiO}_2}\right) \tag{12.4}$$

$$\text{Modified}\quad \text{V ratio} = \left(\frac{\%\ \text{CaO}}{\{\sum \%(\text{SiO}_2 + \text{Al}_2\text{O}_2 + \text{P}_2\text{O}_5)\}}\right) \tag{12.5}$$

Basicity,

$$\text{B} = \left(\frac{\{\sum\ \%(\text{all basic Oxides})\}}{\{\sum\ \%(\text{all acid Oxides})\}}\right) \tag{12.6}$$

Generally,

$$\text{B} = \left(\frac{\{\sum\ \%(\text{CaO} + \text{MgO})\}}{\{\sum\ \%(\text{SiO}_2 + \text{Al}_2\text{O}_2 + \text{P}_2\text{O}_5)\}}\right) \tag{12.7}$$

Acid slag: V or B < 1, basic oxide, CaO may or may not be present at all.
Basic slag: V or B > 1.

The opposite chemical nature results in the formation of complex silicates, phosphates etc.

$$2RO + SiO_2 = 2RO \cdot SiO_2 \tag{12.8}$$

$$4R_2'O + P_2O_5 = 4R_2'O \cdot P_2O_5 \tag{12.9}$$

where RO is basic oxides like CaO, MgO etc. and $R_2'O$ is alkali oxides like Na_2O, K_2O etc.

The basic oxides are by and large ionic solids:

$$RO = R^{2+} + O^{2-} \tag{12.10}$$

$$R_2'O = 2R'^{+} + O^{2-} \tag{12.11}$$

The acid oxides have dominant covalent bonding and hence do not dissociate into respective simple ions as above. In an orthosilicate melt the dissociation appears to be as:

$$2RO \cdot SiO_2 = 2R^{2+} + SiO_4^{4-} \tag{12.12}$$

and that indicates as there is no activity of oxygen ions (O^{2-}) in the slag. This composition can be referred to as a *neutral silicate slag*. On the other hand, the neutral composition has deficient of oxygen ions (O^{2-}), it is termed as *acid slag*.

As basic oxides contain the free oxygen ions (O^{2-}) in the slag and so it is termed as *basic slag*. A slag with V ratio equal to two has the maximum oxidizing power for any FeO content. The basicity and oxidizing power of a slag are independent properties. In practice basicity is generally related to lime concentration and oxidizing power to the FeO content of the slag, although FeO is also a basic oxide. However, slag with much higher V ratio is required to effectively eliminate phosphorus and sulphur from the melt during refining.

The main function of slag is traditionally associated with its action to absorb deoxidation products. During steelmaking, if the slag acts only as a receiver for oxides product, it does not take part in the reactions and it may be very viscous (e.g. acid Bessemer slag); so that it cannot be easily separated from the molten bath; that slag is known as *dry slag*.

If the slag acts as a receiver for oxides product as well as take part in refining reactions, it must be fluid i.e. low viscosity and high fluidity, it can be easily separated from the molten bath; that slag is known as *wet slag*. A fluid slag in general has good thermal conductivity and increases mass transport and thereby accelerate reactions.

The viscosity of any slag can be decreased by increasing the temperature. Highly basic slags are not thin (i.e. fluid) enough even at high temperature (1700 °C). The viscosity of these slags is generally decreased by the addition of fluorspar (i.e. spar, CaF_2) or bauxite (Al_2O_3).

The fluorspar dissociates as:

$$CaF_2 = Ca^{2+} + 2F^{-}. \tag{12.13}$$

The fluorine ions (F^-) act like oxygen ions (O^{2-}) in modifying the complexes and thereby decrease the viscosity. The sag containing fluorspar cannot be used as a phosphatic fertilizer.

Aluminum oxide (Al_2O_3) is a very refractory oxide, its addition does reduce the softening point of some oxides and hence the viscosity of slags decreased. Bauxite (Al_2O_3) is used where fluorspar cannot be used.

12.7 Ternary Diagram

Since slags contain three to four metal oxides; to understand the effect of one oxide to another, ternary diagram should be understood first. The components of the slag are located at the corners of the triangle while the sides represent the three binary systems. Figure 12.4 shows the ternary diagram of $CaO–Al_2O_3–SiO_2$ system. Any point within the triangle specifies the composition of a ternary system.

There are two methods for determining the composition of a point within the triangle as follows:

(i) The length of perpendiculars drawn from the sides of the equilateral composition triangle to the point are proportional to the quantities of *A*, *B* and *C* in the sample (Fig. 12.5).
e.g. Composition at *x* is composed of xa% *A*, xb% *B* and xc% *C*.
Therefore, xa + xb + xc = 100, and *A* = 100 – (*B* + *C*).

(ii) (a) Lines are drawn through the composition point *y*, parallel to the sides of the equilateral composition triangle. The intersection of these lines with any side of the triangle is proportion to the opposite component (Fig. 12.6).
(b) To find the composition of the system at the point y, drawing three lines from this point parallel to the sides of the triangle. The corresponding length of the lines show the % of components *A*, *B* and *C* in the slag (Fig. 12.7).

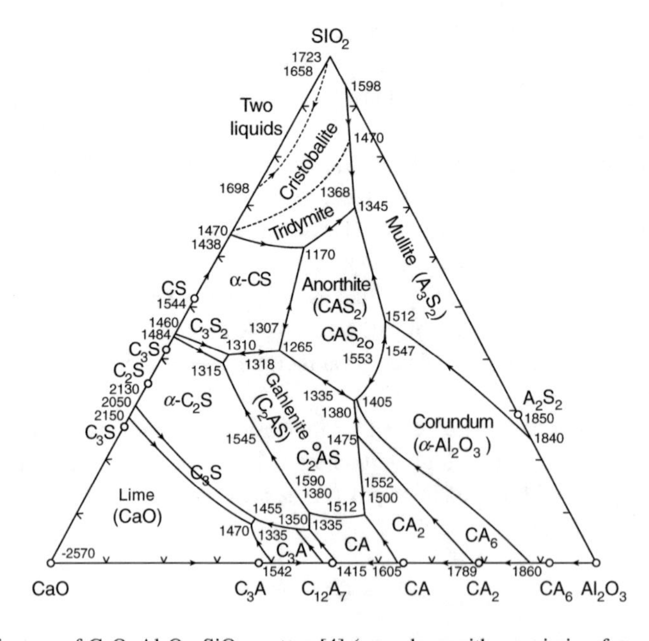

Fig. 12.4 Ternary diagram of $CaO–Al_2O_3–SiO_2$ system [4] (reproduce with permission from Author)

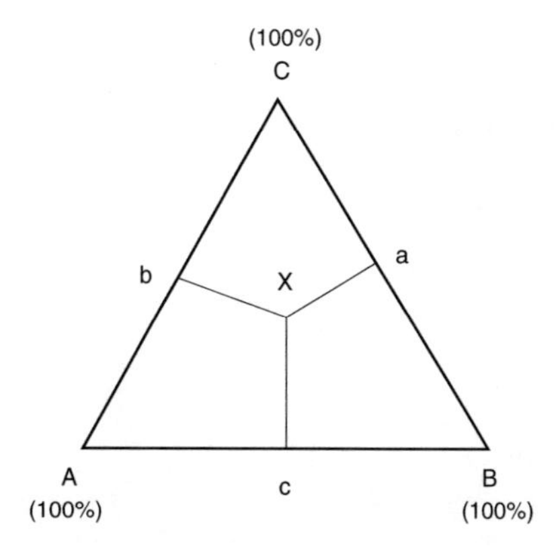

Fig. 12.5 Ternary diagram of A-B-C (perpendicular method)

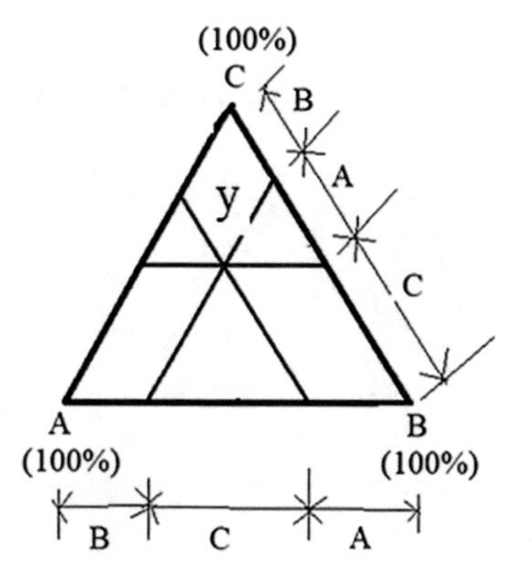

Fig. 12.6 Ternary diagram of A-B-C (parallel method)

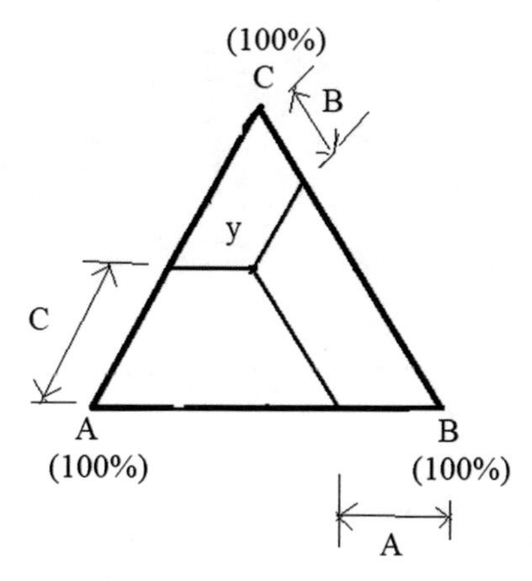

Fig. 12.7 Ternary diagram of A-B-C (parallel method)

12.8 Basic Differences

12.8.1 Difference Between Cast Iron and Steel

Cast Iron	Steel
1. It is an alloy of Fe–C–Si	1. It is an alloy of Fe–C–Mn
2. Theoretically maximum C content 6.67% (by weight)	2. Theoretically maximum C content 2% (by weight)
3. Practically, maximum C content about 3–4% due to present of other alloying elements	3. Practically, maximum C content (for high carbon steel) about 1.5–1.7%
4. On cast condition, it cannot be rolled (i.e. it is brittle nature)	4. On cast condition, it can be rolled (i.e. it is ductile nature)
5. Melting point is lower than steel (i.e. around 1150–1300 °C)	5. Melting point is higher (i.e. >1375 °C)

12.8.2 Difference Between Plain Carbon Steel and Alloy Steel

Plain carbon steel	Alloy steel
1. Total alloying elements content less than 5% (mostly <3%)	1. Total alloying elements (TAE) content more than 5%, but less than 25%
2. Low carbon or mild steel: → 0.25–0.30% C	2. Low alloy steel: → less than 5% TAE.
3. Medium carbon steel: → 0.25–0.30% to 0.65–0.70% C	3. Medium alloy steel: → 5–10% TAE
4. High carbon steels: → 0.65–0.70% to 1.6–1.7% C	4. High alloy steels: → 10–25% TAE

12.8.3 Difference Between Alloy Steel and Ferro-Alloy

Alloy steel	Ferro-alloy
1. Total alloying elements content not more than 25% and iron content not less than 75% i.e. Iron is more and alloying elements are less	1. Total alloying elements content not less than 50% and iron content not more than 50% i.e. Iron is less and alloying elements are more
2. Alloying elements are more than one	2. Alloying elements are at least one

Probable Questions

1. Discuss the earlier processes of steelmaking.
2. What is 'wrought iron'?
3. "During process of refining of pig iron for wrought ironmaking, melt become more and more viscous." Why?
4. What do you understand by 'blister steel'?
5. How steelmaking processes are classified?

6. Discuss the three routes of steelmaking from iron ore.
7. What are the sources of heat in steelmaking processes?
8. What do you understand by autogeneous Processes? How are classified of that processes?

References

1. R.D. Pehlke, *Unit Processes of Extractive Metallurgy* (Elsevier, New York, 1973)
2. *World Steel in Figures 2019* (World Steel Association)
3. S.K. Dutta and R. Sah, *Alternate Methods of Ironmaking (Direct Reduction and Smelting Reduction Processes)* (S. Chand & Company Ltd., New Delhi, 2012)
4. A.K. Singh, Ph.D. Thesis *Study on The Effect of Different Sols on High Alumina Castable Refractory*, NIT Rourkela, 2017. (https://www.researchgate.net/publication/318122611_Study_on_The_Effect_of_Different_Sols_on_High_Alumina_Castable_Refractory)

13 Raw Materials for Steelmaking

Main raw material for pneumatic or autogenous steelmaking processes is hot metal and pig iron and/or scrap/sponge iron is for non-autogenous steelmaking processes. The oxidizing agents used for steelmaking are air, oxygen and iron oxide. Lime is generally added during steelmaking to make the slag basic enough to retain phosphorus and sulphur in the slag phase. The pre-treatment of hot metal is carried out for improving cost-effectiveness in steelmaking by reducing the volume of slag. Pre-treatment of hot metal is done for de-siliconization (up to $Si < 0.15\%$), de-sulphurization and de-phosphorization before charging to the steelmaking furnace.

13.1 Raw Materials

The main raw materials for steelmaking are as follows:

- Sources of metallic iron,
- Oxidizing agents,
- Fluxes,
- Deoxidizers and alloy additions,
- Furnace refractory and
- Sources of heat.

13.2 Sources of Metallic Iron

There are mainly two sources of metallic iron for steelmaking: (i) primary and (ii) secondary.

13.2.1 Primary Sources of Metallic Iron

Primary source of metallic iron is hot metal from blast furnace. Hot metal, which is molten form of impure iron, is the product of blast furnace after the reduction of iron ore. Pig iron is a solid form of impure iron after cast in a small piece. Hot metal/pig iron can be classified according to silicon and phosphorus contents.

S. K. Dutta and Y. B. Chokshi, *Basic Concepts of Iron and Steel Making*,
https://doi.org/10.1007/978-981-15-2437-0_13

Acid pig iron contains low sulphur and phosphorus (<0.04% each), but high in silicon (2.0–2.5%). *Swedish iron* contains 3–4% C, 2.0–2.5% Si, 1.0–1.5% Mn, <0.04% P and <0.04% S.

Basic pig iron contains low silicon (<1.0%), but high in phosphorus (>1.5%). *Thomas iron* contains 3–4% C, 0.5–1.0% Si, 0.5–1.0% Mn, <0.04% S and >1.5% P.

13.2.2 Secondary Sources of Metallic Iron

Secondary sources of metallic iron are (i) steel scrap, (ii) sponge iron/DRI/HBI and (iii) iron carbide.

13.2.2.1 Scrap

Steel plants itself generate a lot of scrap in the form of rejected material, chop-out of ingot/concast head, etc. These are variously called as home scrap, home returns, plant returns, circulating scrap, etc., which are about 25–30% of ingot steel production. Due to the adoption of continuous casting (concast) in place of ingot casting, the circulating scrap generation is decreased to 5–10%. Although it is feasible to use 100% hot metal charge in a steelmaking furnace, it is also necessary to use the home scrap, which is generated within the plant, as part of the charge to utilize some of the scrap.

(a) Scrap is an energy-intensive, valuable commodity, and price of scrap is dictated by market demand and supply. Scraps are generated primarily from three main sources:

   (i) Reclaimed scrap (also known as obsolete scrap) is obtained from old cars, demolished buildings, discarded machineries and domestic objects.
   (ii) Industrial scrap (also known as prompt scrap) is generated by industries using steel within their manufacturing processes.
   (iii) Revert scrap (also known as home scrap) is generated within the steelmaking and forming processes in industries, e.g. chop-out from ingot/concast, rolling operation; metallic losses in slag; defected or rejected portion of ingot/concast, etc.

   The last two types of scrap are good quality, since they are near to chemical composition of the desired molten steel composition, and thus, these are ideal for recycling. Alloy steel scraps are the cheap sources of alloying elements for the charge.

   Reclaimed/obsolete scraps frequently have a quite variable composition and quite often contain contaminant elements, such as Zn and Sn that are undesirable for steelmaking. Levels of residual (or tramp) elements such as Cu, Sn, Sb, As, Ni, Cr and Mo are high (0.13–0.73%) in obsolete scrap (as shown in Table 13.1) and can affect the quality of product, i.e. steel, if they are not diluted during the melting. Since these tramp elements do not oxidize during the refining of steelmaking, they remain as residual elements in steel. The presence of residual elements is above

Table 13.1 Residual (or tramp) elements for different raw materials [1]

S. No.	Raw material	Residual element % (Cu + Sn + Ni + Cr + Mo)
1	Sponge iron/DRI	0.02
2	Pig iron	0.06
3	No. 1 Bundle (steel scrap)	0.13
4	No. 1 Heavy melting scrap	0.20
5	Automobile shredded scrap	0.51
6	No. 2 Heavy melting scrap	0.73

certain specified levels that effect the properties of steel. As the scrap gets recycled again and again, the residuals may increase beyond safe levels.
Thus, there is a need for very low levels of residual (or tramp) elements in the steel that will be forced to use good-quality prompt scrap but cost is much higher. The alternative is to use a combination of the contaminated obsolete scrap along with good-quality iron units or virgin iron unit. These are materials which contain little or no residual elements. Good-quality iron units are typically in the form of iron carbide (Fe_3C), sponge iron/DRI (0.02%), pig iron and hot metal (0.06%) (as shown in Table 13.1).

(b) Scrap can be further classified into four broader groups based on its chemistry from the point of view of steelmaking:

  (i) Scrap contains volatile elements like Zn, Cd, Pb, etc.,
  (ii) Scrap contains non-oxidizable elements (during refining) like Cu, Ni, Sn, Mo, As, W, etc. (This category is known as the problem of residuals.),
  (iii) Scrap contains partially oxidizable elements like P, Cr, Mn, etc.,
  (iv) Scrap contains completely oxidizable elements like Al, Si, Ti, V, Zr, etc.

(c) Scrap is also classified based on its physical size, its source and the way in which it is prepared:

  (i) Number (No.) 1 bundles, No. 1 factory bundles, No. 1 shredded, No. 1 heavy melt.
  (ii) No. 2 heavy melt, No. 2 bundles, No. 2 shredded.
  (iii) Turnings, shredded auto, rail, rail wheels, etc.

China, Korea, Taiwan, India, Middle East, and Turkey heavily depend upon imported scrap. This high demand caused scrap prices increased to 400% between 2003 and 2008. Although the scrap price slumped during the 2008–2009 recession, the price strongly recovered in 2010.

Raw materials selection is very important to assure low nitrogen content. In steels produced in an electric arc furnace (EAF), nitrogen gas comes from two main sources:

- From the scrap (mainly from purchased scraps) which contain a lot of voids (filled with air),
- By air infiltration into the furnace.

To assure low nitrogen content of the steel and to avoid sudden changes in nitrogen from the raw materials, all raw material inputs to steelmaking should be sampled and analyzed (Table 13.2). Based upon this analysis, care should be taken to minimize the addition of any materials that might represent an important nitrogen contribution.

13.2.2.2 Sponge Iron/DRI/HBI

Another secondary source of metallic iron is sponge iron/DRI/HBI. Basic oxygen furnaces (LD/BOF) are normally operated with plant return scrap as coolant, and scrap is the main raw material for electric arc furnaces (EAF)/induction melting furnace (IMF). However, with the adoption of continuous casting (concast) method and improvement in steel yield, the generation and availability of plant return (i.e. scrap) is decreased and resulted in a shortage of scrap throughout the world in 1980 onwards. The scrap shortage can be overcome by alternative sponge iron/DRI/HBI which acts as a coolant for steelmaking in LD/BOF, as well as raw material for EAF/IMF. The use of sponge iron as a replacement of scrap in steelmaking is now established.

Table 13.2 Nitrogen gas content in raw materials for steelmaking [2]

S. No.	Materials	Nitrogen content, ppm
1.	Shredded scrap	100
2.	Internal scrap	50
3.	Sponge iron/DRI	20
4.	Coke	9000
5.	Anthracite coal	10,000
6.	Graphite for injection	13,000
7.	Fe–Si	320
8.	Si–Mn	85
9.	Fe–Nb	270
10.	Low carbon Fe–Mn	710
11.	High carbon Fe–Mn	140
12.	Fe–Mn (electrolytic, flakes)	45
13.	CaF_2	3060
14.	Fluxes	450

Advantages of sponge iron/DRI/HBI are given as follows:

- Uniform size
- Consistent composition
- Low of tramp element (0.02%)
- Low impurity level (S, P)
- Stable price (since it is a product) than scrap (which is by-product, price vary with market demand and supply).

Disadvantages of sponge iron/DRI/HBI:

- Higher slag volume due to high gangue content
- Increased flux consumption due to high SiO_2 content
- Increased hot metal consumption because of higher cooling effect and lower Fe input
- Lower yield because of lower Fe input and high Fe loss through slag.

13.2.2.3 Iron Carbide

Another alternative material to scrap is iron carbide (Fe_3C). Iron carbide is a stable and product of metallic iron with a bonus carbon (6.0–6.5%). Besides the natural advantage of low content of tramp elements (like that of sponge iron), iron carbide provides an additional benefit of saving energy by carbon–oxygen exothermic reaction. The iron carbide is an easily injectable and instantly dissolving charge in steel bath and supporting foamy slag practice in EAF by carbon boil and an effective material to produce low hydrogen, nitrogen and sulphur steels.

Iron carbide is far more effective and less costly than any other materials for producing high-quality steel. A report for the US Department of Energy Technology Roadmap Program [3] identified iron carbide as the preferred material for nitrogen control in EAF steelmaking. Total CO_2 emissions from the process are reported to be 2.17 t CO_2/t-steel production. Table 13.3 shows the chemical analysis of iron carbide compared to DRI (i.e. sponge iron) and HBI.

Table 13.3 Compared chemical analysis of iron carbide with DRI (i.e. sponge iron) and HBI [4]

	Iron carbide	Midrex DRI	Midrex HBI
Total iron, Fe_T	89–93%	90–94%	90–94%
Metallic iron, Fe_m	0.5–3.0%	83–90%	83–90%
Carbon	6.0–6.5%	1.0–2.5%	0.5–1.5%
Iron carbide, Fe_3C	90–96%	–	–
Magnetite, Fe_3O_4	0.5–2.0%	–	–
Gangue, $SiO_2 + Al_2O_3$	1–4%	3–6%	3–6%

Being hard, dense, chemically stable, and granular size, iron carbide is easy to handle and safe to transport by ship [4]. Being fine and heavy, steelmakers can easily inject it into electric arc furnaces (EAFs) using submerged lances. Iron carbide is far more effective and less costly than any other means for removing nitrogen and producing high-quality steel. After carrying the iron carbide grains into the bath, the injection gas (nitrogen or air) rises to the bath surface without significantly reacting with the metal. This explains why injecting inert gas into an EAF fails to remove dissolved nitrogen and hydrogen and why vacuum degassing is so slow and expensive. In contrast, iron carbide forms swarms of fine bubbles through a different mechanism.

When iron carbide enters an EAF, it dissolves instantly. Next, the dissolved carbon reacts with the small amount of iron oxide left in the iron carbide product. The carbon and iron oxides form carbon monoxide. The reaction occurs on a minuscule scale, but extensively. This generates an immense quantity of very fine carbon monoxide bubbles [4]. The tiny bubbles create a vigorous metal boil, rapidly homogenizing the bath, removing nitrogen and hydrogen gases and creating a foaming slag. These properties are extremely beneficial. Steel produced with 15% iron carbide can meet stringent quality standards. This steel is suitable for deep-drawn products. The liquid steel contains 30 ppm nitrogen and 3 ppm hydrogen during tapping time.

Injecting iron carbide directly before the completion of the EAF batch provides the best nitrogen and hydrogen removal [4]. If iron carbide provides a large portion of iron units to an EAF batch, injection of iron carbide can commence as soon as the EAF has enough molten steel to submerge the injecting lance. The furnace heat does not damage the injecting pipe because the transport gas adequately cools the lance. Dust losses are not evident. The widespread generation of tiny carbon monoxide bubbles thoroughly mixes the bath. The mixing is far more effective than argon injection mixing, with iron carbide bath attaining full mixing in 1 min, than 4 min with high flow rates of argon injection. DRI, HBI, and pig iron fail to provide proper mixing and reduction of nitrogen and hydrogen gases from the bath.

Iron carbide is a revolutionary feed material for steelmaking. This new feed material for electric arc furnaces (EAFs) will profoundly impact the steel industry during the coming decade because of its outstanding metallurgical, economic, and environmental benefits. Iron carbide generates the lowest carbon emissions [4] of all processes to produce virgin steel, emitting one-half to one-third the carbon dioxide of other production routes. Iron carbide is ideal raw material for EAF steelmaking. It is granular, non-pyrophoric and dissolves instantly in molten steel. This makes it easy to ship and simple to inject into EAFs.

13.3 Oxidizing Agents

The oxidizing agents used for steelmaking are air, oxygen and iron oxide.

- Air was used for the Bessemer processes; nitrogen present in air was dissolved in liquid steel and that made steel brittle.
- By the development of production of high-purity (99.5%) oxygen at cheaper rate, after World War II, oxygen is used as oxidizing agent for the oxygen steelmaking processes, but nitrogen content of refined steel increase with decreasing the purity of oxygen.
- Iron oxide is used in the form of (a) lump iron ore (Fe_2O_3) and (b) mill scale (Fe_3O_4, Fe_2O_3).

  (a) Iron ore contains 60–67% iron and 25–29% oxygen.
  (b) Mill scale is nearly pure form of oxides (93–95% Fe_3O_4), produced during hot deforming (i.e. rolling, forging, etc.) of steel.

  The use of iron oxide as an oxidizing agent results in improving the iron yield of the process, but it needs thermal energy to dissociate itself and make oxygen available for refining reactions:

$$[\mathrm{Fe}] + (\mathrm{Fe_2O_3}) = 3(\mathrm{FeO}),\ \Delta H^0_{298} = 28.03\,\mathrm{kJ/mol\,Fe_2O_3} \quad (13.1)$$

Steel scrap acts as cooling material in steelmaking. Iron ore also acts as cooling material in steelmaking. Cooling effect of iron ore is nearly four times more than steel scrap. Difference between scrap and iron ore are shown in Table 13.4.

13.3.1 Deoxidizers and Alloy Additions

Since steelmaking is an oxidizing process, some oxygen must be dissolved in liquid steel. To remove the excess oxygen in liquid steel, deoxidizers are used. Various deoxidizers are used in steelmaking as follows:

(i) Elements like Si (as Fe–Si), Mn (as Fe–Mn), Al, etc., are added primarily as common deoxidizers.
(ii) Elements like Zr, B, Ti, etc., are added for deoxidation in special cases.

Table 13.4 Difference between scrap and iron ore

S. No.	Scrap	Iron ore
1.	Heat requires raising the temperature from room temperature to melting point	Heat requires to dissolve iron ore (Fe_2O_3) in slag
2.	Latent heat of fusion for melting	Dissociation of Fe_2O_3 to FeO, [Fe] + (Fe_2O_3) = 3(FeO)
3.	Heat requires melting point to bath temperature	Endothermic reaction ($\Delta H^0_{298} = 28.03\,\mathrm{kJ/mol\,Fe_2O_3}$),
4.	Scrap require about 4.4 kg	Equivalent iron ore require 1 kg
5.	1% scrap reduce the metal temperature by 8 °C	1% iron ore reduced temperature by 33 °C

(iii) Elements like Cr, W, Mo, Ni, V, Nb, etc., are added generally as alloying additions.

Excess additions of deoxidizer, after consuming for deoxidation, can remain in the melt as alloying additions. Compositions of common ferro-alloys are as follows:

Fe–Si: 1–2% C, 50–78% Si;
Fe–Mn: 0.07–0.5% C, 6.5% Si, 85–90% Mn (low carbon);
Fe–Mn: 7% C, 78–82% Mn (high carbon);
Si–Mn: 0.08% max C, 28–32% Si, 56–61% Mn;
Fe–Cr: 0.025–0.75% max C, 2% max Si, 67–73% Cr (low carbon);
Fe–Cr: 7% max C, 3% max Si, 58–65% Cr (high carbon).

13.4 Fluxes

A flux is a substance which is added during smelting and refining to bring down the softening point of the gangue materials, to lower down the viscosity of slag and to decrease the activity of some component to make it stable in the slag phase.

Lime (CaO) is generally added during steelmaking to make the slag basic enough to retain phosphorus and sulphur in the slag phase. Fluorspar (CaF_2) and bauxite (Al_2O_3) are also added to decrease the viscosity of slags. Limestone is not used at all, except for open-hearth furnace, due to a lot of heat absorbed during calcination of limestone as well as that takes time to dissociate; hence, the availability of CaO become late for slag formation.

Typical analyses of fluxes are as follows:

(i) Lime: 90–95% CaO, 2–3% MgO, 1.5% SiO_2, 0.1–0.2% S.
(ii) Calcined dolomite: 55% CaO, 34–38% MgO, 3–4% SiO_2, 0.01% S.
(iii) Bauxite: 54–56% Al_2O_3, 11–14% Fe_2O_3, 1–2% SiO_2, 1–2% TiO_2.
(iv) Fluorspar (spar): 75–85% CaF_2, 10% max. SiO_2, 0.8% max. S.

13.5 Furnace Refractory

Steelmaking furnace is lined by suitable refractory materials. The lining is eroded during steelmaking due to chemical attract by the slag; hence, the material of lining is also required as a recurring consumable raw material.

(a) Acid furnace: fireclay, silica sand.
(b) Basic furnace: dolomite, dolomite enriched with magnesite, chrome-magnesite, etc. Typical composition of stabilized dolomite brick is 40.0% CaO, 40.3% MgO, 14.4% SiO_2, 3.5% Fe_2O_3 and 1.5% Al_2O_3.

13.6 Sources of Heat

In autogenous or pneumatic processes, heat of oxidations due to the refining of impurities is enough to meet the heat requirement of steelmaking; that is, no extra heats are required from outside. The amount of heat generated is always more than necessary so that scrap and/or iron ore is charged to keep the bath temperature within required limits.

External heats are required for non-autogenous processes. The sources of heats are as follows:

(a) chemical: (i) liquid: oils, tar, etc., (ii) gas: producer gas, water gas, coke oven gas, natural gas and blast furnace gas. Chemical fuels should be burnt with excess air.
(b) Electrical: (i) induction heating, (ii) arc heating: AC and DC.

13.6.1 Heat Balance of Steelmaking Process

(I) Input (i.e. source of heat):

(i) Hot metal from blast furnace: The sensible heat carrying the hot metal is given by the linear relationship:

$$q_{\mathrm{HM}} = 0.22\,T + 17. \quad (13.2)$$

where q_{HM} is the sensible heat, kWh/t; and T is the hot metal temperature, °C.

(ii) Heat of refining reactions:

$$[\mathrm{Si}] + \{\mathrm{O_2}\} \rightarrow (\mathrm{SiO_2}),\ \Delta H^0_{298} = -871.53\,\mathrm{kJ/mol\ of\ Si} \quad (13.3)$$

$$[\mathrm{Mn}] + 1/2\{\mathrm{O_2}\} \rightarrow (\mathrm{MnO}),\ \Delta H^0_{298} = -384.72\,\mathrm{kJ/mol\ of\ Mn} \quad (13.4)$$

$$[\mathrm{C}] + 1/2\{\mathrm{O_2}\} \rightarrow \{\mathrm{CO}\},\ \Delta H^0_{298} = -111.72\,\mathrm{kJ/mol\ of\ C} \quad (13.5)$$

$$2[\mathrm{P}] + 5/2\{\mathrm{O_2}\} \rightarrow (\mathrm{P_2O_5}),\ \Delta H^0_{298} = -1585.73\,\mathrm{kJ/mol\ of\ P} \quad (13.6)$$

(iii) Extra heat from:

(a) fuel (liquid or gaseous),
(b) electrical energy,
(c) CO combustion:

$$2\{\mathrm{CO}\} + \{\mathrm{O_2}\} \rightarrow 2\{\mathrm{CO_2}\},\ \Delta H^0_{298} = -282.42\,\mathrm{kJ/mol\ CO} \quad (13.7)$$

The above (i) and (ii) are only sources for autogenous processes; and (iii) is the source for non-autogenous processes.

(II) Output (i.e. consumption of heat):

(i) To heat up the hot metal charge from 1300 to >1500 °C.
(ii) To melt scrap from room temperature to melting point and melting point to higher temperature.

(iii) To make the slag and to keep the slag in molten condition.
(iv) To heat up the refractory lining.
(v) To compensate the losses through radiation, conduction and in the form of sensible heats of liquid steel, liquid slag and exit gases.

13.7 Pre-treatment of Hot Metal

The pre-treatment of hot metal was developed by the Japanese steel industry to improve cost-effectiveness in steelmaking by reducing the volume of slag. Pre-treatment of hot metal is carried out in two or three steps; the first step is de-siliconization (up to Si < 0.15%). After de-slagging, the hot metal is de-sulphurized and de-phosphorized of the hot metal, prior to charging the hot metal in steelmaking furnaces. By injecting with nitrogen plus oxygen, a mixture of sinter fines, burnt lime, calcium fluoride and some calcium chloride are charged. In some practices, sodium carbonate is also injected alone.

The objectives of hot metal pre-treatment have bearing on economics and have environmental advantages as follows:

1. Increasing recycling of steelmaking slag (low in phosphorus) to blast furnace (BF) for the recovery of iron and manganese,
2. Developing a process with low slag volume by using hot metal containing low silicon, sulphur and phosphorus.

13.7.1 De-siliconization

To improve steelmaking process in open-hearth furnaces, attempts were made in the 1940s in the European and North American steel plants to lower the silicon content of hot metal from more than 1% Si to 0.4–0.6% Si. This was done simply by adding dried mill scale either onto the blast furnace runner or into the hot metal transfer ladle. The efficiency of de-siliconization can be increased by going in for injection, and in that case, a Si level of around 0.35% would be possible.

Depending on the operating conditions, there is a temperature rise or loss of about ±20 °C when mill scale or sinter fines are used for de-siliconization. Removal of Si in hot metal is normally accomplished by oxidation using gaseous oxygen or iron oxides. The oxidation of Si with oxygen gas raises the temperature due to exothermic reaction:

$$[\mathrm{Si}] + \{\mathrm{O_2}\} = (\mathrm{SiO_2}),\ \Delta H^0_{298} = -871.53\,\mathrm{kJ/mol} \tag{13.8}$$

A temperature loss in hot metal may occur because of slag removal, using moist reagents, extra fluxes or carrier gases. With oxygen lancing, the hot metal temperature will increase to 120–150 °C.

As hot metal is treated with various de-siliconization agents (viz. mill scale, steelmaking dust, iron ore, oxygen), both the Si and Mn contents of the metal are decreased. The use of iron oxides generally leads to similar decrease in both Si and Mn contents. Special practices like submerged oxygen injection with a controlled flow-rate into the ladle have been developed to retard some Mn loss.

The slag produced during de-siliconization contains primarily SiO_2, FeO, MnO and CaO. Since this slag contains low phosphorus and sulphur, it can be recycled to the sinter plants.

13.7.2 De-sulphurization

In most steel plants outside Japan, pre-treatment of hot metal is confined to de-sulphurization in the transfer ladle with various injected materials such as lime plus spar, lime plus magnesium, calcium carbide plus magnesium or calcium carbide plus limestone.

In hot metal, the de-sulphurizing agent combines with sulphur present in liquid iron as a sulphide and is then transferred to the slag. The agent should be thoroughly mixed in hot metal, and the reaction product should separate out from liquid iron. With the introduction of powder injection technology, the de-sulphurization processes based on the use of powdered reagents have become more effective in attaining low sulphur levels in hot metal while reducing de-sulphurization cost. There are four reagents which are considered for external de-sulphurization of hot metal. These reagents are: (i) soda ash (Na_2CO_3), (ii) powdered lime (CaO), (iii) calcium carbide mix (CaC_2-based powder mixture), and (iv) magnesium-based powders (Mg and inert material).

The quantity of magnesium (Mg) to be injected together with CaO or CaC_2 is adjusted according to initial sulphur content of the hot metal. De-sulphurization occurs primarily by the following reactions:

$$\{Mg\} + [S] = MgS(s) \tag{13.9}$$

$$CaC_2(s) = \{Ca\} + 2[C] \tag{13.10}$$

$$\{Ca\} + [S] = CaS(s) \tag{13.11}$$

Most of the de-sulphurization is done by magnesium. For the removal of 0.04% S in the hot metal, with about 70% efficiency of magnesium usage, the amount of the reagent to be injected would be 1.45 kg/thm for the 30% Mg plus 70% CaO mixture. For this quantity of solid injection with flowing nitrogen at the rate of 0.035 Nm^3/kg solids, the heat absorbed from the melt would be about 4310 kJ/thm. On the other hand, the heat generated by the reaction (13.9) would be about 5060 kJ/thm for 0.04% ΔS which compensates for the thermal energy absorbed in heating the injected material to the bath temperature.

Soda ash is a common reagent. Rourkela Steel Plant, India, had adopted CaC_2-based process which injected CaC_2 mixture (which consisted of CaC_2 and an additive of limestone and graphite). The limestone in the mixture generated enough gases which helped in stirring the hot metal and the graphite prevented explosion by mixing with air and generated heat. The reaction proceeded according to the equation:

$$(CaC_2) + 2(CaO) + 3[S] = 3(CaS) + 2\{CO\} \tag{13.12}$$

The process consists of pneumatically injecting the calcium mixture through a lance into the hot metal. Dry air is used as a carrier gas. The calcium carbide mixture is injected at the rate of 40–60 kg/min.

The variables affecting the de-sulphurization with carbide mixture injection are given as follows:

(i) Initial sulphur in hot metal,
(ii) Amount of carbide consumed (kg/t),
(iii) Initial hot metal temperature,
(iv) Ladle treatment time,
(v) Slag carry-over,
(vi) Lance depth and angle of dip.

The average sulphur removal is around 0.05%, and the final sulphur in the ladle is in the range of 0.01–0.015%. Calcium carbide-based compounds are cheaper as compared to magnesium/ magnesium- based compound. Lime and soda ash are used as de-sulphurizing agents. These materials are cheaper but are less effective de-sulphurizers. However, cost is much less as compared to above reagents.

13.7.3 De-phosphorization

The properties of steel are strongly influenced by phosphorus content in steel. With an increase in the demand of high-grade steel and with the larger uses of continuous casting process, the demand for low phosphorus steel is expected to increase in future. Steel is used for extra deep drawing quality and other grades: 0.015% S (max), 0.015% P (max), 0.0035% N (max) and 0.002% O (max) (preferable 0.0015% max).

De-phosphorization is favoured at low temperature and with high slag basicity. Phosphorus is primarily removed through slag-metal reaction. De-phosphorization capability of slag is measured by the following two parameters:

1. The phosphate capacity of the slag which directly relates to the ability of the slag to absorb phosphorus,
2. The equilibrium phosphorus partition ratio which is the ratio of the phosphorus content of slag to the phosphorus content in metal on a weight percentage basis.

Transfer of phosphorus from metal to slag is an oxidation process given by:

$$[P] + 5/2[O] + 3/2\left(O^{2-}\right) = \left(PO_4^{3-}\right) \tag{13.13}$$

Although the reaction (13.13) as formulated above is conceptually correct, the equilibrium constant cannot be evaluated because the thermodynamic activity or activity coefficient of phosphorus ion cannot be determined experimentally.

For phosphorus reaction between metal and slag, the system may be described by the following six components: [Fe], [P], [O], (CaO), (SiO_2) and (MO), where MO represents MgO plus other minor slag components, e.g. MnO, Al_2O_3 and CaF_2. With the assumption that phosphorus has virtually no effect on its activity coefficient in the metal and slag phase, the equilibrium distribution ratio {(%P)/[%P]} between slag and metal may be described by the following functional relations, depending on the number of phases present.

(I) Two phases: slag and metal

$$\left(\frac{(\%P)}{[\%P]}\right)_{P,T} = \varphi\{[\%O], (\%SiO_2), (\%MgO), B\} \tag{13.14}$$

where

$$B = \text{Basicity} = \frac{(\%\ CaO)}{(\%\ SiO_2)} \tag{13.15}$$

(II) Three phases: MgO-saturated slag and metal

$$\left(\frac{(\%P)}{[\%P]}\right)_{P,T} = \varphi\{[\%O], (\%SiO_2), B\} \quad (13.16)$$

(III) Four phases: MO and CaO-saturated slag and metal

$$\left(\frac{(\%P)}{[\%P]}\right)_{P,T} = \varphi\{[\%O], (\%SiO_2)\} \quad (13.17)$$

Physico-chemical conditions for de-phosphorization are summarized as follows:

1. Phosphorus distribution ratio in $CaO–FeO–SiO_2$ slags can be given as:

$$\log\left[\frac{(\%P)}{[\%P]}\right] = 0.071\{(\%CaO) + 0.1(\%MgO)\} + 2.5\log(\%T_{Fe}) + 8.26T - 8.56 \quad (13.18)$$

2. Na_2O-based fluxes are more effective for de-phosphorization than CaO-based fluxes.
3. Na_2CO_3 is unstable at high temperatures and reacts with carbon in the hot metal to form CO gas. De-phosphorization reaction with soda ash is given by:

$$Na_2CO_3 + 2/3[P] = 2/3(Na_3PO_4) + 1/3\{CO\} + 2/3C \quad (13.19)$$

4. Addition of as low as 2.5% Na_2O in these slag gives high phosphorus partition ratio at FeO less than 5% in slag.
5. The rate and degree of de-phosphorization with soda-based slag increase with higher Na_2O/SiO_2 ratio.
6. One of the reasons for increased efficiency of de-phosphorization with Na_2O is the decrease in the activity coefficient of phosphate ions in the slag.
7. Soda ash addition provides both Na_2O in slag as well as oxidizing agent through the decomposition product CO_2.
8. When Na_2O is replaced by MgO, CaO, BaO at constant SiO_2 level in slag, the phosphate is in the order Ba > Ca > Mg.
9. Addition of fluxes such as CaF_2 and $CaCl_2$ increases de-phosphorization rate by reducing the melting point of the fluxes.
10. In the absence of iron oxide in slag, the addition of CaF_2 is not effective and decreases the phosphate capacity. With a small amount of FeO in slag (up to 10%), the addition of CaF_2 increases the phosphate capacity as CaF_2 increases the activity coefficient of FeO in slag.
11. Phosphorus partition ratios are temperature dependent, which is a function of slag composition.
12. Presence of MnO in slag decreases its de-phosphorization ability.
13. Highly basic slag has high phosphorus partition ratio. Beyond CaO/SiO_2 ratio of 6, distribution ratio becomes constant. Partial replacement of CaO by BaO increases phosphorus partition ratio.

13.7.4 Advantages of Pre-treatment to Hot Metal

1. By reducing the silicon content of hot metal (up to <0.2%), a substantial saving is in lime consumption and the extent of de-phosphorization became more stable.
2. A decrease in slag volume also lowers the extent of refractory wear and increases the steel yield.
3. At the minimum slag practice decreases scrap melting in BOF which to be an advantage for easier control of the steelmaking operation.
4. The overall economics is improved.

Probable Questions

1. What is the difference between Swedish iron and Thomas iron?
2. What are the main sources of steel scrap?
3. What are the secondary sources of iron in steelmaking processes? Discuss.
4. What are the merit and demerit of sponge iron?
5. 'Iron ore has more cooling power than scrap in steelmaking'. Explain.
6. 'Scrap contains high amount of residual elements'. What problems it may pose during steelmaking? How these elements can be reduced in steel product?
7. What are the sources of oxygen for steelmaking processes?
8. What are the roles of slag in steelmaking processes?
9. What do you understand by 'oxidizing power of slag' and 'wet and dry slag'?
10. Why pre-treatment of hot metal is done? Discuss the (i) de-siliconization and (ii) de-sulphurization processes.
11. What are the physico-chemical conditions for de-phosphorization? Discuss the advantages of pre-treatment to hot metal.
12. Why do you need to carry out external de-sulphurizaion of hot metal prior to steelmaking? What kind of flux is used for such de-sulphurizaion?

References

1. *Intitution of Engineers (India) MM,* 69, 23 (Sept 1988)
2. S.K. Dutta, A.B. Lele, Iron & Steel Review **55**(3), 162 (2011)
3. *American Iron & Steel Institute and Department of Energy (AISE/DOE),* Technology Roadmap Program, 31 Mar 2004
4. http://iicarbide.com/archives/IIC_iron_carbide_for_EAFs_rev_02.pdf
5. R.H. Tupkary, *Introduction to Modern Steelmaking* (Khanna Publishers, Delhi, 1991)

14 Steelmaking Processes

Bessemer proposed a production method of steelmaking within a very short time (15–20 min) by blowing air through molten pig iron without using fuel. Siemens brothers, Germany, developed temperature of the furnace around 1600 °C for the first time, on the principle of regeneration of heat from exhaust gases. Regeneration of heat was done by passing cold air through the hot checker work which was heated previously by the passing of hot exhaust gases. Steel scrap/pig iron could be melted and kept molten even at the end of refining, which is popularly known as open-hearth process. Open-hearth process is a slow process, due to the diffusion of oxygen from gas phase to metal bath phase. Diffusion of oxygen can be faster by lancing oxygen to the molten bath. Submerged injection process (SIP) and twin-hearth process are developed by the modification of open-hearth furnace.

14.1 Acid Bessemer Process

Major advance in steelmaking occurred in 1855, when Henry Bessemer took out his first patent for the production of malleable iron and steel without using fuel. Bessemer proposed a production method of steelmaking within a very short time (15–20 min) by blowing air through molten pig iron. After many years of research and hard work, during which he overcame many practical difficulties, Bessemer started production of steel by his process, in Sheffield, UK. The process consisted of the production of tool and high carbon steel by the partial de-carburization of Swedish pig iron (low sulphur and phosphorous). Later, by the complete de-carburization of pig iron, he succeeded in producing mild steels which were partly recarburized by the addition of ferromanganese. In 1856, R. Mushet had observed that a certain percentage of manganese was necessary in the finished steel to make it workable. Since the time of Bessemer, the process which bears his name has been modified very little. Main developments have been in the design of various items of the plant and in the methods of control. Process was the first to produce a large-scale production where hot metal could rapidly and cheaply be refined and converted into liquid steel.

Bessemer originally used *acidic Swedish pig iron* which had low sulphur and phosphorous (<0.04% each), but high in silicon (2.0–2.5%). *Swedish iron* contains 3–4% C, 2.0–2.5% Si, 1.0–1.5% Mn, <0.04% P and <0.04% S. The vessel, in which liquid pig iron/hot metal is converted to liquid steel by oxidation, is known as *converter*. The converter was lined with an acidic material (i.e. silica bricks), and slag formed from the products of oxidation was acidic in nature, that is why the process was known as *acid Bessemer process*. It is an autogenous or pneumatic process, so the oxidation of impurities, specially the silicon, produces a large amount of heat, which not only

S. K. Dutta and Y. B. Chokshi, *Basic Concepts of Iron and Steel Making*,
https://doi.org/10.1007/978-981-15-2437-0_14

maintains the temperature and fluidity of the molten pig iron and slag, but also increases metal temperature.

In acid Bessemer process, the silicon should be in the range of 2.0–2.5%.

- Case I: if the silicon falls below 1.0%, the metal is liable to cool down, i.e. chilled, owing to insufficient heat being produced by the oxidation of silicon; this is known as *blow-cold* condition; that is, insufficient heat is produced by the oxidation of silicon.
- Case II: when the silicon exceeds 2.5% (i.e. about 3.0%), the heat produced by the oxidation of silicon may be excessive and cause the metal to blow too hot or overheated; this is known as *blow-hot* condition. If the blow is too hot, i.e. temperature of the molten bath is high, (i) there is a danger of the carbon being eliminated before all the silicon is removed (as shown in Fig. 14.1), and high residual silicon in the metal renders the resultant steel unserviceable; (ii) the silicon passes into the slag as a mixed silicate of Fe and Mn; therefore, higher the silicon, the greater the loss of iron in the slag; and (iii) finally, volume of slag is also increased, i.e. more heat losses.

When flow of air is blown through molten pig iron which contains 3–4% C, 2.0–2.5% Si, 1.0–1.5% Mn, <0.04% P and <0.04% S, these elements together with some iron are oxidized; the carbon escapes as a gas (i.e. CO), while the silicon and manganese, along with some iron, may be oxidized to form a mixed ferrous-manganese silicate slag which is acidic in nature.

14.1.1 Refining

The refining of molten pig iron takes place by the oxidation of impurities (i.e. Si, Mn and C), which cannot be oxidized directly with the oxygen in air. As per Ellingham diagram (Fig. 14.1), the silicon has a good affinity for oxygen than iron, but due to *mass effect*, iron first reacts with oxygen to form FeO. Because the amount of iron (93–94%) present in the metal is much more than the silicon (only 1.0–2.5%), this product FeO now reacts with the impurities. All reactions take place in the bulk of metal and are known as *volume zone reactions*. Since acidic pig iron contains low sulphur and phosphorous (<0.04% each), then there is no need to add lime (as flux) for removal of sulphur and phosphorous to make basic slag.

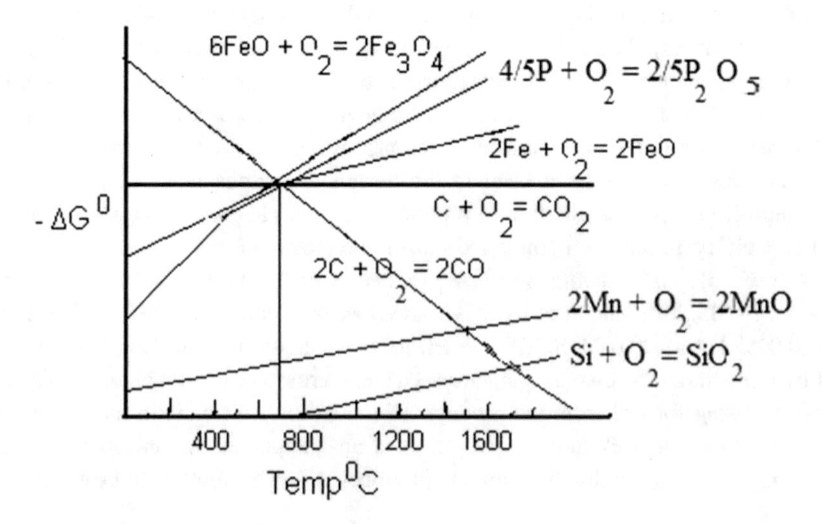

Fig. 14.1 A portion of Ellingham diagram

$$2[\text{Fe}] + \{\text{O}_2\} = 2(\text{FeO}), \Delta H^0_{298} = -264.4 \text{ kJ/mol of Fe} \quad (14.1)$$

$$[\text{Si}] + 2(\text{FeO}) = 2[\text{Fe}] + (\text{SiO}_2),\ \Delta H^0_{298} = -342.67 \text{ kJ/mol of Si} \quad (14.2)$$

$$[\text{Mn}] + (\text{FeO}) = [\text{Fe}] + (\text{MnO}),\ \Delta H^0_{298} = -120.29 \text{ kJ/mol of Mn} \quad (14.3)$$

$$[\text{C}] + (\text{FeO}) = [\text{Fe}] + \{\text{CO}\}, \Delta H^0_{298} = 152.71 \text{ kJ/mol of C} \quad (14.4)$$

CO gas burns to CO_2 at the mouth of the converter and produces a luminous flame:

$$2\{\text{CO}\} + \{\text{O}_2\} = 2\{\text{CO}_2\}, \Delta H^0_{298} = -282.42 \text{ kJ/mol of CO} \quad (14.5)$$

Some of the higher oxide of iron is also formed:

$$6(\text{FeO}) + \{\text{O}_2\} = 2(\text{Fe}_3\text{O}_4),\ \Delta H^0_{298} = -104.08 \text{ kJ/mol of FeO} \quad (14.6)$$

$$4(\text{Fe}_3\text{O}_4) + \{\text{O}_2\} = 6(\text{Fe}_2\text{O}_3),\ \Delta H^0_{298} = -124.74 \text{ kJ/mol of Fe}_3\text{O}_4 \quad (14.7)$$

and these higher oxides are partially reduced:

$$2(\text{Fe}_3\text{O}_4) + [\text{Si}] = 6(\text{FeO}) + (\text{SiO}_2),\ \Delta H^0_{298} = -224.69 \text{ kJ/mol of Si} \quad (14.8)$$

$$(\text{Fe}_2\text{O}_3) + [\text{Fe}] = 3(\text{FeO}),\ \Delta H^0_{298} = 28.03 \text{ kJ/ mol of Fe}_2\text{O}_3 \quad (14.9)$$

$$(\text{SiO}_2)(\text{partially}) + (\text{MnO}) = (\text{MnO} \cdot \text{SiO}_2), \Delta H^0_{298} = -29.71 \text{ kJ/mol} \quad (14.10)$$

$$2(\text{FeO}) + (\text{SiO}_2)(\text{rest}) = (2\text{FeO} \cdot \text{SiO}_2),\ \Delta H^0_{1773} = -27.07 \text{ kJ/mol} \quad (14.11)$$

Hence, the slag in acid Bessemer process can be referred to as FeO–MnO–SiO_2 type dry slag. FeO, together with MnO and SiO_2, forms a highly siliceous slag, i.e. ferrous-manganese silicate slag. The composition of slag is 55–65% SiO_2, 12–18% MnO, 12–18% FeO, 1–3% Fe_2O_3 and other oxides balance, which is chemically acidic in nature, and *dry slag* (i.e. high viscosity and low fluidity) is formed due to high amount of SiO_2 present in slag. When ratio of Si/Mn in hot metal is 2.0–2.5, then dry slag is formed; when the value of ratio is less than 2.0–2.5, then *wet slag* is formed. If high amount of FeO and MnO is present in slag, fluidity of slag increases, and so wet slag is formed.

Hence, the materials used for lining of the acid converter must also be acidic in nature. Body of the converter is made of soft silica bricks and bottom tuyeres of fireclay (Fig. 14.2). The life of the body lining is 800–1000 heats[1], bottom lasts only 25–40 heats; hence, detachable bottom is used.

All the reactions taking place in the bath of converter are reflected in the form of flame at the mouth of converter. As soon as the molten hot metal has been poured into the converter, while in horizontal position the air blowing is turned on (at a pressure of about 60–70 cm of Hg) and the converter is rotated into a vertical position. Immediately a small flame and a large quantity of sparks come out from the mouth. After two or three minutes, the flame begins to increase, and the quantity of sparks decrease. This indicates the removal of Si and Mn, and it continues until the carbon begins to burn off, when the luminous flame increases in size (5–8 m) and become brighter. When most of the carbon has been removed, the flame diminishes and becomes less bright until the flame drops, which indicates the complete removal of carbon, then blowing of air should be stopped; otherwise, metal would be oxidized, i.e. loss of metallic iron.

[1]Heat means tap-to-tap time.

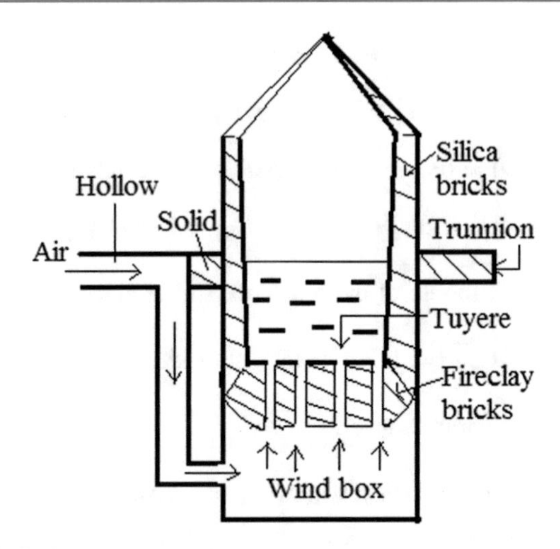

Fig. 14.2 Acid Bessemer furnace

14.1.2 Air Blowing

All the reactions taking place in the converter are reflected in the flame which comes out from the mouth of the converter as follows:

1. First period: Small flame and large quantity of sparking, then a light brown flame at the mouth due to Si, Mn part of Fe oxidized and slag formation.
2. Second period: Flame gradually increases and brighter (5–8 m height), C → CO which is burn in the air at mouth to convert CO_2.
3. Third period: Flame becomes short and reddish brown for Fe and Mn further oxidation.

14.1.3 Heat Balance

When a thermal balance sheet of acid Bessemer process was prepared, the following headings should be considered:

Heat Input:

1. Heat content in the hot converter (6%),
2. Heat content in the molten metal (33%),
3. Heat develop by the oxidation of impurities (e.g. Si, Mn, C) (60%),
4. Heat develop by slag formation (1%).

Heat Output:

1. Heat supply to make finish molten steel (45%),
2. Heat supply to make molten slag (10%),
3. Heat carries away by gases (19%),
4. Heat content in the hot converter (6%),
5. Heat lost due to radiation and conduction (20%).

14.2 Basic Bessemer Process

The availability of large deposits of iron ore rich in phosphorous in some European countries promoted the development of a basic-lined converter in which basic pig iron could be refined. In 1877, P. C. Gilchrist and his cousin Sidney Gilchrist Thomas completed their research which led to the use of calcined dolomite (35–36% MgO and 55–58% CaO) as a basic lining. Magnesite is also used as basic lining. By introducing dolomite lining for converters rendered possible use of phosphoric pig iron (>1.5% P), that led to develop *basic Bessemer process,* or it is also known as *Thomas process. Basic pig iron/Thomas iron* contains 3–4% C, 0.5–1.0% Si, 0.5–1.0% Mn, <0.04% S and >1.5% P.

The removal of phosphorous requires a basic slag and a basic-lined converter. The basic slag is formed from lime (12–15% of weight of hot metal) which is added to the converter just before the hot metal is poured into it.

14.2.1 Refining

Apart from phosphorous, the removals of impurities are similar to the acid Bessemer process. The basic Bessemer process is slightly different than that of acid Bessemer process. At the initial stages, the basic process is like that of acid process, but at later stage CaO comes into solution with the slag which helps in removal of phosphorous. CaO dissolves in solution and acts actively as a base. In basic process, CaO replaces some of FeO in slag, since CaO is a strong basic oxide and its affinity for acidic oxides (e.g. SiO_2, P_2O_5, etc.) is greater than that of FeO.

Phosphorous cannot remove until all the other impurities (i.e. Si, Mn and C) have been eliminated (as shown in Fig. 14.1). The flame drops, indicating the end of de-carburization, but blowing of air is continue for removal of phosphorous, because the free energy of formation for CO is lower than P_2O_5 at that temperature. The removal of phosphorous could not reflected to the flame and blowing continue; hence, this blowing is known as *after-blow period.* The dense brown fumes come out from the mouth of the converter; these fumes are typical of iron oxidation.

$$2[\mathrm{P}] + 5(\mathrm{FeO}) = 5[\mathrm{Fe}] + [\mathrm{P_2O_5}],\ \Delta H^0_{298} = -135.44\ \mathrm{kJ/mol\ of\ P} \tag{14.12}$$

$$[\mathrm{P_2O_5}] + 4(\mathrm{CaO}) = (4\mathrm{CaO}\cdot\mathrm{P_2O_5}),\ \Delta H^0_{298} = -1738.79\,\mathrm{kJ/mol} \tag{14.13}$$

P_2O_5, which is formed by oxidation of phosphorous, is unstable at that temperature; so P_2O_5 should be combined with dissolve CaO to form tetra-basic phosphate of lime (i.e. basic slag). The volume zone reactions are occurred at the initial stage, and then, the dissolved CaO makes it a basic slag. The basic slag in basic Bessemer process is CaO–FeO–MnO–SiO_2–P_2O_5. Slag contains 45–55% CaO, 5–9% SiO_2, 7–12% FeO, 2–8% Fe_2O_3, 18–22% P_2O_5, 2–6% MnO and other oxides balance.

Total iron oxide content in the acid Bessemer slag is greater (12–18% FeO) than the basic Bessemer slag (7–12% FeO), but the activity of FeO in the acid slag is lower than the basic slag. Why? Because in acid Bessemer slag is in the form of FeO–MnO–SiO_2 type. Since SiO_2 is a strong acidic oxide which strongly bonded with FeO (here, FeO acts as basic oxide); hence, there are no free FeO available in the acid slag. On the other hand, basic Bessemer slag is in the form of CaO–FeO–MnO–SiO_2–P_2O_5 type. Since CaO is a strong basic oxide and SiO_2 is a strong acidic oxide; hence, they make strong bond. Whatever FeO present in the basic slag is freely available, and hence, the activity of FeO is increased. Since activity depends on the freely available element or compound in the solution.

14.2.2 Air Blowing

All the reactions taking place in the converter are reflected in the flame which comes out from the mouth of the converter as follows:

1. First period: Sparking and short flame with reddish colour, burning of lime particles; Si, Mn and part of Fe oxidized.
2. Second period: It is called *fore-blow*, elongated flame, C oxidized.
3. Third period: After drop of the big flame, air blowing still continue which is known as *after-blow period*, it form very short flame; P oxidized.

14.2.3 Heat Balance

Thermal balance sheet of basic Bessemer process is as follows:

Heat Input:

1. Heat content in the hot converter (6%),
2. Heat content in the molten metal (33%),
3. Heat develops by the oxidation of impurities (e.g. Si, Mn, C) (55%),
4. Heat develops by slag formation (6%).

Heat Output:

1. Heat supply to make finish molten steel (45%),
2. Heat supply to make molten slag (15%),
3. Heat carries away by gases (14%),
4. Heat lost due to radiation and conduction (20%),
5. Heat content in the hot converter (6%).

14.2.4 Limitations

Bessemer steelmaking suffered from some limitations as follows:

(i) Nearly 15–20% of the heat was lost in general by the flue gases and in particular sensible heat content of the nitrogen gas, and so thermal efficiency of Bessemer process is very poor. If this heat was utilized, the steel become overheated, which could be avoided by the addition of scrap, iron ore or mill scale as coolants.

(ii) One of the disadvantages of Bessemer process was the resulting high nitrogen (0.012–0.015%, i.e. 120–150 ppm) content in the Bessemer steel, compared to desired nitrogen level which was 50–60 ppm.

(iii) Very frequent change of the bottom of the converter was needed.

(iv) Bessemer process could not refine in one stage for high silicon (1.0–1.25%), medium phosphorus (0.3–0.4%) hot metal (like Indian hot metal). A duplex process (i.e. acid Bessemer process and basic open-hearth process) was necessary to treat that hot metal.

Hence, nowadays nowhere in world used Bessemer processes for the production of steel, after oxygen steelmaking processes are developed.

14.3 Open-Hearth Process

Before 1850s the furnace temperature was raised up to 1400–1450 °C, pig iron could be readily melted and refined; but the product was pasty mass as purity increased the melting point of metal was also increased. In 1861, Siemens brothers, Germany, developed temperature of the furnace around 1600 °C for the first time, on the principle of regeneration of heat from exhaust gases. A part of the heat in the hot exhaust gases is somehow stored and returned to the furnace as sensible heat in a reversible cycle. Regeneration of heat was done by passing cold air through the hot checker work which was heated previously by the passing of hot exhaust gases. Steel scrap/pig iron could be melted and kept molten even at the end of refining. Martin brothers, France, designed the furnace. Therefore, it is known as *Siemens-Martin Process* which is popularly known as *Open-Hearth Process*, because the hearth can be seen by raising the door of the furnace (Fig. 14.3). Hear heat requirement is supplied by the combustion of gases or liquid fuel. It is a slow process. Initially, the furnace was used to melt steel scrap alone, but for economic reasons it was soon developed to take a mixture of scrap and hot metal/pig iron as charge materials. It could be operated either as an acid or a basic process depending upon the charge composition and with or without hot metal charge. Lining for acid process was fireclay bricks, and for basic process, lining was dolomite or chrome-magnesite bricks. Basic process was most popular. Basic open-hearth process is suitable for refining of medium phosphorus hot metal.

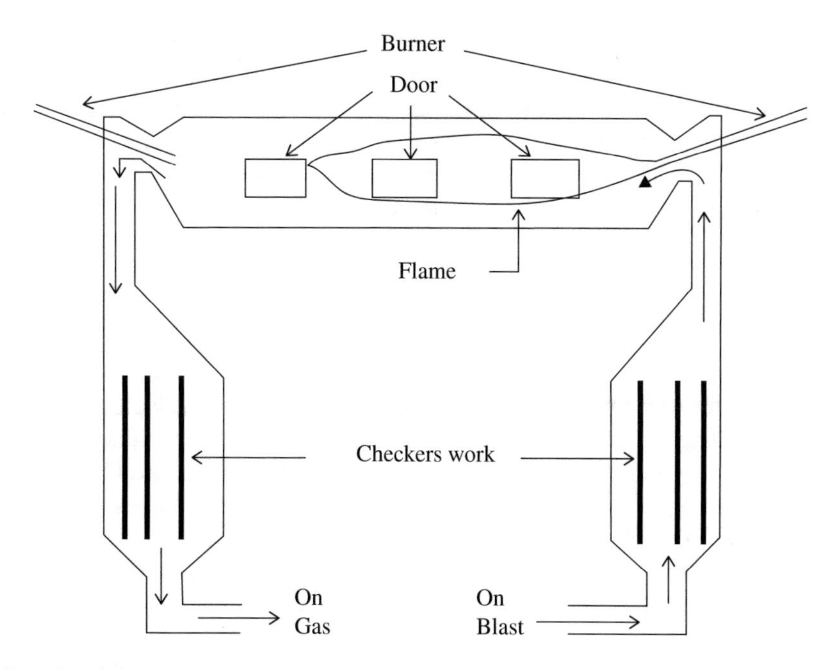

Fig. 14.3 Open-hearth furnace

The principle is regeneration of heat from exhaust gases; the blast and gaseous fuel are made hot by passing through the hot checker work which is heated previously by the passage of hot exhaust gases that is known as *on blast condition or up-takes*. There are two checker works on both the sides: one is supplied heat to the blast and fuel gas and another is pre-heated itself by hot exhaust gases that is known as *on gas condition or down-takes*. Gaseous/liquid fuel burnt inside of the furnace and hot flame is passed on the top of the charging materials.

Open-hearth furnace process can be operated on the following charge combination:

1. 100% solid pig iron or steel scrap (or 60–85% scrap and 40–15% pig iron) → Means too much thermal load is required to melt 100% solid material.
2. 100% hot metal → Metallurgical load is high; i.e., amounts of impurities in metal are high.
3. 50% hot metal + 50% solid scrap → Best option, if hot metal is available.
4. 100% blown metal → Initially impurities (3.5% C, 1.2–1.5% Si and 0.4% P), particularly Si and C, are removed in acid Bessemer furnace, then that metal (1% C, $<$0.1% Si and 0.4% P; known as *blown metal*) is charged into open-hearth furnace. Carbon cannot be less than 1%, because that is required for boiling action in open-hearth furnace.

Operational steps involved are as follows:

1. Charging,
2. Melting down the charge,
3. Refining,
4. Finishing.

14.3.1 Charging

Charge materials consist of (i) hot metal (50–75%), (ii) scrap (25–50%), (iii) limestone (5–7% of metallic charge) as flux, and (iv) iron ore (6% of metallic charge) as oxygen supplier (if 3rd charging option is adopted). Initially, solid scrap, limestone and iron ore can be charged by mechanically charging box to the furnace through the door. The materials should be charged in such a way that it is uniformly distributed over the entire hearth and that should not come in the way of free passage of the top hot flame from one end to the other end of the hearth.

14.3.2 Melting

In an integrated steel plant, a mixture of blast furnace and coke oven gases were used as a fuel. After solid charging to the furnace, fuel supply is switched on to burn the fuel at high rate to heat the solid charge for quick melting. After partially melting down, then burning rate of fuel should be lower down, because: (i) to prevent overheating the refractory lining, since metal would not take much heat for melting; (ii) efficiency of heat would not be utilized properly, and (iii) metal would be oxidized, i.e. more metal would be lost due to high temperature. If oxygen was lancing, that helped in the rapid melting of scrap. After semi-molten of scrap, hot metal was charged to the furnace. If hot metal was charged before semi-molten of scrap, hot metal was chilled into the furnace, due to which the temperature difference is high.

Since the charge is in contact with an oxidizing atmosphere during melting:

$$2[\mathrm{Fe}] + \{\mathrm{O_2}\} = 2(\mathrm{FeO}),\ \Delta H^0_{298} = -264.43\ \mathrm{kJ/mol\ of\ Fe} \quad (14.14)$$

After charging hot metal, enough heat is available to the bath and then iron ore reacted with iron in metal:

$$\mathrm{Fe_2O_3(s)} + [\mathrm{Fe}] = 3(\mathrm{FeO}),\ \Delta H^0_{298} = 28.03\ \mathrm{kJ/mol\ of\ Fe_2O_3} \quad (14.15)$$

Reaction (14.15) is an endothermic reaction that is why enough heat is required. Then, FeO reacts with impurities of hot metal.

$$[\mathrm{Si}] + 2(\mathrm{FeO}) = 2[\mathrm{Fe}] + (\mathrm{SiO_2}),\ \Delta H^0_{298} = -342.67\ \mathrm{kJ/mol\ of\ Si} \quad (14.16)$$

$$[\mathrm{Mn}] + (\mathrm{FeO}) = [\mathrm{Fe}] + (\mathrm{MnO}),\ \Delta H^0_{298} = -120.29\ \mathrm{kJ/mol\ of\ Mn} \quad (14.17)$$

$$[\mathrm{C}] + (\mathrm{FeO}) = [\mathrm{Fe}] + \{\mathrm{CO}\},\ \Delta H^0_{298} = 152.72\ \mathrm{kJ/mol\ of\ C} \quad (14.18)$$

$$(\mathrm{SiO_2})(\mathrm{partially}) + (\mathrm{MnO}) = (\mathrm{MnO \cdot SiO_2}),\ \Delta H^0_{298} = -29.71\ \mathrm{kJ/mol} \quad (14.19)$$

$$2(\mathrm{FeO}) + (\mathrm{SiO_2})(\mathrm{rest}) = (\mathrm{2FeO \cdot SiO_2}),\ \Delta H^0_{1773} = -27.07\ \mathrm{kJ/mol} \quad (14.20)$$

The initial slag is formed as FeO–MnO–SiO_2 type acidic slag, which react with basic lining (i.e. CaO and MgO) at the early stage. Carbon in metal reacts with oxygen to form CO gas evolution which create boiling action to the bath, that is known as *carbon boil*; since this boiling is due to oxygen supply from the iron ore, so this boiling action is known as *ore boil*. Due to that, part of the slag comes out from the furnace through the middle door automatically and drops into a slag pocket. By the ore boil, the slag is flushing out from the furnace through the front doors, and this is known as *front flushing of slag*. This is a good practice to reduce the volume of the slag; since the slag act as an insulating blanket and non-conductor of heat, so the thickness of the slag blanket decreases the heat transfer from gas phase to metal phase through slag phase. Since heat is supplied from the gas phase as hot flame (i.e. top of the bath) to metal phase through slag phase, only one-fifth, i.e. 20%, heat can be transferred through the slag layer. By flushing slag from the doors, volume of slag is decreased; so heat transfer takes place faster from flame to the bottom of the bath. Due to boiling action, the advantages are: (i) good slag–metal contact, (ii) good oxygen transfer to the bath, (iii) good heat transfer to the bath, (iv) good homonization of the bath and (v) help to separate inclusions from the bath by floating. Only disadvantage of flushing slag is lot of iron loss (as FeO) in slag.

After reaching heat to the bottom of hearth where limestones are lying, now limestone at the bottom gets heat for dissociation:

$$\mathrm{CaCO_3(s)} = (\mathrm{CaO}) + \{\mathrm{CO_2}\}, \Delta H^0_{298} = 161.3\,\mathrm{kJ/mol} \quad (14.21)$$

CaO is lighter than metal, so CaO is floated up on the metal and slag; and CO_2 gas comes out as baubles from the bath and again creates boiling action of the bath; this boiling action is known as *lime boil*, since this is due to the calcinations of limestone.

CO_2 gas again acts as oxidizing agent to react with carbon in the metal:

$$\{CO_2\} + [C] = 2\{CO\},\ \Delta H^0_{298} = -617.57\ \text{kJ/mol of}\ CO_2 \quad (14.22)$$

Lime boil is much more vigorous than ore boil. Why?

Because (i) at a time all limestones are dissociated and suddenly release a lot of CO_2 gas; (ii) one mol of CO_2 gas produces two mols of CO gas by oxidizing carbon in metal, that is why lime boil is more vigorous than ore boil.

14.3.3 Refining

Objects of refining period are: (i) to lower down phosphorous and sulphur content to a level of safety below the maximum level specified, (ii) to lower down remaining carbon to a specific level, and (iii) to rise the bath temperature to the desired level.

Phosphorous and sulphur can be removed at refining period only, because the condition of phosphorous removal is not satisfied earlier, i.e. need a basic and fluid slag. Earlier slag contains high FeO (30–35%) and low CaO (10–15%). As the temperature is increased, more lime dissolves in slag after carbon-boiling and increasing basicity of the slag.

CaO forms the basic slag with other oxides:

$$(CaO) + (SiO_2) + (MnO) + (FeO) = (CaO \cdot SiO_2 \cdot MnO \cdot FeO) \quad (14.23)$$

$$2[P] + 5(FeO) = 5[Fe] + [P_2O_5],\ \Delta H^0_{298} = -135.44\ \text{kJ/mol of P} \quad (14.24)$$

$$[P_2O_5] + 4(CaO) = (4CaO \cdot P_2O_5),\ \Delta H^0_{298} = -1738.79\ \text{kJ/mol} \quad (14.25)$$

To produce 0.4% C steel, carbon should be lower down to below 0.4%, i.e. about 0.39%, or still lower at the final stage. So final adjustment of carbon per cent (in steel) can be done by adding ferro-alloy, which contains small amount of carbon. If tapped steel contains higher per cent carbon (e.g. 0.42% C) than required (i.e. 0.4% C), then the extra carbon (i.e. 0.02%) cannot removed after tapping the steel. The temperature of the bath depends on the carbon content in steel; the lower the carbon content, the higher the melting point of steel. Hence, tapping (i.e. furnace to ladle) and pouring (i.e. ladle to mould) temperatures should be 100 °C more than melting point of steel.

14.3.4 Finishing

At the end of refining, lumpy Fe–Mn and/or Fe–Si are added to the bath to remove excess oxygen from the bath. *Blocking of heat* means to stop the C–O reaction, and this is done by the addition of ferro-alloy. Fe–Mn and/or Fe–Si are now reacted with excess oxygen present into the bath.

$$[O] + Fe{-}Mn/Fe{-}Si \rightarrow (MnO)/(SiO_2) + [Fe] \quad (14.26)$$

Manganese as well as silicon content in metal is also adjusted during finishing.

How to control the following reaction?

$$[C] + [O] = \{CO\} \quad (14.27)$$

$$\text{Equilibrium constant, } k_{14.27} = \frac{p_{CO}}{[\text{wt\% C}] \cdot [\text{wt\% O}]} \quad (14.28)$$

$$[\text{wt\% O}] = \left(\frac{p_{CO}}{k_{14.27} \cdot [\text{wt\% C}]}\right) = \left(\frac{1}{k_{14.27} \cdot [\text{wt\% C}]}\right) \quad (\text{Since } p_{CO} = 1 \text{ atm}) \quad (14.29)$$

The equilibrium relationship between carbon and oxygen is dependent upon p_{CO} but not upon the temperature.

$$\text{The extra, i.e. surplus oxygen, } \Delta O = (\%O_a - \%O_e) \quad (14.30)$$

where O_a is the actual oxygen at the bath, and O_e is the equilibrium oxygen of the reaction.

As long as surplus oxygen is present, the C–O reaction will continue. To stop the C–O reaction, the surplus excess oxygen should be lowered down below the equilibrium oxygen content. This is known as *blocking of heat.*

The oxygen content of the bath is just below the equilibrium oxygen content, but bath does not fully deoxidized. The final deoxidation is done out of the furnace in the ladle, and final deoxidation is done by the addition of Al shots. The final deoxidation is not done in the furnace, because the oxygen content of the slag will be decreased, so oxygen potential of the slag will be reduced; hence, ability to hold P_2O_5 in slag is reduced. Therefore, there will be chance that P will again go back into the metal. This is known as *P-reversion*.

$$\text{At a constant temperature, } \frac{(\text{FeO})}{[\text{FeO}]} = \text{constant} \quad (14.31)$$

When Fe–Si is added to the bath, Si dissolves in metal and reacts with [FeO] in metal:

$$[\text{FeO}] + [\text{Si}] = [\text{Fe}] + (\text{SiO}_2) \quad (14.32)$$

Because of this reason, FeO content in metal phase decreases but at a particular temperature; there is an equilibrium, i.e. $\frac{(\text{FeO})}{[\text{FeO}]}$ is constant. Hence, the FeO in slag phase, to maintain equilibrium, will diffuse into the metal phase, and therefore FeO content of the slag is also decreased; that is, the oxidizing power of the slag is reduced. But for effective de-phosphorization, high oxygen potential of the slag is required; hence, the slag cannot hold so much of P_2O_5, and P_2O_5 is decomposed ($2P_2O_5 \rightarrow 4P + 5O_2$) into phosphorous which goes back to the metal again. This reaction is known as *P-reversion*. This can be avoided by the addition of calculated amount of deoxidizer. But the molten bath still contains dissolved oxygen which can be removed at the ladle with Al shots, which act as a deoxidizer as well as a grain refiner.

Bessemer converter had certain limitation: (i) it required hot metal of 2.0% Si or >1.5% P; (ii) Bessemer steel contained high nitrogen (0.012–0.015%, i.e. 120–150 ppm). But open-hearth furnace had big advantages: (i) it could take different charging material [i.e. solid (scrap/pig iron) as well as liquid (hot metal)]; (ii) open-hearth steel contained low nitrogen (<0.005%, i.e. 50 ppm), which was low enough to provide good ductility for deep drawing steel.

Nowadays, only two countries (e.g. Ukraine and Russia) produced steel in open-hearth furnace (4.8 and 1.7 Mt, respectively, in 2018) [1]. Byelorussia and Uzbekistan stopped production in open-hearth furnace from 2012; India stopped production in open-hearth furnace from 2015. Hence, open-hearth furnace process is also outdated like Bessemer processes.

14.3.5 Modification of Open-Hearth Furnace

Open-hearth process is a slow process, due to the diffusion of oxygen from gas phase to metal bath phase. Diffusion of oxygen can be faster by lancing oxygen to the molten bath. Modification of open-hearth furnace is performed by: (i) submerged injection process (SIP) and (ii) twin-hearth process.

14.3.5.1 Submerged Injection Process (SIP)

Submerged injection process is a process where pure oxygen is supplied instead of iron ore. Oxygen is injected by lance directly to the metal bath through slag layer from the door or roof to reduce total heat time due to faster melting of scrap and fuel consumption. The productivity of the process can be increased by lancing oxygen directly into the metal phase. The advantages are as follows:

(i) The endothermic dissociation of iron ore is no more required:

$$Fe_2O_3(s) + [Fe] = 3(FeO),\ \Delta H^0_{298} = 28\ \text{kJ/mol of}\ Fe_2O_3 \quad (14.33)$$

(ii) The exothermic oxidation of iron:

$$2[Fe] + \{O_2\} = 2(FeO),\ \Delta H^0_{298} = -262.97\ \text{kJ/mol of Fe} \quad (14.34)$$

(iii) The exothermic oxidation of carbon:

$$2[C] + \{O_2\} = 2\{CO\},\ \Delta H^0_{298} = -111.72\ \text{kJ/mol of C} \quad (14.35)$$

Along with oxygen some hydrocarbon gases (as coolant) are also introduced which cracks:

$$\{C_nH_m\} \rightarrow nC(s) + m/2\{H_2\} \quad (14.36)$$

and lower down the temperature safely and thus protects the tuyeres. The oxygen lancing is done through tuyeres. The stirring of the bath due to the ore boil is replaced by the turbulence caused by the oxygen lancing. The rate of supply of oxygen is increased several folds by using gaseous oxygen instead of iron ore during refining. Pure oxygen is also oxidized the scrap [as per exothermic reaction (14.34)], and faster melting of scrap takes place before hot metal is charged to the furnace. The tap-to tap time has been reduced from 8 to 3 h and even sometime 1.5 h.

14.3.5.2 Twin-Hearth Process

Twin-hearth furnace is an autogenous process in which a stationary open-hearth furnace is partitioned into two compartments by creating a refractory wall at the tap hole line to a level below the roof to allow flue gases to flow from one compartment to the other. The two compartments are run out of phase with each other such that while refining of the molten charge is going on in the first compartment, CO of flue gases enters the second and is burned to CO_2 with oxygen injected through a

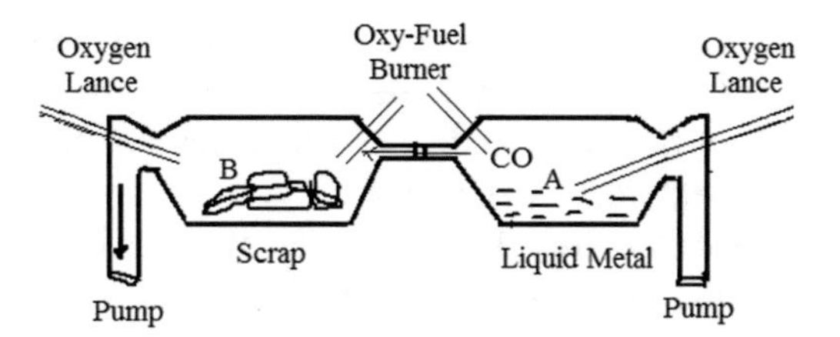

Fig. 14.4 Twin-hearth furnace

lance. The exothermic heat released burning of CO gas preheats the solid scrap in that compartment. The cycle is so phased that by the time refining is over in one, it commences in the other. The process thus permits total heat utilization within the furnace. Twin-hearth process was developed in USSR [2] in 1965 and had gained popularity.

Twin-hearth furnace is a double bath furnace without regenerators, designed for operation with intensive oxygen blowing into the bath through lances from the roof. In addition to these, two oxy-fuel burners are also provided in each hearth for heating. There are two hearths: one acts as scrap melting with oxy-fuel burner and other acts as refining of molten metal by lancing oxygen to the bath. By oxygen lancing, it can utilize the physical and chemical heats from the gases formed during lancing. The hearths are connected to each other with a passage to enable the waste gases to flow from one hearth to other. The area of each hearth is smaller than open-hearth furnace of similar capacity, but the bath depth is more. Both hearths have a common roof. In open-hearth furnace, the roof works as a reflector of heat to the bath, but in twin-hearth furnace, such function is greatly reduced. This has helped to raise the roof height and therefore increased roof life. Since there are no regenerators, the slag pockets are directly connected to the flues through single uptakes. Oxygen lances are provided on roof, along with fuel burners on roof or in back wall. Figure 14.4 shows the schematic diagram of a twin-hearth furnace. The consumption of external fuel and refractory in twin-hearth furnace is reported to be only 10% and 50%, respectively, compared to open-hearth furnace.

The technology of twin-hearth furnace consists of two periods, namely cold and hot period.

(i) The hot period consists of melting and refining, and
(ii) The cold period covers tapping, felting, charging, heating and pouring of hot metals.

During hot period, CO gas is evolved when a large amount of oxygen is blown into the bath:

$$2[\mathrm{C}] + \{\mathrm{O_2}\} = 2\{\mathrm{CO}\},\ \Delta H^0_{298} = -111.72\ \mathrm{kJ/mol\ of\ C} \tag{14.37}$$

If the CO gas is completely burned over the bath by oxygen jet, the heat evolved will be substantially higher than which is needed for the process:

$$2\{\mathrm{CO}\} + \{\mathrm{O_2}\} = 2\{\mathrm{CO_2}\},\ \Delta H^0_{298} = -131.17\ \mathrm{kJ/mol\ of\ CO} \tag{14.38}$$

This heat can be utilized for melting of the scrap by developing a new way of steelmaking in a twin-hearth furnace. There are two baths, A and B, with a common roof, and be connected by a passage through which gases can moved. Each of the baths has a tap hole, slag hole and charging door. While the hot metal in bath A is being blown with oxygen, the bath B is charged with solid scrap which is heated up with the flame of burning CO gas from bath A. The fuel for bath B is mainly by the burning of CO gas coming through the passage from bath A. Motion of the gas is being provided by the draught of the stack, by applying pump.

As a rule, no fuel is supplied, when bath A is blown with oxygen. If fuel is supplied, it is burned in the presence of oxygen, so that fumes passing to the bath B have quite high amount of CO_2 gas. In addition, there are oxygen lances from the roof above bath B and fuel-oxygen burner to supply more fuel to heat up the solid scrap and fluxes when needed.

As the refined metal is tapped from bath A, the direction of fumes is reversed from bath B to bath A, after hot metal is charged into the bath B on the partial melted scrap. Then, oxygen blowing is started in bath B, while bath A is being charged with solid scrap, a new cycle of operation is started.

There is a twin-hearth furnace of a capacity of 300 tonnes in each of the baths, operating 60% hot metal and 40% scrap. The whole cycle is 3½ h, with tapping every 105 min. Small quantity of fuel is required, and total time is reduced to half. This twin-hearth process has many advantages over the conventional open-hearth process. It not only increases productivity by 3–7 times, but also results in considerable saving of heat.

Advantages of twin-hearth furnace are as follows:

1. Higher productivity,
2. Reduced energy consumption,
3. Lower refractory consumption.

Probable Questions

1. Hot metal contains 3.0% C, 1.25% Si, 0.8% Mn, 0.04% P and 0.03% S. Do you need a basic slag? Why?
2. 'Si, Mn and C have good affinity for oxygen than Fe at steelmaking temperature; still Fe reacts first with oxygen'. Why?
3. What do you mean by 'blow-hot condition' in Bessemer converter? What will be the effect on steelmaking?
4. What happens in Bessemer steelmaking, if silicon content is less than 1.0% in hot metal?
5. What are the main sources of heat in acid and basic autogenous steelmaking processes?
6. 'After drop of the big flame in basic Bessemer process, still air blowing is continued' why? What is that known as?
7. What do you understand by 'blown metal'?
8. 'Total FeO content in acid slag is more than basic slag, but the activity of FeO in acid slag is lower than the basic slag'. Why?
9. What were the limitations of Bessemer steelmaking process?
10. What are the problems of using 100% solid pig iron or 100% hot metal as the charge in open-hearth furnace? What is the best combination of charge and why?
11. What is the basic concept of open-hearth process?
12. What is the other name of open-hearth process? Why it is called open-hearth furnace?
13. 'After partly melt down of the solid charge in open-hearth furnace, the burning rate of fuel should be lower down'. Why?
14. Why hot metal is not charged in open-hearth furnace before the partial melting of scrap?
15. 'Front flushing of slag, in open-hearth furnace, is advantages for steelmaking'. Explain. Why this is happened?
16. What are the difference between lime boil and ore boil? Which one is more vigorous than other?
17. How 'blocking of heat' is done in open-hearth furnace?
18. What are the basic concept of SIP and twin-hearth processes?

Examples

Example 14.1 Steel is being made in a Bessemer converter, using air to oxidize the impurities. The converter charge 20 tonnes of hot metal of composition: 3.5% C, 1.5% Si, 1.2% Mn. 3/4th of the carbon form CO and 1/4th form CO_2.

Calculate: (i) the volume of air used and (ii) the composition and total volume of product gases.

Solution

Assume: there is no loss of iron in slag.

$$\text{Reactions}: \quad \underset{28}{Si} + \underset{32}{O_2} = \underset{60}{SiO_2} \qquad \underset{55}{Mn} + \underset{16}{1/2\,O_2} = \underset{71}{MnO}$$

$$\underset{12}{C} + \underset{32}{O_2} = \underset{44}{CO_2} \qquad \& \qquad \underset{12}{C} + \underset{16}{1/2\,O_2} = \underset{28}{CO}$$

Element	Amount in kg		oxygen require in kg
Si	$\rightarrow 20000 \times \left(\frac{1.5}{100}\right) = 300$ kg		$\left(\frac{32}{28}\right) \times 300 = 342.86$ kg
Mn	$\rightarrow 20000 \times \left(\frac{1.2}{100}\right) = 240$ kg		$\left(\frac{16}{55}\right) \times 240 = 69.82$ kg
C	$\rightarrow 20000 \times \left(\frac{3.5}{100}\right) = 700$ k	(i) 3/4 CO	$\left(\frac{16}{12}\right) \times 3/4 \times 700 = 700.0$ kg
		(ii) 1/4 CO_2	$\left(\frac{32}{12}\right) \times 1/4 \times 700 = 466.67$ kg

Total oxygen required = 342.86 + 69.82 + 700.0 + 466.67 = 1579.35 kg
32 kg oxygen occupied volume 22.4 Nm^3 at STP

$$\text{Therefore, 1579.35 kg oxygen occupied volume} = \left(\frac{22.4}{32}\right) \times 1579.35$$
$$= 1105.55\ Nm^3 \text{at STP}$$

Since air contains 21% oxygen by volume
So total volume of air $= 1105.55 \times \left(\frac{100}{21}\right) = 5264.5\ Nm^3$at STP
Nitrogen contain $= 5264.5 - 1105.55 = 4158.95\ Nm^3$at STP (**76.09**%)
CO form $= \times 700 = 980.0\ Nm^3$at STP (**17.93**%)
CO_2 form $= \times 466.67 = 326.67\ Nm^3$at STP (**5.98**%)
Total volume of product gases = **5465.62 Nm^3**at STP (**100**%)

Example 14.2 Chemistry of input and output materials for steel is being made in a basic Bessemer converter process which is as follows:

Element, %	Hot metal	Steel (to be produced)
C	3.40	0.15
Si	1.10	0.01
Mn	0.75	0.25
P	0.40	0.03
S	0.04	0.03

Calculate: (i) Amount of hot metal to be charged per tonne of steel production.
(ii) Amount of slag produced and composition of slag.

Given data: (i) Weight of lime is 50 kg per tonne of steel production (lime contain 94.5% CaO, 2.5% MgO, 1.5% SiO_2 and 1.5% Al_2O_3)
(iii) 1.0% Fe loss with respect to steel production.

Solution

Assumptions: (a) one tonne steel production,
By balancing the chemistry of hot metal and produced steel, we get 94.31% and 99.53% Fe respectively.

Fe balance: Fe input = Fe output
Fe from HM = Fe goes to steel produce + Fe losses in slag

Suppose weight of hot metal = W_{HM}
$0.9431\,W_{HM} = 0.9953 \times 1000 + 0.01 \times 1000$

Therefore, W_{HM} = **1065.95kg**

$$Fe + 1/2\ O_2 = FeO$$
56 72
56 kg of Fe to form 72 kg of FeO

10 kg $\left(\frac{72}{56}\right) \times 10 = 12.86$ kg of FeO

Si balance : Si input = Si output
Si from HM + Si from lime = Si goes to steel produce + Si losses in slag (Suppose weight of Si losses in slag = W_{Si})

$$0.011 \times 1065.95 + 0.015 \times 50 \times \left(\frac{28}{60}\right) = 0.0001 \times 1000 + W_{Si}$$

Therefore, W_{Si} = 11.98 kg

$$Si + O_2 = SiO_2$$
28 60

28 kg Si to form 60 kg of SiO_2
11.98 kg $\left(\frac{60}{28}\right) \times 11.98 = 25.67$ kg of SiO_2 in slag

Mn balance: Mn input = Mn output
Mn from HM = Mn in steel + Mn losses in slag
$0.0075 \times 1065.95 = 0.0025 \times 1000 + W_{Mn}$

Therefore, W_{Mn} = 7.74 kg

$$Mn + 1/2\ O_2 = MnO$$
55 71
55 kg Mn to form 71 kg of MnO
7.74 kg $\left(\frac{71}{55}\right) \times 7.74 = 10.0$ kg of MnO in slag.

P balance: P input = P output
P from HM = P in steel + P goes to slag
$0.004 \times 1065.95 = 0.0003 \times 1000 + W_P$
$W_P = 3.96$ kg
$2P + 5/2\ O_2 = P_2O_5$
$2 \times 31 \quad\quad 142$

62 kg P to form 142 kg of P_2O_5
3.96 kg $\left(\frac{142}{62}\right) \times 3.96 = 9.07$ kg of P_2O_5

S balance: S input = S output
S from HM = S in steel + S goes to slag
$0.0004 \times 1065.95 = 0.0003 \times 1000 + W_S$
Therefore, $W_S = 0.126$ kg
$CaO + S = CaS$
56 32 72

32 kg S to form 72 kg of CaS
0.126 kg $\left(\frac{72}{32}\right) \times 0.126 = 0.28$ kg of CaS
72 kg CaS formation CaO required 56 kg
0.28 kg $\left(\frac{56}{72}\right) \times 0.28 = 0.22$ kg

CaO balance: CaO input = CaO output
CaO from lime = CaO in CaS + CaO in Slag
$0.945 \text{ x } 50 = 0.22 + W_{CaO}$
$W_{CaO} = 47.03\text{kg}$

MgO balance: MgO from lime = MgO in Slag $W_{MgO} = 0.025 \times 50 = 1.25$ kg
Al_2O_3 balance: Al_2O_3 in lime = Al_2O_3 in slag

$$
\begin{aligned}
W_{Al_2O_3} &= 0.015 \times 50 = 0.75 \\
\text{Amount of slag} &= W_{FeO} + W_{SiO_2} + W_{MnO} + W_{P_2O_5} + W_{CaS} + W_{CaO} + W_{MgO} + W_{Al_2O_3} \\
&= 12.86 + 25.67 + 10.0 + 9.07 + 0.28 + 47.03 + 1.25 + 0.75 = \mathbf{106.91\ kg}
\end{aligned}
$$

Composition of slag: 12.03% FeO, 24.01% SiO_2, 9.35% MnO, 8.48% P_2O_3, 0.26% CaS, 43.99% CaO, 1.17% MgO and 0.7% Al_2O_3.

Example 14.3 400 kg scrap and 400 kg hot metal are charged into a basic open-hearth furnace. The composition of charge materials and product is as follows:

Materials	Composition (%)			
	C	Mn	Si	P
Hot metal	3.6	1.9	0.9	0.15
Scrap	0.5	0.3	0.1	0.05
Steel produce	1.0	0.5	0.2	0.04
Iron ore	73% Fe_2O_3	12% MnO	15% SiO_2	–

Slag contains 45% CaO, 20% FeO and 35% SiO_2 and MnO.

Find out: (1) weight of iron ore to be added, (2) weight of limestone to be added and (3) weight of slag produce.

Solution

Basis of calculation: (1) one tonne of steel production and (2) limestone is pure.

Fe balance:

Fe from HM + Fe from scrap + Fe from ore = Fe in steel + Fe loss in slag

$$\left(\frac{93.45}{100}\right) \times 400 + \left(\frac{99.05}{100}\right) \times 400 + \left(\frac{112}{160}\right) \times \left(\frac{73}{100}\right) \times W_{\text{slag}} \times W_{\text{ore}} = \left(\frac{98.26}{100}\right) \times 1000 + \left(\frac{20}{100}\right) \times \left(\frac{56}{72}\right) \times W_{\text{slag}}$$

Or 373.8 + 396.2 + 0.511 W_{ore} = 982.6 + 0.156 W_{slag}

$$\text{Or}\, 0.511\ W_{\text{ore}} + 0.156\ W_{\text{slag}} = 212.6 \tag{1}$$

Si balance:

Si from HM + Si from scrap + Si from ore = Si in steel + Si loss in slag

$$\left(\frac{0.9}{100}\right) \times 400 + \left(\frac{0.1}{100}\right) \times 400 + \left(\frac{15}{100}\right) \times \left(\frac{28}{60}\right) \times W_{\text{ore}} = \left(\frac{0.2}{100}\right) \times 1000 + W_{\text{Si in slag}}$$

Or 3.6 + 0.4 + 0.07 W_{ore} = 2.0 + $W_{\text{Si in slag}}$

Or $W_{\text{Si in slag}}$ = 2.0 + 0.07 W_{ore}

$$\text{Therefore, } W_{\text{SiO2 in slag}} = (2.0 + 0.07\ W_{\text{ore}}) \times \left(\frac{60}{28}\right) = 4.29 + 0.15\ W_{\text{ore}} \tag{2}$$

Mn balance:

Mn from HM + Mn from scrap + Mn from ore = Mn in steel + Mn loss in slag

$$\left(\frac{1.9}{100}\right) \times 400 + \left(\frac{0.3}{100}\right) \times 400 + \left(\frac{12}{100}\right) \times \left(\frac{55}{71}\right) \times W_{\text{ore}} = \left(\frac{0.5}{100}\right) \times 1000 + W_{\text{Mi in slag}}$$

Or 7.6 + 1.2 + 0.093 W_{ore} = 5.0 + $W_{\text{Mn in slag}}$

Or $W_{\text{Mn in slag}}$ = 3.8 + 0.093 W_{ore}

$$\text{Therefore, } W_{\text{MnO in slag}} = (3.8 + 0.093\ W_{\text{ore}}) \times \left(\frac{71}{55}\right) = 4.91 + 0.12\ W_{\text{ore}} \tag{3}$$

CaO balance:
CaO from limestone = CaO in slag
(56/100) × W_{LS} = $W_{CaO\ in\ slag}$
Or

$$W_{CaO\ in\ slag} = 0.56\ W_{LS} \tag{4}$$

Weight of slag = $W_{FeO\ in\ slag}$ + $W_{SiO_2\ in\ slag}$ + $W_{MnO\ in\ slag}$ + $W_{CaO\ in\ slag}$
W_{slag} = 0.2 W_{slag} + (4.29 + 0.15 W_{ore}) + (4.91 + 0.12 W_{ore}) + 0.56 W_{LS}
Or

$$0.8\ W_{slag} = 0.27\ W_{ore} + 0.56\ W_{LS} + 9.2 \tag{5}$$

Again, $W_{CaO\ in\ slag}$ = 0.56 W_{LS} = 0.45 W_{slag}

$$\text{Therefore, } W_{LS} = 0.8\ W_{slag} \tag{6}$$

Again ($W_{SiO_2\ in\ slag}$ + $W_{MnO\ in\ slag}$) = 0.35 W_{slag}
Or (4.29 + 0.15 W_{ore}) + (4.91 + 0.12 W_{ore}) = 0.35 W_{slag} (from Eqs. 2 to 3)
Or

$$0.27\ W_{ore} + 9.2 = 0.35\ W_{slag} \text{ or } W_{ore} = 1.3\ W_{slag} - 9.2 \tag{7}$$

From Eq. (1): 0.511 W_{ore} + 0.156 W_{slag} = 212.6
Or 0.511 (1.3 W_{slag} – 9.2) + 0.156 W_{slag} = 212.6 (from Eq. 7)
Or 0.66 W_{slag} – 4.7 + 0.156 W_{slag} = 212.6 or 0.82 W_{slag} = 212.6 + 4.7 = 217.3
Therefore, $\mathbf{W_{slag}}$ = **265 kg**.
From Eq. 7: $\mathbf{W_{ore}}$ = 1.3 W_{slag} − 9.2 = (1.3 × 265) − 9.2 = **335.3 kg**.
From Eq. 6: $\mathbf{W_{LS}}$ = 0.8 W_{slag} = 0.8 × 265 = **212 kg**

Example 14.4 Composition of combustion gas burned at the open-hearth furnace is 39.8% H_2, 36.9% N_2, 7.3% CH_4, 8.2% CO, 5.5% CO_2 and 2.3% O_2. Composition of exit gas is 11.6% H_2, 77.5% N_2, 7.1% CO, 3.1% CO_2 and 0.7% O_2. How much heat is utilized by the furnace and how much air required for combustion of m^3 gas?

Given: Calorific values of gases (J/m^3): 10,932.79 for H_2, 12,811.41 for CO, 36,091.18 for CH_4.

Solution

Heat generate from burning gases : $H_2 : \left(\frac{39.8}{100}\right) \text{x } 10{,}932.79 = 4{,}351.25 \text{ J}$

$$CO : \left(\frac{8.2}{100}\right) \text{x } 12{,}811.41 = 1{,}050.54 \text{ J}$$

$$CH_4 : \left(\frac{7.3}{100}\right) \text{x } 36{,}091.18 = 2{,}634.66 \text{ J}$$

Total heat generated by burning gases = 4,351.25 + 1,050.54 + 2,634.66 = 8,036.45 J

$$\text{Heat carries away by exit gases : } H_2 : \left(\frac{11.6}{100}\right) \text{x } 10{,}932.79 = 1{,}268.20 \text{ J}$$

$$CO : \left(\frac{7.1}{100}\right) \text{x } 12{,}811.41 = 909.61 \text{ J}$$

$$\text{Total heat carries away by exit gases} = 1{,}268.20 + 909.61 = 2{,}177.81 \text{ J}$$

Therefore, heat utilized by the furnace = 8,036.45 – 2,177.81 = **5,858.64 J**

$$H_2 + 1/2\ O_2 = H_2O \rightarrow O_2\text{require} = \left(\frac{39.8}{100}\right)\text{x}1/2 = 0.199 \text{ m}^3$$

$$CO + 1/2\ O_2 = CO_2 \rightarrow O_2\text{require} = \left(\frac{8.2}{100}\right)\text{x}1/2 = 0.041 \text{ m}^3$$

$$CH_4 + 2O_2 = CO_2 + 2H_2O \rightarrow O_2\text{require} = \left(\frac{7.3}{100}\right)\text{x}2 = 0.146\text{m}^3$$

$$\text{Total } O_2\text{require} = 0.199 + 0.041 + 0.146 = 0.386 \text{ m}^3$$

$$O_2 \text{ present in the combustion gas} = \left(\frac{2.3}{100}\right) = 0.023\text{m}^3$$

Total O_2 require $= 0.199 + 0.041 + 0.146 = 0.386$ m^3

O_2 present in the combustion gas $= \left(\frac{2.3}{100}\right) = 0.023$m^3

Hence, actual O_2 required for combustion of gas = 0.386 − 0.023 = 0.363 m^3
Air contains 21% O_2,
So, air required for combustion of gas = **1.73 m^3/m^{3}of gas**.

Problems

Problem 14.1 Steel is being made in a Bessemer converter, using air to oxidize the impurities. The converter charges 10 tonnes of hot metal of composition: 4.0% C, 1.4% Si, 1.2% Mn. 3/4th of the carbon form CO and 1/4th form CO_2.

Calculate: (i) the volume of air used, (ii) the composition and total volume of product gases and (iii) weight of slag produce.

[Ans: (i) Volume of air: **2871.93 Nm^3** at STP, (ii) Total volume of product gases: **3015.49 Nm^3** at STP; **75.24% nitrogen,18.57% CO and 6.19% CO_2 and (iii) Wt of slag: 454.91 kg**].

Problem 14.2 800 kg hot metal (4.0% C, 1.2% Si, 1.0% Mn, 0.4% P) is charged with scrap (0.25% C, 0.25% Si, 0.5% Mn, 0.04% P) into a basic open-hearth furnace. Limestone is added to maintain slag basicity 2.5. Find out (i) amount of scrap, and (ii) amount of limestone to be charged also and (iii) amount of slag produced with composition.

Given: (i) Product steel contains 0.75% C, 0.2% Si, 0.4% Mn, 0.04% P; (ii) Limestone contains 2% SiO_2, 3% Al_2O_3; and (iii) 1.5% Fe loss in slag.

[Ans: (i) **256.57 kg, (ii) 127.44 kg, (iii) 131.36 kg** (14.68% FeO, 20.55% SiO_2, 5.19% MnO, 5.05% P_2O_3, 51.61% CaO, and 2.91% Al_2O_3)].

References

1. *World Steel In Figures 2019* (World Steel Association)
2. S.R. Prabhakar, U. Batra, *Institution of Engineers (India) MM,* 76, 36 (May 1995)
3. R.H. Tupkary, *Introduction to Modern Steelmaking* (Khanna Publishers, Delhi, 1991)

15 Oxygen Steelmaking Processes

Instate of blowing air from the bottom of the Bessemer converter, in LD process, blowing oxygen from top of the converter. Hot metal, contains high or medium phosphorus, is used for LD process, to remove phosphorus from the hot metal, lining of converter is always basic. In LD converter, reactions are basically slag-metal reactions. The mass transfers are occurring by force convection which is caused the turbulence of the bath. The discovery of Maxshutte tuyere has given birth of new oxygen steelmaking process (i.e. OBM). By bottom blowing with oxygen, the intensity of mixing of the metal bath is increased by about ten times compared to the LD process.

High-phosphorous hot metal created difficulty for refining in top blown process that leads to develop rotary oxygen process of steelmaking. The basic principles are (i) CO gas is burned inside the vessel, generating heat transfer to the lining of vessel and (ii) rotation of converter to get heat transfer from lining of vessel to molten bath.

Although LD/BOF process is now old due to lack of proper bath agitation, slow rate of the slag-metal reactions and over oxidation of the bath. By increasing pressure of oxygen or lowering the lance height contribute in increasing the splashing and slopping, and over oxidation of the bath. To overcome these draw backs, bottom blown processes are developed. The combined blowing process where oxygen alone or oxygen and an inert gas, like argon or nitrogen, are blown simultaneously from top and bottom to achieve relatively accelerated slag-metal reactions and suppressed over oxidation of the bath.

15.1 LD Process

The refining of hot metal by oxygen lancing was first tried by Professor R. Duren in his laboratory in Switzerland. First pilot plant trials on 2–5 tonnes converters were successfully carried out to cast refined steel at Linz and Donawitz in Austria on 1949. First commercial LD plant, with 10–15 tonnes capacity converter, was established at Linz and first commercial casting was produced on 1952. Soon after second plant was installed at Donawitz, Austria. In 1954 outside of Austria, first plant was installed at Hamilton, in Canada. LD process was first adopted in India at Rourkela Steel Plant in 1956 and production was started in 1960. LD stands for ***L**inz* and ***D**onawitz* (two cities of Austria) where the process was developed. This is also known as basic oxygen process (BOP) or basic oxygen furnace (BOF) process.

S. K. Dutta and Y. B. Chokshi, *Basic Concepts of Iron and Steel Making*,
https://doi.org/10.1007/978-981-15-2437-0_15

Main features of this process are as follows:

1. Instate of blowing air in Bessemer processes, here use 99.5% oxygen,
2. Instate of blowing air from the bottom of the Bessemer converter, here blowing oxygen from top of the converter,
3. Hot metal contains high or medium phosphorus is used for LD process,
4. To remove phosphorus from the hot metal, lining of converter is always basic.

15.1.1 Design of Converter

It is either referred as LD converter or vessel. The converter is divided into three segments: (1) spherical bottom, (2) cylindrical body (or shell) and (3) conical throat (as shown in Fig. 15.1). Each of these segments is made of welded steel plates. Earlier design, the spherical bottom was detachable to help cooling and relining. But modern vessels are without joints, i.e. the segments are welded to form one single piece. Vessels without joints are much safer. The throat may be eccentric or concentric with the rest of the body.

Advantages of eccentric throat:

1. It is easier to remove the slag with low metal loss, because in the horizontal position, the metal level is much lower than slag level.
2. Any splashing of metal and slag from the mouth are thrown in one side of the converter only, hence cleaning-up problem apply only to one side.
3. The construction of hood and insertion of a lance are easy.
4. The lance is protected and controlled in a better way since hood is inclined.

Disadvantages of eccentric throat:

1. The shop floor is congested, since the charging and tapping are carried out on the same side.
2. Relining of converter is difficult.

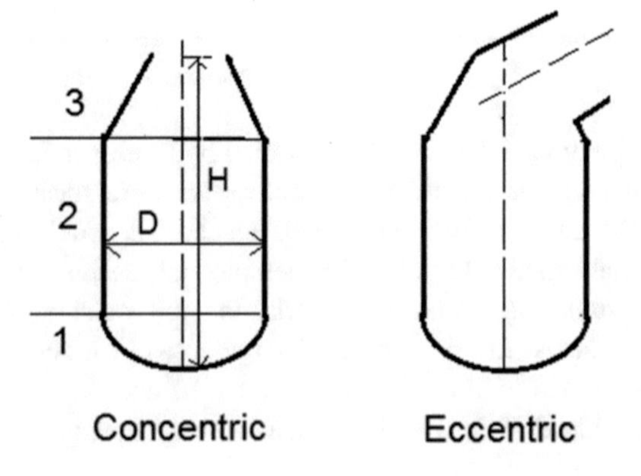

Fig. 15.1 LD converter

Advantages of concentric throat:

1. The shop floor is less congested, since the charging and tapping are carried out on the opposite sides, i.e. it provides greater flexibility.
2. Relining of converter is easier.

Disadvantages of concentric throat:

1. The splashing of metal and slag from the mouth are thrown throughout the throat, so cleaning-up is difficult.
2. It is more difficult to protect lance from radiation heat of molten metal.

Despite above disadvantages, the concentric throat is almost universally adopted. LD converter is mounted on trunnion ring ribs situated near the middle of its height at the cylindrical portion, which help movement of the converter more freely. The vessel is capable to rotate through 360°, but practically rotation is rarely exceeding 220°, which allows charging from overhead hoppers on one side of the vertical axis; and tapping, slagging and other operations on the other side.

Height (*H*) to diameter (*D*) ratio varies from 1.06 to 1.85; 1.5 is most preferred ratio. Capacity increases vessel diameter is also increased; since height has some limitation, hence converters of large capacities have a greater inner diameter and a lower *H*:*D* ratio (Table 15.1). Lining thickness depends on quality of lining material. Thickness of lining varies between 0.6 and 1.0 m.

Detachable bottom facilitates relining of the vessel, since the vessel can be cooled very quickly, brick laying is organized more conveniently with materials being supplied from below, and the whole work is done in a shorter time, because bottom may be simultaneously prepared in a separate place. Converters of a capacity of more than 100 tonnes, however, have a solid non-detachable bottom made integral with a solid, this is done to make the structure more rigid and to eliminate break-out (i.e. leak-out) of liquid metal through the joint.

The conical throat portion of converter is made detachable to reduce repair time, since the lining of this portion is subject to rapid wear by splashed metal and slag, high temperature, and the metallic structure is subject to cracking. Specific volume of the vessel was earlier just over 1 m^3/t capacity, but it was reduced to 0.75 m^3/t capacity. Again, trend is 1 m^3/t capacity. Figure 15.2 shows LD converter.

The amount of splashing is mainly determined by the distance from the bath surface to the throat, i.e. by the height of the free space in the converter. The diameter of the throat is determined from the weight of scrap to be charged for melting. The more scrap is to be charged during the short time specified, the charging scoop must be wider, and therefore the wider must be the throat of the converter. On the other hand, too large a diameter of the throat may be the cause of suction of atmospheric air into the vessel. This can increase concentration of nitrogen in the final steel.

Table 15.1 Dimensions of the converters [1]

Parameter	Capacity, tonnes		
	75	200	300
Height (*H*), m	7.5	9.0	9.0
Diameter (*D*), m	5.5	6.7	8.5
H:*D* Ratio	1.36	1.34	1.06
Bath diameter, m	4.0	5.0	6.5
Bath depth, m	1.3	1.5	1.8
Nose diameter, m	1.65	2.3	3.5

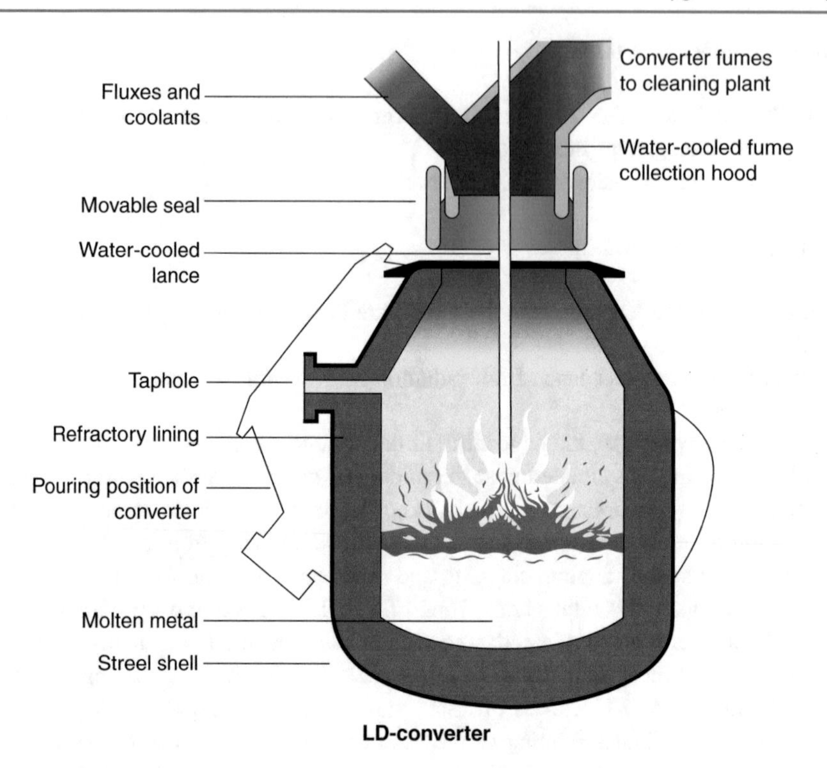

Fig. 15.2 LD converter [2]

15.1.2 Lance

Oxygen is introduced into the LD converter from the top through a water cooled lance. The lance is made by three concentric stainless steel tubes to circulate water around the central tube and pass oxygen through the most inner tube (as shown in Fig. 15.3). The tip of the lance is called nozzle which is made copper (melting point is 1085 °C). The nozzle is always exposed to a very high temperature, due to hot gases and heat of radiations from molten bath. But nozzle is not water cooled, still it does not melt at the exposed of high temperature. Why? This is due to (i) *Joule-Thomson Effect,* i.e. since oxygen comes out from the nozzle at very high pressure (10–12 kg/cm^2, where 1 atm = 1.033 kg/cm^2) and due to sudden change of pressure (from 10–12 kg/cm^2 to 1.033 kg/cm^2) large expansion of volume take place due to that tremendous cooling effect occur which prevent the nozzle to melt. (ii) Since nozzle is made of copper which is good conductor of heat, so whatever some heat come in contract, immediately that heat is dispersed to the stainless steel tube which is water cooled; and there is no concentration of heat is created, hence there is no possibility of melting of the nozzle.

Lance is nearly 8–10 m long and its diameter varies with vessel capacity in the range of 20–25 cm. Rate of cooling water is 50–70 m^3/h at a pressure of 5–7 kg/cm^2. The lance height (h) varies from 90 to 200 cm from the surface of molten bath.

15.1.3 Oxygen Lancing

As soon as the oxygen jet strikes the molten bath, a distinct crater is formed, and the peripheral liquid is considerably splashed to form two lips (as shown in Fig. 15.4). Turbulence is formed, impure metal

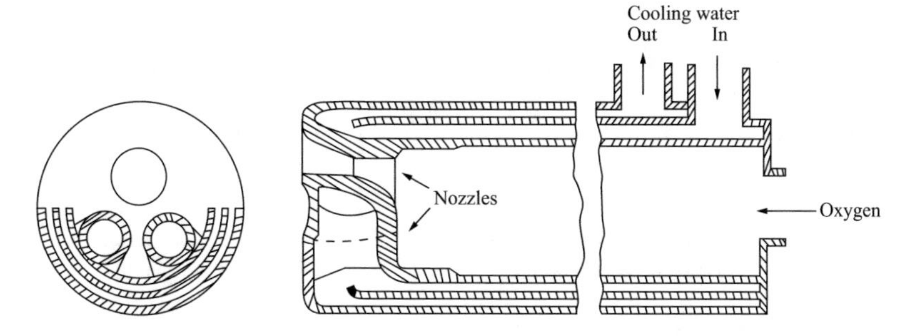

Fig. 15.3 Oxygen lance [1]

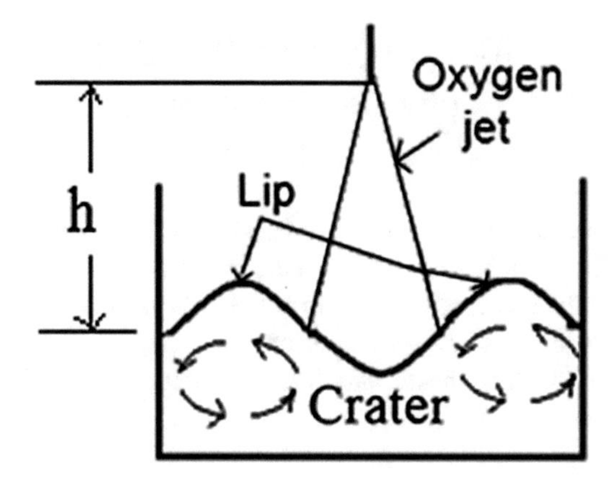

Fig. 15.4 Oxygen jet strikes the molten bath

droplet from the interior goes to the crater for refining and then ultimately falls of pure metal droplet due to gravity and its own weight. The advantage is that new impure metal droplets, whose densities are lower than pure metal droplets, are continuously exposed to oxygen jet.

As the nozzle angle or height of lance increase, the oxygen jet cannot penetrate more into the bath, so on the surface FeO or Fe_2O_3 are formed (exothermic reaction) they react with SiO_2 forming $2FeO \cdot SiO_2$ (i.e. slag). This exothermic reaction supplies the heat for dissolution of CaO in slag phase which helps for de-phosphorization.

$$2[\mathrm{Fe}] + \{\mathrm{O_2}\} = 2(\mathrm{FeO}), \quad \Delta H^0_{298} = -264.43 \text{ kJ/mol of FeO} \tag{15.1}$$

$$4[\mathrm{Fe}] + 3\{\mathrm{O_2}\} = 2(\mathrm{Fe_2O_3}), \quad \Delta H^0_{298} = -821.32 \text{ kJ/mol of } \mathrm{Fe_2O_3} \tag{15.2}$$

$$2(\mathrm{FeO}) + (\mathrm{SiO_2}) = (2\mathrm{FeO} \cdot \mathrm{SiO_2}), \quad \Delta H^0_{1773} = -27.07 \text{ kJ/mol} \tag{15.3}$$

Towards the end of the oxygen lancing period, carbon is less in the bath and so the lance is taken up at an elevated position and if lancing of oxygen is continued for a very long time from the elevated position, then FeO in slag is converted to Fe_2O_3 in slag. So, Fe_2O_3 in slag is increased (as shown in Fig. 15.5).

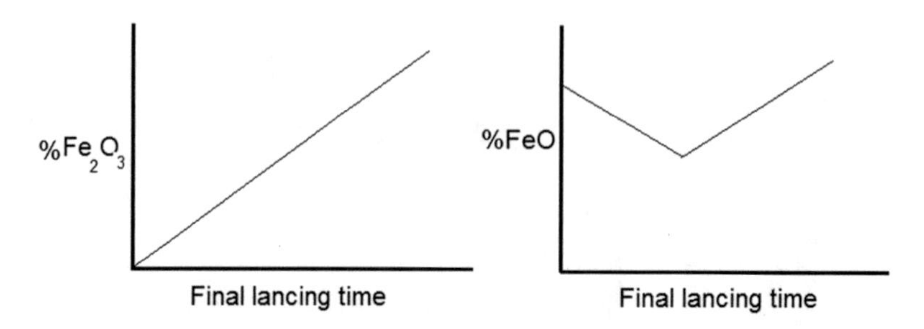

Fig. 15.5 Final lancing time versus Fe_2O_3 and FeO in slag

$$2(\mathrm{FeO}) + n\{\mathrm{O_2}\} = (\mathrm{Fe_2O_3}) + m\{\mathrm{O_2}\} \tag{15.4}$$

$$2(\mathrm{FeO}) + \{\mathrm{CO_2}\} = (\mathrm{Fe_2O_3}) + \{\mathrm{CO}\} \tag{15.5}$$

$$(\mathrm{Fe_2O_3}) + [\mathrm{Fe}] = 3(\mathrm{FeO}) \tag{15.6}$$

$$2[\mathrm{Fe}] + \{\mathrm{O_2}\} = 2(\mathrm{FeO}) \tag{15.1}$$

The net effect is the percentage of both FeO and Fe_2O_3 in slag increased, so Fe content of the slag increased, i.e. Fe losses. This is a bad practice. Hence, lancing of oxygen should be stopped at right moment, i.e. to control the final lancing time to protect Fe loss as FeO in slag. In the final position of lance, oxygen is absorbed in the slag. Therefore, oxygen potential of the slag increases that help dissolution of more lime in slag phase, and hence basicity of slag increases that favour the de-phosphorization reaction.

15.1.3.1 Oxygen Jet

Oxygen is blown at pressures of 10–12 kg/cm^2 through a nozzle, so that the jet formed at the nozzle exit is supersonic and generally has a speed between 2.0 and 2.5 times the speed of sound (i.e. 331.45 m/s in dry air). A supersonic jet is characterized by a supersonic core in which the jet velocity is higher than the speed of sound (Fig. 15.6). As the jet travels away from the nozzle, it is retarded by the converter atmosphere so that the supersonic core shrinks radially, and the axial velocity gradually decreases until at some distance away from the nozzle, the jet becomes fully subsonic. The main factor affecting the length of the supersonic core is the blowing speed and the ratio of the densities of the jet gas and the ambient medium. This ratio would very depend on the flow rate and the lance height.

The average flow rate of oxygen is 8000 N m^3/h for 80–100 tonnes furnace. During the lancing of oxygen, the jet should be expanded to obtain maximum impact area at the bath surface. At the same time, it should also penetrate the bath surface to a maximum extent.

The depth of penetration (d) of a jet in the bath depends on as follows:

$$\text{(i) Pressure of oxygen jet } (P), \quad d \propto P \tag{15.7}$$

$$\text{(ii) Diameter of the nozzle throat } (D_t), \quad d \propto D_t \tag{15.8}$$

$$\text{(iii) Height of the lance from the molten bath } (h), \quad d \propto (1/\sqrt{h}) \tag{15.9}$$

Therefore,

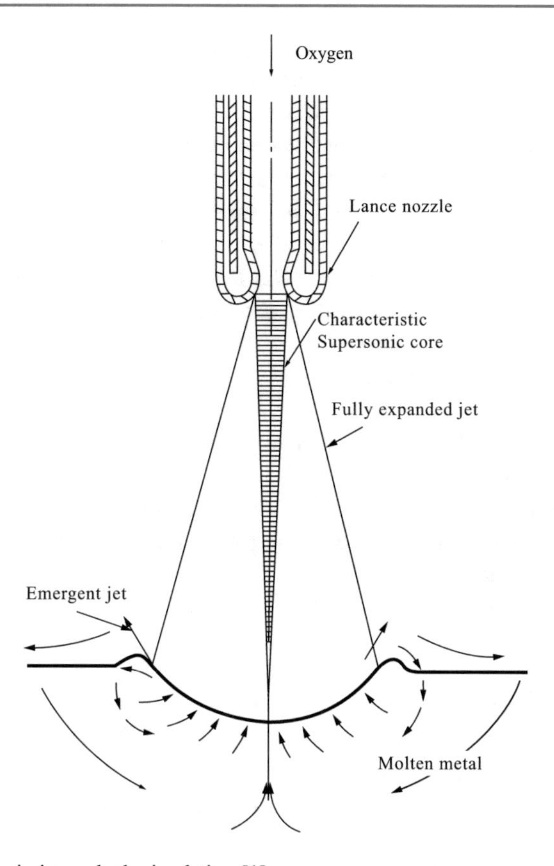

Fig. 15.6 Effect of supersonic jet on bath circulation [1]

$$d \propto \left(\frac{PD_t}{\sqrt{h}}\right) \tag{15.10}$$

Hence,

$$d = \text{Constant} \times \left(\frac{PD_t}{\sqrt{h}}\right) \tag{15.11}$$

As the height of the lance is raised and the pressure of the oxygen jet is reduced, the zone of contact of the oxygen jet with the metal increases and correspondingly the depth of penetration of oxygen into metal decreases. Under this condition, FeO content of the slag increases, which favours dissolution of lime.

The depth of penetration (*d*) controls the related rate of de-carbonization and de-phosphorization reactions. If the depth of penetration is more, then the rate of carbon–oxygen reaction will be more in metal phase due to oxygen potential in metal phase will be more. If the depth of penetration is less, then the rate of phosphorus–oxygen reaction will be more in slag phase due to oxygen potential in slag phase will be more.

Previously, a single nozzle lance is used for small size converter. Modern big converter used multiple nozzles lance. 6–8 nozzles are used for 350–400 t vessels. By using multiple nozzles lance, increase the surface area of jet impact area. For the same jet impact area, the lance height will be more for single nozzle lance than multiple nozzles lance (as shown in Fig. 15.7). If h_1 and h_2 are the lance heights for three and single nozzle lances for same jet impact area; for same impact area, single nozzle lance requires more height than three nozzle lances. If the lance height increases, the depth of penetration decreases; hence, rate of de-carbonization decreases. The lance height shorter means the depth of penetration is more, the jet impact area is more, and so the rate of de-carbonization and de-phosphorization reactions will be increased. For same impact area, there will be more penetration in case of multiple nozzles lance, so there will be less amount of FeO formation at the surface.

The axis of nozzle in a multiple nozzles lance is inclined (θ) to the vertical axis of the lance by nearly 10°. When (i) $\theta > 10° \rightarrow$ then depth of penetration in the bath decreases; (ii) $\theta < 10° \rightarrow$ the jets are interacting and overlap each other.

Advantages of multiple nozzles lance:

1. Reduce oxygen consumption per tonne of steel production,
2. Remelting of higher amount of scrap,
3. Production rate is increased because of faster rate of reaction and good slag-metal contact,
4. Reduce the splashing sufficiently, so increasing total yield by about 3%,
5. Cleaning under the converter is decreased,
6. Decrease formation of FeO in slag up to 3%,
7. Higher residual Mn is left in the bath, so that the consumption of Fe–Mn is reduced.

Splashing of metal is reduced subsequently by lower down the level of the bath. When carbon oxidation starts, the foaming slag is formed due to evolving of CO gas and slag swells up. By using multiple nozzle lances, the surface area of oxygen jet is increased, and foams are eventually distributed throughout the blowing time and foams are not coming out vigorously at any time and ejection of slag (i.e. slopping) is decreased.

Modern LD converters, multiple nozzles lance (up to 8 nozzles) are used. Main advantages in increasing the number of outlets on a lance are to allow the total oxygen throughput to be increased without effective increase the pressure exerted by the jets on the bath surface. This pressure determines the depth of penetration and amount of liquid metal splashed from the bath. By increasing the number of nozzle openings, to increase the rate of oxygen supply without decreasing the metal yield and phosphorous content can be lower down to 0.015%.

Comparative evaluation of three nozzles lance (TNL) vis-a-vis the six nozzles lance (SNL) reveals the following [3]:

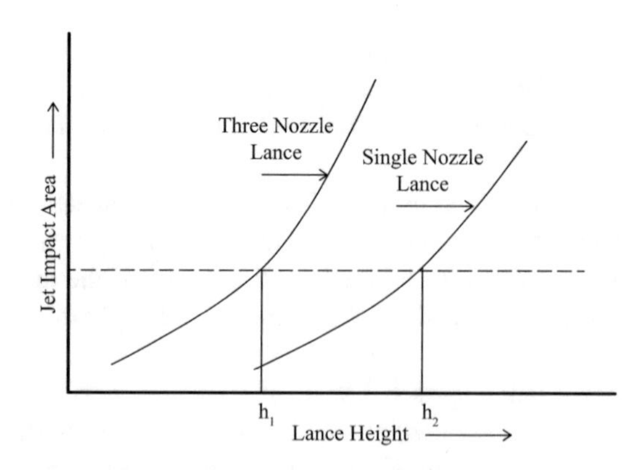

Fig. 15.7 Jet impact area for single and three nozzle lances

1. The consumption of lime is increased slightly from 75 to 85 kg/tonne liquid steel (kg/tls) and at the same time dolomite addition (23 kg/tls) was eliminated for the SNL heats.
2. Distinctly higher iron ore use is possible with the SNL, thereby reducing the cost of steel.
3. Average turn-down phosphorus is 0.012% in the SNL case compared with 0.024% for the TNL and the average phosphorus partition coefficient increases from 60–65 to 100–130. Therefore, there is a significant improvement in de-phosphorization due to SNL.
4. De-sulphurization remains practically unaffected.
5. Slopping/splashing is completely eliminated by adopting bath agitation process (BAP) along with SNL.
6. The yield is improved from 89.4 to 89.9% (i.e. 5 kg/t).
7. The lining life (with tar dolomite bricks) is increased from 300–400 heats for TNL to 600–900 heats for SNL.
8. Average tapping temperature in the SNL case is higher than TNL.

15.1.3.2 Oxygen Jet Momentum

Total energy of the jet from the multiple nozzles lance is distributed over a larger surface area of the bath. This improves slag-metal interaction. The important parameters are jet momentum, jet height and fluid properties. Jet momentum, means quantity of motion, can be calculated as follows [4]:

$$\begin{aligned} &\text{Momentum} = \text{mass} \times \text{velocity} \\ \text{Or} \quad &M = m \times u \end{aligned} \tag{15.12}$$

Again mass flow rate of gas (kg s^{-1}),

$$m = \rho_G \cdot A \cdot u \tag{15.13}$$

where

ρ_G density of gas (kg m^{-3}),
A cross-sectional area of nozzle opening $(m^2) = \left\{\frac{\pi d^2}{4}\right\}$,
d throat diameter of each nozzle (m),
u gas velocity (m s^{-1}).

For a lance with n nozzles and total gas flow rate (Q, m^3 s^{-1}),
Then, the flow rate of gas through each nozzle:

$$Q_n^* = \frac{Q}{n} \tag{15.14}$$

Therefore, mass flow rate of gas (kg s^{-1}) through each nozzle:

$$m_n^* = \left\{\frac{(Q \cdot \rho_G)}{n}\right\} \tag{15.15}$$

Again mass flow rate of gas (kg s^{-1}) through each nozzle:

$$m_n^* = \rho_G \cdot A \cdot u_n^* \tag{15.16}$$

Now combined Eqs. (15.15) and (15.16):
Velocity of gas at each nozzle:

$$u_n^* = \left\{\frac{Q}{(n \cdot A)}\right\} = \left\{\frac{4Q}{(n \cdot \pi d^2)}\right\} \tag{15.17}$$

When throat diameter of each nozzle (d) and total flow rate (Q) are constant, then, Eq. (15.15) become:

$$m_n^* = \frac{K_1}{n} \tag{15.18}$$

and Eq. (15.17) become:

$$u_n^* = \frac{K_2}{n} \tag{15.19}$$

where K is the constant, $K_1 = Q \cdot \rho_G$ and $K_2 = \left\{\frac{4Q}{(\pi d^2)}\right\}$.

Therefore, momentum of jet in each nozzle:

$$M_n^* = m_n^* \cdot u_n^* = \frac{K_1}{n} \cdot \frac{K_2}{n} = \frac{K}{n^2} \tag{15.20}$$

where $K = K_1 \cdot K_2$.

Hence, the momentum of jet in each nozzle is inversely proportional to the square of the number of nozzles in the lance.

Therefore, total momentum for all the jets is:

$$M = \sum M_n^* = n \cdot M_n^* = \frac{K}{n} \tag{15.21}$$

Total momentum for all the jets is inversely proportional to the number of nozzle.

Less momentum or force of jet means less splashing of metal, i.e. less loss of iron or more iron yield.

15.1.4 Mechanism of Refining

A turbulent recirculation zone forms directly under and around the jet impact area. In this region, the gas jets coming out from a multi-hole nozzle hit the liquid metal at a high impact pressure and penetrate the metal phase; this penetration helps fast dissolution of oxygen in metal phase.

First iron is oxidized by oxygen to form FeO, then FeO reacts with Si, Mn in metal to form oxides and again these oxides react each other to form slag.

$$2[Fe] + \{O_2\} = 2(FeO) \tag{15.1}$$

$$[Si] + 2(FeO) = (SiO_2) + 2[Fe] \tag{15.22}$$

$$[Mn] + (FeO) = (MnO) + [Fe] \tag{15.23}$$

$$(MnO) + (SiO_2) = (MnO \cdot SiO_2) \tag{15.24}$$

$$2(FeO) + (SiO_2) = (2FeO \cdot SiO_2) \tag{15.3}$$

In LD converter, reactions are basically slag-metal reactions. As the oxygen jet strikes the bath, initially a slag of FeO–MnO–SiO_2 is formed, it is known as silico-ferrite slag. The temperature of impact zone is very high (~3000 °C) and presence of high pure FeO in the slag which help the quick dissolution of lime in slag phase. After dissolution of lime in slag that becomes calci-ferrite slag (CaO–SiO_2–MnO–FeO). For continuous reactions, the impurities must be brought continuously to the slag-metal interface. If these mass transfers are done only by diffusion, then the rate of reactions would be very slow; which is opposite to experience that the rate of reactions is very fast. So, what is the other mechanism of mass transfers?

The mechanism of mass transfers is mainly due to turbulence of the bath. The mass transfers are occurring by force convection which is caused the turbulence of the bath. There are four factors for making turbulence:

1. Oxygen jet impact,
2. Density difference,
3. Evolution of CO gas,
4. Temperature difference.

(1) Oxygen jet impact: When the oxygen jet strikes the metal surface, turbulence is created; the metal droplets go up to the lips where the oxygen is very easily available and then the oxygen-impurity (in metal) reactions take place rapidly. As the impurities are oxidized and get out from the metal droplets, it became heavier and came down from the lips. New metal droplets (i.e. lighter) occupy their position and by this way a continuous contact between new metal droplets and oxygen jet are maintained (as shown in Fig. 15.8a).

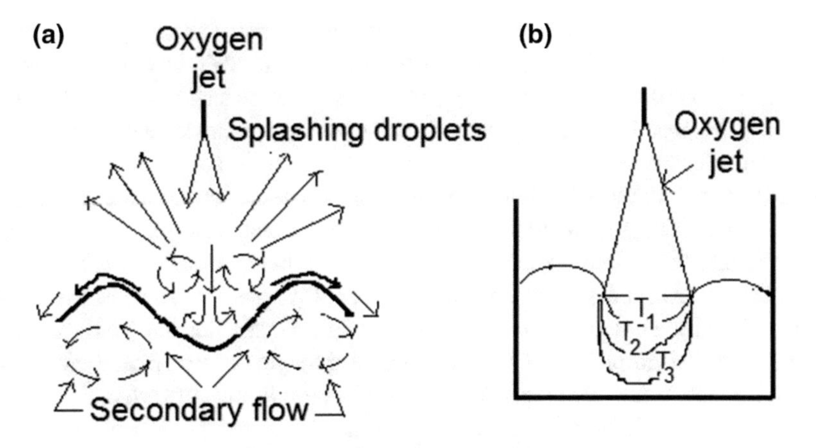

Fig. 15.8 Turbulence of the bath by **a** oxygen jet impact, **b** temperature difference

(2) Density difference: When the metal droplets contain impurities (like Si, Mn, C, P in Fe) their density is lowered than that of the pure metal (Fe) droplets, i.e. pure metal (Fe) has a higher density and so it is heavier than that of the impure metal which has a lower density and hence lighter. When impurities of metal droplets are removed at slag-metal interface and that became pure metal, its density increases. Hence, metal droplet becomes heavier, which are came down. New metal droplets, which are lighter, go up to the interface; by this way, turbulence is created.
(3) Evolution of CO gas:

$$2[\mathrm{C}] + \{\mathrm{O_2}\} = 2\{\mathrm{CO}\} \quad (15.25)$$

Evolution of CO gas bubbles makes boiling the bath, due to that good turbulence take place in the bath.
(4) Temperature difference: Temperature is also responsible for causing the turbulence. The temperature at first layer, below the jet impact, is highest (i.e. $T_1 = \sim 3000\ ^\circ\mathrm{C}$) and at the subsequent layer the temperatures are somewhat low ($T_1 > T_2 > T_3$; as shown in Fig. 15.8b) and so there is tendency for attaining the equilibrium temperature of the bath which causing the turbulence.

15.1.5 Characteristic of Slag

Slag formed in a LD converter is a foaming slag, i.e. large numbers of gas bubbles are entrapped in the slag phase. Due to this foaming nature of slag, the slag-metal contact is increased, and this ensures good slag-metal reactions and decrease of heat loss by liquid metal. In LD process, this foaming slag is an advantage, but in open-hearth process, this foaming slag is a disadvantage because the slag opposes the heat conduction as well as heat transfer in open-hearth process, since slag is a bad conductor of heat.

In LD process, 2/3rd of the lime is added with the charge. The rest 1/3rd of the lime is added in step wise. When adding lime, the oxygen flow rate is decreased and after completing the addition of lime, the oxygen flow rate is again increased step by step.

15.1.6 Mechanism of Carbon Reaction

In oxygen steelmaking, large numbers of metal droplets are generated and ejected into the slag phase due to impact of the high speed oxygen jet on the metal bath. This leads to the formation of a slag-metal emulsion. In this emulsion, large area of contact between slag and metal droplets enhances the kinetics of refining reactions, among them decarburization is a prominent reaction.

The bulk of the de-carburization takes place in the turbulent region of jet impact irrespective of the fact whether the slag is solid, liquid or foamy. Metal droplets are ejected from the jet impact zone, this leads to the formation of a large interfacial area between the metal droplets and the slag. But it is difficult to distinguish the de-carburization occurring in the bulk metal from that occurring in the metal droplets. Slag in LD process is heterogeneous and always contains some entrained gas bubbles and solid material (either undissolved or precipitated). At no stage the slag is 100% liquid. A significant part of the metal droplets falls back and travel through the semi-liquid slag. Through this mechanism, the droplets can cause slag foaming and slopping in the converter.

During oxygen blowing, large amount of very fine metal droplets are thrown into the slag phase, since slag in the LD converter is foaming nature; so a large number of very fine metal droplets are

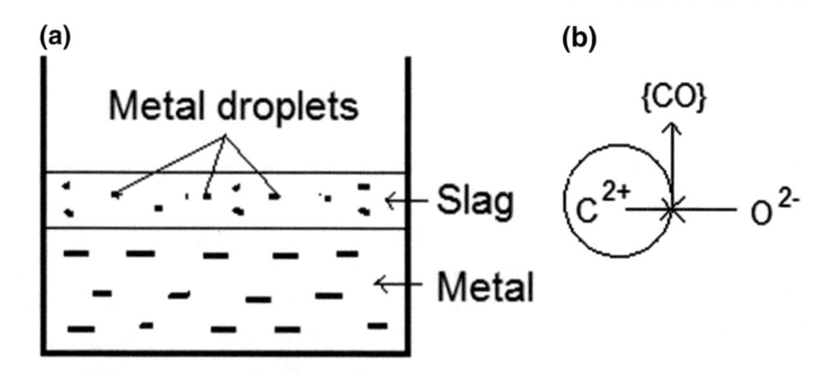

Fig. 15.9 **a** Slag-metal emulsion, **b** single metal droplet in slag phase

entrapped or suspended in the slag phase and a *slag-metal emulsion* is formed (as shown in Fig. 15.9). These metal droplets may remain suspended in the slag phase. Due to presence of impurities (particularly carbon) within the metal droplet and causing decrease in its density and suspension of metal droplet in the emulsion zone. With the carbon-oxygen reaction, there is simultaneous escape of CO gas from the metal droplet. This will lead to increase in density of metal droplets. Then the density of metal droplet is become greater than emulsion density, the metal droplet is return back to the metal bath. But the retention time of these metal droplets within the slag phase is very small, since the metal droplets are heavier than slag. Bulk of the carbon–oxygen reaction occurs in the slag-metal emulsion.

In the emulsion zone, the decarburization of metal droplet occurs by the following reaction, wherein the oxygen supplied from slag (dissociation of FeO) reacts with carbon within the metal droplet and forms carbon monoxide (CO) gas at the surface of the metal droplet (as shown in Fig. 15.9b).

Because of emulsification, the slag is saturated with FeO, i.e. oxygen potential of slag is very high. The smaller the droplet, greater will be the surface area. As soon as a metal droplet comes within the slag phase, oxygen starts diffusing into it due to concentration gradient. This droplet during its short residence time in slag phase becomes loaded with oxygen. Carbon in metal droplet reacts with that loaded oxygen. Hence, a bubble of CO is formed within the emulsion at metal-slag interface.

Dissociation of FeO:

$$(\mathrm{FeO}) = \left[\mathrm{Fe}^{2+}\right] + \left[\mathrm{O}^{2-}\right] \tag{15.26}$$

Carbon ionize within the metal droplet:

$$[\mathrm{C}] = \left[\mathrm{C}^{2+}\right] + 2\mathrm{e} \tag{15.27}$$

Carbon–oxygen reaction:

$$\left[\mathrm{C}^{2+}\right] + \left[\mathrm{O}^{2-}\right] = \{\mathrm{CO}\} \tag{15.28}$$

Iron back to metal:

$$\left[\mathrm{Fe}^{2+}\right] + 2\mathrm{e} = [\mathrm{Fe}] \tag{15.29}$$

Overall reaction:

$$(FeO) + [C] = [Fe] + \{CO\} \quad (15.30)$$

Hence, carbon–oxygen reaction starts within the droplet at the interface, it may not require for nucleation. Since there are large numbers of droplets, so there will be large numbers of interfaces. Hence, homogeneous nucleations are occurred at the emulsion. During peak-period of carbon–oxygen reaction, 30% of metal may be at the emulsion, these droplets are very fine which mean high rate of mass transfer.

There are also some other sites which are favourable to heterogeneous nucleation, during oxygen blowing, some oxygen bubbles are not dissolved in the bath. These undissolve oxygen bubbles may act as sites for carbon–oxygen reaction. The undissolve lime particles, which are floating on the slag; as well as steel scraps and lining which have cracks and crevices containing entrapped air. All these acts as nucleating sites. The cracks and crevices of the converter lining may also act as a site for nucleation.

The major part of carbon–oxygen reaction may have occurred in LD vessel at the slag phase, so the reaction is very fast. The de-carburization reaction will not be completed in a single cycle since the retention time of metal droplet is a fraction of second. The metal droplet goes to the slag phase, subsequently, droplet comes back to the metal phase and again goes to the slag phase; this type of cycle is going on until major amount of carbon is removed from the metal.

The overall de-carburization rate can be written as:

$$W_f \cdot \left(\frac{dC}{dt}\right)_f = W_e \cdot \left(\frac{dC}{dt}\right)_e + W_m \cdot \left(\frac{dC}{dt}\right)_m \quad (15.31)$$

where

W_f the total weight of metal in the furnace,
W_e the weight of metal in the emulsion,
W_m the weight of metal in the metal phase,
$\left(\frac{dC}{dt}\right)$ the de-carburization rate.

Again

$$\left(\frac{dC}{dt}\right)_e = \left[\frac{(\%C_m - \%C_e)}{t_r}\right] \quad (15.32)$$

where $\%C_m$ = % of C in metal phase, $\%C_e$ = % of C in emulsion, t_r = retention time of the droplet within the emulsion.

After carbon–oxygen reaction is over and then percentage of carbon in the metal phase is equal to the percentage of carbon in emulsion.

15.1.7 Manganese Reaction

Manganese in metal is oxidized by FeO in slag to form MnO in slag:

$$[Mn] + (FeO) = (MnO) + [Fe] \quad (15.23)$$

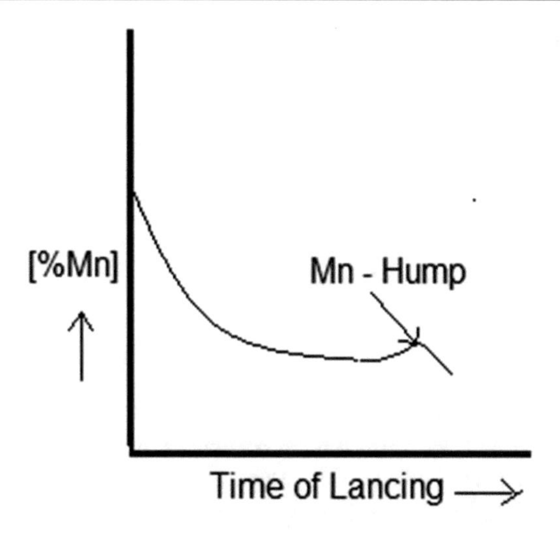

Fig. 15.10 Mn reaction

This reaction is a slag-metal reaction, FeO content of the bath is controlled by Mn. FeO is formed due to oxygen enrichment of the bath by oxygen lancing. When the activity of FeO in slag is decreased, i.e. the concentration of FeO in the bath gradually falls below the equilibrium limit, and then the reaction will start in the back-ward direction. Therefore, equilibrium is disturbed. MnO in slag is decomposed and Mn gets back into the metal. This is known as *Mn-reversion* that creates the *Mn-hump* (as shown in Fig. 15.10), which take place at the pick period of carbon–oxygen reaction:

$$(\mathrm{FeO}) + [\mathrm{C}] = [\mathrm{Fe}] + \{\mathrm{CO}\} \tag{15.30}$$

When the rate of carbon–oxygen reaction decreases, enough FeO will be present in the bath, then again, the Mn reaction processed to the forward direction.

During the initial period of slag formation, when the temperature is not yet high enough to dissolve the lime completely, so MnO combines with SiO_2 to form slag.

$$(\mathrm{MnO}) + (\mathrm{SiO_2}) = (\mathrm{MnO} \cdot \mathrm{SiO_2}) \tag{15.24}$$

With an increase of the temperature, lime is dissolved in slag phase; hence, basicity of the slag is increased; high acidic oxide, SiO_2 is now strongly bonded with high basic oxide, CaO.

$$(\mathrm{MnO} \cdot \mathrm{SiO_2}) + (\mathrm{CaO}) = (\mathrm{CaO} \cdot \mathrm{SiO_2}) + (\mathrm{MnO}) \tag{15.33}$$

By the above reaction, the activity of MnO in the slag is increased. The content of FeO in the slag phase is drastically decreased on that time due to reduction reactions of FeO by Si, P and particularly with C. MnO is reduced by C in iron or iron itself:

$$(\mathrm{MnO}) + [\mathrm{C}]/[\mathrm{Fe}] = [\mathrm{Mn}] + \{\mathrm{CO}\}/(\mathrm{FeO}) \tag{15.34}$$

This behaviour of Mn is responsible for the characteristic of Mn-hump. The content of FeO in the slag increases, once again reaction (15.23) goes forward direction; hence, again curve goes downward.

15.1.8 Phosphorous Reaction

The conditions for de-phosphorization as follows:

1. Basic slag,
2. Low temperature,
3. Oxidizing atmosphere,
4. Low viscosity and high fluidity of the slag,
5. Good slag-metal contact.

Oxidation of phosphorous begins from the beginning of the blowing, due to de-phosphorization condition prevails, when the carbon content of the bath is still higher. This is an essential feature of the LD process as compared with the basic Bessemer process, in which the oxidation of phosphorous starts only at the after-blow period when 0.04–0.06% C is left in the bath.

The phosphorous is oxidized by the following reactions:

$$2[P] + 5(FeO) = (P_2O_5) + 5[Fe] \tag{15.35}$$

$$(P_2O_5) + 4(CaO) = (4CaO \cdot P_2O_5) \tag{15.36}$$

In LD converter, the initial slag is acidic (FeO–MnO–SiO_2) in nature. Due to blowing 100% oxygen, FeO content in the slag is higher and temperature is also increased; so CaO is dissolved in slag very quickly from the beginning. Since the dissolution of CaO in the slag is an endothermic process; so that high temperature is required and presence of high amount of FeO is also helped for quick dissolution of CaO in the slag. Now, slag becomes basic (FeO–MnO–SiO_2–CaO) in nature, which helps phosphorous oxidation reaction at the early stage. For the removal of phosphorous from the metal to slag is proceed with success, require the amount of free lime dissolved in the slag must be sufficiently high.

The iron and phosphorous lines, on the Ellingham diagram (Fig. 15.11), are so close to each other that the entire phosphorous in the charge materials gets reduced along with iron in the blast furnace for ironmaking. This has posed a serious problem in the oxidation of phosphorous in steelmaking. Two lines (FeO and P_2O_5) are widely separated during LD steelmaking process by decreasing the activity of P_2O_5 in slag using a strong and excess basic flux like lime. The standard and non-standard lines on the Ellingham diagram are shown in Fig. 15.11. Slag with a high basicity (2–4) decreases the activity of P_2O_5 (10^{-15} to 10^{-20}) in slag. In general, steelmaking slag contains up to 25% P_2O_5, but even then, the activity of P_2O_5 in slag remains extremely low at high basic slag. For effective removal of phosphorous, high basicity of slag must be required.

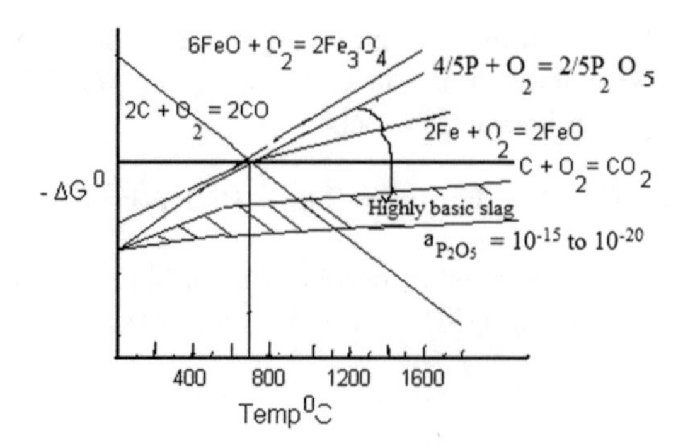

Fig. 15.11 Portion of Ellingham diagram

A high content of FeO in slag speeds up the dissolution of lime. This process occurs with success in the initial and in the final period of blow. For that reason, in LD process, phosphorous can be removed from the metal without much difficulty; provided that the condition of slag must be maintained properly. The best results, for de-phosphorization, are obtained when the final slag of the melt contains 40–45% CaO, 7–14% MnO, 15–18% SiO_2 and 8–12% FeO.

For elimination of phosphorous requires the lime (CaO) should be dissolved in slag, when CaO is dissolved in the slag then P_2O_5 reacts with it and the activity of P_2O_5 is lowered down. If the slag does not contain dissolved CaO, then product P_2O_5 (by P–O reaction) is not entrapped by CaO; hence, P_2O_5 is decomposed into phosphorous and oxygen ($P_2O_5 \rightarrow P + O_2$), since the free energy of P_2O_5 is high. This newly formed phosphorous again absorbed by the metal. This is known as *reversion of phosphorous*. The free energy of formation of P_2O_5 is higher at the steelmaking temperature and so it will be unstable and decomposes to get back into the metal if the slag does not contain dissolved CaO.

15.1.9 Sulphur Reaction

The sources of sulphur in the metal are hot metal and lime. So the contents of sulphur in lime and hot metal should not exceed the specified limits. Small amount sulphur (6–8%) passes from the metal into the gaseous phase (SO_2 gas) under the action of blowing oxygen by the reactions:

$$[\mathrm{FeS}] + 3/2\{\mathrm{O_2}\} = \{\mathrm{SO_2}\} + (\mathrm{FeO}) \tag{15.37}$$

$$[\mathrm{MnS}] + 3/2\{\mathrm{O_2}\} = \{\mathrm{SO_2}\} + (\mathrm{MnO}) \tag{15.38}$$

But the major part of sulphur transfer from metal to slag phase by the following reactions:

$$[\mathrm{FeS}] \rightarrow (\mathrm{FeS}) \tag{15.39}$$

$$(\mathrm{FeS}) + (\mathrm{CaO}) = (\mathrm{CaS}) + (\mathrm{FeO}) \tag{15.40}$$

$$(\mathrm{FeO}) + [\mathrm{C}]/[\mathrm{Si}] \rightarrow [\mathrm{Fe}] + \{\mathrm{CO}\}/(\mathrm{SiO_2}) \tag{15.41}$$

The rate of the reaction (15.40) is closely linked with the rate of formation of an active fluid and basic slag. In LD process, 60% S is removed, whereas in open-hearth process this value is half of the LD process.

The condition for de-sulphurization:

1. Basic slag,
2. High temperature,
3. Reducing atmosphere,
4. Low viscosity and high fluidity of the slag,
5. Good slag-metal contact.

According to the distribution law:

Sulphur partition coefficient,

$$L_S = \left(\frac{(\mathrm{wt\%S})}{[\mathrm{wt\%S}]}\right) \tag{15.42}$$

For LD process, this value is 12–14; whereas for open-hearth process value is 8–10. High sulphur partition coefficient means more de-sulphurization. 50–60% of the initial sulphur is removed in LD process. In LD converter, the slag is basic right from the beginning, the basicity also of the order of 2.3–2.8 (whereas the basicity of BF slag is 1.3), it is much higher (i.e. viscosity is low and fluidity is good); so sulphur removal is good. Since temperature, at the oxygen penetration zone, is higher and turbulence in the bath is more which ensure a good slag-metal contact. Because of these reasons, although the condition of (3), i.e. reducing atmosphere, is not fulfilled; hence, the sulphur reaction occurred nicely in LD converter. Here, sulphur reaction is kinetically controlled rather than thermodynamically controlled.

Metal de-sulphurization reaction proceeds slowly because it is a diffusion process. It may be faster by improving the bath mixing, increasing temperature, fluidity and basicity of the slag, as well as the activity of sulphur. At the initial stage of the heat, when the metal is rich in carbon and silicon, the activity of sulphur is high. A part of sulphur is removed at the initial stage of the process when the temperature of the melt is still relatively low through its reaction with manganese in metal:

$$[Mn] + [S] = (MnS) \tag{15.43}$$

Increasing the concentration of FeO in slag promotes dissolution of lime, and therefore favours de-sulphurization. Most intensive de-sulphurization occurs at the end of the heat when the lime dissolves completely in slag phase and slag basicity increases to 2.8 and more. With an increase of slag basicity, the residual concentration of sulphur in the metal becomes lower, so that the coefficient of sulphur distribution between slag and metal can be increased.

15.1.10 Control of Carbon and Phosphorus Reactions

In case of basic Bessemer process, the carbon cannot be hold at any particular level since all carbon must be removed before phosphorus removal. But this is not the case in LD process. The relative rates of de-carburization and de-phosphorization can be controlled: (i) by controlling the oxygen flow rate and (ii) by adjusting the lance height.

When oxygen jet strikes the bath a crater is formed, if the flow rate is high the crater will be deeper, and it will touch the metal phase also. Hence, the rate of carbon–oxygen reaction is increased. That means more carbon will be removed when the crater will penetrate the metal phase, i.e. more metal droplets will go to slag phase and formed slag-metal emulsion.

If the oxygen jet penetrates only on the slag phase due to higher lance height:

$$Fe \rightarrow FeO \rightarrow Fe_2O_3.$$

Since the rate of de-phosphorization is directly proportional to oxygen potential of slag. So high oxygen potential of slag favour de-phosphorization because: (i) P-reaction is a slag-metal type of reaction and needs oxidizing atmosphere, (ii) lime dissolves faster in a slag containing high FeO. Hence, to control the de-phosphorization and de-carbonization reactions, the lance height and the oxygen flow rate should be controlled.

15.1.11 Process Controlling Factors

The refining process in the LD converter is influenced by the following inputs:

1. Hot metal: (i) composition and (ii) temperature,
   → The amount of heat, which become available to the process, is determined by the composition and temperature of the hot metal.
2. Rate of oxygen input in relation to: (i) nozzle diameter, (ii) nozzle shape, (iii) distance of nozzle from molten bath surface, (iv) oxygen pressure and (v) weight of charge,
   → The rate of oxygen input determines the refining time. The rate of oxygen is depending on diameter and shape of nozzle. Oxygen is almost completely absorbed by the bath; the refining time is limited only by the process of the metal-slag reactions. The reactions are also influenced by the oxygen pressure and as well as the distance between nozzle and the bath surface.

   (a) At constant pressure of oxygen, the greater the distance, greater the area of the impact circle of the oxygen on the bath, and smaller the degree of penetration.
   (b) Increase the oxygen pressure with a constant distance raises the impact pressure and penetration.

3. Slag forming additions: (i) quantity, (ii) time of addition, (iii) chemical and physical characteristics,
   → Thermal energy available during the refining is best utilized for melting of scrap, not to be utilized for calcinations of limestone which may be added as fluxing material. Hence, the use of limestone, as a flux, is therefore not recommended. There is another reason to prefer lime, as a flux, instead of limestone. Lime is straightway available for fluxing and formation of slag at early stage. If limestone is added as a flux, lime will be available only after calcinations, i.e. the slag formation will be delayed.
   Efficiency and economy of the process very much depends upon the formation of thin, oxidizing and basic slag as early as possible during the oxygen blowing. Use of bauxite, silica-sand and fluorspar, as fluxes, are helped to dissolve lime quickly in solution. Lumpy lime is slower to dissolve in slag than fine powdered lime. But fine powder lime tends to fly off. A small granular form of lime is the best choice. Lime consumption varies around 2–5% of the weight of metal charged.
4. Other charging materials: (i) scrap, (ii) iron ore and (iii) sponge iron/DRI,
   → Scrap, iron ore and sponge iron/DRI are used as coolants to best utilize the excess heat energy available during refining. The use of steel scrap as a coolant is a must since the home scrap generated in the plant must be recycled back into the process. LD converter can take up to 25% of the metal charge as scrap.
   Steel scrap has some disadvantages:

   (i) Steel scraps are non-uniform size,
   (ii) Wide variation of composition, and
   (iii) Contain high amount of tramp elements (0.13–0.73%).

Since oxygen gas is used as an oxidizing agent, so iron ore is not required for the same purpose. Iron ore can act as a coolant but is not preferred because the available excess thermal energy can be economically used in melting maximum load of circulating scrap in an integrated steel plant. If scrap is not available iron ore can be used as coolant, 100 kg/t of steel. In its cooling effect, 1% ore in the charge is nearly equivalent to 2.5% steel scrap.

Sponge iron/DRI can be substituted of scrap, due to shortage of scrap, sponge iron can be used as coolant. Sponge iron has some advantages:

(i) Uniform size,
(ii) Known composition, and
(iii) Contain less tramp elements (>0.02%).

15.1.12 Economics of Process

Several factors that have a bearing on the economics of the process are given below:

1. If the slag contains too high FeO which causes slopping (i.e. over flowing) of slag from the mouth of the converter, so extra erosion of the lining takes place. This is due to the lance height is maximum from the bath, so more oxygen dissolve at the slag phase.
2. Lime and fluxes should be added in such quantity that would give the correct basicity of the slag.
3. Finishing temperature should not be higher than necessary temperature; otherwise lining life would be decreased.
4. At the end of heat for dead soft steels, surplus oxygen should be avoided, otherwise, CO would burn to CO_2 and unduly heated the vessel hood.
5. Excessive addition of fluxes (spar or bauxite) should be avoided; otherwise, they react with the lining whose life would be decreased.

15.1.13 Operating Results/Performance

The production of quality steel through LD has become predominant to meet the bulk demands of various grades of steel. Quality requirements for steels like extra deep drawing (EDD), deep drawing (DD) grades for automobile, food processing and packaging industries, low sulphur steel for pipe-lines and thick plates; low nitrogen steel for boiler plates or applications requiring high-strength low-alloy (HSLA) steels for transformers, etc. There are demand an achievement of the composition with respect to alloying elements as well as control on the residuals such as sulphur, phosphorous, nitrogen, hydrogen and inclusions.

LD process can make a wide range of products including carbon and low alloy steels, soft steel for sheet formation of all kinds including extra deep-drawing qualities. It is an advantage in deep drawing steels to have a low nitrogen and phosphorous content. At the end of the blow, 0.002% N can be obtained without difficulty and it will probably rise by a further 0.001% during tapping and teeming, i.e. $0.003\%_{max}$ N. Phosphorous can be reduced to 0.015–0.02%, carbon can be reduced to 0.04% and up to 65% of the charged sulphur is removed. Metallic yield of steel depends largely upon the charge itself, i.e. hot metal, scrap/ore/sponge iron. Steps of LD operation are shown in Fig. 15.12.

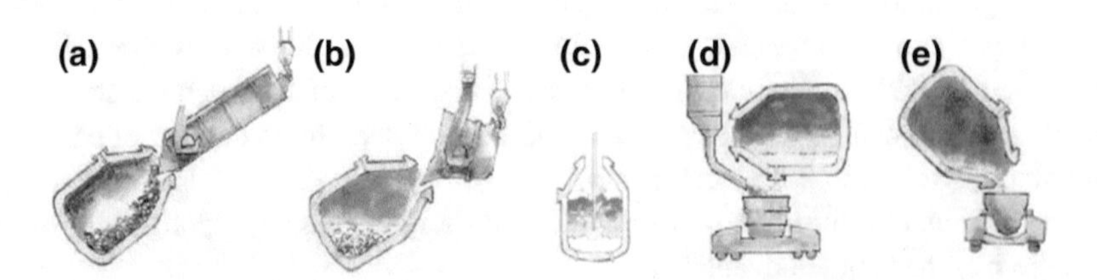

Fig. 15.12 Steps of LD operation [2]: **a** charging scrap, **b** charging hot metal, **c** oxygen blowing, **d** tapping and addition of ferro-alloys and **e** slag tapping (reproduced with permission from Ispatguru)

Operational losses can be arose from the following:

1. Slopping and splashing from the converter mouth,
2. Formation of skull at the mouth,
3. Foaming slag which carries some of the iron droplets,
4. High combined iron (as FeO) in the slag,
5. Loss of metal during taping of slag.

These losses can be minimized to a considerable degree by the experienced operator.

15.1.14 Lining of Converter

The property requirements for LD converter lining are as follows:

1. *Thermal shock resistance* which is characterized by the number of successive cycles of heating to 850 °C and cooling with running water until the material loses 20% of its mass.
2. *Corrosion resistance,* i.e. resistance of its material to the eroding action of metal, slag and dust.
3. *Abrasion resistance,* i.e. its impact strength to resist heavy impacts of charged scrap.
4. *Refractoriness,* i.e. ability of the material to withstand high temperature retaining its initial shape.

At the disposal of raw material and depending on steelmaking tradition, various lining techniques for LD/BOF were developed in different countries. In Austria, magnesite is used which is available in enough quantities. Other European countries use dolomite as lining material. USA, Japan and Australia again use magnesite as lining material.

It is possible to produce synthetic raw material with every CaO/MgO ratio required, modern practice is a zonal lining of the LD/BOF vessel. The qualities of lining are chosen according to the wear in the different zones. The range of qualities is graduated from 100% dolomite, passing magnesite enriched dolomite to 100% magnesite, i.e. the proper CaO/MgO ratio maintain to the right spot. Since the price of magnesite brick is 2.5–3.0 times more than dolomite brick. In any case, the vital point is the performance of the lining, i.e. high rate of disposal with best utilization of the bricks at lowest costs per tonne produced steel. This can be realized using an adopted zonal lining.

The earlier refractory lining for LD converter was based on dolomite, magnesite or chrome-magnesite. The high basicity of the LD slag and the high temperature of the bath promoted rapid wear of the refractories in general, and in particular at the slag-metal interface, i.e. around trunnions. Modern LD converters are lined with zonal lining by magnesite refractory bricks. The magnesite–carbon bricks are also used. The total Al_2O_3 and Fe_2O_3 content in the magnesite refractory should be less than 4%, to improve its resistance to slag attack. The resistance of slag for magnesite is improved by its high bulk density and low impurity content.

(1) **Zonal Lining**: Generally, lining of the LD converter was done with pitch bonded dolomite bricks which can last up to 570 heats and specific consumption of refractory was 3.64 kg/t of steel production. By zonal lining, life of converter was increased to 930 heats and consumption of refractory was decreased to 2.1 kg/t.

Since erosion in all areas of LD converter is not uniform, so LD converter can be divided into three different zones: (i) high erosion zone: around trunnions (9% of total area), (ii) medium erosion zones: above and lower portion of trunnions (16%) and (iii) low erosion zone: remaining portion (75%) (as

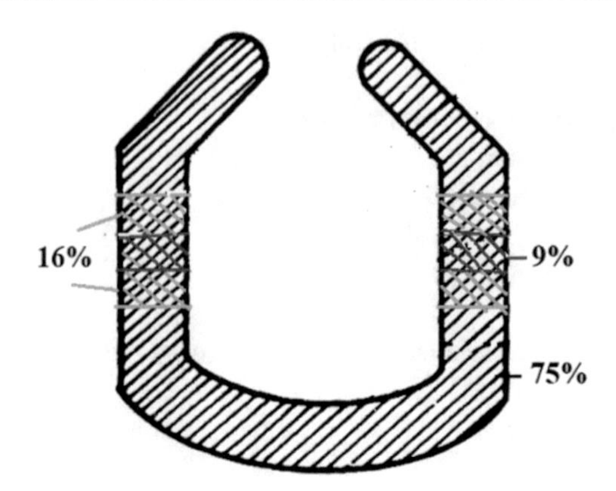

Fig. 15.13 Zonal lining of LD converter

Table 15.2 Properties of some basic bricks for LD converter

	Dolomite [5]	Magnesite enriched dolomite [5]	100% magnesite [5]	MgO–C bricks [6]
Chemical analysis				
MgO, %	>36	77	>94	>97
CaO, %	<61	20	<3	
SiO_2, %	<1.5	<1.5	<2	
(Al_2O_3 + Mn_3O_4), %	<1			
Fe_2O_3, %	<1	<1	<1	
Pitch content, %	4–5	4–5	>4	
Carbon, %				10–12
Physical properties				
Bulk density, g/cm^3	2.89	2.99	3.09	3.0–3.1
Total porosity, %	8–12	7–12	6–10	8–10
Cold crushing strength, kg/cm^2	>300	>300	>300	500–700

shown in Fig. 15.13). Depending upon the converter brick lining pattern and operating practice, some zones erode faster than others. Zonal lining is, therefore, practiced with different qualities of enriched magnesite bricks.

Zonal lining was based on installation of magnesite bricks having higher percentage of MgO content, in areas of excess wear, i.e. at the slag-metal interface. The high-grade bricks cover only 25% of the whole lining area; out of that 9% area (i.e. around trunnions) is covered by high-grade magnesite bricks, and 16% area (i.e. above and lower portion of trunnions) magnesite enriched pitch bonded dolomite bricks (as shown in Table 15.2).

(i) 100% magnesite (>94% MgO) bricks in the trunnion section where maximum wear took place (9% of the whole lining area).
ii) Above and below of the trunnion section (8% area each side), magnesite enriched (77% MgO) pitch bonded dolomite bricks.
(iii) The rest 75% area of the lining exist pitch boned dolomite bricks.

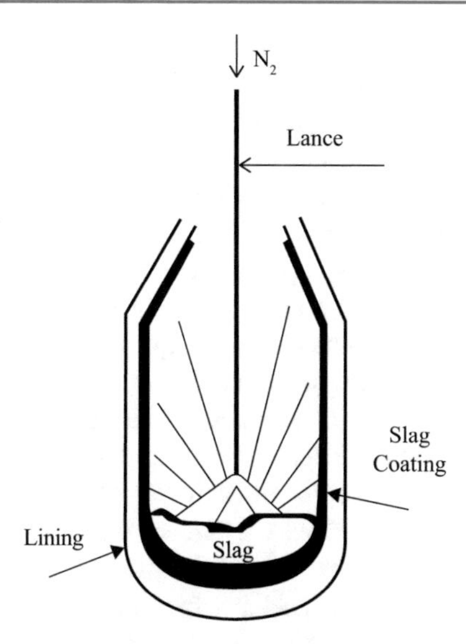

Fig. 15.14 Schematic diagram of slag splashing in LD converter

(2) **Magnesite–Carbon Bricks**: By using different qualities of high duty MgO–C bricks, lining life of converter had reached to the level of 4500 heats or more in all plants of Steel Authority of India Ltd (SAIL), India. In SAIL, highest lining life of 12,235 heats with an average lining life of 9499 heats had been achieved by Bhilai Steel Plant, India during 2011–2012 and that of India is 13,771 heats achieved by JSW, India during 2006–2007. An increase in lining life does not only decrease refractory consumption (0.4–1.1 kg/t) as well as melting cost, but also contribute to efficient steelmaking. The properties of improved quality MgO–C bricks having MgO content >97% and graphite of fixed carbon (96–98%) for LD converter are given in Table 15.2.

(3) **Lining by Slag Splashing**: Slag contains low FeO and a high MgO, should be desirable for slag splashing. Such slag condition is achieved through addition of calcined dolomite after metal tapping. A portion of the slag is retained in the converter after tapping. Slag splashing is done by injecting nitrogen gas through lance with control flow rates for 1–4 min. Slag can be deposited into the inner surface of the converter with a layer formation (Fig. 15.14). Lining life is increased to 8000 heats by slag splashing technique. This technique is used in Japan.

15.1.15 Pollution Control

Air pollution is the major issues in a LD/BOF shop. There are two broad areas of air pollution: (i) undesirable gases such as carbon monoxide, fluorides or metal vapours and (ii) solid particles such as oxide dusts.

Hot metal reloading: Pouring of hot metal from the mixing car into the transfer ladle results in plumes of fine iron oxide and carbon flakes. This mixture is known as *kish*, which is the major source of dust and dirt inside the iron and steelmaking shop. Hot metal is saturated with carbon and when the temperature drops during pouring into the ladle, the carbon precipitates out as tiny graphite flakes due to solubility of carbon in hot metal decreases with decreasing the temperature. The method of control for that pouring should be inside an enclosure (i.e. hood) and collection of the fume in a bag-house.

Charging the BOF: There is some fume generated when the scrap hits the bottom of the furnace. However, the major emission is generated while pouring hot metal into the furnace. Here, very dense oxide clouds, kish and heavy flames rise quickly. Some shops have suction hoods on the charging side above the furnace that collects fume and diverts the heat away from the crane. However, many shops are not so equipped and must rely on slow pouring to limit the fume emission from the roof monitors to comply with regulations. Pouring too fast results in heavy flame generation that has been known to anneal the crane cables, causing spillage accidents.

Blowing during melting and refining: During the main blow, the high concentration of oxygen results in the evolution of iron oxide fumes in higher quantities. Approximately 90% by weight of total fumes is generated at this point. The fumes consist of hot gases at high temperature and heavy concentrations of solid particles. The solid particles contain heavy metal oxides such as chromium, zinc, lead, cadmium, copper and other, depending on the scrap mixture; and also contain dusts from flux additions. The gas composition is approximately 80–90% CO and rest is CO_2. In open combustion hood, where air is induced just above the furnace into the cleaning system, the temperature can raise as high as 1925 °C due to further combustion of CO.

De-sulphurization and skimming of hot metal: Often, the reloading hood is designed to accommodate the de-sulphurization operation. During slag skimming, the splash of slag and metal is felled in a vessel (usually a slag pot) which also generates fine oxides fume.

Sampling and testing: Turning the furnace down for testing and temperature measurement, sampling generates fine oxide fumes at a relatively low rate.

Tapping: Some fume comes directly from the furnace, but most comes from liquid steel colliding with the bottom of the ladle or other liquid steel. The addition of alloys also increases the fume during tapping. Some fume is collected in the main hood, but much escapes through the roof monitors.

Materials handling: The handling of fluxes, alloys and treatment reagents can be a significant source of fume. Materials used in small amounts per heat can be transported in bags or super sacks to eliminate dust during transfer.

15.2 Oxygen Bottom Blowing Processes

There is some drawback in LD process likes:

(i) more splashing and slopping,
(ii) slow slag-metal reaction,
(iii) poor bath agitation,
(iv) over oxidation of slag.

To overcome the above problems of LD process, oxygen bottom blowing processes are developed in Europe in late 1960s. Pure oxygen is introduced from the bottom through tuyeres with a peripheral shield of protective fluid (hydrocarbon gas or oil).

Main advantages of these processes are as follows:

(i) No splashing and slopping,
(ii) Quick slag-metal reaction,
(iii) Vigorous bath agitation,
(iv) Decreasing over oxidation of slag.

There are two processes: (i) OBM and (ii) LWS.

15.2.1 OBM

The discovery of *Maxshutte tuyere* has given birth of new oxygen steelmaking process. The tuyeres are made of two concentric tubes, oxygen being blown through the inner tube, made of copper, and protective gas through the annular space between the tubes; outer tube made of stainless steel (as shown in Fig. 15.15b). The elliptical bottom is detachable. By bottom blowing with oxygen, the intensity of mixing of the metal bath is increased by about ten times compared to the LD process. Oxygen is introduced into the bath through the tuyeres at the bottom of the furnace within a peripheral shield of a gaseous hydrocarbon (act as coolant). The endothermic decomposition of this hydrocarbon:

$$2\{C_nH_m\} \rightarrow 2nC(s) + m\{H_2\} \tag{15.44}$$

This reaction is endothermic, i.e. absorbs the intense heat generated by the reactions of oxidations which would otherwise damage the tuyeres.

The process was developed by the Maximillianshutte Iron and steel Company, West Germany in 1968 and was named as ***O***xygen ***B***ottom Blown ***M***axshutte Process (OBM). The US Steel developed their own version of similar process named as ***Q***uiet or ***Q***uick ***B***asic ***O***xygen ***P***rocess (Q-BOP). This process does not splash the metal (i.e. quiet the bath), since oxygen is blown from the bottom of the liquid metal; this process has also faster slag-metal reaction (i.e. quick process).

The tuyeres are inserted from one side of the elliptical detachable bottom of the converter in such a way that the oxygen (60–65 N m^3/t or 200 N m^3/min at 10 atm) would be surrounded by a protective hydrocarbon gas (like propane, butane or natural gas at 0.5 N m^3/min at 6 atm). On entry to high temperature, propane (C_3H_8) cracks down in an endothermic reaction and deposition of carbon which helps to protect the tip of the tuyere from high temperature of the bath due to the heat generated by the oxidation reactions.

$$\{C_3H_8\} \rightarrow 3C(s) + 4\{H_2\} \tag{15.45}$$

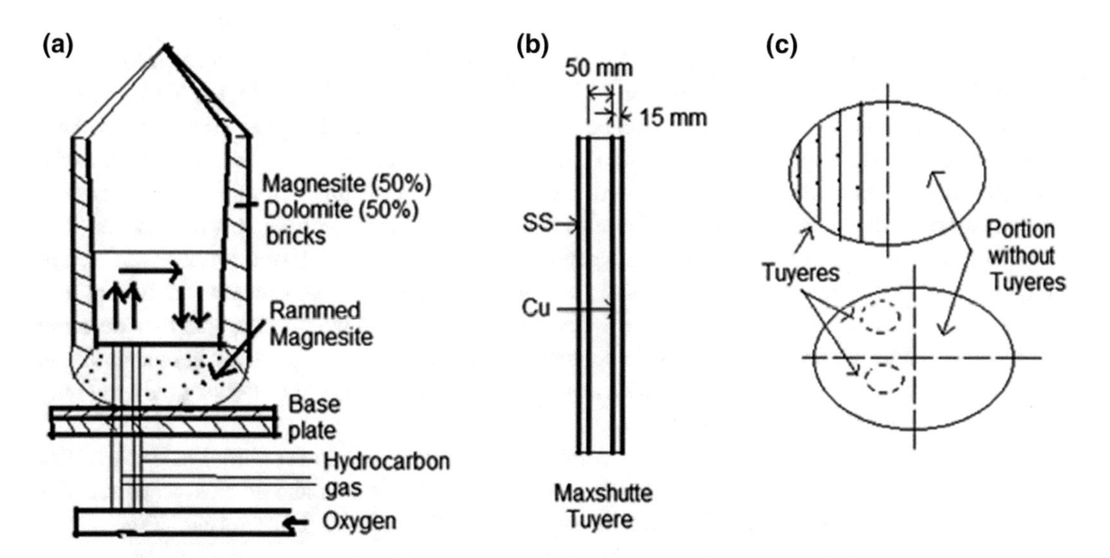

Fig. 15.15 **a** OBM converter, **b** Maxshutte tuyere and **c** arrangement of tuyeres

Sensible heat requires increasing the temperature of product to steelmaking temperature. Other product of cracking is hydrogen, 50% of that burns with dissolved oxygen and generates heat:

$$\{H_2\} + [O] = \{H_2O\} \qquad (15.46)$$

Remaining hydrogen is passed through the liquid metal with some of the amount is absorbed by the liquid metal. The amount of hydrogen dissolved increases particularly towards the end of the blowing (2.5–3.5 cm^3 of H_2 per 100 g of metal). The inert gas (nitrogen or argon) is blown through the inner tube of tuyere instead of oxygen at the end of refining and at that time no hydrocarbon is used, to desorb dissolved hydrogen from liquid metal. Hydrogen comes down to 1.2–1.8 cm^3 of H_2 per 100 g of metal and nitrogen content in steel increases to 0.004–0.006%. By blowing inert gas help to control de-carburization and decreased carbon up to 0.03% or even 0.01%.

There are 5–15 tuyeres. To promote good turbulence in the bath (i.e. to ensure good slag-metal contact), as well as to avoid the damage of tuyeres during charging of scrap; the tuyeres are arranged only on half the portion of the converter bottom (Fig. 15.15c). With this type of arrangement, it is possible to ensure that the direction of metal flow is upwards in the tuyere portion and downwards in the other portion of the vessel (as shown in Fig. 15.15a). A continuous circulation of the bath and intimate slag-metal contact is, thereby, maintained to prevent erosion of lining.

The converter is lined with 50% magnesite and 50% dolomite bricks (0.35 m thick) which can last 450 heats. Rammed magnesite is used for detachable bottom, fixed with base steel plates, whose life is 250–300 heats.

The converter is tilted towards other portion of tuyeres for charging. Solid scrap (35%), high phosphorus hot metal (4.5% C, 1.8% P) and lime are charged into the converter on the other portion where no tuyeres are present. The converter can be filled with hot metal until the level comes up to the first row of tuyeres. The oxygen blowing is started and converter move to vertical position. After 16–17 min converter is again turned down for sampling and initial siliceous slag raked off. Fresh lime is added to make slag basicity (3.4) to remove phosphorus and sulphur. Total lime consumption is about 125 kg/t of liquid steel. Total oxygen blowing time has been brought down by 25% by injection lime powder (90% < 0.1 mm) with the oxygen jet, but that gives some wear of pipes. The distance from the equilibrium between slag and steel is, thereby, substantially reduced. Fluxes such as lime powder are added to the oxygen, resulting in an entirely different mechanism of slag formation than in the LD process. It bottom blowing with oxygen and lime powder injection, the reactions occur directly in the metal bath. As the concentration of the iron in the bath is more than 90%, and thus, higher than all other incidental elements together, and it can be expected that the injected oxygen preferably oxidizes the iron first. The oxide droplets forming above the tuyere tip react with the carbon dissolved in the iron. The carbon monoxide bubbles generated, thereby contain, probably still during the ascension, constantly decreasing oxide amounts. The injected lime will react with such oxide layers. The ascending slag particles have their highest oxygen potential in the beginning of the ascension through the metal bath. Due to the reaction with silicon, manganese, carbon and phosphorus, equivalent iron oxide amounts are consumed. Only after the increase of the oxygen content in the bath by a decrease of the carbon concentration, the iron oxide content in the ascending slag particles is increased and thus also the slag floating on the bath. This increase of the iron oxide content of the slag bulk can only be observed at the end of oxygen blowing.

De-sulphurization capability in the OBM converter is higher due to lower partial pressure of CO gas, lower iron oxide content in the slag, better gaseous de-sulphurization and efficient agitation. The de-sulphurization capability in the OBM is superior to that of the LD process (Fig. 15.16). De-sulphurization extends to the low sulphur range in the OBM process, and consequently the

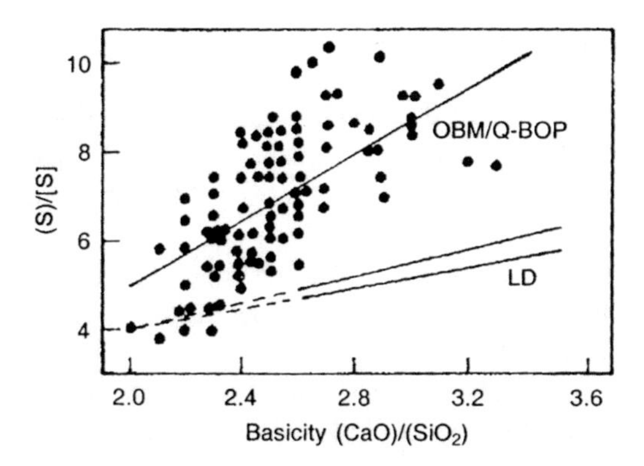

Fig. 15.16 Effect of basicity on sulphur distribution [7] (reproduced with permission from ironmaking and steelmaking)

amount of de-sulphurization agent can be reduced. Tap-to-tap time for OBM is 40 min against 50 min for LD.

After first blow (for 16–17 min) → Metal composition become: 0.3% C, 0.08% P,
→ Slag: 18–22% P_2O_5, 12–15% FeO; Fresh lime is added.
Second blow (for 1–2 min) → Metal composition become: 0.1% C, 0.025% P, <0.03% S,
Third blow (only nitrogen at 2 N m^3/min) → Metal composition become: 0.0004% H and 0.005% N.

Main advantages of OBM process:

(i) More efficient agitation of the bath, as a result there are no temperature and concentration gradients that causes slopping,
(ii) Oxygen utilization is so efficient that very little oxygen is available for post-combustion of CO gas above the bath,
(iii) Since there is no emulsification of slag and metal, it is necessary to inject lime as a fine powder along with oxygen,
(iv) Reaction proceed more efficiently,
(v) Better iron yield,
(vi) Better phosphorus and sulphur partition between metal and slag,
(vii) Higher manganese content and lower oxygen content, hence reduce ferro-alloy consumption;
(viii) Operation of the process is very near to equilibrium.

Disadvantages of OBM process:

(i) Bottom of the converter should have arrangement for injection of at least three gases (oxygen, hydrocarbon and nitrogen),
(ii) OBM converter can only be titled in one direction due to the position of tuyeres in one side of the vessel,
(iii) Nitrogen content of OBM steel is higher than LD steel,
(iv) Cost of refractory per tonne of steel in OBM is higher than LD because bottom does not give the same life of the vessel.

Comparison of OBM with LD process:

1. Moderate capital investment, overall operation costs lower than that of the LD process.
2. Due to quiet blowing, OBM vessel can be filled up to 40% compared with LD only 20–25% filled. In LD, the vessel space required is about 0.8–1.0 m^3/t of steel, in OBM that is about 0.6 m^3/t.
3. The blowing time of OBM is 25% less than LD process. The blow itself is much quiet so that splashing is completely eliminated, and slopping is drastically reduced. As a result, iron yield increases 91–93%, compared with 90% in LD.
4. It is possible to charge up to 35% scrap in OBM process with respect to 25–30% in LD process.
5. Average bottom life of OBM converter (250–300 heats) is five times more than basic Bessemer converter (50–60 heats).
6. FeO content in slag is lower (12–14%) as against 18–20% in LD. Hence, slag of OBM is less aggressive, but slag of LD is over-oxidizing. For producing 0.1% C steel, FeO content 5% in OBM compared with 18–20% in LD and producing 0.03% C steel, FeO content 13% in OBM compared with >25% in LD.
7. Dust content of OBM (3 kg/t) of the exit gases, which is less than LD process (15 kg/t).
8. Higher rate of de-carburization and ability to refine to a much lower carbon levels without much loss of iron in the slag.
9. A more effective de-phosphorization and de-sulphurization with lesser consumption of lime.
10. A higher efficiency of oxygen utilization and therefore relatively lower specific oxygen consumption.
11. A very homogeneous bath chemistry and temperature than in LD, due to the blowing end finally nitrogen is blowing into the bath.
12. Quick (i.e. faster) slag-metal reaction and vigorous bath agitation in OBM than LD. Stirring intensity is ten times more in OBM than LD that gives better partition of P and S.
13. In LD >70% Mn in metal is oxidized, but in OBM only 30–40% Mn in metal is oxidized, as well as lower oxygen content in steel, resulting 30–50% less ferro-alloy consumption.
14. Reaction mechanism is different from that in LD.

15.2.2 LWS

A process similar in design using fuel oil as a protective media was developed in France in 1969 and was named as Creusot ***L***oire, ***W***endel-*Sidelor* and Establishments ***S***prunck process (LWS). The LWS process is not used outside of France. The oxygen is injected through the bottom by means of double tuyeres. The protection of the tuyeres is ensured by liquid fuel oil injected by pumps in the gap between inside and outside pipes. The bottom life reaches more than 500 heats.

There are three main characteristics of the LWS process as follows [8]:

(i) The agitation of the bath is excellent during the largest part of the main blow. Only at the end, during de-phosphorization, the agitation decreases because there is no carbon boil due to no formation of CO gas.
The agitation is also very efficient during the very short stirring by inert gas after the main oxygen blow. Therefore, quick exchanges between phases and equilibria are obtained.

(ii) The reactions take place in the sequence governed by thermodynamic laws, successively remove silicon, carbon, phosphorus and the slag oxidation take place towards the final stage of the blow.
(iii) Due to lime powder injection, the silica (SiO_2) formed at the beginning of the blow can be quickly neutralized into calcium silicate ($CaO \cdot SiO_2$); the slopping of lime particles and viscous silica slag is thus avoided.
(iv) The de-phosphorization and de-sulphurization occur earlier, if fluorspar powder (i.e. flux) is added along with lime powder. Both reactions take place at the tip of the tuyeres and just above, at the impact of oxygen flow containing powder materials.

15.3 LD-AC/OLP

LD process was originally designed to refine hot metal containing less than 0.4% phosphorus. LD process was modified by the CNRM in Belgium to refine high phosphorus hot metal (>1.5% P), by introducing major portion of lime charge as powder through the lance along with oxygen; and this process was put into commercial practice at the works of ARBED in Luxembourg. Hence, the process is known as incorporating ***L***inz-***D***onawitz with ***A***RBED and ***C***NRM (LD-AC). This process was developed for high phosphorus (1.5–2.0% P) hot metal particularly for UK and European countries; it is not applicable for Indian hot metal (0.4% P). Aim of the process is that lime shall be available at the point of high temperature, where the oxygen jet strikes the metal and so immediate fluxing and slag formation take place quickly. So, the slag is foaming in nature, oxygen bubbles are entrapped in the slag; lime is dissolved quickly and increasing the basicity of slag which helps de-phosphorization.

Shape of vessel changes by increasing the diameter, but not height and height: diameter ratio is decreased (H:D is >1.4 for LD and 1.0–1.1 for LD-AC). There are a lot of similarity between LD-AC and LD converters in design and operation. Vessel shape is altered to a tulip shape (i.e. like tulip flower) with two tap holes at opposite sides (as shown in Fig. 15.17a). Vessel volume is more (20%) than LD to accommodate extra slag volume generated due to high phosphorous in hot metal. The design of lance is shown in Fig. 15.17b.

High phosphorous content hot metal is charged along with scrap, ore, bauxite and lump lime in the vessel. Oxygen is lancing at high velocity to the metal bath to oxidize the impurities and raising the temperature of metal and slag. Much of the lime (2/3rd) in finely powder (0.01–2.0 mm) form is charged along with oxygen lancing (after 5 min of oxygen lancing) to form slag quickly and to carry away the oxides from the metal. Rest 1/3rd of lime is charged as lump (40–60 mm) condition. Basic aim is that lime should be available at the place where the high temperature and jet of oxygen strikes the metal, so lime is dissolved quickly. Thus, resulting in immediate fluxing and foaming slag is formation. But it is necessary to control the extent of foaming slag and avoid it over flowing from the vessel mouth. For this reason (i.e. for foaming) slag-metal contact becomes good. Consumption of lime will be less when it is powder form because lump lime could not dissolve fully due to the rapid speed of the process. The oxygen is lanced at high velocity on the metal surface, oxidizing the impurities and raising the temperature of metal and slag. Lime is necessary to form slag and removal of impurities from the bath. So, lime is injected along with the oxygen.

Fundamentally, the rate of de-phosphorization in LD-AC is increased for high phosphorous hot metal, and over well before the de-carburization of the bath. This is possible by introducing major portion of lime charge as powder through the lance along with oxygen, due to high-temperature lime immediately goes to slag phase and de-phosphorization reaction starts from the beginning. Basically, in LD-AC, if hot metal containing more than 0.4% P is refined to produce at least one intermediate

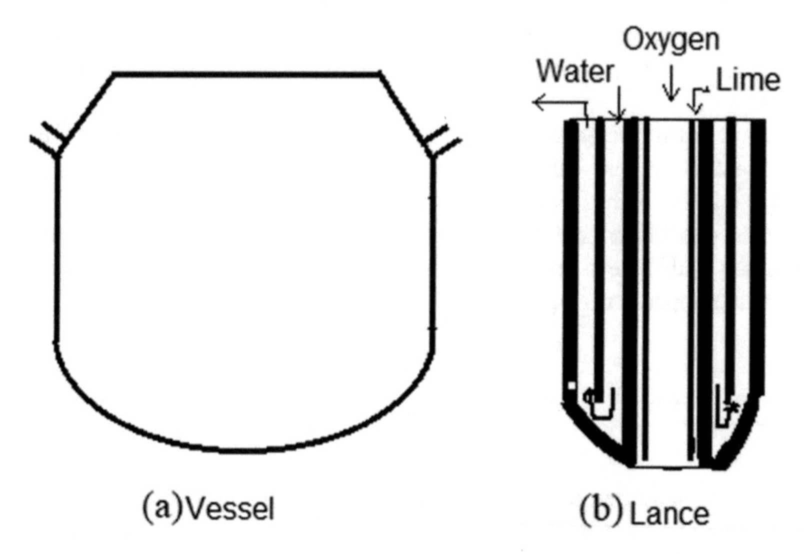

Fig. 15.17 **a** LD-AC vessel, **b** lance

phosphoric slag that is removed from the vessel and the heat is finished by further additions of lime and oxygen blowing.

LD-AC process charges less scrap than LD process, because a large excess of lime is to be added to form slag to accommodate the extra per cent of phosphorous in hot metal. Initially, iron ore and dolomite are charged to help dissolution of lumpy lime. Previous heat vessel is hot and contains some slag; scrap, hot metal and 1/3rd lime are added initially. If silicon content in hot metal is low, some silica may be charged to form foamy slag quickly. Blowing is started at its highest position. Powder lime (1–2 mm) is added through lance after 5 min of blowing. Lance is then carefully lowered to control foam of the slag and de-carburization reaction.

Phosphorous content in hot metal too high, so phosphorous cannot remove in one stage. Phosphorous can remove in two or three stages:

Hot metal: 3.8–3.5% C, >1.5% P

→ First slag off (45–50% CaO, 18–20% P_2O_5, 5–8% SiO_2 and 10–11% FeO) and composition of bath: 1.0–1.5% C, 0.2–0.3% P,
→ Second slag off (50% CaO, 10% P_2O_5, 2–4% SiO_2, 20–25% FeO, few% of MnO and MgO) and composition of bath: 0.2% C, 0.04% P, 0.04% S,
→ Third slag off and composition of bath: 0.05–0.1% C, 0.01% P, 0.01% S.

Lance is then carefully lowered to control the foam of the slag, after 15 min blowing stopped and first slag is drained out (not completely), some slag keeps on the bath (this slag is known as *preformed slag*) for formation of second slag quickly after adding powder lime. After first slag off, carbon in metal is brought down to 1.0–1.5% and 0.2–0.3% P. The remaining C and P are removed by the second slag. This is known as *double slag practice*. Lime powder helps to form the second slag and shortens the lancing time. Oxygen is blown again for 5–6 min; at the end of refining sample is taken and analysed. Temperature is measured, and heat is tapped leaving second slag in the vessel for the next heat.

Removal of sulphur can be take place up to 70%. For extra deep drawing quality steel should have P and S of the order of 0.01% each; and nitrogen content 0.001–0.002% for 99.5% pure oxygen lancing. So, these types of steel cannot be produced by even two slags practice, it is necessary to make third slag.

Fresh addition of coolant (i.e. scrap) and lime; oxygen blowing again start for 5–8 min. Heat is tapped leaving most of all second slag in the vessel for the next heat. Tap-to-tap time is 55–60 min. Life of dolomite lining vessel is 250–280 heats; refractory consumption is 8–10 kg/t of liquid steel.

To control the process, the following points to be considered:

1. Calculation of the lime requirement to give an appropriate slag basicity,
2. Rate of lime charge, which is most important factor;
3. Variations of the oxygen flow rate,
4. Adjust height of lance to make foaming of slag,
5. Temperature can be adjusted by adding scrap and iron ore (as coolants),
6. Adjust endpoint of low carbon heats by the size and colour of the flame at the vessel mouth.

Similar process was developed by IRSID in France at the same time as the LD-AC process. Here, also lime powder is injected with the oxygen lancing to the molten bath. Hence, the process is called ***O***xygen-***L***ime ***P***owder Process, i.e. OLP process.

15.4 Rotary Oxygen Processes

High phosphorous hot metal created difficulty for refining in top blown process, that lead to develop rotary oxygen process of steelmaking. Rotary melting furnaces are used earlier, and fuel burner was replaced by oxygen lance. This development had two distinct roots: (1) Kaldo process and (2) Rotor Process. In both processes, the fundamental different with respect to top blown process are: (a) CO gas is burned inside the vessel, generating heat transfer to the lining of vessel, (b) rotation of converter to get heat transfer from lining of vessel to molten bath.

The temperature of the bath is high (>1400 °C), since above 1000 °C, CO gas is stable, so CO gas is form within the bath due to oxidation of carbon in metal; CO gas comes out of the bath and oxidize to CO_2 gas within the converter and produce heat.

$$\{CO\} + 1/2\{O_2\} = \{CO_2\}, \quad \Delta H^0_{298} = -282.97\,\text{kJ/mol of CO} \tag{15.47}$$

But the problem is upper portion of converter is over-heated and damaged of the lining. This problem can be solved by rotating the converter and burning the CO gas with extra amount of oxygen within the vessel. Molten bath continuously absorbed heat from the lining and there is no harm of the lining life. The yield of the processes is also increased. Since more heat is available and so more scrap can be charged.

Main advantages for rotary oxygen processes are as follows:

(i) Better slag-metal contact,
(ii) Quick dissolution of lime and formation of slag; so early de-phosphorization and de-sulphurization reactions take place,
(iii) More precise control of temperature and slag,
(iv) Generating heat by burning of CO gas inside vessel, which is utilized for steelmaking,
(v) Good heat transfer from gas phase to metal phase via the lining of the vessel.

15.4.1 Kaldo Process

Kaldo process was developed by Professor B. Kalling in Sweden (1954). The largest converter (8.5 m length and 5.5 m diameter, 135 t capacity) can be tilted at least 180° and usually 360°, to permit charging, tapping, etc. (as shown in Fig. 15.18). Converter is inclined during operation at 16–20° with horizontal and speed of rotation up to 25–30 rpm to its own axis. Metal slides back along the lining wall. Oxygen is injected through water-cooled lance which inclined at 22–30° to the horizontal and oscillates about 15–20 times per minutes. High speed of rotation causes intimate mixing of the slag and metal and so facilitates the steelmaking reactions. Due to slag-metal contact is very good, so reactions will be accelerated, i.e. faster. Hence, by controlling speed of rotation, the kinetics of the reactions can be controlled; also controlled the relative ratio of C and P elimination. It is possible to reduce P very low level (i.e. 0.025%) while C is 0.5% or more. At very high speed (>30 rpm), formation of cascading occurs, i.e. metal droplets fall through gaseous atmosphere and oxidized.

Initial hot metal composition is 3.5–4.0% C, 1.8–2.0% P, 0.2–0.3% Si, 0.5% Mn, 0.05–0.06% S; after lower down speed of rotation to 10–12 rpm and decrease the angle of lance; C and P comes down to 0.4% and 0.11%, respectively. First slag (52% CaO, 8% SiO_2, 18% P_2O_5, 3.5% MgO and 9% FeO) is taken out. Fresh lime and iron ore are charged at low speed of rotation (6–8 rpm) and low oxygen flow rate to form second slag (50% CaO, 6% SiO_2, 10% P_2O_5, 4% MgO and 18% FeO); C and P further comes down to 0.1% and 0.03%, respectively. After second slag off again fresh lime and iron ore are charged to form third slag (43% CaO, 4% SiO_2, 7% P_2O_5, 6.5% MgO and 27% FeO); C and P further comes down to 0.05% and 0.018%, respectively.

By lower down the lance height, the de-carburization reaction can be increased and the de-phosphorization reactions should be decreased. Again, high rotation speed of converter, there will be less buildup of FeO in slag and high rate of de-carburization reaction; and low rotation speed of converter, there will be high buildup of FeO in slag and high rate of de-phosphorization reaction.

Fluxes are added as and when required. Excess heat is absorbed by addition of iron ore and scrap. Iron ore is preferable than scrap as cooling agent. Steel scrap can be used but it required large

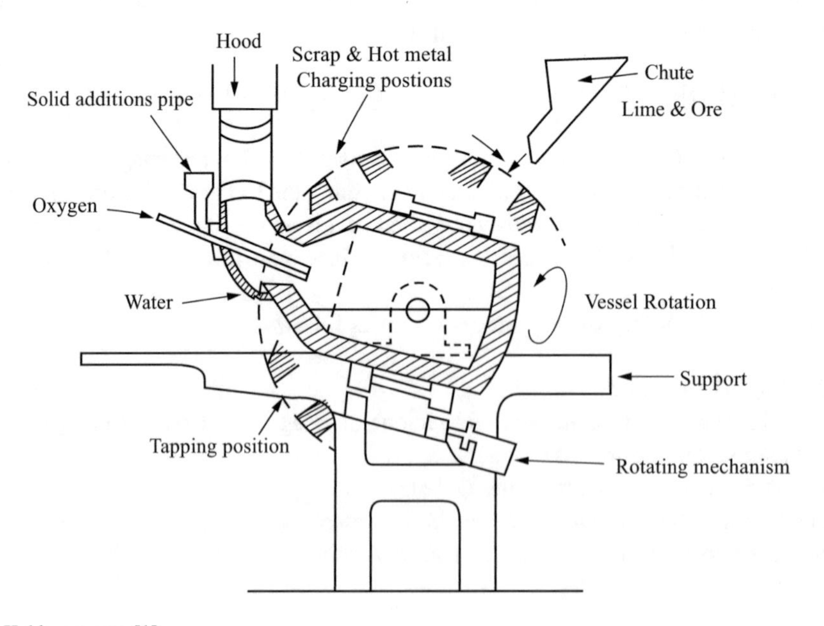

Fig. 15.18 Kaldo process [1]

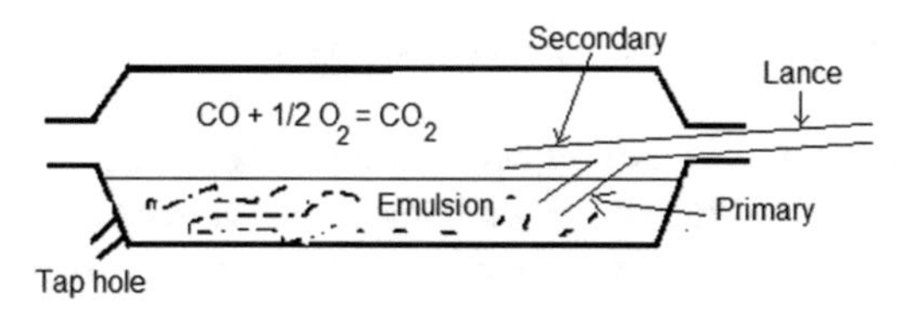

Fig. 15.19 Rotor furnace

amount (about 40%), but ore addition is only about 12–14%, since cooling effect of iron ore is three times more than steel scrap. Lime addition is 13–14% of hot metal. Lancing time is 35–40 min. Tap-to-tap time is 40–60 min.

Converter is lined with tar bonded dolomite bricks (0.35–0.45 m thick) and lining life is only 50 heats. Combustion of 80% CO gas within the converter is super-heated the refractory lining, due to extra heat generated by CO burning within the converter, so the life of lining is decreased.

Advantages:

1. High iron yield,
2. Less iron loss in the slag,
3. Low sulphur, low nitrogen and low oxygen content in steel,
4. Less blowing time, i.e. tap-to-tap time is less,
5. High amount of scrap/iron ore can be charged as coolant.

Problems: (i) lining life is relatively short (50 heats only) and (ii) mechanical problem to rotate large furnace.

15.4.2 Rotor Process

Rotor process was developed in Germany (1952). 100–120 t capacity vessel has 4.5 m diameter and 15 m long. The speed of rotation is low (0.2–4.0 rpm). There are two lances: (i) water-cooled primary lance (made of copper), incline 27–45° to the bath surface, oxygen is passed at the rate of 70–85 N m^3/min at 6–12 atm pressure; and (ii) secondary lance with 2–4 nozzles to supply oxygen for burning CO gas within the vessel and also supply lime powder (50%) (as shown in Fig. 15.19). Dolomite/magnesite bricks are used for vessel lining, life is 400–500 heats. 40–45% scrap can be charged as coolant. Tap-to-tap time is 120–150 min (including hot repair of the furnace).

Lining materials used for lining: (i) rammed tar dolomite, (ii) unburnt tar dolomite brick, (iii) burnt dolomite brick and (iv) magnesite ramming mass.

15.5 New Developments in Oxygen Steelmaking Processes

Although LD/BOF process is now old due to lack of proper bath agitation, slow rate of the slag-metal reactions and over oxidation of the bath. By increasing pressure of oxygen or lowering the lance height contribute in increasing the splashing and slopping, and over oxidation of the bath. To overcome these draw backs, bottom blown processes are developed. Advantages of bottom blown process as compared to top blowing process:

1. Faster slag-metal reaction, as well as vigorous bath agitation,
2. Better iron yield,
3. Better phosphorous and sulphur partition between metal and slag,
4. Higher manganese content and lower oxygen content, hence decreasing ferro-alloy consumption,
5. Possibility of production of extra low carbon steels without over oxidation of metal and slag,
6. Low nitrogen content in steel.

The gas injection under the metal surface, which is realized in the bottom blowing processes, leads to a better equilibrium between metal and slag than in the top blowing processes. This observation has led the steelmakers who operate top blowing converters along with bottom blowing adequate means to improving the mixing of metal and slag. Two important advantages are expected from mixing action:

(i) Better control of the oxygen partition between metal and slag during the lancing, which decreases the slopping and consequently improves the productivity and the metallic yield.
(ii) Reducing the disequilibrium between metal and slag at the end of the lancing, which leads to decrease more phosphorous and sulphur.

The combined blowing process where oxygen alone or oxygen and an inert gas like argon or nitrogen are blown simultaneously from top and bottom to achieve relatively accelerated slag-metal reactions and suppressed over oxidation of the bath. The top, bottom and combined blowing processes are compared in Fig. 15.20.

There are two types of processes have been developed:

(I) Mixing by neutral or inert gas which can be introduced under the metal bath surface either through (a) porous bricks/plugs (e.g. LBE) or (b) tuyeres (e.g. LD-KG, LD-AB, LD-OTB).
(II) Mixing by oxygen injection under the metal bath surface and as well as from top of the bath, i.e. combined top and bottom blowing (e.g. BSC, STB, LD-OB, K-BOP, KMS, LD-HC).

Metallurgical improvement for combined blowing processes are as follows:

(i) Improve blowing behaviour (e.g. no slopping and more reproducible de-carburization reaction),
(ii) Better iron yield,
(iii) More favourable steel analysis (lower phosphorous and sulphur contents) and more predictable results,
(iv) Better control of bath oxidation leading to higher manganese content and saving ferro-alloys.

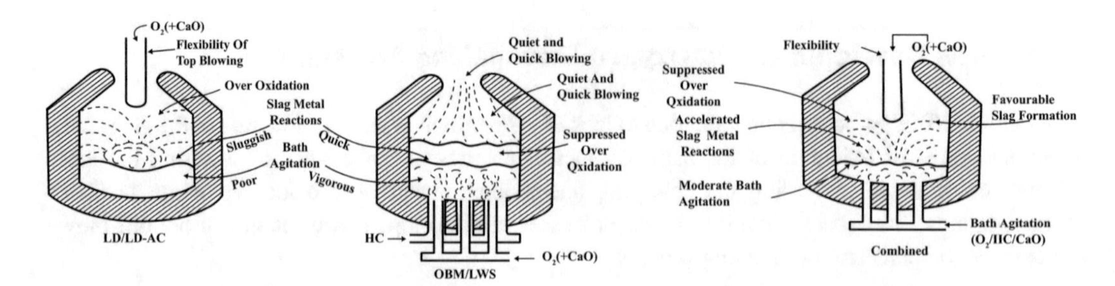

Fig. 15.20 Top, bottom and combined blowing processes [1]

15.5.1 Mixing by Inert Gas Through Porous Bricks

15.5.1.1 LBE Process

LBE process was developed by ARBED (Luxembourg) and IRSID (France). In the LBE process, an inert gas is injected through permeable or porous elements, i.e. plugs set in the bottom of the converter, during and after the oxygen blowing; stirs thoroughly metal and slag, as well as decreases the disequilibrium between metal and slag. Since in the process lancing of oxygen, bubbling of the bath and attending the equilibrium are take place; hence, this process is known as ***L***ance-***B***ubbling-***E***quilibrium process, in short LBE process. An advantage of porous bricks/plugs is the fact that they can be used intermittently, i.e. they can be cut off or put into operation whenever necessary.

The most important requirement for the industrial application of the LBE process is to dispose of permeable elements which keep during their life, a time-independent high permeability to the gases, which have a life time at least equal to that of the converter lining and work in good safe conditions. Improvements in the technology of the permeable elements brought their life to 500 heats from their earlier life 130 heats, and finally improvement of the slag practice increased their life to a point where they last as long as the converter.

Magnesia or chrome-magnesia plugs (i.e. bricks of pyramid shape) have 450–500 mm thickness and the dimension of the surface exposed to the bath is about $100 \times 150\ \text{mm}^2$ (as shown in Fig. 15.21b). The inert gas is fed to the permeable elements (number four) through separate lines, each of them having a valve, a pressure gauge and a flow metre. The flow rate of gas varies and depending upon refractory wear. The maximum flow rate is $8\ \text{m}^3/\text{min}$ and the wear rate is less than 1 mm/heat.

The injection of inert gas during refining has allowed a marked improvement of the blowing behaviour with high phosphorus hot metal, an increase of 1.4% in iron yield was observed. A stirring by inert gas through porous plugs, after the blow, has two main effects: (i) de-carburization of the metal and (ii) a decrease of the disequilibrium.

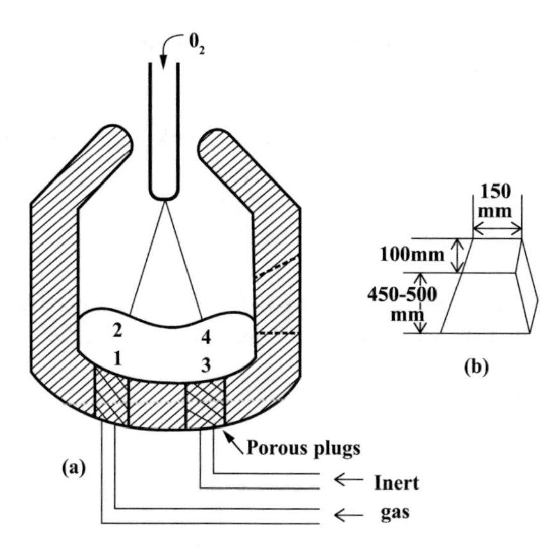

Fig. 15.21 **a** LBE converter, **b** shape of porous plug

The major advantages are as follows:

1. To improve the rate of post-combustion of CO into CO_2 in the converter, increasing its scrap melting capacity without supply any heat source and without any loss in productivity.
2. To control the blow by the inert gas flow rate, the lance height and the oxygen flow rate remaining constant.
3. A decrease of the phosphorus, sulphur and dissolved oxygen content of the metal.
4. To obtain a deep de-carburization of the steel without excessive oxidation of the bath.
5. The mean iron content of the first slag has decreased from 8 to 5%, whereas slopping has completely disappeared, and iron yield has been improved by 1.4% (Table 15.3).
6. Unmelted scrap at the end of the blow has completely vanished due to the better bath movement. Lime dissolution is complete for all grain size.

These allow a greater flexibility in the scrap and hot metal consumption, an improvement of the steel yield and quality, a reduction in aluminium and ferro-alloys consumption, as well as a better control of the oxygen partition between the slag, the metal and the gas in the converter. Injection of inert gas is also performed after the end of the oxygen blow. This bubbling ensures a further de-phosphorization, de-sulphurization and de-carburization of the metal.

Carbon in the metal is oxidized by dissolved oxygen or iron oxides in the slag depending upon the dissolved oxygen content. Figure 15.22 shows the carbon–oxygen relationship before and after stirring by inert gas. By long stirring times, very low carbon contents can be achieved (0.02%). As shown in Fig. 15.22, the value of $\%C \times \%O$ product is equal to 0.0018 as an average after gas injection compared to 0.0036 after the oxygen lancing, a result similar to the one obtained with OBM.

The metallurgical results of this process are compared with the conventional LD process on Table 15.4 for low phosphorus hot metal. It shows the mean phosphorus content of the metal reduced from 0.01 to 0.007% after the inert gas injection to the melt.

Advantages:

1. Better flexibility of the scrap/hot metal ratio, 39–44% scrap can be charged;
2. Decrease of the blowing time,
3. Iron yield is increased by 1.4%,
4. Aluminum/ferro-alloy consumption is decreased by 0.33 kg/t of liquid steel,
5. Phosphorus and sulphur contents are lower down,
6. Decrease of iron content of the slag, 14% instead of 18%,
7. Decrease lime consumption about 3 kg/t of steel.

By long stirring times, very low carbon contents (0.02%) can be achieved with less oxygen content in the bath. Phosphorus in the bath also comes down from 0.021 to 0.011%, but nitrogen consumption increases by 0.2 m^3/t.

Table 15.3 Scrap input and iron yield [9]

	LD-AC	LBE
Hot metal, kg/t	786	698
Pig iron, kg/t	54	13
Scrap, kg/t	271	390
Iron ore, kg/t	6	4
Iron yield, %	95.1	96.5

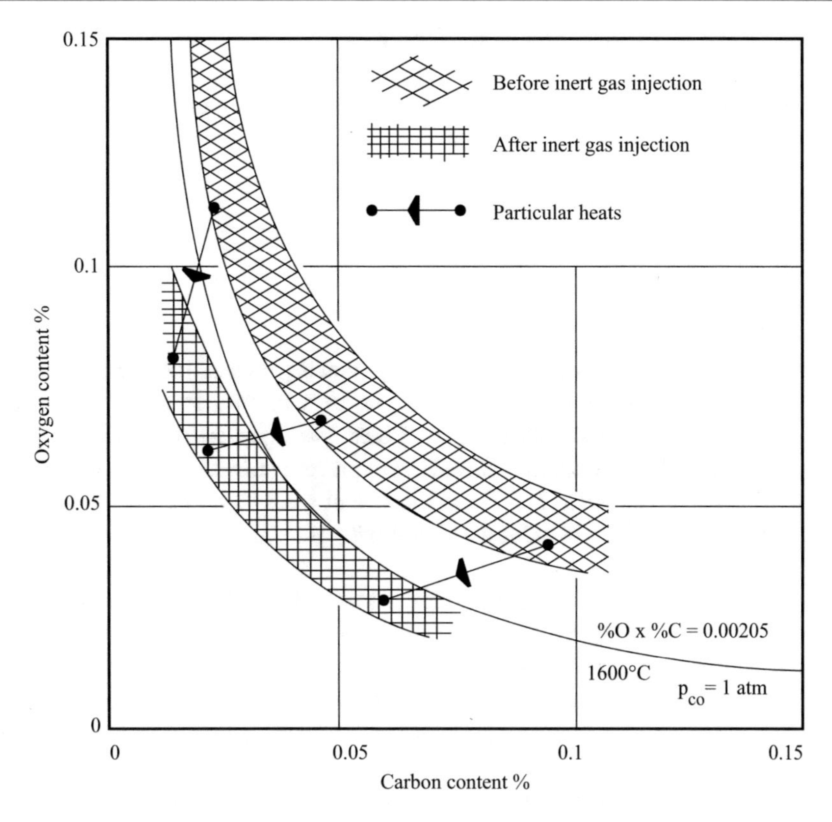

Fig. 15.22 Effect of inert gas injection on the carbon and oxygen contents of the metal [9]

Table 15.4 Comparison LBE process versus LD process for mild steel production [9]

Chemical composition of metal (%)	Conventional LD process	LBE process		Difference LBE-LD
		Before injection	After injection	
Carbon	0.06	0.095	0.06	0
Manganese	0.22	0.24	0.215	−0.005
Phosphorus	0.01	0.01	0.007	−0.003
Sulphur	0.017	0.017	0.016	−0.001
Oxygen	0.06	0.04	0.03	−0.03
Nitrogen	0.0022	0.0022	0.0027	+0.0005
Iron in slag	20	18	17.5	−2.5
Nitrogen consumption	0		0.17 m^3/t	+0.17 m^3/t

Introduce of inert gas under the surface of the metal, steelmakers are preferred, through the porous plugs than the tuyeres; because porous plugs have advantage of intermediate passing of inert gas, i.e. whenever inert gas is required that can be passed. But for tuyeres, inert gas is passed from the beginning to end of the process; otherwise, liquid metal is penetrated the tuyeres and after solidification of liquid metal that will be jammed the tuyeres. The consumption of inert gas is much more in tuyeres than through porous plugs.

15.5.2 Mixing by Inert Gas Through Tuyeres

Introduce of inert gas under the surface of the metal bath through submerged tubes, i.e. tuyeres, on this line the following processes are developed:

(i) LD-KG: developed by Kawasaki Steel, Japan
(ii) LD-AB: developed by Nippon Steel, Japan
(iii) LD-OTB: developed by Kobe Steel, Japan.

15.5.2.1 LD-KG Process

Various advantages of bottom blown process over top blowing process are attributed to the stronger agitation of the bath. To improve the agitation of the bath in the LD converter, Kawasaki Steel Corporation, Japan developed a simple process of inert gas injection from the bottom of the LD vessel. This process called ***LD-Kawasaki Gas Stirring*** process (LD-KG) (as shown in Fig. 15.23).

A flow rate of approximate 0.02–0.05 N m^3/t min inert gas is used in this process which is less than OBM process (7–8 N m^3/t min). But uniform mixing time is a little shorter than that of the LD by 20 s.

As shown in Fig. 15.24, total iron (T. Fe) in slag is lower in the LD-KG than in the LD; and this is due to reduction of FeO in slag by C. Figure 15.25 shows the relationship between phosphorus distribution and total iron in slag. Higher phosphorus distribution ratio is obtained in case of the LD-KG with respect to the LD process. However, in OBM/Q-BOP, this distribution is much higher than that of LD-KG. This is due to high degree of bath agitation which has a large influence on the phosphorus distribution.

Using LD-KG process, slopping has been markedly diminished. The reason supposed is that bubbling gas which disperses slag from the surface of the metal, diminishes CO generation in the slag layer. Two phenomena mentioned above; less generation of slopping and less total iron content in

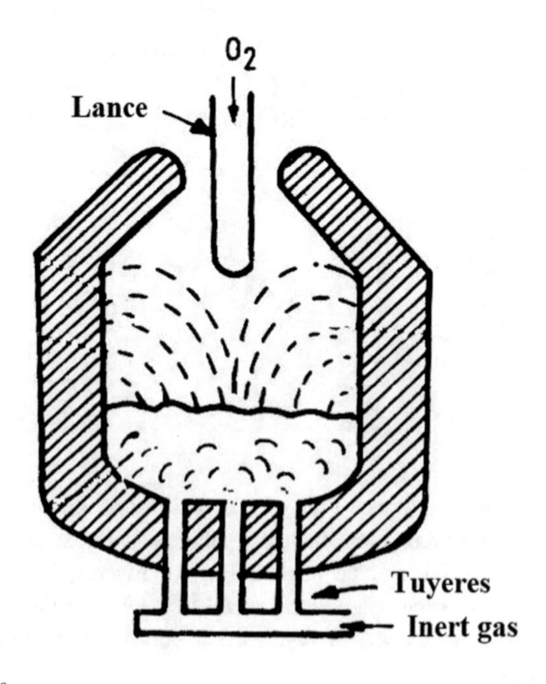

Fig. 15.23 LD-KG process

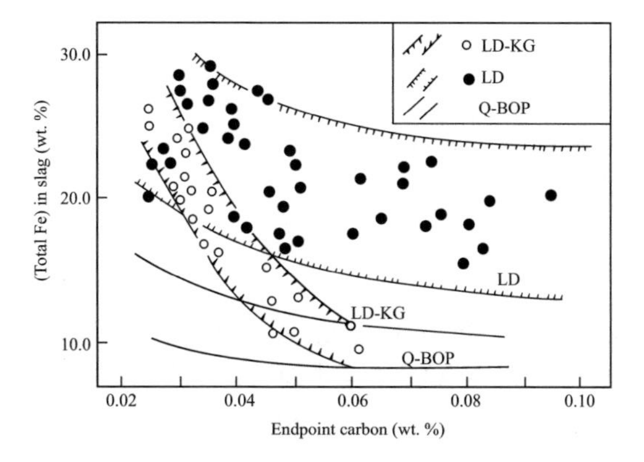

Fig. 15.24 (Total Fe) in slag versus endpoint carbon [10]

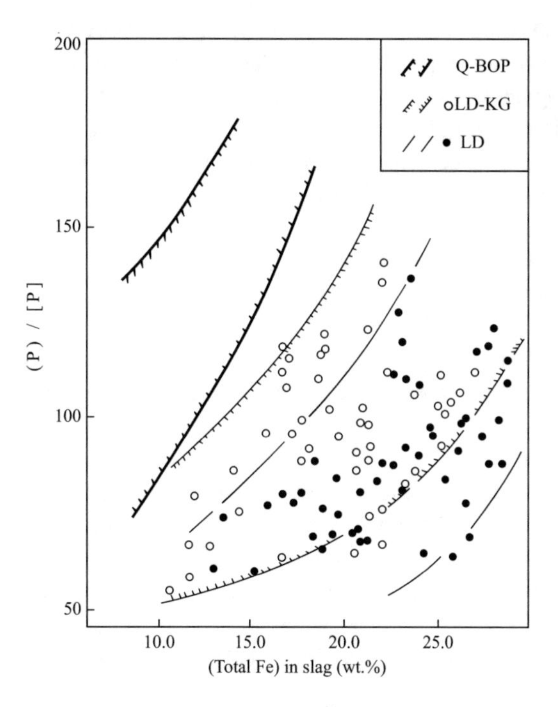

Fig. 15.25 Phosphorus distribution {(P)/[P]} versus (total Fe) in slag [10]

slag, increased iron yield by about 0.5%. Oxygen consumption also has been reduced by 1.7 m^3/t of hot metal. This indicates that de-carburization reaction proceeds in higher efficiency by bottom blowing of inert gases.

Kobe Steel, Japan had examined blowing by mixing with neutral gas, when producing high carbon steels. Increasing the amount of stirring gas steadily improves de-phosphorization for low carbon steels. 0.05 N m^3/t min of inert gas is used for high carbon steels.

Advantages:

(i) Slopping has been drastically reduced,
(ii) Decrease of total Fe content in slag,
(iii) Increase of iron yield by 0.5%.

Disadvantages:

(i) Excessive cost of the mixing gas when argon is used as a stirring gas,
(ii) Excessive nitrogen content in steel when nitrogen is used for stirring.

15.5.3 Combined Blowing Processes

Combined blowing means those processes where oxygen is introduced from top as well as from the bottom. It was particularly developed to increase the consumption of scrap rate. Main advantages: (i) due to good bath agitation, higher Mn recovery, i.e. less ferro-alloy consumption; (ii) better approach of P equilibrium, (iii) quick scrap dissolution, (iv) less slopping and (iv) higher iron yield (0.6%). Disadvantage: nitrogen content in steel is higher when nitrogen is used for stirring.

Different processes are developed as follows:

1. STB → SUMITOMO at Kashima, Japan
2. LD-OB → Nippon steel at Yawata, Japan
3. K-BOP → Kawasaki Steel at Mizushima, Japan
4. KMS → Klockner, West Germany
5. LD-HC → Hainaut-Sambre, Marchienne and CRM, Liege (Belgium).

15.5.3.1 KMS Process

Around 1980, the cost of hot metal production was drastically increased in North America and Europe; so high amount of scrap was charged to the steelmaking furnace, instead of hot metal which was higher cost (nearly double). There was a need to increase more scrap melting capabilities of the process. Technical development was done by Klockner, West Germany for using Maxshutte tuyeres to increase the scrap melting capability of the standard OBM process by modification as ***Klockner–Maxshutte Steelmaking*** or ***Klockner–Maxshutte Scrapmelting*** (KMS) process. Figure 15.26 shows KMS process.

Aim of the process:

1. Good heat transfer and energy utilization by the action of the burners to the scrap charged, by using bottom tuyeres as oxy-fuel burners for scrap pre-heating,
2. Protection of the refractory lining against high flame temperatures of burners by the scrap,
3. A low-cost method for steel production.

This process is developed to provide increased scrap melting capability through:

(i) Post-combustion of converter off-gases,
(ii) Pre-heating of scrap within the converter using the bottom tuyeres as oxy-fuel burners,
(iii) Introducing external energy by carbon injection during the process blowing,
(iv) Introducing the blowing simultaneously from top and bottom of the converter.

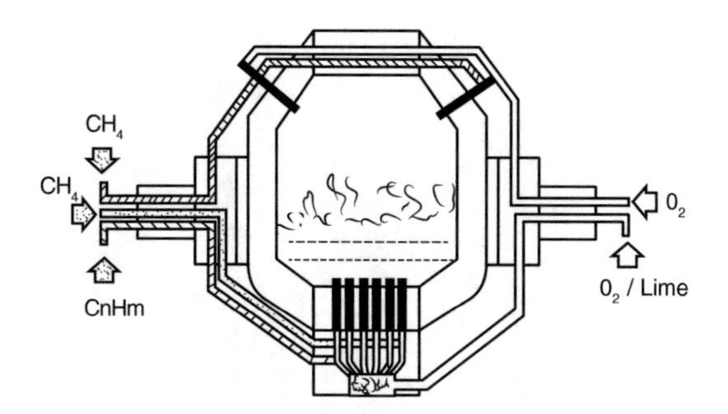

Fig. 15.26 KMS process

These developments can be divided into three types:

(i) Simultaneous top and bottom blowing → partial combustion of CO gas inside the furnace $\left(CO + 1/2\,O_2 = CO_2, \Delta H^0_{298} = -282.42\,\text{kJ/mol}\right)$ → heat transfer to the melt → 30% scrap can be charged (as coolant) → lower iron oxidation.
(ii) Injection of fuel oil through bottom tuyeres (which act as burners) for pre-heating of scrap before charging hot metal to the converter → scrap pre-heated through the bottom tuyeres, the fuel efficiency can increased more than twice → without much oxidizing of iron in the scrap → 40% scrap can be charged → this can be achieved by pre-heating of the scrap and partial combustion of CO gas inside the converter.
(iii) Injection of carbon in the form of coke or low volatile coal into the melt during oxygen lancing → combined with pre-heating of the scrap, partial combustion of CO gas and injection of carbon → 50% scrap can be charged.

It provides a low-cost method for steel production. By bottom injecting oxygen, the intensity of mixing of the metal bath is increased by about ten times compared to the LD process. The distance from the equilibrium between slag and metal is substantially reduced. Lime powder is added along with oxygen (which is reduced by 30%), resulting in an entirely different mechanism of slag formation than LD process. Comparison of characteristic values of LD, OBM and KMS processes are shown in Table 15.5.

The steps for KMS process are as follows [12]:

1. Furnace is empty, refractory is hot. All tuyeres are protected from over-heating by low flow rate of air or nitrogen gas.
2. Scrap is charged in one or two baskets, furnace is turned vertically.
3. Pre-heating starts with oil and oxygen passing through the bottom tuyeres, which act as burners.
4. Pre-heating is stopped according to the calculated time (by computer) based on heat and material balance.
5. Furnace is turned horizontal to receive hot metal charge and then turned vertically after bottom tuyeres get high flow rate of nitrogen gas.
6. Blowing starts by switching over from nitrogen to oxygen gas through bottom and through side tuyeres. Natural gas (or propane) is used as protective gas for tuyeres. Powder lime is added as required with oxygen gas through the bottom tuyeres.
7. Blowing is stopped, according to computer calculated amount of total oxygen required, by switching over again to nitrogen gas and furnace is turned down for sampling.

Table 15.5 Comparison LD, OBM and KMS processes [11]

	LD	OBM	KMS
Top blowing	Oxygen	–	Oxygen
Bottom blowing	–	Oxygen	Oxygen/coal/oil
Scrap rate in % of total metallic charge	Standard (25)	30–35	40–50
Total O_2 blowing rate, N m^3/t min	3	3.0–5.5	3.0–5.5
O_2 consumption	Standard	−4 N m^3/t	0.1–0.15 N m^3/kg additional scrap
Blowing behaviour	Heavy slopping	No slopping	No slopping
Perfect mixing time of bath (s)	120	10	20–50
(P)/[P] at 0.04% C	70	120	120
% [Mn] at 0.04% C	Standard	+0.1–0.2	+0.1–0.2
%(Fe) in slag at 0.04% C	Standard (22%)	10	10
Fe loss via dust (kg Fe/t)	10	2	3–8
%[O] at 0.04% C	0.06–0.09	0.04	0.04
[H] in ppm at 0.04% C	2–3	3–5	3–5
[N] in ppm at 0.04% C	20–40	10–25	20–40
Minimum [C] in %	0.03–0.04	0.005–0.01	0.01
Yield of liquid steel%	Standard	+17–20	+1–2 depending on scrap quality

Advantages:

- It can continue steel production at a high level with scrap melting when hot metal production is reduced by unforeseen mishaps.
- It is the ability to reduce production costs, substituting usually more expensive hot metal with less costly scrap and oxy-fuel.
- It is the ability to increase steel production for a given amount of hot metal.

15.5.3.2 LD-HC Process

LD-HC process is developed jointly by ***Hainaut-Sambre***, Marchienne and ***CRM***, Liege (Belgium) in combination of ***LD*** process in 1980. LD-HC combined blowing process is a process where a small part of the total oxygen is blown from the bottom tuyeres. This development due to the following reasons:

- To assure low carbon steels production with optimized metallurgical results at turndown, enough stirring of the bath is necessary,
- The bottom blowing with a high rate of oxygen (>10 to 20%) needs more investments and particularly an injection of lime powder through the tuyeres.

For these reasons, it was decided to blow 2–10% of total oxygen and a part of the lime through the bottom OBM (i.e. Maxshutte) tuyeres; and natural gas act as the shielding gas. With these types of tuyeres, the blowing of other gases or mixture of gases (Ar, N_2, air, etc.) is also possible. The LD-HC process is, therefore, one of the moderate stirring intensity processes.

During refining, the blowing pattern remains constant as well as the bottom flow rate (0.08–0.26 N m^3/t min) and the top lance oxygen flow rate (2.6 N m^3/t min). With the adopted working conditions, the blowing is stable without any slopping. The important stirring of the bath provokes a thickening of the slag which can lead to splitting problems during refining of low phosphorus hot metal. To avoid this problem, the bottom oxygen flow rate must be reduced, or powdered lime must be used. Another advantage of the process, heats having a hot metal containing 1% Si or 1% P, has been blown successfully without changing the standard practice. A short argon stirring is generally performed before tapping.

As it is now well established, the carbon content at tapping and the bath oxidation are decreased in the combined blowing processes compared with top blowing process (i.e. LD/BOF). In the case of the LD-HC process, the improvement is closely related to the stirring intensity of the bath and thus to the amount of oxygen blown through the bottom. Figure 15.27 shows the relationships between the carbon content and the dissolved oxygen at turndown in the case of several bottom oxygen blowing percentages (2.5–10%). Increasing flow rates through the bottom lead to a substantial reduction of the oxygen potential of the bath and permit to achieve carbon contents, nearly 0.02% without excessive slag and bath oxidation.

Another important aspect is the marked decrease of the iron oxidation of the slag (as shown in Fig. 15.28). For a carbon content of 0.03%, the slag iron content is equal to 27% with the LD process, and only 17% with the LD-HC process. A short stirring (1–2 min) with inert gas after the end of blowing further decreases the carbon and oxygen contents of the steel.

The sharp decrease of the slag content leads as a consequence to an increase of the residual steel manganese content at turndown. The initial manganese level of the hot metal is around 0.5%. For carbon content of 0.05%, the saving in manganese for the LD-HC process reaches 0.12% compared to the LD situation. The gain is still more important for lower carbon levels.

The relationship between the sulphur partition ratio and the basicity index is illustrated in Fig. 15.29 for both LD and LD-HC processes. The results presented in Fig. 15.29 show that, in the presence of powdered lime, a change of the bottom oxygen blowing percentage from 4 to 10% does not modify the de-sulphurization results. The improvement of the de-sulphurization reaction for the LD-HC is particularly significant for normal and high slag basicity. With a basicity ratio of 3, the sulphur partition ratio in LD is 4.5, while it reaches 7.5 in LD-HC; with a basicity index of 3.6, the sulphur partition in LD-HC is higher than 10, thus nearly the value in LD.

Due to the important rate of stirring in the LD-HC process, de-phosphorization is good, even though the iron content in slag is far lower than that in top blown converters. The effect of the stirring intensity (from 4 to 8%) is to accelerate the reaction between the slag and the molten steel. Phosphorus come down after first turndown from 0.018 to 0.017% and iron content in the slag also come down from 18 to 16.3% at the basicity 2.56.

As a direct consequence of the better metallurgical results achieved at turndown, the number of heats which need a correction for temperature or chemical analysis adjustment are reduced with the

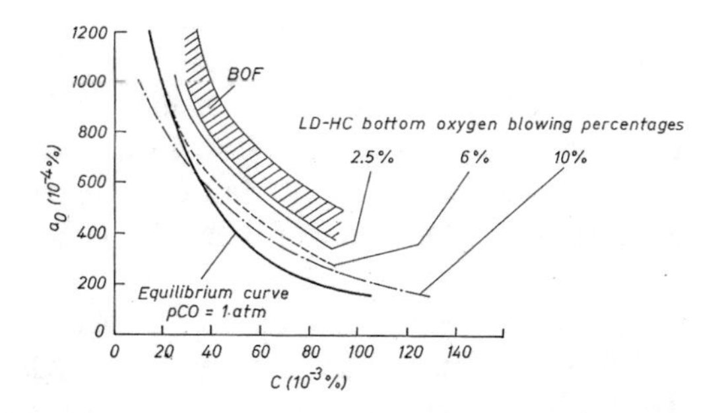

Fig. 15.27 Relationships between the carbon content and the dissolved oxygen [13]

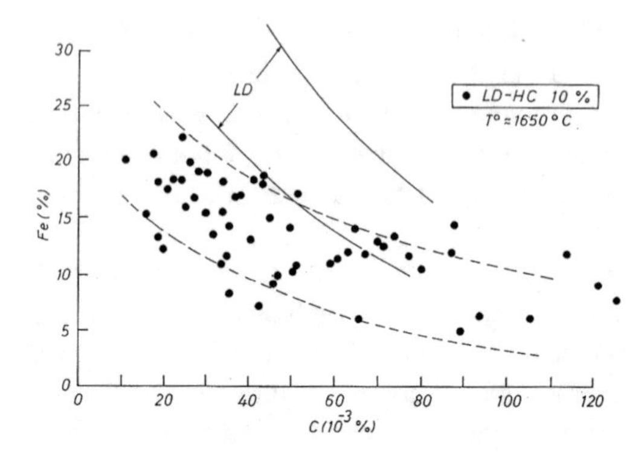

Fig. 15.28 Iron content in the slag relate with carbon content in melt [13]

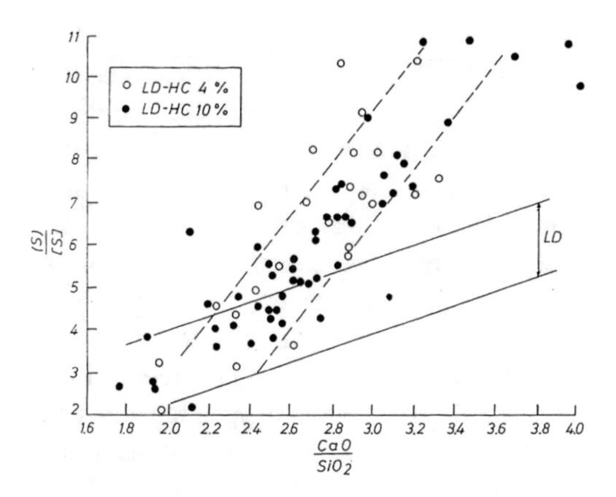

Fig. 15.29 Relationship between the sulphur partition ratio and slag basicity index [13]

LD-HC process compared to the LD process. Hydrogen content 2–3 ppm and 5–6 ppm for 0.05% C and 0.04% C steels baths, respectively. To decrease the hydrogen content of the bath, a short stirring with argon gas (0.5 m^3/t of liquid steel) is performed just before tapping.

Probable Questions

1. What are the main features of LD process? What is LD stands for?
2. Discuss the advantages and disadvantages of eccentric throat of LD converter.
3. 'Nozzles of lance in LD converter are made of copper and exposed to high temperature, and that are also not water cooled; but still that are not melted at all' Why?
4. What are the advantages of multi-nozzle lance over a single nozzle lance in LD process?
5. What do you understand by 'Joule-Thomson effect' in LD converter?
6. Discuss the mechanism of jet-metal interaction in LD process. How lance height can affect the sequence of refining?
7. What do you mean by 'depth of penetration of oxygen jet'? What are the factors on which depth of penetration depend? What is the co-relation between them? What are the conditions for de-carburization and de-phosphorization in terms of depth of penetration?

8. What is the mechanism of mass transfer in LD process?
9. Mechanism of refining in LD process depends on 'forced convection', not by 'diffusion'. Explain.
10. What do you understand by 'slopping' and 'splashing'?
11. 'Steelmaking is an oxidizing process, de-sulfurization required reducing atmosphere, still 60% S can be removed in LD converter'. How?
12. 'Foaming slag is an advantage in LD process, but it is disadvantage for open-hearth process'. Why?
13. Discuss the mechanism of refining in LD converter. How can you control the relative rates of de-carbonization and de-phosphorization in LD converter?
14. What are the factors that causes turbulence in LD converter? Discuss.
15. Explain the terms 'foam' and 'emulsion' with reference to LD process. Do they help refining in LD process? If so, how? Explain.
16. What do you mean by 'slag-metal emulsion'? What is the mechanism of de-carburization in LD converter? Write the equation for overall de-carburization rate.
17. What do you understand by Mn-hump? How it occurs?
18. 'The iron and phosphorous lines, on the Ellingham diagram, are so close to each other that the oxidation of phosphorous in steelmaking cannot be done without loss of iron'. How this problem can be overcome at the LD process?
19. How you can control of the carbon and phosphorus reactions in LD process?
20. How 'blocking of heat' is done in LD process?
21. Discuss the zone-lining and lining by slag splashing in LD converter.
22. '100% magnesite bricks are used at trunnion region in LD converter'. Why?
23. What do you mean by 'Maxshutte tuyere'? How it functions?
24. 'The outer tube of the 'Maxshutte tuyere' is made of stainless steel and the inner tube is made of copper'. Why?
25. 'In OBM converter, tuyeres are placed in one side of the converter'. Why?
26. Discuss the basic principle of OBM process and state the main advantages of OBM process. Why iron loss in OBM process is less compared to LD process?
27. Explain the LD-AC process and its merits over that of conventional LD process.
28. Explain the basic principle of Kaldo and Rotor processes.
29. What advantage you get by rotating the Kaldo converter?
30. Why rotary steelmaking processes are not popular?
31. What is the function of secondary lance in Rotor converter?
32. Discuss the advantages of bottom blown process over that of top blowing process. How many types of modern oxygen steelmaking processes are developed?
33. 'For purging of inert gas, porous plugs are preferred instead of tuyeres at the bottom of a converter/ladle'. Why?
34. What are the metallurgical improvements achieved by combined blowing processes?
35. What is 'OLP' stand for? Explain the basis for the development of OLP process of steelmaking. State the major operational steps that can enhance the refining rate.
36. What is KMS stands for? Discuss the basic principle of KMS process of steelmaking.
37. Discuss the LD-HC process.

Examples

Example 15.1 Chemistry of input and output materials for LD process is as follows:

Element, %	Hot metal	Scrap	Steel (to be produced)
C	3.50	0.20	0.15
Si	1.20	0.02	0.01
Mn	0.75	0.40	0.25
P	0.40	0.04	0.03
S	0.04	0.04	0.03

Calculate: (i) Amount of hot metal to be charged per tonne of steel production.
(ii) Amount of slag produced and composition of slag.

Given data: (i) Weight of steel scrap is 200 kg per tonne of steel production,
(ii) Weight of lime is 50 kg per ton of steel production (lime contain 94.5% CaO, 2.5% MgO, 1.5% SiO_2 and 1.5% Al_2O_3).
(iii)1.5% Fe loss with respect to steel production.

Solution

Assumptions: (a) One tonne steel production,
By balancing we get, 94.11%, 99.30% and 99.53% Fe contain in hot metal, steel scrap and produced steel, respectively.

Fe Balance: Fe input = Fe output
Fe from HM + Fe from steel scrap = Fe goes to steel produce + Fe losses in slag
Suppose weight of hot metal = W_{HM}
$0.9411\ W_{HM} + 0.993 \times 200 = 0.9953 \times 1000 + 0.015 \times 1000$
Therefore, $W_{HM} = 862.5$ kg

$$\mathrm{Fe} + 1/2\,\mathrm{O_2} = \mathrm{FeO}$$

56 72

56 kg of Fe to form 72 kg of FeO

15 kg $\left(\frac{72}{56}\right) \times 15 = 19.29$ kg of FeO

Si Balance: Si input = Si output
Si from HM + Si from steel scrap + Si from lime
= Si goes to steel produce + Si losses in slag (suppose weight of Si losses in slag = W_{Si})
$0.012 \times 862.5 + 0.0002 \times 200 + 0.015 \times 50 \times \left(\frac{28}{60}\right) = 0.0001 \times 1000 + W_{Si}$

Therefore, W_{Si} = 10.64 kg

$$Si + O_2 = SiO_2$$

28 60

28 kg Si to form 60 kg of SiO_2

10.64 kg $\left(\frac{60}{28}\right) \times 10.64 = 22.8$ kg of SiO_2 in slag

Mn Balance: Mn input = Mn output

Mn from HM + Mn from scrap = Mn in steel + Mn losses in slag

$0.0075 \times 862.5 + 0.004 \times 200 = 0.0025 \times 1000 + W_{Mn}$

Therefore, W_{Mn} = 4.77 kg

$$Mn + 1/2\,O_2 = MnO$$

55 71

55 kg Mn to form 71 kg of MnO

4.77 kg $\left(\frac{71}{55}\right) \times 4.77 = 6.16$ kg of MnO in slag.

P Balance: P input = P output

P from HM + P from scrap = P in steel + P goes to slag

$0.004 \times 862.5 + 0.0004 \times 200 = 0.0003 \times 1000 + W_P$

$W_P = 3.23$ kg

$$2P + 5/2\,O_2 = P_2O_5$$

2 × 31 142

62 kg P to form 142 kg of P_2O_5

3.23 kg $\left(\frac{142}{62}\right) \times 3.23 = 7.4$ kg of P_2O_5

S Balance: S input = S output

S from HM + S from scrap = S in steel + S goes to slag

$0.0004 \times 862.4 + 0.0004 \times 200 = 0.0003 \times 1000 + W_S$

Therefore, W_S = 0.125 kg

$$CaO + S = CaS$$

56 32 72

32 kg S to form 72 kg of CaS

0.125 kg $\left(\frac{72}{32}\right) \times 0.125 = 0.28$ kg of CaS

72 kg CaS formation CaO required 56 kg

0.28 kg $\left(\frac{56}{72}\right) \times 0.28 = 0.22$ kg

CaO Balance: CaO input = CaO output
CaO from lime = CaO in CaS + CaO in slag

$$0.945 \times 50 = 0.22 + W_{CaO}$$

$$W_{CaO} = 47.03\,\text{kg}$$

MgO Balance: MgO from lime = MgO in slag

$$W_{MgO} = 0.025 \times 50 = 1.25\,\text{kg}$$

Al_2O_3 Balance: Al_2O_3 in lime = Al_2O_3 in slag

$$W_{Al_2O_3} = 0.015 \times 50 = 0.75$$

$$\text{Amount of slag} = W_{FeO} + W_{SiO_2} + W_{MnO} + W_{P_2O_5} + W_{CaS} + W_{CaO} + W_{MgO} + W_{Al_2O_3}$$

$$= 19.29 + 22.8 + 6.16 + 7.4 + 0.28 + 47.03 + 1.25 + 0.75 = \mathbf{104.96\,kg}$$

Composition of slag: 18.38% FeO, 21.72% SiO_2, 5.87% MnO, 7.05% P_2O_5, 0.27% CaS, 44.81% CaO, 1.19% MgO and 0.71% Al_2O_3.

Example 15.2 Chemistry of input and output materials for LD steelmaking as follows:

Element/compound %	Hot metal	Scrap	DRI	Steel produced	Lime	Fe–Mn	Fe–Si
C	4.0	0.37	1.75	0.25		0.1	1.5
Si	1.0	0.27		0.2		2.0	70.0
Mn	0.5	0.61		0.25		72.5	
P	0.4	0.05	0.03	0.03		0.1	
S	0.04	0.05	0.01	0.03		0.03	
CaO			1.1		95.0		
MgO			0.8		2.5		
SiO_2			1.0		1.5		
Al_2O_3			1.45		1.0		
FeO			8.36				
Fe_{Tol}			92.0				
Amount charged, kg	?	200	325		?	5	5

Calculate (a) amount of *H:M* charged, (b) amount of lime charged, (c) amount of oxygen consumption and (d) amount of slag produced and composition of slag.
Given: Fe loss in slag and fumes are 1.5 and 0.6% of liquid steel.
Basicity of slag = 3.0

Solution

Assumptions: One tonne steel production.
By balancing we get, 94.06%, 98.65%, 99.24%, 28.5% and 25.27% Fe contain in hot metal, steel scrap, produced steel, Fe–Si and Fe–Mn, respectively.

Fe Balance: Fe input = Fe output
Fe from HM + Fe from steel scrap + Fe from DRI + Fe from Fe–Si + Fe from Fe–Mn = Fe goes to steel produce + Fe losses in slag + Fe losses in fumes
Suppose weight of hot metal = W_{HM}

$$\text{So, } 0.9406 \times W_{HM} + 0.9865 \times 200 + 0.92 \times 325 + 0.285 \times 5 + 0.2527 \times 5$$
$$= 0.9924 \times 1000 + 0.015 \times 1000 + 0.006 \times 1000$$

Therefore, W_{HM} = 546.91 = **547 kg**

$$Fe + 1/2\,O_2 = FeO$$
$$56 \quad 16 \quad 72$$

56 kg of Fe to form 72 kg of FeO

$$15\,\text{kg} \qquad \left(\frac{72}{56}\right) \times 15 = 19.29\,\text{kg of FeO in slag}$$

For 56 kg Fe oxygen required 16 kg

$$15\,\text{kg} \qquad \left(\frac{16}{56}\right) \times 15 = 4.29\,\text{kg of oxygen}$$

Si Balance: Si input = Si output
Si from HM + Si from scrap + Si from Fe–Si + Si from Fe–Mn + Si from gange material in DRI + Si from lime = Si in steel + Si goes to slag

$$0.01 \times 547 + 0.0027 \times 200 + 0.7 \times 5 + 0.02 \times 5 + 0.01 \times 325 \times \left(\frac{28}{60}\right) + 0.015 \times W_L \times \left(\frac{28}{60}\right)$$
$$= 0.002 \times 1000 + W_{Si}$$

Therefore, $W_{Si} = 9.13 + 7 \times 10^{-3} W_L$

$$Si + O_2 = SiO_2$$
$$28 \quad 32 \quad 60$$

28 part Si form 60 part of SiO_2

$$(9.13 + 7 \times 10^{-3} W_L)\ldots\left[(9.13 + 7 \times 10^{-3} W_L) \times \left(\frac{28}{60}\right)\text{part of } SiO_2 \right. \tag{1}$$

CaO Balance: CaO input = CaO output
CaO from lime + CaO from gange material in DRI = CaO goes to slag

$$\begin{aligned} & 0.95 \times W_L + 0.011 \times 325 = W_{CaO} \\ & \text{or, } 0.95 \times W_L + 3.575 = W_{CaO} \end{aligned} \tag{2}$$

$$\begin{aligned} \text{Since Basicity} = 3 &= \left(\frac{\Sigma\, \text{CaO in slag}}{\Sigma\, \text{SiO}_2\, \text{in slag}} \right) \\ &= \left[(0.95 \times W_L + 3.575) / \left\{ (9.13 + 7 \times 10^{-3} W_L) \times \left(\frac{60}{28} \right) \right\} \right] \end{aligned}$$

(from Eqs. 1 and 2)
Therefore, W_L = 60.9 kg
From Eq. 1: W_{Si} = 9.556 kg
SiO_2 in slag = $W_{Si} \times \left(\frac{60}{28}\right) = 20.48\,\text{kg}$,
Out of this SiO_2 coming from oxidation $= [20.48 - (0.01 \times 325) - (0.015 \times 60.9)]$
$= 16.32\,\text{kg}$
Therefore, oxygen required for SiO_2 formation by oxidation $= \left(\frac{(32 \times 16.32)}{60}\right) = 8.7\,\text{kg}$
From Eq. 2: W_{CaO} = 61.43 kg.

Mn Balance: Mn input = Mn output
Mn from HM + Mn from scrap + Mn from Fe–Mn = Mn in steel + Mn goes to slag
$0.005 \times 547 + 0.0061 \times 200 + 0.725 \times 5 = 0.0025 \times 1000 + W_{Mn}$
Therefore, W_{Mn} = 5.08 kg
$[\; Mn + 1/2\, O_2 = MnO$
$\quad 55 \quad\quad 16 \quad\quad\quad 71 \;]$
MnO in slag $= 5.08 \times \left(\frac{71}{55}\right) = \mathbf{6.56\,kg}$
Oxygen require $= 5.08 \times \left(\frac{16}{55}\right) = \mathbf{1.48\,kg}$

C Balance: C input = C output
C from HM + C from scrap + C from Fe–Si + C from Fe–Mn + C from DRI
= C in steel + C goes as gases
$0.04 \times 547 + 0.0037 \times 200 + 0.015 \times 5 + 0.001 \times 5 + 0.0175 \times 325$
$= 0.0025 \times 1000 + W_C$
W_C = **25.89 kg**
Assume that C form gases 90% CO and 10% CO_2
$C + 1/2\, O_2 = CO$
$12 \quad 16 \quad\quad 28$

Oxygen required for CO formation = $\left(\frac{16}{12}\right) \times 0.9 \times 25.89 = \mathbf{30.06\,kg}$

$C + O_2 = CO_2$

12 32 44

Oxygen required for CO_2 formation = $\left(\frac{32}{12}\right) \times 0.1 \times 25.89 = \mathbf{6.90\,kg}$

P Balance: P input = P output

P from HM + P from scrap + P from Fe–Mn + P from DRI

= P in steel + P goes to slag

$0.004 \times 547 + 0.0005 \times 200 + 0.001 \times 5 + 0.0003 \times 325$

$= 0.0003 \times 1000 + W_P$

Therefore, W_P = 2.09 kg

$2P + 5/2\,O_2 = P_2O_5$

62 80 142

62 kg P to form P_2O_5 oxygen require = 80 kg

2.09 kg $= \left(\frac{80}{62}\right) \times 2.09 = \mathbf{2.70\,kg}$

P_2O_5 in slag = $\left(\frac{142}{62}\right) \times 2.09 = \mathbf{4.79\,kg}$

S Balance: S input = S output

S from HM + S from scrap + S from Fe–Mn + S from DRI

= S in steel + S goes to slag

$0.0004 \times 547 + 0.0005 \times 200 + 0.0003 \times 5 + 0.0001 \times 325$

$= 0.0003 \times 1000 + W_S$

Therefore, W_S = 0.053 kg

$CaO + S = CaS$

32 72

CaS in slag = $\left(\frac{72}{32}\right) \times 0.053 = 0.1188 = \mathbf{0.12\,kg}$

MgO Balance: MgO input = MgO output

MgO from DRI + MgO from lime = MgO goes to slag

$0.008 \times 325 + 0.025 \times 60.9 = W_{MgO}$

$\mathbf{W_{MgO} = 4.12\,kg}$

Al_2O_3 Balance: Al_2O_3 input = Al_2O_3 output

Al_2O_3 from DRI + Al_2O_3 from lime = Al_2O_3 goes to slag

$0.0145 \times 325 + 0.01 \times 60.9 = W_{Al_2O_3}$

$W_{Al_2O_3} = \mathbf{5.32\,kg}$

$\textbf{Weight of slag} = W_{FeO} + W_{SiO_2} + W_{CaO} + W_{MnO} + W_{P_2O_5} + W_{CaS} + W_{MgO} + W_{Al_2O_3}$

$= 19.29 + 20.48 + 61.43 + 6.56 + 4.75 + 0.12 + 4.12 + 5.32$

$= \mathbf{122.07\,kg}$

Composition of slag: 15.8% FeO, 16.78% SiO_2, 50.32% CaO, 5.37% MnO, 3.89% P_2O_5, 0.1% CaS, 3.37% MgO and 4.36% Al_2O_3.

$\textbf{Oxygen consumption} = O_{FeO} + O_{SiO_2} + O_{MnO} + O_{CO} + O_{CO_2} + O_{P_2O_5}$

$= 4.29 + 8.70 + 1.48 + 30.06 + 6.90 + 2.70 = \mathbf{54.13\,kg}$

Example 15.3 LD converter is charged with 10 tonne of hot metal containing 3.5% C, 1.25% Si, 0.75% Mn and rest Fe. The impurities are oxidized and removed from HM by lancing pure oxygen into the converter. 40% C oxidize to CO_2 and 60% C to form CO. Find out (i) total amount of oxygen gas require (in N m^3), (ii) the weight of slag produce.

Mention clearly, assumptions made. [Given: 1% Fe (w.r.t. HM) is lost.]

Solution

Assumptions: (a) Steel contains 100% Fe,
(b) Oxygen used is pure form.

HM 10 tonne = 10,000 kg

3.5% C in HM = $\left(\frac{3.5}{100}\right) \times 10,000 = 350\,\text{kg}$

1.25% Si in HM = $\left(\frac{1.25}{100}\right) \times 10,000 = 125\,\text{kg}$

0.75% Mn in HM = $\left(\frac{0.75}{100}\right) \times 10,000 = 75\,\text{kg}$

Fe loss = $\left(\frac{1}{100}\right) \times 10,000 = 100\,\text{kg}$

$$\text{Fe} + 1/2\,\text{O}_2 = \text{FeO}$$

56 kg of Fe to form 72 kg of FeO

100 kg $\left(\frac{72}{56}\right) \times 100 = \mathbf{128.57\,kg}$ of FeO goes in slag

For 56 kg of Fe to form FeO oxygen require 11.2 N m^3

100 kg $\left(\frac{11.2}{56}\right) \times 100 = \mathbf{20\,N\,m^3}$

$$Si + O_2 = SiO_2$$
$$28 \quad 22.4 \quad 60$$

28 kg of Si to form 60 kg of SiO_2

125 kg $\left(\frac{60}{28}\right) \times 125 = \mathbf{267.86\,kg}$ of SiO_2 goes to slag

28 kg of Si to form SiO_2 oxygen require 22.4 N m^3

125 kg $\left(\frac{22.4}{28}\right) \times 125 = \mathbf{100\,N\,m^3}$

$$Mn + 1/2\,O_2 = MnO$$
$$55 \quad 11.2 \quad 71$$

55 kg of Mn to form 71 kg of MnO

75 kg $\left(\frac{71}{55}\right) \times 75 = \mathbf{96.82\,kg}$ of MnO goes to slag

For 55 kg of Mn to form MnO oxygen require 11.2 N m^3

75 kg $\left(\frac{11.2}{55}\right) \times 75 = \mathbf{15.27\,N\,m^3}$

$$\text{Out of 350 kg C, 40\% C to form } CO_2 = \left(\frac{40}{100}\right) \times 350 = 140\,\text{kg}$$
$$\text{60\% C to form CO} = \left(\frac{60}{100}\right) \times 350 = 210\,\text{kg}$$

$$C + O_2 = CO_2$$
$$12 \quad 22.4 \quad 44$$

Oxygen required for CO_2 formation = $\left(\frac{22.4}{12}\right) \times 140 = \mathbf{261.33\,N\,m^3}$

$$C + 1/2\,O_2 = CO$$
$$12 \quad 11.2 \quad 28$$

Oxygen required for CO formation = $\left(\frac{11.2}{12}\right) \times 210 = \mathbf{196\,N\,m^3}$

Weight of slag produce = weight of FeO + weight of SiO_2 + weight of MnO
= 128.57 + 267.86 + 96.82 = **493.25 kg**

Total oxygen require = for FeO + for SiO_2 + for MnO + for CO_2 + for CO
= 20 + 100 + 15.27 + 261.33 + 196 = $\mathbf{592.60\,N\,m^3}$

Example 15.4 LD converter is charged with 20 tonnes of hot metal containing 4.0% C, 1.2% Mn, 1.4% Si and rest Fe. The impurities are oxidized and removed from hot metal by lancing oxygen. Three-fourths of carbon goes as CO and one-fourth goes as CO_2. Calculate: (i) oxygen required (in cubic metres), (ii) the composition and total volume of product gases. Mention clearly, the assumptions made in the above case.

Solution

Assumptions: (i) Steel is almost pure Fe and impurities are completely oxidized,
(ii) No loss of Fe during the process.

Si content in HM: $\left(\frac{1.4}{100}\right) \times 20,000 = 280\,\text{kg}$

$Si + O_2 = SiO_2$

28 22.4

28 kg of Si react with $22.4\,\text{N m}^3\,O_2$

280 $\left(\frac{22.4}{28}\right) \times 280 = \mathbf{224\,N\,m^3\,O_2}$

Mn content in HM: $\left(\frac{1.2}{100}\right) \times 20,000 = 240\,\text{kg}$

$2Mn + O_2 = 2MnO$

2×55 22.4

110 kg of Mn react with $22.4\,\text{N m}^3\,O_2$

240 $\left(\frac{22.4}{110}\right) \times 240 = \mathbf{48.87\,N\,m^3\,O_2}$

Carbon content in HM: $\left(\frac{4.0}{100}\right) \times 20,000 = 800\,\text{kg}$

C form CO: $800 \times \left(\frac{3}{4}\right) = 600\,\text{kg}$

C form CO_2: $800 \times \left(\frac{1}{4}\right) = 200\,\text{kg}$

$2C + O_2 = 2CO$

2×12 22.4 2×22.4

24 kg of C react with $22.4\,\text{N m}^3\,O_2$ to form CO

600 $\left(\frac{22.4}{24}\right) \times 600 = \mathbf{560\,N\,m^3\,O_2\,for\,CO}$

600 $\left(\frac{2 \times 22.4}{24}\right) \times 600 = \mathbf{1120\,N\,m^3\,CO}$

$C + O_2 = CO_2$

12 22.4 22.4

12 kg of C react with $22.4\,\text{N m}^3\,O_2$ to form CO_2

200 $\left(\frac{22.4}{12}\right) \times 200 = \mathbf{373.33\,N\,m^3\,O_2\,for\,CO_2}$

200 $\left(\frac{22.4}{24}\right) \times 200 = \mathbf{373.33\,N\,m^3\,CO_2}$

Total oxygen required: 224 + 48.87 + 560 + 373.33 = 1206.2 N m^3 O$_2$

Total volume of gases: 373.33 + 1120 = **1493.33** N **m^3**

%CO = $\left(\frac{1120}{1493.33}\right) \times \mathbf{100 = 75\%}$

%CO$_2$ = $\left(\frac{373.33}{1493.33}\right) \times \mathbf{100 = 24.9998 = 25\%}$.

Example 15.5 Calculate the temperature of the liquid steel rise if 20 kg of ferrosilicon is added to 50 t ladle at 1600 °C. Ferrosilicon contains 80% Si. Assuming half of silicon added reacts with dissolved oxygen and rest half simply dissolved in molten steel.

Given: Specific heat of molten steel 44 J/mol K, $H_{1873} - H_{298}$ for silicon is 33.5 kJ/mol, heat of solution of silicon in liquid steel is −119.3 kJ/mol and heat of reaction for [Si] + 2[O] = SiO_2 (s) is −594.6 kJ/mol. All data are at 1873 K.

Solution

Si in ferrosilicon = 20 × 0.8 = 16 kg.

$$\text{Si (s) at 298} \rightarrow \text{Si (s) at 1873}, H_{1873} - H_{298} = 33.5\,\text{kJ/mol} = (33.5/28) \times 1000 = 1196.43\,\text{kJ/kg Si}$$

$$\text{Si (s) at 1873} \rightarrow [\text{Si}]\text{ at 1873}, \Delta H = -119.3\,\text{kJ/mol} = (-119.3/28) \times 1000 = -4260.71\,\text{kJ/kg Si}$$

$$\text{Si (s) at 298} \rightarrow [\text{Si}]\text{ at 1873}, \Delta H_s = -3064.28\,\text{kJ/kg Si}$$

Heat generated for 16 kg Si dissolved in molten steel = 16 × −3064.28 = −49,028.48 kJ

$$[\text{Si}] + 2[\text{O}] = \text{SiO}_2\,(\text{s})\ \Delta H_r = -594.6\,\text{kJ/mol} = -21{,}235.71\,\text{kJ/kg Si}$$

Now half of silicon added reacts with dissolved oxygen = 16/2 = 8 kg
So, heat generated for 8 kg Si reacts with dissolved oxygen = −169,885.71 kJ
Total heat generated for 16 kg Si dissolved in molten steel and 8 kg Si reacts with dissolved oxygen = −49,028.48 + (−169,885.71) = −218,914.19 kJ
Specific heat of molten steel 44 J/mol K = 44/56 = 0.786 kJ/kg K
Now heat required for 50 t molten steel to rise 1 K = 0.786 × 50,000 = 39,300 kJ
39,300 kJ heat required for 50 t molten steel to rise temperature 1 K

$$\text{Therefore, } 218{,}914.19\,\text{kJ heat rise temperature for } 50\,\text{t molten steel} = (218{,}914.19/39{,}300) = 5.57\,\text{K}$$

The temperature of the liquid steel rise by 5.6 K.

Problems

Problem 15.1 Chemistry of input and output materials for LD process is as follows:

Element, %	Hot metal	Scrap	Steel (to be produced)
C	3.25	0.22	0.15
Si	1.15	0.02	0.01

(continued)

Element, %	Hot metal	Scrap	Steel (to be produced)
Mn	0.80	0.40	0.30
P	0.40	0.04	0.03
S	0.04	0.04	0.03

Calculate: (i) Amount of hot metal to be charged per tonne of steel production.
(ii) Amount of slag produced and composition of slag.

Given data: (i) Weight of steel scrap is 250 kg per tonne of steel production,
(ii) Weight of lime is 50 kg per tonne of steel production (lime contains 95% CaO, 2.0% MgO, 1.5% SiO_2 and 1.5% Al_2O_3),
(iii) 1.5% Fe loss with respect to steel production.

[Ans: (i) **807.12 kg**, (ii) **101.85 kg** (18.94% FeO, 20.16% SiO_2, 5.68% MnO, 6.81% P_2O_5, 0.27% CaS, 46.43% CaO, 0.98% MgO and 0.74% Al_2O_3)]

Problem 15.2 LD converter is charged with 10 tonne of hot metal containing 4.0% C, 1.5% Si, 1.0% Mn and rest Fe. The impurities are oxidized and removed from HM by lancing pure oxygen into the converter. One-quarter of the carbon is oxidized to CO_2 and three quarters to form CO. Find out (i) total amount of oxygen gas require (in N m^3), (ii) the volume of product gases.

Mention clearly, assumptions made.
[Ans: (i) 607 N m^3, (ii) 747 N m^3 (25% CO_2 + 75% CO)]

Problem 15.3 Chemistry of input and output materials for LD steelmaking is as follows:

Element/compound %	Hot metal	Scrap	DRI	Steel produced	Lime	Fe–Mn	Fe–Si
C	3.5	0.25	1.7	0.2		0.1	1.5
Si	1.1	0.27		0.2		2.0	70.0
Mn	0.5	0.61		0.25		72.5	
P	0.4	0.05	0.03	0.03		0.1	
S	0.04	0.05	0.01	0.03		0.03	
CaO			1.1		95.0		
MgO			0.8		2.5		
SiO_2			1.0		1.5		
Al_2O_3			1.45		1.0		
FeO			8.36				
Fe_{Tol}			92.0			25.27	28.5
Amount charged, kg	550	200	?		55	5	5

Calculate (a) amount of DRI charged, (b) amount of oxygen consumption, and (c) amount of slag produced and composition of slag.
Given: Fe loss in slag is 1.5 of liquid steel

[Ans: (a) Wt of DRI = **313.37 kg**, (b) Amount of oxygen = **51.82 kg**, (c) Amount of slag = **116.9 kg** (16.5% FeO, 18.4% SiO_2, 47.6% CaO, 5.6% MnO, 4.1% P_2O_5, 0.1% CaS, 3.3% MgO and 4.4% Al_2O_3.)]

References

1. R.H. Tupkary, *Introduction to Modern Steelmaking* (Khanna Publishers, Delhi, 1991)
2. https://images.search.yahoo.com/search/images
3. *Tata Tech*, May 1996
4. A.K. Chakrabarti, *Steel Making* (Prentice-Hall of India Pvt. Ltd., New Delhi, 2007)
5. J. Stradtmann, W. Munchberg, C. Metzger, *Proceedings of International Symposium on Modern Developments in Steelmaking*, vol. 2, Jamshedpur, India, 1981, p. 4.1.1
6. A. Dasgupta, A.K. Ganguly, JPC Bull. Iron Steel **XIII**(7), 28 (2013)
7. M. Saigusa et al., Ironmaking Steelmaking **7**(5), 242 (1980)
8. A. Maubon, *Conference Proceedings of International Symposium on Modern Developments in Steelmaking*, vol. 2, Jamshedpur, India, 1981, p. 5.4.1
9. F. Schleimer, R. Henrion, F. Goedert, G. Denier, J. Grosjean, *Conference Proceedings of International Symposium on Modern Developments in Steelmaking*, vol. 1, Jamshedpur, India, 1981, p. 5.5.1
10. T. Kosukegawa et al., *Conference Proceedings of International Symposium on Modern Developments in Steelmaking*, vol. 1, Jamshedpur, India, 1981, p. 5.1.1
11. E. Fritz, *Conference Proceedings of International Symposium on Modern Developments in Steelmaking*, vol. 2, Jamshedpur, India, 1981, p. 5.2.1
12. J. Pearce, E.G. Schempp, *Conference Proceedings of International Symposium on Modern Developments in Steelmaking*, vol. 2, Jamshedpur, India, 1981, p. 2.4.1
13. H. Jacobs, M. Dutrieux, C. Marique, C. Coessens, Met. Plant Tech. **2**, 20 (1986)

16 Electric Furnace Processes

The main furnaces that are used to produce steel through secondary route are electric arc furnaces (EAF) and induction melting furnaces (IMF). Steelmaking by EAF, their development, charge modification and design aspect are discussed. DC-EAF and IMF are also discussed. The quality steels are efficiently produced in EAF using steel scrap. Major problem faced by steelmakers was short supply, fluctuating prices and extremely heterogeneous nature of scrap. Sponge iron/HBI has uniform chemical and physical characteristics and it hardly contains any tramp metallic elements and has low sulphur content. The partial/total replacement of scrap by sponge iron/HBI is used as feed materials in EAF for quality steels production. EAF steelmakers also prefer to charge hot metal from a mini-blast furnace directly in the EAF to improve productivity, decrease consumption of electric power and cost of production. The new process, CONARC, which is half BOF and half EAF, i.e. combines oxygen converter and electric arc furnace technologies. It employs a twin shell EAF which can economically handle raw material input of solid scrap and hot metal in varying proportions. The EOF, ECOARC, FASTEEL processes and shaft furnace technology are also discussed. Different methods of stainless steel production are described in details.

16.1 Introduction of Electric Furnaces

Modern steelmaking has been broadly divided into two categories (i) primary route and (ii) secondary route. The steel produced from hot metal, using iron ore as a raw material in its initial stage (i.e. reduction stage) in blast furnace, is considered as primary route of steelmaking. The process in which steel is produced using scrap/sponge iron/pig iron is known as secondary route of steelmaking. The main furnaces that are used to produce steel through secondary route are (1) electric arc furnaces (EAF) and (2) induction melting furnaces (IMF). The families of electric furnaces are shown in Fig. 16.1.

(1) Arc furnaces: Arc furnaces are two types: (i) direct arc and (ii) indirect arc.

   (i) Direct arc furnace: In a direct arc furnace, current flows from the electrode to the charge and heat is transferred directly from the arc to the charge and part of heat is also generated in the charge itself. Three electrodes arc furnace was developed by Paul Heroult (in France) for steelmaking in 1899. In an electric arc furnace, scrap and or sponge iron/DRI is melted and

S. K. Dutta and Y. B. Chokshi, *Basic Concepts of Iron and Steel Making*,
https://doi.org/10.1007/978-981-15-2437-0_16

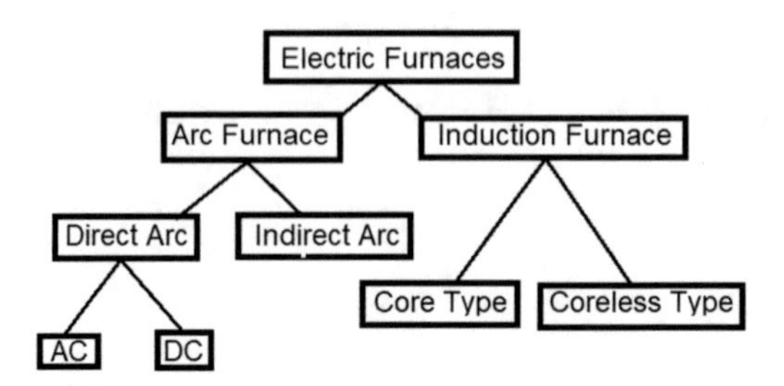

Fig. 16.1 Families of electric furnaces

converted into high-quality steel by using power electric arcs formed between a cathode and anode. There are two types of direct arc furnaces: (a) AC arc furnace and (b) DC arc furnace.

(a) Alternate current (AC) arc furnaces has three electrodes and they are fed from a three-phase supply input, one phase connected to each electrode and the charge is the neutral point (as shown in Fig. 16.2). There are acid lining EAF and basic lining EAF; basic furnace is most popular due to removal of sulphur and phosphorous from the melt.

(b) Direct current (DC) arc furnace has only a single electrode which acts as cathode and the bottom of the vessel act as anode (as shown in Fig. 16.2). These furnaces are applicable only in large sizes.

(ii) Indirect arc furnace: In an indirect arc furnace, arc is struck between two carbon electrodes and heat is transferred to the charge by radiations (as shown in Fig. 16.3). Indirect arc furnaces are of small capacities and do not developed steelmaking temperature readily. These furnaces are generally used in non-ferrous foundries for low-temperature melting.

(2) Induction furnace: In induction furnace, heat energy for melting metallic charge is obtained from induced current produced by the principle of electro-magnetic induction. Furnace is comprised of lined refractory crucible in a cylindrical steel shell. The coil lined with refractory material act as a primary coil. When electric current is passed through this coil, induced current is produced in metallic charge. The heat produced from the electric resistance, melts the solid charge. Due to

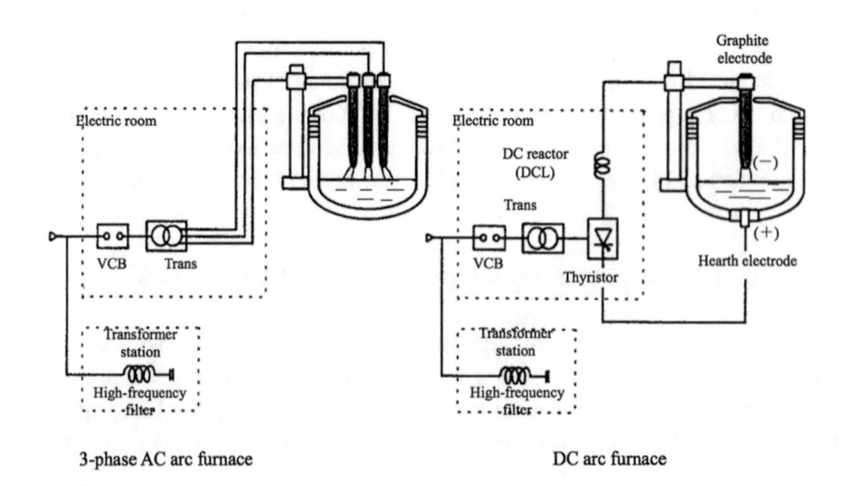

Fig. 16.2 Electric arc furnace [1]

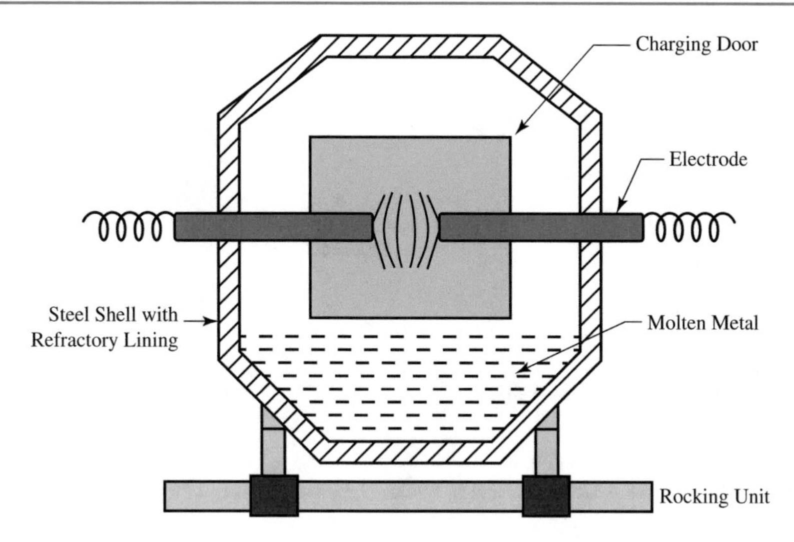

Fig. 16.3 Indirect arc furnace [1]

electro-magnetic action, induction heating produces circular eddy currents within the molten mass creating a stirring effect ensuring uniformly mixed homogeneous molten metal. Electromagnetic induction is used to heat the metal. An alternating current is supplied to a primary water coil tube which sets up a variable magnetic field around that water coil. The variable magnetic flux in term induces an electromotive force in the secondary circuit (i.e. metallic charge materials), so that the metal is melted by the alternating currents formed in it. Induction furnaces are made in a wide range of sizes. The induction furnaces are classified into the following three types based on their frequency of the current supplied:

- High-frequency furnaces (200–1000 kHz) supplied from valve generators.
- Medium-frequency furnaces (500–10,000 Hz) supplied from rotary or thyristor convertors.
- Low-frequency furnaces (50 Hz) which are fed directly from the electrical mains.

As molten metal is excited by current opposite to current flowing in induction coil, molten metal is agitated to raise its surface in the centre. Surface of molten metal is risen higher as frequency becomes lower, i.e. agitation of molten metal occurs stronger in low-frequency furnace than in high-frequency furnace. This effect of agitation makes it possible to ensure uniform temperature of molten metal and its uniform quality as well as to promote entrapment of material charged and fusion of chemical composition adjusting agents, especially carbon addition. On the other hand, excessive agitation may cause such troubles as oxidative wearing of molten metal and fusing out of refractories.

There are two types of induction furnaces: (i) core type (where always liquid metal keeps in the channel) and (ii) coreless type (therc are no channel). In induction furnace category, coreless induction furnaces are very popular for steelmaking.

(i) Induction furnace core type: This type of induction furnace is also called *induction channel furnace*. In this furnace, induction heating takes place in the channel which is a small and narrow area at the bottom of the main molten bath. Heat is transfer from the channel to main melting metal in crucible. The channel passes through a steel core and the coil assembly (as shown in Fig. 16.4a). This type of furnace is mostly used for non-ferrous metals.

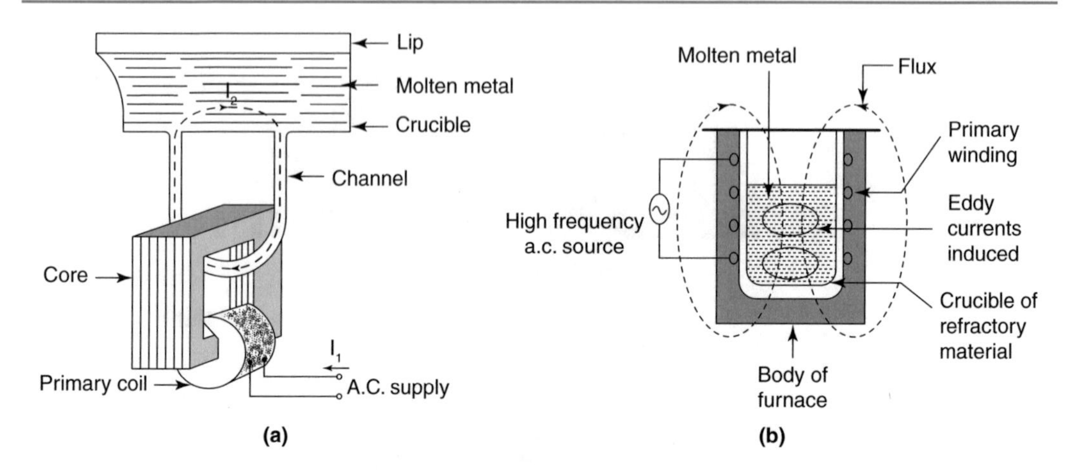

Fig. 16.4 Induction furnace [2]. **a** Core type and **b** coreless type (reproduced with permission from Your electrical guide)

(ii) Induction furnace coreless type: This type of the furnace is also called *induction melting furnace* (IMF) or *induction crucible furnace*. Coreless induction furnaces (medium or high frequency) are used for steelmaking and steel foundries. The coreless induction furnace has a simple construction and basically consists of a refractory-lined crucible. The furnace consists of a crucible, inductor coil and shell and stirring mechanism (as shown in Fig. 16.4b). The refractory crucible is surrounded by primary water coil tube. Metallic materials are charged into the crucible. An alternating current supplied to a primary water coil tube sets up a variable magnetic field around that water coil. The variable magnetic flux in term induces an electromotive force in the secondary circuit (i.e. metallic charge materials), so that the metal is melted by the alternating currents formed in it and generated heat.

16.2 Electric Arc Furnace

Earlier, electric arc furnace (EAF) was used mainly for special alloy steels production. Nowadays, plain carbon and low alloy steels are also produced in EAF. Earlier tap-to-tap times of over three hours were quite common, and specific power consumption was often well over 700 kWh/t, nearly twice the thermodynamic requirement. On that time, EAF steelmaking was an expensive and slow process. High metallic yields coupled with high quality of the product obtainable in this process were fully exploited in the production of costly alloy steels.

This technology had undergone rapid development and became the second largest steelmaking technology behind basic oxygen steelmaking technology. Total theoretical energy required to melt the scrap and to super-heat it to the typical tapping temperatures requires around 350–370 kWh/t of steel. These energies are provided by the electric arc, fuel injection and oxidation of impurities present in scrap. In practice, the energy use is highly dependent on product mix, local scrap and energy costs. Factors such as composition of scrap, power input rates and operational practices (such as post-combustion, scrap pre-heating, etc.) can greatly influence the overall energy consumption. Presently, power consumption in EAFs is reported in the range between 300 and 550 kWh/t throughout the world.

The total energy required to produce one tonne of liquid steel in an electric arc furnace (EAF) by melting scrap is only one-third of that required to produce a tonne of steel from iron ore using the blast furnace and basic oxygen furnace methods of the integrated steel producer [3]. Because of the energy efficiency, high productivity and comparatively low capital cost of new modern EAFs, the per cent of steel produced in world by this process [4] reached 28.8% in 2018.

EAF process is more popular due to the following:

1. It requires less capital investment.
2. It requires less installation period.
3. Any grade and superior quality of steels can be produced from scrap/sponge iron.
4. Increasing availability of alternate iron sources, like sponge iron, hot metal, etc.
5. Alloying elements can be added directly to the furnace with minimum loss, and hence, composition of bath can be controlled easily.
6. Temperature of the molten bath can be controlled within narrow limits.
7. Improvement in the operation, control, efficiency and high metallic yield and
8. It has readymade market and easily available of main raw material from local market.

Capacities of EAFs vary between 1 and 150 t, but 5 t or 10 t are more popular for alloy steel production. Main raw material for electric arc furnace (EAF) is steel scrap, which is 50–60% of production cost and 60–80% cast of raw materials. Nowadays, there is shortage of scrap due to increase of continuous casting yield; scrap is substituted by sponge iron/DRI/HBI partially or fully.

16.2.1 Main Parts of EAF

An electric arc furnace used for steelmaking consists of a refractory-lined vessel, usually water-cooled wall in larger sizes, covered with a refractory roof and through which three graphite electrodes introduce into the furnace. The furnace is primarily consisting of the main parts as follows:

1. Furnace body, i.e. shell, hearth, wall and spout, door, etc.,
2. Roof and roof-lifting arrangements,
3. Electrodes, their holders and supports,
4. Gears for furnace movements and
5. Electrical equipments, i.e. the transformer, cables, electrode control mechanism, etc.

16.2.1.1 Furnace Body

It consists of shell, sidewall, tap hole, spout and door.

(a) Furnace shell: The furnace shell must have sufficient strength to hold the weight of the metal and the lining, as well as withstand the pressure of the lining which expands on heating. The outer shell should not be heated to more than 100–150 °C during operation. The shell is made by welded carbon steel sheets of 15–50 mm thick, depending on the furnace capacity. The top of the shell is open and is covered with a tight fitting but removable dome-shaped roof structure. Shape of the shell determines shape of reaction chamber. Hearth is the bottom portion of the furnace. Next to the shell is a layer of fireclay bricks at the bottom and is followed by dolomite or magnesite bricks. The working hearth is made by ramming tarred magnesite. Hearth can be cylindrical, conical or cylindro-conical (as shown in Fig. 16.5). The hearth may be hemispherical in shape, or in an eccentric bottom tapping furnace. The hearth has the shape of a halved egg (as shown in Fig. 16.6a). The shell bottom is usually spherical which ensures maximum strength. The spout for tapping the metal is welded with the bottom shell, and door is situated directly opposite to the spout. Other openings are provided in the furnace shell for oxygen, lime and carbon injections, oxy-fuel burners, etc.

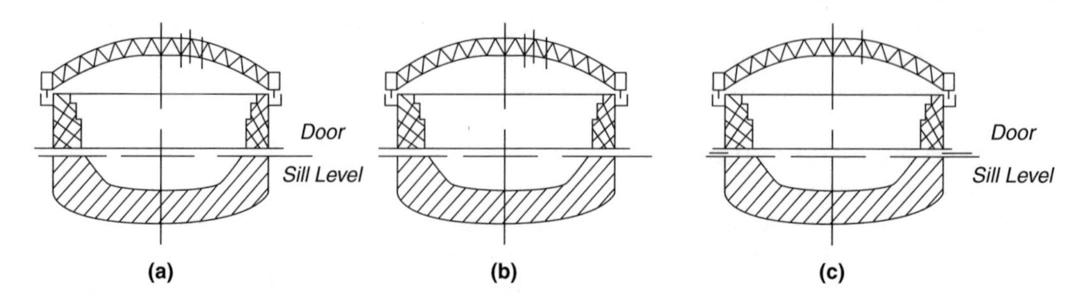

Fig. 16.5 Various forms of shell and hearth of EAF [5]: **a** cylindrical, **b** conical, **c** cylindro-conical (reproduced with permission from Mirtitles.org)

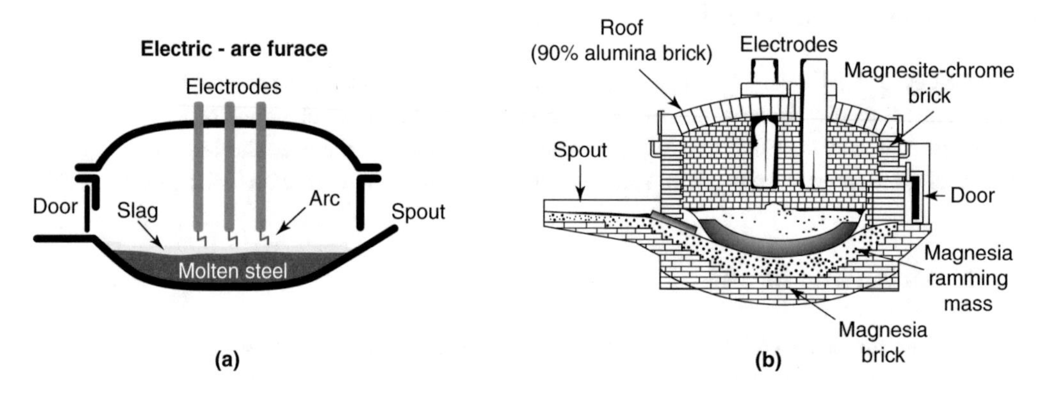

Fig. 16.6 Basic layout of an EAF [6]. **a** Halved egg shape hearth, **b** schematic diagram. (a) Permission from Dr. Dmitri Kopeliovich)

(b) Sidewall: The sidewall is the cylindrical part of the shell and extends vertically from the slag line to the top of the shell. Height of the wall is mainly governed by the nature of the raw materials and the practice of refining. For bulky scrap (i.e. lighter), or oxygen lancing, the height of the wall should be more. Dolomite, magnesite or chrome–magnesite are used for sidewall. Magnesite is better if the furnace is worked hard. Thickness of lining at the sidewall is usually in the range of 35–50 mm.

(c) Tap hole: Gap is left in the wall during lining of wall for putting of tap hole. A round shape refractory is inserted, and the space left around is rammed to make the tap hole.

(d) Spout: Metal is tapping from the furnace through the spout. The spout opening is cut out of the shell, and the spout frame attachment steel plate is welded onto the shell. Tapping spout is a U-shaped chute lined inside with fireclay brick. The rest is rammed along with the hearth as a part of the hearth. It is attached to furnace shell at an angle of 10–12°. New designs and changing the locations of tap holes may reduce the physical and mechanical problems encountered with the spout. One of these alternatives is the side slide gate spout, and another is eccentric bottom tapping with the use of a slide gate (Fig. 16.7).

(e) Door: The shell has door openings cut into it and a cooling frame welded around the door. The door-lifting supports are also welded on the shell. Charging door for smaller furnaces is diametrically opposite to spout. Any additions during refining are made through this door. The size of charging door should be so chosen that free access for visual inspection and fettling of the hearth and free entry of charging boxes should be easily done. The door width is about 0.25 of the diameter of reaction chamber, and height is 0.8 of the width. Door is lined with basic bricks and

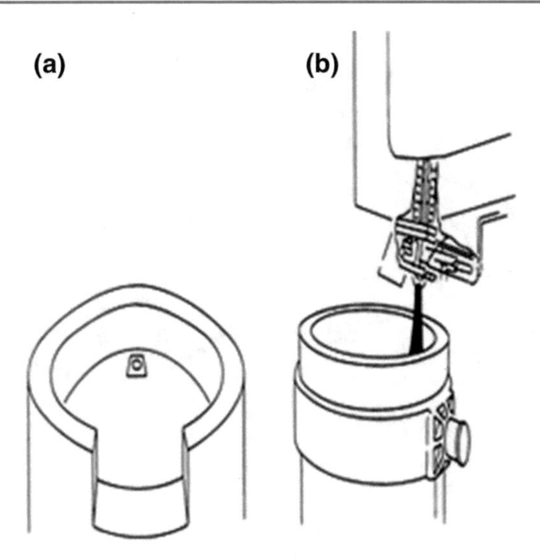

Fig. 16.7 **a** Eccentric bottom tap hole furnace, **b** with slide gate attached to furnace bottom [7]

Table 16.1 General dimensions of EAF [8]

Furnace capacity, t	5	10	20	80	150
Shell dia., m	2.7	3.3	4.5	6.0	7.2
Depth of hearth, m	0.45	0.52	0.72	1.0	1.07
Thickness of hearth, m	0.45	0.52	0.72	0.72	0.73
Roof thickness, m	0.22	0.30	0.30	0.35	0.35

water cooled. This is the main door and used for slag–discharge. Door should be at a proper level to help remove the slag by tilting the furnace on this side. Charging (i.e. scrap) for small furnaces is also done manually through the door. Bigger furnaces have another additional door at right angles to this door. Table 16.1 shows dimensions of EAF.

16.2.1.2 Roof

The roof is a domed shape and rises about one per cent in span towards the centre. Earlier, the roof has three holes located symmetrically to allow insertion of the electrodes, but nowadays there are five holes on the roof; 4th hole for exhaust gases to control the air pollution, and 5th hole for continuous charging of sponge iron/DRI as charging material (as shown in Fig. 16.8). The holes are made from ring-type bricks. The roof is built by high alumina (70–80%) bricks; thickness varies from 25 to 45 cm. Silica bricks (i.e. fireclay bricks) are not used at all, due to formation of $FeO.SiO_2$, by fumes of FeO. The melting point of $FeO.SiO_2$ is very low (<1200 °C), so it melts easily by radiation of the bath and comes out drop by drop to erode the roof for which life of roof will be decreased.

It is a general practice nowadays, and charging is done from the top of the furnace. This requires the roof to be removed from the furnace. There are three types of roof for EAF (Fig. 16.9):

(i) *Fixed roof*: For small furnace, charging is done manually through the door.
(ii) *Swing roof*: Roof along with the electrodes swings clearly off the body of the furnace to allow charging from the top. This type of roof is very common and popular.

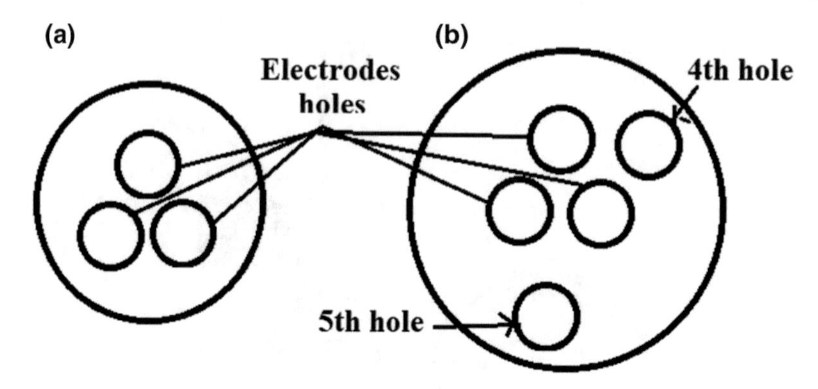

Fig. 16.8 EAF roof: **a** old type, **b** new type

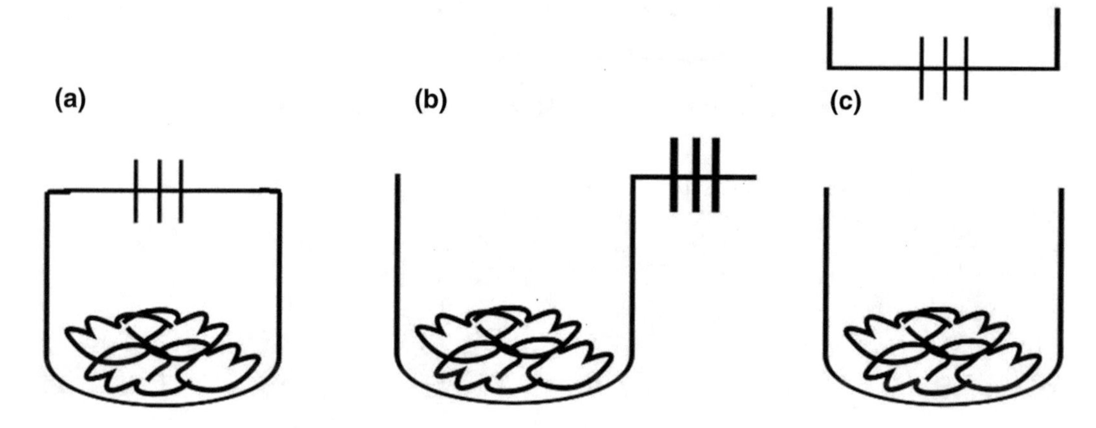

Fig. 16.9 Types of roofs: **a** fixed roof, **b** swing roof, **c** lifting roof

(iii) *Lifting roof*: Roof is lifted a little, and the furnace body moves to one side to facilitate charging from the top.

16.2.1.3 Electrode

For a typical alternating current (AC) electric arc furnace has three electrodes, and for direct current (DC) electric arc furnace has one electrode. Electrodes are round in cross section and typically in segments with threaded couplings, so that as the electrodes wear, new segments can be added easily with the help of nipple. The electrodes are made from either of carbon or graphite that can carry current at high density. Graphite has better electrical conductivity than carbon. Diameters of electrodes are varying from almost a few cm to nearly 100–110 cm, depending upon the furnace capacity; and each piece is about 1–3 m long. The electrodes are supported on three separate arms and held to the electrode arm by a spring-clamped, pneumatically released electrode holder. The electrode arms support the electrodes, holders, bus bars and transformer power cables. The electrodes are raised and lowered by manual or automatic controls. These controls can be operated manually when the bottoms of the electrodes are raised to the roof or when electrodes are shifted. The mechanism of raising and lowering the electrodes is shown in Fig. 16.10. For efficient melting, the arc must be stabilized to supply uniform amount of energy. The electrodes should be vertical from all sightings around the furnace. The electrode is automatically adjusted to maintain a stable arc. Each electrode is provided

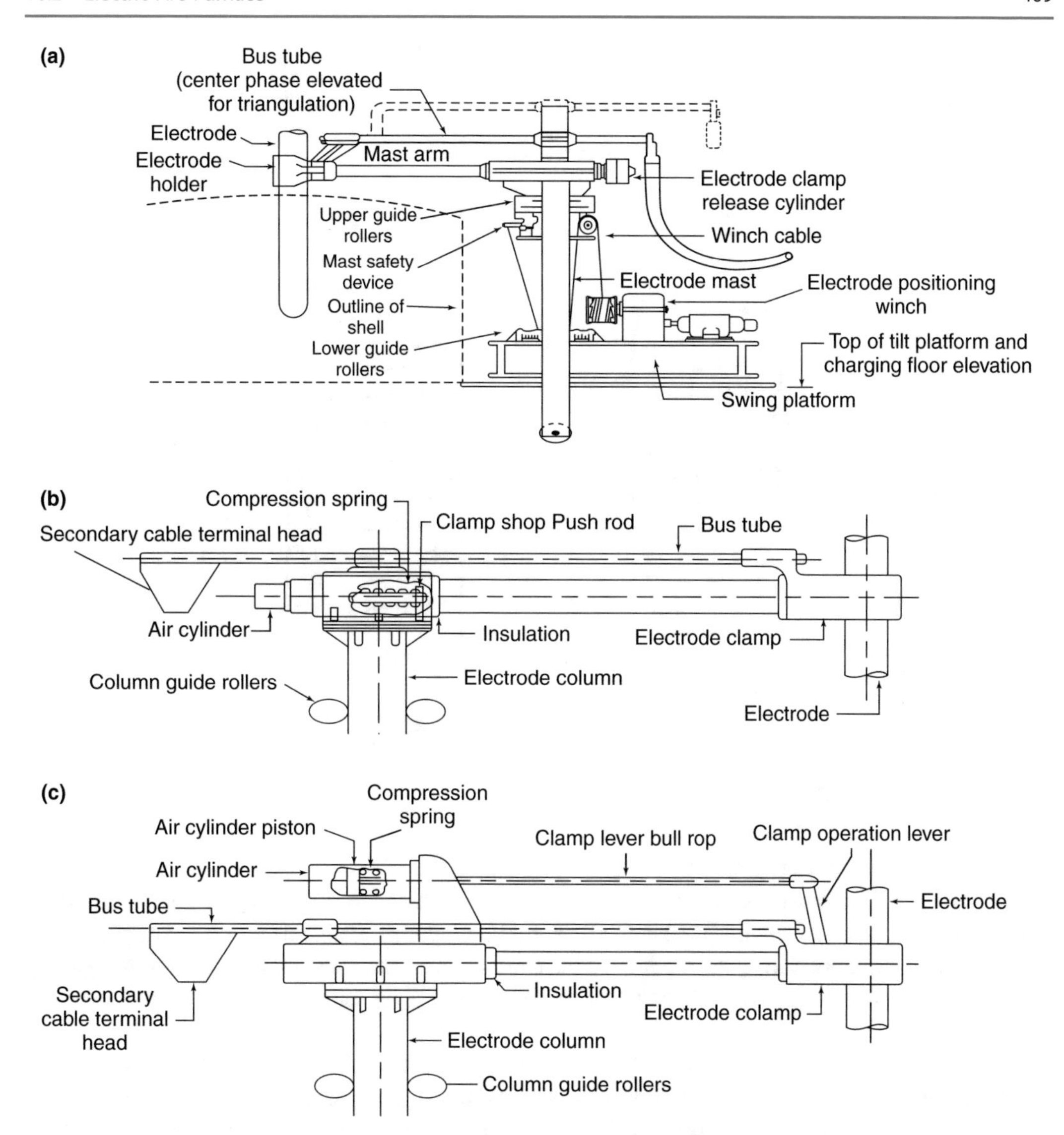

Fig. 16.10 Apparatuses for raising and lowering electrodes [7]. **a** Arrangement of electrode, **b** electrode arm arrangement with power-operated electrode holder and **c** arrangement of electrode arm using power-operated electrode holders using piston

with an independent system of control. Electrodes are costly, and consumption is 5–6 kg/t of liquid steel production. Electrodes are joined with the help of nipple (as shown in Fig. 16.11).

UHP furnace carries heavy current through electrodes and hence needs water-cooled electrode holders. For larger-capacity furnaces, hollow electrodes have been recommended for sponge iron/HBI charging.

16.2.1.4 Gears for Furnace Body Movements

The furnace body needs to be tilted nearly 45° on the tapping side and 15° on the slag–discharge side. The tilting gear is operated either by hydraulic or electrically. The hydraulic tilting arrangement is smooth. Some locking arrangement is necessary for the horizontal position of the furnace.

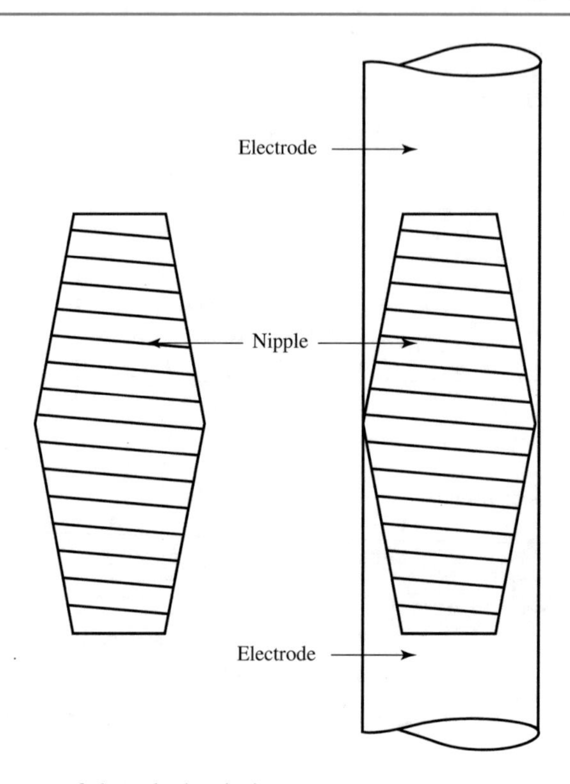

Fig. 16.11 Joining arrangement of electrodes by nipple

16.2.1.5 Electrical Equipment

The electric arc furnace of secondary electrical circuit consists of several components including the delta closure, power cables and bus bars. The regulating system maintains approximately constant current and power input during the melting of the charge. The most arms holding the electrodes carry heavy bus bars, which may be hollow water-cooled copper pipes carrying current to the electrode holders.

Transformer: large transformers are required to run electric arc furnaces. Heavy water-cooled cables connect the bus bars/arms with the transformer located adjacent to the furnace. To protect the transformer from heat, it is installed in a vault. A mid-sized modern steelmaking furnace would have a transformer rated about 60 MVA, with a secondary voltage between 400 and 900 volts and secondary current more than 44 kA. During melting, more power is required than during refining. The transformer capacity, which is designed to suit melting requirements, is usually 170–650 kVA/t furnace capacity. The most powerful furnaces approach a power to tapped steel weight ratio of 1 MVA/t. Many of the furnaces are using foamy slag practice by injecting carbon powder into the furnace. Globally today furnaces with a power to tapped steel weight of 1.5 MVA/t are also available. Power of transformer can be found by means of energy balance for a given capacity and tap-to-tap time. Table 16.2 shows some recommendations on transformer power for furnaces of different size, which are based on experiences in many countries.

16.2.1.6 Power Rating and Consumption

The arc voltage as well as power factor decreases with increasing current in the circuit. Power consumption in the circuit and in the arc increases with increase in the current in the circuit up to certain limit and then decreases. The furnace is best operated at the amperage where maximum

Table 16.2 Recommended transformer power for furnaces [5]

Furnace capacity (t) →	25	50	75	100	150	200	250	400
Transformer power for alloy steel (MVA)	15–18	20–25	–	30–35	–	55–70	–	–
Transformer power for carbon steel (MVA)	18–22	28–32	30–45	40–50	45–60	60–80	90	120
Transformer power for super-powerful furnace (MVA)	–	40	–	60	–	125	–	200

possible power is available in the arc at maximum possible power factor as shown in Fig. 16.12. Electrode ratings should also be considered in deciding the working conditions. Electrode is fully utilized by drawing maximum allowable current through it at maximum possible voltage. Excess current unnecessarily overheats the electrode and lower current results in poor utilization of the electrode.

The power consumption for melting the charge in small furnace is about 600 kWh/t, and it falls to 450 kWh/t for big furnaces. Additional power is required during refining and which varies considerably with practice between 150 and 400 kWh/t.

Total power input is spent approximately in various ways as follows:

1. To the charge for melting and refining: 50–60%,
2. Electrical losses: 8–12%,
3. Water-cooling losses: 3–5%,
4. Radiation losses: 20–30%.

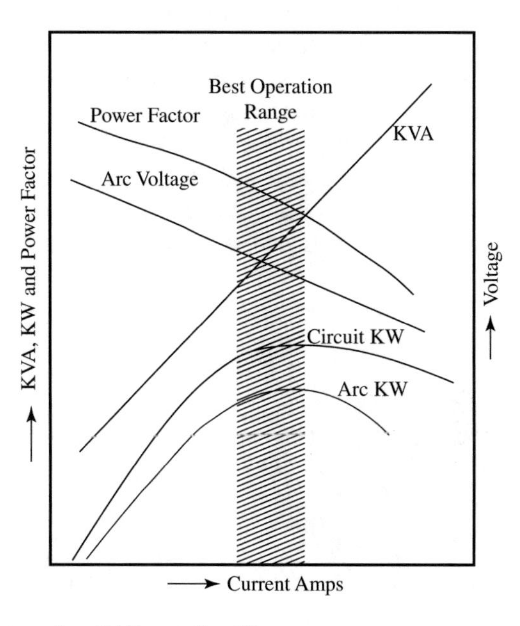

Fig. 16.12 Electrical parameters of an EAF operation [8]

16.2.2 Process

EAF process may be divided into the following stages:

1. Fettling of the furnace,
2. Charging,
3. Melting,
4. Refining,
5. Finishing and tapping of single slag heat/slag-off and making reducing slag,
6. Finishing and tapping of double slag heat.

16.2.2.1 Fettling of the Furnace

After tapping of earlier heat, slag is completely taken out, and lining is inspected by naked eye. (i) Chemically active slag erodes the furnace banks at the slag line due to high temperature. (ii) Lining of hearth may be damaged due to charging of heavy pieces of scrap. (iii) Hearth lining may suffer due to lumps of iron ore charged during melting period. (iv) Local ramming mass may be wash off during boiling of the bath.

Fettling of the furnace means repair of the furnace damage before charging. Eroded portion usually the slag and slag-metal interface line, tap hole, spout and damaged area of the hearth are to be repaired in hot condition by granular dolomite or magnesite refractory materials.

16.2.2.2 Charging

Charging is invariably carried out from the top of the big furnace. A drop bottom basket/pan is used for charging. Roof along with the electrodes swings clearly off the body to allow scrap, lime, spar which are charged by the basket (as shown in Fig. 16.13). Quick meltdown of the charge depends to a large extent on proper proportioning of light, medium and heavy scrap and on the sequence of charging. The basket is filled with 20% light, 40% medium and 40% heavy scrap in the sequence of light, medium, heavy scrap. First basket should be unloaded from as low height as possible to diminish the impact force on the hearth. Again, light scrap along with lime (2–3% as flux) and spar (to increase fluidity of slag) are charged at the top. During melting lump iron ore/mill scale (6%) is also is also charged. Since light scrap and small pieces are filled up the spaces between larger ones, this ensures better electric conductivity of the charge and therefore shorter time to melt down.

Sometime broken electrode/coke is also added as extra carbon to the furnace during charging, so that good carbon boil should be occurred. Hence, opening carbon content in the charge should be

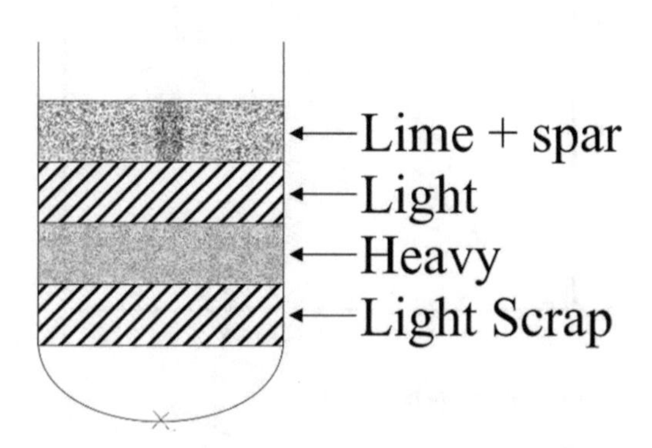

Fig. 16.13 Drop bottom basket/pan

(a) 0.3% above the lower limit for high carbon steels, (b) 0.4% above the lower limit for medium carbon steels and (c) 0.5% above the lower limit for low carbon steels for making carbon boil to remove the inclusions from the bath. De-carburization (i.e. oxidization of carbon) gives boiling action of the bath due to bubbling of CO gas through the molten bath, this is known as carbon boil. The carbon boil helps the following main roles in cleanliness of metal:

(i) Good slag-metal contact to speed up the reactions,
(ii) Good heat transfer,
(iii) Homogenization of the bath,
(iv) Cleanliness of the bath improve by bringing the retained oxides (i.e. inclusions) to the slag phase,
(v) Dissolved hydrogen and nitrogen gases diffuse into the CO bubbles and are flushed out of the molten bath.

16.2.2.3 Melting

After complete charging, roof is replaced again in position and electrodes are lowered down manually and current is supplied to them; arc is struck, then electrodes are put on automatic control. Due to the electric arcing, high temperature is created which melts the charge below the electrodes. Pool of liquid metal is formed under the electrodes by cutting wells in the charge; diameter of liquid well is 30–40% larger than electrode diameter (Fig. 16.14). Important factor is the electric conditions of melting: (i) melting occurs near the electric arc; (ii) greater the voltage used, the larger the arc and correspondingly the lower the current, i.e. lower will be the electric losses and higher the electric efficiency; and (iii) electric losses are proportional to the square of the current. Most advantageous operation is with highest voltage taken from the transformer. A larger arc radiates over a large area. But this arc cannot do any harm to the furnace lining, since it is covered by solid charge, and lining is cooled during charging; hence, lining is not overheated. During lowering of the electrodes, the addition of lime and mill scale/iron ore helps to form slag easily on the metal surface. Slag covers the molten metal to prevent cooling, oxidation of metal, entrapment gases to the metal and carburization of metal by electrodes. The composition of slag is 30–45% CaO, 10–15% SiO_2, 13–20% FeO, 10–15% MnO; basicity is 1.7–3.0. Maximum power is consumed during this melting period.

In the AC furnace, the arc is struck between the electrodes (3 in numbers) and the charge. The current passes from the electrodes through the charge and returns via a second arc to the electrodes. Initially, if the charge is cold and largely scraps, there is inherent arc length variability and accompanying noise which lessens as the charge compacts, arc length stabilizes and melting begins. The

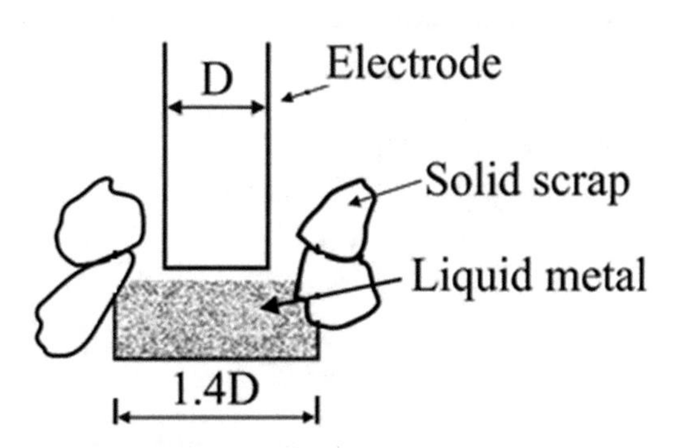

Fig. 16.14 Liquid well formation under the electrode

passage of the current through the charge creates resistance heating, and once a molten pool of steel is formed, the current path is from the electrode to the molten bath.

Another important factor is the electric conditions of melting. The melting occurs near the electric arcs. The greater the voltage used, the longer the arcs, and correspondingly, the lower the current, the lower will be the electric losses (which are proportional to the square of the current) and the higher will be the electric efficiency. A longer arc radiates over a greater area. Thus, the most advantageous operation is with the highest voltage tap of the transformer. Long powerful arcs cannot harm to the lining since this is covered by solid charge. After being cooled during charging, the furnace lining can absorb a considerable amount of heat without overheating.

The pool of molten metal below the electrodes is heated to a high temperature, and at the same time, the entire charge and the furnace lining are being heated. As the line of molten metal rises, the electrodes are lifted by automatic controls to maintain a constant arc length. The time of the melting period depends on: (a) power of the transformer, (b) furnace parameters. The time of melting is shorter with a high useful power of the transformer and with low heat losses through the furnace as well as by blowing oxygen during melting and after partial meltdown. The oxidation of Fe, Si, C and other elements releases a large amount of heat, so by blowing oxygen can shorten the melting time by 10–20 min.

During melting, the silicon, manganese and phosphorus of the charge are oxidized. One way to intensify the process of electric steelmaking is to carry out the de-phosphorization also at the same time of melting. De-phosphorization requires the slag of specific basicity (2.0–3.0), and for that purpose, lime and iron ore/mill scale are added to the furnace together along with the charge. At the end of melting, first to control is phosphorous. De-phosphorization can be easily done due to slag oxidizing and basicity of the order of 2.2–2.5. After complete melting of scrap, a sample of metal is taken for rapid analysis for C, Mn, Si, P, S.

At the second stage of the melting period, part of the slag is allowed to flow from the furnace, and a new slag is formed by adding lime and ore. For a given oxidizability of the slag, the content of phosphorus in metal samples taken after meltdown diminishes with increasing slag basicity and decreasing temperature (as shown in Fig. 16.15). To obtain the same content of phosphorus in the metal at a higher temperature, the basicity of slag should be increased.

16.2.2.4 Refining

Sampling of metal is completed during melting period; after that refining period begins. Now the furnace transformer is switched over to a medium voltage, since high power is no longer needed, because the wall and roof lining have been heated to a high temperature during melting period.

Carbon and phosphorous are adjusted in this period. Since phosphorous has tendency to reversion at high temperature, so before bath pickup high temperature, phosphorous should be fully eliminated by increasing the basicity of slag and decreasing temperature. To lower down the temperature, some

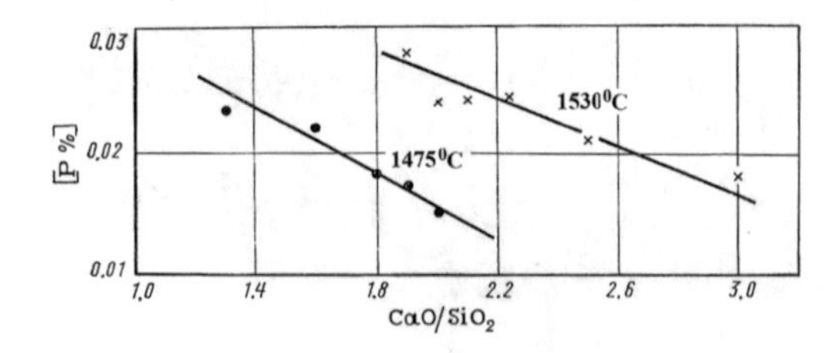

Fig. 16.15 Effect of slag basicity and temperature on phosphorus content in the metal [5] (reproduced with permission from Mirtitles.org)

scrap is charged. By the time all the phosphorous is eliminated, carbon boil should have attained vigorously to clean up the bath. Finishing and tapping of single slag heat: alloying additions for Mn, Cr are done, to lower oxygen content of the bath, to protect them to oxidized and arrest the carbon content by catch carbon technique. Final de-oxidation is done at the ladle by addition of Fe–Si and Al.

16.2.2.5 Slag-off and Making Reducing Slag

The main function of slag is as follows:

(i) Traditionally to absorb de-oxidation products and inclusions in ladle,
(ii) Covering the electrodes from oxidation and protecting the refractories from thermal radiation,
(iii) Removing phosphorus and sulphur from the molten bath,
(iv) Protecting the metal from oxidation, as well as prevention of nitrogen and hydrogen pickup,
(v) Minimizing heat loss and hence increase thermal efficiency of the electric arc furnace.

Addition of iron ore/mill scale to the bath during melting causes the bath to boil intensively due to CO gas formation. The foam formation of slag is made due to the gas evolution, and the level of slag line thus rises.

$$Fe_2O_3(s) + [Fe] = 3(FeO) \tag{16.1}$$

$$[C] + (FeO) = [Fe] + \{CO\} \tag{16.2}$$

The temperature of the bath is raised quickly to start the boil, and iron ore/mill scale is added. The slag should not fall below a (CaO/SiO_2) ratio of 2.5 in order that there should be enough free lime in the slag to fix P_2O_5 as calcium phosphate. Many operators prefer to work work with 3 to 1 ratio. On the other hand, the slag should not be too limey, as this will affect its consistency. The FeO content of the oxidizing slag is normally found to be in the range of 14–20%.

By tilting at an angle of 4–6° to the furnace towards the door side, 60–70% of initial (i.e. oxidizing) slag is skimmed off from the furnace without switching off the current. Alloy steels are produced by double slag practice to achieve the desired level of quality of steel with respect to cleanliness. After removal of oxidizing slag, bath should be carburized by addition of crushed electrodes or coke to form reducing period. Aim of reducing period: (i) de-oxidation of the metal, (ii) formation of reducing slag, (iii) removal of sulphur and (iv) adjustment of steel composition and control of bath temperature.

(i) De-oxidation of the metal: Dissolved oxygen in metal is removed. There are two methods: (a) adding de-oxidants directly to the metal, (b) adding them over the slag to deoxidize the slag.

  (a) Lumpy de-oxidants (i.e. Fe–Mn, Fe–Si, Al, etc.) are added to the bath. Reaction occurs at a depth in the metal bath (i.e. *precipitation de-oxidation*) which depends on density and size of the de-oxidant lumps. That takes little time, but products of reaction (SiO_2, MnO, Al_2O_3) can remain in the metal phase as non-metallic inclusions.
  (b) It involves de-oxidation of the metal via slag (i.e. *diffusion de-oxidation*) which based on the law of distribution of oxygen between two phases, i.e. (FeO)/[FeO] is constant at a given temperature and slag basicity. If the amount of FeO in slag phase is lowered down, corresponding amount of oxygen should pass from metal phase to slag phase to maintain constant of (FeO)/[FeO]. In EAF, it is possible to form slag with only 0.5% FeO or even less, therefore, to produce metal with less than 0.005% oxygen is possible. In this diffusion de-oxidation method, reaction takes place at slag phase or slag-metal interface, which

decreases the contamination of metal by inclusions (i.e. products of reaction, SiO_2, MnO, Al_2O_3, etc.). Hence, quality of metal will be good. Main disadvantages are (i) it takes much more time (i.e. slow process), (ii) metal should be hold for a considerable long time, and hence, temperature may go down and pickup of gases from the surrounding to the metal. Most common de-oxidants in EAF are C, CaC_2, Fe–Mn, Fe–Si, Al, Fe–Ti and complex alloys (e.g. Si–Mn, Si–Ca, Si–Al–Mn, etc.). When carbon acts as de-oxidizer, it not only removes oxygen from slag phase, but it simultaneously diffuses to metal phase and carburized the metal at a rate of 0.02–0.10% C/h. To avoid C–O_2 reaction in the mould, more powerful de-oxidants (such as Si or Al) are used to reduce oxygen below its equilibrium value ($\%O_{eq}$) with carbon [since $\Delta O_{excess} = (\%O_{actual} - \%O_{eqilibrium})$].

(ii) Formation of reducing slag: Reducing slag is formed by adding fresh charge of lime and spar in which a little sand may be added to easy formation of slag. By adding coke (lime: coke = 6:1–12:1), strongly reducing carbide slag is formed:

$$(CaO) + 3C(s) = (CaC_2) + \{CO\} \quad (16.3)$$

Carbide slag breaks into bright grey powder on cooling. In open air, calcium carbide is decomposed by atmospheric moisture or putting carbide slag in water to form acetylene (C_2H_2) gas, which has plangent smell. Due to plangent smell, steelmaker can easily understand formation of carbide slag.

$$(CaC_2) + 2\{H_2O\} = Ca(OH)_2 + \{C_2H_2\} \quad (16.4)$$

(iii) De-sulphurization: There is no other process of steelmaking which can compare with a basic EAF in its ability to remove sulphur from metal. Sulphur can be introduced from the charging materials like lime, spar and ferro-alloys. It can be also introduced during the reducing period of the bath by coke and lime. For low sulphur steelmaking, petroleum coke (contain low sulphur) can be used.
Production of plain carbon steel by EAF contains only 0.012–0.021% S. Removal of sulphur from metal in EAF is affected by: (a) degree of de-oxidation of the metal and slag, (b) carbon content of the metal, (c) slag basicity, (d) temperature, (e) stirring action of metal and slag, (f) holding time of metal under reducing slag and (g) content of sulphur in lime and other feed materials for reducing slag. Amount of sulphur in the metal begins to drop appreciable only after the metal and slag have been de-oxidized to a certain degree, i.e. rate of de-sulphurization increases with decreasing amount of oxygen in slag. The slag basicity can also affect the rate and degree of de-sulphurization. Higher basicity ensures more complete de-sulphurization and greater rate of sulphur transfer from metal to slag phase. High heating of the metal and slag, and active stirring of the bath help good sulphur removal.

$$3(FeS) + (CaC_2) + 2(CaO) = 3[Fe] + 3(CaS) + 2\{CO\} \quad (16.5)$$

Although the most favourable conditions for sulphur removal exist in the reducing period, it is possible to remove sulphur partially during the oxidizing period. This requires a high temperature and high basicity (2.5–3.0) of slag. Figure 16.16 shows how the coefficient of distribution of sulphur between metal and slag depends on slag basicity in the oxidizing period. Figure 16.17 demonstrates the effect that de-oxidation of the slag has on the rate of de-sulphurization in melts for structural steel. It can be seen that the rate of de-sulphurization

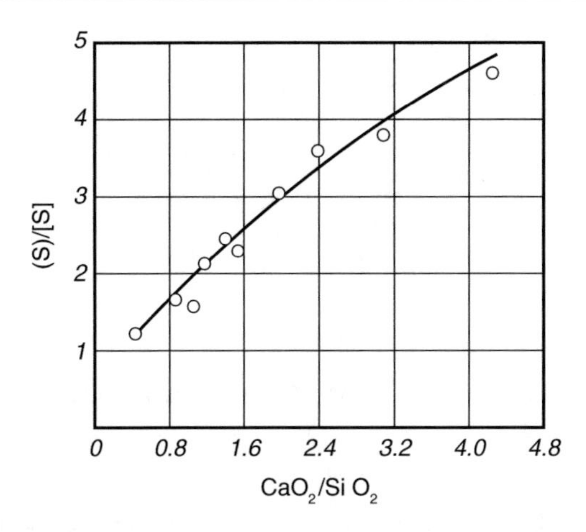

Fig. 16.16 Effect of slag basicity on sulphur distribution [5] (reproduced with permission from Mirtitles.org)

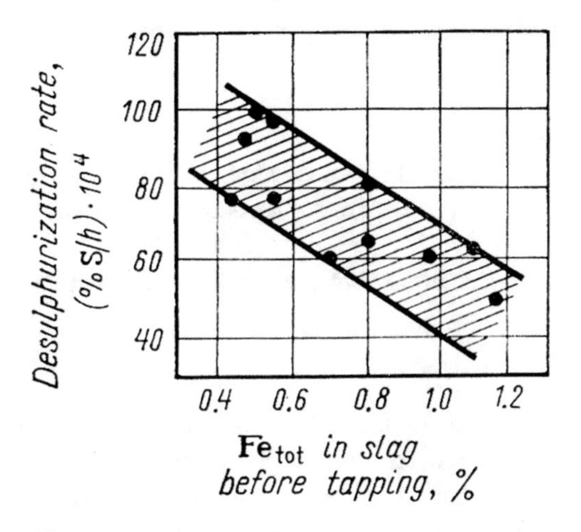

Fig. 16.17 Effect of de-oxidation of the slag on the rate of de-sulphurization in melts [5] (reproduced with permission from Mirtitles.org)

increases with a diminishing amount of oxygen of the metal. Slag basicity can affect the rate and degree of de-sulphurization. Higher basicity ensures more complete de-sulphurization and a greater rate of passage of sulphur from metal to slag phase.

Slag, without carbon or little carbon, which is white colour, is known as *lime slag*. For production of low C (<0.15%) steel production, carbon is not added; instead of that Fe–Si, Al are added as de-oxidizer. Slag, with high per cent of carbon, which is grey colour, is known as *carbide slag*. Carbide slag breaks into bright grey powder on cooling. Chemical compositions of carbide and lime slags are shown in Table 16.3.

(iv) Adjustment of steel composition and control of bath temperature: Again, final sampling of metal is done to adjust the final composition of the metal and control of bath temperature. If the bath temperature is high, then scrap is added to lower down the temperature; and for low-temperature, electric arcing is done to increase the temperature. Steel composition can be adjusted by addition of ferro-alloys.

Table 16.3 Chemical compositions of carbide and lime slags

	Carbide slag	Lime slag
CaO, %	65–70	55–60
SiO_2, %	20–25	25–30
FeO, %	0.5 (max)	1.0 (max)
MgO, %	5–10	5–10

16.2.2.6 Energy Balance

A typical energy balance for a modern EAF is shown as follows [3]:

Input:

- Depending upon the melting operation, about 60–65% of the total energy is coming from electrical energy,
- The remaining is chemical energy arising from the oxidation of elements such as carbon, iron and silicon and the burning of fuel (gas or oil) in oxy-fuel burners.

Output:

- About 53% of the total energy left the furnace along with the liquid steel.
- While the remaining is lost to the slag (10%), waste gas (20%) and cooling losses (17%).

The 20% normally leaving the furnace in the waste gas represents about 130 kWh/t of steel produced.

16.2.2.7 GHG Emissions

The greenhouse gas (GHG) emissions, i.e. CO_2 emissions, are generated during the melting and refining process when carbon is removed from the charge material and carbon electrodes as CO and CO_2 gases. These emissions are captured and sent to a bag house for removal of dust before discharge into the atmosphere. The AOD vessels are small contributors to CO_2 emissions. The CO_2 emissions estimate of 4.6 Mt of CO_2 for EAF is based on the IPCC Guidelines emission factor of 0.08 t of CO_2/t of steel [10]. Emissions of CO_2 are also generated from the use of oxy-fuel burners in EAF. These burners increase the effective capacity of the EAF by increasing the speed of the melting and reducing the consumption of electricity and electrode material, which reduces energy-related GHG emissions. Oxy-fuel burners also increase heat transfer while reducing heat losses and reduce tap-to-tap time. These burners are often designed to minimize the increase in NOx emissions that is a known by-product of the technology by deliberately operating the burners at less than their maximum combustion efficiency; however, this practice increases CO emissions to some extent but in turn lowers CO_2 emissions.

The concentration of solid particles in EAF ventilation air will range from 2 to 15 kg/t of metal poured. The emission rate is 15 kg fume per tonne charged. Iron oxide is in the form of Fe_2O_3. The type of melt and the quality of the scrap charged to the furnace will influence the composition of the fume emitted from the furnace. The quantity of fume will depend more on the melting rate, quality of scrap, the vapour pressure of minor constituents and the degree of oxidation affected by air infiltration and oxygen lancing. Table 16.4 shows typical data on charge and dust emission of a 50–75 t EAF.

Table 16.4 Data on charge and dust emission of EAF [11]

Typical charge, %		Typical dust emission, %	
Home scrap	20	Silica	2
No. 2 baled scrap	25	Alumina	3
Auto-scrap	43	Iron	25
Turning and boring scrap	7	Lime	6
Fluxes, carbon and ore	5	MgO	2
		ZnO	37
		MnO	4
		CuO	0.2
		P_2O_3	0.2
		SO_3	3

16.2.3 Design of Furnace

The calculations for design of different parts of an electric arc furnace are based [12] in Fig. 16.18.

- **Shape and dimensions of the bath**

The hearth is usually conical-spherical with the banks inclined at an angle of 45°. This form ensures quick melting.

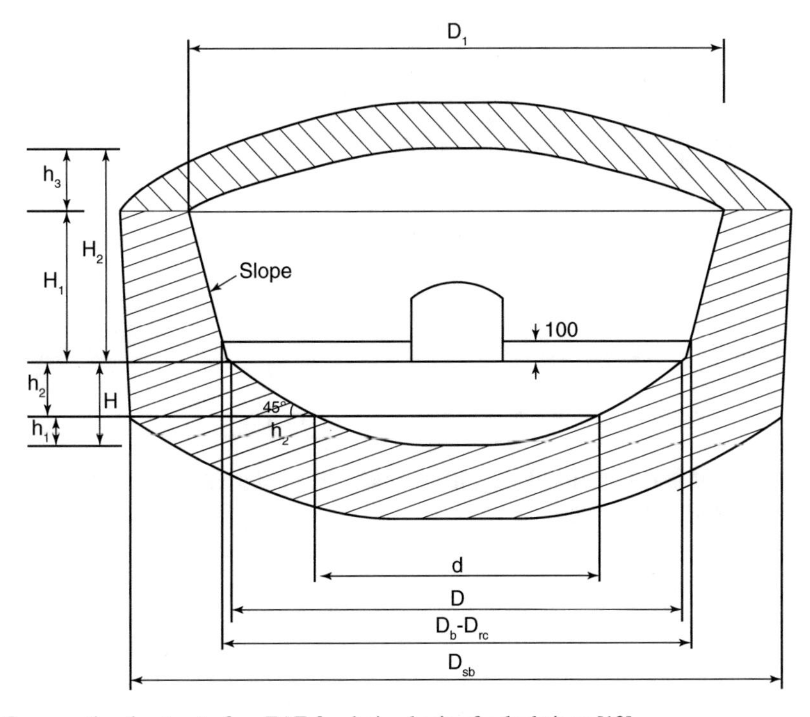

Fig. 16.18 Cross-sectional cut-out of an EAF for design basis of calculations [12]

Since density of steel at melting point is 6.88×10^3 kg/m^3, hence one tonne of molten steel occupies 0.145 m^3, i.e. $V_m = 0.145$ m^3, where V_m is volume of one tonne of liquid steel.

The volume of slag, V_{slag} = 15% volume of liquid steel = 0.15 V_m.

Therefore, total volume of the liquid bath (V_b) will be:

$$V_b = V_m + V_{slag} \tag{16.6}$$

- **Height of reaction chamber**

The diameter-to-depth ratio of the bath operating by the basic process is usually taken as:

$$D = 5H \tag{16.7}$$

where D = bath diameter, m; and H = total depth of the bath, m.

In existing furnaces, the height (h_1) of the spherical portion is roughly 1/5 of the total bath depth:

$$h_1 = 0.2H \tag{16.8}$$

and the depth of the conical portion:

$$h_2 = 0.8H \tag{16.9}$$

Again, the total volume of the bath, V_b, is the sum of the volumes of the truncated cone (V_c) and that of the spherical segment (V_{sp}).

By virtue of the relationships given above:

$$V_b = V_c + V_{sp} = \left\{\frac{1}{3} \cdot \frac{\pi D^2}{4} \cdot h_2\right\} + \left\{\frac{4}{3} \cdot \frac{\pi D^2}{4} \cdot h_1\right\} = \pi \cdot \frac{D^2}{12} \cdot \{h_2 + 4h_1\}$$
$$= \pi \cdot \frac{D^2}{12} \cdot \{1.6H\} = \pi \cdot \frac{D^2}{12} \cdot \left\{1.6\left(\frac{D}{5}\right)\right\}$$

Therefore,

$$V_b = 0.0837D^3\left(\text{m}^3\right) \tag{16.10}$$

Therefore, from Eq. (16.6):

$$0.0837D^3 = V_m + V_{slag}\left(\text{m}^3\right) \tag{16.11}$$

From the volume of liquid steel and slag, the diameter of the furnace (D) can be easily found out.

Diameter of spherical bottom,

$$d = D - 2h_2 \tag{16.12}$$

- **Dimensions of the reaction chamber**

The banks of a furnace are usually made 0.1–0.2 m above the door sill level or the bath surface, to ensure that the slag does not contact the brickwork or reach the joint between the wall blocks and banks.

The diameter of the reaction chamber is:

$$D_{rc} = D + 0.2(\text{m}) \tag{16.13}$$

It is recommended that the following relationship be used in determining the height of the reaction chamber (H_1),

$$H_1 = (0.4 \text{ to } 0.6)D_{rc} \tag{16.14}$$

For large-capacity furnaces, the lower value (i.e. 0.4) is taking.

The height of the furnace roof (h_3) is:

$$h_3 = 0.15D_1 \tag{16.15}$$

where D_1 is the diameter of the furnace roof:

$$D_1 = D_{rc} + 2h_1 \tag{16.16}$$

D_1 is roughly taken 1.0 m wider than D_{rc}.

i.e.

$$\begin{aligned} D_1 &= D_{rc} + 1.0 \\ h_3 &= 0.15D_1 = 0.15(D_{rc} + 1.0) \end{aligned} \tag{16.17}$$

The total height of the roof above the bath level is given as:

$$H_2 = H_1 + h_3 \tag{16.18}$$

$$\begin{aligned} &= 0.5D_{rc} + 0.15(D_{rc} + 1.0) \\ &= 0.65D_{rc} + 0.15 \end{aligned} \tag{16.19}$$

The recommended slope (S) for the inclination of the wall is roughly 10% of the height from the line of banks to roof skewbacks:

$$\text{Slope (S)} = 0.1 \times H_1 \text{ m} \tag{16.20}$$

The diameter of the reaction chamber at the level of the roof skewbacks (i.e. at the level of the upper edge of the furnace shell) is:

$$D_l = D_{rc} + 2S \tag{16.21}$$

$$\begin{aligned} &= D_{rc} + 2(0.1 \times H_1) = D_{rc} + 2\{0.1(0.5D_{rc})\} \\ &= 1.1D_{rc} \end{aligned} \tag{16.22}$$

The thickness of the lining is found by thermal analysis from the condition that the furnace shell should not be heated above 200 °C at the end of the furnace campaign. With the refractory lining δ m thick, the furnace will have the inside diameter of the shell ($D_{i.sh}$):

$$D_{i.sh} = D_{rc} + 2\delta \tag{16.23}$$

where $\delta = 0.6$ m for 100–200 t capacity furnace.

Therefore, $D_{i.sh} = D_{rc} + 2 \times 0.6 = (D_{rc} + 1.2)$ m.

Thickness of shell is kept around 0.05 m.

Therefore, outer side diameter of the shell:

$$D_{o{\cdot}sh} = (D_{i.sh} + 0.1)\mathrm{m}. \tag{16.24}$$

- **Optimum Electric conditions**

The output of an arc furnace and the consumption of energy per tonne of steel production depend to a large extent on how accurately the optimum electric conditions of furnace operation have been determined and set up. The transformer has several voltage taps. What power and what current are optimum for a given tap?

This above question can be answered by using electric performance curve of a furnace which is plotted for each voltage tapping of the transformer based on its no-load and short-circuit characteristics. The curves are plotted as a function of current.

If P_a is the active power taken off the mains and P_l is the loss of electric power:

$$P_a = 1.73\,\mathrm{VI}\cos\varphi\, 10^{-3}\ \mathrm{kW} \tag{16.25}$$

and

$$P_l = 3I^2R\, 10^{-3}\,\mathrm{kW} \tag{16.26}$$

where cos φ is the power factor of the plant, R is phase resistances of the secondary circuit, V is the voltage and I is current.

Therefore, useful power (P_{us}), i.e. power of arcs,

$$P_{us} = P_a - P_l = \left(1.73\,\mathrm{VI}\,\cos\varphi - 3I^2R\right)10^{-3}\ \mathrm{kW}. \tag{16.27}$$

The electric efficiency (η_{el}), i.e. the ratio of the useful power to that taken off the mains:

$$\eta_{el} = \frac{P_{us}}{P_a} \tag{16.28}$$

Apart from active power (P_a), reactive power (P_r) is also taken off the mains; if the active and reactive powers are added together, getting apparent power or rated power (P_{ap}):

$$P_{ap} = \sqrt{P_a^2 + P_r^2} \tag{16.29}$$

$$\cos\varphi = \frac{P_a}{P_{ap}} \tag{16.30}$$

At higher values of reactive power, the power factor is lower and electric energy is utilized less efficiently.

By analogy with the existing furnaces, the transformer power (P_{ap}) may be taken as 35 MVA. Noting the inevitable switching off the furnace during melting required to push the scrap from the banks, partial operation of the furnace at a reduced voltage when the arcs are open and radiation of much heat onto the walls and roof, the average power consumed during the melting period (P_{av}) can be found by using a factor of 0.8–0.9:

$$P_{av} = 0.8 P_{ap}\,\text{kVA} \tag{16.31}$$

Useful power consumed during the melting period:

$$P_{us} = P_{av}\cos\varphi\,\eta_{el}\ \text{kW} \tag{16.32}$$

- **Voltage taps**

For normal course of a melting process, it is necessary to vary the power and the length of arc during various periods of a heat. This is achieved by tapping the high voltage winding of the furnace transformer. The melting period is carried out at full power of the transformer and long arcs (the highest voltage taps), while refining can be done at a reduced power and short arcs (a lower tap). High power can be supplied to furnaces more easily by increasing the secondary voltage.

The upper voltage tap of the secondary voltage for small furnaces can be selected by using the following empirical formulae.

For basic furnaces:

$$V = 15\sqrt[3]{P_{ap}}\ \text{V} \tag{16.33}$$

The largest furnaces have the upper voltage tap within a range of 450–840 V. The lower voltage tap, which is resorted to during the reducing period of a heat, should not exceed 120–163 V (greater values relating to the largest furnaces). A high voltage, therefore longer arcs, might cause overheating of the lining of the roof and walls; that make it difficult to maintain optimum slag conditions in the furnace.

16.2.3.1 Electrode Diameter

The electrodes possess an appreciable electric resistivity and are heated by the passing current, which results in up to 8% of the energy supplied being lost as heat. Electrical losses could be reduced by using electrodes of a larger diameter, i.e. by lowering the current density.

The diameter of electrodes can be found by the formula:

$$d = \sqrt[3]{\frac{0.406\rho\, I^2}{K}}\ (\mathrm{cm}) \tag{16.34}$$

where I linear current, (A),

$$I = \frac{P_{\mathrm{ap}}}{V_{\mathrm{max}}}\frac{10^3}{\sqrt{3}} \tag{16.35}$$

ρ electrode resistivity at 500 °C (for graphitized electrodes, ρ = 10 Ω mm^2/m),
K coefficient (for graphitized electrodes, K = 2.1 W/cm^2).

16.2.3.2 Energy for Melting

The energy required for melting of metal (kJ):

$$q_{\mathrm{melt}} = W \cdot q_{\mathrm{S_c}}(T_2 - T_1) + WL_{\mathrm{f}} \tag{16.36}$$

where

W Weight of the metal to be melted (kg),
$q_{\mathrm{S_c}}$ Specific heat capacity of metal to be melted (kJ/kg K),
T_2 Melting point of the metal (K),
T_1 Room temperature (K) and
L_{f} Latent heat of fusion of the metal (kJ/kg K).

Some thermal data for charge materials are shown in Table 16.5, and calculated enthalpy values of oxides at 1750 °C (2023 K) are given in Table 16.6.

16.2.3.3 Heat Losses by Radiation and Other

(i) The heat losses by radiation from the furnace:

$$q_{\mathrm{r}} = €.\sigma.A.\left(T_2^4 - T_1^4\right) \tag{16.37}$$

Table 16.5 Thermal data for charge materials

Material	Specific heat, kJ/kg. K × 10^3	Latent heat, kJ/kg.K	Material	Heat energy for melting at (1750 °C) 2023 K, kJ/kg
Steel scrap	681.97	271.95	Ni	1062.0
HC–Fe–Cr	670.0	324.52	Slag	1600.0
HC–Fe–Mn	700.0	534.65		
LC–Si–Mn	628.0	578.78		
Fe–Mo	1005.0	486.78		
DRI	837.0	271.95		
HM	510.0	–		

Table 16.6 Calculated enthalpy values of oxides at 2023 K

Oxide	ΔH^0_{298} (kJ/mol)	ΔH^0_{2023} (kJ/mol)	ΔH^0_{2023} (kJ/g)	ΔH^0_{2023} (kJ/kg)
CO	−110.54	−53.47	−1.9096	−1909.64 of CO
CO_2	−393.51	−301.71	−6.8570	−6857.05 of CO_2
FeO	−264.43	−146.77	−2.04	−2038.54 of FeO
MnO	−384.93	−289.53	−4.08	−4077.93 of MnO
P_2O_5	−1492.01	−1042.32	−7.34	−7340.27 of P_2O_5
SiO_2	−910.44	−794.17	−13.24	−13,236.08 of SiO_2
Cr_2O_3	−1129.68	−909.81	−5.99	−5985.60 of Cr_2O_3

where

q_r Heat losses by radiation in watts,
ϵ Emissivity of liquid steel (0.28 for steel at 1725 °C),
σ Stephen Boltzman's constant (5.67×10^{-8} W/m^{2}K),
A Cross-sectional area of furnace, m^2,
T_1 Room temperature, K,
T_2 Temperature of the molten bath, K.

(ii) Other heat losses from the furnace = @ 4% of total heat input (average).

It is found by calculation (Example 16.7) for EAF of capacity 180 t and 7.6 m diameter, the radiation loss per tonne of liquid steel is 67 kWh.

Other heat losses from the furnace = 0.04 × 825 = 33 kWh.

(If total heat input for EAF is 825 kWh/t).

Therefore, total heat loss for radiation and other heat losses at EAF = 67 + 33 = **100 kWh/t**.

16.3 Further Developments in EAF

Developments in EAF take place with the aim as follows:

(i) Reducing operational cost,
(ii) Improving efficiency of the furnace and
(iii) Quality of products.

Developments in EAF can be subdivided into three heads as follows:

1. Design aspect,
2. Process modifications and
3. Charge modifications.

16.3.1 Design Aspect

16.3.1.1 Rapid Melting Technology

The rate of melting of solid charge determines the total heat time and hence productivity of the furnace. It is directly proportional to the power input (i.e. kVA/t capacity of furnace). Increasing size of EAFs has been the application of high power (HP) and ultra-high-power (UHP) transformers to feed these furnaces. This change in place of regular power transformers is not only rapid melting but it results in more stable arc in terms of uniform and consistent power input. Long and thin arc is inherently unstable and unsteady in compared to short and fat arc. The use of lower power factors and higher currents allows high-power inputs to be achieved with short arcs and shorter tap-to-tap times; that are the concept of UHP.

The transformers can be classified as:

(i) Low power → 100–200 kVA/t,
(ii) Medium power → 200–400 kVA/t,
(iii) High power (HP) → 400–700 kVA/t and
(iv) Ultra-high power (UHP) → >700 kVA/t.

HP and UHP transformers must be adopted along with water-cooled cables of special design, special electrode and so on for increasing the production rate. The UHP operation may lead to heat fluxes and increased refractory wear, so cooling of the furnace panels is necessary. This results in heat losses that partially offset the power savings. Total energy savings were estimated to be 15 kWh/t (i.e. 0.054 GJ/t). Use of UHP in place of regular power supply results in saving time of melting and refining of the order of 30% and 10%, respectively. Many EAF operators have installed new transformers and electric systems to increase the power of the furnaces.

Single-electrode direct current arc furnace (DC-EAF) is also used for faster melting. Single electrode (as cathode) is used in a DC arc furnace, and the bottom of the vessel serves as the anode. Electric current flows down from the electrode to an anode mounted in the bottom of the furnace. Due to conductive bottom of the DC-EAF, the arc burns vertically from the electrode to the bath. The favourable behaviour of the arc is also utilized in the electrical layout of the furnace so that it can be operated with a higher voltage and consequently with a longer arc. The arc is generally submerged in the foaming slag. Melting takes place very rapidly and gives good yield of the DC-EAF. Powerful long arc operation promoting better and efficient heat transfer, decreasing power consumption. In addition, DC-EAF also has other features including higher melting efficiency and extended hearth life. Net energy savings are limited to 9–22 kWh/t. Net energy savings over older AC furnaces are estimated to be 0.32GJ/t of steel. Compared to new AC furnaces, the savings are limited to 0.032–0.079 GJ/t of steel [13] and emissions reduction potential is 52.9 kg CO_2/t of steel.

DC-EAF also reduces noise and electrical flicker, increases efficiency and reduces electrode and refractory consumption. Electrode consumption is about half of that with AC furnace. Since the volume of exhaust gases and dust is 2–3 times lower in case of DC-EAF than that of in AC-EAF. DC furnace apart from its low energy and electrode consumption ensures excellent thermal and metallurgical homogenization of the liquid steel. Ultra-low carbon steel can be easily produced since carbon pickup does not exceed 0.05%.

16.3.1.2 Water-Cooled Panels

Use of water-cooling systems for walls and roof becomes absolutely necessary for economical operation. It is so because the furnace diameter cannot be increased at liberty and hence the refractory consumption, otherwise exceeds the acceptable values for large-capacity UHP EAFs. The water-cooled panels are of tube-on-tube design of steel pipes and cover up to 85% of the wall above the slag line. The advantage of the external type water-cooled panels is greater distance between electrodes and sidewall, as well as larger volume for receiving light scrap; and water leakage will not penetrate the furnace lining. Refractory and electrode consumption are reduced drastically, and more scrap can be accommodated in the furnace, yielding about 20% more liquid metal. The roof is also water-cooled, except for the delta region around the electrodes. The dome is 60% lower than that for a brick-lined roof resulting in lower electrode consumption. Water-cooled panels are reducing power-on time by 13% and increasing productivity by 10%.

16.3.1.3 Blowing Oxygen and Carbon Injection

Blowing oxygen in molten steel releases heats because the reactions of oxygen with iron, carbon, silicon, in the molten metal are exothermic in nature; hence, time of melting reduces, the productivity is increased and the consumption of electricity decreases as well as decreasing the electrode and refractory consumption. In the past, lancing operations were carried out manually using a consumable pipe lance. Modern operations use automatic non-consumable, water-cooled lances for injecting oxygen into the steel. Many of these lances also have the capability to inject carbon as well.

Injection of carbon brings the benefits of (i) for 100% scrap practice or when carbon content of the bath is insufficient to produce CO gas for foaming slag, and (ii) oxidation of carbon produces CO gas and generates heat. It is to be noted that carbon injection requires oxygen injection for carbon oxidation.

16.3.1.4 Oxy-Fuel Burners

Oxygen–fuel burners are installed for the cold spots of the furnace, so that the time taken to melt scrap in these areas is the same as that taken by the arc at the hot spots. Hence, lower electrical power consumption results together with an improved heat balance. Oxygen/natural gas burners can help faster melting the charge and reduce electrical consumption and electrode usage while shortening the melting time from tap to tap. Improper use of these burners can cause excessive oxidation of the melt and the electrodes, as well as that can reduce the life of the refractory lining.

Oxy-fuel burners are now used on most of the EAFs. These burners increase the effective capacity of the furnace by increasing the speed of the melt and reducing the consumption of electricity and electrode, which ultimately reduces greenhouse gas (GHG) emissions. Energy can be saved 2–3 kWh/t of steel and reduction of heating time. Electrical energy savings 0.14 GJ/t of steel, with typical oxygen injection rates of 18 Nm^3/t of steel and emissions, can be reduced by 23.5 kg CO_2/t of steel [13]. The use of oxy-fuel burners has several other beneficial effects; it increases heat transfer and reduces heat losses. Steelmakers are now making wide use of stationary wall-mounted oxygen gas burners that combination of burner and lance, which operate as a burner mode during the initial period of the melting, and the burner change over to a mode of oxygen lance when a liquid bath is formed. Moreover, by lancing oxygen helps to remove impurities from the molten steel bath. Electricity savings of 35–40 kWh/t of steel can be achieved with typical oxygen injection rates of 18 Nm^3/t of steel. Annual cost saving is done due to reduce tap-to-tap times.

16.3.1.5 Hot Heel Operation

Some liquid steel (20–25%) is kept in the bottom of the furnace during tapping, which helps in the faster melting of fresh scrap or sponge iron/HBI feed into the furnace for the next heat. Since sponge iron/HBI is non-conductive material, that required hot heel for initial melting in EAF.

16.3.1.6 Bottom Stirring

In convectional arc furnaces, there is little natural electrical turbulence within the bath. Due to the absence of stirring, large piece of scrap can take a long time to melt and may require oxygen lancing. Bottom stirring is accomplished by injecting an inert gas through the porous plug in the bottom of the EAF to increase the heat transfer in the melt. In addition, increased interaction between slag and metal leads to an increased liquid metal yield of 0.5%. Inert gas stirring eliminates temperature and concentration gradients, shortens tap-to-tap times, reduces power, electrode and refractory consumption and improves yield of iron and alloys. Furnace with oxygen injection is sufficiently turbulent, reducing the need for inert gas stirring. The increased stirring can lead to electricity savings of 10–20 kWh/t (0.036–0.072 GJ/t).

16.3.1.7 Eccentric Bottom Tapping

Figure 16.7 shows the eccentric bottom tapping with the use of a slide gate. Eccentric bottom tapping ensures almost slag-free liquid metal tapping in the teeming ladle. Eccentric bottom tapping reduces tap times, temperature losses and slag carry-over into ladle. Power saving of 12–15 kWh/t of steel is estimated by using this technology.

16.3.1.8 Automatic Alloy Feeder

With increasing size of EAF, the net addition has increased in proportion. For meeting specifications of products, the additions must be made more accurately. For these reasons, automatic weighting of additions as per the auto-control directions is made.

16.3.1.9 Coated Electrode

Electrodes are consumed in three ways:

(i) Tip consumption → 50%,
(ii) Surface consumption → 40% and
(iii) Mechanical breakage → 10%.

The tip consumption is directly dependent on electrode current and by dissolution in liquid steel when electrode is lower down for arcing. Surface consumption decreases due to oxidation which can be reduced by coating on electrodes. Mechanical breakage is done by tilting of heavy scrap. Mechanical breakage can be reduced by: (a) automatic control of electrode and (b) use of sponge iron/HBI as feed material.

16.3.1.10 Fifth Hole

Nowadays fifth hole in the roof is common for bigger furnaces for continuous charging of sponge iron/HBI. Due to uniform size of sponge, iron/HBI can be charged continuously. Fifth hole may be: (i) at the delta region of three electrodes or (ii) at in between electrode and periphery of the roof.

16.3.1.11 Switch Gear

EAF is switched on and off frequently than any other electrical appliance. These repetitive switching of up to 150 times a day or 45,000 switching per year is a sizable task for switch gear. The design and material for switch gear must be developed to minimize the wear and breakage.

16.3.1.12 Emission and Noise Control

For bigger furnaces, fourth-hole system is common, where a duct hood is made directly into the roof to evacuate smoke and fumes. The complexity of the fume and dust collection equipment varies from a simple fourth-hole connector to an entire separate room for the melting furnace in which the fumes and dust are collected and where the liquid metal is tapped into a ladle that is brought to the furnace and then removed through a dust protection door. Noise can be easily controlled by DC-EAF.

16.3.2 Process Modifications

16.3.2.1 Foams Slag Practice

Slag foaming has been used in electric arc furnaces (EAF) to protect the refractory materials from the radiation of heat generated by electrodes, decrease the noise level and improve productivity and the energy efficiency of the process. Slag foaming has an important impact on the thermal efficiency, tapping time and refractory lining/electrodes consumption of the EAF. Foamy slag covers the arc and metal surface to reduce radiation heat losses. Foamy slag can be produced by injection carbon (granular coal) and oxygen or by lancing oxygen only when charging of sponge iron/HBI (which contains high carbon) is done. Foamy slag prevents radiation heating to the furnace wall from arc and help to penetrate lighter sponge iron charge in slag-metal interface. By covering the arc in a layer of slag, the arc is shielded, and more energy is transferred to the bath; due to that rate of melting of scrap increases. The effectiveness of slag foaming depends on slag basicity, FeO content of slag, slag temperature and availability of carbon to react with either oxygen or FeO of slag. Foaming slag reduces refractory damage and heat loss from the arc region. Slag foaming increases the power efficiency by at least 20% in spite of a higher arc voltage. The energy savings (accounting for energy use for oxygen production) are estimated at 5–7 kWh/t (0.018–0.025 GJ/t). Foamy slag practice increases productivity as decreasing the tap-to-tap times. Emission reduction of 10.6 kg CO_2/t steel is estimated.

16.3.2.2 Blowing Oxygen

Blowing oxygen in molten steel releases heat because the reactions of oxygen with silicon, carbon and iron in molten metal are exothermic and produces heat. Melting time reduces, and the productivity is increased. Power, electrode and refractory consumption are decreased. Earlier lancing operations were carried out manually using consumable lance pipe. Modern operations are carried out by using of automatic, non-consumable, water-cooled lance for injecting oxygen into the steel. These lances also have the capability to inject carbon as well.

16.3.2.3 Carbon Injection

Injection of carbon has benefits of (i) for 100% scrap practice or when carbon content of the bath is insufficient to produce CO gas for foaming slag, and (ii) carbon oxidation produces CO gas which on post-combustion generates thermal energy.

It is to be noted that carbon injection also requires oxygen injection for carbon oxidation.

16.3.2.4 Scrap Pre-heating

Scrap pre-heating is performed by hot flue gases either in the scrap charging baskets, in a charging shaft before added to the EAF, or in a specially designed scrap conveying system allowing continuous charging during the melting process. The twenty per cent of EAF's energy normally leaving the furnace in the hot waste gas represents about 130 kWh/t of steel produced. Using this hot gas to pre-heat the scrap being charged to the EAF can result in recovering some of this energy and to reduce some of the electrical energy required to melt steel scrap. The heat content of pre-heated scrap (in equivalent kWh/t) is shown in Table 16.7. Additional advantages for scrap pre-heating include: (i) increased productivity, (ii) removal of moisture from the scrap, (iii) reduced electrode consumption and (iv) reduced refractory consumption.

Table 16.7 Heat content in pre-heated steel scrap [3]

Temperature of scrap (°C)	Heat content (kWh/t)
150	22
260	40
370	57
540	81

Pre-heating scrap reduces the power consumption of the EAF by using the waste heat of the EAF as the energy source for the pre-heating operation. Scrap pre-heating can save 0.016–0.2 GJ/t-steel. Scrap pre-heating can be saved energy consumption of melting by 4–50 kWh/t and reduces tap-to-tap times by 8–10 min. Emissions are reduced by 35 kg CO_2/t of steel.

16.3.2.5 Hot Tundish and Ladle

By pre-heating tundish and ladle, chilling effect of liquid metal as well as drastic temperature drop of liquid metal can be avoided. Temperature of liquid metal can be maintained properly.

16.3.2.6 Coupling with Ladle Furnace

EAF-LF combination gives very good quality of products. Very low amount of gases are present in the products. Productivity of EAF is also increased due to EAF which acts as melting unit only and refining of molten metal takes place at ladle furnace (LF). After melting of charges, molten metal is transferred to LF for refining, and again furnace (EAF) is ready for taking the new charges; hence, tap-to-tap time is drastically decreased and number of heats per day is subsequently increased.

16.3.2.7 Process Automation

For meeting specifications of products, process automation is must. For proper heat and mass calculations, automation is essential for charging and control the furnace properly. Process control can optimize operations and thereby significantly reduce power consumption as demonstrated by worldwide. Control and monitoring systems for EAF are moving towards integration of real-time monitoring of process variables, such as steel bath temperature, carbon levels and lance oxygen practice. Neural networks systems analyse data and emulate the best controller and can thus help to reduce power consumption beyond that achieved through classical control systems. For EAF, average power savings were estimated to be 8%, productivity increased by 9–12%, and electrode consumption was reduced by 25%. Electricity savings of 30 kWh/t of steel are estimated, and values may change based on scrap and furnace characteristics [13]. Emissions reduction potential of the technology is 17.6 kg CO_2/t of steel.

16.3.3 Charge Modifications

16.3.3.1 Use of Sponge Iron

Scrap is an extremely heterogeneous by-product, and its quality varies from region to region, type to type and lot to lot. To match demand for high-quality product, steelmakers must start with high quality of scrap. The non-availability of scrap of consistent quality and at a reasonable price necessitated the search for a suitable alternative to scrap. In this context, sponge iron is the best alternative to scrap as charging material in EAF. Sponge iron has uniform chemical and physical characteristics.

Main advantages of sponge iron are:

(i) Uniform composition,
(ii) Uniform size,
(iii) Very low tramp elements (0.02%) with respect to scrap (0.13–0.73%) and
(iv) Low sulphur.

This promotes the use of sponge iron in the charge of EAF, as partial replacement to scrap, which ultimately improves the mechanical and metallurgical properties of the product. The uniform size of sponge iron facilitates continuous charging. Hence, good quality of steels can be produced by addition of sponge iron as charge materials. Since sponge iron has poor thermal conductivity, it is always charged after initially getting molten pool, i.e. producing hot heel by initial melting of steel scrap. Melting of sponge iron in EAF is greatly influenced by factors like carbon content and degree of metallization of sponge iron.

Sponge iron is charged to the furnace time to time to the hot heel by basket in batch charging process, where build-up of sponge iron takes place, which has an adverse effect during the melting period. Creating agglomerates of sponge iron into the furnace is difficult to melt further and requires oxygen lancing. Charging of very large quantity of sponge iron in a batch process poses operational problems. When greater percentages of sponge iron are to be charged, it is necessary to use continuous charging method. The usual practice is to charge a basket of scrap which representing from 20 to 50% of total charge, to melt that scrap partially to create a molten pool and then to feed sponge iron at a particular rate. 100% sponge iron charge is achieved by keeping a heel of molten metal (about 20–25%) in the furnace after tapping and then feeding sponge iron for a new heat. Tramp elements can be controlled in the steel bath by 100% sponge iron charge. The melting and refining are done simultaneously and leaving only an adjusting period for de-oxidizing additions (i.e. blocking of heat) and a temperature check before tapping.

Continuously, charging of sponge iron with metallization of 90–95% and carbon content of 1.8% leads to a foamy slag without addition of further carbon. Increasing carbon level to the sponge iron above 1.8% is not economical since the additional costs in the sponge iron do not fulfil any technical need and are not offset by any cost savings in the melt shop. The carbon reacts with unreduced iron oxide (i.e. FeO) giving CO gas evolution from liquid bath, i.e. carbon boil:

$$(\mathrm{FeO}) + [\mathrm{C}] = [\mathrm{Fe}] + \{\mathrm{CO}\} \qquad (16.2)$$

This carbon boil results into subsequent removal of hydrogen and nitrogen gases, as well as slag inclusion and ultimately producing cleaned steel. The carbon boil also makes foamy slag which prevents radiation loss as well as absorption of nitrogen from atmosphere. Nitrogen pickup to the molten bath depends on dissolved oxygen content in the bath; less oxygen content in the bath, more nitrogen picks up to the molten bath. Therefore, higher carbon content in sponge iron is always desired by steelmakers. If carbon is less than required amount, it must be externally added to the melt. The amount of carbon required (in kg) to reduce the FeO content of the sponge iron is as follows [14].

$$\mathrm{C} = 1.67\left[100 - \%M - \left\{\left(\frac{\%\mathrm{Sl}}{100}\right) \times \%\mathrm{Fe}\right\}\right] \qquad (16.38)$$

where M is the degree of metallization, Sl is the amount of slag and Fe is the amount of iron in the slag.

Metallization of sponge iron has an important factor in steelmaking, and increasing metallization of sponge iron decreased the energy consumption for melting of sponge iron in EAF (as shown in Fig. 16.19). Sponge iron having lower metallization value has relatively higher unreduced iron oxide content. Since FeO reduction is an endothermic reaction, and hence, extra energy is consumed. As a thump rule, loss of 1% of metallization of sponge iron consumes extra energy of 15 kWh/t of liquid steel [15]. Electrode consumption decreases due to fewer electrode breakages, due to the absence of heavy scrap which is titled during melting and breaks the electrode.

As the gangue materials, present in the ore, remain in the sponge iron, the volume of slag is usually larger when a part of scrap is substituted by sponge iron (as shown in Fig. 16.20). As the percentage of gangue increases, slag volume is also increased with fixed per cent of sponge iron charged in EAF.

16.3.3.2 Use of Hot Sponge Iron

Using less energy (in the case of electricity) at the steel plant can reduce emissions at the power plant. With the use of hot sponge iron in an EAF, energy efficiency is highly improved. Charging hot sponge iron into the EAF provides a source of additional energy as well as low residual iron units, leading to

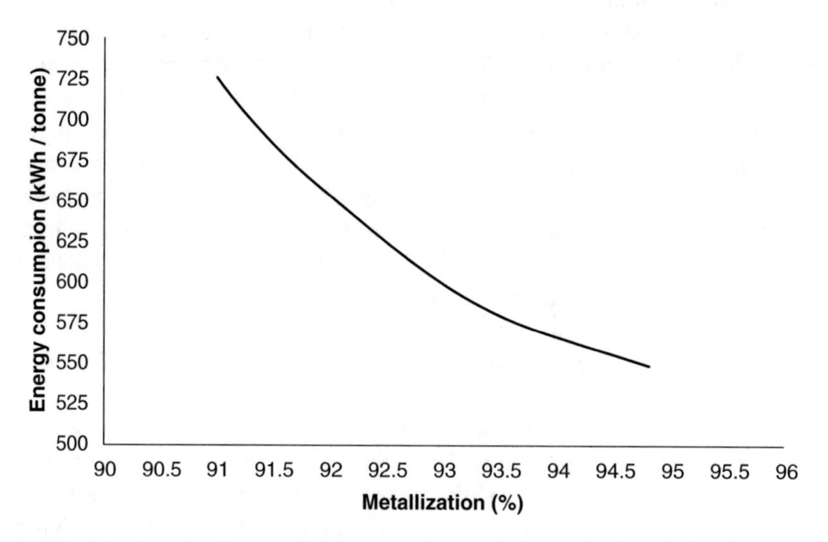

Fig. 16.19 Metallization of sponge iron versus energy consumption in EAF [16]

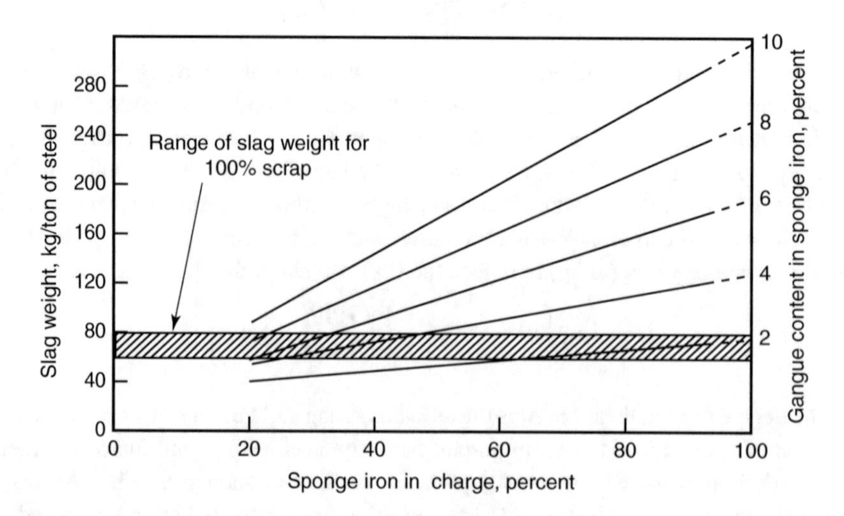

Fig. 16.20 Effect of gangue in sponge iron on slag weight [17]

lower costs and improved furnace productivity. Transport of the hot sponge iron in an inert atmosphere is a key technology to the achievement of these benefits. The difficulty in hot sponge iron transport is not just that the material is hot, but also that is must be kept in an inert atmosphere to avoid oxidation of sponge iron. Midrex has developed three methods [18] [i.e. (i) HOTLINK, (ii) hot transport conveyor and (iii) hot transport vessels] for discharging the sponge iron at elevated temperature, transporting it in hot condition to the melting shop, and charging it to the EAF at 600–700 °C.

Hot charging of sponge iron at 600–700 °C is an effective means of lowering the cost of production per tonne of liquid steel due to the reduction of power (25%) and electrode consumption. These methods lower the electrical energy required per tonne of steel produced, which also reduces CO_2 gas emissions from the power plant. The electrical energy savings occur because less energy is required in the EAF to heat the sponge iron to melting temperature. The rule of thumb is that power consumption can be reduced about 20 kWh/t liquid steel for each 100 °C (373 K) increase in temperature of hot sponge iron [15]. Thus, the savings of electrical energy consumption, when charging of sponge iron at over 600 °C, are 120 kWh/t or more. An additional benefit of the power savings is a reduction in electrode consumption since there is a linear relationship with power consumption. Generally, electrode consumption is 0.004 kg/kWh [19]. Electrode consumption decreases 0.3 kg/t. In addition to the power and electrode savings, hot charging will also increase EAF productivity. Use of hot sponge iron reduces the tap-to-tap time, allowing a productivity increase of up to 20% versus charging at ambient temperature. Hot sponge iron is transferred from the vessel into the furnace through the chute.

The benefits of charging hot sponge iron have been known for many years. Essar Steel, India began experimenting with hot sponge iron transport in 1999 on a small scale. Since mid-1999, the production and use of hot sponge iron have increased continuously up to 100% hot sponge iron charges in DC-EAF.

Main advantages of hot charging of sponge iron to the EAF as follows:

1. Decrease moisture in feed materials,
2. Energy saving by lowering electrical energy consumption, because less energy is required in the EAF to heat the sponge iron up to melting temperature,
3. Decrease in electrode consumption,
4. Increase productivity and yield, due to shortening overall melting time and tap-to-tap time.

There are also environmental benefits of hot sponge iron charging. Retaining the sensible heat in the sponge iron rather than dissipating it to the atmosphere lowers overall emissions two ways. First, the lower electricity demand reduces power plant emissions per tonne of steel produced. Second, for those plants employing carbon injection, reduced energy requirements in the EAF result in less CO_2 given off. With the use of hot sponge iron charging, the DR-EAF route for steelmaking becomes more attractive with respect to CO_2 emissions. Use of 80% hot charged of sponge iron in the EAF results in 46% lower carbon emissions per tonne of steel produced than the BF-BOF route.

Ultimately, the most important consideration for the steelmaker is that hot sponge iron (HDRI) charging can enhance profitability comparing the use of cold sponge iron (CDRI), as shown in Table 16.8. The assumption is that the limiting factor is the size of the EAF at 200 t and the associated electrical system. Use of hot sponge iron at 600 °C enables an increase of 20% in liquid steel output, plus a savings in operating cost [15].

16.3.3.3 Use of Hot Metal in EAF

Steelmakers had been trying to look at methods of steelmaking which could use its existing EAF, LF, caster route to increase the productivity significantly, and reduce dependence on electric power. Hot

Table 16.8 Economics of hot sponge iron charging [18]

Type of sponge iron feed	CDRI	HDRI
DRI volume (Mtpy)	1.40	1.68
EAF heat size (t)	200	200
Feed mix (DRI/scrap)	90/10	90/10
DRI charge temp (°C)	25	600
Tap-to-tap time (min)	65	54
Steelmaking capacity (Mtpy)	1.37	1.64
Increased production (%)		19.7

metal fed to EAF was fitting the above boundary conditions extremely well. EAF process has high flexibility in various types of charge mix, flexibility in producing steel grades, sponge iron/DRI friendliness, better control of cost of metallic, better control of phosphorus, oxygen and carbon and can accept wide percentage of hot metal (30–70% of total charge mix). Hot metal charging into EAF provides additional sensible heat to the furnace, thus leading to power saving and improving productivity. Hot metal can be charged from the roof as well as through launder from slag door. Hot metal contains high carbon and silicon which are the supplier of chemical energies due to that power consumption of EAF drastically reduced. Uses of 50% hot metal in the charge of EAF have saved power of 300 kWh/t, and heat time comes down to 80 min with oxygen blowing of about 40 Nm^3/t.

By using hot metal in EAF:

(i) Energy consumption can be reduced,
(ii) Shorter power-on times,
(iii) Scrap shortage can be overcome.
(iv) Tap-to-tap can be decreased.
(v) Reduced electrode consumption and
(vi) Productivity can be increased drastically, i.e. more number heats can be made per day.

Benefits of using hot metal (3.8–4.2% C, 0.6–0.8% Si, 0.11% max P and 0.08% max S) in EAF are shown in Table 16.9.

As per report, a furnace charge containing 10% pig iron can generate energy as much as 30 kWh/t, but then there are economic limits to oxygen injection in EAF. However, use of hot metal directly from MBF into EAF can improve productivity, lowering the cost of steelmaking because of the lower power consumption and melting time due to the sensible heat content of hot metal.

Table 16.9 Benefits of using hot metal in EAF [20]

Consumption	Without HM (40% pig iron, 30% DRI and 30% scrap)	With HM (45% hot metal, 15% pig iron, 30% DRI and 10% ccrap)	Saving
Power (kWh/t)	500	280	220
Electrode (kg/t)	4.0	2.6	1.4
Oxygen (Nm^3/t)	35	40	−5
Productivity (ratio)	1.0	1.6	0.6
Tap-to-tap time (min)	90	55	35

Table 16.10 Performance and specific consumption of alternate iron sources in EAF [22]

	100% Scrap	50% Iron carbide + 50% scrap	50% Hot metal + 50% scrap	70% Scrap + 30% pig iron	80% DRI + 20% scrap
Scrap, kg/t	1055	547	554	780	215
Alternate iron source, kg/t	–	593 (carbide, 6.5% C)	554 (HM at 1330 °C)	325 (pig iron)	915 (DRI, 1.1%C)
Power consumption, kWh/t	450	425	280	430	650
Electrode consumption, kg/t	4.0	3.9	3.3	3.9	4.6
Lime, kg/t	25	33	37	35	47
Oxygen, Nm^3/t	20	54	42	30	10
Carbon, kg/t	20	–	–	–	15
Metallic yield %	95	91.5	92.5	93.5	93
Tap-to-tap time, min	75	89	69	89	96

16.3.3.4 Use of Iron Carbide in EAF

The iron carbide is injected with carrier gas (nitrogen/argon) through pipes submerged in steel bath at rates ranging from 90 to 450 kg/min. The carbide injection with oxygen is advantaged of the heat generation and thus lowers the electricity consumption. The results so far have indicated satisfactory and trouble-free operation. With 11% carbide, yield rises from 93 to 95.5%, while 35% carbide, total melting time decreased from 47 to 40 min. Power consumption decreased from 450 to 380 kWh/t with rise in carbide proportion from 0 to 35%, due to present of high carbon content (6.0–6.5%) in carbide. No additional carbon charge was needed for carbon boil, and no pre-heating of carbide was necessary. Good bath boil resulted in foaming slag with decreasing FeO content from 24 to 15% and nitrogen level from 82 to 34 ppm.

Higher carbide proportions in the charge for steelmaking and faster carbide injection rates are possible in EAF but up to a certain practical limit only, due to furnace overheating and oxygen cost respectively. Post-combustion of CO from reactor helps in power saving and enhancing productivity. Performance indices of some alternate iron sources in EAF are furnished in Table 16.10.

16.4 DC Arc Furnace

Single-electrode direct current (DC) arc furnace (DC-EAF) is used for faster melting. DC-EAF was pioneered in Europe, and these single-electrode furnaces with direct current (DC) rather than alternating current (AC) have been used in North America for over 30 years. Essar Steel, India, also used DC-EAF for more than 20 years. The main feature of the DC arc furnace is a single graphite electrode acting as cathode, with a conductive bottom together with the charge acting as anode (as shown in Fig. 16.21). Electric current flows down from the electrode to an anode mounted in the bottom of the furnace. Due to conductive bottom of the DC-EAF, the arc burns vertically from the electrode to the bath. The favourable behaviour of the arc is also utilized in the electrical layout of the furnace so that it can be operated with a higher voltage and consequently with a longer arc. The arc is generally

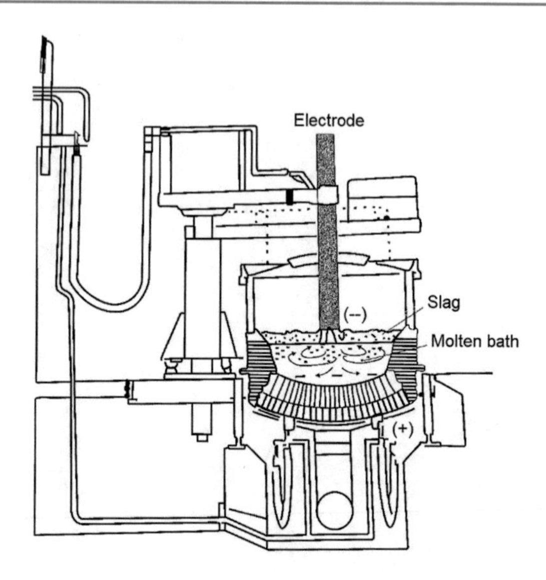

Fig. 16.21 DC arc furnace [23]

submerged in the foaming slag. Melting takes place very rapidly and gives good yield of the DC-EAF. Powerful long arc operation is promoting better and efficient heat transfer, decreasing power and electrode consumption.

16.4.1 Principle

The DC arc acts as a jet pump, drawing surrounding gases and particles into the arc plasma, which has a defined velocity movement from the electrode towards the melt and thus represents a very efficient means of heat transfer from the electrode to the melt. In DC-EAF, centrally located electrode projecting through the roof is connected to the negative pole of the rectifier (i.e. act as cathode). The positive pole is connected to a bottom electrode which is in direct contact with the iron melts or steel billets which act as anode. Since DC-EAF requires only one vertically adjustable electrode, the construction in the roof area is considerably simpler than that of an AC-EAF. Furthermore, the hydraulics and electrode control are also considerably reduced. In the AC-EAF, heavy arcing puts heavy pressures on the grid, while the DC-EAF substantially reduces the load on the electrical grid. The upper and lower parts of the DC-EAF shell are flanged together with a layer of insulating material to electrically insulate the bottom electrode.

Due to conductive bottom of the DC-EAF, the arc burns vertically from the electrode to the bath. The favourable behaviour of the arc is also utilized in the electrical layout of the furnace so that it can be operated with a higher voltage and consequently with a longer arc. The arc is generally submerged in the foaming slag. Melting takes place very rapidly and gives good yield of the DC-EAF.

Based on the distinctive feature of using the heat and magnetic force generated by the current in melting, this arc furnace (DC-EAF) achieves an energy saving of approximately 5% in terms of power consumption in comparison with the three-phase AC arc furnace (AC-EAF). In addition, DC-EAF also has other features including higher melting efficiency and extended hearth life [23]. Net energy savings are limited to 9–22 kWh/t. DC-EAF also reduces noise and electrical flicker, increases efficiency and reduces electrode and refractory consumption. Electrode consumption is about half of that with AC furnace since the volume of exhaust gases and dust is 2–3 times lower in case of DC-EAF than that of in AC-EAF. DC furnace apart from its low energy and electrode consumption ensures excellent thermal and metallurgical homogenization of the liquid steel. Ultra-low carbon steel can be easily produced since carbon pickup does not exceed 0.05%.

16.4.2 Charging

The method of charging of raw materials in DC-EAF greatly influences the melting and refining operations, power, time utilization and productivity of the furnace. There are two types of charging methods: (i) batch charging and (ii) continuous charging.

(i) Batch charging is done by buckets and mainly adopted in case of swing-roof type furnace. Charging materials consist of generally scrap, but from time to time sponge iron/HBI is also preferred.
(ii) Continuous charging can be adopted only for sponge iron/HBI which has uniform sizes. The main advantages of continuous charging are substantial saving in time and avoid radiation losses. It can be continuously charged from the charging hole in the roof of the furnace as follows:

    (a) Single hole in the roof between the electrode and sidewall and
    (b) Centrally hollow electrode.

16.4.3 Operation

A 150 t DC-EAF (160 MVA transformer capacity) with a 7.3 m shell diameter operates with rated current 110–130 kA. The shell of DC-EAF is made of a large copper plate bolted to the furnace bottom and serving as the contact plate conducting the direct current to the bottom. On the top of the copper plate is the electrically conductive hearth consisting of magnesite–graphite bricks lining of one-meter thickness. To ensure a good electrical contact between the charge and the bottom, the furnace is operated with a liquid heel. There are two types of arc produced between electrode and scrap: (i) short and fat arc and (ii) long and thin arc (as shown in Fig. 16.22).

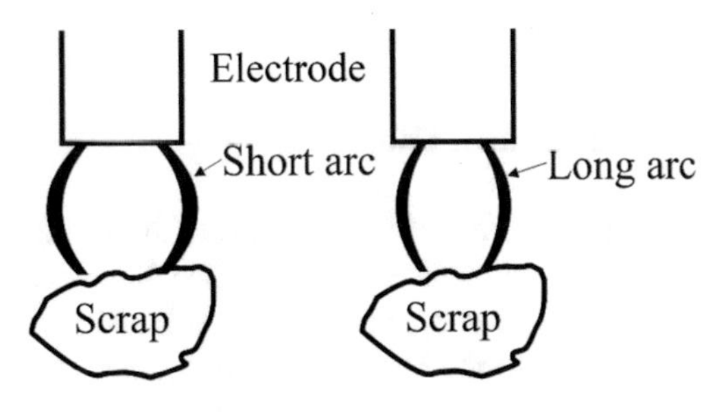

Fig. 16.22 Types of arc

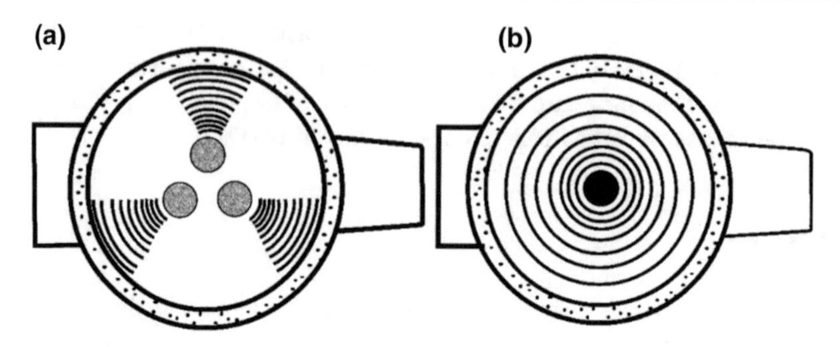

Fig. 16.23 Heat distribution pattern [23]. **a** AC, **b** DC arc furnaces

Long arc operation with foaming slag to shield the arc and to improve heat transfer to the charge had been adopted by AC-EAF melting shops. DC-EAF melting involves the same technique. Longer arcs can be used in DC-EAF operation since the single electrode is centrally located in the furnace and the DC arc is burning vertically towards the bath [23]. The result is a more uniform heating pattern in a DC-EAF than in an AC-EAF, where the three arcs repel each other by magnetic forces creating a concentration of arc heat outside each electrode (as shown in Fig. 16.23).

Reduced electrode consumption is perhaps the most important advantage of DC-EAF. Electrode consumption of DC-EAF is half of AC-EAF. Electrode consumption may be accounted as follows:

- Electrode tip wear (50%), which is proportional to the square of the electrode current,
- Sidewall oxidation (40%), which is proportional to the exposed electrode surface and
- Breakage (10%), which is directly related to the number of electrodes.

The electrode tip wear depends mainly on the temperature. The anode temperature in an arc is always considerably higher than that of the cathode. The suspended electrodes in AC-EAF are anodes in the first half cycle, while it acts as cathodes in the second half cycle. Consequently, the average temperature of AC-EAF is higher than in DC-EAF operation, where the single suspended electrode is always the cathode. Thus, the electrode tip wear is reduced in case of DC-EAF. With the development of electrode saving system, the oxidation loss of the electrode is further reduced. It consists of electrode cooling by means of a water film on the electrode surface that is exposed to atmosphere.

DC-EAF has a centrally positioned electrode, so that the distance between the arc and the refractory lining is constant over the complete circumference of the furnace as shown in Fig. 16.23. The electrode to wall distance is also large and with a vertical arc, and this reduces the refractory wear.

16.4.4 Difference Between AC and DC

A comparison of AC versus DC furnaces led to a lengthy list of differences. On closer inspection, the differences were not AC/DC related but rather due to shop-specific practices. Ultimately, the major differences are method of power transfer and control, arc length and foamy slag practice.

The mechanism for power transfer to the charge is quite different. In the AC furnace, the arc is struck between the electrodes (3 in numbers) and the charge. The current passes through the charge and returns via a second arc to the electrodes. Initially, if the charge is cold and largely scraps, there is inherent arc length variability and accompanying noise which lessens as the charge compacts, arc

Table 16.11 Comparison DC-EAF versus AC-EAF [23]

	AC-EAF	DC-EAF	Difference (DC-AC)
Furnace capacity, t	80	80	–
Shell diameter, m	5.5	5.5	–
Graphite electrode, Number	3	1	−2
Investment cost, %	100	175	+75
Electrode consumption, kg/t	3.0	1.5	−1.5
Energy consumption, kWh/t	447	425	−22

length stabilizes and melting begins. The passage of the current through the charge creates resistance heating and once a molten pool of steel is formed, the current path is from the electrode to the bath. In the case of DC furnace, the electrode (single) acts as the cathode and the anode, a conductive bottom, is a bottom electrode which is billet, metal fin or metal pin type. Current flows from the electrode through the cold scrap, to the cathode. As with AC EAFs, heating is resistance in nature.

The control scheme is quite different, being current controlled in the AC furnace and voltage controlled in the DC furnace. The arc length is 30–50 cm in an AC furnace and 90–120 cm in the DC furnace through the maximum power input which is the same in each case. For batch charging of sponge iron, there is no significant difference between AC and DC operations. For continuous charging, there are some important considerations. AC furnaces have a built-in path to the molten bath. With a DC furnace, however, it is necessary to create a path to the molten bath by ensuring there is a good foamy slag operation to insulate and direct the arc. While the foamy slag practice is critical for good operations in either case, slag height is more critical for DC furnaces due to the longer arc length cited above. The allowable sponge iron feed rate is the same for both, being between 27 and 45 kg/min/MW [28]. Table 16.11 compares the important parameters of DC-EAF with that of AC-EAF for the 80 t furnace.

16.5 Induction Melting Furnace (IMF)

Technology of melting of metals and alloys by electric induction system is fairly old in the world. In India also, high-frequency induction melting furnaces were installed by many steel plants and ordnance factories to produce high alloy and tool steels. Medium-frequency induction melting furnace gets higher temperatures for steelmaking and to cut down power consumption. There was also improvement in electronic circuit by using thyristors and capacitors which took care for higher temperature for melting. In early 70s, this development took place in European countries and USA. Steel foundries in these countries started installing medium-frequency induction melting furnaces of various capacities to manufacture low and high alloy steel castings.

Induction furnaces are producing mild steel of structural quality in considerable quantities at a competitive cost. Improvements are taking place in the panel circuit design. By suitable incorporation of capacitors, companies have increased power factor as much as unity. By improving panel circuit design, power consumption can be achieved below 600 kWh/t. While many induction furnace units have started processing steel melting scrap so that it not only improves the quality of production but also consumes less power. Nowadays, the medium-frequency induction melting furnaces become popular in almost all parts of India, due to low power consumption, low capital investment, low dissolved gases in steel and economic refractory consumption. The steel produced by medium-frequency induction furnaces was found to be cheaper than the steel produced by the electric arc furnaces.

16.5.1 Principle

An induction furnace is an electrical furnace in which the heat is applied by induction heating of metal. Induction furnace is operated by utilizing a strong magnetic field created by passing of an electric current through a coil wrapped around the furnace (as shown in Fig. 16.24). This electric current creates an electromagnetic field that passes through the refractory material and couples with the conductive metallic charge inside the furnace. The magnetic field in turn creates a voltage across, the subsequently secondary current through the metal to be melted. The electrical resistance is given by the metal that produces heat to melt the metal and helps it to reach the high temperature.

The primary circuit of the induction furnace is formed by the coil, and the secondary circuit is the crucible or, rather, the metallic charge in it. The lines of magnetic force link through the metallic charge and induce eddy current in it, and the latter generates heat. The magnetic field and the electro dynamic forces are acting in the crucible of an induction furnace. Main components of an induction furnace are power supply unit consisting of transformer, inverter and capacitor bank, the charging arrangement, the cooling system for the power supply and furnace coil, process control system and the fume extraction equipment.

An induction furnace consists of a non-conductive crucible holding the charge of metal to be melted, surrounded by a coil of copper wire. A powerful alternating current flows through the wire. The coil creates a rapidly reversing magnetic field that penetrates the metal. The magnetic field induces eddy currents, circular electric currents, inside the metal, by electromagnetic induction. The eddy currents, flowing through the electrical resistance of the bulk metal, heat it by Joule heating. In ferromagnetic materials like iron, the material may also be heated by magnetic hysteresis, the reversal of the molecular magnetic dipoles in the metal. Once melted, the eddy currents cause vigorous stirring of the melt, assuring good mixing.

Induction current flows in a concentrated way in the surface of material to be melted. This concentration of current becomes more remarkable as the frequency becomes higher, resulting in better heating efficiency. This is known as *skin effect* and the depth of the surface layer of the metal (charge) where the density of induced current is large, is called *penetration depth*. The heat required to melt the charge is developed mainly in this layer.

16.5.2 Raw Materials

The main raw material for induction furnaces is steel scrap, cast iron and sponge iron/HBI. India is the only country where use of induction furnaces contributes a large share (more than 30%) in annual crude steel production. The amount of sponge iron in the charge mix varies from 0 to 90% depending on its availability and economics of production. Majority of the steel produced through induction furnace route is plain carbon steel and construction quality steel, i.e. structural steel.

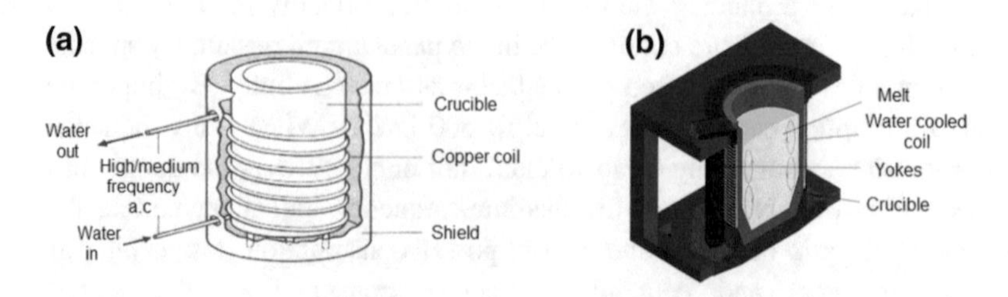

Fig. 16.24 **a** Induction melting furnace, **b** vertical section of IMF [24]

It is recommended that mild steel scrap should be cleaned and properly sized before charging in the furnaces. Many medium-frequency induction melting furnaces units are pre-heating the charge before charging. Sponge iron or hot briquetted iron (HBI) is being used in considerable quantities by the induction furnace units nowadays. Sponge iron contains very low tramp elements. The steel produced by using sponge iron is preferred by rerolling industry because it can roll very well. In case carbon, phosphorous and sulphur are high in scrap, and the percentage of sponge iron used in the charge should be increased to dilute these elements. Higher carbon content in bath can also be removed by using sponge iron because it contains certain amount of unreduced iron oxide (i.e. FeO) that oxygen of FeO reacts with the carbon from the bath to form CO gas which creates boiling action of the bath. Due to the boiling action: (i) homogenized the bath, (ii) easy to remove dissolved gases (like nitrogen, hydrogen, etc.) and (iii) inclusions are separated out. The chemistry of making mild steel ingots can be controlled easily in the induction furnaces by judicious use of scrap, sponge iron and certain ferro-alloys. At present, the structural steel of good quality is produced by the induction furnace. In India, majority of the induction furnaces are operated using acidic lining of silica-based ramming mass to produce structural steel where phosphorous and sulphur are not removed.

The choice of refractory materials is normally monolithic materials and depends on the type of charge, i.e. acidic (silica based), basic (magnesia based) or neutral (alumina based). The durability of the crucible depends on the grain size, ramming technique, charge analysis and rate of heating and cooling of the furnace.

16.5.3 Electromotive Force

The electromotive force of induction is as follows:

$$E = 4.44\,\Phi_{\text{max}}.\text{f.n.}10^{-8}\ \text{V} \tag{16.39}$$

where

Φ_{max} magnetic flux density,
f alternating current frequency,
n number of the inductor turns.

If Φ_{max} drops, the required electromotive force of induction can be retained by increasing current frequency. In that case, the magnetic flux will change more frequently. The electromotive force induced in the metallic charge forms circular currents in the plane of the winding turns, i.e. perpendicular to the magnetic flux axis. The density of induced currents attains its maximum at the surface of the charge and lowers towards the middle. The voltage across two parallel cross sections of a conductor is the same, and the current and current density in an inner element are lower than in a surface element.

The depth of the surface layer of a metallic charge, where the density of induced current is large, is called penetration depth. The heat required to melt the charge is developed mainly in that layer. The penetration depth may be as follows:

$$h_{\text{p}} = 5.03 \times 10^3 \sqrt{\left(\frac{\rho}{\mu f}\right)}\ \text{cm} \tag{16.40}$$

where

ρ resistivity of the charge, Ω cm,
μ magnetic permeability,
f frequency, Hz.

For iron at 20 °C, $\rho = 10^{-5}$ Ω cm and $\mu = 100$; at around 750 °C iron loses its magnetic properties, then $\rho = 1.1 \times 10^{-4}$ Ω cm and $\mu = 1.0$.

The lowest frequency for a given metal at its curic point is determined as follows:

$$f_{\min} \geq 2.5 \times 10^{9}\left(\frac{\rho}{d^{2}}\right) \tag{16.41}$$

where d is the mean diameter of the crucible, cm.

Large furnaces require lower frequency than smaller ones. The energy that is transformed into heat in the charge is as follows:

$$w = I^{2}n^{2}2\pi^{2}\left(\frac{d}{h}\right)\sqrt{(\rho\mu f \cdot 10^{-9})}\ \mathrm{W} \tag{16.42}$$

where

I current in the inductor, A,
d mean diameter of crucible, cm,
h depth of metal in the crucible, cm.

The product (In) is called *ampere-turns*. The energy that is transformed into heat in the charge is proportional to the square ampere-turns and to the square root of resistivity and frequency.

Practical relationships between the dimensions of the inductor and the melting crucible are as follows:

(1) Ratio of the height of the inductor (H) to the depth of metal in the crucible (h):

$$\left(\frac{H}{h}\right) = 1.1\ \text{to}\ 1.3 \tag{16.43}$$

(2) Ratio of the crucible diameter (d) to the depth of metal in the crucible (h) for furnace of various capacity:

Ratio	Upto 500 kg	500–1500 kg	1.5–3.0 t	Above 3 t
$\left(\frac{d}{h}\right)$	$\left(\frac{1}{2}\right)$ to $\left(\frac{2}{3}\right)$	$\left(\frac{2}{3}\right)$ to $\left(\frac{3}{4}\right)$	$\left(\frac{3}{4}\right)$ to $\left(\frac{4}{5}\right)$	$\left(\frac{4}{5}\right)$ to 1

16.5.4 Cooling System

Induction furnaces have two separate electrical systems: one for the cooling system, furnace tilting and instrumentation, and the second for the induction coil power. The power for the induction coil is fed from a 3-phase, high voltage, high amperage electrical line. When AC current flows through the coil, it creates an electromagnetic field which in turn induces eddy currents in the charged material. This charge material gets heated up as per Joule's law, and with further heat, the charge material melts.

The cooling system is normally a through one-way flow system with the copper coils connected to water source through flexible rubber hoses. The inlet is from the top, while the outlet is at the bottom. Loss of heat conducted from the charge through the refractory crucible requires the coil to be cooled with water as the cooling medium to prevent undue temperature rise of the copper coils.

Efficiency of induction furnace is expressed as electrical and heat transfer losses. Electrical losses consist in transformer, frequency converter, condenser, wiring, cable, coil, etc. Loss in coil is essential factor, on which the furnace capacity depends. Heat losses in induction furnace consist of conduction loss of heat escaping from furnace wall to coil side, radiation loss of heat released from melt surface, absorption loss in ring hood, slag melting loss, etc., and the coils of furnace are water-cooled which also result in heat loss.

16.5.5 Operation

In induction furnace steelmaking, no direct oxygen is introduced in the bath. The impurities are oxidized by introducing FeO in the bath. The requirement of FeO for slag formation and oxidation of various elements is fulfilled by addition of sponge iron or mill scale. Thus, the presence of FeO is important as it corresponds to oxygen potential of the slag. Removal of phosphorous takes place by oxidation. The product, phosphorous pentoxide is being held by basic constituents, like CaO, present in the slag. In steelmaking processes, basicity of the slag is maintained by addition of calcined lime. Bedarkar and Singh [29] concluded that development of melting practices with basic or neutral linings is very important from the point of removal of phosphorous or sulphur from the steel. Stirring of slag and metal in IMF is also important. Mixture of sponge iron, cast iron and steel scrap can be successfully melted in the IMF lined with MgO-based lining. Phosphorous can be removed in IMF by maintaining required basicity and FeO in the slag.

Unnecessary super-heating of liquid steel to high temperature costs to energy significantly. Depending on steel specification and temperature loss during transfer of liquid steel to continuous casting machine, super-heat temperature is to be decided. The composition of the slag varies depending on the specific process being used and the type of steel being produced. The composition of furnace and ladle slags is often very complex. Slag generated during melting has tendency to stick on the furnace wall. This reduces volume of furnace hence reduces metal output per heat. Super-heating of metal is done at higher temperature and held for few minutes. This inhibits slag to deposit on the furnace lining keeping furnace clean with full volume.

Tilting of the furnace is to affect pouring of the melt as a last operational activity before casting. The furnace is usually tilted to achieve a maximum angle of 90° or greater for complete pouring of the liquid steel.

16.5.6 Merit and Limitation

The greatest advantage of the induction furnace is its low capital cost when compared with other types of melting furnaces i.e. EAF. Its installation is relatively easier, and its operation is simpler. Among other advantages, there is very little heat loss from the furnace as the bath is constantly covered and there is practically no noise during its operation. The liquid metal in the induction furnace is circulated automatically by electromagnetic action so that when ferro-alloy additions are made, a homogeneous product is ensured in minimum time. The time between tapping and charging, the charging time, power delays, etc., which are items of utmost importance, are meeting the objective of maximum output in tonnes/hour at a low operational cost.

The advantage of the induction furnace is a clean, energy-efficient and well-controllable melting process compared to most other means of metal melting. Since no arc or combustion is used, the temperature of the material is no higher than required to melt it; this can prevent loss of valuable alloying elements. The one major drawback to induction furnace usage in a foundry is the lack of refining capacity; charge materials must be clean of oxidation products and of a known composition and some alloying elements may be lost due to oxidation (and must be re-added to the melt).

An advantage of induction heating is that the heat is generated within the furnace's charge itself rather than applied by a burning fuel or other external heat source, which can be important in applications where contamination is an issue. Operating frequencies range from utility frequency (50 or 60 Hz) to 400 kHz or higher, usually depending on the material being melted, the capacity (volume) of the furnace and the melting speed required. Generally, the smaller the volume of the melts, the higher the frequency of the furnace used; this is due to the skin depth which is a measure of the distance an alternating current can penetrate beneath the surface of a conductor.

One of the most critical problems with the induction furnaces is its limitation to refine steel to reduce phosphorous content below the desired limits. Higher phosphorous and pickup of nitrogen during induction melting make the final steel product hard and brittle and unstable for many critical applications.

The main limitation in maintaining quality of structural steel is controlling the quantum of phosphorus in steel produced through induction furnace route. In this route, the steel retains phosphorous in the range of 0.045% to as high as 0.09% depending on the quality of raw materials. The main source of phosphorous in induction furnace is sponge iron and cast iron, the quality of which is directly related to quality of iron ore. The removal of phosphorous takes place by oxidation. The product of the oxidation is held in combination with basic constituents in the slag. The extent of the removal is governed by equilibrium condition which is characterized by the metal and slag compositions. Effect of temperature is also important.

In other primary steelmaking furnaces, phosphorous is removed using direct oxygen lancing in the bath. Refining the steel with oxygen lancing in induction furnace is difficult as the furnace is operated under full volume condition. The next major limitation with induction furnace steel production in India is that almost all the furnaces are operated with acidic or silica lining, in which it is difficult to maintain the basicity of the slag. Magnesia-based basic lining and alumina-based neutral lining have also been used in induction furnaces. Basic linings are more popular in foundry-based induction furnaces with the heat size less than 5 t. Alumina-based lining in induction furnaces have been tried in a few foundries. The main limitation with alumina lining is its cost which is almost 10–15 times higher compared to silica. Both linings, magnesia based and alumina based, may become popular on acceptance of refining of steel in induction furnace in terms of phosphorous and sulphur removal. Without the regular use, it is difficult to comment on their lining life and economics of steel production.

Depending on the installed power density and the melting practice, the thermal efficiency of the induction furnace can exceed 80%, but usually it ranges from 60 to 78%. The theoretical requirement of energy for melting iron is only 340 kWh/t, whereas the actual power required is around 600 kWh/t. This difference is due to two factors: (i) one inherent in the principle of melting in an induction furnace and (ii) other operation. The inherent reasons in the induction furnace include the inefficiency in (i) electrical bus bar losses, (ii) eddy current losses, (iii) refractory losses and (iv) cooling water losses, etc. The operational losses are largely due to unnecessary and excessive holding of liquid steel in the induction furnace.

Merits of using sponge iron in induction furnace can be summarized as follows [31]:

(i) No additional de-sulphurization is required, and at the same time, one can get product with sulphur content as low as 0.012–0.015%.
(ii) Final product contains low amount of residual metals like chromium, copper, molybdenum, tin, etc.
(iii) Charging time decreased which also reduces overall heat loss.
(iv) It improves the product quality consistency.

16.5.7 Difference Between EAF Versus IMF

Induction furnace differs from electric arc furnaces in the following ways. As compared to arc furnaces, induction melting furnaces (IMF) do possess the following characteristics:

- High, relatively narrow melting vessel (i.e. low d/h ratio),
- Low crucible wall thickness,
- Relatively small area of liquid metal in contact with slag,
- Lower slag temperature,
- No carburizing during melting down,
- Powerful bath agitation.

The comparison of the operating parameters of IMF with those of EAF is given in Table 16.12. The IMF has the following technical advantages over EAF:

- It is possible to melt very low carbon steel, since there are no electrodes.
- Liquid steel contains very low in gases, due to absence of arcing.
- Furnace productivity is high, since very low oxidation of alloying elements.
- Temperature of the process can be controlled quite accurately.
- Metallic yields are higher.
- Capital expenditure and space requirement are lower.
- Induction furnace is suitable for charging addition at any time due to the characteristics of the bath agitation.
- Process is relatively cleaner and lesser environment related expenditure.

Table 16.12 Comparison of EAF and IMF [30]

Consumption	EAF	IMF
Electrical energy, kWh/t	490–510	580–600
Electrode, kg/t	2.4–2.6	Nil
Refractory, kg/t	4.1–4.2	3.4–3.6
Oxygen, Nm^3/t	15–25	Nil
Flux, kg/t	25–28	Nil
Slag generation, kg/t	65–72	11–15
Dust generation, kg/t	6–10	1–2
Noise level, dB(A)	95–120	82–86

Disadvantages of induction melting furnace are as follows:

- Slag is heated only by the bath, and this may not be enough for it to melt.
- Requirement of minimal wall thickness of the refractory lining is having risk of crack formation, resulting in stoppage of operations.
- It requires more quality of scrap.
- Refining of liquid metal is restricted in general and in particular de-phosphorization and de-sulphurization are restricted due to refractory wear.
- Induction furnaces of very high capacities are not available.

16.5.8 Applications

Induction furnaces are made in a wide range of sizes from less than one kg to one hundred tonnes and are used to melt pig iron and steel, non-ferrous metals (i.e. copper, aluminium and precious metals, etc). Products made with the induction furnace melting by the industry include mild steel ingots/billets for structural purposes, stainless steel ingots/billets for making utensils, wire rods and wires, low alloy steel castings, etc.

16.6 Quality Steel Production by Using Sponge Iron

The high-quality steels essentially mean lower and lower levels of residuals (in ppm) such as sulphur, phosphorous, oxygen, hydrogen and nitrogen; combined content of carbon, sulphur, nitrogen and hydrogen are being produced in the levels of 100–150 ppm (i.e. 0.01–0.015%). The high-quality steels with very low levels of carbon, sulphur, tramp metallic elements (e.g. Cu, Sn, Ni, Cr, Mo, etc.), and non-metallic inclusions are required to meet the demands of space, defence, power plants, automobile, food processing, gas or oil pipe lines and allied fields. The following attributes are generally expected in clean steel:

- Low content of residual impurity elements, such as S, P, O, N and H,
- Absence or restricted amount of trace elements, like As, Sn, Cu, Pb, Bi, etc.,
- Controlled amount and size of non-metallic inclusions (NMI) of mainly oxides, sulphides, nitrides, etc.

The quality steels are efficiently produced in EAF using steel scrap. Major problem faced by steelmakers was short supply, fluctuating prices and extremely heterogeneous nature of scrap in 1980s. Over and above, tramp metallic elements (0.13–0.73%) are present in the scrap which is highly undesirable for production of high-quality steels. Sponge iron/HBI has uniform chemical and physical characteristics, and it hardly contains any tramp metallic elements (0.02%) and has low sulphur content. The partial/total replacement of scrap by sponge iron/HBI is used as feed materials in EAF for quality steels production, which ultimately improves the mechanical and metallurgical properties of the product.

16.6.1 Charging

Charging of sponge iron/HBI in EAF is carried out by (i) batch and (ii) continuous mode.

(i) Batch charging: When maximum 20–30% sponge iron/HBI to be charged, this can be done by manually through the door of the furnace; or sometime from the basket. Sponge iron/HBI, sandwiched between layers of scrap in the basket, is charged in molten steel bath prepared initially with scrap.
(ii) Continuous charging: It is done from the roof with the special hole (i.e. fifth hole) on that, leads to a substantial saving in time and avoids radiation losses. As and when needed, sponge iron/HBI along with other necessary alloys and slag forming constituents can be fed directly into the furnace at the point where it is needed most.

Continuous charging mode has several significant benefits in the operation of the EAF as follows [14]:

- Less power-off time,
- Lower heat loss,
- Higher power efficiency,
- Improved bath heat transfer and faster metallurgical reactions,
- Combined melting and refining steps,
- Reduced charge to tap time.

The combined effect of above these is increased furnace productivity and reduced both electrode and power consumption related to batch charging mode. Attaining the best melting results also requires control of the feeding rate of sponge iron. The feeding rate is primarily dependent upon the effective power input but is also influence by the sponge iron/HBI composition, bath temperature and heat losses. For a given sponge iron/HBI, EAF is usually calibrated by plotting the change in bath temperature in unit time as a function of the sponge iron/HBI feeding rate, as shown in Fig. 16.25. For most furnaces, constant bath temperature is maintained at the feed rates of sponge iron/HBI in the range of 27–35 kg/min MW of applied power. The critical feeding rate of sponge iron/HBI is 33 kg/min MW [14]. Feeding rate is very crucial, if the rate is fast, then melt will be chilled; if rate is slow, then melt is heated up.

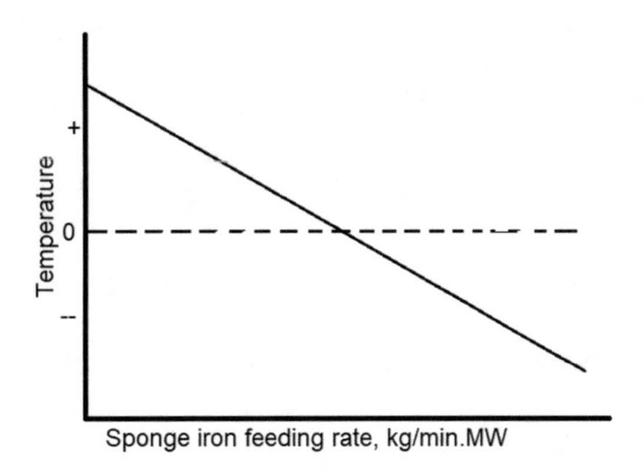

Fig. 16.25 Effect of sponge iron feeding rate on bath temperature [14]

16.6.2 Melting

After fettling of the furnace is completed, initially scrap is charged by basket, and maximum power is applied to melt the initial charge. When the charge is melted and getting molten pool or hot heel, a temperature of 1570 °C is attained. Continuous feeding of sponge iron from the fifth hole to the EAF is started. This also requires full power to be supplied during the melting. The slag at the beginning of sponge iron feeding is viscous and acidic; later, it become more fluid and basic. Irrespective of charging mode (i.e. continuous or batch), sponge iron is always charged after initially getting molten pool, i.e. hot heel by melting of steel scrap or remaining liquid metal (20–25%) of earlier heat.

The greatest benefits of continuous charging are generally attained when the melting of sponge iron takes place at the slag-metal interface, a condition which necessitates the sponge iron quickly penetrating the slag layer. Failure of a substantial portion of the sponge iron to quickly penetrate the slag layer results in excessive cooling of the slag with subsequent build-up sponge iron both in and on the slag. These build-ups, commonly called *icebergs* or *ferrobergs* just like iceberg in sea, result in erratic furnace operation, several refractory lining attack and potentially wild furnace boil. Due to porous nature of sponge iron, sometimes penetration to the slag is difficult by the lower specific gravity of porous sponge iron (2.4–3.0) than that of furnace slag (3.1–3.3), and this creates a tendency for the sponge iron to float on the slag. This penetration problem can be solved by two ways: (a) to increase the density of the sponge iron, density of the sponge iron can be increased by making compact to form HBI / briquetted sponge iron (specific gravity 3.5); or (b) to decrease the density of the slag, density of the slag can be decreased by making foaming slag (specific gravity 1.0).

Melting of sponge iron in EAF is also greatly influenced by factors like (i) carbon content and (ii) degree of metallization of sponge iron.

(i) Carbon content of sponge iron reacts with unreduced iron oxide content of sponge iron leading to CO gas evolution from liquid bath that causes: (a) boiling action in the molten bath to remove of hydrogen and nitrogen gases, homogenization of the bath and producing cleaned steel by helping to float the inclusions; (b) as well as producing foaming slag.

Carbon boil occurs at slag-metal interface by the reaction:

$$(\mathrm{FeO}) + [\mathrm{C}] = [\mathrm{Fe}] + \{\mathrm{CO}\} \tag{16.2}$$

Foaming slag, which prevents the heat loss of the bath and overheated the refractory wall of the EAF, is formed due to carbon boil. Therefore, higher carbon content in sponge iron is always desired, that is why steelmakers always prefer gas-based sponge iron (where carbon content 1.0–2.5%) instead of coal-based sponge iron (less than 1% C). If carbon is less than required amount, it must be externally added to the melt.

The amount of carbon (in Kg) requires for reducing FeO content of the sponge iron [14]:

$$\mathrm{C} = 1.67\left[100 - \%M - \left\{\left(\frac{\%\mathrm{Sl}}{100}\right) \times \%\mathrm{Fe}\right\}\right] \tag{16.38}$$

where M is the degree of metallization, Sl is the amount of slag and Fe is the amount of iron in the slag.

The combination of vigorous carbon boil and deep foaming slag has the following benefits for EAF operation:

- Lower power consumption,
- Faster melting rate,

- Decrease electrode consumption,
- Longer refractory life and
- Enhance steel quality.

(ii) The degree of metallization varies from 85 to 95% depending on the process adopted for sponge iron production. Low degree of metallization leads to economic disruption such as higher energy consumption, higher slag volume, consumes more heat and lower yield during steelmaking. As a thumb rule, 1% decrease in degree of metallization of sponge iron consumes extra power 15 kWh/t of liquid steel [14].

Although 10–20% scrap is usually charged to the EAF, the use of 100% sponge iron is advantageous. Hot heel practice is common in which 10% or more of the molten metal from the prior heat, plus some slag are retained in the furnace after tapping previous heat. Sponge iron is then continuously charged under full power to this molten pool, and a deep foaming slag is established which minimizes destructive arc radiation to the refractory lining. The furnace productivity is increased by continuous charging sponge iron late in the melting cycle, thereby combining melting and refining. The usual procedure is to melt the scrap in the furnace with one-third to one-half the normal rate of full power. As the bath temperature reaches around 1540 °C, a good carbon boil and foaming slag are established; the feeding rate of sponge iron is increased at full power to maintain the bath temperature in the range of 1540–1580 °C. At the end of melting period, the feeding rate of sponge iron is decreased while maintaining full power, thus raising the bath temperature to the taping temperature (which is 100–150 °C above the melting point of steel produced).

By continuous charging of sponge iron/HBI, increase the furnace productivity due to combining melting and refining periods. The high carbon in the sponge iron could decrease the nitrogen content in the steel. High carbon content in the sponge iron is preferred to decrease the energy consumption in the EAF. Besides the benefit of decreasing the energy consumption, another important improvement was observed, namely the fact that the nitrogen content at meltdown was lower as the carbon per cent increase in the sponge iron. High carbon per cent in the sponge iron can be important to decrease the energy consumption of the EAF by means of increasing oxygen injection (the oxidation of carbon representing an important chemical energy contribution). The final nitrogen content in steel would be very low (7–12 ppm) when sponge iron/HBI with 2.90% C was applied. Closed door practice (EAF) is also very important to avoid nitrogen absorption from the atmosphere.

16.6.2.1 Rate of Melting [31]

(a) Based on initial weight:
Sponge iron/HBI dissolves in molten bath,

$$z = x - y \tag{16.44}$$

where x and y are the initial weight (g) and final weight (g) of sponge iron/HBI before melting and after melting at time t (sec), respectively.
Fraction of sponge iron/HBI dissolution, at time t (sec),

$$f = \frac{z}{x} \tag{16.45}$$

Therefore, the rate of sponge iron/HBI dissolution (per sec),

$$R = \frac{\mathrm{d}f}{\mathrm{d}t} \tag{16.46}$$

(b) Based on amount dissolved:
Laboratory melting rate (R_1, g/sec) of sponge iron/HBI can be calculated by:

$$R_1 = \frac{z}{t} \tag{16.47}$$

Equation (16.47) is not valid for industrial scale, so Eq. (16.47) can be modified by R_2, kg/min:

$$R_2 = 0.06R_1 \tag{16.48}$$

(c) Based on overall melting of sponge iron/HBI:
If total weight of sponge iron/HBI (W_n, kg) melts in t_n minutes, then net melting rate (R_n, kg/min):

$$R_n = \frac{W_n}{t_n} \tag{16.49}$$

16.6.2.2 Iron Yield

Iron yield (%) can be calculated:

$$\left[\frac{\{(F_2 \cdot W_2) \cdot 100\}}{\{(F_1 \cdot W_1) + (F_T \cdot W_0)\}}\right] \tag{16.50}$$

where F_1, F_T and F_2 are fraction of total iron present in steel scrap, sponge iron/HBI and steel product, respectively. W_1, W_0 and W_2 weight of steel scrap, sponge iron/HBI and steel product, respectively.

16.6.3 Mechanism of Nitrogen Removal

It was established that nitrogen removal from molten bath was possible due to the CO gas evolution by the carbon oxidation, and additionally that by applying big proportions of sponge iron/HBI, low nitrogen levels are possible via dilution. CO gas from the reduction of FeO in sponge iron is evolved rapidly, while the sponge iron remains buoyant in the slag phase. The rate of CO gas evolution from the reduction of FeO in sponge iron/HBI at steelmaking temperatures is fast, controlled by heat transfer and occurs at relatively low temperatures ranging between 800 and 1200 °C. CO gas bubbleshelp to remove nitrogen as well as hydrogen gases from the molten bath [19]. They also help to avoid nitrogen pickup of the molten bath from the air. By means of maintain of reducing atmosphere in the furnace and continuous slag flushing, the dissolved gases collected in the slag are removed. Sponge iron/HBI containing high carbon is the key to maintain of proper slag foaming by means of CO gas evolution throughout the heat.

16.6.4 Product Characteristics

By using sponge iron/HBI, the highest quality flat products can be produced with excellent formability and ageing characteristics. Sponge iron/HBI is also ideally suited to produce very low sulphur steels which are growing in worldwide importance. Since sponge iron/HBI contains very low sulphur, the final sulphur concentration in the product steel can be reduced from 0.02 to 0.004% by increasing sponge iron/HBI in the feed material from 35 to 95% [14]. At such low sulphur levels, problems with segregation and hot shortness are non-existent. The steel of 0.035% S and P levels gave impact value of 0.4 kg cm and for same steel when S and P are controlled to 0.015% level, impact value was improved to 3.0 kg cm.

The purging effect of the prolonged carbon boil occurring during melting of sponge iron/HBI also results in very low hydrogen and nitrogen gases content in steel compared to that of scrap practice as shown in Table 16.13. Addition of fluxes, carbon and ferromanganese are minimized to avoid their nitrogen contributions. Table 16.14 shows the nitrogen gas content for different processes for steel production. 15 ppm nitrogen content in steel can be achieved by using 100% HBI (2.7% C) charge in EAF.

With increased quality consciousness on the part of consumers, it is important that input scrap quality should be controlled. The most troublesome tramp metallic elements (i.e. Cu, Co, Sn, As, Sb, Ni, Mo, etc.) from scrap are ultimately concentrated in product steel. Their presence has been found to induce undesirable resistance to deformation, hot shortness and mechanical defects. Hence, a charge mix with sponge iron/**HBI** has helped in lowering their menace considerably. Figure 16.26 shows the dilution effect of the addition of sponge iron upon the bath residual metal levels. By using sponge iron, the highest quality flat products can be produced with excellent formability and ageing characteristics. On the other hand, low P, C and Si grades of steel can be produced only by using sponge iron to the extent of 40% of the charge.

It is observed [31] that with increase in the sponge iron/HBI proportion in the charge, carbon content of the product increases but tramp elements and sulphur are decreased. With increase in the sponge iron/HBI percentage in the charge, net melting rate and iron yield are increased.

With increase in the demand of special quality steel throughout the world, product should contain low amount of tramp elements as well as low sulphur and it should have less segregation in the products. Mechanical properties of steel produced are improved by using sponge iron/HBI in the charge material. Segregation of carbon in product steel is also decreased. By using sponge iron/HBI in the charge, high iron yield is achieved, and tap-to-tap time of the heats is subsequently decreased.

16.6.5 Advantages of Using Sponge Iron in EAF

(i) Low power consumption: Melting is achieved with longer arc using a higher voltage. Since melting and refining are done simultaneously, overall power consumption decreases.
(ii) Faster melting time: Foaming (low density) slag allows rapid penetration of sponge iron to the slag-metal interface and enhances heat transfer.

Table 16.13 Comparison of gas content in steel [14]

Gas	Gas content, ppm	
	100% scrap heat	75% sponge iron heat
Hydrogen	2.0–5.0	1.5–3.0
Nitrogen	40–60	20–25

Table 16.14 Nitrogen content for processes of steel production [19]

Process	BOF	EAF (100% Scrap)	EAF (100% HBI with 2.4% C)	EAF (100% HBI with 2.7% C)
Nitrogen content in steel, ppm	30	60–100	25	15

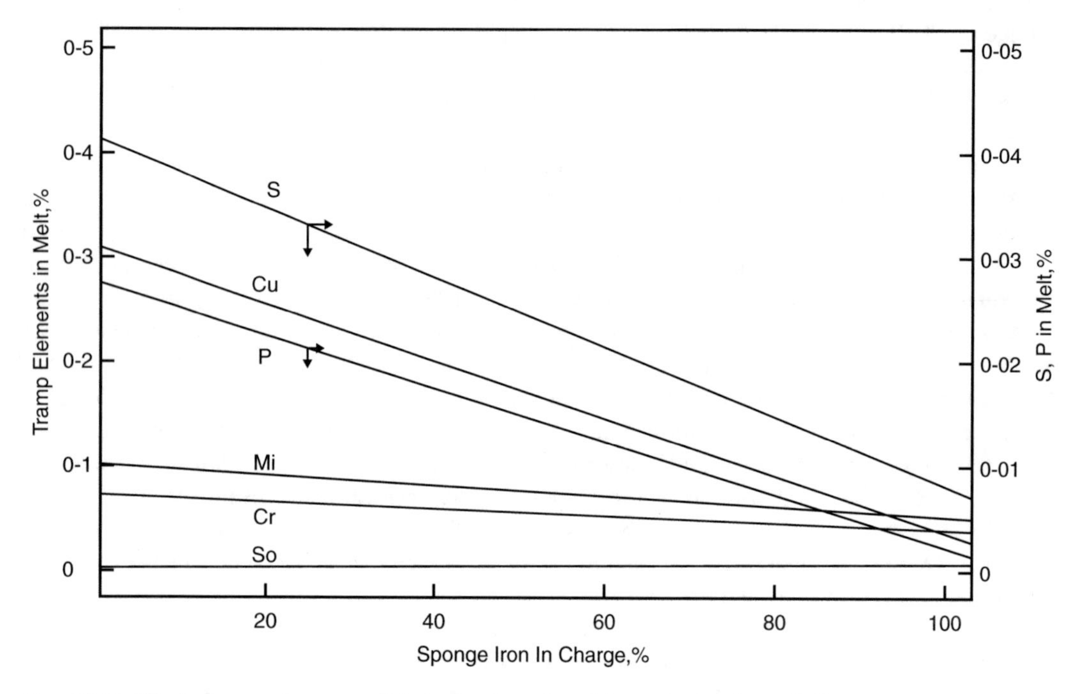

Fig. 16.26 Effect of sponge iron on melt in tramp elements, sulphur and phosphorus [14]

(iii) Decrease electrode consumption: Furnace atmosphere is rich in CO gas, which decreases oxidation of electrode, as well as decrease in electrode breakage due to tilting of scrap.
(iv) Longer refractory life: EAF can be operated at full power without serious refractory wear because arc radiation is absorbed by the foamy slag.
(v) Enhance steel quality: CO gas evolved at the slag-metal interface (as carbon boil) removes nitrogen and hydrogen gases from the bath and improves slag-metal mixing to help removal of sulphur and phosphorus from the bath.

16.7 Use of Hot Metal in EAF

A growing concern about the dwindling energy reserves and the immediate problem of spiralling energy costs have led to technological innovations which have resulted in substantial reductions in the specific energy consumption in the world steel industries.

The rising cost of electricity and its difficult availability in India has led to the development of the mini-blast furnace ironmaking to complement of EAF route, both to reduce costs and to increase overall production. With the availability of hot metal/pig iron from various mini-blast furnace, the use of hot metal/pig iron in the EAF has shown a dramatic increase among the major producers. With door lances blowing oxygen up to 20 Nm^3/min and up to 30% solid pig iron has been successfully utilized in the EAF.

Hot metal contains carbon and silicon, which are inherent energetic advantages, allow for significant reduction in electric power cost for EAF operation. EAF steelmakers prefer to charge hot metal from a mini-blast furnace directly in the EAF to improve productivity and decrease cost of production. The energy content of hot metal consists of both sensible and chemical heat. The sensible heat of hot metal is given by the linear relationship [33]:

$$q_{HM} = 0.22T + 17 \quad (16.51)$$

where q_{HM} is the sensible heat of hot metal, kWh/t; and T is the hot metal temperature, °C.

The chemical heat of hot metal, depends on the amount of the elements present in hot metal, is produced due to highly exothermic oxidation reactions. At a charge ratio of 30% hot metal, power savings up to 220 kWh/t can be achieved. It has been reported that every 1% addition of chemical energy replaces 5 kWh/t of electrical energy [15]. To achieve the desired steel quality of high standard, better quality iron charging material must be used. Since hot metal contains only about 0.06% residual elements as compared to >0.25% in steel scrap, substituting scrap with hot metal improves the quality of steel. Overall residual elements of steel decrease, and hence, hot metal acts as diluent to the molten bath and increases the number of heats per day from 8 to 16, i.e. double the production. Overall benefits obtained by using hot metal as charge material to the EAF are shown in Table 16.15.

There are two methods of hot metal charging into the EAF:

(a) Charging through roof: Initially, scrap is melted in the EAF with closed roof. After liquid pool of metal is formed, melting is halted, and the roof opened; the hot metal is quickly poured into the furnace. The disadvantage is that the melting process must be interrupted, and radiation losses occur due to opening of the roof.
(b) Charging using a launder: Initially, scrap is melted in the EAF with closed roof. Oxygen lancing is performed to cut the heavy scrap which is near the door as to permit the insertion of the launder through the door. The hot metal is poured smoothly with the help of launder into the EAF, while scrap melting continues without interruption. This is the most preferred technique as it does not increase the tap-to-tap time and no heat losses by radiation occur. It is essential that the charging time should be keep as short as possible (about 10 t/min) to minimize cooling of the hot metal.

Table 16.15 Benefits of using hot metal in EAF [33]

Parameter	Charge		Saving
	100% Scrap	55% Hot metal + scrap[a]	
Energy consumption, kWh/t	670	295	375
Electrode consumption, kg/t	4.66	2.95	1.71
Oxygen, Nm^3/t	20	35	−15
Power-on/melting time, min	120	57	63
Tap-to-tap time, min	149	90	59
Number of heats/day	8	16	8

[a]At Usha Martin, Jamshedpur (India)

Advantages offer by use of hot metal in EAF:

(i) Decrease electrical power consumption,
(ii) Shorter power-on times and tap-to-tap times,
(iii) Increase in furnace productivity, by increasing number of heats per day,
(iv) Reduction in electrode consumption,
(v) Decrease in scrap consumption,
(vi) Less dependence on electric power for steelmaking.

16.7.1 CONARC Process

This process is developed in 1990s due to: (i) shortage of steel scrap, (ii) rising costs as well as unreliable supply of electric power and (iii) low productivity and higher power consumption with respect to developed countries. These problems can be overcome by using hot metal [which has heat content, $q_{HM} = (0.22T + 17)$ kWh/t (Eq. 16.51)] as well as chemical energy content [33] in EAF to produce steel at lower costs, because: (a) major electric power saving (i.e. cost saving), (b) increase productivity with high quality of product and (c) independence from scrap shortage and its price fluctuation.

Nowadays, world steel production is shared by two major processes: LD/BOF and EAF. Basic chemistry of steelmaking is nearly similar in both these processes [33], but they differ from one another in several ways.

Characterization of LD/BOF process:

(i) Ability to generate heat energy (i.e. heat of reaction) within itself for meeting process requirement,
(ii) Hot metal and scrap (20–25%) as the common burden materials and
(iii) Short tap-to-tap time resulting in high productivity.

Characterization of EAF process:

(i) Solid scrap and/or sponge iron/HBI as the common burden materials,
(ii) Require external energy (i.e. electric energy) for the melting,
(iii) Productivity slightly lower than LD/BOF process.

These two dissimilar technologies have merged under a set of special circumstances develops a new technology. This new process which is half BOF and half EAF, i.e. combines oxygen converter and electric arc furnace technologies, termed as *Electroxy/CONARC* process. It employs a twin shell EAF which can economically handle raw material input of solid scrap and hot metal in varying proportions. The high energy content ($q_{HM} = 325$ kWh/t at 1400 °C and $q_{HM} = 314$ kWh/t at 1350 °C) of hot metal is used to produce chemical energy in form of heat of reactions and hence reduce consumption of electric power and increase productivity. Quality and cost of steel produced depend largely on the quality of hot metal. Hence, quality of hot metal must be up to the mark and consistency of composition. Silicon in hot metal is oxidized with highly exothermic reaction:

$$[\mathrm{Si}] + 2[\mathrm{O}] = (\mathrm{SiO_2}), \Delta \mathrm{H}^{\circ}_{298} = -871.53\,\mathrm{kJ/mol\ of\ Si} \quad (16.52)$$

Increasing Si content in hot metal, oxygen requirement also increased. Sulphur and phosphorous content in steel depend on their content in hot metal, so sulphur and phosphorous content in hot metal should be as low as possible.

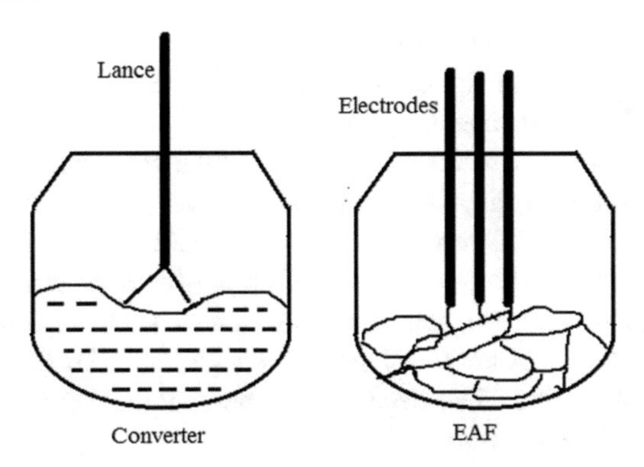

Fig. 16.27 CONARC process

CONARC is combining the conventional ***con**verter process* with *electric* ***arc*** *steelmaking* in a furnace unit having two identical vessels (as shown in Fig. 16.27). One vessel operates in the converter mode using the top lance, while the other vessel operates in the arc furnace mode with graphite electrodes. CONARC process is a twin vessel of EAF, which economically handles raw materials input of solids and hot metal in varying proportions. Figure 16.27 shows the left vessel is working as a converter and the right one as EAF.

The furnace is equipped with graphite electrodes connected with transformer, which can be moved from one vessel to the other. Oxygen is guided into the melt by water-cooled top lance system, which also can be moved from one vessel to the other.

16.7.1.1 Charging

- Through roof: Initially, scrap is melted in the vessel with closed roof by electric arcing. After liquid pool is formed, melting is halted, the roof opened, and hot metal is quickly poured in to the vessel. This minimizes the temperature drop. Major disadvantages: (i) melting of scrap must be interrupted and (ii) radiation losses due to the opening of the roof.
- Using launder: Scrap is first charged, roof is closed and electrodes are introduced to start melting. For faster melting, oxygen lancing is performed to cut the heavy scrap. After liquid pool is formed, insertion of launder takes place through the door. Hot metal is then poured smoothly at the rate of 10 t/min into the vessel, while scrap melting continues without interruption. It takes 5–10 min; pouring time should be kept as short as possible to minimize cooling of hot metal. Advantages: (i) melting of scrap has not interrupted, (ii) no heat losses by radiation, since no roof opening during pouring and (iii) no increase in tap-to-tap time.

16.7.1.2 Operation

CONARC process operation can be divided into two stages:

(i) EAF step: Electric power is used for melting of cold charge (i.e. scrap and/or sponge iron) to form liquid heel and adjusting the bath temperature (which is now acted as EAF),
(ii) Converter step: Hot metal is poured over to the liquid heel (which is now acted as converter), lance is introduced and injecting oxygen to oxidize the impurities of the bath. De-carburization of bath is also done.

The process starts with charging of scrap, and the solid charge is melted by electric energy with the help of graphite electrodes to form a liquid heel. Now this vessel acts as EAF. The electrodes are removed, and hot metal is poured into that vessel. As soon as the top lance is being brought into position, the oxygen blowing is started; now vessel acts as converter. The impurities of hot metal (i.e. C, Si, Mn and P) are oxidized and generating large amount of heat (i.e. chemical energy). Then, cold materials like sponge iron or scrap are added as coolant to the furnace to utilize these enormous heat energies and thus to avoid overheating of the molten bath. After bath temperature and composition adjustment, the top lance is removed away and tapping of liquid melt takes place. The tap-to-tap time was reduced by optimizing the furnace operation to 60 min in average (for 170 t furnace), so that the capability to produce more than 20 heats per day. The energy consumption was dropped to approximately 310 kWh/t (for 50% hot metal + 50% sponge iron). Small part of the melt kept in the furnace for charging of sponge iron. The electrodes are brought back into operation positions to melt the solid charge (i.e. scrap/sponge iron).

When one vessel act as converter to remove impurities from melt, then on the same time other vessel act as EAF to melt the solid charge. The CONARC process fulfils a maximum flexibility in a raw material charging. The furnace can be operated in the range of 100% hot metal to 100% sponge iron/scrap. The CONARC process is used to produce good quality of steel and increase the productivity. Due to that, large numbers of EAF (AC and DC), in the world, are working with hot metal. The plants of Russia, Belgium, France, Poland, India, Japan and South Africa are used hot metal as feed material [34] in EAF.

Advantages of the process:

- Wide range of charging can be done: 100% scrap, 100% sponge iron, 100% hot metal, or 50% scrap/sponge iron +50% hot metal, etc.
- Decrease tap-to-tap time: 57–64 min, i.e. increase productivity,
- More than 20 heats/day is possible.
- Electric energy consumption drops from 450–500 kWh/t to 310 kWh/t (for 50% hot metal +50% sponge iron).
- Independence from scrap shortage and its price fluctuations,
- By off-gas heat, scrap can be pre-heated.
- Major cost saving,
- Low P (<0.01%), low S (0.013%) and low C (0.003–0.04%) steel can be produced.

16.7.2 EOF Process

Developed in Brazil for replacing small, uneconomic open-hearth furnaces about 35 years ago, the ***E**nergy **O**ptimizing **F**urnace (EOF)* was conceived to utilize the sensible heat of small and medium sized steel converters in an effective way. The EOF is essentially an oxygen steelmaking vessel which employs extensively side blowing of oxygen and also provides options for auxiliary fuel (solid or liquid) injection. Oxygen is injected horizontally into the molten bath from the sides through specially designed, submerged tuyeres; and oxygen is blown above the bath through water-cooled injectors. The process is capable of handling solid, liquid or mixed charges; and there is no requirement of electric energy for melting.

The principle of EOF lies in the injection of oxygen into the bath to react with carbon of the charge producing CO gas. Carbon required for this reaction is come either from hot metal or added as a carbonaceous material. Oxygen is introduced through submerged tuyeres and atmospheric injectors.

$$2[C] + \{O_2\} = 2\{CO\}, \Delta H^\circ_{298} = -111.72\,\text{kJ/mol of C} \qquad (16.53)$$

By picking up additional oxygen, above the bath inside the melting chamber, the CO gas is burnt to CO_2.

$$2\{CO\} + \{O_2\} = 2\{CO_2\}, \Delta H^{\circ}_{298} = -282.42 \text{kJ/mol of CO} \quad (16.54)$$

The extra heat thus generated is used to melt the pre-heated scrap.

The EOF process has provisions for post-combustion, coal addition and scrap pre-heating. Figure 16.28 shows the EOF with a two-stage scrap pre-heater located above the furnace where the enthalpy of the hot flue gases is used to pre-heat the scrap around 850 °C, thereby contributing to energy optimization by recovery of heat. The pre-heated scrap is dropped into the lower chamber and melted. The process neither requires any high-power distribution system nor electrodes; rather it operates with minimum auxiliary equipment. The tap-to-tap time is ideally suited for subsequent refining the melted and partially refined metal in a secondary process of steelmaking.

The process theoretically can be charged up to 100% scrap but typically 40–60% is used. It is a combined blowing basic oxygen steelmaking process where a mix of hot metal (55–85%), scrap (15–40%) and sponge iron (0–15%) acts as charging materials. About 20 kg/t of coal is also used. Hot metal is charged into the vessel followed by scrap from lower pre-heating chamber. The scrap is pre-heated in a series of pre-heating chambers to about 800–1200 °C. Oxygen is blown through two submerged tuyeres and one or two lances. A second scrap is charged after five minutes of oxygen blowing. Post-combustion of the emerging gases above the molten bath is done by using four atmospheric injectors, thus generating heat, a part of the heat is used to the metallic bath and rest heat for scrap pre-heating.

About 70 m^3/t of oxygen is used for 50:50 scrap to hot metal charged mix, out of that one-third is blown into the molten metal bath and remain two-third is blowing above the bath for post-combustion. The post-combustion gas is used to pre-heat the scrap, and therefore, the energy from the post-combustion does not have to be completely transferred to the molten bath.

The submerged injected oxygen reacts with the carbon of the bath, formation of CO gas which promotes intense bath agitation (carbon boil), beneficial for reaction kinetics and temperature homogenization. The CO gas is burnt with the oxygen from the atmospheric injectors and generating heat. Due to carbon boil, bath surface swell and increasing the exposure to oxygen from the supersonic lances and capturing part of the heat generated by after burning. The combination of these, the extremely fast decarburizing and temperature increase of the bath, resulting in blowing times similar to that of the BOF. The continuous slag-free tapping as well as the instantaneous release of scrap from the scrap pre-heater allows tap-to-tap times decreases less than 30 min.

The EOF owes its thermal efficiency to the optimized use of energy derived from the following sources:

- Chemical energy released within the extended bath surface by the reactions between the injected oxygen and the oxidizable elements including carbon which can be added through the injectors,
- Chemical energy derived from the oxidation reactions in the furnace atmosphere mainly involving the conversion of CO to CO_2,
- Sensible heat transferred from the hot gases to the cold scrap charged into the pre-heater.

Advantages of EOF process:

1. Low investment and operating cost,
2. Flexibility in terms of metallic charge,
3. Good metallurgical control of carbon, phosphorous and sulphur,

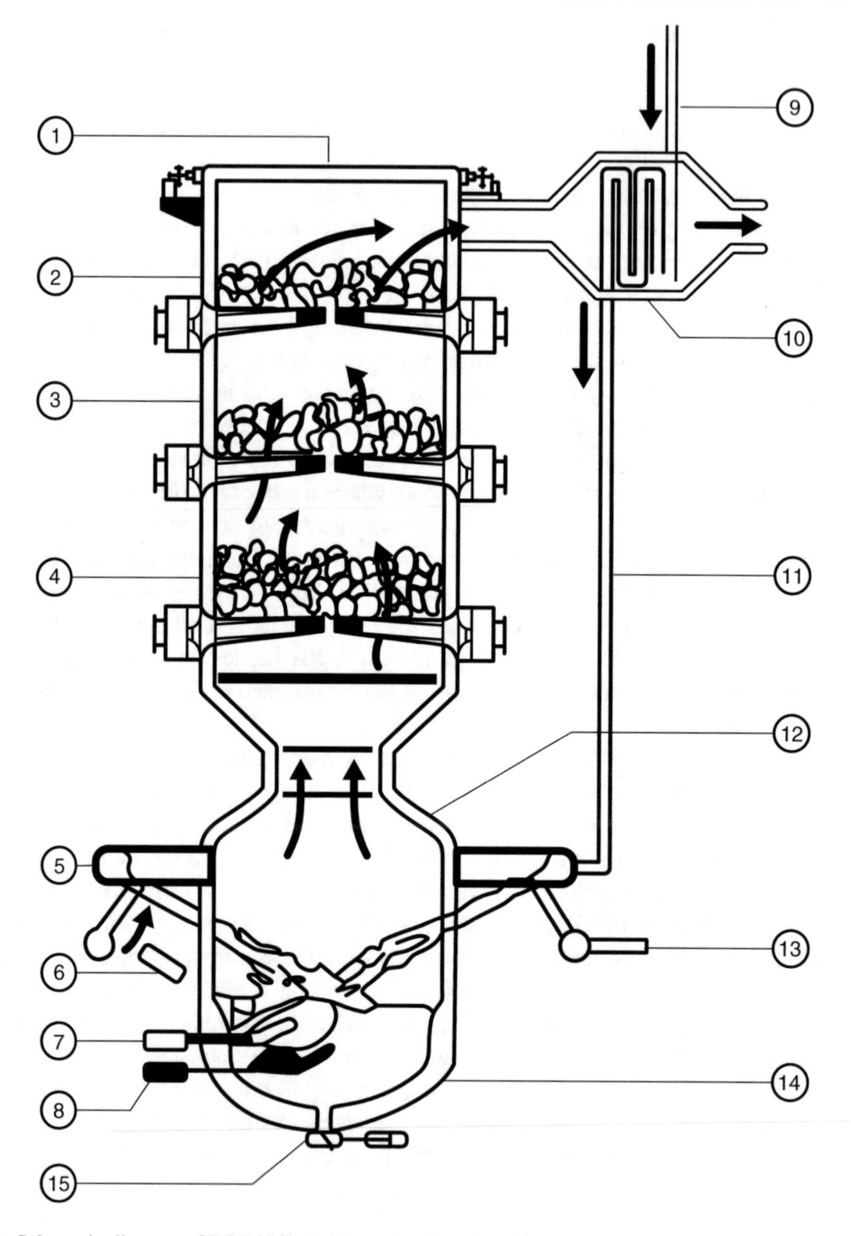

Fig. 16.28 Schematic diagram of EOF [36]. 1: Scrap charging, 2: cold scrap, 3: scrap pre-heater, 4: pre-heated scrap, 5: pre-heated air, 6: oxy-fuel burner, 7: submerged O_2 tuyere, 8: coal injection, 9: cold air, 10: recuperator, 11: pre-heated air, 12: water-cooled elements, 13: additional O_2, 14: furnace vessel, 15: bottom tapping (reproduced with permission from *Joint Plant Committee* (JPC), India)

4. High productivity and furnace availability,
5. Low inclusions due to continuous flushing of slag during blowing and slag-free tapping,
6. Decreasing electric energy consumption (25–45 kWh/t) and
7. Very low noise level and emissions.

Economic considerations for EOF on operating cost as follows:

- High productivity optimizes fixed cost,
- Low energy consumption,
- Low cost of operating consumables,
- Low maintenance cost and
- Waste heat recovery options.

The EOF experienced a strong evolution and became a refence also in terms of quality steels used for forging, seamless pipes, cold heading, roller bearings, etc. approved by the strictest clients, regarding quality control. One of the most remarkable aspects in this regard is the ultra-low phosphorous (<0.01%) content in steel which may be reached consistently. Table 16.16 shows the data of EOF on different charge mixed. 41 heats/day tapped at Hospet Steels Ltd, Karnataka, India; and 48 heats/day tapped at GERDAU Divinopolis, Brazil [37]. Tata Steel also commissioned 80 t EOF in 1990.

16.7.3 ECOARC Process

ECOARC is considered a few steps ahead of the several EAF scrap pre-heating systems, which have been developed by NKK, Japan, in 1998 on a 5 t pilot plant built at NKK in Toyama keeping in mind the renewed emphasis of the Government on environment where regulation for control or reduction of fume, dust, odours and dioxins has been legislated. With this background, NKK developed ECOARC (***ECO**logically* friendly and economically ***ARC** furnace*), as a shaft-type furnace to meet future environmental requirements and increase energy saving to below 200 kWh/t, with a target of 150 kWh/t. The schematic diagram of ECOARC is shown in Fig. 16.29.

ECOARC consists of melting chamber and scrap pre-heating shaft. The shaft is directly connected to the melting chamber and that should be avoided air infiltration. Scrap is continuously fed into the pre-heating shaft and is in constant feeding to the molten bath at the melting chamber. During the

Table 16.16 Data for different charge mixed in EOF [39]

	A kg/t	%	B kg/t	%	C kg/t	%	D kg/t	%
Metallic charge								
Hot metal	675	59.5	575.7	50.5	335	30.0	–	–
Iron scrap	329	29.0	–	–	–	–	–	–
Steel scrap	131	11.5	563	49.5	781	70.0	1100	100
Total	1135	100	1138.7	100	1116	100	1100	100
Consumption								
Fuel oil, kg/t	3		3–5		3–5		3–5	
Carbon, kg/t	0		12–18		50–60		85–100	
Oxygen, Nm^3/t	75		75–85		90–100		110–135	
Refractories, kg/t	2.8		4–5		4–5		5–6	
Blowing time, min	35		35–40		40–45		40–50	
Heats/day	21.5		20–21		19–20		18–20	

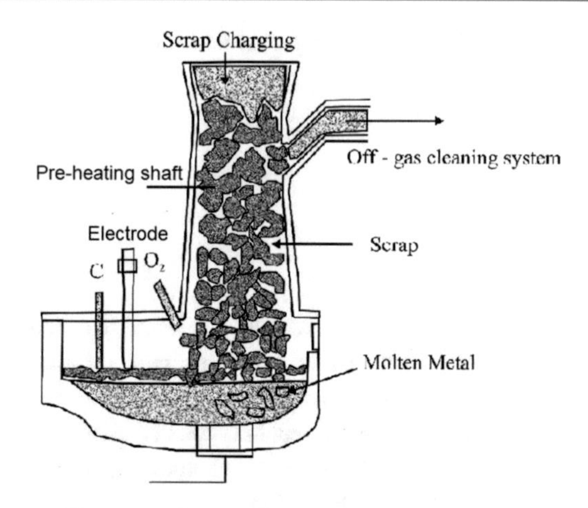

Fig. 16.29 ECOARC process [40] (reproduced with permission from *Joint Plant Committee* (JPC), India)

Table 16.17 Benefit of ECOARC compared with conventional EAF [40]

	EAF (conventional)	ECOARC (Version I)	ECOARC (Version II)
Electrical power consumption, kWh/t	380	210	150
Oxygen consumption, Nm^3/t	33	33	45
Electrode consumption, kg/t	2	1	1
Carbon consumption, kg/t	25	25	36
Consumption for the fuel for the burner, l/t	20	5	5
Dust generation, kg/t	18	9	9

melting period, pre-heated scrap constantly sinks into the molten bath, and the scrap descends the shaft. So new scrap can be charged continuously on the top of the shaft. Temperature of molten metal is about 1500–1530 °C during the melting period, due to the present of solid scrap in contact with the molten metal. The furnace is tilted about 15° towards the pouring side, and the pre-heated solid scrap is still in the melting chamber and shaft; now the refining is started. During the refining period, gap between the molten metal and solid scrap decreases due to tilting of the furnace. This allows the molten metal to be heated to 1600 °C or more. After reaching the proper temperature of the molten metal, the furnace is again tilted more to increase the gap between the molten metal and solid scrap; and one heat of molten steel is tapped. After tapping (70–75%), the furnace is turned down to the horizontal position with some molten metal remaining, and the next melting period begins.

In the ECOARC process, the flue gases (i.e. CO and CO_2), which are generated in the melting chamber, are passed to the pre-heating shaft to heat the solid scrap efficiently. The temperature reached to the pre-heated solid scrap is about 900–1000 °C. ECOARC process consumes 210 kWh/t of electrical power and 33 Nm^3/t of oxygen; if the oxygen consumption rate is increased to 45 Nm^3/t, then electrical power consumption rate decreases to 150 kWh/t (as shown in Table 16.17).

16.7.4 CONSTEEL Process

CONSTEEL process is the solution for the constant challenges faced by EAF operators. The CONSTEEL process has been successfully developed at Shaoguan Iron and Steel, China. The efficient use of available energy, meeting strict environmental regulations, flicker reduction, harmonic disturbance reduction and increasing productivity are some of the features of CONSTEEL process. In the CONSTEEL process, pre-heated solid scrap is continuously charged through a side opening in the upper shall. Hot gases from EAF are conveyed from the scrap feeding port of the furnace and through a pre-heater tunnel, flowing in counter-current to the scrap for pre-heating. Coal is injected along with oxygen into the EAF to form CO and CO_2 gas mixture as well as generating heat energy. The mixture of CO and CO_2 gases is conveyed through the tunnel, where CO is post-combusted by locally added air. 28% hot metal is directly charged in the CONSTEEL process at Shaoguan, China. Various units of CONSTEEL process have 67–250 t/h (production rate) per unit.

Advantages of CONSTEEL process:

1. Increased production (about 30%),
2. Decreased electrical energy consumption,
3. Decreased electrode consumption,
4. Less manpower requirement,
5. Decreased refractory and maintenance cost,
6. Decreased production cost,
7. Decreased environmental impact and noise level,
8. Reduced electrical disturbances,
9. 30% less dust produced.

16.7.5 FASTEEL Process

The success of Rotary Hearth Furnace (RHF)-based DRI technology (i.e. FASTMET process) has led to development of *FASTEEL* Process by Midrex and Kobe Steel. The RHF is a flat, refractory hearth rotating inside a high temperature, low pressure and circular tunnel kiln. The composite agglomerate (either pellets or briquettes) is produced from a mixture of iron oxide fines (such as ore fines, mill scale) and fine carbon sources, such as coal, coke fines, charcoal or other carbon bearing solid. The agglomerated feed is placed in the hearth, one layer thick, on FASTMET process. Burners are located above the hearth which provides heat required to raise the temperature of the feed agglomerates for reduction. The burners are fired with natural gas, fuel oil or pulverized coal. The agglomerates are fed, stay in the hearth for one revolution (typically 6–12 min) and discharged continuously depending on the reactivity of feed mixture and target product quality.

The DRI (hot or cold), produced by FASTMET process, is charged to an electric melter known as electric ironmaking furnace (EIF), developed by EMC International, a sister company of Midrex. This has led to the development of FASTMELT process. The EIF produces high-quality liquid iron known as Fastiron.

The FASTEEL process for making steel developed by Techint, Kobe Steel and Midrex merges the hot metal (i.e. Fastiron) benefits of FASTMELT process with continuous pre-heated scrap feeding in CONSTEEL process. These processes working together provide a new method of production of high-quality steel that offers significant environmental improvements, higher quality product, lower capital cost and less expensive operation than traditional method of production. FASTEEL process combines 1/3 Fastiron hot metal from FASTMELT process with 2/3 pre-heated scrap from CON-STEEL process.

16.7.6 Shaft Furnace Technology

The 20% heat energy normally leaving the EAF as the waste gas represents about 130 kWh/t of steel produced. Using this gas to pre-heat the scrap being charged to the EAF can result in recovering some of this energy and reduced of the electrical energy required to melt steel scrap.

The advantages for scrap pre-heating are as follows [3]:

- Increased productivity,
- Energy saving potential up to 77 kWh/t,
- Removal of moisture from the scrap,
- Reduced electrode consumption,
- Reduced refractory consumption.

The scrap pre-heating has been used for over 35 years primarily in countries with high electricity costs such as Japan and Europe. Conventional scrap pre-heating involves the use of hot gases from furnace to heat scrap in the bucket prior to charging the scrap into the EAF. Conventional scrap pre-heating can be done by the hot furnace gases from the fourth hole in the EAF to a special hood over the charging bucket. Typically, the gases leave the EAF at about 1200 °C, enter the bucket at 815 °C and leave at around 200 °C. The amount of pre-heating depends on the heat transfer to the scrap which is a function of scrap size, time and temperature. Typically, the scrap is pre-heated to a range of 315–450 °C. This amount of pre-heating will typically reduce energy consumption by 40–60 kWh/t, electrode consumption by 0.3–0.36 kg/t and tap-to-tap time by 5–8 min.

Disadvantages to conventional scrap pre-heating are as follows [3]:

- Scrap sticking to bucket and short bucket life,
- Poor controllability of pre-heating,
- For tap-to-tap times less than 70 min, lead to minimal energy savings that cannot justify the capital expense of a pre-heating system.

The shaft furnace technology has the outcome of the efforts to overcome the shortfalls of the bucket pre-heating techniques. It uses the energy obtained by the post-combustion to heat the scrap. This technology has evolved in stages. With a single shaft furnace, at least 50% of the scrap can be pre-heated, whereas a finger shaft furnace (involving a shaft with a scrap retaining system) allows pre-heating of the total scrap amount. A further modification is the double shaft furnace which consists of two identical shaft furnaces positioned next to one another and which is serviced by a single set of electrode arms. These technologies [42] require close control of carbon monoxide and oxygen concentrations to reduce the risk of explosions.

16.7.6.1 Single Shaft Furnace

The *single shaft furnace* (SSF) design was developed based on a conventional EAF with scrap pre-heating facilities (Fig. 16.30). The shaft serves the purpose of holding the scrap for pre-heating to an average temperature of 800 °C by means of the furnace off-gases before it is melted in the furnace

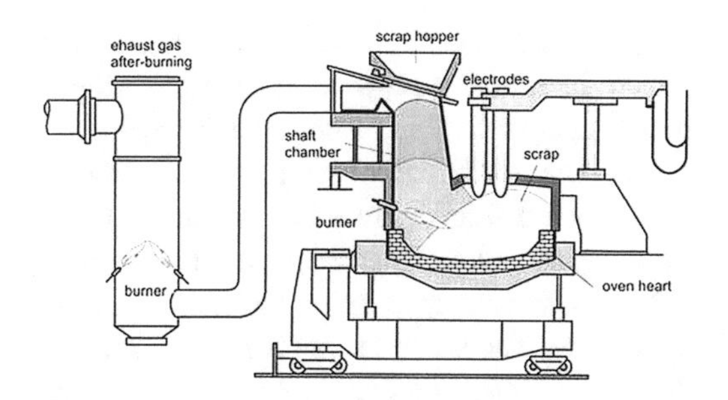

Fig. 16.30 Single shaft furnace (SSF) on the top of EAF [1]

Table 16.18 Operating parameters and consumptions for SSF and DSF at 95 t capacity furnace [41]

Consumptions/parameters	SSF	DSF
Electrical energy, kWh/t	130	340–360
Electrodes, kg/t	1.8	1.6
Oxygen, Nm^3/t	30	25–30
Fuel, Nm^3/t	7	6–8
Charge carbon, kg/t	15	8–10
Carbon powder, kg/t	5	3–8
Power-on time, min	33–40	35–37
Tap-to-tap time, min	51–60	38–43
Productivity, t/h	96–112	130–153

vessel. The SSF has been successful started in UK, Turkey, China and USA. The main performance data are shown in Table 16.18.

With single shaft system, main benefits:

(i) up to 100% scrap pre-heating,
(ii) electric energy saving up to 120 kWh/t of liquid steel,
(iii) up to 15% reduced electrode consumption,
(iv) up to 2% increase in metallic yield,
(v) up to 25% lower dust emissions and
(vi) up to 20% productivity increase.

16.7.6.2 Double Shaft Furnace

Double shaft furnace (DSF) technology is one of the logical solutions which substantially reduce the idle time, operating based on twin shell philosophy. Two shaft furnaces of equal size are positioned adjacent to each other and serviced by one set of electrode arms. Melting and refining are carried out in the active furnace, during which time the inactive furnace is tapping and subsequently charged with scrap. The consumption values are given in Table 16.18.

16.7.6.3 Finger Shaft Furnace

The full potential of the hot off-gases cannot be utilized in the single shaft furnace (SSF), since no scrap is placed in the shaft during refining; therefore, the finger shaft furnace was developed. It is not only capable of serving steel plants based on 100% scrap but is also suitable for processing sponge iron or hot metal in cases where high quality, low residual steels are produced.

The furnace is equipped with scrap retaining, water-cooled fingers in the shaft. These fingers are closed during the refining period. The first basket of scrap to be pre-heated for the next heat is charged on to these fingers. The water-cooled fingers are fabricated from heat resistant steel and reach a lifetime exceeding that of any water-cooled panel fabricated from steel. After tapping the previous heat, the furnace is brought back into melting position and the fingers are opened to allow the pre-heated scrap to drop into the furnace. The second scrap basket is charged immediately into the shaft. Charging time and energy losses are substantially reduced in this way since the furnace and shaft do not have to be removed for first basket charging. Power input is promoted by the hot scrap conditions from the beginning of the melting process. The melting rate is accelerated by using sidewall oxy-fuel burners until the heat has been evenly distributed throughout the scrap column. Once all the scrap has descended into the melting vessel, the fingers can be closed and are ready to receive the first basket of the next heat for pre-heating. Pre-heating the first scrap load retained on the fingers takes less than 15 min. The exact energy savings depend on the scrap used and the degree of post-combustion (oxygen levels). For the finger shaft furnace, tap-to-tap times of about 35 min are achieved, which is about 10–15 min less compared to EAF without efficient scrap pre-heating. The process [42] may reduce electrode consumption, improve yield by 0.25–2%, increase productivity by 20% and decrease flue gas dust emissions by 25%.

Advantages: [41, 42]

- The maximum scrap pre-heating capacity of the off-gas during the EAF process is used by charging the scrap directly into the off-gas stream.
- The scrap, while being pre-heated in the off-gas stream, serves as the first filter for the off-gas, which means less contaminated dust, and the metallic yield in the furnace is increased.
- Energy saving potential up to 110 kWh/t is approximately 25% of the electricity input.
- CO_2 emission reduction potential reduced by 35 kg CO_2/t.
- Cost savings in production and the payback within one year.

16.8 Stainless Steel Production

Steels contain 12–30% Cr with or without other alloying elements are known as *stainless steels* or *rust-free steels*, carbon contain below 0.12% (average 0.07–0.08%). Common stainless steel is 18:8 → 300 series contains 18% Cr, 8% Ni and 0.03% C. Stainless steel displays unique engineering properties in terms of strength, toughness and overall corrosion resistance over a wide range of temperature.

16.8.1 Earlier Method

Stainless steel was first produced in the USA at least on commercial scale by what came to be known as the *dilution process* by using low carbon ferrochromium as the main source of chromium. Initially, mild or plain carbon steel scraps were melted in electric arc furnace. Stainless steel was produced by addition of low C–Fe–Cr to plain carbon steel melt to make up the entire Cr; thus, it was very costly method. On that time, stainless steel scrap could not be reused economically, without chromium losses and carbon pickup.

16.8.2 Rustless Process

Stainless steel scrap accumulated in the plants and the question of its eventual utilization is to be a very serious problem. By 1931, the stainless steel scrap problem became more serious matter; not only because piled up of the scrap, but also due to the popularity of austenitic chromium–nickel grade (18:8) which became potential for valuable alloying elements.

The first significant commercial use of stainless steel scrap as a charge constituent in the electric arc furnace started in 1931 at the Rustless Iron and Steel Corporation, USA. Large amount of iron ore or mill scale was used as the oxidizing agent, and the chromium is oxidized from the metal along with carbon. By oxidation, chromium reduces large amount of iron oxide to metallic iron. This chromium oxide makes the slag thick and later chromium was taken back from slag to the metal bath by reduction. This process was known as *Rustless process*. This process used still oxygen was not available chiefly for refining of steel. Process took 8–12 h for stainless steel production.

Sources of chromium for stainless steel production:

(i) 1/3 Cr coming from high C–Fe–Cr,
(ii) 1/3 Cr coming from chrome ore and
(iii) 1/3 Cr coming from stainless steel scrap.

The charging basket was filled with first light stainless steel scrap, chrome ore, then heavy scrap followed by iron ore (15–20%). High C–Fe–Cr was charged above iron ore followed by a little light scrap.

Iron ore supplied oxygen to the bath, not only oxidized carbon during refining, but it also oxidized a lot of chromium.

$$Fe_2O_3(s) + [Fe] = 3(FeO), \quad \Delta H^0_{298} = 28.03\,kJ/mol\ of\ Fe_2O_3 \tag{16.55}$$

$$[C] + (FeO) = [Fe] + \{CO\}, \quad \Delta H^0_{298} = 152.72\,kJ/mol\ of\ C \tag{16.56}$$

$$2[Cr] + 3(FeO) = 3[Fe] + (Cr_2O_3), \quad \Delta H^0_{298} = -165.67\,kJ/mol\ of\ Cr \tag{16.57}$$

Reaction (16.57) gave the necessary heats to raise the bath temperature. When carbon is lower down below the specification level, the oxidizing conditions were replaced by reducing conditions to get back chromium from slag phase by addition of Fe–Si or Al to the bath.

$$(Cr_2O_3) + 3/2[Si] = 2[Cr] + 3/2(SiO_2), \quad \Delta H^0_{298} = -187.04\,kJ/mol\ of\ Cr_2O_3 \tag{16.58}$$

The equilibrium constant for reaction (16.58),

$$k_{16.58} = \frac{\left\{a_{Cr^2} \cdot a_{(SiO_2)^{3/2}}\right\}}{\left\{a_{(Cr_2O_3)} \cdot a_{Si^{3/2}}\right\}} \tag{16.59}$$

Therefore,

$$\left\{\frac{a_{(Cr_2O_3)}}{a_{Cr^2}}\right\} = (1/k_{16.58}) \cdot \left\{\frac{a_{(SiO_2)^{3/2}}}{a_{Si^{3/2}}}\right\} \tag{16.60}$$

In basic slag, activity of acidic SiO_2 in slag $\{a_{(SiO_2)}\}$ is very low due to strong bonding with basic oxide CaO; and activity of Cr_2O_3 in slag $\{a_{(Cr_2O_3)}\}$ is much higher, since Cr_2O_3 is a basic oxide, it is remained free in basic slag. But that value should be lower by addition of ferrosilicon. Fe–Si is very effective for basic slag. Hence, Cr transfer from slag to metal phase is favoured and Cr from slag is going back to the metal phase by reduction. Higher the basicity of slag, better is the Cr recovery in metal phase.

In acid slag, the activity of Cr_2O_3 in slag $\{a_{(Cr_2O_3)}\}$ is lower than basic slag, since Cr_2O_3 is bonded with SiO_2, so activity of Cr_2O_3 is less. Hence, more stronger oxidizing agent is required to recover Cr from Cr_2O_3 in acid slag. Cr transfer from slag to metal is much favourable if a stronger de-oxidizer like Al is used (instead of Fe–Si).

After chemical analysis, final adjustment of the chemistry of the bath is done by adding low C–Fe–Cr. The heat is tapped after final temperature adjustment. This is very cheaper process as compared to earlier process to produce stainless steel because cheaper sources of Cr are used for obtaining the necessary alloying content.

16.8.3 Rapid Process

This process was developed, after developing the cheaper process of oxygen production, in 1939. The development of the technique of lancing oxygen into EAF bath after the solid charge melting down revolutionized stainless steelmaking practice. Oxygen was lancing to the molten bath to refine in general and de-carburization in particular to the bath. A consumable lance is used for small furnaces and a water-cooled lance for bigger furnaces.

The chromium–carbon equilibrium curves of molten metal under atmospheric pressure and at the different temperatures are shown in Fig. 16.31. It is clearly seen that the equilibrium of the system for any given chromium content is shifted to lower carbon as the temperature is increased (line 1 in Fig. 16.31); and carbon content is increasing with increased chromium at constant temperature (line 2). Oxygen lancing helps to increase the temperature of the molten bath from its usual temperature of 1600–1800 °C at the expense of nearly 3% Cr. Sometimes, oxygen lancing is started before the heat is fully molten condition.

$$4/3[\mathrm{Cr}] + \{\mathrm{O_2}\} = 2/3(\mathrm{Cr_2O_3}) \tag{16.61}$$

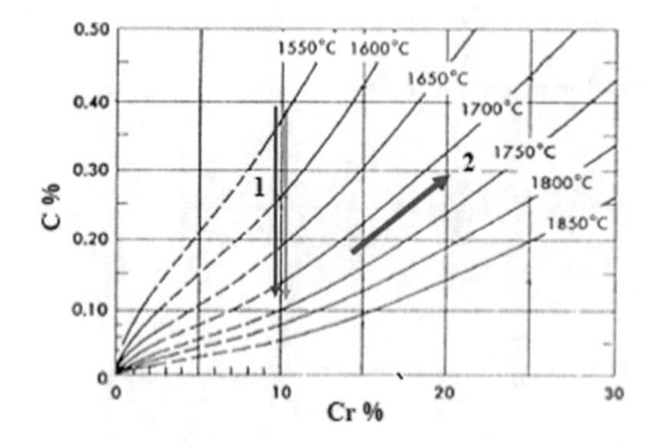

Fig. 16.31 Carbon–chromium temperature equilibrium diagram [1]

$$2[C] + \{O_2\} = 2\{CO\} \tag{16.62}$$

Reverse of reaction (16.62):

$$2\{CO\} = 2[C] + \{O_2\} \tag{16.63}$$

Addition of reactions (16.61) and (16.63):

$$4/3[Cr] + 2\{CO\} = 2/3(Cr_2O_3) + 2[C] \tag{16.64}$$

Multiple by 3/2 of reaction (16.64):

$$2[Cr] + 3\{CO\} = (Cr_2O_3) + 3[C], \quad \Delta G^0_{16.65} = -785,127.6 + 522.78T \text{ J/molCr}_2\text{O}_3 \tag{16.65}$$

Change of free energy of the reaction is zero at equilibrium, i.e. $\Delta G^0_{16.65} = 0$.
So,

$$785,127.6 + 522.78T = 0$$

Hence,

$$T = 1501.83\,\text{K i.e.}1502\,\text{K i.e.}1229\,°\text{C}.$$

i.e. above 1230 °C, change of free energy of the reaction become positive, i.e. $\Delta G^0_{16.65} = +$.
Hence, reaction (16.65) goes to backward direction that means (Cr_2O_3) is reduced by C in metal.
The equilibrium constant for reaction (16.65),

$$k_{16.65} = \left\{ \frac{a_{(Cr_2O_3)} \cdot a_{C^3}}{a_{Cr^2} \cdot P_{CO^3}} \right\} \tag{16.66}$$

To maintain the equilibrium constant ($k_{16.65}$) value remain constant, to maintain low carbon in metal, the chromium in metal is also become low; i.e. for removal of carbon, the chromium in metal is also oxidized.

Figure 16.31 shows C–Cr equilibrium diagram, low carbon is equilibrium with fixed chromium as increase the temperature. Hence, the carbon in the melt can, therefore, be preferentially oxidized from a Fe–C–Cr bath only at much higher temperature, which can be created by oxygen lancing, without much oxidation of chromium in metal. If any chromium in metal oxidized, that would come back to the metal phase later, i.e. in reducing period.

When stainless steel bath is de-carburized by using oxygen lancing, the atmosphere of the above the bath is essentially pure CO gas at one atmosphere and increase the temperature. This decides the maximum level of Cr can exist in the bath in equilibrium at a given level of C and temperature of the bath.

At 1600 °C → 10% Cr is equilibrium with 0.25% C.
At 1800 °C → 10% Cr is equilibrium with 0.07% C.

Excess Cr would be oxidized and lost in slag. So as %C decreases in metal, %Cr is also decreased in metal; i.e. de-carburization of metal without Cr oxidation is not possible. The chromium oxidation can be prevented by increasing the temperature. Carbon in metal can be oxidized below the required level without much oxidation of chromium in metal: (a) at high temperature (1800 °C), or (b) lower down the partial pressure of CO gas.

The rate of oxidation of Cr, Mn, and Fe is increased with increased lancing rate of oxygen up to 30 Nm^3/hr. Higher initial Cr content in melt, higher final temperature of the bath at the end of refining.

16.8.4 New De-carburization Techniques

Using oxygen lancing in EAF has the following disadvantages:

1. Utilization factor of power is low.
2. Although cheaper high C–Fe–Cr is used to a large extent, but low C–Fe–Cr is still needed for final adjustment and in term of cost it is not negligible amount.
3. Oxygen lancing develops high bath temperature (1800 °C) which affects the lining life and it causes discontinuity of production.
4. Heterogeneity in composition of the bath.
5. Recovery of chromium is low and more Fe losses.

De-carburization of steel bath in the presence of chromium in metal is an important reaction. The basic equilibrium reaction governing stainless steelmaking:

$$2[\mathrm{Cr}] + 3\{\mathrm{CO}\} = (\mathrm{Cr_2O_3}) + 3[\mathrm{C}] \tag{16.65}$$

The equilibrium constant for reaction (16.65),

$$k_{16.65} = \left\{ \frac{a_{(\mathrm{Cr_2O_3})} \cdot a_{\mathrm{C}^3}}{a_{\mathrm{Cr}^2} \cdot P_{\mathrm{CO}^3}} \right\} \tag{16.66}$$

By oxygen lancing, atmosphere of the bath is essentially pure CO gas at one atmosphere; that decides the maximum level of Cr can exist in the bath in equilibrium with a given level of C in metal at a particular temperature of the bath. Any excess Cr in metal would be oxidized and go to slag that can be recovered later stage.

Hence, C in metal decreases along with Cr in metal is also decreased; i.e. de-carburization without Cr oxidation is not possible. This can be prevented by (a) increasing the temperature at the cost of lining life, and (b) thermodynamic alternative is to reduce the equilibrium partial pressure of CO gas in the ambient atmosphere of the melt.

Figure 16.32 shows C–Cr equilibrium diagram with different pressures of CO gas. At the same temperature (e.g. 1700 °C), higher per cent chromium is equilibrium with higher per cent of carbon at 1 atm pressure; when the equilibrium partial pressure of CO gas is decreased, then higher per cent chromium is equilibrium with lower per cent of carbon. That means lower down the per cent of carbon at very low level without much loss of chromium. This has been achieved on commercial level process: (i) AOD: argon–oxygen–de-carburization technique → the partial pressure of CO in the ambient atmosphere is decreased by injecting argon–oxygen gas mixture; and (ii) VOD: vacuum–oxygen–de-carburization technique → oxygen is lancing for carbon oxidation under vacuum, this process is specially effective for extra low carbon and nitrogen grade stainless steels.

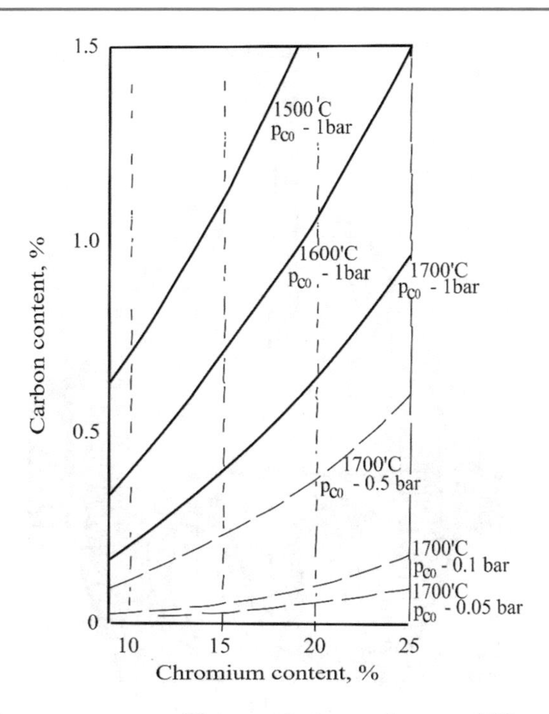

Fig. 16.32 Carbon–chromium temperature equilibrium with change of pressure [43]

Two types of processes, developed for stainless steelmaking, are as follows:

1. Duplex process (i.e. EAF/IMF-AOD/VOD),
2. Triplex process (i.e. EAF/IMF-Converter-VOD/LF).

16.8.4.1 Duplex Process

In duplex process, after melting of charge in primary unit, refining and de-carburization are carried out in a single secondary vessel. Thus, melting of scrap and ferro-alloys takes place in EAF/IMF, followed by refining in AOD/VOD with a downstream facility using either an inert gas and/or vacuum method, depending on the particular technique employed to reduce CO gas partial pressure.

Argon–Oxygen–De-carburization (AOD)

Argon–oxygen–de-carburization (AOD) process is developed at Linde Division of Union Carbide, USA, on the principle reducing the partial pressure of CO by injecting inert gas argon. More than 75% of stainless steel produced worldwide by this process. The development of argon–oxygen–de-carburization (AOD) process revolutionized stainless steelmaking. The process based on the dilution principle, i.e. decreasing partial pressure of CO gas by argon–oxygen gas mixture, allowed steelmakers to use lower-cost raw materials and to shorter the steelmaking times compared to the other processes. Stainless steel production process mainly has two stages: (i) melting and (ii) refining.

(i) Meltdown of the charge materials, such as scrap and process alloy, takes place in an EAF which achieves highest energy density and short meltdown time. Carbon in melt should be maintained 0.7–0.8% for good carbon boil during refining.
(ii) Refining is carried out in a converter (Fig. 16.33) which has solid bottom with side tuyeres for blowing argon–oxygen gas mixture. In the AOD process, the heat is treated under atmospheric

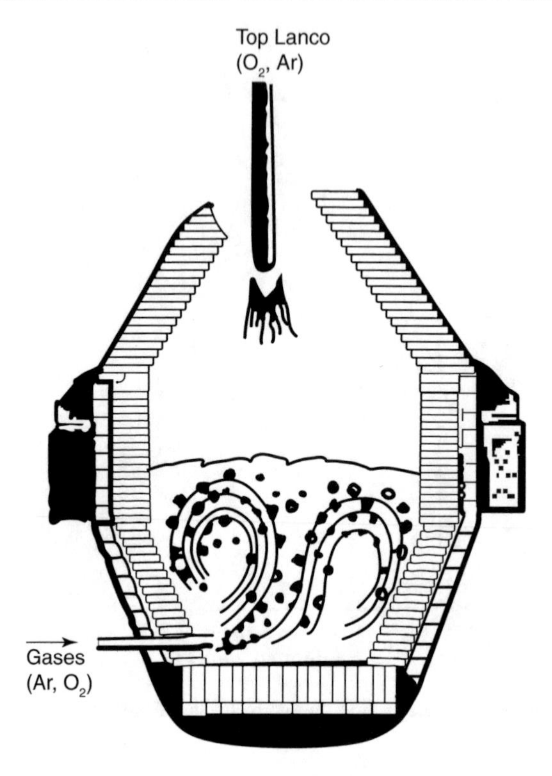

Fig. 16.33 AOD converter [44]

conditions, top and side/bottom oxygen blowing techniques are applied. Inert gas is injected into the melt through side/bottom tuyeres to reduce the system partial pressure. This step drastically reduces chromium oxidation during de-carburization. An argon–oxygen ratio (1:3–3:1) is blowing directly into the melt through tuyeres at the side of the vessel. The inert (Ar) gas dilutes the percentage of CO gas in the rising gas bubble, thus lowering partial pressure of CO ($p_{Ar} + p_{CO} = 1$ atm). The required partial pressure of CO is obtained by adjusting the oxygen to argon gas ratio, which largely depends on carbon content in melt during the refining of steel. The argon–oxygen ratio is 1:3 at the beginning and 3:1 at the final stage when temperature raises to 1710 °C. Nickel, stainless steel scrap and high C–Fe–Cr are added as coolants.

Oxygen blowing from top in AOD has the greatest impact on reducing process cost and increasing productivity. 75–90% of available heat energy from the combustion reactions is transferred to the molten bath, which decreases the consumption of ferrosilicon, increases the amount of cold charge and/or to reduce the EAF tap temperature.

The chromium is getting back from slag to metal by Fe–Si, 97% recovery take placed. The consumption of argon gas is 18–20 Nm^3/t. Argon gas can be substitute by nitrogen gas in the initial stage of refining, due to higher cost of argon; at the end of refining, nitrogen can be replaced by argon. The consumption of argon gas comes down to 8–10 Nm^3/t. But nitrogen in stainless steel is increased from 50 to 1000 ppm. Nitrogen is austranite (γ) stabilizer, to replace nickel, it is used in 200 series (which contain: 17–19% Cr, 4–6% Ni, 9% Mn, 0.15% C and 0.25% N). Good de-sulphurization occurs by using Ar gas stirring at the end of refining. Total time taken for AOD is two hours, and lining life is 80 heats.

Major attributes of AOD process are as follows:

(i) Raw materials and processing flexibility, using low cost alloys,
(ii) Rapid de-carburization rate, i.e. high productivity,
(iii) Slag handling capability,
(iv) Scrap melting capability,
(v) Optimization of the melting furnace operation and cost,
(vi) Ease maintenance,
(vii) High product quality, low residual gas content.

Vacuum–Oxygen–De-carburization (VOD)

In the vacuum process, the partial pressure of CO is kept low during refining at a markedly reduced pressure. Initially, charge is melted in an EAF, carbon in melt should be maintained 0.7–0.8% for good carbon boil during refining. Refining of the melt and subsequent reduction of chromium oxide takes place in a VOD vessel inside a vacuum chamber (Fig. 16.34). Oxygen is introduced by a lance that passes through a vacuum seal at the top of the vacuum vessel. At the same time, argon is injected through a porous plug from the bottom of the vessel to stir the bath. Refining is controlled by varying the distance between oxygen lance and top of bath, evacuation and purging rates. By continually lowering the chamber pressure down to less than 1μ bar, the CO gas partial pressure can be reduced to a great extent than AOD process. 18% Cr is equilibrium with 0.45% C at 1600 °C and 1 atm pressure. Hence, lower carbon levels up to 0.02% can be successfully achieved, with 15–18% Cr at 1600 °C and low pressure, a comparatively reduced loss of chromium to the slag. Argon stirring is necessary; otherwise de-carburization reaction would be delayed due to lack of mass transfer of carbon from bottom to the surface where C–O reaction would be feasible. With the help of synthetic slag (2–3% weight of metal charge) and argon purging, 80% sulphur in metal is removed; and finally, metal contains 0.01% S. Argon gas consumption is 1 Nm^3/t and time required for VOD is 2–2.5 h. Chromium recovery is 97%. End of refining, vacuum is broken, and the bath is de-oxidized with Al or Fe–Si.

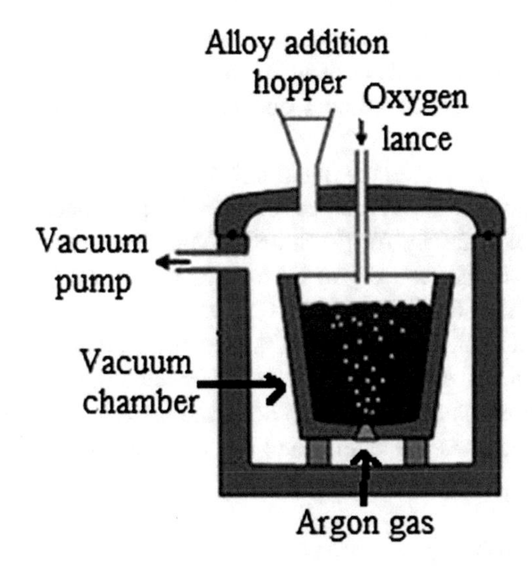

Fig. 16.34 VOD process

16.8.4.2 Triplex Process

Duplex process has certain disadvantages with respect to the ratio of ferro-alloys to scrap in the case of VOD and higher argon, ferrosilicon consumption and shorter converter life in the case of AOD. Moreover, duplex process involves lengthy process time in the secondary vessel/converter. This leads to a bottle neck during the process which will ultimately slow down or even stop EAF/IMF production. However, to overcome these above disadvantages and produce different grades including the ultra-low carbon, nitrogen content varieties with greater economy and better quality, triplex process is developed. Even steel quality can be improved by adopting the triplex process during the final stages of de-carburization in VOD/LF vessel.

In triplex process, after melting in primary unit, refining is carried out in two different vessels. In first vessel, de-carburization and major refining take place and final stages of de-carburization and de-gassing take place in second vessel. There are four commercially viable routes of triplex process, each consisting of three units, in which successive steps of the process are carried out, as follows:

- EAF-ASM-LF,
- UHPF-OTBC-VOD,
- IMF-MRK-LF,
- KMS.S-K.OBM.S-VOD.

Compare to duplex process, triplex process has the following merits:

1. Increase productivity,
2. Increase number of heats per day,
3. Increase scrap to liquid metal yield,
4. Improve quality of the product,
5. High operational flexibility,
6. Comparatively lower cost of production.

EAF-ASM-LF

Stainless steel is produced by a combined set up of electric arc furnace (EAF), argon secondary melting (ASM) converter and ladle furnace (LF). This triplex process is developed to produce stainless steel with the aimed at certain technical improvements in refining like the lowering of carbon, silicon and sulphur contents. This leads to better utilization of facility, higher reproducibility, decrease in costs and improvement in product quality. This route is commercially exploited by Panchmahal Steel, India.

Initially, scrap is melted in EAF, then the liquid metal is transferred to the ASM converter and is blown by using various proportion of oxygen and nitrogen. Heat is generated by oxidation of carbon and silicon. The bath is de-oxidized by Al, Fe–Si or Fe–Mn. Temperature of the bath is maintained above 1700 °C, which favours the recovery of Cr from slag to metal. At the end of refining, only nitrogen/argon is blown to the bath.

Liquid steel is tapped from converter into the LF. Necessary alloy additions are done to control final chemistry of the bath. Dissolved nitrogen is removed by bubbling of argon gas from porous plug at the bottom. In addition, de-sulphurization is also done and helps to floating up of smaller inclusion particles. Moreover, it homogenizes the temperature and composition of the bath.

UHPF-OTBC-VOD

The other route of triplex process constitutes production of stainless through the ultra-high-power furnace (UHPF), oxygen top and bottom converter (OTBC) and vacuum oxygen de-carburization

Table 16.19 Chemistry changes during process [45]

	Per cent								Ppm		
	C	Mn	Si	P	S	Cr	Ni	Cu	O_2	N_2	H_2
Production of 309 L by UHPF-OTBC-VOD											
Before VOD	0.3	0.40	0.05	0.018	0.015	24.5	13.0				
After De-C	0.015	0.38	0.02	0.018	0.01	24.0	13.0				
After reduction and alloying	0.015	1.6	0.4	0.019	0.008	23.5	13.5				
Production of 200 series by IMF-MRK-LF											
After melting	2.5	4.0		0.045	0.035	12.0	1.0	1.5			
After MRK	0.08	8.0		0.045	0.025	14.5	1.0	1.5	70	1800	4
After LF	0.08	8.5		0.045	0.015	14.5	1.0	1.5	30	1800	3

(VOD) process. This route is adopted by Mukund Limited, India. Scrap, high carbon ferrochrome and ferronickel are initially melted in UHPF. The liquid melt is then transferred to the OTBC where de-carburization (up to 0.3% C) and refining are performed. After de-slagging, the melt is tapped into ladle before taking to VOD vessel for further de-carburization to achieve less than 0.03% C. VOD also contributes to inclusion control and de-gassing of nitrogen. Increased productivity of low carbon grades, better quality and greater economy are achieved by this route. Table 16.19 shows the progress of chemistry during production of 309L grade of stainless steel.

IMF-MRK-LF

The third route of triplex process makes use of the combination of induction furnace, converter and ladle furnace. Initially, scrap and high carbon ferrochrome are melted in the induction furnace, and then, molten metal is charged to metal refining converter (MRK). Either oxygen is blown from top lance or mixture of oxygen–nitrogen/argon is blown through the bottom tuyere of the converter for de-carburization. Proper gas mixing promotes faster de-carburization and improves recoveries of alloying elements. MRK operation is highly predictable, reproducible and refined steel contain low hydrogen and oxygen. The refined steel is then transferred to ladle furnace (LF) for de-sulphurization of the melt. Gentle stirring with nitrogen/argon in the LF removes inclusions and thus improves the quality. LF can be operated with single-electrode working on DC power supply. The single electrode reduces the refractory erosion and carbon pickup of the melt. This route is adopted for 200 series stainless steel production in India. Table 16.19 shows the progress of chemistry during 200 series stainless steel production.

KMS.S-K.OBM.S-VOD

The fourth route of triplex process makes use of two separate converters and a VOD unit. This route is adopted by Kawasaki Steel Corporation, Japan. The production of stainless steel based on hot metal without EAF operation can be possible by this route. Initially, de-phosphorization of the hot metal is carried out in a mixing car or charging ladle and then subsequently, molten metal is charged to the KMS.S converter. Oxygen is bottom blown into the bath of the KMS.S converter through hydrocarbon shielded tuyeres vertical positioned within the converter bottom. Because of the cracking of the hydrocarbon gas a protective mushroom formation covers the tuyeres and adjacent refractories, protecting them against the aggressive and corrosive bath environment, prolonging tuyeres and refractory life. In comparison with AOD and its characteristic oxygen side blowing technique, the KMS.S converter allows for faster melting of large solid charges (i.e. scrap, ferrochrome, chromite ore, etc.) due to the improved bath movement from oxygen bottom blowing. In this converter, smelting reduction of chromite ore and scrap melting are carried out. The process heat is generated by the combustion of coke addition and by the partial post-combustion of the off-gas (i.e. CO) due to oxygen lancing from the top of the converter. The generation of additional energy is utilized for the melting of the extra solid charge. The carbon content of the bath is increased during the smelting reduction reactions by adding coke in accordance with the equation [46]:

$$[\%C] = 4.04 + 0.084[\%Cr]. \tag{16.67}$$

The chromium saturated molten metal (i.e. 9–13% Cr, 5.5% C) is transferred to the second converter [i.e. OBM stainless converter (K.OBM.S)] for de-carburization of the bath with combined oxygen blowing from top and bottom, the chromium content is adjusted and the bath temperature is increased from 1470 to 1700 °C. The de-carburization reaction should be stopped at about 0.12% C in melt to limit chromium oxidation. The chromium oxidation can also be prevented by using mixed gas blowing. With the injection of lime powder into the bath, improves de-oxidation and de-sulphurization during the reduction period. CaO rich slag from K.OBM.S converter can be charged to the KMS.S converter for slag formation.

Final de-carburization and reduction reactions are performed in the subsequent VOD vessel under vacuum conditions. Total time requires for this process which is less than 70 min, i.e. high productivity up to 20 heats per day is possible.

16.8.5 Stainless Steel Production by IMF

Stainless steel production by induction melting furnace (IMF): The requirement of stainless steel in India was high. Stainless steel sheets were imported from Japan in early seventies for the manufacture of utensils. There is a good response from entrepreneurs to manufacture stainless steels. Many EAFs started producing stainless steels, but due to carbon pickup from electrodes, it was difficult to maintain the carbon below 0.08%. A thick layer of fluxes was put on the surface of bath and some nitrogen blown to control carbon pickup and absorption of hydrogen. But stainless steel so produced was not good quality. In early eighties, small IMFs were used for manufacture of stainless steels by using low carbon ferrochrome and low carbon ferromanganese [48]. Stainless steel scrap of 304 and 302 qualities were imported at concessional customs duty rates. Thus, IMFs started producing stainless steels since early eighties by using imported stainless steel scrap. At present, stainless steel is produced by melting stainless steel scrap and casting ingots by IMFs and adding ferro-alloys, then transferring the liquid metal to AOD vessels for refining. After AOD treatment, some units carry out further refining in LF and then producing billet by Concast process.

Probable Questions

1. What are the function of fourth and fifth holes at the EAF roof?
2. Why EAF roof cannot be made by fireclay bricks?
3. Why light scrap is put at the bottom of the basket?
4. Draw a neat schematic sketch and show the details of EAF.
5. How electrodes are joining without loss of material?
6. Discuss step by step about double slag practice in EAF steelmaking.
7. Why carbide slag is not suitable for producing low carbon steels? Justify.

8. Why reducing sag is required for sulphur removal EAF?
9. How do you understand carbide slag formation?
10. How do you prepare a reducing slag in an EAF? How does it help in de-sulphurization?
11. What are the modern developments of EAF steelmaking process?
12. What are the advantages of hot DRI and hot metal as a feed material in EAF steelmaking?
13. 'Hot metal is the cost-effective charge substitute in EAF'. Discuss.
14. What are the methods of charging sponge iron in EAF? Which one is the best?
15. 'For quality steel production in EAF, sponge iron is used as charge substitute'. Discuss.
16. Densities of sponge iron and EAF slag are 2.7 and 3.3 g/cm^3, respectively. How sponge iron can be penetrate in the slag phase for melting at EAF?
17. If DRI contain 90% metallization and 8.5% FeO, how much Fe goes to liquid steel per 100 kg DRI?
18. 'Steelmakers prefer high carbon sponge iron for steelmaking'. Why?
19. Why hot heel is required to melt the sponge iron?
20. Discuss the advantages of continuous charging mode in EAF.
21. Discuss use of iron carbide in EAF.
22. Discuss the basic principles of IMF and its merits.
23. Discuss the difference between EAF versus IMF.
24. Discuss the basic principles of CONARC process and its merits.
25. Discuss the basic principles of DC-EAF and its merits.
26. Discuss the rustless process of stainless steelmaking.
27. Explain how stainless steel is produced in an EAF, starting from virgin metal scrap.
28. State the recent methods of stainless steel production. How that is better than older process?
29. How loss of chromium is prevented during de-carburization in stainless steelmaking? How chromium is recovered from slag phase in stainless steel production?
30. How can you refine a high carbon Fe–Cr–C melt in an AOD converter? Explain the physico-chemical principle involved.
31. 'In basic slag of stainless steelmaking, Cr_2O_3 is easily reduced by Fe–Si; but in acid slag, that requires strong oxidizing agent'. Why?
32. What are the basic principles of AOD and VOD processes?
33. What is EOF stands for? Discuss the basic principles of EOF Process.

Examples

Example 16.1

(a) Calculate amount of DRI required along with 500 kg steel scrap to produce one tonne steel in EAF. (b) Calculate amount of slag produce. (c) Amount of lime required for making slag (V ratio 2.5). Given: (i) 1.5% Fe loss in slag w.r.t liquid steel (ii) spar and Fe–Mn charged 5 kg each.

Element/compound %	Steel produced	Scrap	DRI	Spar	Lime	Fe–Mn
C	0.15	0.26	1.25			0.2
Si	0.20	0.25				1.5
Mn	0.35	0.25				72.5
P	0.03	0.04	0.05			0.4

(continued)

Element/compound %	Steel produced	Scrap	DRI	Spar	Lime	Fe–Mn
S	0.03	0.04	0.01	0.3	0.15	0.05
CaO			1.0	0.5	95.85	
MgO			1.0		2.5	
SiO_2			1.5	1.5	1.5	
Al_2O_3			0.5	0.7		
FeO			9.64			
Fe_{Tol}			92.5			
CaF_2				97.0		
Amount charged, kg		500	?	5	?	5

Solution

Assumptions: One tonne of liquid steel production.

By balancing, we get 99.16, 99.24 and 25.35% Fe contain in steel scrap, produced steel and Fe–Mn, respectively.

Fe balance: Fe input = Fe output
Fe from steel scrap + Fe from DRI + Fe from Fe–Mn = Fe in steel produce + Fe losses in slag
Suppose weight of DRI = W_{DRI}.

$$\left[\left(\frac{99.16}{100}\right) \times 500\right] + \left[\left(\frac{92.5}{100}\right) \times W_{DRI}\right] + \left[\left(\frac{25.35}{100}\right) \times 5\right]$$
$$= \left[\left(\frac{99.24}{100}\right) \times 1000\right] + \left[\left(\frac{1.5}{100}\right) \times 1000\right]$$

Or $0.925\ W_{DRI} = 1007.4 - 497.07 = 510.33$
Therefore, W_{DRI} = **551.71 kg**

$$\mathrm{Fe} + 1/2\mathrm{O_2} = \mathrm{FeO}$$
$$56 \quad 11.2 \quad 72$$

56 kg of Fe to form 72 kg of FeO
15 kg... $\left(\frac{72}{56}\right) \times 15 =$ **19.29 kg** of FeO in slag

Si balance: Si input = Si output
Si from scrap + Si from Fe–Mn + Si from gange material in DRI + Si from lime + Si from spar = Si in steel + Si goes to slag

$$\left[\left(\frac{0.25}{100}\right) \times 500\right] + \left[\left(\frac{1.5}{100}\right) \times 5\right] + \left[\left(\frac{1.5}{100}\right) \times \left(\frac{28}{60}\right) \times 551.7\right] + \left[\left(\frac{1.5}{100}\right) \times \left(\frac{28}{60}\right) \times W_L\right]$$
$$+ \left[\left(\frac{1.5}{100}\right) \times \left(\frac{28}{60}\right) \times 5\right] = \left[\left(\frac{0.2}{100}\right) \times 1000\right] + W_{Si}$$

$$1.25 + 0.075 + 3.86 + 7 \times 10^{-3} W_L + 0.035 = 2 + W_{Si}$$

Or $5.2219 + 7 \times 10^{-3}\, W_L = 2 + W_{Si}$
Therefore,

$$W_{Si} = 3.222 + 7 \times 10^{-3} W_L \tag{1}$$

$$\begin{array}{ccc} Si + O_2 = SiO_2 \\ 28 \quad 32 \quad 60 \end{array}$$

28 part Si form 60 part of SiO_2

$$(3.222 + 7 \times 10^{-3} W_L)\ldots\left[(3.222 + 7\text{x}10^{-3} W_L)\text{x}\left(\frac{60}{28}\right)\right]\text{part of } SiO_2 \tag{2}$$

CaO balance: CaO input = CaO output
CaO from lime + CaO from gange material in DRI + CaO from spar = CaO in slag
$0.9585 \times W_L + 0.01 \times 551.7 + 0.005 \times 5 = W_{CaO}$
or,

$$0.9585 \times W_L + 5.542 = W_{CaO} \tag{3}$$

$$\text{Since Basicity} = 2.5 = \left(\frac{\Sigma\text{CaO in slag}}{\Sigma SiO_2 \text{ in slag}}\right) = \left(\frac{(0.9585 \times W_L + 5.542)}{\left[(3.222 + 7 \times 10^{-3}.W_L) \times \left(\frac{60}{28}\right)\right]}\right)$$

Therefore, $(0.9585 \times W_L + 5.542) = (17.26 + 37.5 \times 10^{-3} W_L)$

$$0.921 \times W_L = 11.718 \text{ or } W_L = \mathbf{12.72\,kg}$$

From Eq. (3): $W_{CaO} = 0.9585 \times W_L + 5.542 = \mathbf{17.74\,kg}$
From Eq. (2): SiO_2in slag $= \left[(3.222 + 7 \times 10^{-3} W_L) \times \left(\frac{60}{28}\right)\right] = \mathbf{7.09\,kg}$

Mn balance: Mn input = Mn output
Mn from scrap + Mn from Fe–Mn = Mn in steel + Mn goes to slag
$0.0025 \times 500 + 0.725 \times 5 = 0.0035 \times 1000 + W_{Mn}$
Therefore, $W_{Mn} = 1.375$ kg

$$\begin{bmatrix} Mn & + & 1/2O_2 & = & MnO \\ 55 & & 16 & & 71 \end{bmatrix}$$

MnO in slag $= 1.375 \times \left(\frac{71}{55}\right) = \mathbf{1.78\,kg}$

P balance: P input = P output

$$\begin{aligned} &\text{P from scrap} + \text{P from Fe–Mn} + \text{P from DRI} \\ &\qquad = \text{P in steel} + \text{P goes to slag} \\ &0.0004 \times 500 + 0.004 \times 5 + 0.0005 \times 551.7 \\ &\qquad = 0.0003 \times 1000 + W_p \end{aligned}$$

Therefore, $W_P = 0.1958$ kg

$$\begin{array}{ccccc} 2P & + & 5/2O_2 & = & P_2O_5 \\ 62 & & 80 & & 142 \end{array}$$

P_2O_5 in slag $= 0.1958 \times \left(\frac{142}{62}\right) = \mathbf{0.45\,kg}$

MgO balance: MgO input = MgO output
MgO from DRI + MgO from lime = MgO goes to slag
$0.01 \times 551.7 + 0.025 \times 12.72 = W_{MgO}$
W_{MgO} = **5.84 kg**

Al_2O_3 balance: Al_2O_3 input = Al_2O_3 output
Al_2O_3 from DRI + Al_2O_3 from spar = Al_2O_3 goes to slag
$0.005 \times 551.7 + 0.007 \times 5 = W_{Al_2O_3}$
$W_{Al_2O_3}$= **2.79 kg**

CaF_2 balance: CaF_2 input = CaF_2 output
CaF_2 from spar = CaF_2 goes to slag
$0.97 \times 5 = W_{CaF2}$
W_{CaF2} = **4.85 kg**

$$\begin{aligned} \text{Weight of slag} &= W_{FeQ} + W_{Sio_2} + W_{CaQ} + W_{MnQ} + W_{P2QS} + W_{MgQ} + W_{Al_2O_3} + W_{CaF_2} \\ &= 19.29 + 7.09 + 17.74 + 1.78 + 0.45 + 5.84 + 2.79 + 4.85 = \mathbf{59.83\,kg}. \end{aligned}$$

Example 16.2 Calculate amount of scrap required along with 500 kg DRI to produce one tonne steel in EAF. Find out amount of slag produce and amount of lime required for making slag. Fluorspar and Fe–Mn are charged 5 kg each.

	C	Mn	Si	S	P	Fe_T	FeO	CaO	MgO	SiO_2	Al_2O_3	CaF_2
Steel	0.15	0.35	0.2	0.03	0.03							
Scrap	0.26	0.25	0.25	0.04	0.04							
Fe–Mn	0.2	72.5	1.5	0.05	0.4							
DRI	1.25			0.01	0.05	92.5	9.64	1.0	1.0	1.5	0.5	
Lime				0.15				95.85	2.5	1.5		
Fluorspar				0.3				0.5		1.5	0.7	97.0

Given: 1.5% Fe loss in slag w.r.t. liquid steel and basicity of slag (B) is 3.0.

Solution

Assumptions: One tonne steel production.
By balancing, we get 99.24, 99.16 and 25.35% Fe contain in produced steel, scrap and Fe–Mn, respectively.

Fe balance: Fe input = Fe output
Fe from scrap + Fe from DRI + Fe from Fe–Mn = Fe goes to steel produce + Fe losses in slag

Suppose weight of Scrap = W_{Sc}

$$\left\{\left(\frac{99.16}{100}\right) \times W_{SC}\right\} + \left\{\left(\frac{92.5}{100}\right) \times 500\right\} + \left\{\left(\frac{25.35}{100}\right) \times 5\right\} = \left\{\left(\frac{99.24}{100}\right) \times 1000 + \left\{\left(\frac{1.5}{100}\right) \times 1000\right\}\right.$$

$$0.9916W_{SC} + 462.5 + 1.27 = 992.4 + 15$$

$$0.9916W_{Sc} + 463.77 = 1007.4$$

Therefore, $\mathbf{W_{Sc}}$ **= 548.24 kg**

$$Fe + 1/2O_2 = FeO$$

56 kg Fe form 72 kg FeO

Therefore, 15 kg Fe form $\left\{\left(\frac{72}{56}\right) \times 15\right\}\text{kg FeO} = 19.29\,\text{kg FeO}$ goes to slag.

Si balance: Si input = Si output

Si from scrap + Si from gange material in DRI + Si from Fe–Mn + Si from lime + Si from spar = Si in steel + Si goes to slag

$$\left\{\left(\frac{0.25}{100}\right) \times 548.24\right\} + \left\{\left(\frac{1.5}{100}\right) \times \left(\frac{28}{60}\right) \times 500\right\} + \left\{\left(\frac{1.5}{100}\right) \times 5\right\} + \left\{\left(\frac{1.5}{100}\right) \times \left(\frac{28}{60}\right) \times W_L\right.$$

$$+\left\{\left(\frac{1.5}{100}\right) \times \left(\frac{28}{60}\right) \times 5\right\} = \left\{\left(\frac{0.2}{100}\right) \times 1000 + W_{Si}\right.$$

$$1.37 + 3.5 + 0.075 + 7 \times 10^{-3}W_L + 0.035 = 2.0 + W_{Si}$$

Therefore, $W_{Si} = 4.98 + 7 \times 10^{-3}\ W_L$

$$Si + O_2 = SiO_2$$

28 32 60

28 part Si form 60 part of SiO_2

$$(4.98 + 7 \times 10^{-3}W_L)\ldots\left[(4.98 + 7 \times 10^{-3}W_L) \times \left(\frac{60}{28}\right)\right]\text{part of } SiO_2 \text{ goes to slag} \quad (1)$$

CaO balance: CaO input = CaO output

CaO from lime + CaO from gange material in DRI + CaO from spar = CaO goes to slag

$\left\{\left(\frac{95.85}{100}\right) \times W_L + \left\{\left(\frac{1.0}{100}\right) \times 500\right\} + \left\{\left(\frac{0.5}{100}\right) \times 5\right\} = W_{CaO}\right.$

or,

$$0.9585 \times W_L + 5.025 = W_{CaO} \quad (2)$$

$$B = 3.0 = \left(\frac{W_{CaO}}{W_{SiO_2}}\right) = \left(\frac{(0.9585 \times W_L + 5.025)}{[(4.98 + 7 \times 10{-}3W_L) \times 2.14]}\right)$$

Therefore, $(0.9585 \times W_L + 5.025) = (31.97 + 0.045\ W_L)$

$0.9135W_L = 26.945$, hence, W_L = **29.5 kg**

Therefore, SiO_2 goes to slag $= [(4.98 + 7 \times 10^{-3}W_L) \times 2.14)]$

$= 10.65 + 0.015W_L =$ **11.09 kg**

From Eq. 2: $W_{CaO} = 0.9585 \times W_L + 5.025 = 0.9585 \times 29.5 + 5.025 =$ **33.3 kg**

Mn balance: Mn input = Mn output

Mn from scrap + Mn from Fe–Mn = Mn in steel + Mn goes to slag

$$\left\{\left(\frac{0.25}{100}\right)\times 548.24\right\}+\left\{\left(\frac{72.5}{100}\right)\times 5=\left\{\left(\frac{0.35}{100}\right)\times 1000\right\}+W_{Ma}\right.$$

$1.37+3.63=3.5+W_{Mn}$

Therefore, W_{Mn} = 1.5 kg

$$\begin{matrix}[Mn + 1/2\ O_2 = MnO \\ 55 \quad 16 \quad 71]\end{matrix}$$

MnO in slag $= 1.5\times\left(\frac{71}{55}\right)=\mathbf{1.94\,kg}$

P balance: P input = P output

P from scrap + P from Fe–Mn + P from DRI

= P in steel + P goes to slag

$\left\{\left(\frac{0.04}{100}\right)\times 548.24\right\}+\left\{\left(\frac{0.4}{100}\right)\times 5\right\}+\left\{\left(\frac{0.05}{100}\right)\times 500=\left\{\left(\frac{0.03}{100}\right)\times 1000\right\}+W_P\right.$

$0.22+0.02+0.25=0.3+W_P$

Therefore, W_P = 0.19 kg

$$\begin{matrix}2P + 5/2\,O_2 = P_2O_5 \\ 62 \quad 80 \quad 142\end{matrix}$$

P_2O_5 goes to slag $=\left(\frac{142}{62}\right)\times 0.19=\mathbf{0.44\,kg}$

MgO balance: MgO input = MgO output

MgO from DRI + MgO from lime = MgO goes to slag

$$\left\{\left(\frac{1.0}{100}\right)\times 500\right\}+\left\{\left(\frac{2.5}{100}\right)\times 29.5\right\}=W_{MgQ}$$

$\mathbf{W_{MgO}=5.74\,kg}$

Al_2O_3 balance: Al_2O_3 input = Al_2O_3 output

Al_2O_3 from DRI + Al_2O_3 from spar = Al_2O_3 goes to slag

$\left\{\left(\frac{0.5}{100}\right)\times 500\right\}+\left\{\left(\frac{0.7}{100}\right)\times 5\right\}=W_{Al_2O_3}$

$\mathbf{W_{Al2O_3}=2.54\,kg}$

CaF_2 balance: CaF_2 input = CaF_2 output

CaF_2 from spar = CaF_2 goes to slag

$\left\{\left(\frac{97}{100}\right)\times 5\right\}=W_{CaF_2}=\mathbf{4.85\,kg}$

Weight of slag $=W_{FeQ}+W_{SiQ_2}+W_{CaQ}+W_{MaQ}+W_{P_2O\Sigma}+W_{MgQ}+W_{Al_2O_3}+W_{CaF_2}$

$=19.29+11.09+33.3+1.94+0.44+5.74+2.54+4.85=\mathbf{79.19\,kg}.$

Example 16.3 Calculate amount of scrap required along with 600 kg DRI to produce one tonne steel in EAF. Find out amount of lime required for make basicity of slag 3.5. Fe–Mn is charged 5 kg. Also find out amount of slag produce and yield of Fe.

	C	Mn	Si	S	P	Fe_{Met}	FeO	CaO	MgO	SiO_2	Al_2O_3
Steel	0.1	0.25	0.1	0.03	0.03						
Scrap	0.25	0.3	0.2	0.04	0.04						
DRI	2.5				0.05	85.6	9.5	1.0	1.0	2.0	0.5
Lime				0.15				96	1.5	2.5	
Fe–Mn	0.2	72.5	1.5	0.05	0.4						

Given: 1.0% Fe loss in slag w.r.t. liquid steel.

Solution

Assumptions: One tonne steel production.
By balancing, we get 99.49%, 99.17% and 25.35% Fe contain in produced steel, scrap and Fe–Mn, respectively.

Fe balance: Fe input = Fe output
Fe from scrap + Fe from DRI (Fe from Fe–Metallization in DRI + Fe from FeO in DRI) + Fe from Fe–Mn = Fe in steel produce + Fe losses in slag
Suppose weight of scrap = W_{Sc}

$$\left\{\left(\frac{99.17}{100}\right) \times W_{Sc}\right\} + \left[\left\{\left(\frac{85.6}{100}\right) 600\right\} + \left\{\left(\frac{9.5}{100}\right) 600 \left(\frac{56}{72}\right)\right\} + \left\{\left(\frac{25.35}{100}\right) \times 5\right\}\right.$$
$$= \left\{\left(\frac{99.49}{100}\right) \times 1000 + \left\{\left(\frac{1.0}{100}\right) \times 1000\right\}\right.$$
$$0.9917 W_{Sc} + [513.6 + 44.33] + 1.27 = 994.9 + 10 = 1004.9$$
$$0.9917 W_{Sc} = 1004.9 - 559.2 = 445.7$$

Therefore, $\mathbf{W_{SC} = 449.43\,kg}$

$$Fe + 1/2O_2 = FeO$$
56 kg Fe form 72 kg FeO

Therefore, 10 kg Fe form $\left\{\left(\frac{72}{56}\right) \times 10\right\}$kg FeO = **12.86 kg** FeO goes to slag.

Si balance: Si input = Si output
Si from scrap + Si from gange material in DRI + Si from Fe–Mn + Si from lime = Si in steel + Si goes to slag

$$\left[\left(\frac{0.2}{100}\right) \times \mathbf{449.43}\right] + \left[\left(\frac{2.0}{100}\right) \times \left(\frac{28}{60}\right) \times 600\right] + \left\{\left(\frac{1.5}{100}\right) \times 5\right\} + \left[\left(\frac{2.5}{100}\right) \times \left(\frac{28}{60}\right) \times W_L\right]$$
$$= \left[\left(\frac{0.1}{100}\right) \times 1000\right] + W_{Si}$$

Or 0.9 + 5.6 + 0.075 + 0.012 W_L = 6.575 + 0.012 W_L = 1 + W_{Si}
Therefore, W_{Si} = 5.575 + 0.012 W_L

$$SiO_2 \text{ in slag } = (5.575 + 0.012 W_L) \times \left(\frac{60}{28}\right) = 11.95 + 0.026\, W_L \quad (1)$$

Since

$$\begin{array}{ccc} Si + O_2 = SiO_2 \\ 28 \quad 32 \quad 60 \end{array}$$

28 part Si form 60 part of SiO_2

CaO balance: CaO input = CaO output
CaO from lime + CaO from gange material in DRI = CaO in slag
$0.96 \times W_L + 0.01 \times 600 = W_{CaO}$
or,

$$0.96W_L + 6.0 = W_{CaO} \tag{2}$$

Since Basicity $= 3.5 = \left(\frac{W_{Cao}}{W_{sio_2}}\right) = \left(\frac{(0.96W_L + 6.0)}{(11.95 + 0.026\, W_L)}\right)$
Therefore, $(0.96\, W_L + 6.0) = 41.825 + 0.09\, W_L$
or $0.87\, W_L = 35.825$ or $W_L =$ **41.18 kg**
From Eq. (2): $W_{CaO} = 0.96\, W_L + 6.0 =$ **45.53 kg**
From Eq. (1): SiO_2 in slag $= [11.95 + 0.026\, W_L] =$ **13.02 kg**

Mn balance: Mn input = Mn output
Mn from scrap + Mn from Fe–Mn = Mn in steel + Mn goes to slag
$\left\{\left(\frac{0.3}{100}\right) \times + \mathbf{449.43}\right\} + \left\{\left(\frac{72.5}{100}\right) \times 5\right\} = \left\{\left(\frac{0.25}{100}\right) \times 1000\right\} + W_{Mn}$
$1.35 + 3.625 = 2.5 + W_{Mn}$
Therefore, $W_{Mn} = 2.475\,\text{kg}$

$$\begin{array}{lll} [Mn + 1/2O_2 = & & MnO \\ 55 & 16 & 71] \end{array}$$

MnO in slag $= 2.475 \times 2.475 \times \left(\frac{71}{55}\right) = \mathbf{3.2\,kg}$

P balance: P input = P output
P from scrap + P from Fe–Mn + P from DRI
= P in steel + P goes to slag
$\left\{\left(\frac{0.04}{100}\right) \times \mathbf{449.43} + \left\{\left(\frac{0.4}{100}\right) \times 5\right\} + \left\{\left(\frac{0.05}{100}\right) \times 600\right\} = \left\{\left(\frac{0.03}{100}\right) \times 1000\right\} + W_P\right.$
$0.18 + 0.02 + 0.3 = 0.3 + W_P$
Therefore, $W_P = 0.2$ kg

$$\begin{array}{cccc} 2P + & 5/2O_2 & = & P_2O_5 \\ 62 & 80 & & 142 \end{array}$$

P_2O_5 goes to slag $= \left(\frac{142}{62}\right) \times 0.2 = \mathbf{0.46\,kg}$

S balance: S input = S output
S from scrap + S from Fe–Mn + S from lime
= Sin steel + S goes to slag

$$\left\{\left(\frac{0.04}{100}\right)\times \mathbf{449.43}\right\}+\left\{\left(\frac{0.05}{100}\right)\times 5\right\}+\left\{\left(\frac{0.15}{100}\right)\times \mathbf{41.18}\right\}=\left\{\left(\frac{0.03}{100}\right)\times 1000\right\}+W_S$$

$0.18+0.0025+0.062=0.3+W_S$

$W_S = 0.244 - 0.3 = -0.056$ i.e. no loss of S

Hence, S remains in Steel = $\left(\frac{0.244}{1000}\right)\times 100=\mathbf{0.024}\%$

MgO balance: MgO input = MgO output

MgO from DRI + MgO from lime = MgO goes to slag

$\left\{\left(\frac{1.0}{100}\right)\times 600\right\}+\left\{\left(\frac{1.5}{100}\right)\times \mathbf{41.18}\right\}=W_{MgO}$

$\mathbf{W_{MgO}=6.62\,kg}$

Al_2O_3 balance: Al_2O_3 input = Al_2O_3 output

Al_2O_3 from DRI = Al_2O_3 goes to slag

$\left\{\left(\frac{0.5}{100}\right)\times 600\right\}=W_{Al_2O_3}$

$W_{Al_2O_3}=\mathbf{3.0\,kg}$

Weight of slag $= W_{FeO}+W_{SiO_2}+W_{CaO}+W_{MnO}+W_{P_2O_5}+W_{MgO}+W_{Al_2O_3}$

$= 12.86+13.02+45.53+3.2+0.46+6.62+3.0=\mathbf{84.69\,Kg}$

Iron yield (%) can be calculated:

$$\left[\frac{\{(F_2\cdot W_2)\cdot 100\}}{\{(F_1\cdot W_1)+(F_T\cdot W_0)+(F_3\cdot W_3\}}\right] \qquad \text{(a)}$$

where F_1, F_T, F_3 and F_2 are fraction of total iron present in steel scrap, DRI, Fe–Mn and steel product, respectively. W_1, W_0, W_3 and W_2 weight of steel scrap, DRI, Fe–Mn and steel product, respectively.

99.49%, 99.17% and 25.35% Fe contain in produced steel, scrap and Fe–Mn, respectively. DRI contain 85.6% Fe_M and 9.5% FeO

Wt of scrap = **449.43** kg, DRI = 600 kg, Fe–Mn = 5 kg and steel produce = 1000 kg

$$\text{Total Fe contain in DRI}=\left[\left\{\left(\frac{85.6}{100}\right)\times 600\right\}+\left\{\left(\frac{9.5}{100}\right)\times 600\times\left(\frac{56}{72}\right)\right\}=513.6+44.33=557.93\,\text{kg}$$

Fraction of Fe in DRI = (557.93/600) × 100 = 92.99%

From Eq. (a):

$$\left[\frac{\{(F_2\cdot W_2)\cdot 100\}}{\{(F_1\cdot W_1)+(F_T\cdot W_0)+(F_3\cdot W_3\}}\right]=\left[\frac{\{(0.9949\times 1000)\cdot 100\}}{\{(0.9917\times 449.43)+(0.9299\times 600)+(0.2535\times 5)\}}\right]=\mathbf{99\%}.$$

Example 16.4 Calculate amount of scrap required along with hot metal (at 1350 °C) to produce one tonne duplex stainless steel in CONARC process. Find out (i) amount of slag produce and its composition, (ii) amount of oxygen consumed and (iii) amount of Ni and Fe–Mo used.

	C	Mn	Si	S	P	Fe	Cr	Mo	Ni	Wt. (kg)
Steel	0.03	1.2	1.5	0.03	0.03	71.16	18.5	2.75	4.8	1000
Scrap	0.26	0.25	0.25	0.04	0.03	99.17				?
HM	3.5	0.5	1.0	0.04	0.4	94.56				332
HC Fe–Cr	7.0	0.18	4.0	0.05	0.05	26.22	62.5			306
LC Si–Mn	0.1	57.5	27.5			14.9				22
Fe–Mo	0.15					44.85		55.0		?
	CaO	MgO	SiO_2	Al_2O_3	CaF_2	S				
Lime	95.85	2.5	1.5	–	–	0.15				52
Fluorspar	0.5	–	1.5	0.7	97.0	0.3				5.5

Given: (i) 1.5% Fe loss in slag w.r.t. liquid steel and 0.6% Fe in fumes, (ii) 90% C forms CO and 10% C forms CO_2 and (iii) there is no loss of Ni and Mo.

Solution

Mo balance: Mo input = Mo output
Mo from Fe–Mo = Mo in Steel
$(0.55 \times W_{\text{Fe–Mo}}) = (0.0275 \times 1000)$
Therefore, weight of Fe–Mo, $W_{\text{Fe–Mo}} = 27.5/0.55 =$ **50 kg**

Ni balance: Ni input = Ni output
Ni from Ni metal = Ni in Steel
Therefore, weight of Ni = Ni in Steel = (0.048 × 1000) = **48 kg**

Fe balance: Fe input = Fe output
Fe from HM + Fe from scrap + Fe from HC Fe–Cr + Fe from LC Si–Mn + Fe from Fe–Mo = Fe in steel produce + Fe losses in slag + Fe losses in fumes
Suppose weight of scrap = W_{Sc}
$\{(0.9456 \times 332) + (0.9917 \times W_{sc}) + (0.2622 \times 306) + (0.149 \times 22) + (0.4485 \times 50)\}$
$= \{(0.7116 \times 1000) + (0.015 \times 1000) + 0.006 \times 1000)\}$
$\{313.94 + 0.9917\,W_{Sc} + 80.23 + 3.28 + 22.43\} = \{711.6 + 15 + 6\} = 732.6$
$0.9917 W_{Sc} = 732.6 - 419.88 = 312.72$
Therefore, W_{Sc} = 315.34 i.e. **315 kg**

$$\begin{array}{ccc} \text{Fe} + 1/2\text{O}_2 = \text{FeO} \\ 56 \quad 16 \quad 72 \end{array}$$

56 kg Fe form 72 kg FeO

$$15\,\text{kg Fe form} \left\{\left(\frac{72}{56}\right) \times 15\right\} \text{kg FeO} = \mathbf{19.29\,kg}\ \text{FeO goes to slag.}$$

Therefore, 56 kg Fe react with 16 kg oxygen

$$15\,\text{kg Fe} \quad \left\{\left(\frac{16}{56}\right) \times 15\right\} = \mathbf{4.29\,kg}\ \text{oxygen}$$

Si balance: Si input = Si output

Si from HM + Si from scrap + Si from HC Fe–Cr + Si from LC Si–Mn + Si from lime + Si from fluorspar = Si in steel produce + Si losses in slag

Suppose Si losses in slag = W_{Si}

$\{(0.01 \times 332) + (0.0025 \times 315) + (0.04 \times 306) + (0.275 \times 22) + \left(0.015 \times \left(\frac{28}{60}\right) \times 52\right)$
$+ \left(0.015 \times \left(\frac{28}{60}\right) \times 5.5\right)\} = \{(0.015 \times 1000) + W_{Si} = 15 + W_{Si}$

Therefore, $W_{Si} = \{3.32 + 0.79 + 12.24 + 6.05 + 0.364 + 0.039\} - 15 = 7.80\,\text{kg}$

$$\begin{array}{l} Si + O_2 = SiO_2 \\ 28 \quad 32 \quad 60 \end{array}$$

28 kg Si reacts with 32 kg O_2 to form SiO_2

7.80 kg Si ... $7.80 \times \left(\frac{32}{28}\right) = \mathbf{8.91\,kg}\, O_2$ to form SiO_2

Weight of SiO_2 in slag $= \left\{7.80 \times \left(\frac{60}{28}\right)\right\} = \mathbf{16.72\,kg}$

Mn balance: Mn input = Mn output

Mn from HM + Mn from scrap + Mn from HC Fe–Cr + Mn from LC Si–Mn = Mn in steel + Mn goes to slag

$\{(5 \times 10^{-3} \times 332) + (2.5 \times 10^{-3} \times 315) + (1.8 \times 10^{-3} \times 306) + (0.575 \times 22) = (0.012 \times 1000) + W_{Mn}$

$W_{Mn} = (1.66 + 0.79 + 0.55 + 12.65) - 12 = 15.65 - 12 = 3.65\,\text{kg}$

$$\begin{array}{l} Mn + 1/2O_2 = MnO \\ 55 \qquad 16 \qquad 71 \end{array}$$

MnO in slag = $3.65 \times \left(\frac{71}{55}\right) = \mathbf{4.71\,kg}$

55 kg Mn react with 16 kg oxygen to form MnO

1.65 kg...$\left\{3.65 \times \left(\frac{16}{55}\right)\right\} = \mathbf{1.06\,kg}$ oxygen to form MnO

C balance: **C** input = C output

C from HM + C from scrap + C from HC Fe–Cr + C from LC Si–Mn + C from Fe–Mo = C in steel + C goes to gases

$\{(0.035 \times 332) + (2.6 \times 10^{-3} \times 315) + (0.07 \times 306) + (1.0 \times 10^{-3} \times 22) + (1.5 \times 10^{-3} \times 50)$
$= (3 \times 10^{-4} \times 1000) + W_C$

$W_C = (11.62 + 0.82 + 21.42 + 0.02 + 0.08) - 0.3 = 33.66$ kg

Therefore, C goes to gas: $W_C = \mathbf{33.66\,kg}$

90% C forms CO, i.e. 0.9 × 33.66 = 30.29 kg C forms CO

$$\begin{array}{l} C + 1/2O_2 = CO \\ 12 \qquad 16 \qquad 28 \end{array}$$

12 kg C react with 16 kg oxygen to form CO

30.29 kg...$\left\{30.29 \times \left(\frac{16}{12}\right)\right\} = \mathbf{40.39\,kg}$ oxygen to form CO

CO formed $= \left\{30.29 \times \left(\frac{28}{12}\right)\right\} \doteq \mathbf{70.68\,kg}$

10% C forms CO_2, i.e. 0.1 × 33.66 = 3.366 kg C forms CO_2

$C + O_2 = CO_2$

12 32 44

12 kg C react with 32 kg oxygen to form CO_2

3.366 kg…{3.366×} = **8.98 kg** oxygen to form CO_2
CO_2 formed = {3.366×} = **12.34 kg**

P balance: P input = P output
P from HM + P from scrap + P from HC Fe–Cr = P in steel + P goes to slag
$\{(4 \times 10^{-3} \times 332) + (3 \times 10^{-4} \times 315) + (5 \times 10^{-4} \times 306) = (3 \times 10^{-4} \times 1000) + W_P$
Therefore, $W_P = (1.33 + 0.095 + 0.153) - 0.3 =$ **1.28 kg**

$$2P + 5/2O_2 = P_2O_5$$
$$62 \quad 80 \quad 142$$

P_2O_5 goes to slag $= 1.28 \times \left(\frac{142}{62}\right) =$ **2.93 kg**
62kg P react with 80kg oxygen to form P_2O_5
1.28kg…$\left\{1.28 \times \left(\frac{80}{62}\right)\right\} =$ **1.65 kg** oxygen to form P_2O_5

S balance: S input = S output
S from HM + S from scrap + S from HC Fe–Cr + S from lime + S from fluorspar = S in steel + S goes to slag
$\{(4 \times 10^{-4} \times 332) + (4 \times 10^{-4} \times 315) + (5 \times 10^{-4} \times 306) + (1.5 \times 10^{-3} \times 52) + (3 \times 10^{-3} \times 5.5) = (3 \times 10^{-4} \times 1000) + W_S$
Therefore $W_S = 0.133 + 0.126 + 0.153 + 0.078 + 0.017 = 0.507 - 0.3 =$ **0.207 kg**

$$S + CaO \rightarrow CaS$$
$$32 \quad 56 \quad 72$$

32 kg S form 72 kg CaS
0.207 kg …$\left\{0.207 \times \left(\frac{72}{32}\right)\right\} =$ **0.466 kg** CaS
32 kg react with 56 kg CaO
0.207 kg …$\left\{0.207 \times \left(\frac{56}{32}\right)\right\} = 0.36$ kg CaO to form CaS.

CaO balance: CaO input = CaO output
CaO from lime + CaO from fluorspar = CaO in slag + CaO to form CaS
$\{(0.9585 \times 52) + (5 \times 10^{-3} \times 5.5)\} = W_{CaO} + 0.36$
Therefore $W_{CaO} = (49.842 + 0.0275) - 0.36 =$ **49.54 kg**

MgO balance: MgO input = MgO output
MgO from lime = MgO goes to slag
$(10.025 \times 52) = W_{MgO}$
$\mathbf{W_{MgO} = 1.3}$ **kg**

Al_2O_3 balance: Al_2O_3 input = Al_2O_3 output
Al_2O_3 from fluorspar = Al_2O_3 goes to slag
$(7 \times 10^{-3} \times 5.5\} = W_{Al_2O_3}$
$W_{Al_2O_3} = 0.$**04 kg**

CaF_2 balance: CaF_2 input = CaF_2 output
CaF_2 from fluorspar = CaF_2 goes to slag
$(0.97 \times 5.5) = W_{Flu} =$ **5.34 kg**

Cr balance: Cr input = Cr output
Cr from HC Fe–Cr = Cr in Steel + Cr loss in slag
(0.625 × 306) = (0.185 × 1000) + W_{Cr}
Therefore, W_{Cr} = 191.25 − 185.0 = 6.25 kg

$$2Cr + 3/2O_2 = Cr_2O_3$$
$$2 \times 52 \quad 3 \times 16 \quad 152$$

104 kg Cr form 152 kg of Cr_2O_3
kg Cr … {6.25 × $\left(\frac{152}{104}\right)$} = **9.13 kg of Cr_2O_3**
104 kg Cr react with 48 kg oxygen
6.25 kg Cr … {6.25 × $\left(\frac{48}{104}\right)$} = **2.88 kg of oxygen**

$$\begin{aligned}\textbf{Weightofslag} &= W_{FeO} + W_{SiO_2} + W_{MnO} + W_{PROS} + W_{CaO} + W_{CaS} + W_{MgO} + W_{Al_2O_3} + W_{CaF_2} \\ W_{Cr_2O_3} &= 19.29 + 16.72 + 4.71 + 2.93 + 49.54 + 0.47 + 1.3 + 0.04 + 5.34 + 9.13 \\ &= \mathbf{109.47\,kg}\end{aligned}$$

Composition of slag: 17.62% FeO, 15.27% SiO_2, 4.3% MnO, 2.68% P_2O_5, 45.25% CaO, 0.43% CaS, 1.19% MgO, 0.04% Al_2O_3, 4.88% CaF_2, 8.34% Cr_2O_3.
Weight of oxygen = O for FeO + O for SiO_2 + O for MnO + O for CO + O for CO_2 + O for P_2O_5 + O for Cr_2O_3 = 4.29 + 8.91 + 1.06 + 40.39 + 8.98 + 1.65 + 2.88 = **68.16 kg**.

Example 16.5 Calculate amount of electric power required to produce one tonne duplex stainless steel in CONARC process by addition of 315 kg scrap along with 332 kg hot metal (at 1350 °C) (as given data in Example 16.4).

Given:

(i) Heat content of HM = (0.22T + 17) kWh/t, T is in °C.
(ii) Heat of reactions at 2023 K ($-\Delta H^0_{2023}$, kJ/kg) as the following:

CO → 1909.70, CO_2 → 6857.05, FeO → 2038.53, MnO → 4078.14
P_2O_5 → 7340.20, SiO_2 → 13,236.38, Cr_2O_3 → 5985.63.

Material	Specific heat, kJ/kg K × 10^3	Latent heat, kJ/kg K	Material	Specific heat, kJ/kg K × 10^3	Latent heat, kJ/kg K
Steel scrap	681.97	271.95	Fe–Mo	1005.0	486.78
HC Fe–Cr	670.0	324.52	LC Si–Mn	628.0	578.78

Specific heat for HM is 0.51 kJ/kg K. For melting of 1 t Ni at 2023 K energy requires 295 kWh, and 1 t slag formation at 2023 K energy requires 444.44 kWh.

Radiation and other heat losses = 100 kWh.

Solution

Heat input:

(i) Sensible heat carries by hot metal = $(0.22T + 17)$ kWh/t = $\left(\frac{(0.22T+17)}{1000}\right)$ kWh/kg

at 1350 °C = $\left(\frac{(0.22\times1350+17)}{1000}\right) = 0.314\,\text{kWh/kg}$

So 332 kg HM carries sensible heat = 0.314 × 332 = **104.25 kWh**

(ii) Heat of Reactions:

Heat formation for 19.29 kg FeO = 2038.53 × 19.29 = 39,323.24 kJ
Heat formation for 16.72 kg SiO_2 = 13,236.38 × 16.72 = 221,312.27 kJ
Heat formation for 4.71 kg MnO = 4078.14 × 4.71 = 19,208.04 kJ
Heat formation for 70.68 kg CO = 1909.70 × 70.68 = 134,977.60 kJ
Heat formation for 12.34 kg CO_2 = 6857.05 × 12.34 = 84,616.0 kJ
Heat formation for 2.93 kg P_2O_5 = 7340.20 × 2.93 = 21,506.79 kJ
Heat formation for 9.13 kg Cr_2O_3 = 5985.63 × 9.13 = 54,648.80 kJ

$$\begin{aligned}\text{Total heat of reactions} &= 39{,}323.24 + 221{,}312.27 + 19{,}208.04 + \mathbf{134{,}977.60}\\ &\quad + 84{,}616.0 + 21{,}506.79 + 54{,}648.80\\ &= 575{,}592.74\,\text{kJ} = (575{,}592.74/3600) = \mathbf{159.89\,kWh}\end{aligned}$$

(Since 1 kWh = 3600 kJ)

Total heat input = (i) + (ii) = 104.25 + 159.89 = **264.14 kWh**
Heat Output:

$$\begin{aligned}\text{Melting of scrap} &= \{W_{Sc} \times q_{sc} \times (T_2 - T_1)\} + (W_{Sc} \times L_f)\\ &= \{315 \times 0.68197 \times (2023 - 298)\} + (315 \times 271.95)\\ &= 370{,}565.45 + 85{,}664.25 = 456{,}229.7\text{kJ} = 126.73\,\text{kWh}\end{aligned}$$

$$\begin{aligned}\text{Melting of HC Fe–Cr} &= \{306 \times 0.67 \times (2023 - 298)\} + (306 \times 324.52)\\ &= 353{,}659.5 + 99{,}303.12 = 452{,}962.62\,\text{kJ} = \mathbf{125.82\,kWh}\end{aligned}$$

$$\begin{aligned}\text{Melting of LHC Si–Mn} &= \{22 \times 0.628 \times (2023 - 298)\} + (22 \times 578.78)\\ &= 23{,}832.6 + 12{,}733.16 = 36{,}565.76\,\text{kJ} = \mathbf{10.16\,kWh}\end{aligned}$$

$$\begin{aligned}\text{Melting of Fe–Mo} &= \{50 \times 1.005 \times (2023 - 298)\} + (50 \times 486.78)\\ &= 86{,}681.25 + 24{,}339.0 = 111{,}020.25\,\text{kJ} = \mathbf{30.84\,kWh}\end{aligned}$$

For melting of 1 t Ni at 2023 K requires 295 kWh.
So, 1 kg Ni at 2023 K requires (295/1000) = 0.295 kWh.
Therefore, for melting 48 kg Ni = (48 × 0.295) = **14.16 kWh**
Total heat requires for melting = 126.73 + 125.82 + 10.16 + 30.84 + 14.16 = **307.71 kWh**

$$\begin{aligned}\text{The heat requires for increasing temperature of HM} &= \{332 \times 0.51 \times (2023 - 1623)\}\\ &= 67{,}728\,\text{kJ} = \mathbf{18.81\,kWh}\end{aligned}$$

Heat requires 1 t slag formation at 2023 K = 444.44 kWh
i.e. heat requires for 1 kg slag formation at 2023 K = (444.44/1000) = 0.444 kWh
Therefore, heat requires for 109.47 kg slag formation = 109.47 × 0.444 = **48.60 kWh**

Radiation and other heat losses = **100 kWh**
Total Heat output = 307.71 + 18.81 + 48.60 + 100 = **475.12 kWh**

Therefore, power require = Heat output − Heat input = 475.12–264.14 = **210.98 kWh**.

Example 16.6 The initial weight and final weight of HBI before melting and after melting are 850 g and 455 g, respectively, at 1 min at a laboratory induction furnace. Find out: (i) fraction of HBI dissolution, (ii) rate of HBI dissolution (per sec), (iii) laboratory melting rate and (iv) melting rate for industrial scale.

Solution

(i) Fraction of HBI dissolution, at time 1 min, i.e. 60 s,

$$f = \frac{x-y}{x} = \frac{850-455}{850} = 0.465 \quad (1)$$

(ii) Therefore, the rate of HBI dissolution (per sec),

$$R = \frac{\mathrm{d}f}{\mathrm{d}t} = \frac{0.465}{60} = 7.75 \times 10^{-3}\,\mathrm{s}^{-1} \quad (2)$$

(iii) Laboratory melting rate (R_1, g/s) can be calculated by:

$$R_1 = \frac{z}{t} = \frac{850-455}{60} = 6.58\,\mathrm{g/s} \quad (3)$$

(iv) The melting rate for industrial scale (Kg/ min):

$$R_2 = (R_1/1000) \times 60 = (6.58/1000) \times 60 = 0.395\,\mathrm{kg/min}. \quad (4)$$

Example 16.7 Calculate the heat losses by radiation per tonne from the furnace of 7.6 m diameter at 1750 °C. Capacity of the furnace is 180 t.

Solution

The heat losses by radiation from the furnace:

$$q_r = €.\sigma.A.\left(T_2^4 - T_1^4\right) \quad (1)$$

where

q_r = Heat losses by radiation in watts,
€ = Emissivity of liquid steel (0.28 for steel at 1725 °C),
σ = Stephen Boltzman's constant (5.67×10^{-8} W/m^{2}K),
A = Cross-sectional area of furnace, m^2;
T_1 = Room temperature, K;
T_2 = Temperature of the molten bath, K.

Cross-sectional area of furnace, $A = \left(\frac{\pi d^2}{4}\right) = \left(\frac{\pi (7.6)^2}{4}\right) = 45.34 m^2$
Room temperature, T_1 = 298 K,
Temperature of the molten bath, T_2 = (1750 + 273) = 2023 K
Therefore, from Eq. (1):

$$\begin{aligned} q_r &= 0.28 \times \left(5.67 \times 10^{-8}\right) \times 45.34 \times \left\{(2023)^4 - (298)^4\right\} \\ &= 71.98 \times 10^{-8} \times \left\{\left(1.67 \times 10^{13}\right) - \left(0.79 \times 10^{10}\right)\right\} = 71.98 \times 10^{-8} \times (167000 - 79) \times 10^8 \\ &= 12{,}014.97\,\text{kWh} \end{aligned}$$

Hence, loss of radiation per tonne = (12,014.97/180) = 66.75 kWh/t.

Example 16.8 An EAF discharged dust emission and slag at the rate of 15 kg/t and 60 kg/t, respectively. Dust contains 23.9% Fe_2O_3 and 9.7% FeO; slag contains 23.4% FeO.

Find out how much Fe loss per tonne of liquid steel in dust emission and slag at EAF.

Solution

23.9% Fe_2O_3 content in 15 kg dust
Therefore, total Fe_2O_3 content in dust = 0.239 × 15 = 3.585 kg/t
Similarly, total FeO content in dust = 0.097 × 15 = 1.455 kg/t
Again, 160 part Fe_2O_3 contains 112 part of Fe
So, 3.585 kg Fe_2O_3 contains = (112/160) × 3.585 = 2.51 kg
72 part FeO contains 56 part of Fe
Hence, 1.455 kg FeO contains = (56/72) × 1.455 = 1.13 kg
Therefore, total Fe loss as Fe_2O_3 and FeO in dust emission = 2.51 + 1.13 = **3.64 kg/t**
23.4% FeO content in 60 kg slag,
So, total FeO content in slag = 0.234 × 60 = 14.04 kg/t
Again 72 part FeO contains 56 part of Fe
Hence, 14.04 kg FeO contains = (56/72) × 14.04 = 10.92 kg/t
So, total Fe loss as FeO in slag = **10.92 kg/t**
Hence, total Fe loss as Fe_2O_3 and FeO in dust emission and FeO in slag
= 3.64 + 10.92 = **14.56 kg/t liquid steel.**

Example 16.9 Density of steel at melting point is 6.88×10^3 kg/m^3, volume of slag = 15% of liquid steel. Find out the diameter of the furnace and depth of height; as well as inside and outside diameters of the shell.

Given: (i) ratio of diameter-to-depth = 5, (ii) height (h_1) of the spherical portion, $h_1 = 0.2H$; (iii) depth of the conical portion, $h_2 = 0.8H$;(iv) total volume of the bath, $V_b = 0.0837\ D^3$; and (v) diameter of the reaction chamber, $D_{rc} = D + 0.2$.

Solution

Since density of steel at melting point is 6.88×10^3 kg/m^3, one tonne of molten steel occupies 0.145 m^3 i.e. $V_m = 0.145$ m^3. Therefore, 100 t molten steel will occupy 14.5 m^3 (V_m is volume of liquid steel).
The volume of slag, V_{slag} = 15% of V_m, i.e. $V_{slag} = 0.15 \times 14.5 = 2.175$ m^3.

Therefore, total volume will be 16.675 m^3.

The diameter-to-depth ratio of the bath operating by the basic process is usually taken as:

$$D = 5H \quad (1)$$

where,

D = bath diameter, m,
H = total depth of the bath, m,

Since the height (h_1), of the spherical portion is roughly 1/5 of the total bath depth:

$$h_1 = 0.2H \quad (2)$$

and the depth of the conical portion:

$$h_2 = 0.8H \quad (3)$$

The total volume of the bath, V_b,

$$V_b = 0.0837D^3\left(\mathrm{m}^3\right) \quad (4)$$

Therefore,

$$0.0837D^3 = 16.675\left(\mathrm{m}^3\right).$$

Therefore, from Eqs. (1), (2) and (3), getting the following values:

$$\mathrm{D}^3 = 16.675/0.0837 = 199.22$$

Therefore, **D = 5.84 m**
Since $D = 5H$, so $H = D/5 = 5.84/5 =$ **1.168 m**
Since $h_1 = 0.2H$ and $h_2 = 0.8H$
So, $h_1 = 0.2 \times 1.168 = 0.234$ m and $h_2 = 0.8 \times 1.168 = 0.934$ m

Diameter of spherical bottom,

$$d = D - 2(h_2) = 5.84 - 2(0.934) = 3.972\,\mathrm{m}.$$

The diameter of the reaction chamber is:

$$\begin{aligned} D_{rc} &= D + 0.2(\mathrm{m}) \\ D_{rc} &= 5.84 + 0.2 = 6.04\,\mathrm{m} \end{aligned} \quad (5)$$

It is recommended that the following relationship be used in determining the height of the reaction chamber (H_1),

$$H_1 = (0.4 \text{ to } 0.6)D_{rc} \quad (6)$$

For large-capacity furnaces, taking the lower value (i.e. 0.4).

Therefore, $H_1 = 0.4\ D_{rc} = 0.4 \times 6.04 = 2.416$ m
The height of the roof (h_3) is:

$$h_3 = 0.15D_1 \quad (7)$$

where D_1 is roughly taken 1.0 m wider than D_{rc}.

$$\begin{aligned} D_1 &= D_{rc} + 1.0 \\ h_3 &= 0.15D_1 = 0.15\,(D_{rc} + 1.0) \end{aligned} \quad (8)$$

Therefore, $h_3 = 0.15\ (6.04 + 1.0) = 1.056$ m
The total height of the roof above the bath level is given as:

$$\begin{aligned} H_2 &= H_1 + h_3 \\ H_2 &= 2.416 + 1.056 = 3.472\,\text{m} \end{aligned} \quad (9)$$

The recommended slope (S) for the inclination of the wall is roughly 10% of the height from the line of banks to roof skewbacks:

$$\text{Slope (S)} = H_1 \times 0.1 = 2.416 \times 0.1 = 0.2416\,\text{m}$$

The diameter of the reaction chamber at the level of the roof skewbacks (i.e. at the level of the upper edge of the furnace shell) is:

$$\begin{aligned} D_1 &= D_{rc} + 2S \\ D_1 &= 6.04 + 2 \times 0.2416 = 6.523\,\text{m} \end{aligned} \quad (10)$$

The thickness of the lining is found by thermal analysis from the condition that the furnace shell should not be heated above 200 °C at the end of the furnace campaign. With the refractory lining δ m thick, the furnace will have the following inside diameter of the shell:

$$D_{i.sh} = D_{rc} + 2\delta \quad (11)$$

where $\delta = 0.6$ m for 100–200 t capacity furnace.
Therefore, $D_{i.sh} = 6.04 + 2 \times 0.6 = 7.24$ m.
Thickness of shell is kept around 0.05 m.
Therefore, outer side diameter of the shell:

$$\begin{aligned} \text{D}_{0.sh} &= \text{D}_{i.sh} + 2 \times \text{Thickness of shell} \\ &= 7.24 + 2 \times 0.05 = 7.34\text{m}. \end{aligned}$$

Example 16.10 If transformer power (P_{ap}) is taken as 35 MVA. The relation of the average power consumed during the melting period (P_{av}) and the transformer power (P_{ap}) may be taken as:
$P_{av} = 0.8\ P_{ap}$ (kVA). Find out the upper voltage tap required for melting, when cos φ and η_{el} are 0.85 and 0.9, respectively.

Solution

Since

$$P_{av} = 0.8P_{ap}(\text{kVA})$$
$$= 0.8 \times 35{,}000 = 28{,}000\,\text{KVA}$$

Useful power consumed during the melting period:

$$P_u = P_{av}\cos\varphi\,\eta_{el}(\text{kW})$$
$$= 28{,}000 \times 0.85 \times 0.9 = 21{,}420\text{KW}$$

The upper voltage tap of the secondary voltage for small furnaces can be selected by using the following empirical formulae.

For basic furnaces: $V = 15\sqrt[3]{P_{AP}} = \mathbf{490.66\,V}$.

Problems

Problem 16.1 The initial weight and final weight of HBI before melting and after melting are 750 g and 400 g, respectively, at 1 min at a laboratory induction furnace. Find out: (i) fraction of HBI dissolution, (ii) rate of HBI dissolution (per sec), (iii) laboratory melting rate and (iv) melting rate for industrial scale.

$[\text{Ans: (i) } 0.467,\ \text{(ii) } 7.78 \times 10^{-3}\text{s}^{-1},\ \text{(iii) } 5.83\,\text{g/s and (iv) } 0.35\,\text{kg/min}]$.

Problem 16.2 (a) Calculate amount of DRI required along with 700 kg steel scrap to produce one tonne steel in EAF. (b) Calculate amount of slag produce. (c) Amount of lime required for making slag (V ratio 2.0) and (d) yield of Fe.

Given: (i) 1.5% Fe loss in slag w.r.t liquid steel, (ii) spar and Fe–Mn charged 5 kg each.

Element/compound %	Steel produced	Scrap	DRI	Spar	Lime	Fe–Mn
C	0.10	0.20	1.20			0.2
Si	0.20	0.25				1.5
Mn	0.30	0.25				72.5
P	0.03	0.04	0.05			0.4
S	0.03	0.04	0.01	0.3	0.15	0.05
CaO			1.0	0.5	95.85	
MgO			1.0		2.5	
SiO_2			1.5	1.5	1.5	
Al_2O_3			0.5	0.7		
FeO			9.64			
Fe_{Tol}			92.5			
CaF_2				97.0		
Amount charged, Kg		500	?	5	?	5

[Ans: (a) 338.0 kg, (b) 49.39 kg, (c) 11.96 kg and (d) 98.5%].

Problem 16.3 Calculate amount of scrap required along with 500 kg hot metal (at 1350 °C) to produce one tonne steel in CONARC process. Fe–Mn and lime are charged 5 and 30 kg respectively. How much oxygen required to produce one tonne steel. Find out amount of slag produce.

	C	Mn	Si	S	P	CaO	MgO	SiO_2	Al_2O_3
Steel	0.1	0.25	0.1	0.03	0.03				
Scrap	0.25	0.3	0.2	0.03	0.04				
HM	3.5	1.2	1.5	0.04	0.4				
Lime				0.15		94.5	1.5	2.5	1.5
Fe–Mn	0.2	72.5	1.5	0.05	0.4				

Given: (i) 2.0% Fe loss in slag w.r.t. liquid steel. (ii) 90% carbon forms CO and 10% carbon forms CO_2.
[Ans: 552.6 kg, 46.18 kg, 88 kg].

Problem 16.4 Calculate amount of hot metal (at 1350 °C) required along with 350 kg stainless steel scrap to produce one tonne stainless steel in CONARC process. Find out how much lime is required to get basicity of 3.0. How much oxygen required to produce one tonne steel. Find out amount of slag produce. Also find out how much power required.

Element/Compound %	Steel produced	SS scrap	HM	Spar	Lime	Ni	HC Fe–Cr	HC Fe–Mn	Fe–Mo
C	0.03	0.08	3.5				7.0	7.0	0.15
Si	0.8	1.0	1.0				4.0	1.5	
Mn	2.0	2.0	0.5				0.18	72.5	
P	0.03	0.04	0.4				0.05	0.4	
S	0.02	0.03	0.04	0.3	0.15		0.05	0.05	
Fe	67.12	70.85	94.56				26.22	18.55	44.85
Cr	22.0	18.0					62.5		
Ni	5.0	8.0				100			
Mo	3.0								55.0
CaO				0.5	95.85				
MgO					2.5				
SiO_2				1.5	1.5				
Al_2O_3				0.7					
FeO									
Fe_{Tol}									
CaF_2				97.0					
Amount charged, Kg	1000	350	?	5.5	?	22	261.2	20.0	55.0

[Ans: Wt. of Hm = 367.34 kg, wt of lime = 70.06 kg, total oxygen consumption = 69.99 kg, amount of slag = 136.8 kg; and power require = 134.19 kWh].

Problem 16.5 Calculate amount of DRI required along with 600 kg scrap to produce 1 t steel. Given: Scrap contains 98.5% Fe and steel contains 99.3% Fe. DRI contains 9.64% FeO and 85% metallization. 1.5% Fe loss in slag and 0.5% in exit gases w.r.t. liquid steel production.
[Ans: 488.7 kg].

Problem 16.6 Calculate amount of DRI to be charged along with 300 kg hot metal (at 1350 °C) and 114 kg MS scrap to produce one tonne duplex stainless steel (SAF 2205) in CONARC process. Other materials are charged as follows: HC Fe–Cr → 363 kg/t, HC Fe–Mn → 30 kg/t, Fe–Mo → 55 kg/t, Ni → 50 kg/t, Lime → 82.2 kg/t, and Spar → 5.5 kg/t.

How much oxygen required to produce one tonne steel. Also find out amount of slag produce. Assume: (i) 1.5% and 0.6% Fe loss in slag and flumes, respectively, (ii) 90% C forms CO and 10% C forms CO_2.
Composition of charge materials and stainless steel produce is shown as the following:

Mat.	Composition %											
	C	Mn	Si	S	P	Fe	Cr	Ni	CaO	MgO	SiO_2	Al_2O_3
HM	3.5	0.5	1.0	0.04	0.4	94.56						
MS scrap	0.26	0.25	0.25	0.04	0.03	99.17						
SS scrap	0.08	2.0	1.0	0.03	0.04	70.85	18	8				
DRI	1.25	–	–	0.01	0.05	85.0		FeO 9.64	1.0	1.0	1.5	0.5
HC Fe–Cr	7.0	0.18	4.0	0.05	0.05	26.22	62.5					
HC FeMn	7.0	72.5	1.5	0.05	0.4	18.55						
LC SiMn	0.1	57.5	27.5	–	–	14.90						
FeMo	0.15	–	–	–	–	44.85		Mo 55				
Lime	–	–	–	0.15	–	–	–	–	95.85	2.5	1.5	
Spar	–	–	–	0.3	–	–	–	–	0.5	CaF_2 97.0	1.5	0.7

AISI grade	%C	%Cr	%Ni	%Mo	%Mn	%Si	%S	%P	%Fe
SAF 2205	0.03	22.0	5.0	3.0	2.0	0.8	0.02	0.03	67.12

[Ans: (i) Wt of DRI = 184.84 kg, (ii) amount of oxygen require = 79.11 kg and (iii) amount of slag = 155.12 kg].

References

1. https://images.search.yahoo.com/search/images
2. https://www.yourelectricalguide.com/2017/03/induction-heating-induction-furnace.html
3. http://infohouse.p2ric.org/ref/10/09048.pdf
4. World Steel Association, *World Steel in Figures* (2019)
5. F.P. Edneral, *Electrometallurgy of Steel and Ferro-alloys* (MIR Publishers, Moscow, 1979)
6. https://www.substech.com/dokuwiki/doku.php?id=electric_arc_furnace_eaf
7. ASM, *ASM Metal Handbook*, 9th edn., vol. 15, p. 537 (For Casting).
8. R.H. Tupkary, *Introduction to Modern Steelmaking* (Khanna Publishers, Delhi, 1991)
9. L.B. Singh, IIM Metal News **11**(1), 25 (2008)
10. U.S. Environmental Protection Agency, *Available and Emerging Technologies for Reducing Green House Gas Emissions from Iron and Steel Industry* (U.S. Environmental Protection Agency, North Carolina, 2012)
11. AIMMPE, *Electric Furnace Steelmaking*, vol. 1 (AIMMPE, US, 1962), p. 61
12. R. Oza, S.K. Dutta, Iron Steel Rev. **58**(7), 202 (2014)
13. http://ietd.iipnetwork.org/content/electric-arc-furnace
14. S.K. Dutta, A.B. Lele, Tool Alloy Steels **27**(7), 193 (1993)
15. S.K. Dutta, JPC Bull. Iron Steel **XIII**(1) 61 (2013)
16. P. Tatia, *Paper presented at NMD-ATM of IIM 2014* (Pune, India, 2014) (Privately Collected)
17. A.K. Chakrabarti, *Steel Making* (Prentice-Hall of India Pvt Ltd, New Delhi, 2007), p. 90
18. T. Ames, J. Kopfle, *Direct from Midrex*, 2nd edn. (Quarter 2007), p. 7
19. S.K. Dutta, A.B. Lele, Iron Steel Rev. **55**(3), 162 (2011)
20. S.K. Dutta, JPC Bull. Iron Steel, **XIV**(1), 13(2014) (Sp Issue)
21. S.N. Guha, S. Pal, *Proceedings of the International Seminar on Alternative Routes for Ironmaking in India* (Kolkata, India, 2009), p. 51
22. A. Ganguly et al., IIM Metal News **19**(2), 13 (1997)
23. S.K. Dutta, A.B. Lele, Tool Alloy Steels, **28**(11), 317 (1994)
24. https://en.wikipedia.org/wiki/Induction_furnace
25. S.K. Dutta, Iron Steel Rev. **59**(3), 155 (2015)
26. S.K. Sarna, JPC Bull. Iron Steel **XV**(12), 12 (2015)
27. S.K. Dutta, A.B. Lele, Tool Alloy Steel **32**(12), 8 (1998)
28. C. Farmer, *Direct from Midrex* (4th Quarter, 2000), p. 3
29. S.S. Bedarkar, R. Singh, Trans. Indian Inst. Met. **66**(3), 207 (2013)
30. S.K. Sarna: JPC Bull. Iron Steel, **XV**(4), 12 (2015)
31. S.K. Dutta, A.B. Lele, N.K. Pancholi, Trans. Indian Inst. Met. **57**(5), 467 (2004)
32. S.K. Dutta, A.B. Lele, Indian Foundry J. **47**(3), 29 (2001)
33. S.K. Dutta, A.B. Lele, Inst. Eng. (I) J.-MM **82**, 1 (2001)
34. S.K. Dutta, A.B. Lele, JPC Bull. Iron Steel **VI**(7), 16 (2006)
35. G. Kleinschmidt, G. Walter, S.K. Bose, *International Symposium on Minerals and Metals challenges Beyond 2000*, 56th edn. (ATM of IIM, Bhilai, India, 2000), p. 115
36. Henrique, JPC Bull. Iron Steel **XIV**(9), 11 (2014)
37. J. Singh, A.K. Ray, JPC Bull. Iron Steel **VII**(12), 5 (2007)
38. K. Datta et al., Trans. Indian Inst. Met. **47**(6), 373 (1994)
39. S.R. Prabhakar, U. Batra, Inst. Eng. (I) J.-MM **76**, 36 (1995)
40. S. Dewan, JPC Bull. Iron Steel **11**, 3 (2000)
41. Technical Notes, IIM Metal News **3**(2), 20 (2000)
42. http://ietd.iipnetwork.org/content/shaft-furnace-scrap-preheating
43. S.K. Dutta, A.B. Lele, Trans. Indian Inst. Met. **56**(1), 19 (2003)
44. N.K. Bharal, *Key Note Lecture at the National Seminar on Recent Advances in Making, Shaping and Applications of Stainless steel* (Vadodara, India, 2001), p. 57 (Privately Collected)
45. S.K. Dutta, A.B. Lele, Iron Steel Rev. **43**(6), 56 (1999)
46. E. Fritz, J. Steins, Iron Steel Rev. **41**(1), 54 (1997)
47. A. Ganguly, R.H.G. Rau, IIM Metal News **1**(4), 3 (1996)
48. R.P. Varshney, JPC Bull. Iron Steel **2**(1), 90 (2002)

Secondary Steelmaking 17

The increasing demand for high-quality steel products has led to the continuous improvement of steelmaking practice. These requirements have led to development of various kinds of treatment of liquid steel in ladles which are known as secondary steelmaking or ladle metallurgy. Different methods of ladle furnace process, de-gassing processes and injection ladle metallurgy are discussed in detail. The de-oxidation practice: precipitation de-oxidation and diffusion de-oxidation are described. Different types of inclusion and its control are also described.

17.1 Introduction of Secondary Steelmaking

Competitive and high-quality steelmaking is unthinkable today without secondary metallurgy that perfect link between steelmaking and casting. Customers' demands have led to stringent steel quality with respect to composition in a narrow range, high degree of cleanliness, absence of segregation, low gases content, etc. The increasing demand for high-quality steel products has led to the continuous improvement of steelmaking practice. These requirements have led to development of various kinds of treatment of liquid steel in ladles which are known as *secondary steelmaking* or *ladle metallurgy*. *Secondary steelmaking* or *secondary metallurgy* is defined as any post-steelmaking process performed at a separate station prior to casting. There is also an emerging trend of reducing total non-metallic residuals (i.e. C, S, P, N, O, H) to very low levels [in ppm (parts per million)]. Secondary steelmaking has played a vital role in meeting these requirements. Modern steelmaking of today is therefore characterized by a primary process designed primarily to melt and rough refine hot metal, scrap or sponge iron, by a product-oriented secondary refining stage to finally adjust the melt and by the continuous casting of the steel. Now primary steelmaking is used for rapid scrap melting and gross refining only and leaving further refining and control of gases by secondary steelmaking.

Secondary steelmaking processes are used to achieve one or more of the following requirements:

(i) Improvement in quality,
(ii) Improvement in production rate,
(iii) Decrease in energy consumption.
(iv) Improve economy by better yield and higher availability of the various processes.
(v) Use of restively cheaper raw materials,
(vi) Use of alternate sources of energy and
(vii) Improvement of recovery of alloying elements.

S. K. Dutta and Y. B. Chokshi, *Basic Concepts of Iron and Steel Making*,
https://doi.org/10.1007/978-981-15-2437-0_17

The objectives of secondary steel refining are as following:

1. Homogenization of the steel bath with respect to temperature and chemical composition,
2. Improved alloy yield and composition control within narrow limits,
3. De-carburization,
4. De-phosphorization, de-oxidation, de-sulphurization and improved steel cleanness, i.e. inclusion control,
5. De-gassing, i.e. removal of nitrogen and hydrogen gases,
6. Adjustment of teeming temperature and
7. Providing sufficient holding time during treatment and teeming.

Secondary steel refining is carried out even in a transfer ladle fitted with porous plug for inert gas (i.e. argon) purging. Argon supply line is connected to this plug, just before tapping. Similarly, injection treatment can be carried out in the transfer ladle itself. However, the bath tends to lose temperature in all these treatments, and hence, these cannot be carried out for too long period, even if required. There are three types of processes: (1) ladle furnace, (2) de-gassing processes and (3) injection ladle metallurgy.

17.2 Ladle Furnace (LF)

Main aim of ladle furnace is as follows:

1. To control of chemical composition and temperature in a very narrow level,
2. To make very low level of sulphur (below 0.005%) by efficient de-sulphurization,
3. Addition of micro-alloying elements in proper sequence to have maximum recovery.

Ladle furnace is used to relieve the primary melter of most secondary refining operations, and its primary functions are:

1. Reheating of liquid steel through electric power conducted by graphite electrodes,
2. Homogenization of steel temperature and chemistry through inert gas stirring,
3. Formation of a slag that protects refractory from arc damage; concentrates and transfers of heat to the liquid steel; trap inclusions; and arranges for de-sulphurization.

Secondary functions that can be included with a ladle furnace are:

1. Alloy additions to provide bulk chemical control,
2. Cored wire addition for changing the morphology of inclusions,
3. Separation of inclusions for high cleanliness,
4. Removal of gases like hydrogen and nitrogen,
5. Provide a means for deep de-sulphurization.
6. Provide a means for de-phosphorization.
7. Act as a buffer for downstream steelmaking.

It is a simple ladle like furnace (as shown in Fig. 17.1) provided with (i) bottom porus plug for argon purging and (ii) lid with electrodes to become an arc furnace for heating the bath. Another lid may be provided to connect it to vacuum line, if required.

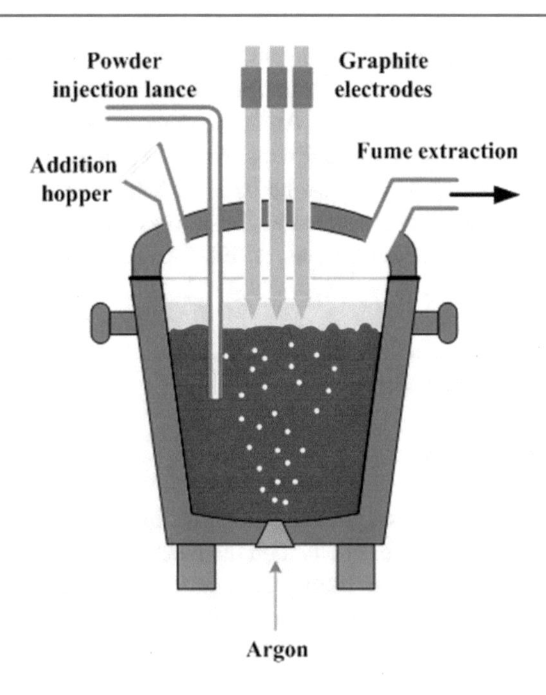

Fig. 17.1 Ladle furnace [1]

The function of the porous plug is to provide gas stirring of the molten metal to promote homogenization of bath temperature and composition. Normal stirring operations are performed by argon gas through a purge plug arrangement in the bottom of the ladle, to help flotation of non-metallic inclusions and faster rate of de-sulphurization. Sulphur level in the bath comes down to 50 ppm after ladle furnace treatment. Argon gas is passed through porous plug at the rate of 0.18 Nm^3/min from beginning to end. Total time required in ladle furnace is 40–50 min, and power consumption is about 50 kWh/t.

It can carry out stirring, vacuum treatment, synthetic slag refining, purging gas, injection, etc. all in one unit without restraint of temperature loss, since it is capable of being heated independently. Fumes and dust generated during heating and alloying operations at the LF will exit the water-cooled ladle roof. Every ladle furnace need not be equipped with all these arrangements.

17.2.1 ASEA-SKF Furnace

ASEA-SKF furnace is a special variety of LF only. This furnace is essentially a teeming ladle for which additional fittings are provided. Metal in the ladle is stirred by an electromagnetic stirrer provided from outside. Ladle shell is made of austenitic stainless steel with refractory lining. Two ladle covers, i.e. lids, are used: (i) for vacuum seal for de-carburization and de-gassing and (ii) lid with electrodes to act as EAF to heat the melt for compensation of heat loss.

When the de-carburization and vacuum de-gassing is over, the first roof is replaced by the second roof which contains three electrodes. Then, final alloying and temperature adjustments are made. De-sulphurization of steel can also be done by preparing a reducing basic slag under the electrode cover.

It is a process which possesses group of treatment units usually consisting of separate de-slagging, arc heating and vacuum treatment units. Slag is removed to prevent reverse of phosphorous in metal, after which ferro-alloy addition is carried out. Arcing is done to increase the temperature for compensating the cooling effect of the alloying additions. After that, de-gassing is done in a vacuum for reducing the oxygen and hydrogen content in the bath for achieving hydrogen contents as low as 1.5 ppm. The electromagnetic stirring is created in the process which helps in floating non-metallic inclusions and result in production of clean steels. ASEA-SKF units have incorporated with inert gas stirring to enable de-sulphurization.

17.3 De-gassing Processes

17.3.1 Gases in Liquid Steel

During the primary steelmaking process, gases like oxygen, hydrogen and nitrogen are dissolved in the liquid steel. These gases have a harmful effect on the mechanical and physical properties of steel. Dissolved oxygen [O] from liquid steel cannot be removed as molecular oxygen $\{O_2\}$, and its removal is termed as de-oxidation. The term de-gassing is used for the removal of dissolved hydrogen [H] and nitrogen [N] gases from liquid steel as molecular hydrogen $\{H_2\}$ and nitrogen $\{N_2\}$. Since the de-gassing process of liquid steel is carried out in teeming ladles under vacuum, it is also known as vacuum de-gassing of liquid steel.

Removal of hydrogen [H] and nitrogen [N] gases from liquid steel is necessary since both of these gases affect the properties of steel. Solubility of hydrogen [H] in steel is low at low temperature. Excess hydrogen [H] is rejected during solidification and results in pinhole formation as well as formation of blow holes or porosity in solidified steel. Few ppm of hydrogen gas causes blistering and loss of ductility. In case of nitrogen [N] gas, maximum solubility of nitrogen [N] in liquid iron is 450 ppm and less than 10 ppm at room temperature. During solidification, excess nitrogen is rejected which can cause formation of either blow holes or nitrides. Excess nitrogen also causes embrittlement of heat affected zone during welding of steels and impairs cold formability of steel.

Removal of gases from molten steel is taken place by the following reaction:

$$2[\mathrm{G}] = \{\mathrm{G}_2\} \tag{17.1}$$

where G is any diatomic gas (i.e. hydrogen or nitrogen).

The equilibrium constant (k'_{G}) for the reaction (17.1) is given by:

$$k'_{17.1} = \frac{p_{\mathrm{G}_2}}{[h_{\mathrm{G}}]^2} \tag{17.2}$$

where p_{G_2} is partial pressure of a gas (in atm) in equilibrium with molten steel and $[h_{\mathrm{G}}]$ is the Henrian activity of gas in molten steel.

$$\text{Since } [h_{\mathrm{G}}] = [f_{\mathrm{G}}\mathrm{wt\%\,G}], \text{for very dilute solution}, f_{\mathrm{G}} = 1;\ \text{so } [h_{\mathrm{G}}] = [\mathrm{wt\%\,G}] \tag{17.3}$$

where [wt% G] denotes wt% of gas dissolved in molten steel.

Therefore, Eq. (17.2) may be written in the form of *Sievert's law*:

$$[\text{wt\% G}] = \sqrt{\frac{p_{G_2}}{k'_{17.1}}} = k_H \cdot \{\sqrt{p_{G_2}}\} \tag{17.4}$$

where $k_G = \sqrt{\frac{1}{k'_{17.1}}}$ = De-gassing constant

According to Sievert's law, the solubility of a gas in liquid metal is proportional to the square root of the partial pressure of that gas. At 1600 °C and under a pressure of 1 atmosphere, nitrogen content of the liquid steel is equal to 0.04% [3]. The solubility of nitrogen in steel also increases with temperature. Alloying elements which form stable nitrides increase the solubility of nitrogen in liquid iron alloys. Vanadium, niobium, chromium and manganese are in this group. Elements like carbon, silicon, nickel, copper and tin reduce the solubility of nitrogen in liquid iron alloys (as shown in Fig. 17.2).

Most of the common alloying elements (e.g. C, B, Si, Al, Sn, etc.), however, reduce the solubility of hydrogen in liquid iron alloys. The solubility of hydrogen in liquid iron alloys increases in the presence of alloying elements (like Ti, Ta, V, Nb, Cr, Mn, etc.) as shown in Fig. 17.3.

Dissolved oxygen from liquid steel cannot be removed as molecular oxygen gas, and its removal is termed as de-oxidation. The first activity after tapping of liquid steel in ladle is to de-oxidize it with the help of suitable elements. The most effective de-oxidizer is aluminium. The different primary steelmaking processes can generate varying content of tap oxygen in the range of 250–1200 ppm. It is therefore expected that the amount of alumina generated through de-oxidation will also vary.

Total oxygen in steel is present in two forms: (i) free oxygen is essentially present in dissolved form and (ii) combined oxygen as non-metallic inclusions. Free or active oxygen is controlled by equilibrium thermodynamics. Steel, after being suitably killed by Al, has a dissolved oxygen content of about 3–5 ppm. Total oxygen generally drops after every processing stage of steelmaking [4]:

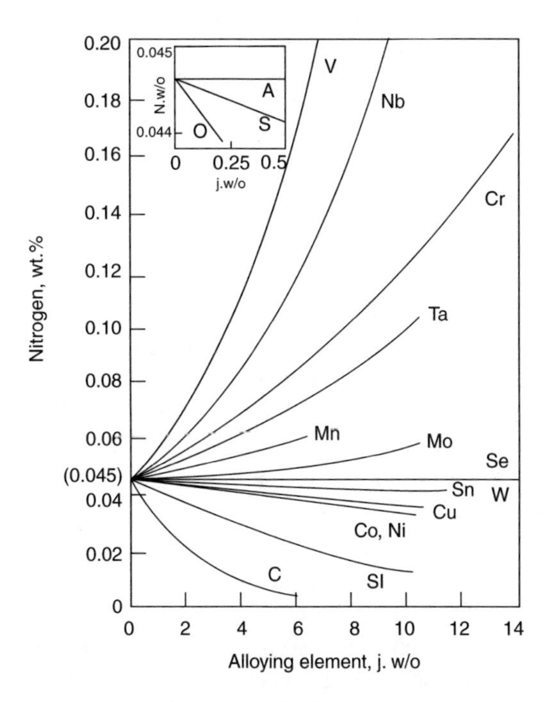

Fig. 17.2 Solubility of nitrogen in liquid iron alloys at 1600 °C and 1 atmosphere pressure of nitrogen gas [2]

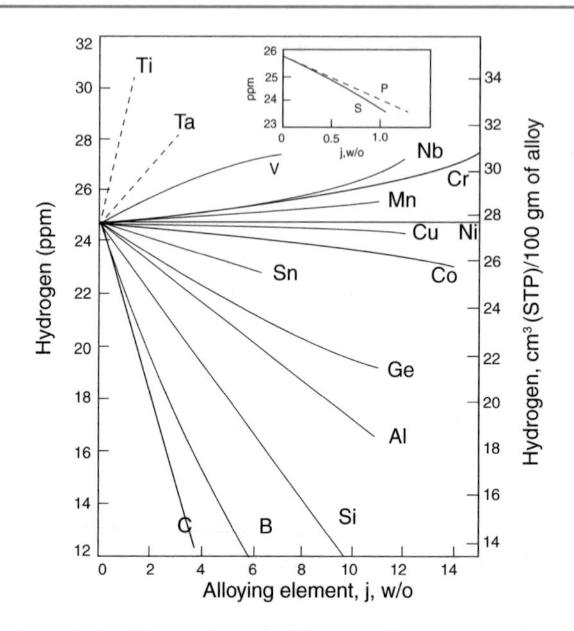

Fig. 17.3 Solubility of hydrogen in liquid iron alloys at 1600 °C and 1 atmosphere pressure of H_2 gas [2]

- Ladle: 40–45 ppm,
- Tundish: 25–30 ppm,
- Mould: 20–25 ppm.

Nitrogen pickup is essentially the difference in nitrogen content measured between different stages of steelmaking. It is taken as an indicator of air entrainment during transfer of molten steel from ladle to tundish and from tundish to mould. The low dissolved oxygen content after de-oxidation enables rapid absorption of air in liquid steel. Nitrogen pickup therefore serves as a crude indirect measure of reoxidation and the consequent deterioration in steel cleanliness. It is interesting to note that sulphur being a surface-active element, very low sulphur content in steel facilitates higher pickup of nitrogen and reoxidation.

17.3.2 Vacuum De-gassing

It was only in the early 1950s that the problem of producing steel with minimum gas content was solved through the development of a method of vacuum treatment of liquid steel in the ladle before teeming. The vacuum treatment of liquid steel in the ladle started first with the ladle to ladle and ladle to ingot mould vacuum de-gassing processes for the removal of hydrogen. Originally, vacuum de-gassing of liquid steel is carried out under reducing condition at a pressure ranging from 0.5 to 10 mbar (1 mbar = 0.75 mm Hg or 0.00102 kg/cm^2) with the objective of reducing the hydrogen content to less than 2 ppm.

Initially, the concept of vacuum de-gassing was used primarily for the removal of hydrogen gas from liquid steel, but sooner it served many other purposes also for the production of clean steels. Since around 1980 or so, there has been an increased use of vacuum de-gassing for the production of ultra-low carbon steels with carbon contents of 30 ppm or less. Furthermore, with the development of interstitial-free steels with carbon and nitrogen contents of 30 ppm or less, a treatment under vacuum has become a necessity. Presently, a vacuum de-gassing treatment has become an essential facility for a steel melting shop producing quality steel.

The general features of vacuum de-gassing are as follows [5]:

- Desorption of gases is a gas–metal interfacial reaction. The atomic [H] or [N] from the liquid steel must diffuse at the gas–metal interface, where it is converted to molecular hydrogen or nitrogen gas which can then be desorbed. The effectiveness of vacuum treatment increases with increase in surface area of liquid exposed to vacuum. The increased surface area of liquid steel exposed to vacuum, e.g. in the form of a thin stream or gas induced stirring accelerates the de-gassing process.
- Temperature of liquid steel drops during vacuum de-gassing process. More is the surface area of stream exposed to vacuum higher is the temperature drop.
- The de-gassing time need to be kept at minimum.
- The degree of de-gassing increases with the degree of vacuum. Vacuum of the order of 1 mm Hg (1 mm Hg = 1 torr) or even less than 1 mm Hg is used in the practice.

Many processes are used in commercial practice for vacuum de-gassing treatment of molten steel after tapping from furnace. There can be broadly classified into three groups: (1) ladle de-gassing, (2) stream de-gassing and (3) circulation de-gassing.

17.3.2.1 Ladle De-gassing

A ladle de-gasser is used to reduce the concentrations of dissolved gases (H_2, N_2, O_2) in the liquid steel; homogenize the liquid steel composition and bath temperature; remove oxide inclusions from the liquid steel; and provide the conditions that are favourable for final de-sulphurization. Vacuum de-gassing is practised in steel works to achieve the following main objectives [6]:

- To remove gases, such as hydrogen, nitrogen and oxygen from the steel,
- To produce steels of low to ultra-low carbon content (<30 ppm C),
- To produce steels within close chemical composition ranges,
- To improve steel cleanliness by removing part of the oxygen through vacuum de-oxidation and by the agglomeration and floating of non-metallic inclusions,
- To control pouring temperatures, especially for continuous casting operation.

The ladle is placed in a vacuum chamber and stirred with an inert gas, while the chamber is evacuated. Alternatively, the ladle may have a sealing arrangement (by Al foil) on its periphery for a lid to be fitted which forms the vacuum chamber. Liquid steel can be treated in a chamber de-gasser without arcing. This can be done with two different stirring systems:

- An inductively stirring of the liquid steel bath and
- The bath stirring by bubbling argon gas through a porous plug located in the ladle bottom.

The removal of sulphur is achieved through slag-metal reactions, which are promoted by strong argon gas flushing (bubbling) within the vacuum envelope. The chamber de-gassing process requires: (a) rapid evacuation of the vacuum chamber, (b) maintenance of vacuum while at the same time sucking out a heavy flow of inert gas, (c) immediate availability, (d) dust resistance and (e) safe operation under harsh conditions.

The top lancing method is used for steel refining in the ladle with powder injection and stirring the melt with high argon flow rates. For gas bubbling at moderate rates, e.g. <0.6 Nm^3/min, a porous

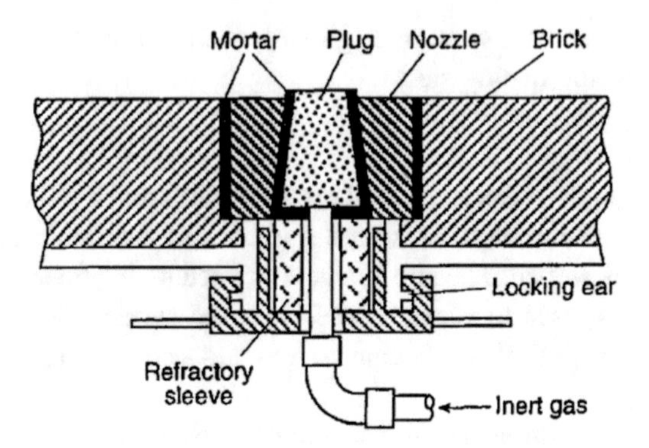

Fig. 17.4 Porous plug assembly at the bottom of a ladle [7]

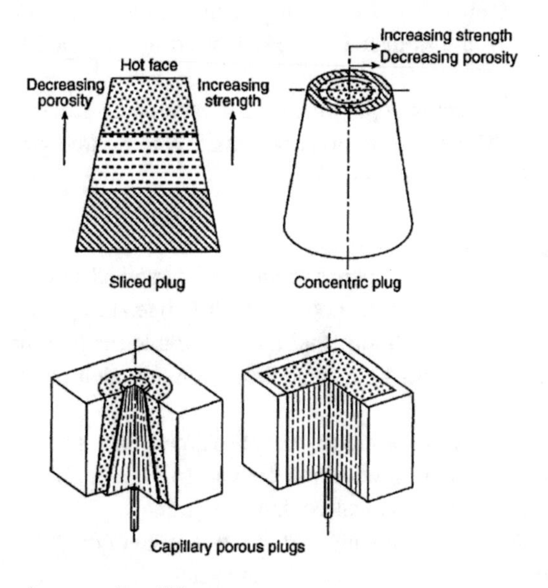

Fig. 17.5 Typical examples of porous plugs [7]

refractory plug is used. This is fitted usually on the bottom of the ladle as shown in Fig. 17.4. There are various designs of porous plugs as shown in Fig. 17.5.

Ladle is provided with a porous plug at its bottom to purge argon gas. The ladle is placed in a vacuum chamber. The vacuum chamber is equipped with a hopper to make additions of elements as and when it is needed. Stirring gas is introduced either from top through the roof by a submerged refractory tube or through the porous plug fitted at the bottom of the ladle. For effective de-gassing of fully killed steel, it is better to purge argon gas through the bottom of the ladle. Stirring the bath enhances rate of gas removal. Vigorous removal of gases also causes splashing of liquid steel. Hence, ladle is not filled completely and around 25% of its height is kept as free board to accommodate the splashed droplets of liquid steel. Pressure is maintained in between 1 and 10 mm Hg for effective de-gassing. During de-gassing, additions are made for de-oxidation and alloying. In certain cases, ladle is heated to compensate for the loss of heat during de-gassing. For the effectiveness of de-gassing, it is necessary that carry-over slag from primary steelmaking furnace is to be as low as possible. Carry-over slag contains FeO, and since oxygen content of steel is in equilibrium with FeO content of slag, oxygen content of steel increases.

The fundamental requirements for the ladle de-gassing process are as follows:

- Sufficient free board in the ladle to contain the vacuum induced slag and steel boil,
- An inert gas percolating through the steel bath for stirring, inclusion separation and enhancement of vacuum de-gassing performance,
- Sufficient super-heat in the steel to avoid skull formation and
- Means to deliver additives while the ladle is inside the vacuum chamber.

The argon gas connection to the ladle is established when the ladle is set in place inside the vacuum chamber. The vacuum chamber is evacuated to the required operating pressures by a vacuum pumping system. Nature of fluid motion and intensity of turbulence during vacuum treatment of liquid steel are of considerable importance due to their significant influence on mixing, mass transfer, inclusion removal, refractory lining wear, entrapment of slag and reaction with atmosphere. Rising gas bubbles are either the only source or the principal source of stirring. The bubbles are gases evolved due to de-gassing (N_2, H_2 and CO) as well as injected argon gas.

In vacuum de-gassing, the chamber pressure is very small as compared to the ferrostatic head of the liquid in vessel. Consequently, the bubbles expand enormously when they rise to free surface. This causes the phenomenon known as *bubble bursting* because of which liquid metal droplets get ejected into the vacuum chamber in large numbers.

Main features of this process are: (i) accelerated reactions under vacuum conditions; (ii) achievement of low contents of carbon, hydrogen and nitrogen; and (iii) improved steel cleanliness, especially with respect to oxides and sulphides.

In some processes, steel is heated by electric arc or induction to compensate for the heat loss during de-gassing and help dissolve large amount of de-oxidizers and alloying additions. These additions are made when de-gassing is nearing completion but well ahead of the point when stirring of the bath due to de-gassing practically stops. Stirring is essential for homogenization of the bath composition. For adequate de-gassing, pressure drop should be 1–10 mm Hg.

Steel Homogenization with Gas Stirring

The mixing of liquid steel with gas injection to achieve homogenization of melt temperature and composition is due primarily to the dissipation of the buoyancy energy of the injected gas. The effective stirring power of gas is given by the following thermodynamic relation, which was derived by Pluschkell [8].

$$\varepsilon' = \left(\frac{n'RT}{M}\right) \ln\left(\frac{P_t}{P_o}\right) \tag{17.5}$$

where

ε' stirring power, W t^{-1},
n' molar gas flow rate,
R gas constant, 8.314 J $mol^{-1}K^{-1}P$,
T temperature, K,
M mass of steel, t (tonne),
P_o gas pressure at the melt surface, atm,
P_t total pressure at injection depth H (m),
$= P_o + \rho g H$

ρ steel density, 6940 kg m^{-3} at 1600 °C,
g 9.81 m s^{-2}.

With these values, Eq. (17.5) is modified to the following:

$$\varepsilon'\left(\mathrm{W\,t^{-1}}\right) = 14.23\left(\frac{V'T}{M}\right)\log\left(1+\frac{H}{1.48P_o}\right) \tag{17.6}$$

where V' = gas flow rate, Nm^3 min^{-1}.

The gas stirring time to achieve 95% homogenization is called the mixing time, t. Several experimental and theoretical studies have been made to formulate the mixing time in terms of the gas stirring power, ladle diameter and depth of gas injection in the melt. The following relation is from the work of Mazumdar and Guthrie [9].

$$t(s) = 116(\varepsilon')^{-1/3}\left(D^{5/3}H^{-1}\right) \tag{17.7}$$

where

D average ladle diameter,
H depth of gas injection.

Calculated mixing times [7] are given in Fig. 17.6 for the simplified case of D = ~H. In a 200 t heat, the melt homogenization will be achieved with argon gas bubbling for 2.0–2.5 min at the rate of 0.2 Nm^3/min.

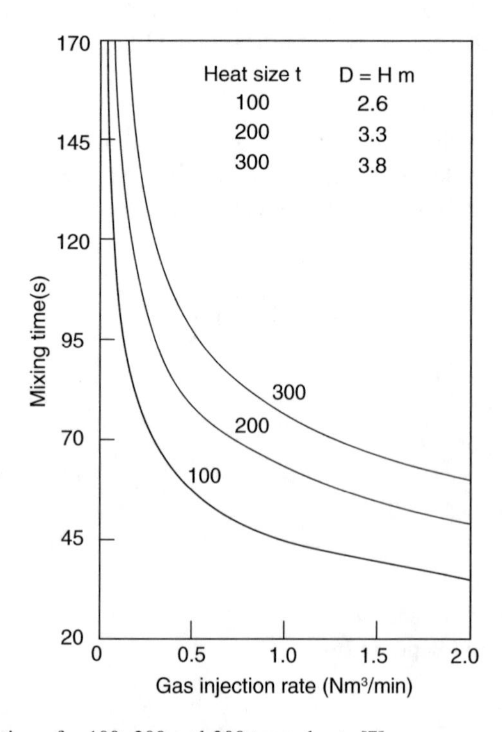

Fig. 17.6 Calculated mixing times for 100, 200 and 300 tonne heats [7]

Vacuum De-gassing (VD) Process

This is a simple ladle de-gassing unit with provisions for alloying additions. Here, vacuum is created through a vacuum pumping system. Pressure as low as 0.5 mm Hg is created. The process is capable of (i) homogenization of liquid steel bath regarding both temperature and composition, (ii) fine adjustment of chemistry and (iii) improved de-oxidation and reduction in [H], [N] and [O] contents.

De-sulphurization is a big problem for heats directly processed through this unit from primary steelmaking furnace. However, the problem can be sorted out through ensuring reduced slag in the ladle before sending the heat to VD unit (Fig. 17.7), and enhanced de-sulphurization is caused by slag-metal mixing.

Vacuum Arc De-gassing (VAD) Process

Vacuum arc de-gassing (VAD) process (Fig. 17.8) is a chamber de-gassing process with electrodes added for reheating the liquid steel. This is a single unit process in which the ladle sits in a vacuum chamber and is stirred by inert gas through porous plug at the bottom with provision for heating through electrodes and alloying additions. After addition of lime to the liquid steel in the ladle, arcing is carried out at a pressure of 250–300 mm Hg to raise the temperature and fuse the lime followed by short duration de-gassing, additions for chemistry adjustment and deep de-gassing to pressures as low as 1 mm Hg. Argon stirring is continued in all the operational steps, and the adjustment of flow rate is done for different operations being carried out during the VAD process. The heating rate is around 3–4 °C/min, and during heating, the rate of argon gas flow is kept on the lower side. In this system, under vacuum, carbon–oxygen reaction under the high-temperature arc is of great help in achieving low [O] content without any solid reaction product. Removal of [H] depends on (i) argon flow rate, (ii) pressure of vacuum chamber and (iii) initial [H] content. H_2 levels as low as 1.5 ppm are achieved caused by intense mass transfer by argon and low partial pressure of hydrogen gas because of dilution of liberated carbon monoxide gas. The main advantage of this process is the high degree of de-sulphurization as high as 80% for production of steels with sulphur levels as low as 0.005%. VAD is now a widely used process to produce clean steel.

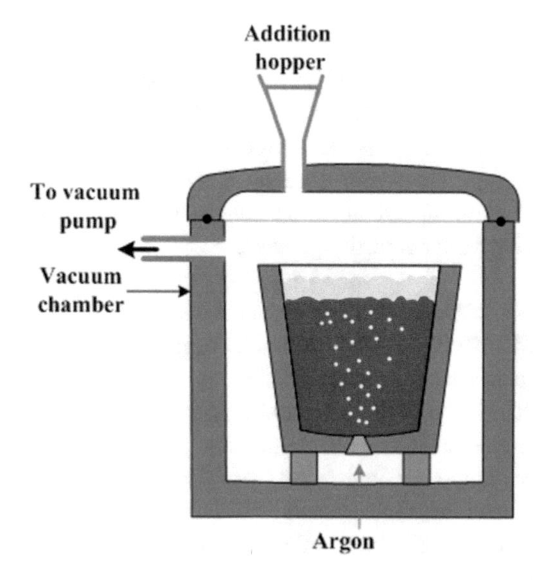

Fig. 17.7 Ladle vacuum de-gassing unit [1]

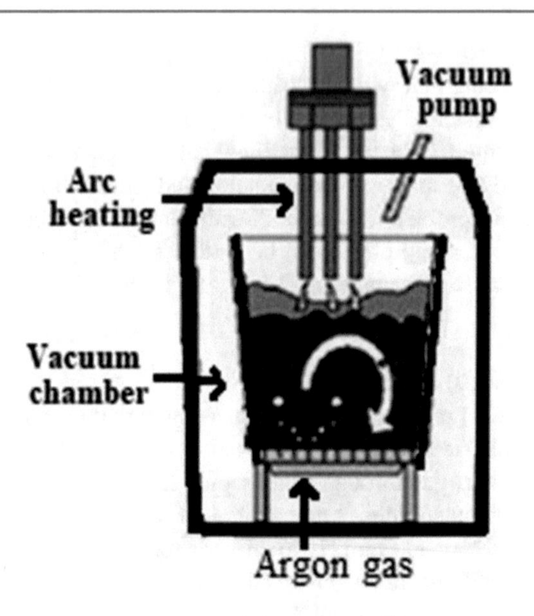

Fig. 17.8 VAD process

Liquid steel is held in a ladle which is put inside a vacuum chamber. Ladle is not filled completely, up to 30% of the total volume (i.e. ¼ height) should be provided free board; else liquid metal may spill over the edges. Steel may be stirred by bubbling an inert gas or by an electromagnetic stirrer while being exposed in vacuum. The teeming ladle, after tapping, is kept in a vacuum chamber which is then evacuated. Metallurgical reactions such as de-gassing, de-oxidation, de-carburization, de-sulphurization as well as alloying take place under vacuum conditions. The ladle is removed after a predetermine time when the correct teeming temperature of steel is reached. Teeming is carried out in open atmosphere. As the pressure fails (in vacuum), the vigorous in the ladle increases and the bath appears as if it is boiling.

17.3.2.2 Stream De-gassing

In stream de-gassing, liquid steel is poured into another vessel which is under vacuum. Sudden exposure of liquid stream in vacuum leads to very rapid de-gassing due to the increased surface area created by breakup of stream into droplets. This process helps the hydrogen dissolved in steel, to be evacuated by a vacuum pump. The major amount of de-gassing occurs during the fall of liquid stream. The height of the pouring stream is an important design parameter. Stream de-gassing technology has the following variants in the practice.

- Ladle-to-mould de-gassing: Pre-heated ingot mould with hot top is placed in vacuum chamber. Above the chamber a tundish is placed. Liquid steel tapped in the ladle is at super-heat equivalent of 30 °C. The ladle is placed above the tundish. Bottom pouring of liquid steel is into the tundish which is desirable. Schematic of ladle-to-mould de-gassing is shown in Fig. 17.9.
- Ladle-to-ladle de-gassing: In ladle-to-ladle de-gassing, a ladle with the stopper rod is placed in a vacuum chamber. Ladle containing liquid steel from primary steelmaking furnace is placed on top of the vacuum chamber, and the gap is sealed by aluminium foil. Alloy additions are made under vacuum. Stream is allowed to fall in the ladle where liquid steel is de-gassed (Fig. 17.10). Alloy additions are made under vacuum.

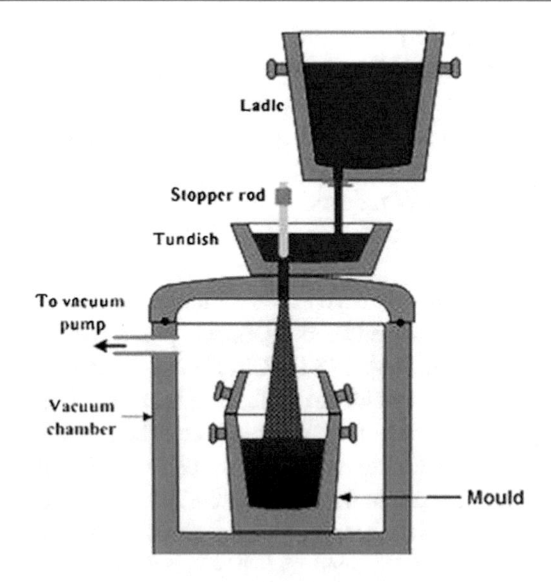

Fig. 17.9 Schematics of ladle-to-mould de-gassing [1]

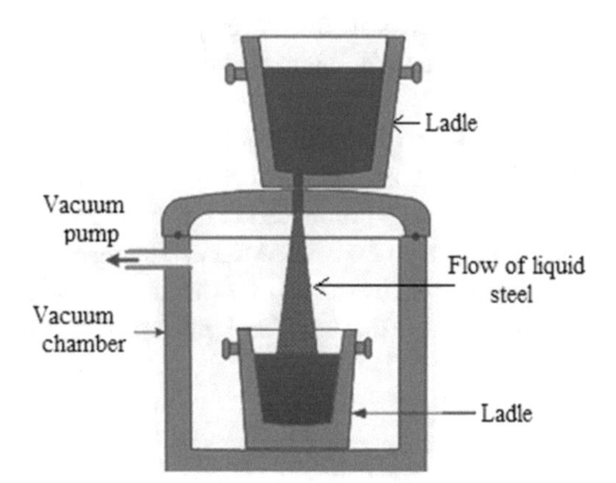

Fig. 17.10 Ladle-to-ladle de-gassing

In some plants, de-gassing is done during tapping. In this arrangement, liquid steel from primary steelmaking furnace is tapped into tundish or small ladle. From the small ladle, liquid stream is allowed to fall into a ladle which is evacuated. Ladle is closed from top with a special cover which contains exhaust opening. Liquid steel (with 25–30 °C super-heat) is tapped into ladle.

Liquid steel flows down in the form of a stream from the furnace or ladle to another ladle or mould during its exposure to vacuum, i.e. molten steel is exposed to vacuum in the form of a stream of metal flowing from one vessel to another. Rapid rate of de-gassing take place due to large increase in surface area of steel in the form of falling droplets. The steel is poured either from a teeming ladle, a pony ladle or a tundish. It is received either in a ladle or in a mould kept inside a vacuum chamber.

Stream de-gassing processes are used: (a) ladle-to-mould de-gassing, (b) ladle-to-ladle de-gassing, (c) tap de-gassing, (d) gero vacuum casting process and (e) Therm-I-Vac process.

17.3.2.3 Circulation De-gassing

In the recirculation de-gassing practice, liquid steel can circulate in the vacuum chamber continuously by special arrangement. In this process, a vacuum chamber is positioned above the ladle possessing a snorkel or snorkels which are dipped into the liquid steel bath. Liquid steel is either continuous or intermittently circulated during its exposure to vacuum. There are two types of circulation de-gassing processes: (a) RH process and (b) DH process.

RH De-gassing Process

RH process was developed by ***Rheinstahl Heinrich*** Shutte at Germany in 1957. The chamber is a cylindrical steel shell with two legs, called snorkels, openings at the top to provide for exhaust, alloy additions, observation and control (Fig. 17.11). Circulation is carried out in a refractory lined chamber equipped with two snorkels which are immersed in the steel bath. By reducing the system pressure and by injecting inert gas into the up-leg snorkel, the liquid steel rises into the vacuum chamber where de-carburization and other reactions take place. The steel then recirculates back into the ladle through the down-leg snorkel. In this way, the entire heat can be rapidly treated. Cylindrical shell is lined with fireclay bricks in the upper portion and alumina (Al_2O_3) bricks in the lower portion to sustain high temperature. The legs are lined with Al_2O_3 refractories. Argon (Ar), the lifter gas, is injected at the inlet snorkel to increase the liquid steel velocity entering inlet snorkel. The diameter of up leg is greater than of the down leg to increase circulation rate [10]. It is lined with fireclay bricks in the upper portion and high alumina bricks in the lower portion which directly meets liquid steel during de-gassing. The snorkels are lined with higher quality of Al_2O_3 bricks from both sides since they are dipped in liquid steel.

The vacuum chamber is heated to 900–1000 °C before use. The chamber is lifted and lowers down to an appropriate level in the ladle containing molten steel. The chamber is evacuated, and liquid steel just raises the chamber. The argon gas is then introduced in the inlet snorkel, and gas expands and rises, thereby raising the velocity of liquid steel in the inlet. After achieving the required immersion depth of the snorkel, the reaction vessel is evacuated by means of a vacuum pump system which is connected to the reaction vessel through off take duct (exhaust). The liquid steel density at 1600 °C is about 6.94

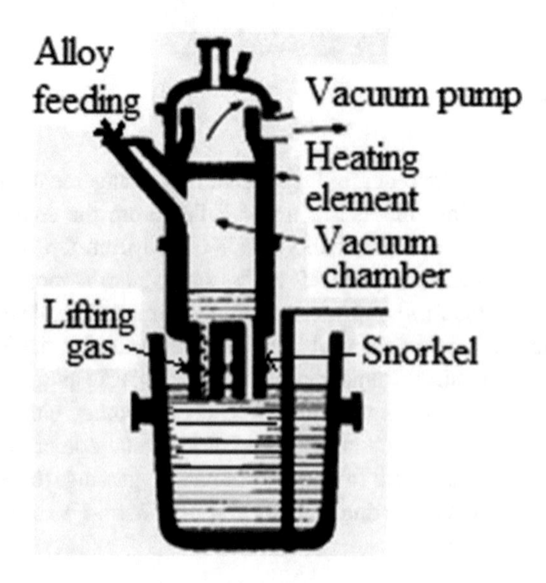

Fig. 17.11 RH de-gassing

tonnes per cubic metre. The atmospheric pressure exerted on the ladle surface causes the steel in the snorkels to rise to a barometric height of approximately 1.45 m under deep vacuum conditions. Now liquid steel exposes the vacuum and de-gassing takes place efficiently. Due to gravity, the liquid steel flow back in the ladle via another snorkel. Since de-gassed steel is slightly cooler and denser compared to the ladle's liquid steel, hence it forces the lighter untreated steel (i.e. without de-gassing) going upwards; thereby this phenomenon is ensuring adequate mixing and homogeneity of the bath. The rate of circulation of steel is controlled by adjusting the vacuum and rate of flow of lifter gas.

The vacuum treatment in RH plants produces steel which fulfils the demand of high steel qualities. To achieve this, the liquid steel is allowed to circulate in a vacuum chamber where a considerable drop in pressure causes it to disintegrate into the smallest of the parts. The increase in the surface area allows the liquid steel to de-gassing to the best possible extent. The process needs reliable vacuum units since it should be able to suck off very large flow rates under very difficult conditions of dusty atmosphere and high temperatures. The speed of de-gassing in an RH unit increases with increase in rate of circulation (R) of liquid steel through the vacuum chamber.

Circulation velocity increases with increasing argon flow rate in the up-leg of de-gasser. The rate of circulation (*R*, t/min) of liquid steel can be correlated as follows [11]:

$$R = A \cdot X \tag{17.8}$$

where A = a constant, and

$$X = \left[Q^{1/3} \cdot D^{4/3} \cdot \left\{\ln\left(\frac{p_1}{p_2}\right)\right\}\right] \tag{17.9}$$

where

Q argon injection rate, Nl/min;
D internal diameter of leg, m;
p_1 pressure of base of down-leg,
p_2 pressure in vacuum chamber.

Since the RH process is based on the exchange of liquid steel between the steel ladle and the RH vessel, the rate of steel recirculation determines the velocity of metallurgical reactions and the duration of the process assuming a defined metallurgical target. Liquid steel circulation depends on the geometry of the equipment such as snorkel diameter, the radius of the equipment and the position and number of lift gas tuyeres. At the end of de-gassing, when nearly 90% gas is removed, alloy addition may be made depending upon the super-heat available. The circulation of steel ensures homogeneous composition of the bath. Temperature of steel drops down about 25–50 °C depending upon the ladle size. Amount of argon gas requires 0.015–0.075 m^3/t of steel treated. End of the treatment, vacuum is broken, the chamber lifted out and liquid steel is teemed in a normal way in open air.

The operation of RH de-gasser includes as the following [10]:

- Heating of the cylindrical chamber to the desired temperature (varies in between 900 and 1500 °C),
- Lowering of the chamber into liquid steel up to a desired level,
- Evacuation of the chamber so that liquid begins to rise in the chamber,
- Introduction of the lifter gas which expands and creates a buoyant force to increase the speed of liquid steel rising into the inlet snorkel,

- De-gassing of the liquid steel in the chamber takes place and it flows back through the other snorkel into the ladle. This de-gassed steel is slightly cooler than steel in the ladle. Buoyancy force created by density difference (density of cooler liquid steel is more than the hot steel) stirs the bath.
- Rate of circulation of liquid steel in cylindrical chamber controls the de-gassing. Circulation rate depends upon amount of lifter gas and the degree of vacuum. Average rate of circulation of steel is nearly 12 t/min, and nearly 20 min are required to treat 100 t of steel to bring down the gas contents to the desired levels. The specific consumption of Ar is around 0.075 m^3/tonne.
- Additions of ferro-alloys can be made at the end of de-gassing depending on the super-heat.

RH process has several advantages which include (i) heat losses are relatively low, (ii) alloy additions can be adjusted more closely and (iii) small vacuum pumping capacity is adequate since smaller volume is to be evacuated as compared with ladle to ladle or stream de-gassing.

The technology of oxygen lancing into the RH vessel is developed by Kawasaki Steel, Japan known as KTB system. Oxygen is blown into the vacuum chamber through a water-cooled top lance, on the surface of the molten steel to accelerate de-carburization as well as to generate heat by post-combustion of the CO gas evolving from the bath.

$$\{CO\} + 1/2\{O_2\} = \{CO_2\} \tag{17.10}$$

For improved de-carburization reaction, argon gas can be additionally blown into the vacuum chamber through porous bricks. The reaction surface area is increased accordingly.

DH De-gassing Process

In ***Dortmund-Horder*** (DH) de-gassing process, a small portion, 10–15% of the total steel in the ladle, is treated at a time under vacuum. DH de-gassing process has a single snorkel and operates by repeatedly sucking the liquid steel into the vacuum chamber and then releasing it back into the ladle. The process is repeated until required de-gassing is achieved. It is also known as a *lifter de-gassing process*. The vacuum chamber has one long leg (i.e. snorkel) to be dipped in the liquid steel. The arrangement of a vessel and the ladle is somewhat like RH process except that in DH de-gassing process the cylindrical vessel has one snorkel. The DH chamber is equipped with heating facility, alloying addition arrangement and exhaust system. Bottom of the cylindrical vessel is provided with a snorkel which can be dipped into the liquid steel (Fig. 17.12). Chamber is lined with fireclay bricks in the upper portion and high alumina bricks in the lower portion. The snorkel is lined with higher quality of Al_2O_3 bricks from both sides since it is dipped in liquid steel. The length of the snorkel is sufficiently large to realize the effect of atmospheric pressure on rise of steel in the snorkel.

The following are the important steps for operation the DH de-gassing process [10]:

- DH chamber is pre-heated and lowered in the ladle so that snorkel tip dips below the liquid steel surface.
- The evacuated chamber is moved up and down so that steel enters the chamber.
- The chamber is moved 50–60 cm with a cycle time of 20 s.
- Adequate de-gassing is possible in 20–30 cycles.
- A layer of slag is kept in the ladle to minimize heat losses.
- The DH de-gassing process can operate with lower super-heats compared with RH de-gassing since DH unit has heating facility.

Since only 10–15% liquid steel is exposed at a time, 7–10 cycles are required to expose entire steel at least once. For adequate de-gassing is possible in about 20–30 cycles, and additional 5–10 cycles may be required depending upon the alloy additions. 15–20 min are required for de-gassing. Now a days, magnesite bricks are used in the chamber. High alumina is used for snorkel that last for 90–100 heats. The main wear is by slag attack, which has been overcome by: (i) use of suitable refractories and (ii) minimizing the entry of slag from the ladle into the DH unit.

17.3.2.4 De-sulphurization During Vacuum De-gassing

Sulphur elimination from liquid steel also takes place during vacuum de-gassing especially in the presence of carbon and silicon. However, injections of powders of Ca–Si, Ca–Ba alloys, rare earth metals and CaO are also carried out during vacuum treatment of steel to get extra low sulphur in steel, and desirable inclusion is also modified.

Calcium (Ca) and barium (Ba) form CaS and BaS, respectively, upon reaction with sulphur, whereas cerium (Ce) forms several sulphides out of which CeS is the stablest one at steelmaking conditions. Ce also forms an oxy-sulphide, Ce_2O_2S. All these compounds are solids at steelmaking temperatures. It should be noted that all these elements form very stable sulphides as well as oxides. Therefore, they are also acted as strong de-oxidizers, as well as de-sulphurizers and both oxides and sulphides are formed [12].

The overall reaction can be represented as:

$$[S] + (MO) = [O] + (MS) \tag{17.11}$$

The equilibrium constant (k) for the reaction (17.11):

$$k_{17.11} = \left\{ \frac{[a_O] \cdot (a_{MS})}{[a_S] \cdot (a_{MO})} \right\} \tag{17.12}$$

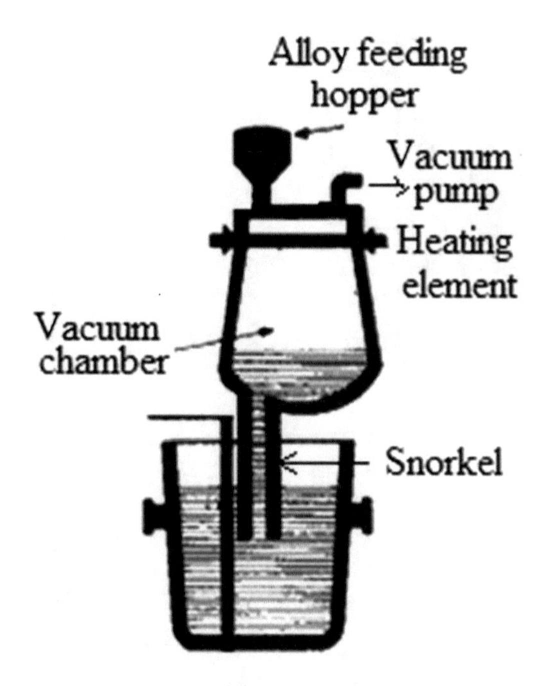

Fig. 17.12 DH de-gassing

Assuming pure MO and MS, so

$$(a_{\rm MQ}) = 1 \text{ and } (a_{\rm MS}) = 1 \tag{17.13}$$

For dilute solution:

$$[a_{\rm O}] = [h_{\rm O}] = [W_{\rm O}] \text{ and } [a_{\rm S}] = [h_{\rm S}] = [W_{\rm S}] \tag{17.14}$$

Since f_i = 1 for dilute solution.
Now combined Eqs. (17.12) to (17.14):

$$k_{17.11} = \frac{[h_{\rm O}]}{[h_{\rm S}]} = \frac{[W_{\rm O}]}{[W_{\rm S}]} \tag{17.15}$$

The values of $k_{17.11}$ for different systems can be calculated from free energies of the reactions. Figure 17.13 shows the variation of equilibrium constant ($k_{17.11}$) for different reactions with temperatures. From Fig. 17.13, it should be noted that Ba is the strongest de-sulphurizer, and Mg is the weakest de-sulphurizer. Ca and Ce are lying in between Ba and Mg. To lower down [W_S], [W_O] should be lower down for maintaining $k_{17.11}$ constant; or other hand [W_O] should be removed by de-oxidizer like C or Si. If the [W_O] is increased, then the reaction (17.11) goes to backward direction that means sulphur will be come back to the metal. To remove or lower down of [W_O] is done under vacuum by carbon to form CO gas.

$$[{\rm O}] + [{\rm C}] = \{{\rm CO}\} \tag{17.16}$$

It has been noted that the slag-metal equilibrium with respect to sulphur is closely attained.

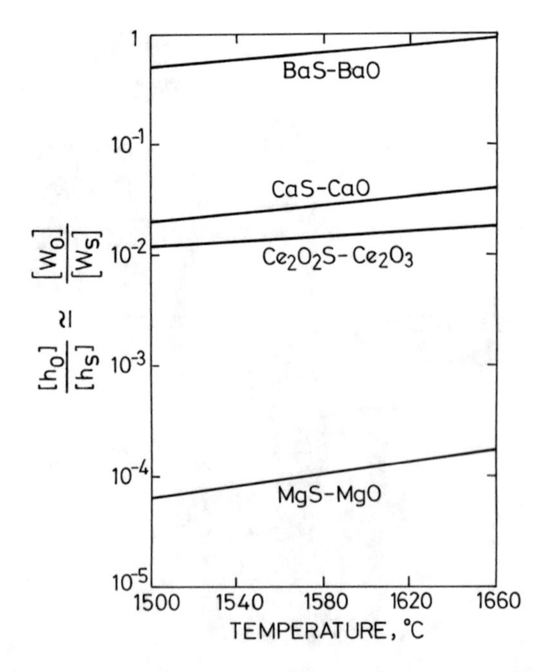

Fig. 17.13 Oxygen/sulphur activity ratio in liquid iron for some sulphide oxide equilibria at 1600 °C [12]

In the presence of carbon or silicon as reducing agents, the most probable reactions are:

$$[S] + (CaO) + [C] = (CaS) + \{CO\} \tag{17.17}$$

$$[S] + 2(CaO) + 1/2[Si] = (CaS) + 1/2(2CaO \cdot SiO_2) \tag{17.18}$$

In Eq. (17.18), the selection of dicalcium silicate is somewhat arbitrary, but it is the most stable lime-silica compound and is therefore likely to be produced during reaction.

The equilibrium constants are as follows:

$$k_{17.17} = \left\{ \frac{(a_{CaS}) \cdot p_{CO}}{[h_S] \cdot (a_{CaO}) \cdot [a_C]} \right\} \tag{17.19}$$

$$k_{17.18} = \left\{ \frac{(a_{CaS}) \cdot (a_{2CaOSiO_2})^{1/2}}{[h_S] \cdot (a_{CaO})^2 \cdot [a_{Si}]^{1/2}} \right\} \tag{17.20}$$

Considering the iron is saturated with carbon and all the solid compounds are pure,
$a_i = 1$; $a_{CaS} = 1$, $a_{CaO} = 1$, and $a_{2CaOSiO2} = 1$.
Since $h_i = f_i$ wt% i
So

$$k_{17.17} = \left\{ \frac{\{p_{CO}\}}{[f_S \cdot \text{wt\% S}] \cdot [f_C \cdot \text{wt\% C}]} \right\} \tag{17.21}$$

$$k_{17.18} = \left\{ \frac{1}{[f_S \cdot \text{wt\% S}] \cdot [f_{Si} \cdot \text{wt\% Si}]^{1/2}} \right\} \tag{17.22}$$

Again,

$$[\text{wt\% S}] = \left\{ \frac{\{p_{CO}\}}{k_{17.17} \cdot f_S \cdot [f_C \cdot \text{wt\% C}]} \right\} \tag{17.23}$$

$$[\text{wt\% S}] = \left\{ \frac{1}{k_{17.18} \cdot f_S \cdot [f_{Si} \cdot \text{wt\% Si}]^{1/2}} \right\} \tag{17.24}$$

Sulphur content in metal (Eq. 17.17) is pressure dependent, and solid lime would therefore be expected to be an effective de-sulphurizing agent under vacuum conditions. From the free energy value of Eq. (17.17): $\Delta G^0_{17.17} = 96{,}608.6 - 105.19T$ J

For the typical vacuum steel-melting temperature of 1560 °C, steel with 0.25 wt% C and 0.30 wt% Si, under a carbon monoxide partial pressure of 10^{-6} atm, the sulphur content should be 2.5×10^{-7} wt% at 1560 °C.

$$\begin{aligned} &\text{Since } \Delta G^0_{17.17} = 96{,}608.6 - 105.19T\,\text{J}, \\ &\text{at } 1560\,^\circ\text{C} = 1833\,\text{K} : \Delta G^0_{17.17} = -96{,}204.67\,\text{J} \end{aligned} \tag{17.25}$$

Again,

$$\Delta G^0_{17.17} = -RT \ln k_{17.17} \tag{17.26}$$

Therefore,

$$\ln k_{17.17} = \frac{\Delta G^0_{17.17}}{-RT} = \frac{(-96204.67)}{(-8.314 \times 1833)} = 6.313 \tag{17.27}$$

Therefore, $k_{17.17} = 551.69$.

$$\begin{aligned}[\text{wt\% S}] &= \left\{\frac{\{p_{CO}\}}{k_{17.17} \cdot f_S \cdot [f_C \cdot \text{wt\% C}]}\right\} \\ &= \frac{10^{-6}}{(551.69 \times 0.25)} \quad (\text{Consider} f_i = 1). \\ &= \frac{10^{-6}}{137.92} = \left(0.07 \times 10^{-7}\right)\end{aligned} \tag{17.28}$$

17.4 Injection Ladle Metallurgy

The objectives, for better quality steels with controlled chemistry and superior and homogeneous mechanical properties, are as follows:

- To bring down the level of basic impurities such as carbon, sulphur, phosphorus, nitrogen, hydrogen, oxygen and tramp elements (like copper, tin, antimony, bismuth, etc.); and
- To reduce the non-metallic inclusions or to modify them so that their harmful effects are counter-balanced.

In this context, injection metallurgy, i.e. treatment of molten steel with powdered metallic materials and fluxes coupled with gas stirring or gas stirring along, is finding growing acceptance in the steel industry. Since it involves less investment, is easy to operate and facilitates quicker processing at reasonable reaction efficiency for a wide span of heat sizes. Adoption of injection metallurgy will thus lead to significant time saving as it relieves the main steelmaking units (LD/BOF or EAF) partly or completely from performing the time-consuming refining and finishing of steel heats.

There is a special interest in the control of non-metallic inclusions because of their harmful effect on the subsequent stages and their influence on the properties of the final product. By controlling of the amount, size and chemical composition of the inclusions, it is possible to obtain a final product of good quality without inclusions.

The need to produce low sulphur steels with control of sulphide inclusions and to control the tundish nozzle clogging during Concast, ladle injection process was developed. Treatment of liquid steel with calcium has become an important means of de-oxidation and de-sulphurization, to very low levels and control of the shape, size and distribution of oxide and sulphide inclusions. Benefits directly attributable to calcium treatment include greater fluidity, smoother continuous casting by reducing nozzle blockage and improved cleanliness, machinability, ductility and impact strength in the final product. Calcium treatment of liquid steel is started for prevention of nozzle clogging in aluminium-killed plain carbon steels, and calcium helps to liquefy the alumina. In this process, nozzle clogging is also reduced; particularly in billet and thin slab casting where the liquid steel must pass through narrow refractory spaces. Calcium treatment is also done extensively for inclusion modification to improve properties of the product.

Calcium is usually added in the alloyed form like Ca–Si, Ca–Fe, Ca–Si-Ba and Ca–Si-Ba–Al to restrict the vapour pressure of calcium in the molten bath. Most commonly used injection powder is Ca–Si (30% Ca), alloying with Si helps to reduce vapour pressure of Ca at steelmaking temperature. To ensure optimum recovery of Ca, it is necessary to inject Ca–Si deep into the bath. Other reagents are: CaO, CaO–CaF_2–Al, CeO–SiO_2, Mg–CaO, CaC_2 and combinations of them.

Main purpose of particle injection is to feed of a fine-grained material deep into the molten steel. Major factors influencing the injection method: (i) nature of the powder, (ii) powder size distribution, (iii) powder/gas ratio and (iv) injection depth.

(i) Nature of the powder: The nature of the powders used is governed primarily by the purpose of the injection.
(ii) Powder size distribution: Optimum powder size must be determined. Fine size powder provides a large reaction surface, and coarse powders create a small reaction surface.
(iii) Powder/gas ratio: Solid particles are mostly injected by a stream of argon gas. The gas bubbles rise through the liquid. The particles trapped in the bubble hit the gas–liquid interface where they will either pass through the interface or remain trapped inside the bubble. The rising solids might vaporize (e.g. Mg), melt and dissolve (Ca, CaC_2, Ca–Si) and then vaporize.
(iv) Injection depth: Reaction zones are:

    (a) Jet zone in front of nozzle where three-phase dispersion liquid metal-solid-gas is formed,
    (b) Bubble plume where bubbles, solid and/or partially molten or reacted particles and liquid steel rise and react;
    (c) Break through zone at the melt surface where there is reaction with air. Reaction products are absorbed into the top slag.

There are two popular methods of injection [13]: (1) submerge injection through lance and (2) cored wire injection.

17.4.1 Submerge Injection Through Lance

Calcium powder is injected along with an inert carrier gas through a submerged lance (as shown in Fig. 17.14). The major problem with the submerged powder injection with inert carrier gas like nitrogen is that around 40% of the particle rises as particle inside the gas bubble, which does not participate much in the heat and mass transfer process. This makes the process less efficient. Liquid slag injection could evolve as a potential technique for enhanced kinetics of impurity transfer from metal to slag.

17.4.2 Cored Wire Injection

Calcium additions can be used for several metallurgical objectives such as de-sulphurization, shape control of sulphide and oxide inclusions. Introduction of calcium in steel has always been a challenge to steelmakers because of the low melting point (848 °C), low boiling point (1484 °C) and extremely low solubility. The solubility of calcium in liquid iron is reported to be 300 ppm at 1600 °C and 1 atm of calcium vapour pressure [14]. Due to high vapour pressure, it should be injected into the

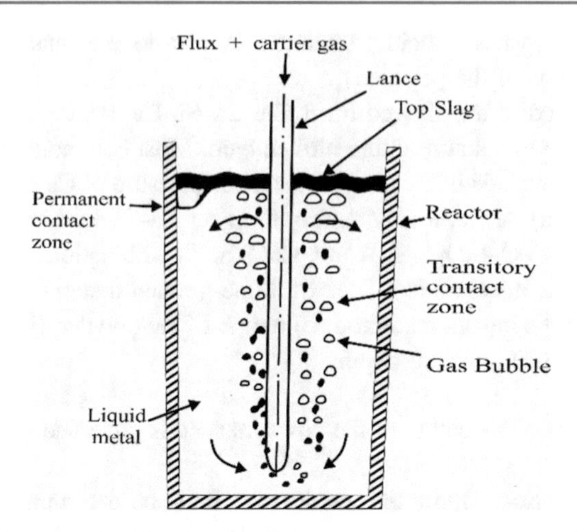

Fig. 17.14 Powder injection through submerged lance [13] (reproduced with permission from ***IIM Metal News***)

bath. There are two methods of calcium injection: powder injection as Ca–Si and Ca wire injection. In this case, calcium powder is compacted in a hollow mild steel casing that appears in the form of a wire. To make this cored wire, alloy addition is cost-effective and metallurgically efficient. It is important that the wire must travel to the bottom of the ladle before it finally melts and releases the powder and the filling material must be consumed by the melt to the maximum extent. When alloy additions are made to steel, a steel shell forms around the injected material. The same is true for wires. The formation of shell allows the wire to be injected deep into the bath before the calcium begins to vaporize. Ca wire goes deep into the bath before vaporization occurs due to faster rate of injection. Calcium wire is injected into the ladle as shown in Fig. 17.15.

When using calcium wires for sulphide shape control, it is necessary to de-oxidize the steel with aluminium and de-sulphurize the steel to less than 0.01% S prior to wire injection. Different types of Ca wires are shown in Table 17.1. Typical consumption of Ca wire is approximately 0.3–0.4 kg/tonne of steel.

$CaO–CaF_2–Al$ powder is very suitable for de-sulphurization of low Si steels. Ca, Ca–Al and Ca–Si may also be injected into the liquid steel bath in the form of flux cored wires. Because of high vapour pressure of Ca, it must be injected deep into the bath. Calcium first react with oxygen, then modifies oxide inclusions and de-sulphurization and finally provides sulphide shape control. Hence, steel melt is usually de-oxidized with Al before flux cored wire injected.

Ca wire injection converts Al_2O_3 inclusions into low melting $CaO–Al_2O_3$. Liquid $CaO–Al_2O_3$ rise more easily than high melting point Al_2O_3 inclusion. $CaO–SiO_2$-based flux may be injected into the steel melt in ladle to lower melting point of oxide inclusion and to promote the coagulation of the inclusions. As a result, flotation of the inclusion is enhanced, and melt becomes clean. $CaO–CaF_2$ and $CaO–Al_2O_3–CaF_2$ fluxes have strong de-sulphurization effects. Low sulphur (<0.001%) steel can be produced by such flux injection.

When calcium is injected deep into the melt as an alloy of Ca–Si, Ca–Al or as pure Ca and mixed with nickel or iron powder, the subsequent reactions in liquid steel will be the same. The series of reactions are expected to occur in Al-killed steels containing alumina inclusions as follows:

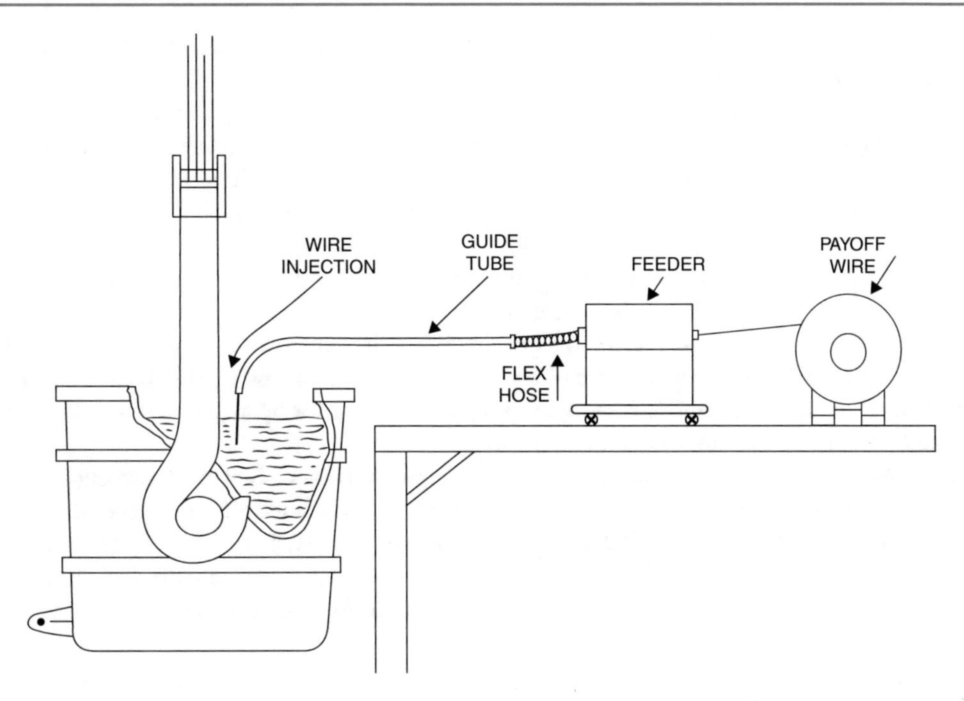

Fig. 17.15 Schematic diagram of Ca cored wire injection into a ladle [13] (reproduced with permission from ***IIM Metal News***)

Table 17.1 Types of Ca wire [15]

Type of wire	Dia. (mm)	Composition
Metallic calcium wire	5.4	38% Ca
Metallic calcium wire	8.0	48% Ca
Ca–Si wires	8.0	47% Ca–Si
Ca–Si wires	12	45% Ca–Si

$$\mathrm{Ca(1)} = \{\mathrm{Ca}\} \tag{17.29}$$

$$\{\mathrm{Ca}\} = [\mathrm{Ca}] \tag{17.30}$$

$$[\mathrm{Ca}] + [\mathrm{O}] = (\mathrm{CaO}) \tag{17.31}$$

$$[\mathrm{Ca}] + [\mathrm{S}] = (\mathrm{CaS}) \tag{17.32}$$

$$[\mathrm{Ca}] + (x + 1/3)(\mathrm{Al_2O_3}) = (\mathrm{CaO} \cdot x\mathrm{Al_2O_3}) + 2/3[\mathrm{Al}] \tag{17.33}$$

$$(\mathrm{CaO}) + 2/3[\mathrm{Al}] + [\mathrm{S}] = (\mathrm{CaS}) + 1/3(\mathrm{Al_2O_3}) \tag{17.34}$$

Thermodynamically, if sulphur or oxygen is dissolved in the steel at moderate levels, or if Al_2O_3 inclusions are present in steel, calcium will react with oxygen or sulphur until the contents of reactants are very low (<2 ppm). The formation of calcium sulphide (reaction 17.32) can occur if the calcium and sulphur contents are sufficiently high. Since calcium has higher affinity for oxygen than for sulphur, the addition of calcium initially results in a more-or-less pronounced conversion of the alumina into calcium aluminates (reaction 17.33), until the formation of calcium sulphides starts as the addition of calcium continues. For steels with low sulphur content, reaction (17.33) will take place first following by reaction (17.34).

Most grades of steel are treated with calcium using either a Ca–Si alloy or a Ca–Fe(Ni) mixture, depending on the silicon specification. This treatment is made after trim additions and argon rinsing. In most melt shops, the cored wire injection system is used in the calcium treatment of steel. The melting and boiling points of calcium are 839 °C and 1500 °C, respectively. The cored wire containing Ca–Si or Ca–Fe(Ni) is injected into the melt at a certain rate such that the steel casing and contents of the cored wire melt deep in the steel bath, e.g. 1.5–2.0 m below the melt surface without hitting the bottom of the ladle or resurfacing without being completely melted. In 200–240 tonnes heats, the calcium treatment was made by the cored wire injection of Ca–Si at the rates of 0.16–0.36 kg calcium/t of steel, the cored wire of 16 × 7 mm cross section or 13 mm diameter is injected at the rate of about 180 m min^{-1}. If the steel is to be de-gassed for hydrogen removal, the calcium treatment should be made after vacuum de-gassing.

By combined use of a synthetic basic top slag cover and injection of Ca-based powders, S levels in liquid steel can be brought down to very low levels. Oxide inclusions also are modified. Top slag usually belongs to $CaO–CaF_2$ or $CaO–Al_2O_3–CaF_2$. Basicity is maintained high value. Aluminium content in steel is decreased primarily by the reaction of Al with unstable oxides in the top slag. Carry-over slag from furnace to the ladle must be minimized to avoid phosphorus reversion.

17.4.3 Efficiency of Calcium

The material balance for calcium consumption is expressed as follows [16]:

$$W_i = W_b + W_o + W'_o + W_s + W_v \tag{17.35}$$

where

W_i Amount of calcium injected,
W_b Amount of calcium dissolved in the bath,
W_o Amount of calcium present in aluminates and sulphides,
W'_o Amount of calcium reacted with alumina subsequently floated out,
W_s Amount of calcium reacted with the slag,
W_v Amount of calcium escaped via the vapour phase and subsequently burnt at the bath surface.

It is generally accepted that $W_b \ll W_o$,
thus, giving for the efficiency of calcium usage [$\eta(\mathrm{Ca})_u$]:

$$\eta(\mathrm{Ca})_u = \frac{W_o + W'_o}{W_i} \times 100\% \tag{17.36}$$

While efficiency of the calcium retention [$\eta(\mathrm{Ca})_r$] in the steel is given by:

$$\eta(Ca)_r = \frac{W_o}{W_i} \times 100\% \tag{17.37}$$

The calcium retention efficiency [$\eta(\mathrm{Ca})_r$] decreases with the increasing quality of calcium injected. The amount of calcium to be injected must be adjusted in accordance with the degree of cleanliness of the steel or its total oxygen content. Obviously, injecting more calcium than can react with the available inclusions leads to a low calcium retention efficiency. Furthermore, it is to be expected that the calcium retention efficiency in the continuous casting mould or in the teeming ladle will be less than the retention efficiency in the ladle because of flotation of calcium-containing inclusions out of the bath in the time interval prior to casting or teeming.

17.5 De-oxidation of Steel

Steelmaking is a process of selective oxidation of impurities in molten iron. During this oxidizing process, oxygen is bound to dissolve in the molten steel (i.e. molten iron); at 1600 °C and 1800 °C, solubility of oxygen in iron are 0.23% and 0.48%, respectively. Pure iron, whose melting point is 1539 °C, becomes more fusible as the oxygen concentration of the solution increases, the minimum melting point at 1524 °C corresponding to an oxygen concentration of 0.16%. With further increase in temperature, iron dissolves more oxygen (in the form of FeO). Maximum oxygen content in iron is given by Chipman and co-workers [17] as follows:

$$\log \%[O]_{max} = -\left(\frac{6320}{T}\right) + 2.734 \tag{17.38}$$

Similarly, relationship for solubility of oxygen in liquid iron:

$$\%[O]_{max} = 0.131 \times 10^{-2} T - 1.77 \quad (\text{where } T \text{ is temperature, } ^\circ\text{C}) \tag{17.39}$$

and in terms of FeO:

$$[\%FeO]_{max} = 0.589 \times 10^{-2}\,\mathrm{T} - 7.96 \quad (\text{where T is temperature, } ^\circ\text{C}) \tag{17.40}$$

Since the quantity of inclusions in steel is related to the oxygen dissolved in molten metal, the steelmaker must carry out the heat to decrease the oxygen content of the liquid bath to a minimum at the end of the refining period. At the end of the refining, a considerable amount of oxygen (0.05–0.1%) contents in liquid low carbon steel.

Since solubility of oxygen in solid steel is negligibly small (0.003%), therefore during solidification of steel in ingot or continuous casting mould, the excess oxygen is evolved in the form of gases causes defects such as blow holes and non-metallic inclusions in casting. Hence, it is necessary to control oxygen content in molten steel, i.e. excess oxygen should be removed from liquid steel, before it is pouring into mould. The removal of residual oxygen content from refined steel is known as *de-oxidation*. Oxygen content in molten bath in furnace is high, and it is necessary and brings it down by carrying out de-oxidation at the end. De-oxidation never takes place at constant temperature. Temperature of molten steel keeps dropping from furnace to mould. Addition of de-oxidizer also causes some temperature change due to heats of reaction.

There are two types of de-oxidation practice: (1) precipitation de-oxidation and (2) diffusion de-oxidation.

17.5.1 Precipitation De-oxidation

A de-oxidant (or de-oxidizer) should have higher affinity for oxygen than iron and interact with liquid iron to form an oxide precipitate, i.e. to convert the homogeneous system (molten steel with dissolved

Table 17.2 Properties of some oxides [18]

S. No.	Oxide	Melting point (°C)	Specific gravity at 20 °C
1.	Al_2O_3	2045	3.85
2.	MnO	1785	5.40
3.	SiO_2	1713	2.40
4.	TiO_2	1840	4.20
5.	ZrO_2	2700	5.75
6.	Steel	1539	> 7.00

oxygen) to a heterogeneous system (molten steel with suspended oxide). As the density of the precipitate (Table 17.2) is lower than that of the molten steel, it separates out from the metal.

Precipitation de-oxidation can be expressed as:

$$x[\mathrm{M}] + y[\mathrm{O}] + = \left(\mathrm{M}_x\mathrm{O}_y\right) \tag{17.41}$$

where [M] and [O] denote de-oxidizer and oxygen dissolved in molten steel, and (M_xO_y) is the oxide product in slag.

The equilibrium constant for reaction (17.41),

$$k'_{17.41} = \frac{(a_{\mathrm{M}_x\mathrm{O}_y})}{\{[h_\mathrm{M}]^x \cdot [h_\mathrm{O}]^y\}} \tag{17.42}$$

where $k'_{17.41}$ is equilibrium constant. $(a_{\mathrm{M}_x\mathrm{O}_y})$ is activity of M_xO_y in slag for the Raoultian standard state. $[h_\mathrm{M}]$ and $[h_\mathrm{O}]$ are activities of dissolved M and O, respectively, in molten steel in the Henrian standard state.

Again

$$h_i = f_i \text{ wt\%}\, i \tag{17.43}$$

If the de-oxidation product is pure, then $a_{\mathrm{M}_x\mathrm{O}_y} = 1$. Also, for very dilute solution, f_i may be taken as one. Equation (17.42) may be rewritten as:

$$[\text{wt\% M}]^x \cdot [\text{wt\% O}]^y = \frac{1}{k'_{17.41}} = k_{17.41} \tag{17.44}$$

where $k_{17.41}$ is known as de-oxidation constant. [wt% M] and [wt% O] are wt% of M and O in liquid metal, respectively, at equilibrium with pure oxide.

Again,

$$k_{17.41} \text{ is varies with temperature: } \log k_{17.41} = -\left(\frac{A}{T}\right) + B \tag{17.45}$$

where T is temperature in Kelvin, and A and B are empirical constants.

If only one de-oxidant (i.e. M) is added, then product is pure M_xO_y, i.e. $a_{M_xO_y} = 1$. On the other hand, simultaneous addition of two de-oxidants (e.g. Fe–Si and Fe–Mn) leads to formation of liquid (e.g. $MnO \cdot SiO_2$).

Consequently, a_{MxOy} is less than 1, and hence, [wt% M]^x. [wt% O]^y is less than that obtained by simple one de-oxidant (e.g. Fe–Si) addition. At a fixed value of [wt% M], therefore [wt% O] would be less in the later case.

More metal (i.e. M, de-oxidizer) is introduced for de-oxidation, decreasing the residual concentration of [O] in the steel. Best results are obtained when the de-oxidizer dissolves quickly and the products of de-oxidation rapidly separate from the metal and float-up to be readily picked up by the slag phase.

De-oxidation of liquid steel is carried out mostly without help of slag, such as in ladle, tundish and mould. Even in furnace lump de-oxidizers are often added into the metal bath directly. In all these cases, the product of de-oxidization, this is an oxide forms of solid as precipitates. Therefore, such a technique is known as *precipitation de-oxidation*. This de-oxidization practice is used all steelmaking processes except EAF process.

The precipitation de-oxidation has also two types: (i) simple de-oxidation and (ii) complex de-oxidation.

(i) Simple de-oxidation: When one type of de-oxidizer is added to the metal phase and the product is solid, then de-oxidation process is known as simple de-oxidation. De-oxidation may be carried out in furnace (minor amount), ladle, tundish as well as in the mould. It consists of addition of elements such as Mn, Si and Al which have stronger affinities for oxygen as compared to iron. These elements are added mostly in the form of ferro-alloys.
Si, Al, Ti, Zr, Ca and rare earth metals act as simple de-oxidizers, and products are in the solid particle due to their higher melting points (Table 17.2). Most common de-oxidizers are Mn, Si and Al; products are in the solid small particles suspended in liquid steel.

$$2[\mathrm{FeO}] + [\mathrm{Fe{-}Si}] = 3[\mathrm{Fe}] + (\mathrm{SiO_2}) \tag{17.46}$$

$$[\mathrm{FeO}] + [\mathrm{Fe{-}Mn}] = 2[\mathrm{Fe}] + (\mathrm{MnO}) \tag{17.47}$$

$$3[\mathrm{FeO}] + 2[\mathrm{Al}] = 3[\mathrm{Fe}] + (\mathrm{Al_2O_3}) \tag{17.48}$$

(ii) Complex de-oxidation: When more than one de-oxidizer are added to the molten steel simultaneously, that is known as *complex de-oxidation* and product is liquid. Some important complex de-oxidizers are: Si–Mn, Ca–Si, Ca–Al, Al–Mn-Si, Ca–Si–Al, etc. Conversion of solid and smaller size to complex inclusions which are occurred in molten conditions at steelmaking temperature; so that they are more readily formed the larger inclusions which are easily floated up on the liquid bath. By this way clean steel or quality steel can be produced. Complex de-oxidation is being used drastically for better quality product and offers the following advantages:

   (a) The dissolved oxygen content in metal is lower in complex de-oxidation as compared to simple de-oxidation from equilibrium considerations. If only Fe–Si is added, then product is pure solid SiO_2, i.e. $a_{SiO_2} = 1$. On the other hand, simultaneous addition of ferrosilicon and ferromanganese in suitable ratio (Mn/Si = 5–7) leads to formation of liquid $MnO \cdot SiO_2$ (melting point is 1200 °C).
   (b) The de-oxidation product, if liquid, agglomerates easily into a larger size and consequently floats up faster and the liquid steel makes cleaner.

When complex de-oxidizers are used, both acidic and basic oxides are formed simultaneously in such concentrations that their resultant product has a sufficient low melting point, and therefore, it remains liquid for a long time. The finally dispersed liquid has got enough time to agglomerate and float up from the liquid metal to the slag.

When ferrosilicon and ferromanganese are simultaneous addition into the bath, their oxides segregate from the solution as a liquid slag phase and make larger inclusions.

$$2[FeO] + [Fe{-}Si] = 3[Fe] + [SiO_2] \tag{17.46}$$

$$[FeO] + [Fe{-}Mn] = 2[Fe] + [MnO] \tag{17.47}$$

$$[MnO] + [SiO_2] = (MnO \cdot SiO_2) \tag{17.49}$$

Large inclusions are floated up due to density difference between liquid steel and inclusions, bubble attachment and fluid transport in the metallurgical vessel like ladle or tundish. The rate of rise, i.e. velocity of floating segregated liquids (i.e. inclusions) through the melt, can be estimated by Stokes' law:

$$v = \frac{\{2(d_s - d_m)r^2 g\}}{(9\mu)} \tag{17.50}$$

where

v the rate of rise, ms^{-1}.
d_s the density of slag particle, kgm^{-3}.
d_m the density of liquid metal, kgm^{-3}.
r the radius of spherical particle, m.
g the acceleration due to gravity, ms^{-2}.
μ the viscosity of liquid metal, kgs^{-1}.

Since $d_m > d_s$, so value of v will be negative, the slag particle will be floated up against gravity. Terminal velocity is not only depending on $(d_s - d_m)$, but also on the particle size (r). If the particles are very small (i.e. micro-inclusions), that would take time to float up, and hence, they get suspended within the metal and later form inclusions. The micro-inclusions are suspended in the liquid steel and are passed on to the next process. If the particles are big in sizes, then they take less time to float up i.e. larger inclusions float up faster than micro-inclusions. The floating larger inclusions are absorbed by the slag at the top of the molten bath.

17.5.2 Diffusion De-oxidation

Distribution of oxygen between slag and metal, i.e. law of distribution:

$$L_{FeO} = \left(\frac{(FeO)}{[FeO]}\right) \tag{17.51}$$

At a given temperature, L_{FeO} is constant. FeO in slag is de-oxidized by Si, Al or C:

$$(FeO) + Si/Al/C \rightarrow [Fe] + (SiO_2)/(Al_2O_3)/\{CO\} \tag{17.52}$$

To maintain constant of L_{FeO}, FeO in metal, i.e. [O] in metal begins to diffuse from metallic phase to slag phase, that is way this de-oxidation process is known as *diffusion de-oxidation*. By this way,

[O] in metal is decreased and metal becomes free from non-metallic inclusions, which is a product of de-oxidation, remains in the slag phase. This technique is applied in EAF process.

In EAF, the major amount of de-oxidation is done at furnace itself. In furnace, one may lower the iron oxide content of slag by de-oxidizer additions. This induces removal of some oxygen from slag phase and oxygen from metal phase diffuses to the slag phase. In this technique, therefore, de-oxidation of metal bath takes place through reaction with existing slag in the furnace.

17.6 Inclusion and Its Control

Inclusions are non-metallic unwanted particles embedded in the matrix of metals and alloys after solidification. Inclusion in steels like oxide, sulphide and silicates is embedded during the process of steelmaking and solidification. The undesirable impurities are present in steel products either in the form of elements in solid solution, or as simple or complex compounds depending upon their content and thermodynamic stability [4]. Undesirable inclusions, if present in more amounts, produce the material impure and are detrimental to the serviceability of the steel. However, some important property like machinability is favour influenced by certain interstitial inclusion. Hence, it is necessary to know amount and distribution of non-metallic inclusion in steel to evaluate its quality. It is true that no steel can be free from the residuals or trace elements. Neither is it possible to eliminate non-metallic inclusions all together. It is, therefore, important to make a realistic judgement on the limits of the impurity contents, and the critical size of non-metallic inclusions, with respect to specific quality requirement of steel products. Clean steel should not only have a relative low content of impurities and inclusions; more importantly, inclusions larger than a critical size must be avoided. The critical size is of course dependent on the specific application requirements of the product.

In general, inclusions have been found to be harmful the mechanical properties and corrosion resistance of steel. This is more so for high strength steels for critical application. As a result, there is moving to produce *clean/quality steel*. However, no steel can be totally free from inclusions. The number of inclusions has been variously estimated to range between 10^{10} and 10^{15} per tonne of steel.

The properties that are adversely affected by inclusions are fracture toughness, impact properties, fatigue strength and hot workability. The factors responsible for these may be classified as follows:

(a) Geometrical factors: size, shape, size distribution and total volume fraction of inclusions,
(b) Property factors: deformability and modulus of elasticity at various temperatures, coefficient of thermal expansion.

The sources of inclusions are as the following:

1. Precipitation due to reaction of molten steel with de-oxidizer,
2. Mechanical and chemical erosion of refractories and other materials that encounter molten steel,
3. Oxygen pickup by molten metal during teeming and consequent oxide formation.

17.6.1 Classification of Non-metallic Inclusions

Non-metallic inclusions are classified by chemical and mineralogical content. By chemical content, non-metallic inclusions are divided into the following groups:

(i) **Oxides**:

(a) Single: FeO, MnO, Cr_2O_3, TiO_2, SiO_2, Al_2O_3, etc.,
(b) Combined: $FeO \cdot Fe_2O_3$, $FeO \cdot Al_2O_3$, $MgO \cdot Al_2O_3$, $FeO \cdot Cr_2O_3$, etc.

(ii) **Sulphides**:

(a) Single: FeS, MnS, CaS, MgS, Al_2S_3, etc.,
(b) Combined: $FeS \cdot FeO$, $MnS \cdot MnO$, etc.

(iii) **Nitrides**:

(a) Single: TiN, AlN, ZrN, CeN, etc.,
(b) Combined: Nb(C, N), V(C, N), etc., which can be found in alloyed steels and has strong nitride-generative elements in its content: e.g. titanium, aluminium, vanadium, cerium, etc.

(iv) **Phosphides**: Fe_3P, Fe_2P, etc.

All steels contain oxide inclusions due to the presence of de-oxidation products. They influence the mechanical properties and machinability of steel. From a solidification point of view, six steps may be isolated in the formation of a primary inclusion. These are as follows:

1. Nucleation and initial growth,
2. Dendritic growth,
3. Dendritic multiplication,
4. Clustering,
5. Coalescence and
6. Flotation.

Figures 17.16 and 17.17 show the formation of silica inclusions and of alumina inclusions, respectively.

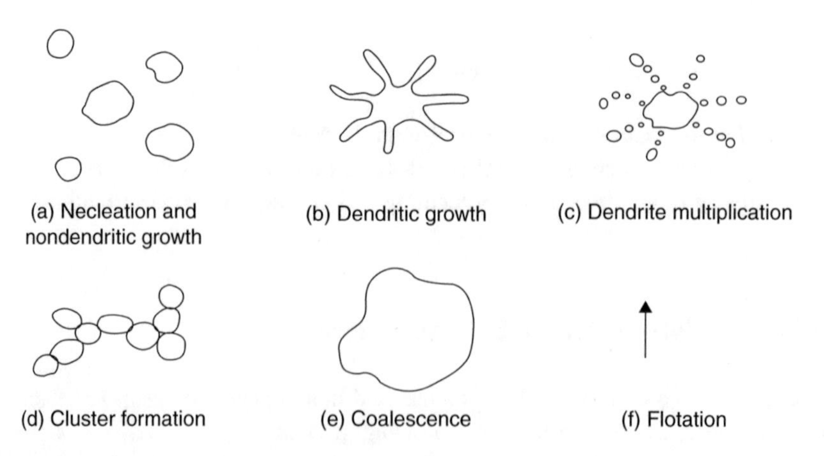

Fig. 17.16 Six possible steps in evaluation of silica inclusions [19]

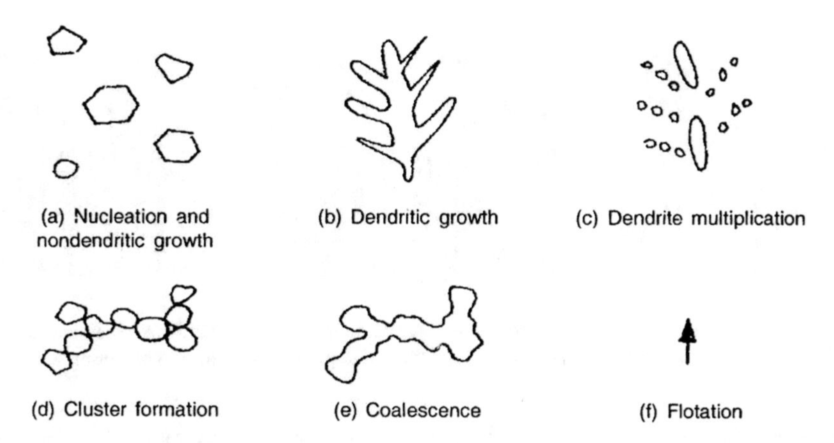

Fig. 17.17 Six possible steps in evaluation of alumina inclusions [19]

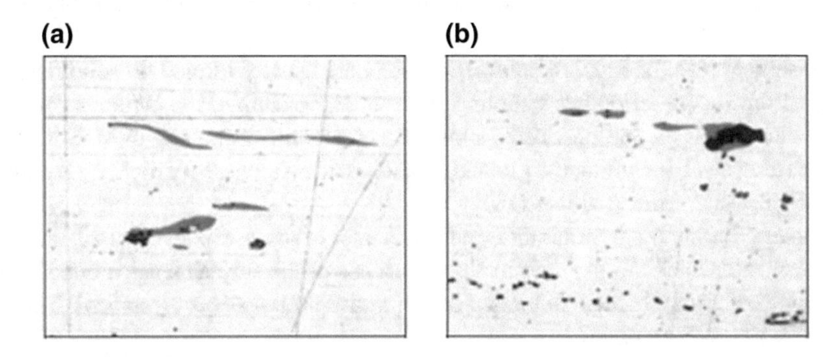

Fig. 17.18 Non-metallic inclusion in steel: oxides-dark grey and sulphides-light grey [20] (reproduced with permission from Total Materia)

Most of inclusions in steels are oxides and sulphides. Among various types of non-metallic inclusions, oxide and sulphide inclusions have been thought harmful for common steels. Usually, nitrides are present in special steels (stainless steels, tool steels, etc.) which have elements with a strong affinity for nitrogen (e.g. chrome, vanadium, etc.) that create nitrides [20]. Figure 17.18 shows sulphides and oxides of non-metallic inclusion in steel.

As mentioned above, most of inclusions in steels are oxides and sulphides. Sulphides in steel have been paid much attention because their presence in steel is hampered the properties of the final products by their deformation during the steel working process; especially their morphology has a significant effect on the steel properties.

17.6.1.1 Inclusion of Manganese Sulphides

According to analysis based on the steel ingots containing 0.01–0.15% S, the morphology of MnS can be broadly classified into three types (as shown in Fig. 17.19):

- **Spherical or globular**: MnS, with a wide range of sizes, is often formed duplex with oxides. The spherical MnS is formed through the two-phase separation of [as shown in Fig. 17.19a].

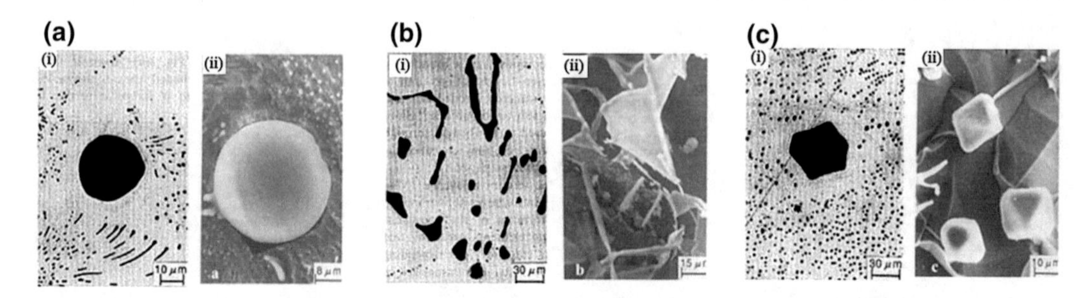

Fig. 17.19 Morphology of MnS depends on steel composition: **a** spherical or globular, **b** irregular eutectic and **c** primary angular [21]. (i) Optical micrograph and (ii) scanning electron micrograph (reproduced with permission from ISIJ)

- **Irregular eutectic**: These sulphides exhibit a complex structure composed of a branch-like part growing along characteristic directions and a plate-like part connecting the branches [as shown in Fig. 17.19b].
- **Primary angular**: This type of MnS shows an octahedral shape [as shown in Fig. 17.19c].

Hence, the sulphides in the modern commercial steel are usually formed on solidification process or in solid steel during the subsequent cooling process. For example, the Widmanstätten plate-like MnS is formed in solid steel, and Fig. 17.20 shows the common morphology of MnS in conventional continuously casting steel, including the globular duplex oxide–sulphide (particle A, B and C) and the Widmanstätten plate-like MnS (particle D).

During calcium treatment, the alumina and silica inclusions are converted to molten calcium aluminate and silicate which are globular in shape because of the surface tension effect. This change in inclusion composition and shape is known as the *inclusion morphology control*. This is demonstrated schematically in Fig. 17.21.

The calcium aluminate inclusions retained in liquid steel suppress the formation of MnS stringers during solidification of steel. This change in the composition and mode of the precipitation of sulphide inclusion during solidification of steel is known as sulphide morphology or sulphide shape control.

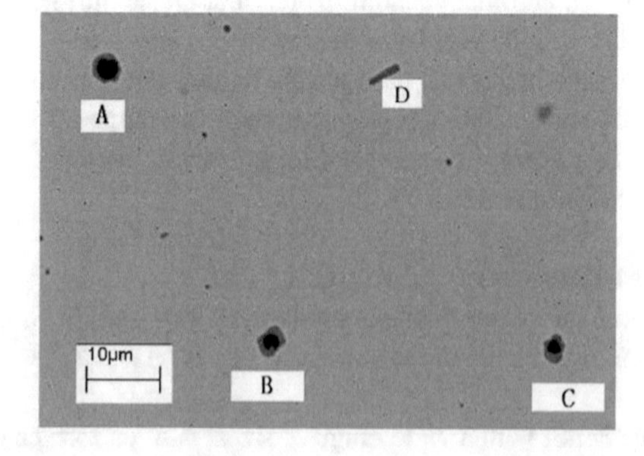

Fig. 17.20 Typical duplex oxide–sulphide inclusion (particle A, B and C) and plate-like MnS (particle D) in conventional continuous casting silicon steel [20] (reproduced with permission from ISIJ)

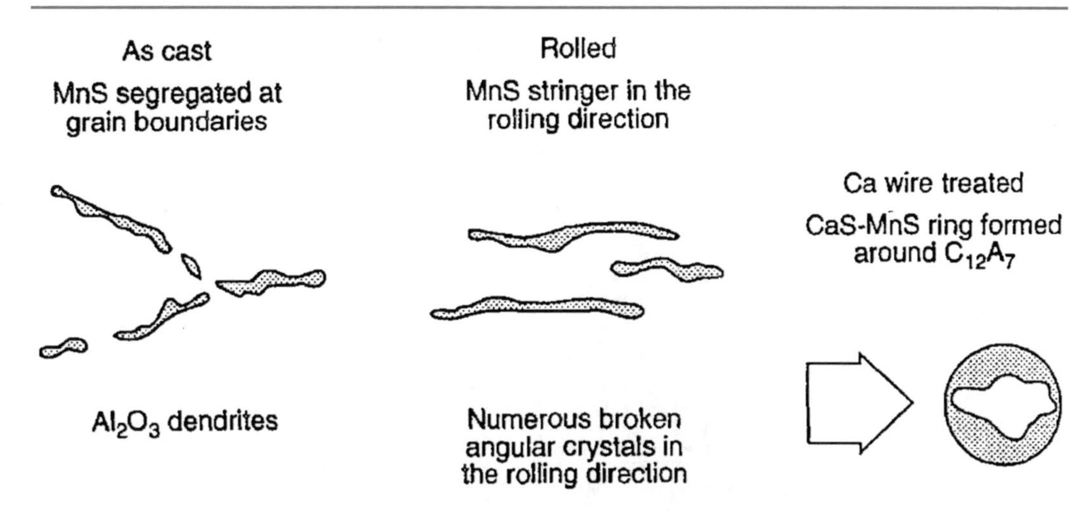

Fig. 17.21 Schematic illustration of modification of inclusion morphology with calcium treatment of steel [7]

17.6.2 Inclusion Control

Inclusion control falls into two categories:

(a) Minimizing the occurrence of inclusions, primarily macro-inclusions,
(b) Modifying the inclusion to an impart globular shape and desirable properties. This is known as *inclusion shape control* or *inclusion modification*.

17.6.2.1 Treatment of Liquid Steel with Calcium

Alumina inclusions occur as de-oxidation products in the aluminium-based de-oxidation of steel. Pure alumina has a melting point above 2000 °C, i.e. these alumina inclusions are present in a solid state in liquid steel. The addition of calcium to steel which contains such inclusions changes the composition of these inclusions from pure alumina to CaO-containing calcium aluminates. Details are shown in Sects. 17.4.1 and 17.4.2.

As it can be seen from Fig. 17.22, the melting point of the calcium aluminates will decrease as the CaO content increases, until liquid oxide phases occur at about 22% of CaO, i.e. when the $CaO \cdot 2Al_2O_3$ compound is first exceeded at 1600 °C. The liquid phase content continues to increase as CaO content rises further and is 100% at 35% of CaO. The minimum melting temperature for the calcium aluminates is around 1400 °C, i.e., such liquid calcium aluminates may be present in liquid form until, or even after, the steel solidifies [22]. Density of liquid calcium aluminates is also lower than density of liquid steel; hence, inclusion of liquid calcium aluminates is floated out from the liquid steel.

In most steel shops, the cored wire containing Ca–Si or Ca–Fe(Ni) injection system is used in the calcium treatment of steel. The melting and boiling points of calcium are 839 °C and 1500 °C, respectively. During calcium treatment, the alumina and silica inclusions are converted to molten calcium aluminates and silicate which are globular in shape because of the surface tension effect. The change in inclusion composition and shape is known as the inclusion morphology control.

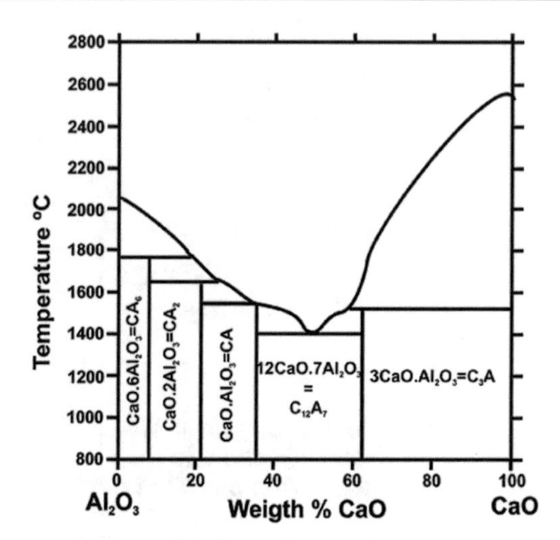

Fig. 17.22 Binary system CaO–Al_2O_3 [22] (reproduced with permission from Total Materia)

The calcium aluminate inclusions retained in liquid steel suppress the formation of MnS stringers during solidification of steel. This change in the composition and mode of precipitation of sulphide inclusion during solidification of steel is known as sulphide morphology or sulphide shape control.

Several metallurgical advantages [7] are brought about with the modification of composition and morphology of oxide and sulphide inclusions by calcium treatment of steel, as follows:

1. To improve steel castability in continuous casting, i.e. minimize nozzle blockage,
2. To minimize inclusion-related surface defects in billet, bloom and slab castings,
3. To improve steel machinability at high cutting speeds and prolong the carbide tool life,
4. To minimize the susceptibility of steel to reheat cracking, as in the heat-affected zones (HAZ) of welds,
5. To prevent lamellar tearing in large restrained welded structures,
6. To minimize the susceptibility of high-strength low alloy (HSLA) line pipe steels to hydrogen-induced cracking in sour gas or sour oil environments. The calcium content in the final product can be controlled within the range of 15–20 ppm,
7. To increase both tensile ductility and impact energy in the transverse and through-thickness directions in steels with tensile strengths below 1400 MPa.

Probable Questions

1. What do you understand by 'secondary steelmaking'? What do you achieve and objectives for that?
2. What are the main aim of ladle furnace? What are the functions of LF? Discuss the basic principles of LF.
3. Why de-gassing of steels are done? Derive *Sievert's law of equation.*
4. How solubility of hydrogen and in liquid iron alloys effect with other alloying elements?
5. Discuss the basic principles of vacuum de-gassing of liquid steel.

6. Why oxygen in liquid steel cannot be removed in elementary form? How dissolved oxygen is removed during vacuum de-gassing?
7. Why ladle is not filled completely during de-gassing of steels? What are the functions of porous plug at the bottom of the ladle?
8. What are the functions of vacuum de-gassing in terms of cleanliness of steels? 'Vacuum is necessary for de-gassing of steel'. Why?
9. What are the objectives of vacuum de-gassing of steel? What are the processes for vacuum de-gassing of steels? Discuss briefly.
10. What is the basic principles involved in RH de-gassing process? Discuss the metallurgical operations carried out in the process.
11. Discuss the basic principles of DH process.
12. Discuss the basic principles of de-sulphurization during vacuum de-gassing.
13. Discuss the basic principles of cored wire injection in liquid steel.
14. Why is it necessary to de-oxidize the bath in the finishing stage of steelmaking?
15. Why de-oxidation is done in molten steel? Explain.
16. 'Complex de-oxidation process is used instead of simple de-oxidation process for quality steel production'. Why?
17. Discuss the thermodynamic of de-oxidation processes.
18. Explain 'precipitation de-oxidation', 'complex de-oxidation' and 'diffusion de-oxidation' processes with example.
19. Why diffusion de-oxidation is better than precipitation de-oxidation process?
20. Differentiate between 'indigenous' and 'exogenous' inclusions in steels. Why inclusions are not welcome to the steel? How that's control?
21. How non-metallic inclusions are classified and discuss the modification of inclusions?

Examples

Example 17.1 Calculate the rising velocity of a 1.5 μm slag particle, rising through stagnant liquid steel at 1600 °C, given slag density is 3000 kg/m^3, the density of liquid steel is 7600 kg/m^3 and viscosity is 7 cP. (Given $g = 9.81$ m/s^2)

Solution

$$T = 1600 + 273 = 1873\,\text{K}$$

According to Stokes' law: $u = \left(\frac{\{2r^2(\rho p - \rho s)g\}}{9\mu}\right)$

where

u velocity of the slag particle, m/s
r radius of the slag particle, m

ρ_p density of the slag particle, kg/m^3
ρ_s density of the liquid steel, kg/m^3
g acceleration due to gravity, m/s^2
μ viscosity, kg/ms

Here

$$\rho_p = 3000\,\text{kg/m}^3 \text{ and } \rho_s = 7600\,\text{kg/m}^3$$
$$r = (1.5/2)\,\mu\text{m} = 0.75 \times 10^{-6}\,\text{m};\ g = 9.81\,\text{m/s}^2$$
$$\mu = 7\,\text{cP} = 7 \times 10^{-2}\,\text{P} = 7 \times 10^{-3}\,\text{kg/ms}$$

Putting these values in the above equation:

$$\begin{aligned} u &= \left[\left\{2 \times \left(0.75 \times 10^{-6}\right)^2 \times (3000 - 7600) \times 9.81\right\} / \left(9 \times 7 \times 10^{-3}\right)\right] \\ &= \left[-\left(50766.75 \times 10^{-12}\right) / \left(63 \times 10^{-3}\right)\right] = -805.82 \times 10^{-9} \\ &= \mathbf{-8.06 \times 10^{-7}\ m/s} \end{aligned}$$

Negative sign of velocity shows that slag particle goes in upward direction.

Example 17.2 Molten steel is termed from a ladle through a 5 cm diameter nozzle at its bottom. If the volume of the molten steel in the initial stage is 2 m^3 and the linear velocity of discharge is 90 cm/s. Calculate the time required to complete empty of ladle.

Solution

According to the law of conservation of mass:

$$[(\text{Amount of fluid entering the reactor per unit time}) - (\text{Amount of fluid leaving the reactor per unit time})] = (\text{Amount of fluid retained in reactor per unit time}) \quad (1)$$

$$\text{Amount of fluid entering the reactor per unit time} = \rho_1 \cdot u_1 \cdot A_1 \quad (2)$$

$$\text{Amount of fluid leaving the reactor per unit time} = \rho_2 \cdot u_2 \cdot A_2 \quad (3)$$

where ρ, u and A are the density, velocity of the fluid and area of the reactor, respectively.

$$\text{Amount of fluid retained in reactor per unit time} = V\left(\frac{d\rho}{dt}\right) \quad (4)$$

where V is the volume of the reactor and ($d\rho/dt$) is the rate of change of density of fluid with time.

Putting the values of Eqs. (2–4) in Eq. (1):

$$\rho_1 \cdot u_1 \cdot A_1 - \rho_2 \cdot u_2 \cdot A_2 = V\left(\frac{d\rho}{dt}\right) \quad (5)$$

$$\text{Since in this case there is no fluid entering to the reactor, i.e. } \rho_1 \cdot u_1 \cdot A_1 = 0 \quad (6)$$

The density of the liquid steel may be taken as constant,

so that Eq. (4) may be written as:

$$\rho\left(\frac{dV}{dt}\right) \tag{7}$$

Hence, Eq. (5) becomes:

$$\rho_2 \cdot u_2 \cdot A_2 = \rho\left(\frac{dV}{dt}\right) \text{ or } dV = -u_2 \cdot A_2 dt \tag{8}$$

Now integrating between t = 0 and t = t, where V = V and V_t = 0, respectively.

$$\int_V^{V_t} dV = -u_2 \cdot A_2 \int_0^t dt$$

Therefore

$$V = u_2 \cdot A_2 t \text{ or } t = \left(\frac{V}{u_2 A_2}\right) \tag{9}$$

Here, V = 2 m^3, u_2 = 90 cm/s = 0.9 m/s and A_2 = [(π/4) $(0.05)^2$] = 1.96 × 10^{-3} m^2.
Hence, t = [2/(0.9 × 1.96 × 10^{-3})] = **1131.77 s = 18.86 min**.

Example 17.3 Consider de-oxidation of molten steel by Al at 1600 °C. The bath contains 1% Mn, 0.1% C. The final oxygen content is to be brought down to 0.001 wt%. Calculate residual Al content of molten steel assuming that [Al]–[O]–(Al_2O_3) equilibrium is attained.

Given:

$$\log k_{Al} = \{(-58473/T) + 17.74\}$$
$$e_{Al}^{Mn} = 0, e_{Al}^{C} = 0.091, e_{Al}^{O} = -6.6, e_{A}^{Al} = 0$$
$$e_{O}^{Mn} = -0.021, e_{O}^{C} = -0.45, e_{O}^{O} = -0.2, e_{O}^{Al} = 0$$

Solution

$$(Al_2O_3) = 2[Al] + 3[O] \tag{1}$$

$$\text{Therefore, } k_{Al} = \left\{[h_{Al}]^2 \cdot [h_O]^3, \downarrow (a_{Al_2O_3})\right\} = \left\{[h_{Al}]^2 \cdot [h_O]^3 \cdot\right\} \quad (\text{since } a_{Al_2O_3} = 1, \text{ for pure } Al_2O_3)$$
$$= [f_{Al} \cdot W_{Al}]^2 [f_O \cdot W_O]^3 \tag{2}$$

$$\text{Now } \log f_{Al} = e_{Al}^{Mn} \times \text{wt\% Mn} + e_{Al}^{c} \times \text{wt\% C} + e_{Al}^{O} \times \text{wt\% O} + e_{Al}^{Al} \times \text{wt\% Al}$$
$$= 0 \times 1 + 0.091 \times 0.1 + (-6.6) \times 0.001 + 0 \times W_{Al}$$
$$= 0.0091 - 0.0066 = 0.0025$$

So f_{Al} = **1.006**

$$\text{Similarly, } \log f_O = e_O^{Mn} \times \text{wt\% Mn} + e_O^{C} \times \text{wt\% C} + e_O^{O} \times \text{wt\% O} + e_O^{Al} \times \text{wt\% Al}$$
$$= (-0.021) \times 1 + (-0.45) \times 0.1 + (-0.2) \times 0.001 + 0 \times W_{Al}$$
$$= -0.021 - 0.045 - 0.0002 = -0.0662$$

So f_O = **0.859**

$$T = 1600 + 273 = 1873\,\text{K}$$
$$\log k_{Al} = \{(-58473/T) + 17.74\} = \{(-58473/1873) + 17.74\} = -13.48$$
Hence $\boldsymbol{k_{Al}}$= **3.32** × **10**$^{\mathbf{-14}}$

From Eq. (2), we get: $k_{Al} == [f_{Al} \cdot W_{Al}]^2 [f_O \cdot W_O]^3$
Therefore, $3.32 \times 10^{-14} = [1.006 \times W_{Al}]^2.[0.859 \times 0.001]^3$
So $[1.006 \times W_{Al}]^2 = 5.237 \times 10^{-5}$
$[W_{Al}]^2 = 5.175 \times \times 10^{-5}$
Therefore $[W_{Al}] = \mathbf{7.193 \times 10^{-3}}$
Hence, residual Al content of molten steel is **7.193** × **10**$^{\mathbf{-3}}$**%**

Example 17.4 Liquid steel contains 0.002 wt% $\underline{O}$. Vanadium is added in the ladle at 1557 °C to make a product containing 1 wt% V.

To what extent, the oxygen content of the liquid steel must be lowered to prevent the oxidation of *V*.

Given: $2[V]_{wt\%} + 3[O]_{wt\%} = \langle V_2O_3 \rangle$, $\Delta G^0 = -780,400 + 267.8T$ J/mol

Solution

$$1557^\circ\text{C} = 1557 + 273 = 1830\,\text{K}$$

Since $\Delta G^0 = -780,400 + 267.8T = -780,400 + 267.8 \times 1830 = -290,326$ J/mol

$$\text{Again, } \Delta G^0 = -RT \ln k = -8.314 \times 1830 \times \ln k$$
$$\ln k = [\Delta G^0/(-8.314 \times 1830)] = 19.08$$

Therefore, $\boldsymbol{k}$ = **1.94** × **10**$^{\mathbf{8}}$
Since $k = \{a_{V_2O_3}/([a_V]^2 \times [a_O]^3)\} = \{1/([\text{wt\% V}]^2 \times [\text{wt\% O}]^3)\} = \{1/([\text{wt\% O}]^3)\}$
[Since $a_{V_2O_3} = 1$ and wt% V = 1 wt%]
Therefore, $[\text{wt\% O}]^3 = \{1/k\} = \{1/(1.94 \times 10^8)\} = 5.161 \times 10^{-9}$
Hence, wt% $\underline{O} = 1.73 \times 10^{-3} =$ **0.00173 wt%**

Thus, 0.00173 wt% $\underline{O}$ is in equilibrium with 1 wt% V. If there is more oxygen than this in the liquid steel, as in the present case (0.002 wt%). So V will start forming V_2O_3. Hence to prevent the loss of V, the oxygen content in liquid steel should be lowered from 0.002 wt% to 0.00173 wt%.

Example 17.5 Calculate the chemical potential of nitrogen gas in liquid steel at 1600 °C. The steel has the following composition: wt% N = 0.01, wt% C = 0.5, wt% P = 0.2, wt%M n = 0.5.

$$\text{Given: } [\text{wt\% N}] \cdot f_{\text{N}} = k_{\text{N}} \cdot P_{\text{N}_2}^{1/2}, \text{ when } \log k_{\text{N}} = [\{(-188.1)/T\} - 1.246]$$
$$e_{\text{N}}^{\text{C}} = 0.25, e_{\text{N}}^{\text{N}} = 0, e_{\text{N}}^{\text{P}} = 0.051, e_{\text{N}}^{\text{Mn}} = -0.02$$

Solution

$$\text{N}_2(\text{g}) = 2[\text{N}]_{\text{wt\%}} \quad (1)$$

Therefore, equilibrium constant, $k = [a_{\text{N}}]^2 / a_{\text{N}_2} = [h_{\text{N}}]^2 / P_{\text{N}_2} = \{[\text{wt\% N}].f_{\text{N}}\}/P_{N_2}$

$$\{[\text{wt\% N}] \cdot f_{\text{N}}\} = \sqrt{(k \cdot P_{\text{N}_2})} = k_{\text{N}} \cdot \sqrt{(\text{P}_{N_2})} \quad (2)$$

$$\text{Since } \log k_{\text{N}} = [\{(-188.1)/T\} - 1.246] = [\{(-188.1)/1873\} - 1.246] = -1.346$$
$$\text{Therefore, } k_{\text{N}} = 0.045 \quad (3)$$

$$\text{Since } \log f_{\text{N}} = e_{\text{N}}^{\text{N}} \text{wt\% N} + e_{\text{N}}^{\text{C}} \text{wt\% C} + e_{\text{N}}^{\text{P}} \text{wt\% P} + e_{\text{N}}^{\text{Mn}} \text{wt \% Mn}$$
$$= 0 \times 0.01 + 0.25 \times 0.5 + 0.051 \times 0.2 - 0.02 \times 0.5 = 0.1252$$
$$\text{Therefore, } f_{\text{N}} = 1.334$$

From Eq. (2), we get:

$$\sqrt{(P_{\text{N}_2})} = \{[\text{wt\% N}] \cdot f_{\text{N}}\}/k_{\text{N}} = [(0.01 \times 1.334)/0.045 = 0.296$$

So $P_{N2} = 0.088$
Therefore, $\mu_{\text{N}_2} = RT \ln P_{N_2} = 8.314 \times 1873 \times \ln(0.088) = 37.86\,\text{kJ/mol}$

Problems

Problem 17.1 Calculate the maximum size of the inclusion that can float up from the bottom of a liquid metal bath of 1 m height in 10 min. Assume Stokes' law to be valid. Viscosity of metal is 5×10^{-3} kg m^{-1} s^{-1}, density of metal is 7200 kg m^{-3} and the density of inclusion is 2700 kg m^{-3}.
[Ans: **1.94×10^{-5} m i.e. 19.4 μm**]

Problem 17.2 Calculate the oxygen potential of liquid steel which is in contact with a molten slag at 1600 °C. Assume equilibrium partitioning of oxygen between slag and metal.
Given:

(i) $[\text{O}]_{\text{wt\%}}$ + Fe (l) = FeO (l), $k_1 = 4.35$ at 1600 °C
(ii) $\text{O}_2(\text{g}) = 2\ [\text{O}]_{\text{wt\%}}$, $k_2 = 6.89 \times 10^6$ at 1600 °C
(iii) $a_{\text{FeO in slag}} = 0.45$
[Ans: **−315.98 kJ/mol**]

Problem 17.3 Estimate the activity of S in hot metal of a blast furnace containing 0.04% S, 4.0% C, 2.0% Si and 1.0% Mn at 1600 °C in 1 wt% standard state.

Given: $e_S^C = 0.24$, $e_S^{Si} = 0.066$, $e_S^S = -0.028$ and $e_S^{Mn} = -0.025$.

[Ans: **0.465]**

References

1. https://images.search.yahoo.com/search/images
2. C.R. Taylor, J.F. Elliot, *Electric Furnace Steel Making* (Physical Chemistry of Liquid Steel, Iron and Steel Society, 1985) p. 291
3. A.K. Chakrabarti, *Steel Making* (Prentice-Hall of India Pvt Ltd, New Delhi, 2007)
4. S.K. Ray, JPC Bull. Iron Steel **IX**(1), 8 (2009)
5. http://ispatguru.com/vacuum-degassing-processes-for-liquid-steel/
6. H.J. Renkens, in *Proceedings of the International Symposium on Quality Steelmaking, Emerging Trends in the Nineties*, Ranchi, India, 1991, p. 107
7. E.T. Turkdogan, *Fundamentals of Steelmaking* (The Institute of Materials, London, 1996, Paperback edition first published in 2010 by Maney Publishing, Leeds, UK), p. 250
8. W. Pluschkell, Grundoperationen pf annenmetallurgischer Prozesse, *Stahl Und Eisen* **101**, 97 (1981) *in German*
9. D. Mazumdar, R.I.L. Guthrie, Metall. Trans. B, **17B**, 725
10. http://ispatguru.com/rh-vacuum-degassing-technology/
11. A. Ghosh, in *Proceedings International Symposium on Quality Steelmaking: Emerging Trends in the Nineties*, Ranchi, India, Nov 1991, p. 85
12. A. Ghosh, *Principles of Secondary Processing and Casting of Liquid Steel* (Oxford & IBH Publishing Co Pvt Ltd, New Delhi, 1990)
13. G.G. Roy, IIM Metal News **17**(1), p18 (2014)
14. A. Kamble et al., Enhancing Calcium Recovery by Introduction of Pure Calcium Cored Wires in Steel Ladles, Steel Tech. **8**(2), 55 (2014)
15. Y. Chokshi, S.K. Dutta, Inclusion and its Control in Steelmaking, Iron Steel Rev. **59**(7), 198 (2015)
16. *The making, shaping, and treating of steel*, 11th edn. (AISE Steel Foundation, Pittsburgh, P.A, USA, 1998), p. 689
17. R.G. Ward, *An Introduction to the Physical Chemistry of Iron & Steel Making* (ELBS, London, 1962)
18. O. Kubeschewski and C.B. Alcock, *Metallurgical Thermo-chemistry*, 5th edn. (Maxwell Macmillan Inter Ed, Pergamon Press, Oxford, 1989)
19. M.C. Flemings, (B) Formation of Oxide Inclusions during Solidification, Int Met Rev. **20**(3), 20 (1977), https://doi.org/10.1179/imtr.1977.22.1.201
20. Z. Liu, Y. Kobayashi, K. Nagai, J. Yang, M. Kuwabara, Morphology Control of Copper Sulfide in Strip Casting of Low Carbon Steel, ISIJ Int. **46**(5), 744 (2006)
21. K. Oikawa, H. Ohtani, K. Ishida, T. Nishizawa, ISIJ Int. **35**(4), 402 (1995)
22. https://www.totalmateria.com/page.aspx?ID=CheckArticle&site=KTS&NM=200

18 Casting Pit Practice

The entire processing of molten steel from the time of tapping until it is solidified including stripping from mould and reconditioning, etc. is known as casting pit practice or pit side practice. Solidification of steel, according to ingot structure, is discussed. Ingot defects are developed during solidification that is described. Different methods of teeming are also discussed.

18.1 Introduction of Casting Pit Practice

After hot metal is refined in a furnace, it is tapped in a special ladle called teeming ladle to which de-oxidizers, recarburizers and alloying elements may be added, while the steel is bringing tapped. A very large bulk of liquid steel is casted into permanent moulds to produce ingots for rolling and forging. The entire processing of molten steel from the time of tapping until it is solidified including stripping from mould and reconditioning, etc. is known as *casting pit practice* or *pit side practice*.

18.2 Teeming Ladle

The ladle is lined with refractory to withstand the tapping temperature and weight of liquid metal. It is a welded or riveted steel shell, refractory lining may be bricked (most commonly adopted) or rammed. The teeming ladle has an opening in its bottom, it equipped with (i) a nozzle, (ii) a stopper rod and (iii) mechanism for raising and lowering the stopper rod to open or close the nozzle [1] (as shown in Fig. 18.1).

Newly lined ladle is thoroughly dried to achieve good life and to eliminate the possibility of hydrogen pickup by the liquid metal. A ladle may be used for nearly 6–8 tapping before it is relined. In each tap, some metal solidifies in the ladle and is known as *skull* which should be carefully removed after the ladle is empty, to increase the ladle life.

(i) Nozzle: The ladle may have one or two nozzles. With the help of nozzle, liquid metal can pour at constant flow rate. Fitting a nozzle in the scaling block is done from either inside or outside the ladle. The choice of the material for making nozzle depends upon the type of steel to be poured. Ordinary fireclay nozzles are used to teem fully de-oxidized or killed steels. High alumina clay nozzles are used for teeming unde-oxidized or rimmed steels. The nozzle diameter decides the

S. K. Dutta and Y. B. Chokshi, *Basic Concepts of Iron and Steel Making*,
https://doi.org/10.1007/978-981-15-2437-0_18

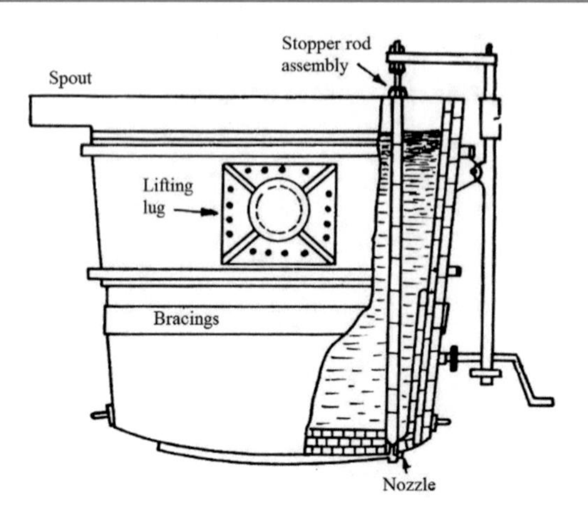

Fig. 18.1 Teeming ladle [1]

time required to pour liquid steel to a given size mould; since diameter increases, pouring rate is also increased.

(ii) Stopper rod: The stopper rod is a refractory covered steel held vertically from the top end by the cross-arm of the stopper lever assembly. A total of 6–12 pieces of refractory sleeves of circular cross section are used to cover the steel rod. The bottom sleeve forms the plug. The plug may be made of ordinary fireclay, high alumina clay, graphite, chrome–magnesite, etc.

Preparation of teeming ladle: After pouring the earlier heat, the ladle must be freed of any solidified metal, i.e. skull and slag. The life of the ladle lining very much depends upon the amount of slag that flows in along with liquid steel during tapping, since slag has the most corrosive effect on fireclay lining. Bad joints allow the skull to grow in these cracks, and the lining is damaged while removing the skull. A newly lined or a repaired ladle must be thoroughly dried before use to increase ladle life and to avoid hydrogen pickup by the liquid steel. Pre-heating of ladle is done by gaseous or liquid fuels. Nowadays, this pre-heating is avoided by insulating blocks which is used for only one heat.

18.3 Ingot Mould

Earlier molten steel is poured into permanent moulds to produce ingots for rolling, forging, etc., except in steel foundry. The moulds are themselves massive cast iron castings of uniform shapes with a cross section like square, rectangular, round, polygonal, etc.

(1) Ingot with a *square cross section* is used for rolling into billets, rails, structural sections, etc.
(2) Ingot with a *rectangular shape*, also known as a slab, is suitable for rolling into flat products like sheets, strips and plates.
(3) Ingot with a *round shape* is not used much but can be used for tube making, rod making, etc.
(4) Ingot with *polygonal shape* is used for wheels, forgings, etc.

Weight of an ingot (from a few 100 kg to 20 t) varies considerably depending upon the rolling and forging mill designs. To help striping the ingot from the mould, the mould walls are tapered from one to the other end.

Types of ingot moulds:

(I) Narrow-end-up (NEU),
(II) Wide-end-up (WEU).

(I) Narrow-end-up (NEU): NEU moulds can be lifted from the lugs to strip the ingot off the mould easily and efficiently. This is commonly used to produce rimming and semi-killed steel ingots for rolling (Fig. 18.2a).
(II) Wide-end-up (WEU): WEU moulds cannot be stripped off without the aid of additional mechanical arrangement. This is now to produce forging ingots of killed steels (Fig. 18.2b).

Cross-sectional shapes of the moulds: The inner walls of the mould may be plane, cambered, corrugated, etc. The plane walled mould has minimum of surface per unit weight of ingot produces, and hence, the cooling rate of steel is least. The specific surface area of the mould walls increases from plane to corrugated which has the effect of increasing the initial ingot skin thickness by promoting faster cooling. Corrugated moulds are more commonly used. The height of corrugation varies from 10 to 20 mm, and the wavelength of corrugated shape varies between 75 and 150 mm.

The mould, in general, should be massive enough to possess sufficient heat capacity, so that the heat evolved during solidification of ingot does not overheat the mould. Small moulds may have walls as thin as 8 cm which increases up to 30 cm for big moulds. The height of the moulds varies between 1.5 and 2.5 m. Forging ingot moulds may be taller than the same size rolling ingot moulds. Since the ingot shrinks on solidification, an extra allowance of 2% in the mould cross section is allowed in the design. Similarly, 1–2% taper is provided in the mould walls. Cast iron has good thermal shock resistance and is hard enough to resist the pressure of feeding. Cast iron has a different coefficient of expansion from that of the steels, and hence, the mould shrinks away from the ingot face on solidification and separating the two readily. The sticking of ingot to the mould is further minimized by applying mould dressings (oil or paint).

18.4 Hot Top

Fully de-oxidized or killed steel shrinks deeply on solidification and may lead to the formation of a pipe. Use of hot top (height 30–60 cm) acts as a reservoir to liquid metal to the main part of the ingot and avoids the formation of such a pipe which otherwise leads to excessive loss of ingot yield during

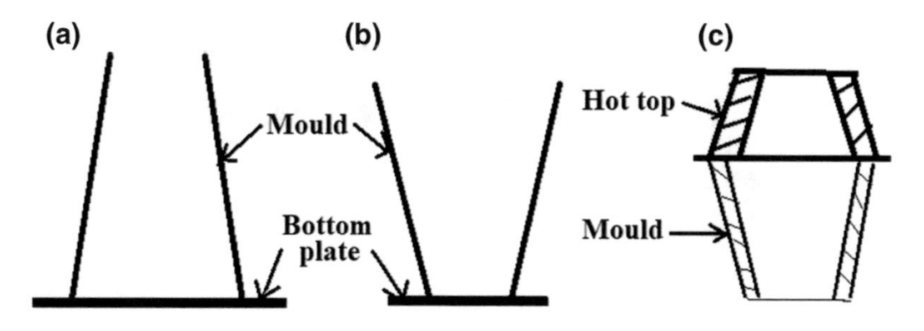

Fig. 18.2 Types of ingot mould: **a** narrow-end-up (NEU), **b** wide-end-up (WEU) and **c** WEU with hot top

working. The shrinkage cavity is than mainly confined to the hot top region. Hot top is also called a feeder head (Fig. 18.2c). Hot top is a cast iron box which is lined from inside with fireclay bricks and is generally placed on top of the wide-end-up (WEU) moulds. The box is stripped first as it has narrow-end-upwards. The ingot may be withdrawn from the wide-end-up mould by holding the head, i.e. portion solidified in the feeder box. Exothermic or insulating materials are often added in the hot top, immediately after the teeming is completed, to avoid solidification in the hot top portion since it must solidify at the end.

18.5 Solidification of Steel

According to ingot structure, steel can be classified into four types (as shown in Fig. 18.3):

(i) Killed steels,
(ii) Semi-killed steels,
(iii) Rimming steels and
(iv) Capped steel.

18.5.1 Killed Steels

When steel is fully de-oxidized by strong de-oxidizers at the furnace or ladle, no activity is observed by way of gas evolution during solidification in the mould. That type of steel is called killed steel. It remains quiet during solidification in the mould as if it is dead and hence it is called killed, i.e. no gas evolution during solidification. Steel contains >0.3% C which are killed steel. Killed steels are always casted in wide-end-up (WEU) moulds with hot tops to avoid the piping due to shrinkage. Structure is quite sound and dense which is suitable for forging. Alloy steels are fully killed steels.

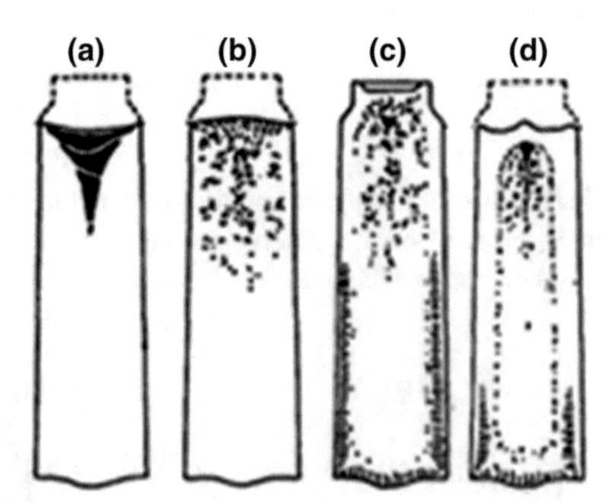

Fig. 18.3 Types of ingot steels **a** fully killed, **b** semi-killed, **c** capped, **d** rimmed

18.5.2 Semi-killed Steels

These are partially de-oxidized steel such that only a small amount of gas is evolved during solidification. Steel contains 0.15–0.3% C. The required de-oxidation may be carried out in the furnace or ladle by addition of Fe–Mn, Fe–Si and Al. Gas is not evolved immediately after pouring of steel in a mould. The top level solidified before gas evolution commences. This is ensured by adding Al in the mould as it is filled. The gas is evolved towards the end of ingot solidification. The shrinkage of steel on solidification relieves the pressure developed in the liquid core of the ingot due to gas evolution. The pipe is thus automatically compensated by the gas evolution. Structural shapes, plates and merchant are the products of semi-killed steel.

18.5.3 Rimming Steels

If the de-oxidation is not carried out at the furnace or ladle, it gives appearance of boiling to liquid steel in the mould. This boiling action is known as rimming, and steel is known as rimming steel. It requires a lot of gas evolution during solidification. Therefore, the steel must contain enough dissolved oxygen and which is possible only in low carbon steel. So carbon content in steel is less than 0.15%.

No de-oxidation is carried out inside the furnace, and only a small amount of de-oxidation, if needed, is carried out itself. Procedure depends on the carbon content of steel: high (0.12–0.15%), low (0.06–0.1%) or <0.06%. Rimming ingot has a smooth surface due to gas evolution at the beginning of solidification. Such a structure is most desirable for rolling of products where surface finish is most important, e.g. flat products. It is cast in narrow-end-up (NEU) moulds.

18.5.4 Capped Steels

This is another variety of rimming steel in which the gas evolution is much less violent than in usual rimming steels, i.e. it is less active steel in cast in bottle-shaped narrow-end-up (NEU) moulds. The carbon contents in steel are about 0.15%. Early gas evolution is prohibited by adding some Al shorts in the mould during teeming.

18.5.5 Mechanism of Solidification

Killed steel is solidified in three zones: (i) chill layer, (ii) middle layer, i.e. columnar crystals and (iii) central portion, i.e. coarse equiaxed grains. The horizontal cross section of a killed ingot is shown in Fig. 18.4.

(i) Chill layer: Liquid metal next to mould walls and bottom is chilled by the cold mould surfaces. This is a thin solid layer and known as a chill, shell or skin of an ingot and has fine equiaxed grains [1]. Rate of solidification is very high in forming the skin. However, rate of solidification soon slows down due to mould expands on heating and skin contracts on solidification, resulting in separation of the two and formation of an air gap in between. It reduces the rate of heat flow and slows down the cooling of ingot.

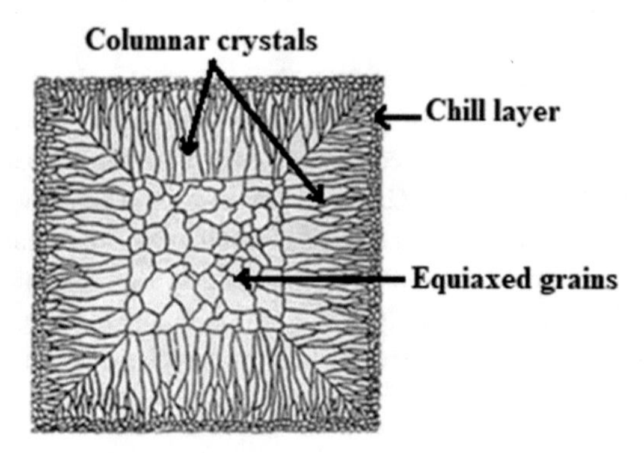

Fig. 18.4 Structure of horizontal cross section of a killed ingot

(ii) Middle layer: Solidification front moves inwards perpendicular to the mould walls, resulting formation of columnar grains form next to chill layer. Their growth is restricted due to adjacent crystals. Columnar crystals rarely extend to the centre of mould.

(iii) Central portion: This portion solidifies at last as equiaxed grains of bigger sizes than those in the chill layer due to slow cooling rate.

18.6 Ingot Defects

Basic aim of the operator is to produce, both physically and chemically a homogeneous ingot which would have a fine equiaxed crystal structure and would be free of chemical segregation, non-metallic inclusions, cavities, etc. and would have a smooth surface finish. Practically, that is not possible. Hence, an ingot, during solidification, develops defects as follows:

1. Pipe formation or piping,
2. Blow holes,
3. Columnar structure,
4. Segregation,
5. Non-metallic inclusions,
6. Internal rupture and hairline cracking and
7. Surface defects.

18.6.1 Pipe Formation

A cavity is formed due to volumetric contraction by the solidification of liquid to solid. That is known as piping, which is about 2.5–3.0% of total volume of ingot. Piping formation mostly found in killed steel. The shape and location of piping in killed steel depends upon the mould type: (a) narrow-end-up (NEU) mould showing long primary and secondary piping (as shown in Fig. 18.5), (b) wide-end-up (WEU) mould showing short pipe.

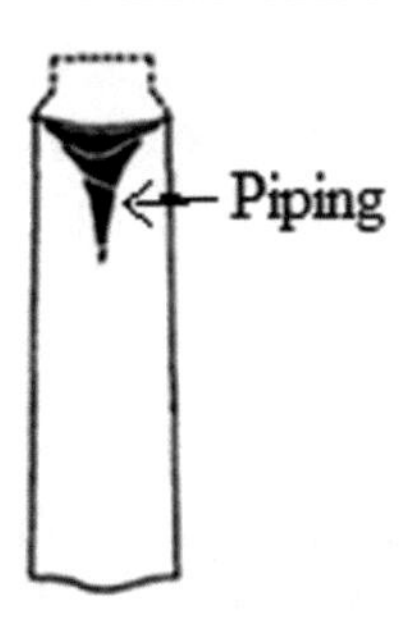

Fig. 18.5 Piping in killed ingot

Rimming and semi-killed steels show slight tendency for piping that can be eliminated by careful practice. Capped steel is practically free of pipe. With hot top feeder head, pipe is confined to the feeder box to pour little more liquid metal after solidification.

Primary or open pipes get oxidized during reheating and do not weld during rolling. As a result, that much of the ingot portion has to be discarded. It decreases the ingot yield. But deep seated the secondary pipe does not get oxidized and is welded up during rolling.

The longer is the primary pipe the lesser is the ingot yield. Wide-end-up moulds are, therefore, preferred to obtain better yield during rolling. Remedies are: (a) by adopting a hot top feeder head, so the pipe is confined to the hot feeder box of a wide-end-up mould. The volume of the feeder box is about 15% of the ingot volume. Using of insulating or exothermic materials on the top to keep the metal in the hot top molten condition for long time. (b) To pour little more metal after partial solidification of the ingot to compensate part of the shrinkage.

18.6.2 Blow Holes

Entrapment of gas evolved during solidification of steel produces cavities (within the solidified steel) known as blow holes in steel. It is formed in all types of steel except killed steel. Blow holes are two types: (a) primary blow holes, and (b) secondary blow holes. Primary blow holes are elongated or like honeycomb, which are located near to skin. Secondary blow holes are more spherical in nature, which are situated further inside.

Formation of blow holes (Fig. 18.6) eliminates partially or fully the pipe formation and thereby increases the ingot yield during rolling. Deeper blow holes do not open up during soaking and rolling operation; hence, that do not get oxidized. Such blow holes are welded up during rolling and do not make any mark in the product. Blow holes that are nearer to the surface often get oxidized due to open-up during rolling. Oxidized blow holes do not weld up during rolling and create surface defects

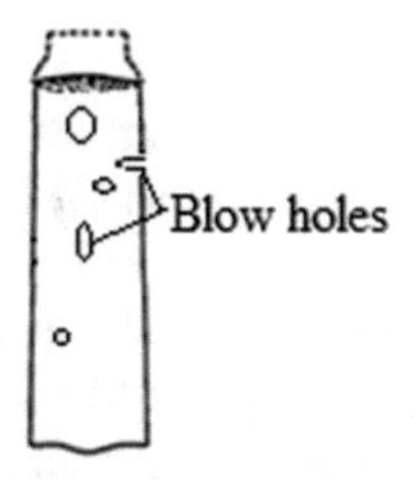

Fig. 18.6 Blow holes in ingot

on the product. Gas evolution during solidification should be controlled to obtain blow holes, by keeping the metal in the molten condition at the top of the mould for long time to avoid thick skin formation.

18.6.3 Columnar Structure

Steel is a crystalline solid, after the formation of initial chill layer further solidification takes place by formation of dendrites which grow along perpendicular to the mould walls. Their lateral growth is restricted due to the growth of adjoining dendrites giving rise to elongated crystals. If the length of these is appreciable, it is known as columnar structure (as shown in Fig. 18.4). Ingot possessing columnar structure tends to crack during rolling, unless in the first few passes with low reduction in cross section.

In general, columnar structure does not extend to the centre in the ingot. The middle portion of the ingot solidifies as equiaxed grains. The relative proportion of columnar and equiaxed grains is adjusted to keep the ingotism to a minimum [1]. The adjustment is carried out with respect to the composition of steel, its temperature while pouring, mould temperature and gas evolution during solidification.

18.6.4 Segregation

Segregation means departure from the average composition; if the concentration is greater than average, it is called positive (+) segregation, and if the concentration is less than average, it is called negative (−) segregation. Segregation is the results of differential solidification of liquid solution. Steel contains C, S, P, Si and Mn; so, it is prone to segregation during solidification.

Initial chill layer of the ingot has practically the same composition of poured liquid steel, i.e. no segregation in the chill layer due to very rapid rate of solidification. Progressive solidification of purer phase (rich in iron) occurs, while the remaining liquid gets richer in impurity contents. Impurity segregated at the top in the shape of pipe that is known as ‘V’ segregation. Side by side inverted ‘V’- or ‘A’-shaped segregation is also observed at the top. It may be due to the sinking of purer crystals down and rising up of the impure liquid in the upper part. The impurities get entrapped in the impure part at the end of solidification. This is the positive (+) segregation. The negative (−) segregation is confined to the lower central portion of the ingot (Fig. 18.7). Segregation is increased with time of solidification required for an ingot, so that large ingots tend to segregate more than small ingots. Segregation increases in the order killed, semi-killed, capped and rimming steels. Segregation can be minimized by prolonged soaking of ingots before working.

18.6.5 Non-metallic Inclusions

The term clean steel is used to refer relative free from the entrapped non-metallic particles in solid ingot. There are two types of non-metallic inclusions: (a) indigenous and (b) exogenous.

(a) Indigenous inclusions: Indigenous inclusions arising during the process of steelmaking, e.g. de-oxidation products like oxides, or precipitate like sulphides (Fe, Mn), carbides (Ti), nitrides (Zr, Ti, V), etc. Oxide inclusions can be kept to a minimum by selecting suitable de-oxidation practice. Enough time is allowed for them to float to the surface in the ladle. To minimize them,

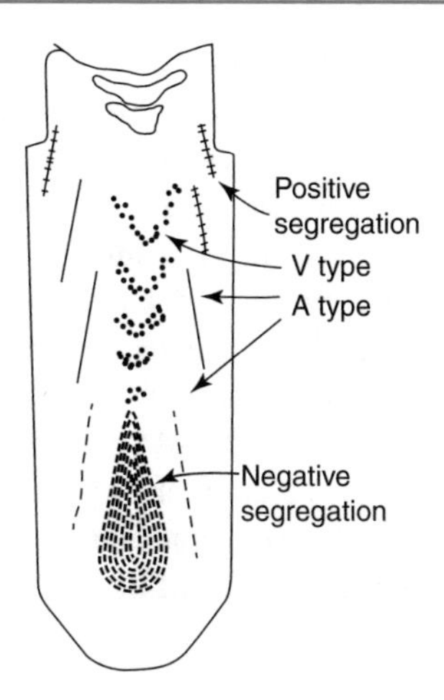

Fig. 18.7 Pattern of macro-segregation in a vertical section of killed steel ingot

de-oxidation carried out in the furnace. Further to decrease the oxygen content of liquid steel, vacuum treatment may be adopted. Almost all sulphur presenting in metal is precipitated as iron or manganese sulphides, to avoid formation of sulphides inclusions, best way is to keep sulphur as low as possible in metal. Nitrides form only if dissolve nitrogen level is high. Addition of Zr, Ti and V is formed stable nitrides.

(b) Exogenous inclusions: Exogenous inclusions arising from mechanical erosion of refectory lining of the furnace or ladle, with this come in contact during its processing. In particular, the erosion of ladle and more so of the refractories used in the assembly of the mould contribute most of the formation of exogenous inclusions. The best way to keep them down is to use minimum of such refractory channels through which metal must flow before solidification. Use of strong refractory for such places is beneficial.

Inclusions are not always undesirable, and these are purposely introduced to gain certain desirable effect, e.g. sulphur is added, to obtain sulphide inclusions, to improve machinability. Exogenous inclusions are used for dispersion strengthening of steels. Stable nitrides are useful since the ill effects of dissolved excess nitrogen are thereby eliminated.

18.6.6 Internal Rupture and Hairline Cracking

Internal rupture is used to denote internally cracked ingot. This rupture due to two causes [1]: (i) too rapid reheating of an ingot such that the outer layers expand more rapidly than the core giving rise to internal rupture. Certain classes of alloy steels are prone to such cracking because of their coarse and weak crystal structure. (ii) Too rapid cooling of an ingot after stripping the mould causes uneven contraction at the surface and in the core, finally resulting in internal rupture.

Internal rupture may extend to the surface and cause surface cracking or may open up during soaking and working. These can be eliminated by preventing too rapid cooling and reheating of an ingot.

Hairline cracking is formed all through the section and is found only after deep etching. These are oriented at random. Hairline cracking is due to hydrogen gas evolution or desorption. Solubility of hydrogen in molten metal is decreased during solidification and further cooling of steel. So excess hydrogen gas is desorbed (i.e. comes out) very slowly even after cooling the steel, for days or even weeks depending upon the type of steel, cross section, the residual stresses and the hydrogen content of steel when it was molten. Alloy steels are more susceptible to hairline cracking. Since the hydrogen content of liquid steel is determined by the type of raw materials used and the steelmaking process adopted, these factors seem to influence hairline cracking appreciably. Hydrogen content less than 2.0–2.5 cc/100 g of liquid is safe to avoid hairline cracking. Hydrogen content can be reduced below this safe level by vacuum treatment of liquid steel. In the absence of such treatment, steel must not be allowed to cool to room temperature unless enough time is allowed for hydrogen to diffuse out. In practice, the steel is held at 600–650 °C to reduce hydrogen content to below the safe limit [1]. The holding time increases with increasing cross section and may even extend to several days or weeks.

Ingot is stripped off the mould very early and is transferred to a soaking pit for prolonged soaking to allow hydrogen to diffuse out. This is also minimizing tendency to form internal ruptures.

18.6.7 Surface Defects

These are the defects that are apparent on the surface of an ingot after solidification or are visible in some form after mechanical working of an ingot. There are two types of surface defects: (i) ingot cracks and (ii) other surface defects.

18.6.7.1 Ingot Cracks

Chilling effect of a mould forms a thin solid layer, i.e. ingot skin. The thickness of skin is important bearing the formation of ingot cracks. Contraction of ingot occurs on skin formation, and expansion of mould on heating tends to separate the two and forms an air gap in between. Ferrostatic pressure of liquid core has to be withstood by skin alone, and if the skin is not thick enough to stand the internal pressure, its ruptures give rise to cracks on the surface.

Thickness of the skin formed depends upon the time of contact of steel with the mould and is calculated by:

$$\text{Thickness}\,\alpha\,\sqrt{\text{time}}\,(t) \tag{18.1}$$

$$\text{i.e. Thickness} = k\sqrt{t}. \tag{18.2}$$

where k is a constant, value varying between 0.9 and 1.2. The value of k depends upon the mould weight, mould design, temperature of liquid steel, temperature of mould, pouring technique and rate of pouring. The factors that tend to decrease the skin thickness, so tend to help formation of cracks: a too high teeming temperature, rapid rate of teeming and too high mould temperature.

(a) Longitudinal cracks: These are more or less parallel to the vertical axis of the ingot and are caused due to the development of lateral tension in the skin. The tendency to form this type of cracks increases if the ratio of cross section to height increases [1]. These are formed more at the bottom position of the ingot. Alloy steels are more prone to form such cracks than mild steels.
(b) Transverse cracks: These are nearly parallel to the base of the ingot and are formed due to longitudinal tension in the skin. The tendency to form this type of cracks increases as the ratio of ingot height to cross section increases. This is the most common type of ingot cracks.
(c) Restriction cracks: These may be longitudinal or transverse in direction and are located at the corners of the ingot. The longitudinal restriction cracks are due to large corner radius of the

ingot. The transverse restriction cracks are due to the friction between the mould and the ingot of a small corner radius.

(d) Subcutaneous cracks: These are internal ruptures close to the surface and are caused due to thermal shocks. These are open up during soaking and/or rolling.

Much of the surface cracks can be eliminated by designing the mould properly. Sufficiently thick mould walls ensure adequate chilling to produce skin strong enough to stand internal ferrostatic pressure. Use of corrugated mould walls in place of plane walls mould increases the specific surface area so that chilling is improved [1]. It also minimizes the transverse tensile stresses in the skin developed due to its solidification.

18.6.7.2 Ingot Cracks Other Surface Defects [1]

1. Scab: A projection on the side surface of an ingot caused by freezing of steel in a cavity in the mould wall or in a mould with uneven wall. A scab produces line marks on the surface during rolling.
2. Lappiness: Lap is a fold in the ingot skin caused by freezing of a slowly rising top surface of the metal in the mould before the pouring is over. Surface gets oxidized due to slowly poured or teeming temperature is low.
3. Splash: Metal drops are thrown off due to impact of metal stream on the mould bottom. These drops stick to the mould wall to form scab which marks on rolling product.
4. Crazing: If a large number of cracks are present in the mould wall, liquid steel may freeze in these cracks and give a network of fins or crazing on the ingot face. This is known as crocodile skin which marks on rolling products.
5. Double skin: Skin formed in the lower part of the ingot shrink and liquid steel flows in the gap between the mould and ingot giving a double skin. This is due to slow rate of pouring and severe chilling effects of the mould.
6. Spongy top: Viscous top tends to rise due to late gas evolution and thereby make the top spongy.
7. Skin holes: These are formed due to entrapment of gas evolved from mould dressing in the skin, which is marked in rolling products.

18.7 Teeming Methods

Teeming means pouring of liquid steel from ladle to an ingot mould. The method of teeming affects the ingot quality. A sound ingot gives high yield and less return scrap during subsequent working operations. The time required for teeming depends upon the teeming practice, i.e. method of pouring, number of ingots simultaneously teemed, the nozzle size, ingot size, ladle size, etc. The teeming temperature should be such that for a given teeming practice the steel must possess adequate super-heat even while teeming the last ingot.

There are three types of teeming methods: (i) direct teeming, (ii) tundish teeming and (iii) bottom teeming (as shown in Fig. 18.8).

(i) Direct teeming: Direct teeming means pouring of liquid steel from ladle to an ingot mould directly (Fig. 18.8a). The rate of teeming can be controlled by the use of different sizes and design of nozzles. The rate of teeming increases as the nozzle diameter increases due to erosion. Magnesite and graphite nozzles are better than fireclay nozzle in this respect. Since the stream of liquid steel directly hits the bottom place of the mould, the wear of the bottom plate is quite severe in direct teeming. This method is used for rolling ingots.

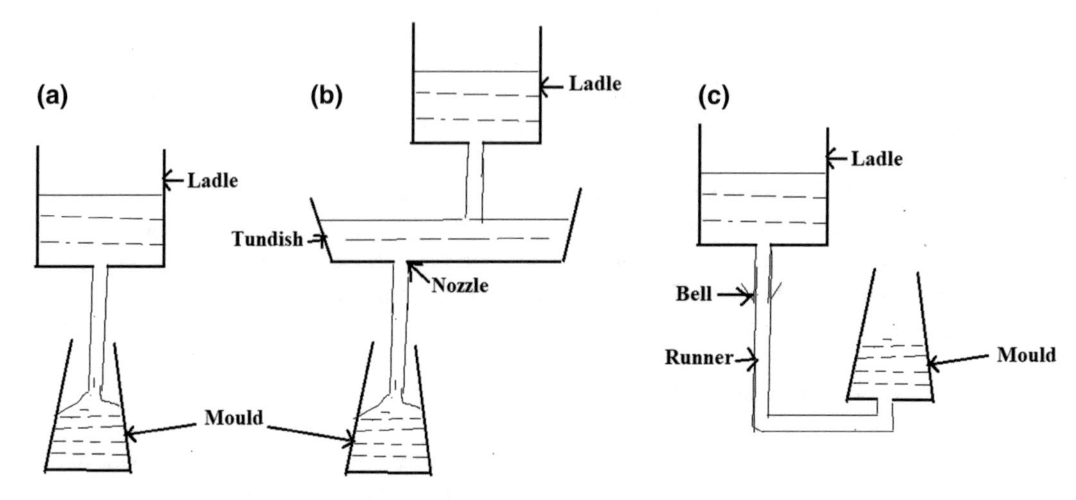

Fig. 18.8 Teeming methods: **a** direct teeming, **b** tundish teeming and **c** bottom teeming

(ii) Tundish teeming: Tundish teeming means pouring of liquid steel from ladle to tundish and then tundish to ingot mould (Fig. 18.8b). A tundish is inserted between the ladle and the ingot mould to ensure uniform rate of metal stream. The tundish has its own nozzle to regulate the flow. The tundish nozzle size is slightly bigger than the nozzle of ladle. Tundish with one or more, up to eight, nozzles are used to distribute the liquid metal evenly for all moulds at a time. This method reduces the total teeming time of a ladle, and the available super-heat in the metal can be utilized fully. This method is used for teeming forging and special alloy steel ingots.

(iii) Bottom teeming: Bottom teeming means pouring of liquid steel from ladle to a riser and at the bottom of the riser is connected with moulds; pouring of liquid steel is from riser to moulds (Fig. 18.8c). This is also known as uphill or indirect teeming. Liquid steel is teemed into a vertical runner which is connected at the bottom to the moulds. The top of the vertical runner is shaped like a bell to make teeming easy. The height of the runner is more than that of the moulds to ensure complete filling of the moulds. In general, one vertical runner is used to feed at least two or as many as twelve moulds (Fig. 18.9). The quality of the bottom teemed ingot is much superior, and the bottom plate wear is much less as compared to top teemed ingots. Use of bottom teeming is economically justified only if the superior quality of the ingot is required.

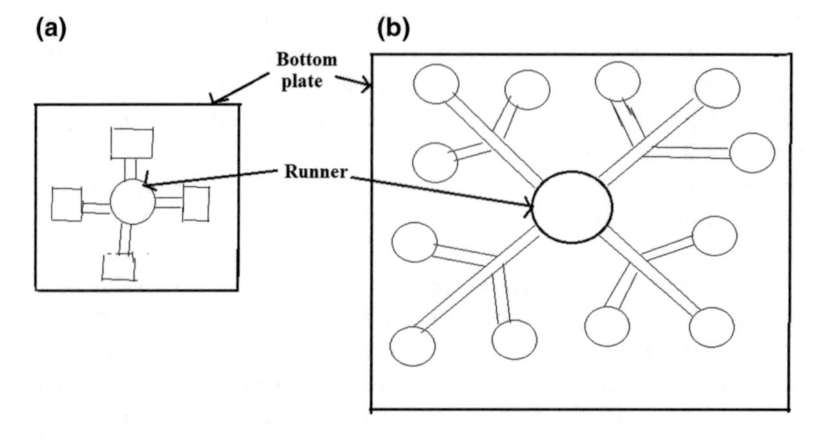

Fig. 18.9 Plan view of bottom teeming: **a** four moulds, **b** twelve moulds

Probable Questions

1. What are the types of ingot?
2. What do you understand by 'Killed', 'Semi-killed' and 'Rimming' steels?
3. Discuss the defects in ingot and their remedies.
4. Discuss the solidification mechanism of killed ingot.
5. Discuss the different types of teeming methods.
6. What do you mean by 'casting pit practice'?
7. With the help of a neat sketch differentiate between the top and bottom pouring technique of ingot making.
8. Differentiate between 'tapping' and 'teeming'. Discuss the factors that affect the tapping temperature.
9. What do you mean by segregation? How can you minimize segregation in steel ingots?

References

1. R.H. Tupkary, *Introduction to Modern Steelmaking* (Khanna Publishers, Delhi, 1991)
2. A. Ghosh, A. Chatterjee, *Ironmaking and Steelmaking* (PHI Learning Private Limited, New Delhi, 2008)

Continuous Casting (CONCAST) 19

Continuous casting may be defined as teeming of liquid steel in a short mould with a false bottom through which partially solidified cast is continuously withdrawn at the same rate at which liquid metal is poured into the mould. The main components for continuous casting, basic principle, mechanism of powder lubrication and types of concast are discussed. Horizontal continuous casting and direct rolling are also discussed.

19.1 Introduction of CONCAST

Earlier ingot route of steel product:

Steelmaking → Ingot casting → Reheating → Primary rolling → Blooms → Billets/Slabs.
Bloom → 100–150 mm thick, 200–500 mm width and 2–3 m long,
Billet → 40–250 mm square and 1–6 m long,
Slab → 100–250 mm thick, 300–2000 mm width and 1.5–5.0 m long,
Strip → 0.1–3.0 mm thick and 4–100 mm width.

The ingot route has the following disadvantages:

1. A large amount of capital must be invested in moulds, bottom plates, transporting equipments, rails, cranes, stripping and reconditioning of moulds, soaking pits, primary mills, etc.
2. Despite all cares taken, defects are occurred in ingot.
3. A certain part particularly the top of the ingots must be discarded in rolling that decreases the yield.
4. If billets are needed, additional rolling are required.

These have led to the development of continuous casting of steel into blooms, slabs or billets, and so that much of the above problems are eliminated.

Hence, nowadays, Steelmaking → Continuous casting → Billets/Blooms/Slabs.

Continuous casting replaced the ingot casting of liquid steel because of the following:

S. K. Dutta and Y. B. Chokshi, *Basic Concepts of Iron and Steel Making*,
https://doi.org/10.1007/978-981-15-2437-0_19

1. Improved yield,
2. Lower capital cost due to elimination of many downstream facilities,
3. Fuel saving for reheating the ingot before rolling,
4. Improve quality of product.

Continuous casting (CONCAST) may be defined as teeming of liquid steel in a short mould with a false bottom through which partially solidified cast is continuously withdrawn at the same rate at which liquid metal is poured into the mould.

19.2 Equipments for CONCAST

The main components for continuous casting of steel consist of the following:

1. Ladle: to hold liquid steel for teeming,
2. Tundish: to regulate the flow of liquid steel into the mould,
3. Water-cooled mould: to allow adequate solidification of the product,
4. Dummy plug bar or false bottom: to temporarily close the bottom of the mould,
5. Withdrawal rolls: to pull out the semi-solid cast continuously from the mould,
6. Cooling sprays: to solidify the cast completely,
7. Bending and/or cutting devices: to obtain handable lengths of the product,
8. Electrical and/or mechanical gears to run the concast machine.

19.2.1 Ladle

Earlier lip poured, or the stopper teeming ladles were used; now, it is bottom poured teeming ladle with side gate system used to supply liquid steel to the tundish. Dolomite or alumina-lined ladles are used to control the dissolved oxygen content of the steel (2–6 ppm oxygen in steel), whereas the dissolved oxygen content of the steel for the silica-lined ladles is 10–20 ppm.

19.2.2 Tundish

A tundish refers to an intermediate vessel placed between the ladle and the mould in continuous casting of steel. The tundish is designed to supply and distribute molten steel to different continuous casting moulds at a constant rate, with continuing emphasis on increased productivity and superior steel quality. Thus, nowadays tundish also acts as a useful reactor for the treatment of liquid steel.

It is necessary for teeming steel from the ladle into the mould via tundish to regulate the flow of liquid steel. Advantages are as follows:

- Flow rate of liquid steel can be controlled more accurately by maintaining the same liquid metal level during entire casting period to minimizing vertexing (i.e. funnel formation). Due to low level of liquid metal, vertexing is formed; on that time, slag also enters the mould.
- It can subdivide the flow rate into several streams for simultaneous cast on multiple stands.
- A single large mould can be teemed with number of streams from the same tundish.
- Even if the slide gate of tundish fails, spare tundish can be used to complete the teeming.

- Tundish is expccted to perform effecting cleaning of the steel. This helps to remove the slag, if that has come along with liquid steel. To effect removal of inclusions, the retention time of liquid steel in tundish is increased by the following means:

  (a) By controlling the flow rate,
  (b) Using suitable fluxes to absorb inclusions,
  (c) Using ceramic filters to trap inclusions.

- Tundish allows final de-oxidation adjustments, addition for de-sulphurization, inclusion modification and temperature control.

Single stand, multiple stands, T-type, V-shape, etc., are used in tundish designs. There are two methods of tundish lining: (a) conventional lining which requires pre-heating, (b) cold castable lining which does not need pre-heating, but it needs insulating powder cover on the liquid steel bath. Lining of tundish is last nearly 40–50 heats.

19.2.3 Mould

Mould is made of copper and water cooled. It has both sides (i.e. top and bottom) opened; hence, bottom is closed by a dummy plug bar in the beginning of casting. Length of the mould should be such that under the conditions liquid steel cooled to adequate skin formation before comes out from the mould. It is nearly 75–140 cm in length and mould is expected to extract 15–20% of the total extractable heat.

Moulds are lubricated to assist stripping. Moisture-free rapeseed oil is mostly used, due to less smoke and flame, no hard and solid residue left on the surface. For smooth operation, a film of 0.025 mm in thickness over mould wall is essential. Oil is supplied continuously from a ring similar in shape to that of the cross section of the mould during casting, and consumption of oil is 50–250 cm^3/t steel.

19.2.4 False Bottom or Dummy Plug Bar

The bottom of the mould is temporarily closed by dummy bar at the beginning. It was similar in cross section to that of product. Head of the bar is supposed to close the bottom of the mould, whatever gap is left is packed with asbestos. Certain amount of scrap is put at the bottom of the mould for quick solidification and form a solid seal against leak out of the liquid. Head of the bar has a bolt or a loop, around which the metal solidifies and thereby ensures a good grip for pulling the casting.

19.2.5 Withdrawal Rolls

There are one or two pairs of rolls to grip the casting and pull it out at a pre-fixed rate, without deforming the product. Pressure exerted by the rolls on the product should neither be excessive to cause its deformation nor less to allow slipping.

19.2.6 Cooling Sprays

Initial (primary) cooling by the water in the mould solidifies only outer portion of the casting (i.e. skin formation), and core of the casting remains in liquid form. Further cooling (secondary) of the casting takes place outside of the mould by high-pressure water sprays. Water sprays are directed from all sides on the casting. Total length of the secondary cooling system is generally of the order of 4–5 times longer as compared to the length of the mould. This cooling must be precise to get a defect-free product. Nearly 80–85% of the total heat is extracted in this zone. Efficiency of secondary cooling depends on: (a) rate of cooling water, (b) temperature of cooling water, (c) direction of spray and (d) water quality and scale formation.

Cast must be fully solidified, be at correct temperature and defect-free, when it emerges from the secondary cooling zone. As the cast comes out from the mould, it is positioned with the help of several guide rolls held horizontally from all sides of the casting. This is known as *roller apron/jacket*. The nozzles of water spray are situated in between these rollers. Roller apron supports the casting and maintains its shape, prevents bulging and helps bending without causing any cracking. Clean water must be used for cooling, dirty water can cause blockage in the nozzles and thereby uneven cooling occur.

19.3 Principle

Figure 19.1 shows schematic diagram of continuous casting process. Since the copper mould is open at both ends, so operation is started with the insertion of the dummy plug bar into the mould to close the bottom temporarily. Initially, liquid steel pours from a ladle to the tundish. Liquid steel with a super-heat of 10–40 °C is poured continuously at control flow rate in the oscillating mould from the tundish. Once the liquid reaches a predetermined level within the mould, the dummy bar is slowly withdrawn by pulling with chains. The rate of withdrawn (i.e. 1.24–1.5 m/min) must exactly match with the rate of pouring for smooth operation of the machine. When the molten steel comes in contact with the water-cooled copper mould, a thin solid skin is formed. By the time metal comes out of the mould, the skin becomes sufficiently thick to withstand the ferrostatic pressure of liquid core, the withdrawn of cast from the mould starts. Liquid steel does not completely solidify in the mould. It is then cooled by secondary cooling zone by direct spaying of water, and finally, it is cooled by radiation heat transfer only until it becomes completely solid. A small area of the cast where the liquid core can press the solid skin against the mould walls maintains a sort of a seal to prevent liquid leaking out from the mould. This acts as a moving seal, when the cast is withdrawn slowly from the mould and an equivalent amount of liquid is poured into the mould. If the cast is withdrawn rapidly, this seal may fracture and may produce cracks in the cast or even leakouts of liquid.

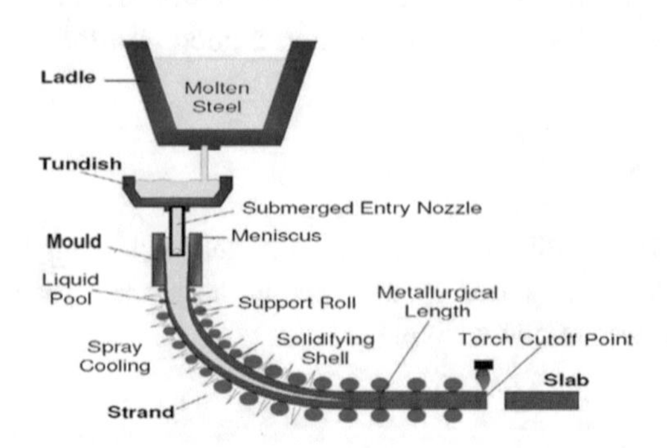

Fig. 19.1 Continuous casting unit of straight mould with progressive bending [10]

The solidification of casting occurs in three stages of cooling. The primary cooling is due to the high rate of flow of cooling water on the mould surface. The cast which comes out of the mould has a thin solid shell and the liquid core. If the solid shell is very thin, then the liquid metal may leak out and this is known as *breakout*. The next stage, secondary cooling, is due to the water sprays directly on the cast, during the downward movement of the casting. Complete solidification occurs in the secondary cooling zone. Uniform cooling is required during this stage. The final stage is the tertiary cooling due to the radiation. This occurs after the stand has become straight. Schematic of cooling processes for continuous casting is shown in Fig. 19.2.

Three major requirements in concast are: (i) liquid core should be bowl shaped; (ii) solidified shell of casting should be strong enough when comes out from the mould, so that it does not crack or break out under pressure of liquid core; and (iii) solidification should be completed before reach to withdrawal rolls. All the above requirements can be met only if the heat extraction from the liquid steel both in the mould as well as in the secondary zone is carried out satisfactorily. Higher the casting speed less is the time available for heat extraction. Therefore, longer would be the length of the liquid core as well as mushy zone and less would be the thickness of the solid shell during coming out from the mould. Hence, under a specific condition, there is maximum permissible or limiting casting speed (v_c). For 50 mm square billet → v_c = 5–6 m/min, and for 2 × 3 m slab → v_c = 0.4–0.5 m/min.

19.4 Types of Casters

There are various types of continuous casters namely: (i) vertical; (ii) vertical with bending; (iii) straight mould with progressive bending; and (iv) curved mould (S-type).

From types (i) to (iv), ferrostatic pressure decreases correspondingly and reduces the danger of internal deformation by bulging and can influence the cleanliness of steel. From the above four types,

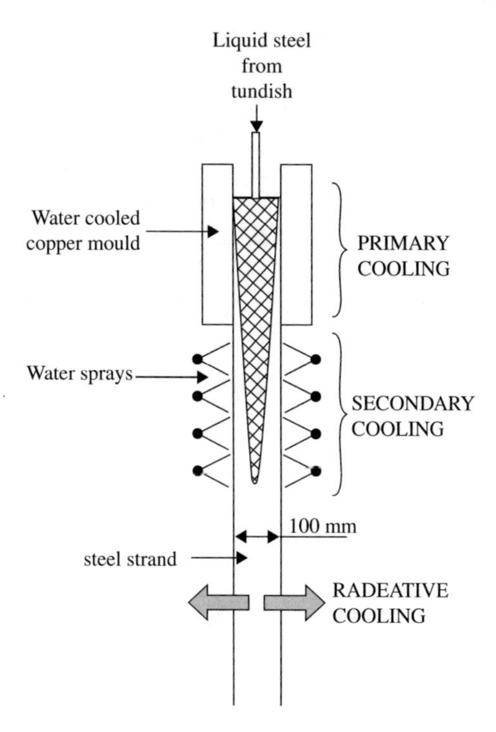

Fig. 19.2 Schematic of cooling processes for continuous casting of steel [11] (reproduced with permission from Springer Nature)

(iii) straight mould with progressive bending and (iv) curved mould are most popular concast machines (Fig. 19.3).

19.5 Mould Powder

To prevent excessive fraction and sticking between the mould and the cast, an oil or lubricant powder is introduced into the mould to help reduce the friction. The lubricant forms a fluid slag on the metal surface preventing the metal to stick on the mould. Rapeseed oil is used for billets and small blooms casting. Solid lubricant powders consist of a mixture of SiO_2–CaO–Al_2O_3–Na_2O–CaF_2 with varying amount of carbon (0.5–20%) are used for slabs and bloom casting.

Mould powder of continuous casting is used primarily to facilitate the smooth passage of casting through the mould. It is also known as *casting powder* and *mould flux* [1]. Mould powder plays an important role in the continuous casting of liquid steels. It is one of the most critical and influential factors for smooth casting of the liquid steel.

The functions of a mould powder in continuous casting are as follows:

(a) To insulate the surface of the mould and stop any freezing,
(b) To prevent re-oxidation of cast surface in the mould,
(c) To achieve a uniform heat transfer between the solidifying cast shell and the water-cooled surface of the mould,
(d) To act as a lubricant and reduce the friction between mould and cast shell and
(e) To absorb inclusions from the molten steel.

Melting rate, sintering rate and viscosity of the mould powder are important parameters. Composition of mould fluxes are: 70% CaO and SiO_2, 0–6% MgO, 2–6% Al_2O_3, 2–10% Na_2O and K_2O, 0–10% F with varying additions of TiO_2, ZrO_2, B_2O_3, Li_2O and MnO. Basicity (CaO/SiO_2) varies in between 0.7 and 1.3. Carbon particles are also used in the form of coke breeze, carbon black or graphite (2–20%), to control the melting rate of flux and to form a reducing atmosphere of CO gas to protect the molten steel from oxidation at the upper mould region.

The flow of the liquid steel in the upper mould region greatly influences the top surface powder layers. The behaviour of the powder layers is very important to steel quality. Time to time mould powder is added to the top surface of the liquid steel. It sinters and melts to form a protective liquid flux layer, which traps impurities and inclusions. This liquid is drawn into the gap between the shell and mould, where it acts as a lubricant and also helps to make heat transfer more uniformly.

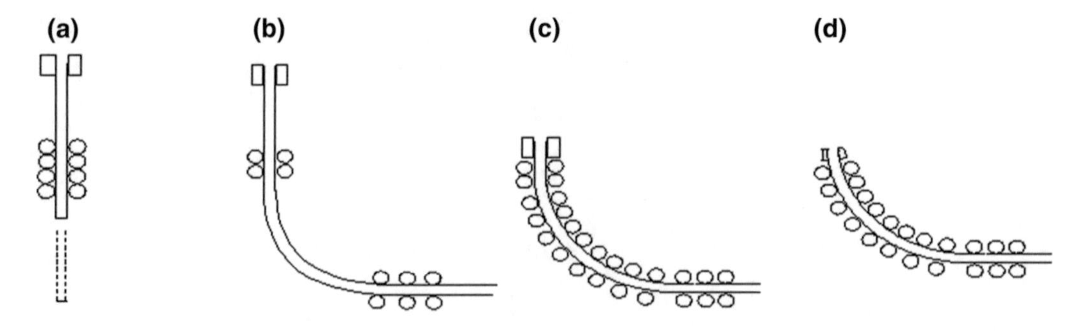

Fig. 19.3 Continuous Casters: **a** vertical; **b** vertical with bending; **c** straight mould with progressive bending; and **d** curved mould

Mould powder is sprayed at the top of the molten steel in the mould either manually or by automatic feeders. Mould powder after additions in the mould acts as follows [1]:

- Heat up and loose some carbon by reaction with oxygen,
- Forms a sintered layer,
- Melts at a definite rate and forms sintered (mushy) and liquid layers. The liquid layer acts as a reservoir to supply liquid slag to the strand. This liquid pool should be deeper than the stroke length to ensure good lubrication;
- Forms a solid slag film through the first infiltration of liquid slag into the mould/strand gap. This slag film is glassy in nature and is typically 2–4 mm thick. This slag subsequently crystallizes in the high-temperature regions adjacent to strand; and
- Forms a liquid slag film typically of 0.1 mm thickness. This liquid slag is down into the gap along the steel shell and lubricates the strand. This lubrication prevents the steel from adhering to the mould thus removing a cause of the strand breakout.

Mould powders are supplied in different forms, namely powder, granulated, extruded and expanding granules. Each type of mould powder has its own advantages and disadvantages related to cost, flowability, thermal insulation, melting rate and health hazards. When the mould powder is applied on the top of molten steel in the continuous casting mould, it must flow over and completely cover the exposed surface of the molten steel, which is particularly important when automated flux feeders are used. If the mould powder does not readily flow, then some cast surface will remain exposed which may oxidize. The result is insufficient thermal insulation and an increased re-oxidation of the steel. There are chance to reduce tendency for the mould powder to absorb non-metallic inclusions.

Mould powders or granules form various layers in the mould as follows [2]:

1. Solid layer or solid flux at the top protects the molten steel against thermal loss and re-oxidation.
2. Molten slag layer or liquid flux, in contact with the liquid steel, can absorb the floating inclusions, which are generated either in the mould, or transferred from tundish.
3. Another important role played by the molten flux is to infiltrate into the gap between the inner surface of mould wall and the cast. This provides lubrication and controls the heat transfer between the solidifying shell and the mould wall.

An illustration of the action of mould powder and meniscus region in the mould of a continuous caster is given in Fig. 19.4.

Mould powders should be stable at all casting speeds for the continuous casting process. Almost all mould powders are mixtures of several mineralogical compounds and carbon. The main function of mould powders is to provide lubrication and to control the mould heat transfer in the horizontal direction between the solidifying cast shell and the water-cooled copper mould. At higher casting speeds associated with thin slab casting, the role of mould powder is more important.

During casting, mould powder melts on the steel surface, forming a layer of liquid mould slag. Subsequently, the mould slag infiltrates between the cast shell and the oscillating mould, creating a thin slag film which solidifies into glassy and crystalline phases. The properties of the slag film dictate the main mould powder functions of strand lubrication and mould heat transfer. During casting, *slag rims* are formed which adhere to the mould walls close by the meniscus. Under stable casting conditions, slag rims are small but play a role during the infiltration of mould slag. However, rims can grow, disturb and even interrupt the casting process [3]. The thickness of the *slag film* ranges between

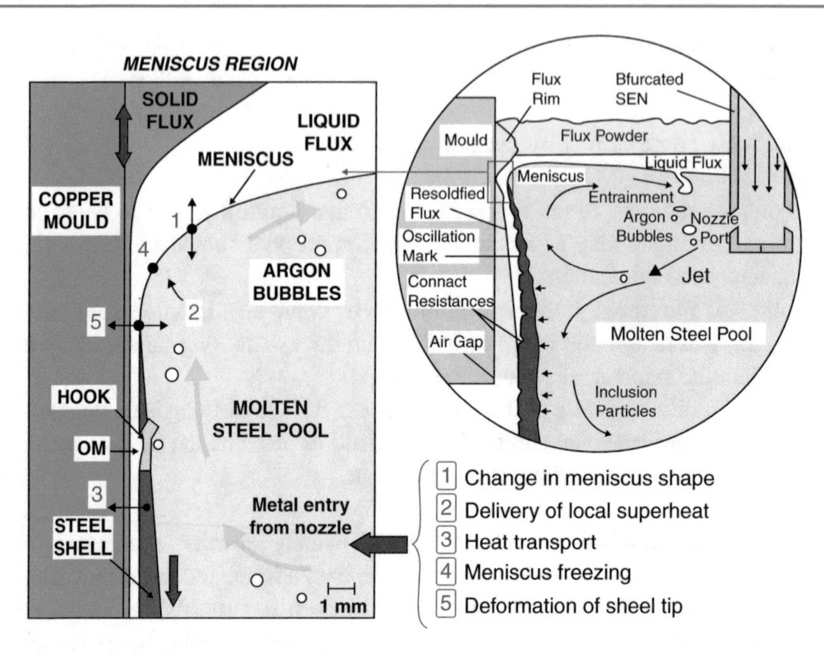

Fig. 19.4 Action of mould powder and meniscus region in the mould [12] (reproduced with permission from author (B. G. Thomas))

0.2 and 0.3 mm. The crystal can be found at the mould side or in the middle of the slag film. Detail of action of flux powder and cooling system is shown in Fig. 19.5.

A bulk chemical analysis of the standard mould powder for thin slab casting is given in Table 19.1. CaO and SiO_2 are responsible for controlling basicity of slag. Na_2O influences the melting range and the viscosity of slag. Amount and type of carbon control the melting rate. Pre-fused granules are preferred by most of the steelmakers because of its advantages of better flowability, homogeneity and less chance of direct entrapment, as compared to normal powder.

It is well known that the role of mould flux in continuous casting can be identified with the terminology [4] *APRIL*. Mould flux can ***absorb*** inclusions, ***protect*** the molten steel from the various gaseous species in the atmosphere, control the ***radiative*** heat transfer to prevent excessive heat removal across the mould, ***insulate*** the steel to inhibit the vertical heat loss and control the effective horizontal heat transfer in the mould and ***lubricate*** between the partially solidified shell and the copper mould to prevent sticking and subsequent breakouts. Thus, the role of mould flux can be identified with APRIL. The control of the radiative heat transfer and the lubrication is of primary focus to the mould flux designer, where the radiative heat is dominated by the type, shape, and number of crystalline phases formed such as cuspidine ($Ca_4Si_2O_7F_2$) and lubrication is dominated by the complex network of the molten flux structure and viscous behaviour of the optimized flux.

Beyond the control of heat transfer in the mould through crystallization, optimizing the viscosity is also essential in maintaining reliable casting operations and decreasing the propensity for caster breakouts. As the typical unoptimized silica-based mould flux is reduced by the Al in the steel, the SiO_2 is substituted with Al_2O_3 and the initial calcium silicate fluxes is modified to a calcium aluminate flux. This change in the composition of the flux changes not only the crystallization behaviour of the flux, but also the viscosity and subsequent lubrication of the mould flux. Al_2O_3 additions lowered the crystallization temperature of the flux and several crystalline phases for fluxes with high concentrations of SiO_2 forms depending on the cooling rate. High Al_2O_3 containing fluxes formed relatively few crystalline phases and were not highly dependent on the cooling rate.

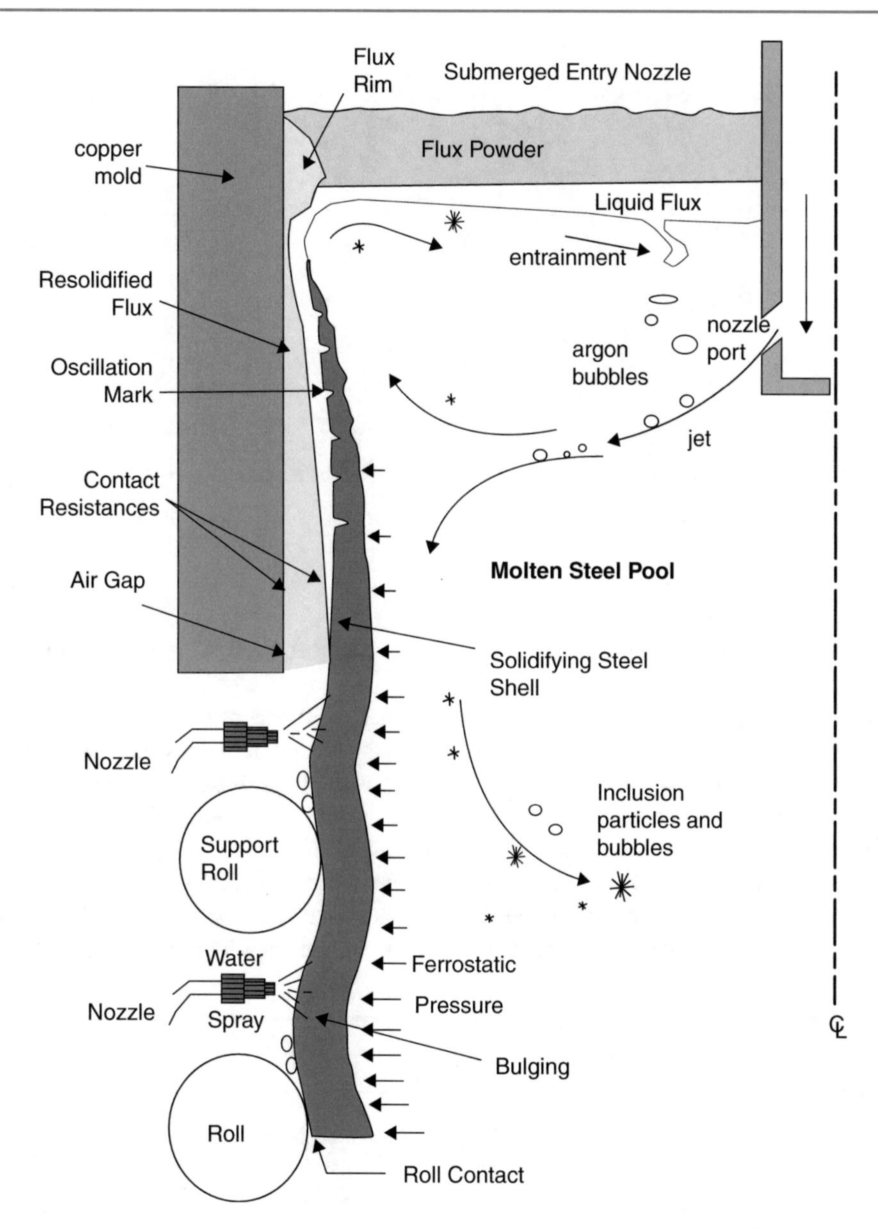

Fig. 19.5 Detail of action of flux powder and cooling system [13] (reproduced with permission from author (B. G. Thomas))

19.5.1 Powder Consumption

Mould powder consumption effects both the lubrication and the horizontal heat transfer during casting. There are various empirical relations which describe the powder consumption as function of the casting speed, the slag viscosity and parameters like break point, mould stroke, oscillation frequency, etc. Initially, a simple and widely applied equation as proposed by Wolf [5] (modified-Wolf equation) was used to describe powder consumption:

Table 19.1 Chemical composition of standard mould powder [3]

Component	Weight, %
SiO_2	20–50
CaO	25–45
MgO	0–10
Al_2O_3	0–10
Na_2O	1–20
FeO	0–6
F	4–10
C	1–25
MnO	0–10
K_2O	1–5

$$Q_s = \left(\frac{0.55}{\eta^{0.5} v_c}\right) \tag{19.1}$$

where Q_s is powder consumption (kg/m^2), η is slag viscosity at 1300 °C (poise or dPa s) and v_c is casting speed (m/min).

At the Tata Steel (Direct Sheet Plant, DSP) in IJmuiden, Netherlands [6], the powder consumption during casting is measured by continuously monitoring the weight of the powder bin. This method is more accurate than other methods like the counting of powder bags. Based on five months of casting operations with casting speeds between 3.5 and 5.8 m/min, the powder consumption data were evaluated and plotted against the casting speed. An illustration is given in Fig. 19.6 where the black line represents the measured powder consumption (kg/m^2). The modified-Wolf equation is plotted in this figure as well, represented by the dotted line. The measured values are about *half* of the predicted values using the modified-Wolf relation. A good fit can be obtained by the following equation [6]:

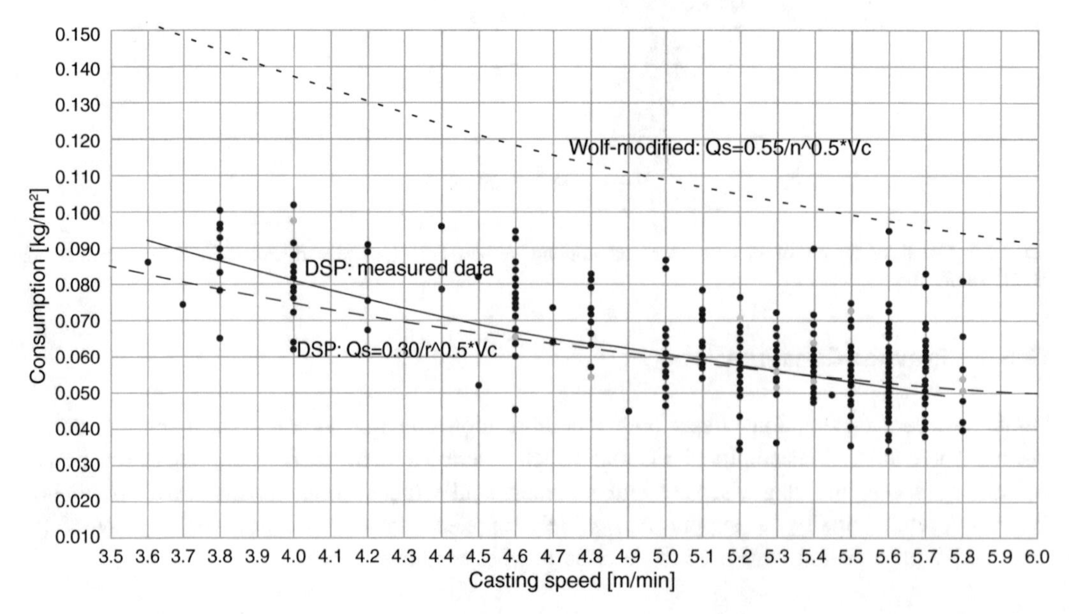

Fig. 19.6 Powder consumption versus casting speed at DSP (black line: measured plant data; dotted line: modified-Wolf equation; dashed line: equation for DSP caster) [6]

$$Q_s = \left(\frac{0.30}{\eta^{0.5} v_c}\right) \tag{19.2}$$

This relation is given by the dashed curve in Fig. 19.6. Note that the indices of v_c and η are similar as those proposed by Wolf.

The view that the actual powder consumption is low is confirmed by some other thin slab casters which report values around 0.1 kg/m^2 at approximately 5 m/min and the QSP process of Sumitomo, reporting a consumption of 0.1 kg/m^2 at a casting speed of 5 m/min and between 0.09 and 0.05 kg/m^2 at a casting speed of 8 m/min.

It is important to realize that the low consumption at the DSP (Netherlands) caster does not cause any operational problems related to strand lubrication or mould heat transfer, i.e. sticking of the shell and the occurrence of surface cracks. Only a minor part of the breakouts at the DSP can be related to the performance of mould powder during casting.

19.5.2 Slag Films

In general, both the mould side and the strand side of the *slag films* showed a smooth surface. This indicates that the control of mould heat transfer during thin slab casting is mainly achieved by the slag film and by the slag film properties themselves. Furthermore, it was found that the residence time of the slag film or at least the part of the film in contact with the mould is very long (up to 10 h or more).

The average thickness of the liquid film can be calculated from the powder consumption during casting and the density of the mould slag [6]:

$$d_l = \left(\frac{fQ_s}{\rho}\right) \tag{19.3}$$

where

d_l Average thickness of the liquid film (m),
f Fraction of the powder forming slag,
Q_s Mould powder consumption (kg/m^2) and
ρ Density of the liquid flux (kg/m^3).

In this approach, any mould slag present in the oscillation marks is neglected.

Given an actual slag consumption of 0.05 kg/m^2 and a slag density of 2600 kg/m^3, the liquid film thickness is approximately 0.0192 mm. As a thumb rule, the average liquid film thickness is at least a tenth of the total film thickness. This indicates an average film thickness of approximately 0.2 mm or more.

19.6 Merits of CONCAST

The advantages of CONCAST are as follows:

1. The yield improves by 10–20% than that of ingot practice.
2. Concast gives saving in energy due to elimination of ingot reheating, primary rolling, etc. Energy requirement for: (a) coke oven → BF → BOF → Concast is 26.14 GJ/t of billet; (b) coke oven → BF → BOF → Ingot → Reheating → Rolling is 29.66 GJ/t of billet; thus, net energy saving is 3.52 GJ/t of billet.

3. The quality of concast product is better than conventionally cast ingots. Concast have favourable grain size, less segregation and more homogeneous chemical composition due to rapid cooling.
4. The caster can be operated most effectively by online computer control.
5. Concast technique has much higher productivity and is comparatively less expensive over a long run.

19.7 Improvements of CONCAST

There are a lot of improvements in CONCAST as follows:

1. Remotely adjustable moulds (RAM),
2. Ladle slag detection system,
3. Submerged entry nozzle (SEN),
4. Electromagnetic stirring (EMS),
5. Electromagnetic breakers (EMBR), and
6. Argon purging through tundish mono-block stopper (MBS).

19.7.1 Remotely Adjustable Moulds (RAM)

The use of RAM is for automation online width change of slabs during casting. Adjustment of slab width during casting is done by moving the mould end plates.

Advantages of RAM:

(i) Increase yield,
(ii) Better caster availability,
(iii) Lower refractory consumption for tundish,
(iv) Lower fuel consumption for pre-heating of tundish and dummy bar head and
(v) Reduce maintenance work.

19.7.2 Ladle Slag Detection System

Electromagnetic slag detection system concentrically arranged sensor coils are placed around the ladle nozzle. One coil is a transmitter and other is a receiver. Conductivity of slag is 1000 times less than that of liquid steel. Difference affects the current in the transmitting coil, the secondary voltage in the receiving coil and the phase difference between current and voltage. This detects the passing slag from ladle to tundish.

19.7.3 Submerged Entry Nozzle (SEN)

Submerged entry nozzle (SEN): The liquid steel is transferred from tundish to mould through the submerged entry nozzle (SEN). This represents a zone of increased turbulence which promotes coalescence of smaller inclusions. This process is desirable in ladle and tundish because larger inclusions easily float up and get absorbed. However, since almost no time is available for separation, coagulation is disadvantageous in SEN. Clogging of the port openings of SEN poses operational problem such as change in flow pattern and has adverse impact on steel cleanliness. The composition of typical clogs has been found to be rich in Al_2O_3. The high frequency of SEN clogging can be taken as an indirect measure of poor steel cleanliness.

The following three important issues are associated with nozzle clogging [2]:

1. Dislodged clogs may either get trapped in steel or change the composition of in-mould slag. Both have been shown to cause defects.
2. Clogs change the flow pattern and jet characteristics of liquid steel coming out of the nozzle ports. This disrupts flow in the mould, resulting in slag entrainment and consequent quality problem.
3. Clogging interferes with mould-level control, as the flow control device tries to compensate for the clog.

The molten metal flows into the oscillating copper mould from the tundish through a submerged entry nozzle (SEN). SEN is used to provide a smooth flow of molten steel into the mould (Fig. 19.7).

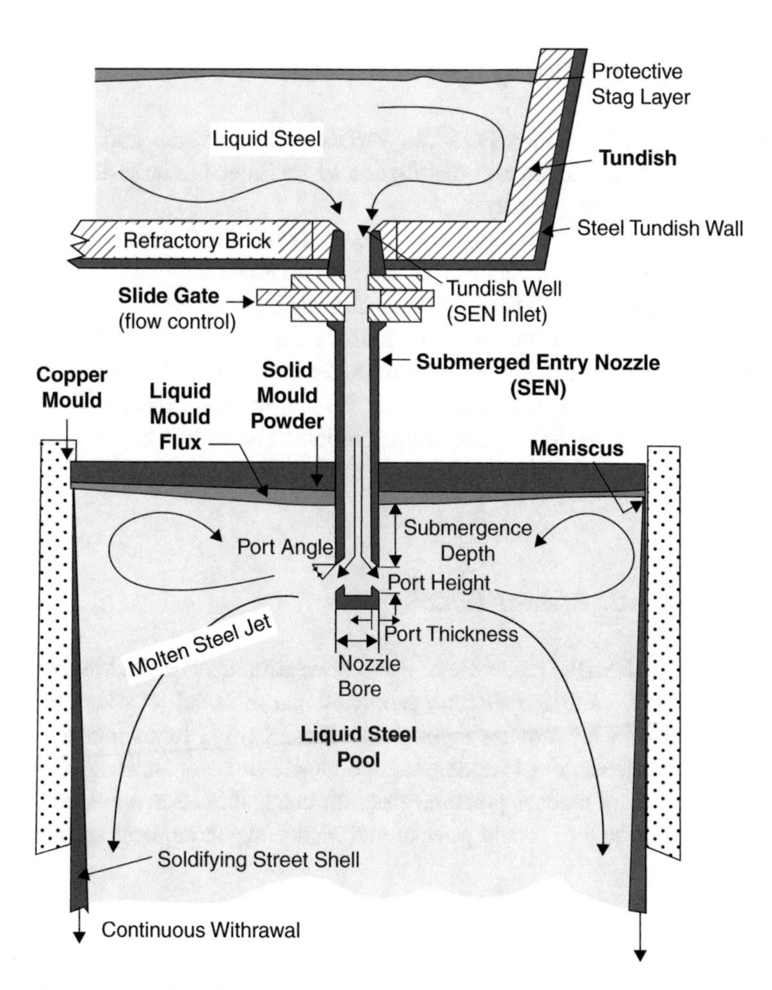

Fig. 19.7 Schematic view of tundish, SEN system and mould [14] (reproduced with permission from Springer Nature)

SEN prevents oxidation of the molten metal. It keeps mould powder, non-metallic inclusions and gases from atmosphere away from the casting. Alumina–graphite has shown excellent erosion resistance to molten steel.

Caster mould is the last refining stage, which offers a possibility of inclusion removal. Inclusions, which are carried into the mould through SEN, may comprise the following varieties [2]:

- De-oxidation products,
- Nozzle clogs,
- Entrained tundish slag,
- Re-oxidation products from air absorption through nozzle leaks,
- Additional sources of inclusions which can be formed inside the mould,
- Mould slag entrained due to excessive velocity of liquid steel at the top near the slag interface.

Inclusions can either float up and get absorbed into the top slag layer or they get entrapped into the solidifying shell to form permanent source of defects in the product. Flow through SEN governs meniscus oscillation, mould powder entrainment and thereby, influence solidified steel quality significantly.

19.7.4 Electromagnetic Stirring (EMS)

Electromagnetic stirring (EMS) influences the surface, super-surface and centreline quality of product. It alters: (i) volume fraction and distribution of non-metallic inclusions, (ii) chemical segregation and (iii) distribution of porosity.

Strong convection velocities in the liquid steel effectively prevent the attachment of gas bubbles to the solidification front. Columnar to equiaxed transaction is certainly the most commonly recognized beneficial effect. EMS suppresses all types of centreline segregations. A number of slag entrapments and a number of pinholes become extremely small. EMS depends on super-heat, stirring parameters and quality of the strand alignment. The stirrer can be placed on any of the four sides of the strand. It is enclosed in a double-shell box of austenite steel. Box allows the magnetic field to pass into the strand. The stirrer is placed behind the supporting rollers. The rollers between the stirrer and the strand are made of austenite steel and thus do not disturb the magnetic field which can penetrate the strand.

19.7.5 Electromagnetic Brakers (EMBR)

Electromagnetic breakers (EMBR) make use of a strong electromagnetic field for breaking the jets of liquid steel emerging from a submerged entry nozzle. When the steel jet enters the magnetic field, current loops are induced in the moving liquid steel. These current loops interact with the applied static magnetic field and generate a breaking force opposite to liquid steel flow. Reduction in the velocity of the liquid steel jet reduces potential risks for crack formation and break outs because of thin solid shell. Defects related to mould powder and inclusions entrapment are also reduced.

19.7.6 Argon Purging Through Tundish Mono-Block Stopper (MBS)

Argon purging through mono-block stoppers of tundish is developed to minimize abnormalities in continuous casting. The problem of alumina clogging in tundish nozzle and submerged entry nozzle, for casting of aluminium-killed steel, has been solved through the implementation of existing anti-clogging concepts of usage of argon gas injection through mono-block stoppers (MBS) and argon gas injection through ladle shroud preventing the adherence of alumina particles by removing them from the inner bore of tundish nozzle and submerged entry nozzle (SEN) (as shown in Fig. 19.8). This method is developed at Visakhapatnam Steel Plant (VSP) [7], India.

Reasons for alumina deposition are: (i) oxidation of aluminium during the killing process of steel (i.e. de-oxidation) and casting process by entrapped air and refractory; (ii) physical and chemical bonding between alumina and inner walls of SEN; and (iii) precipitation of alumina due to drop in

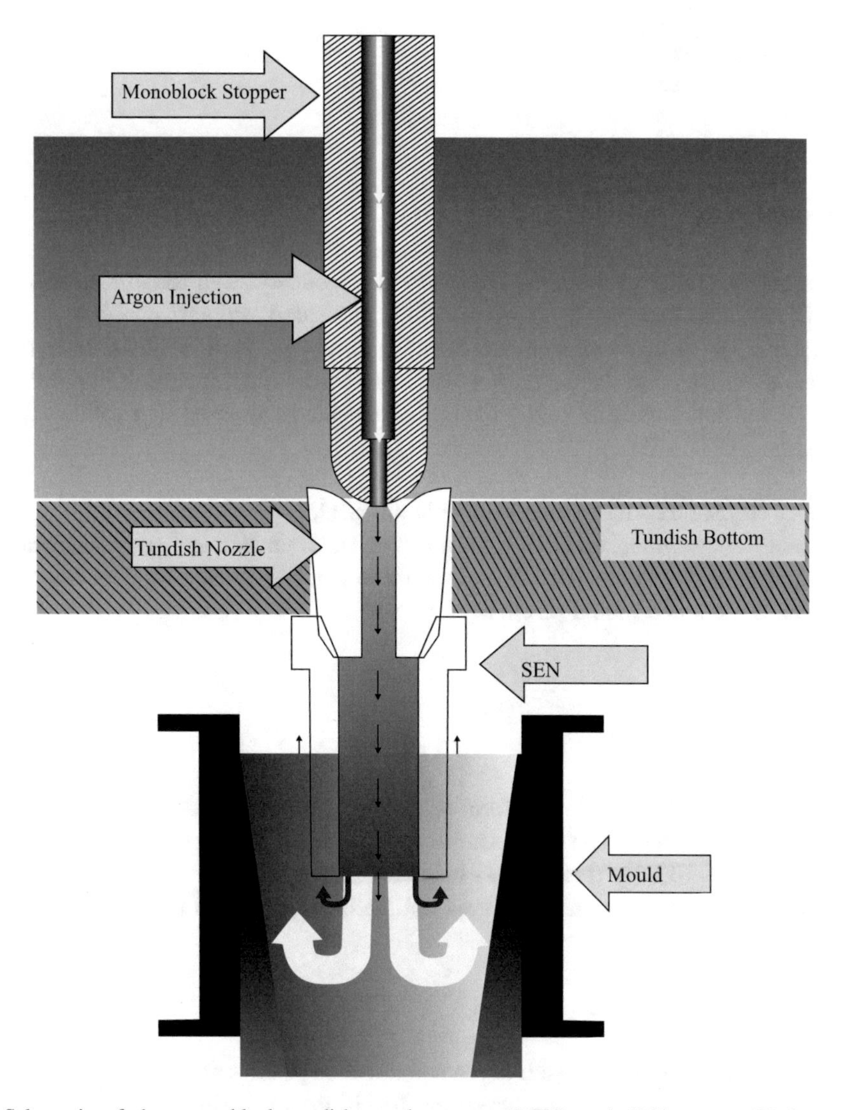

Fig. 19.8 Schematic of the mono-block tundish nozzle stopper (MBS) and SEN system [7] (reproduced with permission from *IIM Metal News*)

liquid steel temperature (as the solubility of Al in liquid steel decreases with temperature) in tundish mono-block and SEN.

Methods to overcome these problems: (a) reduction of excess aluminium in steel, by slag-free tapping and controlled rinsing by argon gas; (b) shroud the steel stream to reduce the interaction with air, by inert gas to avoid the re-oxidation of aluminium in steel during casting; and (c) removal of deposited alumina, by argon gas inside the stopper rods to prevent its continuous growth.

19.8 Quality Control in CONCAST

Quality of steel can be directly attributed to the following technological parameters [8]:

- Cleanliness,
- Chemical homogeneity,
- Free from porosity and cracks and
- Conformance to desired shape.

19.8.1 Cleanliness

Cleanliness is determined by the impurity level in steel. Secondary steelmaking facilities such as VD, LF, RH and DH are used to reduce levels of impurities, particularly sulphur, phosphorus, nitrogen and hydrogen in steels. The objective of continuous casting is not only to maintain the cleanliness level of the liquid steel as delivered to the caster, but also to promote further inclusion removal. To prevent re-oxidation of the liquid steel as well as to get a low inclusion level, the following factors are controlled during CONCAST:

(i) Minimize oxygen transfer from the air to liquid steel,
(ii) Exogenous inclusion pickup from refractories of ladle, tundish and casting powder,
(iii) Control of fluid flow in the tundish and mould to maximize inclusion float out,
(iv) Adoption of optimum mould powder.

19.8.2 Chemical Homogeneity

One of the prerequisites for obtaining uniform mechanical properties and microstructure is chemical homogeneity of the liquid steel. The liquid steel is becoming homogenized by bubbling with inert gas at a high pressure for 3–10 min. This not only helps the alloy elements uniformly distributed, but also attains a uniform temperature and helps to float out the inclusions from the molten bath.

19.8.3 Porosity and Cracks

Microsegregation and porosity is a result of the solidification process and occurs in the axial zone of the slab or billet. The origin of microsegregation lies in the distribution of elements like carbon, manganese, sulphur and phosphorus between solid and liquid during the freezing of steel, as well as the solidification structure and convection forces near the bottom of the liquid pool. The solidification process needs to be controlled to minimize segregation. This can be done by promoting growth of equiaxed crystals in the core of the cast product. The equiaxed zone can be propagated by controlling the factors: (i) degree of super-heat; (ii) steel composition; (iii) section size; and (iv) induced fluid flow.

It is reported that more super-heat promotes columnar growth, and to obtain a reasonable wide equiaxed zone, it is necessary to impose a limit on the degree of super-heat. It is an established practice to permit a maximum degree of super-heat of 20 °C for ordinary grades of steel and 15 °C required for critical applications.

Carbon and phosphorus contents in steel have an influence on the size of the equiaxed zone. Medium carbon (0.17–0.38%) favours an equiaxed structure; similarly, phosphorus (0.02–0.025%) causes the columnar zone to shrink.

There are basically of two types of cracks: (a) surface cracks and (b) internal cracks (as shown in Fig. 19.9).

Surface cracks, which are initiated in the mould, are a serious problem as they oxidize during rolling, giving rise to an oxide rich scam in the rolled product. Internal cracks, which are initiated during solidification, are also contributed to a quality problem, particularly when they do not weld during rolling, left as voids in the plate product. High sulphur contents or excessive super-heat adversely influences the formation of internal cracks. This is related to the fact that high super-heat favours the formation of columnar structure having easy crack paths while a high content reduces the hot ductility of the steel in the temperature range where most cracks form.

19.8.4 Desired Shape

Shape defects are a common problem during continuous casting of billets. Off-squareness is a typical defect seen in billets and is usually accompanied by off-corner, subsurface cracks.

1. Longitudinal cracks: This type of defect varies from small, fine cracks to long, deep ones, usually in the middle half. This is caused by both metallurgical and operational factors. Metallurgical factors are: (i) carbon level and (ii) Mn/S ratio.
2. Carbon level: Longitudinal cracks are most frequently observed around a carbon level of 0.12%. This is due to 4% different in the thermal shrinkage coefficient of δ-Fe and γ- phases that give rise to stresses which are relieved by longitudinal cracking. This is minimized as lower down the casting speed.

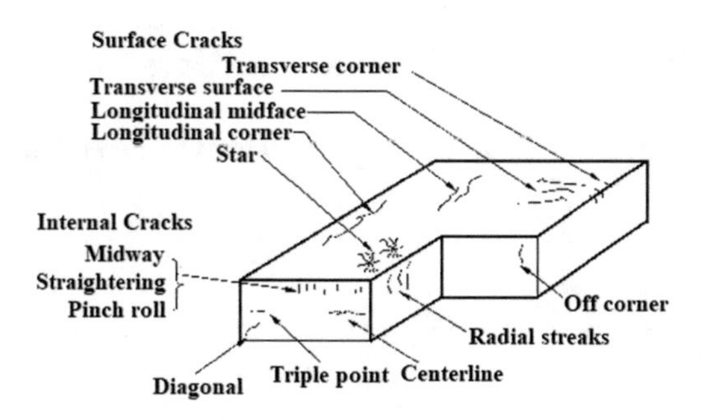

Fig. 19.9 Surface defects of concast [15]

3. Mn/S ratio: Mn/S ratio also influence slab surface quality which improves with increasing Mn/S ratio and decreasing S (in metal) level.
   Operational factors affecting surface quality are: (i) varying or increasing casting speed; (ii) very high temperature in the tundish; (iii) mould condition and mould top zone alignment; and (iv) cooling condition and slab support systems immediately below the mould. A thick slag layer with a significant crystalline fraction reduces the horizontal heat transfer and keeps the shell as thin as possible, for reducing stresses.
4. Transverse cracks: These are rarely seen in round profiles, and they appear due to (i) tensions on the longitudinal direction of strand, (ii) thermal stresses by the uneven solidification of the cast and additional stress by the turbulent flow in the meniscus and (iii) friction of the strand in the mould by the higher casting speed, decreases of liquid metal flow between the mould wall and cast, and increases the edge friction with the increasing viscosity of the mould powder.
5. Centreline cracks: It is largely unaffected by metallurgical factors such as steel chemistry or super-heat. Rolling and or bulging of strand near the point of complete solidification appear to be the main cause of such defect. Strict control of casting speed and extraction roll pressure as well as enhanced secondary cooling may reduce this type of defect.
6. Gas and slag entrapment: Gas and slag entrapment are mainly caused by the turbulence of metal flow. Turbulence may be reduced by adjusting the SEN depth, by reducing the velocity by EMBR and by decreasing viscosity of the flux.
7. Segregation: Principle factors affecting central segregation is the degree of super-heat of liquid steel. Secondary cooling and casting speed indirectly affect segregation.

   (i) Super-heat: Low super-heat lead to a high proportion of equiaxed structure which is effective in minimizing segregation.
   (ii) Secondary cooling: Non-uniformity of cooling will cause uneven solidification, therefore segregation.
   (iii) Section size: As the section size increases, the size of the segregation zone also increases, but the proportion of the cross section occupied by the segregation is decreased.
   (iv) Inclusion accumulation: Higher the casting speed, the number of inclusions close to the inner radius of the slab.

   With EMBR, the amount of inclusions in the slabs cast is significantly reduced due to reduce penetration of liquid steel into the mould.
8. Strand breakout: Strand breakout means liquid steel flows out. This can be prevented by controlling the rate of water flow, casting speed, and proper mould and nozzle design parameters.

19.9 Further Developments of CONCAST Practices

Developments in concast practices have taken place with the invention of three more types of practices, namely:

1. Near-net-shape (NNS) casting,
2. Horizontal continuous casting (HCC) and
3. Direct rolling (ISP and CSP).

19.9.1 Near-Net-Shape (NNS) Casting

The NNS casters are further classified into four different types depending on the thickness of slab or strip produced. These types are as follows [9]:

(i) Thin slab caster: It produces slabs of about 20–70 mm thickness and 1000–2000 mm width which can directly be taken to finishing stand of hot strip mill without any conditioning.
(ii) Strip caster: It produces strips of about 5–20 mm thickness which need some conditioning in the form of limited hot rolling to avoid cracking before further cold rolling.
(iii) Thin strip caster: It produces strips less than 5 mm thickness which can directly be taken to cold rolling mill.
(iv) Foil caster: It produces strips of about 0.02–0.5 mm thickness which can directly be used without further working.

The developments in concast technology to shorten the process route are shown in Fig. 19.10. One of the important types of NNS casters is round continuous caster to produce a cast product of round cross section to produce seamless tubes. The advantages of NNS casters are as follows:

- Comparatively lower capital investment operational and production cost,
- Improved quality of the product (i.e. minimization of segregation and uniform chemical composition of the cast),
- Uniform physical and mechanical properties of the finish product,
- Greater flexibility of production,
- Less energy consumption to produce steel strip or sheet.

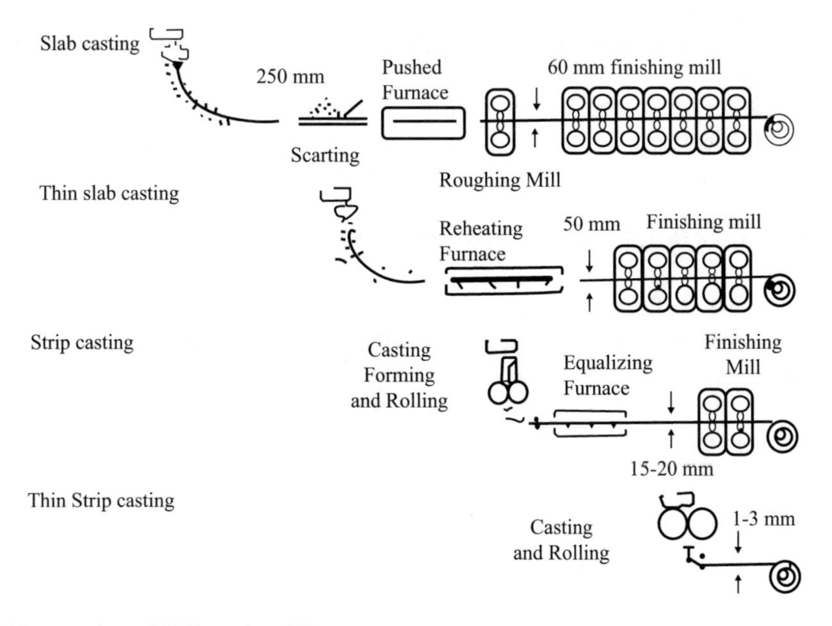

Fig. 19.10 Near-net-shape (NNS) castings [9]

19.9.2 Horizontal Continuous Casting (HCC)

The austenite formed from the delta iron during the cooling of liquid steel or after shaping is extremely coarse is extremely coarse grained. Under these circumstances, the grain boundaries represent nucleation sites for pores which, when exposed to diffusion processes, develop into cracks. This means, the cast strands must be kept hot as far as possible. Particularly, the endangered temperature range between 800 and 1000 °C should be avoided. To get rid of surface defects on modern bloom casters, the machine should satisfy the following requirements [9]:

(i) Largest possible casting radius,
(ii) A long-supported section,
(iii) Operation with dynamically regulated, soft secondary cooling or
(iv) Round sections of the product.

A much simpler solution to this problem is offered in the form of horizontal casters involving the direct rolling of cast steel as shown in Fig. 19.11. Unlike conventional concast, the HCC process is characterized by the direct connection of the tundish with the mould and the absence of bending throughout the entire facilities linearly arranged on the ground level. The HCC process offers the following advantages:

- Lower investment cost compared with conventional casting,
- Lower height and smaller space requirements,
- Reduced manpower requirement to control the process,
- Very low ferrostatic pressure during solidification,
- Electromagnetic stirring is possible by means of permanent magnets,
- No flux powders are used,
- Absence of metal re-oxidation between tundish and mould, an added advantage while casting smaller sections,
- Quick changeover possible to different square or round mould dimensions and
- Elimination of bending.

However, the major drawbacks of HCC are as follows:

- There is no chance to float any inclusion from the stand,
- Gravity segregation of nuclei may cause accumulation of equiaxed crystals in the lower half of the strand and
- The life of the most critical component, the breaking, is quite erratic and imposes serious restrictions on sequence casting.

19.9.3 Direct Rolling (ISP and CSP)

It involves direct linking of the concast caster with the hot rolling process, thereby conserving a substantial amount of energy and maintaining an improved product quality. The hot rolling mill is in line with the cast strand and rolling takes place immediately following casting. It is necessary to use an online heating facility to achieve the required temperature of cast product for its ultimate rolling.

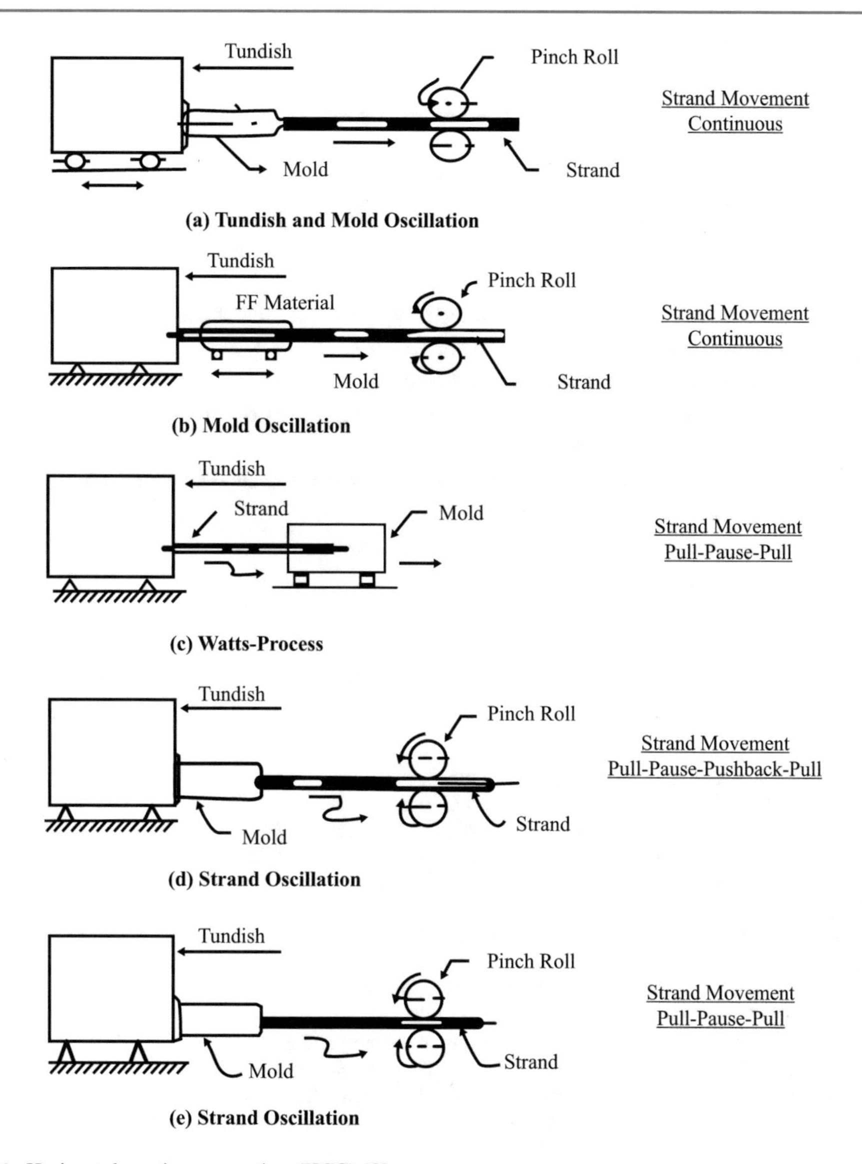

Fig. 19.11 Horizontal continuous casting (HCC) [9]

Online strip production (ISP) is an extension of thin slab casting, directly linking it with the intermediate induction heating and the final inline rolling into hot strips in finishing stands. In this technique, the slab of 60–80 mm thickness is produced by the caster is first reduced to 15–25 mm thickness by hot rolling and finally further rolled to get finish strip of less than 2 mm thickness [9]. The main advantages are: (i) economic production of hot strip/sheet/tube; (ii) energy conservation; and (iii) improved quality product.

In compact strip production (CSP), the caster is directly linked with the rolling mill through the soaking furnace. This technique enables direct rolling of hot strips with several additional advantages.

Probable Questions

1. What are the basic requirements of a continuous casting unit? Why dummy plug bar is required for Concast?
2. Describe the principle of continuous casting of steel with a neat sketch for the solidification characteristics. Is it advantageous as compare to that of ingot casting? How?
3. What are the types of concast machines? Which one is the most popular? What are the new developments in Concast practice?
4. Which material is used for Concast's mould? Why do you select this particular material?
5. Why water-cooled copper mould is used for Concast?
6. What is the function of mould powder in Concast?
7. Discuss the action of mould powder and meniscus region in the mould.
8. What are advantages of Concast?
9. Why do you need to oscillate the mould? What is the advantage of using an electromagnetic stirrer? Where it is placed in Concast machine?
10. Explain the roll of SEN, MBS and tundish in Concast.
11. Discuss about quality control in Concast.

References

1. http://ispatguru.com/continuous-casting-mould-powders/
2. S.K. Ray, *JPC Bulletin on Iron and Steel* **IX**(1) 8 (2009)
3. J. A. Kromhout: *Trans Indian Insti. Matel*, 66(5–6), Oct-Dec 2013, p 587
4. S.S. Jung, G.H. Kim, I. Sohn, *Trans Indian Insti. Matel* **66**(5–6), 577 (2013)
5. M. Wolf, in *Proceedings of METEC Congress 94, 2nd European Continuous Casting Conference* (Steel Institute VDEh, Düsseldorf, Germany, 1994), pp. 78–85
6. J.A. Kromhout, R.C. Schimmel, Ironmak. Steelmak. **45**(3), 249 (2018)
7. M.B.V. Rao et al., IIM Metal News **13**(2), 13 (2010)
8. R. Datta, S. Mishra, Trans. Indian Inst. Matel. **48**(4), 339 (1995)
9. S.K. Dutta, A.B. Lele, Tool Alloy Steels **29**(3), 81 (1995)
10. https://images.search.yahoo.com/search/images/continuouscastingofsteel
11. J. Sengupta, B.G. Thomas, M.A. Wells, Met. Mater. Trans A **36A**, 187 (2005)
12. J. Sengupta, B.G. Thomas, *Journal of Metals* (electronic edition) (Dec 2006)
13. B.G. Thomas, Chapter 15: Modeling for Casting and Solidification Processing, in *Continuous Casting of Steel*, ed. by Yu (Marcel Dekker, New York, 2001) pp. 499–540
14. H. Bai, B.G. Thomas, Met Mater. Trans B **32B**(2), 253–267 (2001)
15. B.G. Thomas, J.K. Brimacombe, I.V. Samarasekera, Trans. Iron Steel Soc. **7**, 7 (1986)
16. T.K. Barat, IIM Metal News **13**(2), 17 (2010)
17. R.H. Tupkary, in *Introduction to Modern Steelmaking* (Khanna Publishers, Delhi, 1991)

Part V

Thermodynamics and Physical Chemistry of Steelmaking

Thermodynamics 20

Success and efficiency of a process is depending on the thermodynamic calculation of the reactions. How much oxygen requires to oxidize impurities present in hot metal should be known before the actual process. Thermodynamics of refining of molten bath, thermodynamics of de-oxidation of liquid steel and thermodynamics of chromium reactions for stainless steel production are discussed. How much de-oxidizer is required for the de-oxidation of the bath and thermodynamics of vacuum degassing are also described.

20.1 Physical Chemistry of Steelmaking

Steelmaking is an oxidation process in which the impurities in the hot metal are oxidized by lancing oxygen into the bath. The products of oxidation like oxides, silicates, phosphates and sulphides are taken by the slag, and the carbon is oxidized to carbon monoxide and goes out as gas. The oxidation reactions occurring are common to all steelmaking processes, regardless of the type of vessel in which it is carried out, e.g. LD/BOF, OBM and EAF.

From the point of view of law of mass action, the required conditions can be achieved by increasing the activities of the reactants, i.e. impurity, and decreasing the activities of the products. For a given composition of iron melt, the activity of the impurity is fixed and hence cannot be increased.

The oxidizing potential of an oxidizing agent can be increased by using atmospheric air (a_O = 0.21) in place of iron oxide ($a_O = 10^{-6}$–10^{-8} in slag phase) and pure oxygen (a_O = 1) instead of air. But once the nature of the oxidizing agent is chosen, it cannot be further increased. The activity of the product can, however, be decreased by combining it with oxide of opposite chemical character, i.e. an acid oxide product is mixed with basic oxide and vice versa.

The silicon and phosphorous form acidic oxides, and hence, a basic flux is needed to form a suitable slag for their effective removal. If the higher the proportion of basic oxide is available, then the lesser will be the chance of backward reaction.

For manganese elimination, since manganese oxide is basic, an acidic flux will be required. During refining, i.e. steelmaking due to mass effect, iron itself gets first oxidized to some extent as FeO which is basic in nature. It is possible to adjust the contents of Si and Mn in hot metal such that the amounts of (FeO and MnO) formed during refining would be able to form a slag essentially of the type FeO–MnO–SiO_2 and fixed-up silicon in it. In FeO–MnO–SiO_2 slag, FeO and MnO have strong bonding with SiO_2; so FeO is not freely available, i.e. oxygen potential of slag becomes low. In such a slag,

S. K. Dutta and Y. B. Chokshi, *Basic Concepts of Iron and Steel Making*,
https://doi.org/10.1007/978-981-15-2437-0_20

P_2O_5 is not stable, because FeO and MnO, basic oxides, together are not strong enough to hold P_2O_5 in slag. To oxidize P in Fe, strong external basic oxides like CaO and/or MgO are needed in enough proportion to form strong basic slag to hold P_2O_5 without any danger of its reversion. Phosphorous is best eliminated by a slag of CaO–FeO–P_2O_5. This basic slag is also capable of removing sulphur from iron melt.

Basically, the steelmaking processes can be divided into two broad categories:

(a) Acid steelmaking process: when Si is the main impurity to be eliminated from hot metal, P and S need not be eliminated at all.
(b) Basic steelmaking process: when P and S (to some extent) are the main impurities to be eliminated from hot metal along with some Si.

Elimination of Mn will take place under both the categories. In the finished steel, P and S must be below 0.04% each. If P is above this limit, steel becomes *cold shortness*[1] and if S is more than that, it becomes *hot shortness*.[2]

20.1.1 Oxidizing Power of Slag

Slag is a part of steelmaking process, since most of the reactions are slag-metal reactions that approach thermodynamic equilibrium. It is then possible to oxidize the impurities of liquid metal through controlling the slag composition. Similarly, it may be possible to control de-phosphorization and de-sulphurization of the metal by controlling the slag. The composition of slag can be controlled by addition of basic or acidic oxides.

If molten iron is in equilibrium with a slag containing FeO:

$$(\mathrm{FeO}) = [\mathrm{Fe}] + [\mathrm{O}] \tag{20.1}$$

Therefore, there is always some dissolved oxygen in the metal phase. This oxygen may be used to oxidize the impurities in iron during refining.

The equilibrium constant of reaction (20.1):

$$k_{20.1} = \frac{\{[a_{\mathrm{Fe}}] \cdot [a_{\mathrm{O}}]\}}{(a_{\mathrm{FeO}})} = \frac{[h_{\mathrm{O}}]}{(a_{\mathrm{FeO}})} = \frac{[\mathrm{wt\%\,O}]}{(a_{\mathrm{FeO}})} \tag{20.2}$$

Since iron is pure, $a_{\mathrm{Fe}} = 1$.

The oxygen content of iron can be reduced by lowering the activity of FeO in slag phase, and that can be increased by increasing the activity of FeO in slag at a constant temperature. Therefore, the activity of FeO in slag can be taken as a measure of oxidizing power of slag in iron and steelmaking.

[1]Cold shortness: Large amount of phosphorus (more than 0.12% P) reduces the ductility, thereby increasing the tendency of the steel to crack when cold worked. This brittle condition at temperatures below the recrystallization temperature is called cold shortness.

[2]Hot shortness: Sulphur in steel is considered injurious except when added to enhance machinability. Sulphur readily combines with iron to form a low-melting iron sulphide. Sulphur causes hot shortness in steel unless sufficient manganese is added. Sulphur has a greater affinity for manganese than iron and forms manganese sulphide which has a melting point above the hot rolling temperature of steel, which prevents hot shortness. The term is used for the character of steel, which becomes brittle at hot-working temperatures above 0.6 Tm (recrystallization temperature, where strain hardening is removed; Tm is melting point of steel, K). Hot shortness hinders hot-working operations.

20.1.2 Sulphide Capacity of Slag

Sulphide or sulphur capacity of a slag (C_S) indicates the capacity to hold the sulphur in slag phase. If a simple equilibrium involving sulphur and oxygen in slag and gas phases,

$$1/2\{S_2\} + (O^{2-}) = 1/2\{O_2\} + (S^{2-}) \tag{20.3}$$

The equilibrium constant can be written as:

$$k_{20.3} = \frac{\left\{(a_{S^{2-}}) \cdot p_{O_2}^{1/2}\right\}}{\left\{(a_{O^{2-}}) \cdot p_{S_2}^{1/2}\right\}} \tag{20.4}$$

If Henrian behaviour is assumed for S^{2-} ions in slag phase, then Eq. (20.4) become:

$$k_{20.3} = \frac{\left\{(\text{wt\% S}) \cdot p_{O_2}^{1/2}\right\}}{\left\{(a_{O^{2-}}) \cdot p_{S_2}^{1/2}\right\}} \tag{20.5}$$

Therefore,

$$\frac{\left\{(\text{wt\% S}) \cdot p_{O_2}^{1/2}\right\}}{\left\{p_{S_2}^{1/2}\right\}} = k_{20.3} \cdot (a_{O^{2-}}) = C_S \tag{20.6}$$

Left-hand side of Eq. (20.6) is defined as sulphur capacity (C_S) of the slag, which depends solely on the composition of slag at a given temperature, and therefore, that is a property of the slag.

$$\text{i.e.} \quad C_S = \frac{\left\{(\text{wt\% S}) \cdot p_{O_2}^{1/2}\right\}}{\left\{p_{S_2}^{1/2}\right\}} \tag{20.7}$$

Hence, the sulphur capacity (C_S) is thus a measure of the ability of slag to absorb sulphur from gases of given oxygen and sulphur partial pressures. The sulphur capacity is used as a measure of de-sulphurizing power.

20.2 Fundamental Thermodynamic Relations

20.2.1 Carbon in Iron–Carbon Alloys

The presence of carbon in iron lowers down the melting temperature of iron from 1539 °C for pure iron to the eutectic temperature of 1157 °C at 4.25 wt% C in iron. The refining of hot metal, saturated with carbon, to low carbon steel with the attendant increase in temperature necessary to maintain fluidity, is the essence of steelmaking.

In iron and steelmaking, iron–carbon alloys contain additional alloying and impurity elements. The concentration of these additional elements usually experienced only a minor effect on the melting relations, but they have a profound influence on the activity of carbon. The effect of the common additional elements had been studied in ternary Fe–C–X solutions, and the activity coefficients, f_C^X, representing the influence of the third elements on the activity of carbon are plotted in Fig. 20.1 as a function of the concentration of the third element. These semi-log plots are linear at low concentration of X, so they can be represented by equations of the form:

$$\log f_C^X = e_C^X[\text{wt\% X}] \tag{20.8}$$

Values of the interaction coefficient of carbon, e_C^X, are given in Table 20.1.

20.2.2 Oxygen in Iron

In comparison with the solubility of carbon in liquid iron, oxygen is relatively small due to high stability of the separating phase, FeO. From the work of Chipman and co-workers, the solubility is known to be given by [1]:

$$\log[\text{wt\% O}]_{sat} = \left(\frac{-6320}{T}\right) + 2.734 \tag{20.9}$$

which indicates saturation by 0.23 wt% O at 1600 °C rising to 0.48% at 1800 °C.

In the unsaturated iron–oxygen solutions, the oxygen activity rises in a linear manner up to the saturation limit; although for refined calculations, a small deviation from ideality may be corrected for dilute solutions using:

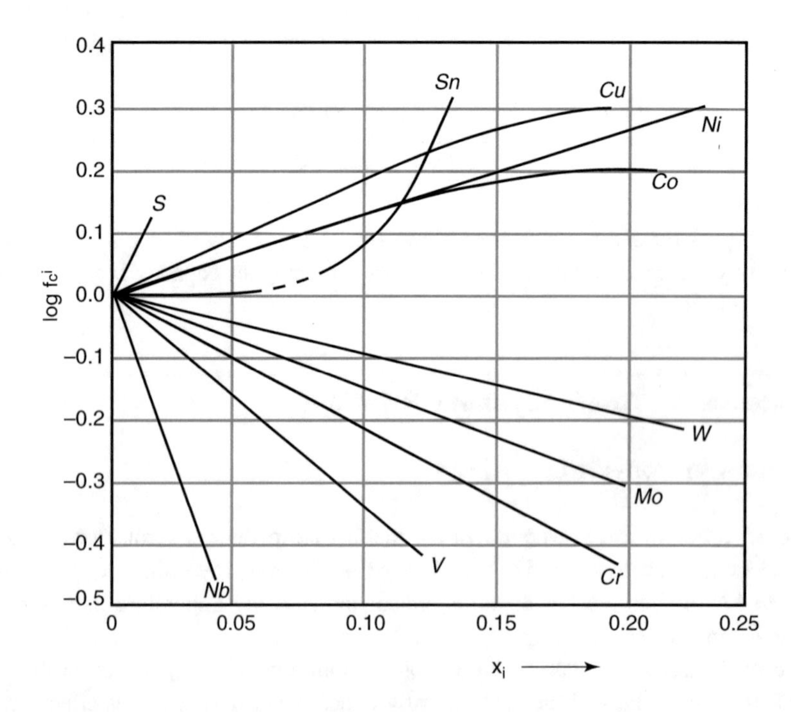

Fig. 20.1 Effect of concentration of a third element on the Henrian activity coefficient of carbon in liquid iron at 1560 °C [1]

Table 20.1 Values of the interaction coefficient, e_C^X, in liquid Fe–C–X alloys at 1560 °C [1]

Element, X	S	Cu	Ni	Co	Sn	W	Mo	Cr	V	Nb
e_C^X	0.09	0.016	0.012	0.012	0	−0.003	−0.009	−0.024	−0.038	−0.06

$$\log f_O^0 = -0.20[\text{wt\% O}] \tag{20.10}$$

The activity of oxygen is influenced by the presence of third alloying elements, the effect of which is shown by the plots of the activity coefficients, f_O^X, in Fig. 20.2. These plots are linear at low concentrations of the third element and follow equations of the type:

$$\log f_O^X = e_O^X[\text{wt\% X}] \tag{20.11}$$

Table 20.2 gives the known values of the interaction coefficient of oxygen, e_O^X.

The free energy change for gaseous oxygen is dissolved to liquid iron:

$$1/2\{O_2\} = [O]_{1\,\text{wt\%}} \tag{20.12}$$

$$\Delta G_{20.12}^0 = -117,300 - 2.89T\,\text{J} \tag{20.13}$$

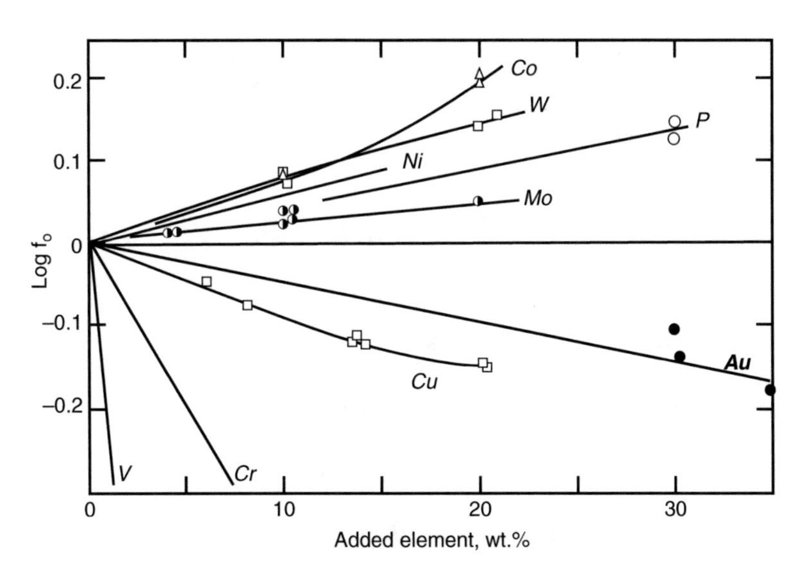

Fig. 20.2 Effect of concentration of a third element on the Henrian activity coefficient of oxygen in liquid iron at 1560 °C [1]

Table 20.2 Values of the interaction coefficient, e_O^X, in liquid Fe–O–X alloys at 1560 °C [1]

Element, X	S	P	W	Co	Ni	Pt	Mo	Mn	Al
e_O^X	1.0	0.07	0.0085	0.007	0.006	0.0045	0.0035	0	12
Element, X	Au	Cu	Si	Cr	V	C			
e_O^X	−0.005	−0.0095	−0.02	−0.041	−0.27	−0.41			

20.3 Thermodynamics of Refining

For the general refining reaction:

$$[\mathrm{X}] + [\mathrm{O}] = (\mathrm{XO}) \tag{20.14}$$

where [X] stands for any impurity element dissolved in molten iron, and [O] is dissolved oxygen in molten iron.

Therefore, equilibrium constant for reaction (20.14):

$$k_{20.14} = \frac{a_{(\mathrm{XO})}}{\{[a_\mathrm{X}] \cdot [a_\mathrm{O}]\}} = \frac{1}{\{[h_\mathrm{X}] \cdot [h_\mathrm{O}]\}} \tag{20.15}$$

where $a_{(\mathrm{XO})} = 1$, for pure compound, h_i = Henrian activity of species i.

To get the value of $k_{20.14}$ needs the value of either free energy change (i.e. ΔG_f^0) of the reaction or the activity coefficient value of [X], [O].

$$\text{Since} \quad h_i = f_i[\mathrm{wt\%}\, i] \tag{20.16}$$

$$\text{Again} \quad \log f_i = \sum e_i^j \cdot \mathrm{wt\%}\, j \quad \text{(for multi-components melt)} \tag{20.17}$$

where e_i^j is known as the interaction coefficient expressing the effect of element, j on element i.

For dilute melt,

$$f_i = 1, \text{therefore}, h_i = [\mathrm{wt\%}\, i] \tag{20.18}$$

Equation (20.15) becomes:

$$k_{20.14} = \frac{1}{\{[\mathrm{wt\%\,X}] \cdot [\mathrm{wt\%\,O}]\}} \tag{20.19}$$

20.3.1 Carbon–Oxygen Equilibrium Reaction

Carbon–oxygen reaction plays a dominant role in steelmaking:

$$[\mathrm{C}] + [\mathrm{O}] = \{\mathrm{CO}\} \tag{20.20}$$

Therefore, equilibrium constant for reaction (20.20):

$$k_{20.20} = \frac{p_\mathrm{CO}}{\{[h_\mathrm{C}] \cdot [h_\mathrm{O}]\}} \tag{20.21}$$

where $[h_\mathrm{C}]$ and $[h_\mathrm{O}]$ are the Henrian activities of carbon and oxygen in liquid iron.

Since $h_i = f_i$ wt% i.

$$\text{Therefore,} \quad [f_\mathrm{C} \cdot \mathrm{wt\%\,C}][f_\mathrm{O} \cdot \mathrm{wt\%\,O}] = \frac{p_\mathrm{CO}}{k_{20.20}} \tag{20.22}$$

$$\text{Or} \quad [\text{wt\% C}][\text{wt\% O}]f_C \cdot f_O = \frac{p_{CO}}{k_{20.20}} \tag{20.23}$$

Since values of f_C and f_O are very small, value of products will be very very small, so that value can be neglected.

$$\text{Therefore,} \quad [\text{wt\% O}] = \frac{p_{CO}}{k_{20.20}[\text{wt\% C}]} \tag{20.24}$$

Equation (20.24) indicates an inverse relationship between [C] and [O] in liquid iron. It may be predicted from Eq. (20.24) that the solubility of [O] in Fe–C melt will increase with an increase in the p_{CO} and vice versa at constant [wt% C]; and at constant pressure (p_{CO} = 1) with decreasing [wt% C], [wt% O] is increased. Figure 20.3 illustrates this phenomenon.

The solubility of carbon in molten iron is raised, and the activity is lowered by manganese, chromium, vanadium and titanium increasingly in that order. These all elements form carbides which are more stable than cementite (Fe_3C), and the free energy of formation of their more stable carbides increases in the same order. Conversely, the solubility of carbon is lowered, and the activity is increased by those elements which form stronger bonds with iron than with carbon.

The difference (Δ%C) between the solubility of carbon in the ternary system Fe–C–X (where X is a solute at low concentration) at a fixed concentration of X and the solubility in a pure iron–carbon alloy is independent of the temperature, and at low concentration of X, the ratio (Δ%C/%X) is also independent of the concentration of X. The effect of alloying element on the solubility of carbon in molten iron is shown in Fig. 20.4.

Activity of carbon in liquid Fe–C melt shows positive and negative deviations from ideality. The presence of carbide forming elements like Nb, V, Cr, W, Mo, etc., decreases the activity of carbon, whereas the presence of non-carbide formers like Cu, Ni, Co, Si, etc., increases the activity coefficient of C in Fe–C melt.

$$\begin{aligned} \log f_C = {} & e_C^{Co}\,\text{wt\% Co} + e_C^{Ni}\,\text{wt\% Ni} + e_C^{Cu}\,\text{wt\% Cu} - e_C^{Nb}\,\text{wt\% Nb} \\ & - e_C^{Cr}\,\text{wt\% Cr} - e_C^{W}\,\text{wt\% W} \end{aligned} \tag{20.25}$$

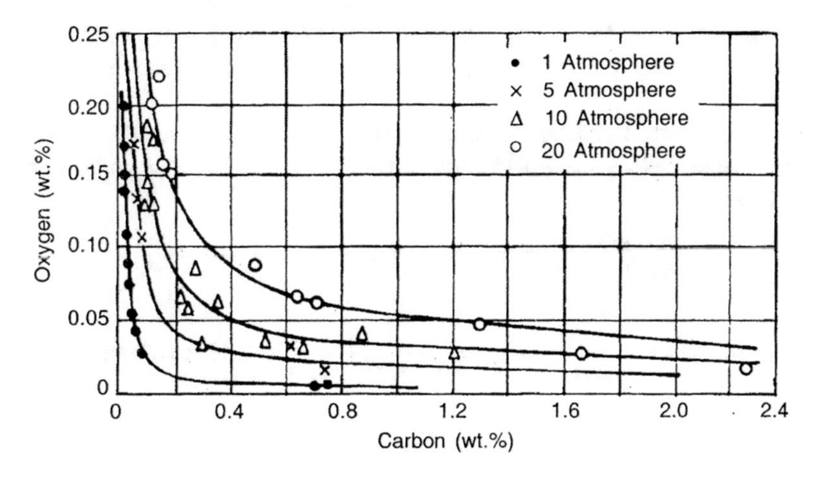

Fig. 20.3 Relation between carbon and oxygen contents in molten iron in equilibrium with carbon monoxide at various pressures [2]

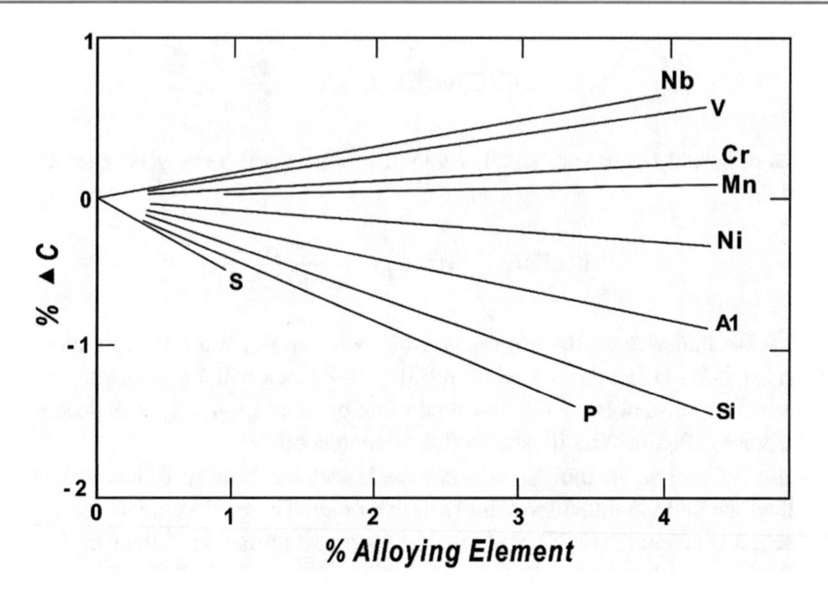

Fig. 20.4 Effect of alloying elements on the solubility of carbon in liquid iron [3]

The product of [wt% C] and [wt% O] is not constant; hence, the activity coefficient must be considered in the calculation of $k_{20.20}$ (Eq. 20.21) which is temperature dependence.

The $k_{20.20}$ correlates with temperature as follows:

$$\log k_{20.20} = \left(\frac{1168}{T}\right) + 2.07 \tag{20.26}$$

At 1600 °C, if wt% C is 0.1%, p_{CO} is 1 atm.

From Eq. (20.26): $\log k_{20.20} = \left(\frac{1168}{1873}\right) + 2.07 = 2.69$

Therefore, $k_{20.20}$ = 489.78.

Now considering dilute solution (i.e. f_i = 1).

Therefore, from Eq. (20.24) became: $[\text{wt.\% O}] = \frac{p_{CO}}{k_{20.20} \cdot [\text{wt\% C}]} = \frac{1}{\{489.78 \times 0.1\}} = 0.02\,\text{wt\%}$.

Hence, it is possible to produce steels with less than 0.1% C using oxygen in metal. The product CO gas passed off into the atmosphere, and reaction (20.20) always tends to go in the forward direction.

At 0.1% C, equilibrium dissolved [O] will be 0.02 wt% at 1 atm pressure. Oxygen content in liquid steel could be decreased much below the above level by treating that under vacuum or under inert gas atmosphere.

At 0.1% C, and $p_{CO} = 0.1\,\text{torr} = \left(\frac{1}{7600}\right)$ atm

From Eq. (20.24) became:

$$[\text{wt.\%O}] = \frac{p_{CO}}{k_{20.20} \cdot [\text{wt\%C}]} = \frac{\left(\frac{1}{7600}\right)}{(489.78 \times 0.1)} = 2.69 \times 10^{-6}\text{wt\%}.$$

20.3.1.1 Mechanism of Carbon–Oxygen Reaction in Pneumatic Processes

In the pneumatic processes, oxygen is readily available at the gas–metal interface. The rate of reaction is affected by the diffusion of carbon to the interface. In most cases, oxygen bearing bubbles pass through the molten metal and stirring of the bath caused in these processes help to faster carbon transfer. The rate of carbon elimination is very high (10 wt% C/h), so the pneumatic processes are the most rapid method of making steel.

If the gas–metal interface is considered on an atomic scale, it is evident that both iron and carbon atoms will be exposed to the oxidizing gas, and the oxidation of both species will occur. This results in the formation of CO, but as the iron atoms predominate over carbon atoms[3], considerable oxidation of iron occurs effectively resulting in the solution of oxygen into the molten iron. In this way, the oxygen content of the iron is raised above that for equilibrium with carbon. Due to the large numbers of bubbles involved in pneumatic processes, there are large numbers of gas–metal contact area at which this reaction may occur. Homogeneous nucleations take place for pneumatic processes.

The degree of oxygen super-saturation is determined by the rate at which carbon and oxygen atoms are transported to a gas–metal interface and react to form CO which then desorbs into the gas phase. Due to the large number of bubbles involved, as, for example, in a OBM converter, there is an enormous metal–gas contact area at which this reaction may occur, and low excess oxygen is found in the metal bath.

20.3.1.2 Mechanism of Carbon–Oxygen Reaction in Open-Hearth Process

In open-hearth process, oxygen from the furnace atmosphere must diffuse across the slag and the metal layers to reach the refractory–metal interface, because heterogeneous reaction requires nucleating sites for nucleation of gas–metal reaction.

During refining in the open-hearth furnace, 20% of the oxygen requirement is derived from the atmosphere, the remaining oxygen supplied by oxide/ore addition.

The chemical absorption of oxygen by the slag to form oxygen ions:

$$1/2\{O_2\} + 2e^- = (O^{2-}) \tag{20.27}$$

Some compensating reaction is required to maintain the overall slag electrically neutral. This is usually considered to be the oxidation of ferrous to ferric iron at the gas–slag interface:

$$2(Fe^{2+}) = 2(Fe^{3+}) + 2e^- \tag{20.28}$$

Hence, the overall reaction at gas–slag interface is as (by adding (20.27) and (20.28)):

$$2(Fe^{2+}) + 1/2\{O_2\} = 2(Fe^{3+}) + (O^{2-}) \tag{20.29}$$

This increases in ferric iron concentration at the upper surface of the slag.

Due to thermal diffusion, transfer of ions from gas–slag interface to slag-metal interface and a reverse reaction takes place:

$$2(Fe^{3+}) + (O^{2-}) = 2(Fe^{2+}) + [O] \tag{20.30}$$

[3]Carbon atoms are strongly surface active and will concentrate at a free surface if equilibrium is attained, but this surface concentration is unlikely to be maintained by the diffusional supply of carbon atoms when these are continuously being removed from the surface by oxidation.

The (Fe^{2+}) ion again goes back to the gas–slag interface and completes the cycle which repeats itself. This results in the transfer of oxygen to the metal phase. Oxygen dissolved in metal diffuses from slag-metal interface to metal–refractory interface and combines with carbon:

$$[C] + [O] = \{CO\} \tag{20.20}$$

The steps in the overall transfer of oxygen from gas phase to the metal phase are represented as follows:

(i) Transfer of oxygen from the furnace gas phase to gas–slag interface,
(ii) Oxygen dissolves in the slag phase,
(iii) Transfer of oxygen from the gas–slag interface to slag-metal interface by diffusion,
(iv) Dissolution of oxygen in the metal phase,
(v) Transfer of oxygen from the slag-metal interface to metal–refractory interface by diffusion,
(vi) Transfer of iron from the slag-metal interface to slag–gas interface,
(vii) Transfer of carbon from metal phase to metal–refractory interface,
(viii) Chemical reaction at metal–refractory interface,
(ix) Transfer of carbon monoxide bubble from metal–refractory interface to metal-slag interface,
(x) Transfer of carbon monoxide bubble from metal-slag interface to slag–gas interface,
(xi) Transfer of carbon monoxide bubble from slag–gas interface to gas phase.

The process of diffusion of oxygen is a very slow process, to metal–refractory interface, that is why open-hearth process is a slow process. Diffusion of oxygen can be faster by eliminating the diffusion path of oxygen across the slag layer by adding lumps of iron ore (Fe_2O_3) which is heavier than slag but lighter than liquid metal, so it settles at the slag-metal interface. It supplies oxygen as:

$$Fe_2O_3(s) = 2(FeO) + [O], \Delta H^\circ_{298} = 292.46\,\text{kJ/mol of } Fe_2O_3 \tag{20.31}$$

Above reaction is an endothermic reaction and known as *oreing of the slag*.

In absent of oreing, rate of de-carburization is only 0.12–0.18% C/h; by oreing rate of de-carburization is 0.6% C/h; and by oxygen lancing, rate will be 3% C/h.

In open-hearth furnace, the oxygen for oxidation of carbon is derived from the slag. Under these conditions, the rate of carbon elimination is an order of magnitude less than the pneumatic processes, i.e. <1 wt% C/h. It is an important factor governing the production rate of a furnace, and this is the reason for open-hearth process which is the slowest process without oxygen injection.

Considering the physics of the formation of a CO bubble, the pressure within a bubble of radius, r in a metal of surface tension, σ is given by [1]:

$$p_{CO} = p_O + \left(\frac{2\sigma}{r}\right) \tag{20.32}$$

where p_O is the static pressure due to atmospheric, slag and metal heads compressing the bubble, and $(2\sigma/r)$ is the additional pressure on the bubble.

For homogeneous nucleation of a CO bubble within the metal phase, Eq. (20.32) indicates that a bubble nucleus of molecular dimensions (r = 6A^0) would have an internal pressure of 5×10^4 atm. This would become very large. For reaction (20.20):

$$[\mathrm{C}] + [\mathrm{O}] = \{\mathrm{CO}\} \quad (20.20)$$

and Eq. (20.21):

$$k_{20.20} = \frac{p_{\mathrm{CO}}}{\{[h_{\mathrm{C}}] \cdot [h_{\mathrm{O}}]\}} \quad (20.21)$$

$$\{[h_{\mathrm{C}}][h_{\mathrm{O}}]\} = \frac{p_{\mathrm{CO}}}{k_{20.20}} \quad (20.33)$$

Since p_{CO} is small value in open-hearth furnace, super-saturation of this nature cannot be attained in practice, and homogeneous nucleation must be disregarded.

Hence, heterogeneous nucleation occurred either on the furnace lining or at the slag-metal interface. Small cavities in the lining are not full of molten metal due to lack of penetration and full of entrapped air which serves as a nucleus for CO bubble. While a suitable nucleus is absorbing carbon monoxide, the bubble grows, and its surface becomes more and more nearly hemispherical. The internal pressure given by Eq. (20.32) consequently rises until the bubble radius equals the crevice radius. Any further growth would then result in a decrease in internal pressure due to the increased bubble radius; the bubble is then unstable and detaches itself from the crevice (Fig. 20.5).

20.3.2 Silicon Reaction

Silica is a very stable oxide, and hence, once silicon is oxidized to silica, the chance of its reversion does not arise in refining of hot metal. Because of the large difference in the free energies of formation of SiO_2 and FeO, as well as the low activity of SiO_2 in basic slag, Si in Fe is readily oxidized during steelmaking.

$$[\mathrm{Si}] + 2[\mathrm{O}] = (\mathrm{SiO_2}) \quad (20.34)$$

Therefore, equilibrium constant for reaction (20.34):

$$k_{20.34} = \frac{a_{(\mathrm{SiO_2})}}{\left\{[h_{\mathrm{Si}}] \cdot [h_{\mathrm{O}}]^2\right\}} \quad (20.35)$$

If Henrian behaviour of silicon and oxygen in iron (for dilute solution) are considered, the product of silica (as pure) for silica saturated slag [then $a_{(\mathrm{SiO2})} = 1$] and considering dilute solution (i.e. $f_i = 1$). (Since $h_i = f_i \times \mathrm{wt\%}\ i$).

By rearrangement of Eq. (20.35):

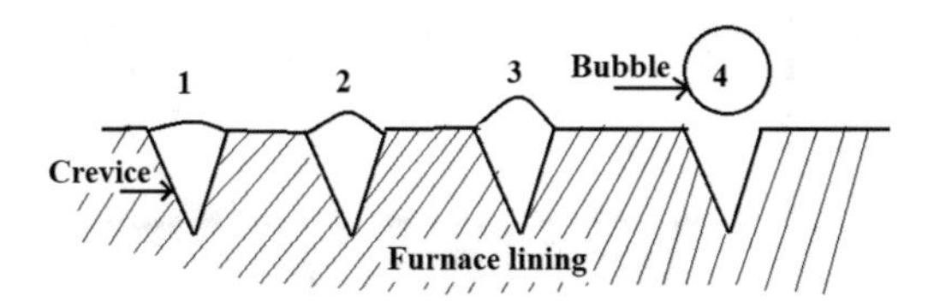

Fig. 20.5 Heterogeneous nucleation at the crevice in the furnace lining: **a** nucleation starts, **b**, **c** growth, **d** bubble detaches from the crevice

$$\left\{[\text{wt\% Si}][\text{wt\% O}]^2\right\} = \frac{1}{k_{20.34}} \tag{20.36}$$

Taking logarithm of Eq. (20.36):

$$\log[\text{wt\% Si}] + 2\{\log[\text{wt\% O}]\} = \log\left(\frac{1}{k_{20.34}}\right) = -\log(k_{20.34}) \tag{20.37}$$

or

$$\log[\text{wt\% Si}] = -[\log(k_{20.34}) + 2\{\log[\text{wt\% O}]\}] \tag{20.38}$$

Experimental plots of log [wt% Si] versus log [wt% O] as shown in Fig. 20.6 also demonstrate a linear relationship. This confirms that substitution of Henrian activities of Si and O by their respective wt% does not involve much error [2]. However, $k_{20.34}$ is constant due to lowering activity of oxygen being compensated by the corresponding increase in the activity of silicon.

$$\log\left(\frac{1}{k_{20.34}}\right) = 11.01 - \left(\frac{29,150}{T}\right) \tag{20.39}$$

Silicon content of melt in a basic process should be as low as possible to decrease lime consumption in maintaining the basicity required for effective phosphorus removal at minimum slag volume. In basic steelmaking, the activity of SiO_2 is so low that its preferential oxidation never poses any problem like that of phosphorus removal. For high silicon in hot metal, external de-siliconization of hot metal should be done.

20.3.3 Manganese Reaction

Next to silica (SiO_2), MnO is most stable oxide product during refining of hot metal. Iron and manganese form ideal solutions, and even in the presence of other solutes, ideal behaviour is assumed. MnO has extensive solubility in slag.

Manganese exchange between metal and slag may be represented by the reaction:

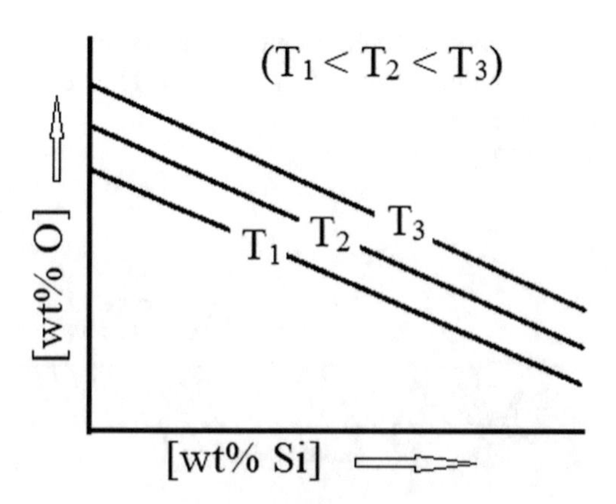

Fig. 20.6 Influence of silicon on the oxygen content of iron in equilibrium with solid silica at the temperature varies from 1550° to 1650 °C

$$[\mathrm{Mn}] + (\mathrm{FeO}) = (\mathrm{MnO}) + [\mathrm{Fe}] \tag{20.40}$$

Therefore, equilibrium constant for reaction (20.40):

$$k_{20.40} = \frac{\{(a_{\mathrm{MnO}}) \cdot [a_{\mathrm{Fe}}]\}}{\{[a_{\mathrm{Mn}}] \cdot (a_{\mathrm{FeO}})\}} \tag{20.41}$$

Considering dilute solution (i.e. f_i = 1) and Fe is pure, so $a_{[\mathrm{Fe}]} = 1$.
So

$$k_{20.40} = \frac{(\mathrm{wt\%\,MnO})}{\{[\mathrm{wt\%\,Mn}] \cdot (\mathrm{wt\%\,FeO})\}} \tag{20.42}$$

Or

$$\left\{\frac{(\mathrm{wt\%\,MnO})}{[\mathrm{wt\%\,Mn}]}\right\} = k_{20.40}(\mathrm{wt\%FeO}) \tag{20.43}$$

Turkdogan [4] derived the following equation for the temperature dependence of the equilibrium constant, $k_{20.40}$:

$$k_{20.40} = \left(\frac{7452}{T}\right) - 3.478 \tag{20.44}$$

The equilibrium constant, $k_{20.40}$, depends on temperature and slag composition. The values of $k_{20.40}$ are derived from Eq. (20.44) and are plotted in Fig. 20.7 against the slag basicity (B). In LD/BOF, OBM/Q-BOP and EAF steelmaking, the slag basicities are in the range 2.5–4.0 and the melt temperature in the vessel at the time of furnace tapping in most practices is between 1600 and 1650 °C, for which the equilibrium constant, $k_{20.40}$, is about 1.8 ± 0.2.

The activity of (MnO) decreases with increasing basicity, and hence, the ratio of {(wt% MnO)/(wt % Mn)} decreases with increase of basicity. At constant basicity, ratio increases with oxidizing power of slag; hence, more Mn recovery takes place. That is, for a given basicity in a basic slag, the equilibrium distribution of Mn between slag and metal is directly proportional to the concentration of the iron oxide.

In low alloy steel, the activity of Mn is proportional to its concentration. The equilibrium constant $k_{20.40}$ decreases with increasing basicity of the slag. Turkdogan gives the equilibrium relation $k_{20.40}$ in basic slag at steelmaking temperature:

$$K_{20.40} = \frac{6}{\mathrm{B}} \tag{20.45}$$

where

$$\mathrm{B} = \left(\frac{\{\%\mathrm{CaO} + 1.4(\%\mathrm{MgO})\}}{\{\%\mathrm{SiO_2} + 0.84(\%\mathrm{P_2O_5})\}}\right) \tag{20.46}$$

This agrees with the operating experience that a higher Mn residual in the metal is obtained with higher temperature. (MnO) and (FeO) are miscible in each other. When Fe–Mn is added to the steel bath, the product is usually a slag (FeO · MnO) which may be solid or liquid at 1600 °C, depending

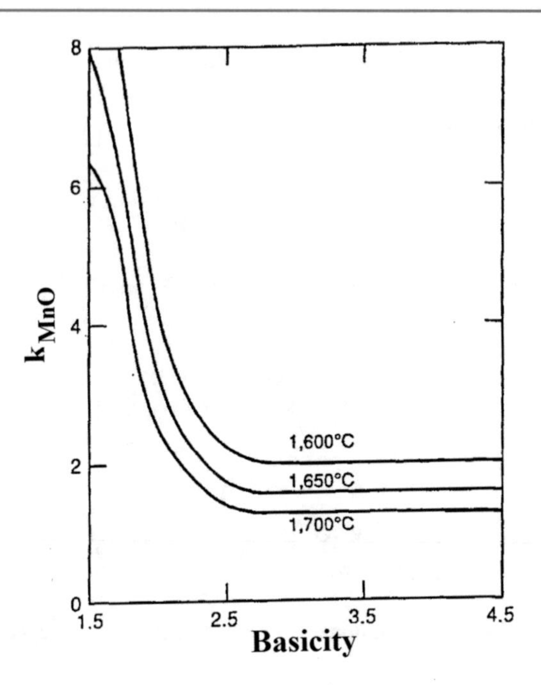

Fig. 20.7 Equilibrium constant, k_{MnO}, related to slag basicity [4]

on the relative proportions of FeO and MnO (as shown in Fig. 20.8). If the (FeO) content is large, it is expected that the slag phase will be liquid and float-up. But more often slag is solid and remains in steel as non-metallic inclusions.

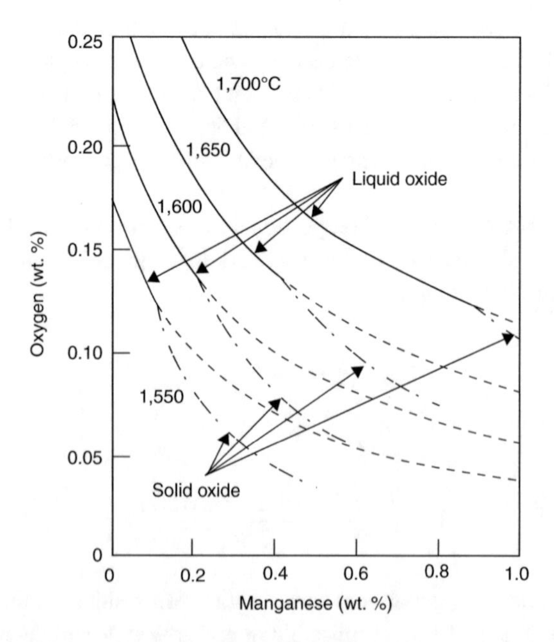

Fig. 20.8 Limiting oxygen contents of iron–manganese alloys in equilibrium with the mixed oxides of iron and manganese [2]

20.3.3.1 Ionic Theory of Manganese Transfer

Since the elements are present in the slag phase only as ions and as neutral species in the metal phase, the reaction can take place only at the slag-metal interface.

The overall reaction:

$$[\text{Mn}] + \left(\text{Fe}^{2+}\right) = [\text{Fe}] + \left(\text{Mn}^{2+}\right) \tag{20.47}$$

The overall reaction consists of several steps, known as kinetic steps. For reaction (20.47), the kinetic steps (as shown in Fig. 20.9) are as follows:

1. Transfer of Mn from the metal phase to the slag-metal interface,
2. Transfer of Fe^{2+} ion from the slag phase to the interface,
3. Chemical reaction (20.47) at the interface,
4. Transfer of Fe from interface to the metal phase,
5. Transfer of Mn^{2+} ion from interface to the slag phase.

Steps (1), (2), (4) and (5) are mass transfer steps. Step (3) is a chemical reaction step, and that is governed by the laws of chemical kinetics. However, the reaction is occurring at the interface and not homogeneously in the bulk; that is why this reaction is heterogeneous reaction.

There are two ways in which the interfacial reaction, i.e. step (3), can take place:

(a) Fe^{2+} ion in slag phase and Mn atom in metal phase can collide with each other at the phase boundary and thus react;

(b) The reaction can proceed via consecutive anodic and cathodic processes with the liquid metal phase acting as the electronic conductor.

$$\text{Anodic reaction}: [\text{Mn}] \rightarrow \left(\text{Mn}^{2+}\right) + 2\text{e}^- \tag{20.48}$$

$$\text{Cathodic reaction}: \left(\text{Fe}^{2+}\right) + 2\text{e}^- \rightarrow [\text{Fe}] \tag{20.49}$$

This is known as the electrochemical mechanism of slag-metal interfacial reaction. This was proposed by Wagner based on such a mechanism operating in corrosion. This is the likely mechanism since it does not require collision of Fe^{2+} ion and Mn atom which is expected to be very slow due to low probability of its occurrence. On the other hand, the anodic and cathodic sites are entirely random in the latter mechanism and independent.

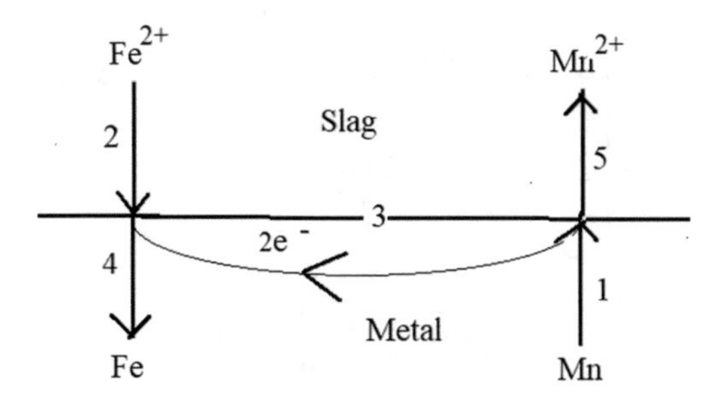

Fig. 20.9 Slag-metal reaction

20.3.4 Phosphorous Reaction

Although phosphorus has a very low boiling point (280 °C), a considerable amount of it remains dissolved in liquid iron because of its strong interaction with iron. Unless silicon and carbon are not removed from the melt, phosphorus cannot be removed. This is a basic problem in steelmaking, because the free energy formation of P_2O_5 is greater than CO and SiO_2. If by some means lowering down the free energy formation of P_2O_5, then it may be possible to remove phosphorus simultaneously with carbon. But this requires decreasing the activity of P_2O_5 in slag which can be achieved by dissolution of P_2O_5 by dissolved CaO in slag.

Lines of iron oxides and phosphorous pentra-oxide in the Ellingham diagram (Fig. 20.10) are so close to each other that entire phosphorous in the charge materials gets reduced along with iron in blast furnace. As all the phosphorus in the iron ore enters the hot metal, this phosphorus must be removed by oxidation in the subsequent conversion to steel. There is a very real risk of oxidizing all the iron in the process. This might have posed a serious problem in the oxidation of phosphorous. This problem is overcome by separating the two lines with decreasing the activity of P_2O_5 in the steelmaking slag using a strong and high basic slag. High basicity slag (2–4) decreases the activity of P_2O_5 (10^{-15}–10^{-20}) in slag by dissolving the P_2O_5 in slag phase. Line of phosphorous pentra-oxide in the Ellingham diagram is shifted from top to far below the iron oxides lines. Although steelmaking slag contains up to 25% P_2O_5, but still activity of P_2O_5 in slag remains extremely low at high basic slag due to stronger bonding of acidic P_2O_5 with basic CaO. The activity of P_2O_5 in slag means freely available of P_2O_5 in slag phase. From Ellingham diagram, it was found that at steelmaking temperature (1600 °C), CO line is much lower than P_2O_5 line; hence, phosphorous removal cannot occur until carbon is oxidized. For high basic slag, activity of P_2O_5 is greatly reduced and line of P_2O_5 is clockwise rotates, i.e. free energy changes involved in the formation of CO gas and P_2O_5 become nearly equal at 1600 °C under high basic slag. Hence, simultaneous removal of carbon and phosphorous is possible in oxygen steelmaking processes. During steelmaking, these two lines (FeO and P_2O_5) are widely separated by decreasing the activity of P_2O_5 using a strong basic slag. Now P_2O_5 becomes more stable than FeO at steelmaking temperature. For effective removal of phosphorous,

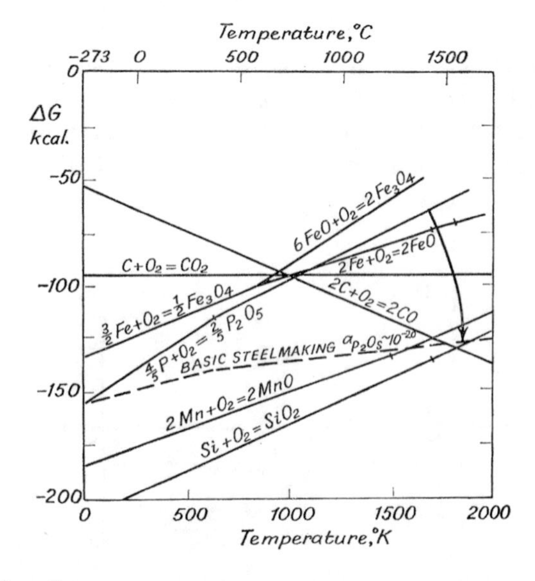

Fig. 20.10 A part of Ellingham diagram

high basic slag must be required. If the basicity of slag falls, then holding capacity of CaO decreases and phosphorous may soon go back to the metal phase.

For elimination of phosphorous requires the CaO to be dissolved in slag, when CaO is dissolved in the slag, then P_2O_5 reacts with it and makes stronger bond with CaO; then, the activity of P_2O_5 is lowered down. If the slag does not contain dissolved CaO, then evolved P_2O_5 (by P–O reaction) is not entrapped by CaO; hence, P_2O_5 is decomposed into phosphorous and oxygen ($P_2O_5 \rightarrow P + O_2$), since the free energy of P_2O_5 is high. This newly formed phosphorous again absorbed by the metal. This is known as *reversion of phosphorous*.

Phosphorous reacted between metal and slag may be represented by:

$$2[P] + 5[O] = (P_2O_5) \tag{20.50}$$

Therefore,

$$k_{20.50} = \frac{(a_{P_2O_5})}{\left\{[a_P]^2 \cdot [a_O]^5\right\}} \tag{20.51}$$

$$\log k_{20.50} = \left(\frac{35880}{T}\right) - 30.31 \tag{20.52}$$

At 1200 °C: $k_{20.50} = 1.12 \times 10^{-6}$, at 1300 °C: $k_{20.50} = 3.16 \times 10^{-8}$ and at 1600 °C: $k_{20.50} = 7.02 \times 10^{-12}$; that means as the temperature increased, the value of equilibrium constant ($k_{20.50}$) decreased, i.e. P transfer from metal to slag phase decreased at high temperature.

Equation (20.51) along with Eq. (20.52) gives the well-known thermodynamic conditions for good de-phosphorization as follows:

- Low temperature,
- High activity of oxygen,
- Low activity of P_2O_5.

Since activity of P_2O_5: $a_{P_2O_5} = \gamma_{P_2O_5} \cdot x_{P_2O_5}$,

where $\gamma_{P_2O_5}$ and $x_{P_2O_5}$ are the activity coefficient and mol fraction of P_2O_5, respectively; it can be reduced by lowering either $\gamma_{P_2O_5}$ and $x_{P_2O_5}$, or both. P_2O_5 is an acidic oxide and so addition of CaO reduces $\gamma_{P_2O_5}$ very significantly; $x_{P_2O_5}$ becomes low.

The rate of P removal from metal bath to slag is given by [5]:

$$-\left(\frac{d[\%P]}{dt}\right) = a_v k_P\left(\%P - \%P_{eq}\right) \tag{20.53}$$

where a_v is slag-metal contact area per unit volume of metal, k_P is the mass transfer coefficient of P in metal, %P is the bath phosphorus content at any time and % P_{eq} is the phosphorus content in equilibrium with slag and is determined by Eq. (20.51). Equation (20.53) suggests that de-phosphorization rate will be higher for large values a_v and k_P. Stirring of bath and slag increases the values of both a_v and k_P. Besides, k_P increases with temperature but that contradicts the requirement of low $\%P_{eq}$ value. Equation (20.53) further shows that the rate of de-phosphorization decreases as bath phosphorus becomes low. As a result, it is always difficult to produce ultra-low P steel.

Thermodynamic analysis given above gives the limiting phosphorus when equilibrium is attained. But in reality, phosphorus partitioning does not attain equilibrium because:

- Slag chemistry changes during the progress of blowing oxygen in LD/BOF, and
- Reaction rate is not sufficiently high for attainment of equilibrium.

Post-stirring brings phosphorus partitioning $\left(\frac{(\%P)}{[\%P]}\right)$ closer to equilibrium and thereby lower phosphorus level could be achieved in most cases.

Phosphorus input in the EAF charge is usually lower than that in LD/BOF due to usage of large amount of scrap and sponge iron. Furthermore, due to presence of gangue in sponge iron, slag volume is also more. So, it is easier to obtain lower phosphorus level in steel. In induction melting furnace (IMF), normally lining is acidic, so phosphorus removal is not possible. But by using basic lining [6], significant amount of phosphorus removal is possible. Mixture of sponge iron, cast iron or/and steel scrap can be successfully melted in the IMF lined with MgO or Al_2O_3-based lining. Phosphorous can be removed in IMF by maintaining required basicity and FeO in the slag.

The values of equilibrium constant, $k_{20.50}$, are increased with decreasing temperature and increasing concentration of CaO in the slag. Instead of the simple reaction expressed in Eq. (20.50), it is possible to assume that the P_2O_5 in the slag is combined with lime to form the compound $4CaO \cdot P_2O_5$.

Thus

$$2[P] + 5[O] + 4(CaO) = (4CaO \cdot P_2O_5) \tag{20.54}$$

$$k_{20.54} = \frac{(a_{4CaO \cdot P_2O_5})}{\left\{[a_P]^2 \cdot [a_O]^5 \cdot (a_{CaO})^4\right\}} \tag{20.55}$$

If the slag is saturated with CaO and with $4CaO \cdot P_2O_5$, the activities of both oxides are unity; so that at equilibrium activity of P in the metal is depended only on activity of O.

$$k_{20.54} = \frac{1}{\left\{[a_P]^2 \cdot [a_O]^5\right\}} \tag{20.56}$$

Hence, phosphorous is oxidized according to the following reactions:

$$2[P] + 5(FeO) = (P_2O_5) + 5[Fe] \tag{20.57}$$

$$(P_2O_5) + 4(CaO) = (4CaO \cdot P_2O_5) \tag{20.58}$$

The factors which influence P transfer from metal to slag are as follows:

(i) High oxygen potential presence in the slag,
(ii) High basicity of slag, i.e. CaO/SiO_2 ratio is more than 2.2, to dissolve (P_2O_5) by (CaO);
(iii) De-phosphorization is more effective at low temperature; increasing the temperature lower the value of phosphorus distribution coefficient.
(iv) Good fluidity of slag.

Combining Eqs. (20.57) and (20.58):

$$2[\mathrm{P}] + 5(\mathrm{FeO}) + 4(\mathrm{CaO}) = (4\mathrm{CaO}\cdot\mathrm{P_2O_5}) + 5[\mathrm{Fe}] \quad (20.59)$$

The equilibrium constant for reaction (20.59):

$$k_{20.59} = \frac{\left\{\left(a_{4\mathrm{CaO}\cdot\mathrm{P_2O_5}}\right)\cdot[Fe]^5\right\}}{\left\{[a_{\mathrm{P}}]^2\cdot(a_{\mathrm{FeO}})^5\cdot(a_{\mathrm{CaO}})^4\right\}} \quad (20.60)$$

Therefore,

$$\left(\frac{(\mathrm{wt\%\,P})}{[\mathrm{wt\%\,P}]}\right) \propto \left\{(\%\mathrm{FeO})^5\cdot(\%\mathrm{CaO})^4\right\} \quad (20.61)$$

Phosphorous distribution coefficient (L_{P}), i.e. $\left(\frac{(\mathrm{wt\%\,P})}{[\mathrm{wt\%\,P}]}\right)$, is influenced by (FeO) and basicity of slag (as shown in Figs. 20.11 and 20.12). Phosphorous distribution coefficient (L_{P}) increases with increasing of (FeO), because of its high oxidizing ability. If (FeO) content increases still further, the distribution line is decreased, and if (FeO) increased, proportion of (CaO) is decreased (according to Eq. 20.61). Hence, the advantage of high oxidizing ability of (FeO) is neutralized by the loss of lime in the slag. Hence, the value of phosphorous distribution coefficient is decreased. Optimum (FeO) for de-phosphorization is around 14–15%.

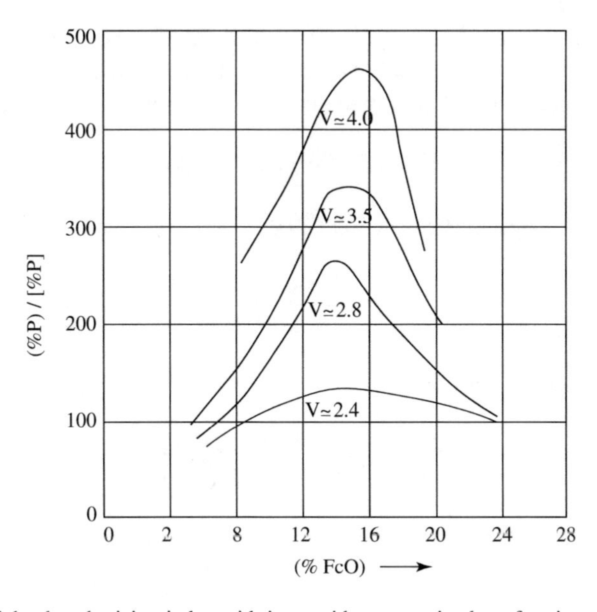

Fig. 20.11 Variation of de-phosphorizing index with iron oxide content in slag of various basicities [7]

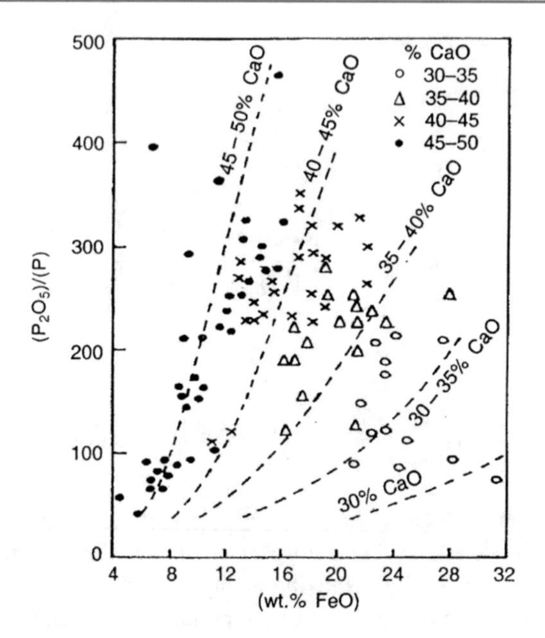

Fig. 20.12 Phosphorous distribution coefficient varies with FeO content in slag at 1600 °C [2]

20.3.5 Sulphur Reaction

20.3.5.1 Molecular Theory

Although sulphur has a very low boiling point (445 °C), like phosphorous (280 °C), a considerable amount of sulphur is found in liquid iron because of its strong interaction with iron and manganese.

Sulphur equilibrium between slag and metal can be expressed by molecular equation such as:

$$[\mathrm{FeS}] + (\mathrm{CaO}) = (\mathrm{CaS}) + (\mathrm{FeO}) \tag{20.62}$$

Therefore, equilibrium constant for reaction (20.62):

$$k_{20.62} = \frac{\{(a_{\mathrm{CaS}}) \cdot (a_{\mathrm{FeO}})\}}{\{[a_{\mathrm{FeS}}] \cdot (a_{\mathrm{CaO}})\}} \tag{20.63}$$

Sulphur distribution coefficient (L_S),

$$\frac{(\%\mathrm{S})}{[\%\mathrm{S}]} = k_{20.62} \cdot \left\{\frac{(\%\mathrm{CaO})}{(\%\mathrm{FeO})}\right\} \tag{20.64}$$

(i) At a fixed basicity of slag,

$$\left\{\frac{(\%\mathrm{S})}{[\%\mathrm{S}]}\right\} \propto \left\{\frac{1}{(\%\mathrm{FeO})}\right\} \tag{20.65}$$

(ii) Similarly, for a fixed (FeO),

$$\left\{\frac{(\%\mathrm{S})}{[\%\mathrm{S}]}\right\} \propto (\%\mathrm{CaO}) \qquad (20.66)$$

For efficient removal of sulphur, high basicity and low oxygen potential are essential. The effect of changing the temperature on sulphur distribution coefficient (L_S) is uncertain.

Condition for de-sulphurization:

1. High basic slag,
2. Low oxygen potential, i.e. low (FeO),
3. Reducing condition,
4. High temperature,
5. Good fluidity of slag.

In basic oxygen process and EAF process, conditions (1), (4) and (5) are easily maintained; but conditions (2) and (3) are not maintained in oxygen processes.

Thermodynamically, lower temperature is bound to improve the de-sulphurization, but high temperature is favoured from kinetic point of view. Since $\left\{\frac{(\%\mathrm{S})}{[\%\mathrm{S}]}\right\}$ is inversely proportional to (FeO) [as per Eq. (20.65)], so at low temperature less (FeO) will be formed. If the concentration of (FeO) increases, then reaction (20.62) goes towards back direction that means sulphur will go back to the metal phase. So (FeO) should be lower down by the following reduction reaction by C or Si (hence reducing condition prevail):

$$(\mathrm{FeO}) + [\mathrm{C}]/[\mathrm{Si}] \rightarrow [\mathrm{Fe}] + \{\mathrm{CO}\}/(\mathrm{SiO_2}) \qquad (20.67)$$

Thermodynamically, lower temperature is bound to improve the sulphur distribution coefficient, but in practice, high temperature is favoured from the kinetic point of view. Hence, de-sulphurization reaction is more kinetically control than thermodynamically.

20.3.5.2 Ionic Theory

Since the slag is an ionic nature, the reaction (20.62) is more suitable for ionic form:

$$[\mathrm{S}] + (\mathrm{O}^{2-}) = (\mathrm{S}^{2-}) + [\mathrm{O}] \qquad (20.68)$$

Therefore, equilibrium constant for reaction (20.68):

$$k_{20.68} = \frac{\{(a_{\mathrm{S}^{2-}}) \cdot [a_\mathrm{O}]\}}{\{[a_\mathrm{S}] \cdot (a_{\mathrm{O}^{2}}\ )\}} \qquad (20.69)$$

Therefore, sulphur distribution coefficient,

$$\left\{\frac{(a_{\mathrm{S}^{2-}})}{[a_\mathrm{S}]}\right\} = k_{20.68} \cdot \left\{\frac{(a_{\mathrm{O}^{2-}})}{[a_\mathrm{O}]}\right\} \qquad (20.70)$$

If the species obey Henrian behaviour, Eq. (20.70) can be rewrite as:

$$\left\{\frac{(\text{wt\% S})}{[\text{wt\% S}]}\right\} = k'_{20.68} \cdot \left\{\frac{(n_{O^{2-}})}{[\text{wt\% O}]}\right\} \tag{20.71}$$

where n_{O2-} is the number of gram ions of oxygen in 100 g of slag, after the oxygen requirements for the formation of SiO_4^{4-}, PO_4^{3-} and AlO_3^{3-} ions have been satisfied.

If plotting $\left\{\frac{(\text{wt\% S})}{[\text{wt\% S}]}\right\}$ versus $\left\{\frac{(n_{O2-})}{[\text{wt\% O}]}\right\}$, the slope of the plot (in Fig. 20.13) is clearly the equilibrium constant, $k'_{20.68}$ and the linear form of the plot passing through origin means value of $k'_{20.68}$ is independent of slag composition.

From Eq. (20.71), it is apparent for any slag-metal equilibrium, and oxygen content in metal is constant; so sulphur distribution coefficient will be increased by increasing (n_{O2-}) value, i.e. by increasing the basicity of slag. If the basicity of slag is constant, i.e. (n_{O2-}) value is also constant; therefore, [wt% O] is assumed to be proportional to the iron oxide content of the slag.

So Eq. (20.71) becomes:

$$\left\{\frac{(wt\% \text{ S})}{[wt\% \text{ S}]}\right\} = \left\{\frac{\text{Constant}}{(\text{mol\% FeO})}\right\} \tag{20.72}$$

Or

$$\left\{\frac{(\text{wt\% S})}{[\text{wt\% S}]}\right\} \propto \left\{\frac{1}{(\text{mol\% FeO})}\right\} \tag{20.73}$$

Hence, low FeO in slag is encouraged de-sulphurization. The effect of slag composition on the activity coefficient of S and iron oxide in the slag is such that the equilibrium constant ($k_{20.68}$) increases with increasing basicity of the slag, i.e. more S transfer from metal phase to slag phase.

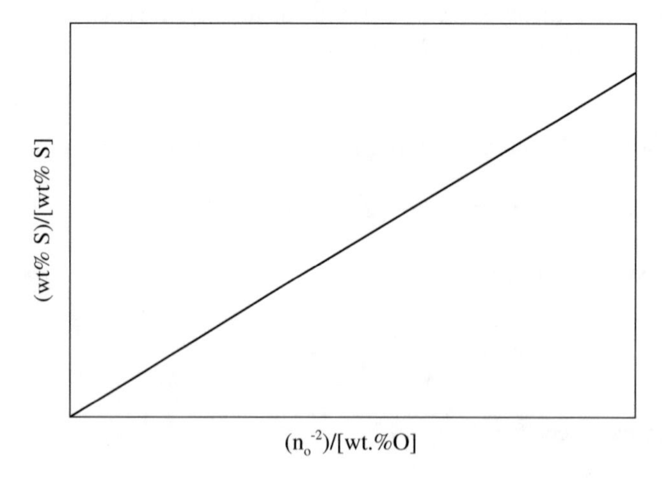

Fig. 20.13 Linear relation between the sulphur and oxygen distribution ratios

$\left\{\frac{(\text{wt\% S})}{[\text{wt\% S}]}\right\}$ is known as de-sulphurizing index or sulphur partitioning coefficient, and its variation with basicity and oxidizing power of slag. For a fixed basicity, the index is inversely proportional to the iron oxide content of the slag or [O] content of the metal, i.e.

$$\left\{\frac{(\text{wt\% S})}{[\text{wt\% S}]}\right\} \propto \left\{\frac{1}{(\text{mol\% FeO})}\right\} \text{or} \propto \left\{\frac{1}{[\text{O}]}\right\} \tag{20.74}$$

Similarly, for a fixed (FeO) content of slag:

$$\left\{\frac{(\text{wt\% S})}{[\text{wt\% S}]}\right\} \propto (\%\text{CaO}) \tag{20.75}$$

For efficient removal of S, high basicity and low oxygen potential are therefore essential. Thermodynamically, lower temperature is bound to improve de-sulphurization index, but high temperature is favoured from the kinetic point of view. From activity point of view, it is easier to remove S from hot metal increase the activity coefficient of S.

Taking logarithm of Eq (20.72):

$$log\left\{\frac{(\text{wt\% S})}{[\text{wt\% S}]}\right\} = \log \text{C} - \log(\text{mol\% FeO}) \tag{20.76}$$

Since steelmaking is carried out under oxidizing conditions (except in EAF reducing period), the efficiency of de-sulphurization is very low. The value of the sulphur distribution coefficient ranges from 50 to 100 under blast furnace conditions and that drops to 5–10 under basic steelmaking conditions.

20.4 Thermodynamics of De-oxidation of Steel

Steelmaking is an oxidation as well as refining process; hence, some oxygen must dissolve in liquid steel during steelmaking process. This excess oxygen should be removed from liquid steel. De-oxidation is the process for removal of excess oxygen from molten steel; by adding metal/ferro-alloy, which have good affinity for oxygen. De-oxidation of steel is carried out before teeming into the mould. De-oxidation of steel is usually performed by adding Mn, Si and Al, or rarely by adding Cr, V, Ti, Zr and B.

20.4.1 Thermodynamics for Oxygen in Molten Steel

The dissolution of oxygen in molten steel:

$$1/2\{\text{O}_2\} = [\text{O}], \Delta G^0_{20.77} = -117{,}300 - 2.89T\,\text{J} \tag{20.77}$$

where [O] is dissolved oxygen in molten steel.

Therefore, equilibrium constant for reaction (20.77):

$$k_{20.77} = \left\{ \frac{[h_O]}{(p_{O_2})^{1/2}} \right\} \quad (20.78)$$

where $[h_O]$ is the Henrian activity of oxygen, since $h_i = f_i \cdot$ wt% i = wt% i (for dilute solution).

Equation (20.78) becomes:

$$k_{20.77} = \left\{ \frac{[wt\%\,O]}{(p_{O_2})^{1/2}} \right\} \quad (20.79)$$

Again

$$\Delta G^0_{20.77} = -RT \ln k_{20.77} = -RT \ln \left\{ \frac{[wt\%\,O]}{(p_{O_2})^{1/2}} \right\} \quad (20.80)$$

Putting the value of $\Delta G^0_{20.77}$ into Eq. (20.80):

$$-117{,}300 - 2.89\,T = -8.314 \times 2.303T \log k_{20.77} = -19.147\,T \log k_{20.77} \quad (20.81)$$

$$\text{Therefore, } \log k_{20.77} = \left(\frac{6126.29}{T} \right) + 0.15 \quad (20.82)$$

Equation (20.82) allows to estimate [wt% O] in liquid steel at any value of p_{O_2} with which the molten steel would be brought to equilibrium. This value of [wt% O] is nothing but solubility of [O] at that p_{O_2}.

At 1600 °C and 1 atm, log $k_{20.77}$ = 3.42 and $k_{20.77}$ = 2,630.26 = [wt% O] (from Eq. 20.79).

This value is physically not possible.

However, oxygen tends to form stable oxide with iron. Therefore, molten iron becomes saturated with oxygen when the oxide starts forming, i.e. when liquid iron and oxide are at equilibrium. This oxide, in the pure form, is denoted as Fe_xO, where $x \cong 0.985$ at 1600 °C. For the sake of simplicity, taking $x = 1$ and designating this oxide become FeO.

For the reaction:

$$(Fe_xO) = x[Fe] + [O] \quad (20.83)$$

Therefore, equilibrium constant for reaction (20.83):

$$k_{20.83} = \frac{\{[h_O] \cdot [a_{Fe}]^x\}}{(a_{Fe_xO})} \quad (20.84)$$

Now, $[a_{Fe}]$ = activity of Fe in metal = 1 (in Raoultian scale), and $(a_{Fe_xO}) = 1$ (when FeO is a pure compound).

So,

$$k_{20.83} = [h_O] \quad (20.85)$$

Hence, $[h_O]$ is in equilibrium with pure Fe and pure FeO.

In case FeO is not pure and it is present in an oxide slag, then $(a_{FeO}) < 1$ and $[h_O]$, i.e. solubility of oxygen in liquid metal in equilibrium with the slag, would be less.

$$k_{20.83} \text{ can be calculated from: } \log k_{20.83} = \left(-\frac{6372}{T}\right) + 2.73 \quad (20.86)$$

At 1600 °C, $\log k_{20.83} = -0.672$ and $k_{20.83} = 0.213 = [h_O] = [\text{wt\% O}]$ (from Eq. 20.85).

Now this value is physically possible. That means oxygen gas cannot go to metal by Eq. (20.77), but through FeO by Eq. (20.83).

20.4.2 De-oxidation Equilibrium

De-oxidation reaction, when the alloying element is added to the steel, can be represented by:

$$(M_xO_y) = x[M] + y[O] \quad (20.87)$$

$$\text{e.g.} \quad (SiO_2) = [Si] + 2[O] \quad (20.88)$$

The de-oxidation constant is given by:

$$k_{20.87} = \frac{\{[h_M]^x \cdot [h_O]^y\}}{(a_{M_xO_y})} \quad (20.89)$$

Now considering M_xO_y as pure, the Henrian activity is considered as 1, i.e. $(a_{M_xO_y}) = 1$

Hence,

$$k_{20.87} = [h_M]^x [h_O]^y \quad (20.90)$$

Therefore, for reaction (20.88),

$$k_{20.88} = [h_{Si}]\,[h_O]^2 \quad (20.91)$$

where $[h_M]$ and $[h_O]$ are Henrian activities of metal and oxygen.

Now, Henrian activity is also defined as:

$$h_i = f_i(\text{wt\%}\,i) \quad (20.92)$$

So,

$$[h_{Si}] = [f_{Si}(\text{wt\% Si})] \quad \text{and} \quad [h_O] = [f_O(\text{wt\% O})] \quad (20.93)$$

where f_i is Henrian activity coefficient of species i, which can be calculated by:

$$\log f_i = e_i^j(\text{wt\%}j) \quad (20.94)$$

Now, for dilute solution, like low alloy steels production, the activity coefficient (f_i) can be taken as unity ($f_i = 1$), and hence,

$$[h_i] = [\text{wt\%}\, i] \tag{20.95}$$

Therefore,

$$k_{20.88} = [\text{wt\% Si}][\text{wt\% O}]^2 \tag{20.96}$$

Now, $k_{20.88} = 2.2 \times 10^{-5}$ and if steel has 0.1% Si.
Then,

$$[\text{wt\% O}] = 0.0148\% = 148\,\text{ppm} \tag{20.97}$$

- Here, for single element de-oxidation, the solubility of oxygen in liquid iron at 1600 °C is given as a function of the concentration of the alloying element.
- As shown in Fig. 20.14, in each case, the melt is in equilibrium with the respective pure oxide.
- It can be clearly seen that aluminium is the strongest of the common de-oxidizers followed by titanium. Rare earth metals are strong as aluminium as de-oxidizers.

The kinetics of a de-oxidation reaction consists of the following steps (or stages) [8]:

(1) Dissolution of de-oxidizer into molten metal,
(2) Chemical reaction between dissolved oxygen and de-oxidizing element at phase boundary,
(3) Nucleation of de-oxidation product and
(4) Growth of nuclei.

20.5 Thermodynamics of De-sulphurization

Another major refining reaction in ladle metallurgy is de-sulphurization. The de-sulphurization reaction is now getting interest in ladle metallurgy. The additives to liquid steel for de-sulphurization are lime and de-oxidizer. The reaction can be expressed as:

$$(\text{CaO}) + [\text{S}] = (\text{CaS}) + [\text{O}] \tag{20.98}$$

The equilibrium constant for reaction (20.98):

$$k_{20.98} = \left\{ \frac{(a_{\text{CaS}}) \cdot [h_{\text{O}}]}{(a_{\text{CaO}}) \cdot [h_{\text{S}}]} \right\} \tag{20.99}$$

At 1600 °C, $k_{20.98} = 4 \times 10^{-2}$.

The value of $k_{20.98}$ decreases with increasing the temperature; therefore, de-sulphurization reaction would be less favourable. However, kinetic of the reaction is favoured by high temperature; this is more important.

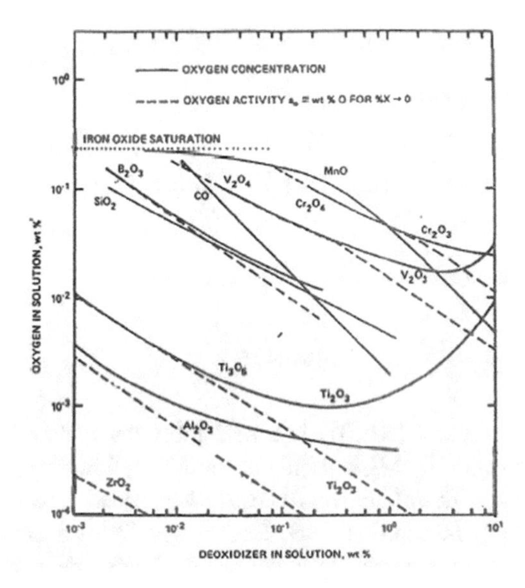

Fig. 20.14 De-oxidation equilibrium in liquid iron alloys at 1600 °C [8]

By assuming unit activities of CaO and CaS:
Then

$$k_{20.98} = \frac{[h_O]}{[h_S]} \tag{20.100}$$

For Si–Mn de-oxidized steel with 50 ppm oxygen in liquid steel, the lowest possible sulphur level is about 0.1%; this steel cannot be effectively de-sulphurized. However, aluminium-killed steel containing 0.04% Al and in equilibrium with CaO saturated calcium aluminate is about 0.0025% S.

Several important factors must be considered when de-sulphurization occurs in liquid steel:

(1) The de-sulphurizing slag can absorb sulphur before CaS is precipitates.
(2) The sulphide capacity (C_S) is a property of the slag which can hold the sulphur in slag phase. Therefore, the higher the sulphide capacity, the greater the ability of the slag to absorb sulphur. The sulphide capacity of $CaO–Al_2O_3$ slag is greater than for $CaO–SiO_2$ slag. The sulphide capacity increases if CaF_2 is present and increase with slag basicity.
(3) It is obvious that the slag can absorb considerable amounts of sulphur before CaS forms. The final sulphur content of the steel will depend on the relative amounts of slag, metal and initial sulphur contents. However, it can be concluded that sulphur levels of less than 0.001% are thermodynamically possible.
(4) These calculations only indicate the thermodynamic possibilities for de-sulphurization. There must be sufficient slag-metal mixing to promote the reaction and extraneous sources of oxygen must be eliminated to insure good de-sulphurization.

Mg and mixtures of Mg and CaO have also been used for steel de-sulphurization. However, the MgS–MgO equilibrium is less favourable than the CaO–CaS equilibrium for de-sulphurization.

$$(MgO) + [S] = (MgS) + [O] \quad (20.101)$$

The equilibrium constant for reaction (20.101):

$$k_{20.101} = \left\{ \frac{(a_{MgS}) \cdot [h_O]}{(a_{MgO}) \cdot [h_S]} \right\} \quad (20.102)$$

Assuming $(a_{MgS}) = 1$ and $(a_{MgO}) = 1$:

$$k_{20.101} = \frac{[h_O]}{[h_S]} \quad (20.103)$$

So value of $k_{20.101}$ for reaction (20.101) is considerably lower than $k_{20.98}$ the reaction (20.98), indicating that de-sulphurization by Mg can only occur at very low oxygen activities. Mg and Mg–CaO mixtures are successful de-sulphurizers because Mg causes local strong de-oxidation, thus promoting reaction (20.101). In addition, such CaO is usually also present; de-sulphurization by reaction (20.98) also occurs. In general, Mg is not as an effective de-sulphurization as Ca in liquid steel. Mg is effective for hot metal de-sulphurization because the oxygen potential and temperature are much lower.

Secondary steelmaking slags consist of CaO, Al_2O_3 and SiO_2 as the major constituents. Among the common de-oxidizers, aluminium is the most powerful. Low oxygen levels can be achieved by de-oxidation (referred to as *killing*) using Al, and these are known as *aluminium-killed steels*.

Based on the above, the overall reaction may be written as:

$$3(CaO) + 3[S] + 2[Al] = 3(CaS) + (Al_2O_3) \quad (20.104)$$

The equilibrium constant for reaction (20.104):

$$k_{20.104} = \left(\frac{\{(a^3_{CaS})(a_{Al_2O_3})\}}{\{(a^3_{CaO})[h^3_S][h^2_{Al}]\}} \right) \quad (20.105)$$

The *partition coefficient for sulphur* (L_S) is defined as:

$$L_S = \left(\frac{(w_S)}{[w_S]} \right) \quad (20.106)$$

Assuming $[h_S] = [w_S]$, $[h_{Al}] = [w_{Al}]$ for dilute solution; and (w_S) in slag proportional to (a_{CaS}). Then, it is possible to write:

$$L_S = \left(\frac{(w_S)}{[w_S]} \right) \alpha\, (a_{CaO}) \left(\frac{\left[w_{Al}^{\frac{2}{3}} \right]}{(a^{1/3}_{Al_2O_3})} \right) \quad (20.107)$$

Increasing wt% of CaO increases a_{CaO} and decreases $a_{Al_2O_3}$, thereby increasing L_S.

Figure 20.15 demonstrates this and shows that L_S increases with increase in wt% CaO in slag and $[w_{Al}]$ in molten steel for CaO–Al_2O_3 slags at 1500 and 1650 °C. $k_{20.104}$ is a function of temperature and is given as [9]:

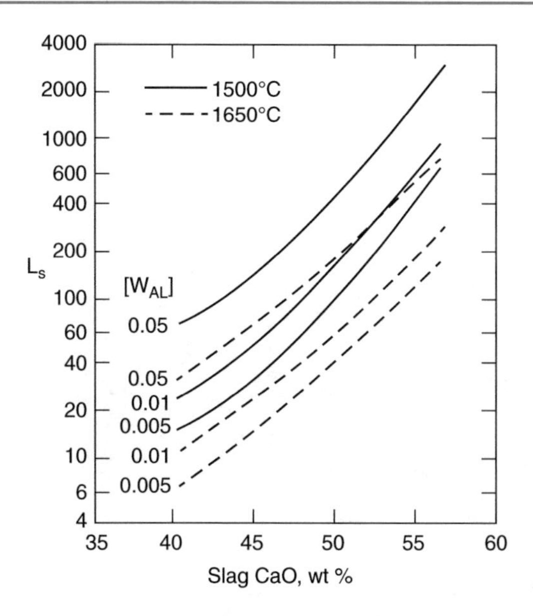

Fig. 20.15 Sulphur partition ratio between slag and metal for Fe–Al alloys in equilibrium with $CaO–Al_2O_3$ slags [8]

$$\log k_{20.104} = \left(\frac{48{,}580}{T}\right) - 16.997 \tag{20.108}$$

Hence, the lower temperature helps de-sulphurization.

If the initial $[w_S]$ in steel is 0.01%, it is to be brought down to 0.002%. The weight of liquid steel is 150 tonnes and that of slag is 2 tonnes.

Hence, the amount of sulphur to be transferred to the slag

$$= \left[150 \times (0.01 - 0.002) \times 10^{-2}\right] \text{tonne} = 0.012\,\text{tonne}$$

Therefore, wt% S in slag$(w_S) = \left(\frac{0.012}{2}\right) x\,10^2 = 0.6\%$

Hence, $L_S = \left(\frac{0.6}{0.002}\right) = 300$.

Since equilibrium may not be attained in the process, L_S should be larger than 300. Plant trials have shown that it is possible to have effective de-sulphurization if L_S is 1000 or above. Such a high value is obtained when the slag is almost saturated with CaO.

20.6 Thermodynamics of Chromium Reactions

Chromium, manganese and silicon occupied a unique position in iron and steelmaking because their oxides show a sequence of progressively increasing stability beyond that of iron oxide, as shown by their free energies of formation in Fig. 20.16. Since their stabilities slightly greater than the oxides of iron and phosphorus, the oxides of chromium, manganese and silicon show behaviour in the blast furnace intermediate between those of iron and phosphorus which are practically completely reduced, and the refractory oxides of aluminium, magnesium and calcium are totally unreduced. The percentage recovery of chromium, manganese and silicon in the blast furnace generally decreases in the

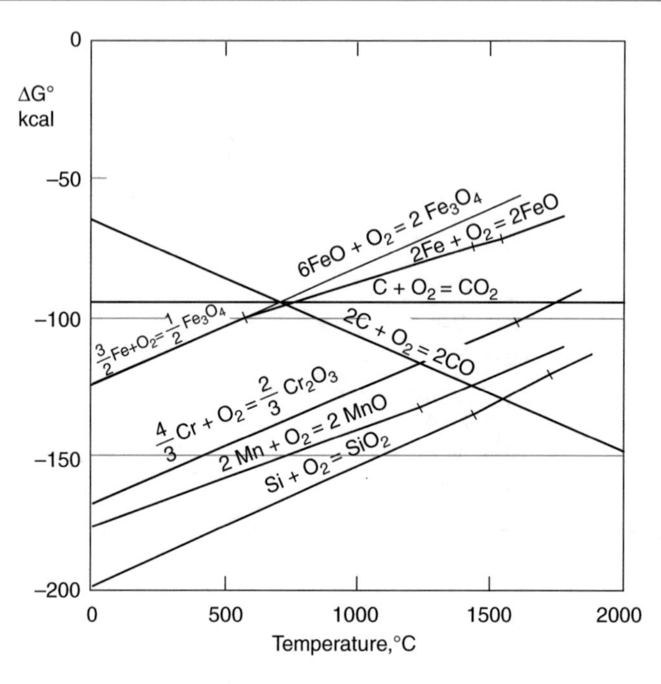

Fig. 20.16 Free energy of formation of the oxides of some selected elements

same order as the increase in the stabilities of their oxides. Their recovery in steelmaking follows the same order.

The controlling factors in the process metallurgy of stainless steelmaking are the thermodynamic equilibria between oxygen, carbon and chromium; and the kinetics of refining processes.

The chromium reactions are important for stainless steel production:

$$2[\mathrm{Cr}] + 3/2\{\mathrm{O_2}\} = (\mathrm{Cr_2O_3}), \Delta G^0_{20.109} = -1120,266 + 259.83T\,\mathrm{J/mol\ of\ Cr_2O_3} \tag{20.109}$$

$$[\mathrm{C}] + 1/2\{\mathrm{O_2}\} = \{\mathrm{CO}\}, \Delta G^0_{20.110} = -111,712.8–87.65T\,\mathrm{J/mol\ of\ CO} \tag{20.110}$$

Revised Eq. (20.110) and multiplication by 3:

$$3\{\mathrm{CO}\} = 3[\mathrm{C}] + 3/2\{\mathrm{O_2}\}, \Delta G^0_{20.111} = -3\Delta G^0_{20.110} = 3(111,712.8 + 87.65T)\mathrm{J} \tag{20.111}$$

Adding Eqs. (20.109) and (20.111):

$$2[\mathrm{Cr}] + 3\{\mathrm{CO}\} = (\mathrm{Cr_2O_3}) + 3[\mathrm{C}] \tag{20.112}$$

Therefore,

$$\Delta G^0_{20.112} = -785,127.6 + 522.78T\,\mathrm{J/mol\ of\ Cr_2O_3} \tag{20.113}$$

At equilibrium, $\Delta G^0_{20.112} = 0$, i.e.—785,127.6 + 522.78T = 0.
Hence, T = 1501.83 K, i.e. 1502 K or 1229 °C.

That means above 1229 °C, for reaction (20.112): $\Delta G^0_{20.112}$ became positive, so reaction (20.112) proceeds in reverse direction. Hence, chromium oxide (Cr_2O_3) is reduced by carbon in metal bath at high temperature.

The equilibrium constant for reaction (20.112) can be written as:

$$k_{20.112} = \left\{ \frac{(a_{Cr_2O_3}) \cdot [a_C]^3}{[a_{Cr}]^2 \cdot p_{CO^3}} \right\} \tag{20.114}$$

$$k_{20.112} = 1 \text{ at } 1502\,\text{K} \tag{20.115}$$

If steels with low carbon and high chromium contents are to be made, value of $k_{20.112} < 1$ is required and temperature should be higher than 1502 K, i.e. 1229 °C. Conditions in the production of stainless steels are largely governed by oxygen reactions with carbon and chromium at different temperatures and partial pressures of CO gas. Figure 20.17 shows the equilibrium conditions for the carbon and chromium contents in molten bath of Fe–C–Cr under the given temperatures and pressures. Carbon oxidation will be varied with temperature at constant chromium content. With rising temperatures and decreasing partial pressures of CO gas, a fixed chromium can be equilibrium with much low carbon.

For example, at higher temperature (e.g. 1800 °C), higher chromium (say 10%) can be equilibrium with much lower carbon (say 0.05%) in the Fe–C–Cr bath, with respect to at temperature, 1600 °C that per cent of chromium (i.e. 10%) can be equilibrium with much higher carbon (say 0.22%) in the melt (as shown in Fig. 20.17). That means carbon in the bath can be preferentially oxidized from Fe–C–Cr bath only at much higher temperature without loss of much chromium. That is the equilibrium of the system for any given chromium content is shifted to lower carbon level as the temperature of the melt increased.

Equation (20.114) can be rewrite as:

$$k_{20.112} = \frac{[a_C]^3}{\left\{[a_{Cr}]^2 \cdot p_{CO^3}\right\}} = \frac{[\text{wt\% C} \cdot f_C]^3}{\left\{[\text{wt\% Cr} \cdot f_{Cr}]^2 \cdot p_{CO^3}\right\}} \tag{20.116}$$

Since $(a_{Cr2O3}) = 1$, for pure and standard state.

Therefore,

$$[\text{wt\% C}] = \frac{\left\{k_{20.112}^{1/3} \cdot [\text{wt\%}\, Cr \cdot f_{Cr}]^{2/3} \cdot p_{CO}\right\}}{[f_C]} \tag{20.117}$$

At constant temperature p_{CO}, and $k_{20.112}$ remains constant. At constant $k_{20.112}$, for de-carburization of the bath, chromium is also oxidized but that is not desirable. That means decreasing carbon in melt, chromium content is also decreased. Hence, de-carburization of the bath is not possible without chromium oxidation.

Carbon content will be a function of both temperatures and partial pressure of CO gas.

So,

$$[\text{wt\% C}] \propto (T, \text{wt\% Cr}, p_{CO}) \tag{20.118}$$

Therefore, thermodynamic alternative is as follows:

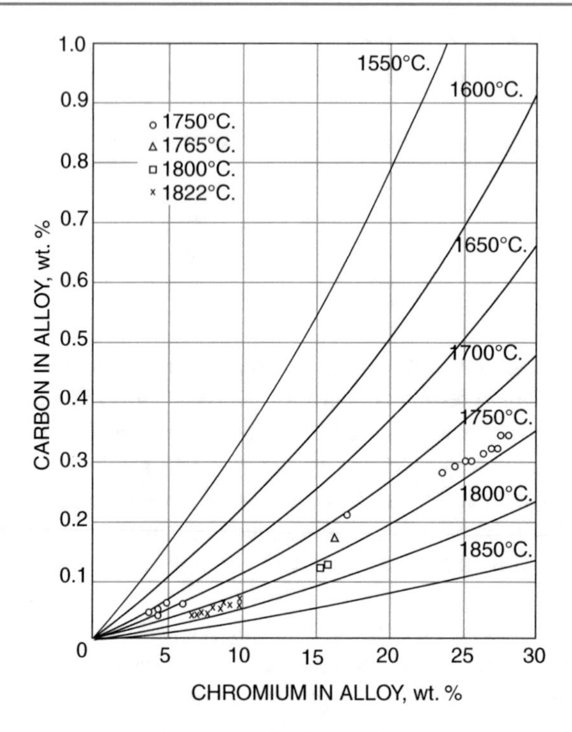

Fig. 20.17 Effect of temperature on the relation between carbon and chromium in liquid steel at 1 atmosphere pressure of CO [10]

1. To increase the temperature, and
2. To lower down the p_{CO} of the bath.

Hence, de-carburization of the bath is possible without chromium oxidation at a specific temperature and low partial pressure of CO (as shown in Fig. 20.18). At low partial pressure of CO, at a particular temperature, much lower carbon is equilibrium with higher chromium.

The partial pressure of CO can be reduced by:

(i) Injecting inert (argon or nitrogen) gas to the bath:
   - AOD (argon oxygen de-carburization) process, injection of inert gas (argon or nitrogen), decreases the partial pressure of CO in the bath, thus allowing higher chromium contents to be in equilibrium with low contents of carbon (as shown in Fig. 20.18).

(ii) Applying vacuum to lower down the partial pressure of CO:
   - VOD (vacuum oxygen de-carburization) process, applying a vacuum to the molten bath to removes CO gas, allows high chromium contents to be in equilibrium with low carbon content. This process is especially effective when carbon content of the product is very low.

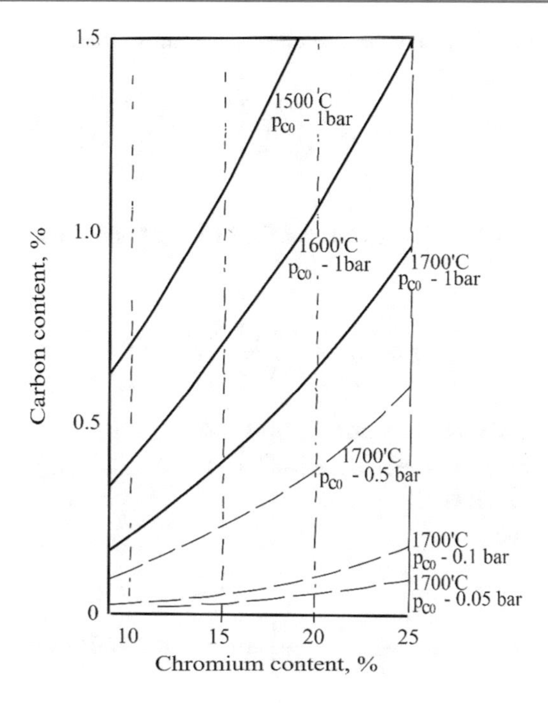

Fig. 20.18 Effect of temperature and pressure on the equilibrium relation between carbon and chromium in liquid steel [11]

20.7 Thermodynamics of Vacuum Degassing

The initial object in vacuum degassing was to lower down hydrogen content of liquid steel to prevent cracks in large forging quality ingots. Later, its objective also included lowering of nitrogen and oxygen contents.

20.7.1 Hydrogen During Vacuum

Removal of hydrogen from molten steel is taken place by the following reaction:

$$2[\mathrm{H}] = \{\mathrm{H_2}\} \tag{20.119}$$

The equilibrium constant (k'_H) for the reaction (20.119) is given by:

$$k'_{20.119} = \frac{p_{H_2}}{[h_H]^2} \tag{20.120}$$

where p_{H_2} is partial pressure of hydrogen (in atm) in equilibrium with molten steel and $[h_\mathrm{H}]$ is Henrian activity of hydrogen.

Since $[h_\mathrm{H}] = [f_\mathrm{H} \cdot \mathrm{wt\%\ H}]$, for very dilute solution, $f_\mathrm{H} = 1$, so $[h_\mathrm{H}] = [\mathrm{wt\%\ H}]$, where [wt% H] denotes wt% of hydrogen dissolved in molten steel. However, the concentration of hydrogen in steel is very low and is usually expressed in parts per million (ppm).

Hence, $[h_\mathrm{H}] = [\mathrm{ppm\ of\ H}]$

Therefore, above Eq. (20.120) may be written in the form of *Sievert's law*:

$$[\text{ppm of H}] = \sqrt{\left\{\frac{p_{H_2}}{k'_{20.119}}\right\}} = k_{20.119} \cdot \left\{\sqrt{p_{\text{H}_2}}\right\} \quad (20.121)$$

where $k_{20.119} = \left\{\sqrt{\frac{1}{k'_{20.119}}}\right\}$ = Degassing constant for hydrogen and it correlates with temperature as follows [8]:

$$\log k_{20.119} = -\left(\frac{1905}{T}\right) + 2.409 \quad (20.122)$$

At 1600 °C, if partial pressure of hydrogen is 0.1 torr.
From Eq. (20.122): $\log \text{k}_{20.119} = -\left(\frac{1905}{1873}\right) + 2.409 = 1.392$
[since T = 1600 + 273 = 1873 K].
Therefore, $k_{20.119}$ = 24.66.
Now

$$p_{H_2} = 0.1\,\text{torr} = \left\{\frac{1}{(760 \times 10)}\right\}\text{atm} \quad [\text{Since } 760\,\text{torr} = 1\,\text{atm}]$$

From Eq. (20.121):

$$[\text{ppm of H}] = k_{20.119} \cdot \left\{\sqrt{p_{H_2}}\right\} = 24.66\sqrt{\left\{\frac{1}{(7600)}\right\}} = 0.28$$

Therefore, thermodynamic point of view, it is possible to bring down the hydrogen content of molten steel below 0.3 ppm by vacuum degassing.

In arriving Eq. (20.121), the activity coefficient of hydrogen (f_{H}) is one. This is a pretty good assumption for mild steel. However, for alloy steels, f_{H} is likely to deviate from one significantly and can be calculated by using Eq. (20.123):

$$\log f_{\text{M}} = \sum \left(\text{e}_M^j \cdot W_j\right) \quad (20.123)$$

20.7.2 Nitrogen During Vacuum

Removal of nitrogen from molten steel takes place by the reaction:

$$2[\text{N}] = \{\text{N}_2\} \quad (20.124)$$

The equilibrium constant (k'_{N}) for the reaction (20.124) is given by:

$$k'_{20.124} = \frac{p_{\text{N}_2}}{[h_{\text{N}}]^2} \quad (20.125)$$

where p_{N_2} is partial pressure of nitrogen (in atm) in equilibrium with molten steel and $[h_N]$ is Henrian activity of nitrogen.

Since $[h_N] = [f_N \cdot \text{wt\% N}]$, for very dilute solution, $f_N = 1$; so $[h_N] = [\text{wt\% N}]$, where [wt% N] denotes wt% of nitrogen dissolved in molten steel. Since the concentration of nitrogen in steel is much larger than that of hydrogen in steel, and former is usually expressed in wt%.

Therefore, above Eq. (20.125) may be written in the form of *Sievert's law*:

$$[\text{wt\% N}] = \sqrt{\left\{\frac{p_{N_2}}{k'_{20.124}}\right\}} = k_{20.124} \cdot \left\{\sqrt{p_{N_2}}\right\} \tag{20.126}$$

where $k_{20.124} = \sqrt{\left\{\frac{1}{k'_{20.124}}\right\}}$ = Degassing constant for nitrogen and it correlates with temperature as follows [8]:

$$\log k_{20.124} = -\left(\frac{187.9}{T}\right) - 1.248 \tag{20.127}$$

At 1600 °C, if partial pressure of nitrogen is 0.1 torr.

From Eq. (20.127): $\log k_{20.124} = -\left(\frac{187.9}{1873}\right) - 1.248 = -1.348$

Therefore, $k_{20.124} = 0.0448$.

Now

$$p_{N_2} = 0.1 \text{ torr} = \left\{\frac{1}{(7600)}\right\} \text{atm} \quad [\text{Since } 760 \text{ torr} = 1 \text{ atm}]$$

From Eq. (20.126):

$$\begin{aligned}[\text{wt\% N}] &= k_{20.124} \cdot \left\{\sqrt{p_{N_2}}\right\} \\ &= 0.0448\left[\sqrt{\left\{\frac{1}{(7600)}\right\}}\right] = 5.14 \times 10^{-4} = 0.0005 \text{ wt\%}\end{aligned}$$

Therefore, it is possible to bring down the nitrogen content of molten steel about 0.0005% by vacuum degassing.

20.7.3 Oxygen During Vacuum

(a) Removal of oxygen from molten steel takes place by the reaction:

$$2[O] = \{O_2\} \tag{20.128}$$

The equilibrium constant ($k'_{20.128}$) for the reaction (20.128) is given by:

$$k'_{20.128} = \frac{p_{O_2}}{[h_O]^2} \tag{20.129}$$

where p_{O_2} is partial pressure of oxygen (in atm) in equilibrium with molten steel and $[h_O]$ is Henrian activity of oxygen.

Since $[h_O] = [f_O \cdot \text{wt\% O}]$, for very dilute solution, $f_O = 1$; so $[h_O] = [\text{wt\% O}]$, where [wt% O] denotes wt% of oxygen dissolved in molten steel.

Therefore, above Eq. (20.129) may be written in the form of *Sievert's law*:

$$[\text{wt\% O}] = \sqrt{\left\{\frac{p_{O_2}}{k'_{20.128}}\right\}} = k_{20.128} \cdot \left\{\sqrt{p_{O_2}}\right\} \tag{20.130}$$

where $k_{20.128} = \left\{\sqrt{\frac{1}{K'_{20.128}}}\right\}$ = Degassing constant for oxygen and it correlates with temperature as follows:

$$\log k_{20.128} = \left(\frac{6120}{T}\right) + 0.15 \tag{20.131}$$

At 1600 °C, if partial pressure of oxygen is 0.1 torr.

From Eq. (20.131): $\log k_{20.128} = \left(\frac{6120}{1873}\right) + 0.15 = 3.417$

Therefore, $k_{20.128} = 2612.16$.

Now from Eq. (20.130):

$$[\text{wt\% O}] = k_{20.128} \cdot \left\{\sqrt{p_{O_2}}\right\} = 2612.16\left[\sqrt{\left\{\frac{1}{(7600)}\right\}}\right] = 29.96\,\text{wt\%}.$$

This value is certainly a high value. It shows that oxygen cannot be removed from liquid steel through reaction (20.128). Hence, oxygen from liquid steel can be removed by other means.

(b) By carbon in liquid steel, removal of oxygen from molten steel takes place by the reaction:

$$[\text{O}] + [\text{C}] = \{\text{CO}\} \tag{20.132}$$

The equilibrium constant (k_{CO}) for the reaction (20.132) is given by:

$$k_{20.132} = \frac{p_{CO}}{\{[h_O] \cdot [h_C]\}} = \frac{p_{CO}}{\{[\text{wt\% O}] \cdot [\text{wt\% C}]\}} \tag{20.133}$$

For very dilute solution, f_O and $f_C = 1$; so $[h_O] = [\text{wt\% O}]$ and $[h_C] = [\text{wt\% C}]$.

The $k_{20.132}$ correlates with temperature as follows:

$$\log k_{20.132} = \left(\frac{1168}{T}\right) + 2.07 \tag{20.134}$$

At 1600 °C, if partial pressure of carbon monoxide is 0.1 torr, wt% C is 0.2.
From Eq. (20.134): $\log k_{20.132} = \left(\frac{1168}{1873}\right) + 2.07 = 2.69$
Therefore, $k_{20.132}$ = 489.78.
From Eq. (20.133):

$$[\text{wt\% O}] = \frac{p_{CO}}{\{k_{20.132} \cdot [\text{wt\% C}]\}} = \frac{\left(\frac{1}{7600}\right)}{\{489.78 \times 0.2\}} = 1.34 \times 10^{-6}\text{wt\%}.$$

This demonstrates that oxygen removal to an extremely low value is possible from a thermodynamic point of view by vacuum degassing in presence of carbon in liquid steel. After ordinary ladle de-oxidation, oxygen content in liquid steel is in the range between 0.003 and 0.04%.

Reaction (20.132) is also the basis for VOD process. A deliberate addition of oxygen to liquid steel under a low pressure (i.e. vacuum) can lower the carbon content of steel significantly.

20.7.4 De-sulphurization During Vacuum

Sulphur elimination takes place during vacuum degassing in the presence of silicon. Nowadays, injection of powders of Ca–Si, Ca–Ba alloy and rare earth metals (Ce, Sc, Ho, etc.) are carried out during vacuum treatment of steel to obtain extra low sulphur in steel.

Calcium and barium form CaS and BaS, respectively, by reaction with S, whereas cerium forms several sulphides, out of which CeS is the stable one at steelmaking condition [8]. Cerium also forms an oxy-sulphide, Ce_2O_2S. All these compounds are solid in steelmaking temperature. From thermodynamic data, it is found that all these elements form very stable sulphides as well as oxides. Hence, they are acted as both strong de-oxidizers and de-sulphurizers.

Overall reaction can be written as:

$$[\text{S}] + (\text{MO}) = [\text{O}] + (\text{MS}) \tag{20.135}$$

The equilibrium constant,

$$k_{20.135} = \frac{[a_O] \cdot (a_{MS})}{[a_S] \cdot (a_{MO})} \tag{20.136}$$

Assuming MO and MS are pure, so (a_{MO}) = 1 and (a_{MS}) = 1.
Since $h_i = f_i \cdot$ wt% i.
For dilute solution, h_i = wt% i (since f_i = 1).
Therefore, for dilute solution:

$$k_{20.135} = \frac{[h_O]}{[h_S]} = \frac{[\text{wt\%}\, O]}{[\text{wt\%}\, S]} \tag{20.137}$$

By plotting $k_{20.135}$ versus T (Fig. 20.19), it is found that barium is the strongest de-sulphurizer and magnesium is the weakest, with calcium and cerium lying in between. Good de-sulphurizer (i.e. low wt% S) calls for a good de-oxidizer (i.e. low wt% O) and as a pre-condition which may be achieved under vacuum by carbon.

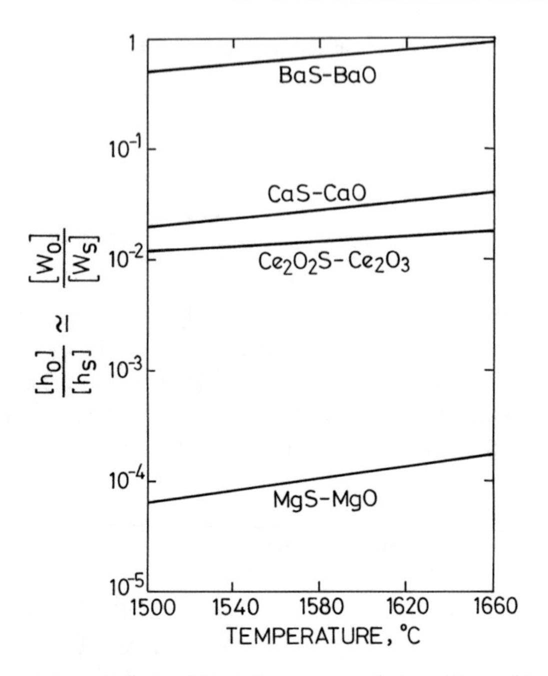

Fig. 20.19 Oxygen/sulphur activity ratio in liquid iron for some sulphide oxide equilibria at 1600 °C [8]

At constant temperature, $k_{20.135} = \frac{[\text{wt\%}\,O]}{[\text{wt\%}\,S]}$ = constant, to lower [wt% S], [wt% O] should be lower down. Otherwise, reaction (20.135) goes backward direction. Hence, reducing atmosphere (by strong de-oxidizer) is required for de-sulphurization as well as high temperature.

On Ca–Si injection in argon-purged ladles showed that the presence of a proper slag at the top of the melt is important to prevent reversion of sulphur during processing. An empirical equation for sulphur distribution between slag and metal has been suggested as:

$$\frac{[\text{wt\%}\,O]}{[\text{wt\%}\,S]} = \left\{\frac{118.4\text{B} - 205}{(\text{wt\%}\,FeO)}\right\} \tag{20.138}$$

where the basicity of the slag,

$$\text{B} = \left(\frac{(\text{wt\%}\,CaO) + 1.4(\text{wt\%}\,MgO)}{(\text{wt\%}\,SiO_2)}\right) \tag{20.139}$$

Since these compounds (CaO, SiO_2) would not be present in pure form, Ca–Si leads to formation of $CaO \cdot SiO_2$.

If (MO) and (MS) are not pure, then it is better to write in ionic form of de-sulphurization reaction:

$$[\text{S}] + (\text{O}^{2-}) = (\text{S}^{2-}) + [\text{O}] \tag{20.140}$$

The equilibrium constant,

$$k_{20.140} = \frac{(a_{S^{2-}}) \cdot [h_O]}{[h_S] \cdot (a_{O^{2-}})} \tag{20.141}$$

Or

$$k_{20.140} \cdot (a_{O^{2-}}) = \left\{ \frac{(a_{S^{2-}}) \cdot [h_O]}{[h_S]} \right\} \tag{20.142}$$

Substitute $(a_{S^{2-}})$ by (wt% S), i.e. (W_S); then, modify $k_{20.140}$ (let that is $k'_{20.140}$):

$$k'_{20.140} \cdot (a_{O^{2-}}) = \left\{ \frac{(w_S) \cdot [h_O]}{[h_S]} \right\} = C'_S \tag{20.143}$$

where C'_S is known as modified sulphur capacity.

Sulphur capacity of slag (C_S), i.e. the ability of a slag to absorb sulphur, according to Richarson [8]:

$$C_S = w_S \cdot \left(\frac{p_{O_2}}{p_{S_2}} \right)^{1/2} \tag{20.144}$$

where (w_S) is the weight per cent sulphur in the slag in equilibrium with a gas having partial pressures of oxygen and sulphur as p_{O_2} and p_{S_2}. Its usefulness stems from the fact that C_S is a property of slag, and at a fixed temperature, it is determined solely by slag composition. The higher the value of C_S, the better the de-sulphurizing ability of the slag. Figure 20.20 shows C_S values for some typical slag systems of interest in secondary steelmaking. The superiority of CaO–CaF_2 slag is obvious. Values of C_S for various slags are available in *Slag Atlas* [12].

For slag-metal reaction, modified sulphur capacity (C'_S) is used.

$$\log C_S = \log C'_S + \left(\frac{936}{T} \right) - 1.375 \tag{20.145}$$

$$\text{At } 1600^\circ\text{C}, \quad C'_S = 7.5\, C_S \tag{20.146}$$

Another parameter, i.e. equation of sulphur partition ratio between slag and metal:

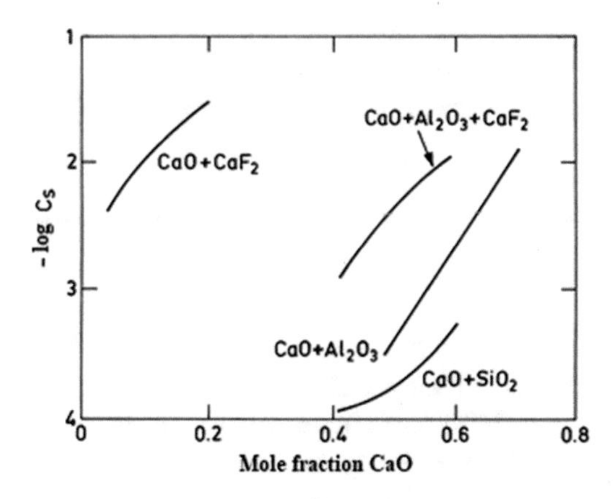

Fig. 20.20 Sulphide capacities of some slags at 1600 °C [8]

$$L_S = \frac{(w_S)}{[w_S]} \quad (20.147)$$

Equation (20.147) is putting in Eq. (20.143) and if $[h_S]$ is taken as $[w_S]$.
Then

$$\mathrm{L_S} = \left(\frac{C'_S}{[h_O]}\right) \quad (20.148)$$

where $[h_O]$ is in liquid steel which is determined by the presence of de-oxidizer, especially by dissolved aluminium. $[h_O]$ may be related to (FeO) content of slag as well. However, it has been found more appropriate to relate it to the former. Figure 20.16 shows L_S as a function of the (CaO) content of slag and aluminium content of metal for CaO–Al_2O_3 slag. Therefore, for good de-sulphurization, Al content of more than 0.02% is generally recommended.

Probable Questions

1. Discuss oxidizing power of slag.
2. What do you mean by sulphide capacity of slag?
3. Discuss thermodynamics of refining and carbon–oxygen equilibrium reaction.
4. Discuss thermodynamics of silicon reaction.
5. What do you mean by 'phosphorous participation'? State the conditions that are required for de-phosphorization of liquid steel bath. Draw a plot showing dependence of phosphorous participation on the FeO content and V-ratio of steelmaking slag.
6. Discuss thermodynamics of phosphorous reaction.
7. Discuss ionic theory of manganese transfer.
8. Discuss ionic theory of sulphur reaction.
9. Discuss thermodynamics for oxygen in molten steel.
10. Discuss thermodynamics of chromium reactions for stainless steel production.
11. 'De-carburization of the bath is not possible without chromium oxidation'. Why? What are the thermodynamic alternatives for that?
12. Discuss thermodynamic of vacuum degassing in steelmaking.
13. Discuss thermodynamics of vacuum degassing for hydrogen.
14. Discuss thermodynamics of degassing for oxygen.
15. Discuss de-sulphurization during vacuum.
16. Find out the correlation for equation of sulphur partition and sulphur capacity.

Examples

Example 20.1 In a steelmaking process to stop C–O_2 reaction at 0.1%C; (i) what is the equilibrium [O] presence in steel at 1600 °C? (ii) If the actual [O] presence is 0.025%, then what is the excess [O] presence in steel? (iii) What is the [C]% retain in steel at that [O] content (if all [O] consume by [C]) ?

Given: $\log k_{CO} = \left(\frac{1056}{T}\right) + 2.131\ (\text{for C} - O_2\ \text{reaction})$

Solution

(i) $[C] + [O] = \{CO\}$,

$$\log k_{CO} = \left(\frac{1056}{T}\right) + 2.131$$

$T = 1600 + 273 = 1873$ K
$\log k_{CO} = \left(\frac{1056}{1873}\right) + 2.131 = 2.6748$
Therefore, $k_{CO} = 495.22 = \left(\frac{p_{CO}}{[h_C][h_O]}\right)$
Since $[h_i] = f_i$ [wt% i].
Therefore, $f_c[\text{wt}\,\%\text{C}] \times f_o[\text{wt}\,\%\text{O}] = \left(\frac{p_{CO}}{k_{CO}}\right)$
Now putting, $f_c = 1$ and $f_o = 1$ [for dilute solution], $p_{CO} = 1$ atm
$[\text{wt}\,\%\text{C}] \times [\text{wt}\,\%\text{O}] = (1/k_{CO}) = (1/495.22) = 2.019 \times 10^{-3}$
Since [wt% C] = 0.1.
Therefore, $[\text{wt\% O}]_{eq} = \{(2.019 \times 10^{-3})/0.1\} =$ **0.02%**.

(ii) Excess $[O] = [O]_{ac} - [O]_{eq} = 0.025 - 0.02 =$ **0.005%**.
(iii) Since [wt% C] x [wt% O] = 2.019×10^{-3} and $[\text{wt\% O}]_{ac} = 0.025\%$.
Therefore, [wt% C] = $\{(2.019 \times 10^{-3})/0.025\}$ = **0.081%**.

Example 20.2 Find out activity of [O] content in liquid steel during reaction of Si at 1600 °C. [Si] content in liquid is 0.8%.
Given $\log k_{SiO_2} = \left(\frac{29{,}700}{T}\right) - 11.24$.

Solution

$$[Si] + 2[O] = (SiO_2) \quad (1)$$

$$k_{SiO_2} = \left(\frac{(a_{SiO_2})}{[a_{Si}][a^2_{O}]}\right) \quad (2)$$

$T = 1600 + 273 = 1873$ K.
Therefore, $\log k_{SiO_2} = [(29{,}700/1873) - 11.24] = 4.617$
$k_{SiO_2} = 41391.77$
Considering formation of SiO_2 is pure form, so $(a_{SiO_2}) = 1$ and
$[a_{Si}] = f_{Si}$ [wt% Si] = [wt% Si] since for dilute solution, $f_{Si} = 1$.
From Eq. 2:
$[a_O]^2 = \left(\frac{(a_{SiO_2})}{k_{SiO_2}[Wt\%\,\text{Si}]}\right) = \left(\frac{1}{(41391.77 \times 0.8)}\right) = 3.0199 \times 10^{-5}$
Therefore, $[a_O]$ = $\mathbf{5.495 \times 10^{-3}}$.

Example 20.3 Calculate the concentration of oxygen in molten iron at 1600^0 C in equilibrium with pure FeO. Given $\log k_{Fe} = -(6372/T) + 2.73$.

Solution

$$\text{FeO(l)} = \text{Fe(l)} + \text{O(wt\%)} \quad (1)$$

$$\text{So}k_{\text{Fe}} = \{[h_{\text{o}}] \cdot [a_{\text{Fe}}]/(a_{\text{FeO}})\} \quad (2)$$

Since a_{FeO} = 1 (for pure FeO) and a_{Fe} = 1 (assume pure Fe).
Again $[h_{\text{o}}] = [f_{\text{o}}]$, $[W_{\text{o}}] = [W_{\text{o}}]$ (if we consider for dilute solution, $f_{\text{o}} = 1$).
Now from Eq. (2), we get:

$$k_{\text{Fe}} = [W_{\text{o}}] \quad (3)$$

T = 1600 + 273 = 1873 K.
log k_{Fe} = −(6372/T) + 2.73 = [−(6372/1873) + 2.73] = −0.672.
Therefore, k_{Fe} = 0.213.
Hence, the concentration of oxygen in molten iron at 1600 °C, $[W_{\text{o}}]$ = **0.213%**.

Example 20.4 Calculate the residual oxygen content of liquid iron containing 0.1 wt% Si in equilibrium with solid silica at 1600 °C.

$$\text{Given}: \quad \text{Si(l)} + \text{O}_2\text{(g)} = \text{SiO}_2\text{(s)}, \Delta G_1^{\text{o}} = -947.68 + 0.199\,T\,\text{kJ/mol}$$
$$\text{O}_2\text{(g)} = 2[\text{O}](1\,\text{wt\% std state}), \Delta G_2^{\text{o}} = -233.47 - 0.006\,T\,\text{kJ/mol}$$
$$\text{Si(l)} = [\text{Si}](1\,\text{wt\% std state}), \Delta G_3^{\text{o}} = -119.24 - 0.053\,T\,\text{kJ/mol}$$
$$e_{\text{Si}}^{\text{Si}} = 0.32, e_{\text{Si}}^{\text{O}} = -0.24, e_{\text{O}}^{\text{O}} = -0.20, \text{and}\, e_{\text{O}}^{\text{Si}} = -0.14.$$

Solution

$$\begin{aligned}
&[\text{Si}](1\,\text{wt\% std state}) = \text{Si(l)} - \Delta G_3^{\text{o}} \\
&2[\text{O}](1\,\text{wt\% std state}) = \text{O}_2\text{(g)} - \Delta G_2^{\text{o}} \\
&\text{Si(l)} + \text{O}_2\text{(g)} = \text{SiO}_2\text{(s)}, \quad \Delta G_1^{\text{o}} \\
&[\text{Si}](1\,\text{wt\% std state}) + 2[\text{O}](1\,\text{wt\% std state}) = \text{SiO}_2\text{(s)}\Delta G_{\text{r}}^{\text{o}}
\end{aligned}$$

$$\begin{aligned}
\text{Therefore}\, \Delta G_{\text{r}}^{\text{o}} &= \Delta G_1^{\text{o}} - \Delta G_2^{\text{o}} - \Delta G_3^{\text{o}} \\
&= (-947.68 + 0.199\,T) - \{(-233.47 - 0.006\,T) + (-119.24 - 0.053\,T)\}] \\
&= -594.97 + 0.258\,T\,\text{kJ/mol} \quad [T = 1600 + 273 = 1873\,\text{K}] \\
&= (-594.97 + 0.258 \times 1873) \times 1000\,\text{J/mol} \\
&= -111736\,\text{J/mol}
\end{aligned}$$

Since $\Delta G_{\text{r}}^{\text{o}}$ = −RT ln k.
Or −111,736 = −8.314 × 1873 × ln k.
Or ln k = 7.175.

$$\begin{aligned}
\log f_{\text{o}} &= e_{\text{o}}^{\text{O}}[\text{wt\% O}] + e_{\text{o}}^{\text{Si}}[\text{wt\% Si}] = (-0.2) \times (\text{wt\% O}) + (-0.14) \times (0.1) \\
&= -0.2\,\text{wt\% O} - 0.014
\end{aligned}$$

$$\log f_{Si} = e_{Si}^{Si}[\text{wt\% Si}] + e_{Si}^{O}[\text{wt\% O}] = 0.32 \times 0.1 + (-0.24) \times (\text{wt\% O})$$
$$= 0.032 - 0.24\text{wt\% O}$$

Equilibrium constant, $k = [a_{SiO2}/(h_{Si} \cdot h_O^2)] = [1/(h_{Si} \cdot h_O^2)]$ (since $a_{SiO2} = 1$).
Again $h_i = f_i \cdot$ wt% i.
Therefore, $k = [1/\{(f_{Si} \cdot \text{wt\% Si}) \cdot (f_O \cdot \text{wt\% O})^2\}]$.
Taking both side logs:

$$\log k = -[\{\log f_{Si} + \log(\text{wt\% Si})\} + 2\{\log f_O + \log(\text{wt\% O})\}]$$
$$= -[\{(0.032 - 0.24\,\text{wt\% O}) + \log(0.1)\} + 2\{(-0.2\,\text{wt\% O} - 0.014) + \log(\text{wt\% O})\}]$$
$$= -[\{(0.032 - 0.24\,\text{wt\% O}) + (-1)\} + \{(-0.4\,\text{wt\% O} - 0.028) + 2\log(\text{wt\% O})\}]$$
$$= [0.996 + 0.64\text{wt\% O} - 2\log(\text{wt\% O})]$$

Since value of wt% O is very small, value of log (wt% O) can be taken as negligible.
[since **ln x = 2.303 log x**; ln k = 7.175 = 2.303 log k,
Therefore, log k = (7.175/2.303) = 3.1155].
So log k = 0.996 + 0.64 wt% O = 3.1155.
Therefore, **wt% O = 3.31**.

Example 20.5 Find out the oxygen concentration in liquid steel after degassing at 1600 °C and 1.0 torr. [wt% C] = 0.2.
Given:

(i) $\log k_O = (6120/T) + 0.15$.
(ii) $\log k_{CO} = (1168/T) + 2.07$.

Solution

$T = 1600 + 273 = 1873\,\text{K}, 760\,\text{torr} = 1\,\text{atm}, \text{so } 1.0\,\text{torr} = (1/760)\,\text{atm}$

$$2[\text{O}] = \text{O}_2 \quad (1)$$

Equilibrium constant,

$$k'_O = \left[p_{O2}/(h_O)^2\right] \quad (2)$$

Since $h_i = f_i \times$ wt% i = wt% i, due to $f_i = 1$.
From Eq. (2):

$$\text{wt\% O} = \sqrt{(p_{O2}/k'_O)} = k_O \times \sqrt{p_{O2}} \quad (3)$$

Since $k_O = \sqrt{(1/k'_O)}$.
Again log k_O = (6120/T) + 0.15 = (6120/1873) + 0.15 = 3.417.
Therefore, k_O = 2615.082; now putting this value to Eq. (3):
wt% O = $k_O \times \sqrt{p_{O2}}$ = 2615.08 × $\sqrt{(1/760)}$ = **94.86%**.
This high value of oxygen content in steel is impossible; hence, oxygen cannot be removed by this way. Therefore, oxygen in steel can be removed by reacting with carbon.

$$[C] + [O] = \{CO\} \tag{4}$$

Equilibrium constant,

$$k_{CO} = [p_{CO}/(h_C \times h_O)] = [p_{CO}/(\text{wt\% C} \times \text{wt\% O})] \tag{5}$$

Again log k_{CO} = (1168/T) + 2.07 = (1168/1873) + 2.07 = 2.693.
Therefore, k_{CO} = 493.853; since [wt% C] = 0.2 and p_{CO} = (1/760).
Putting these values in Eq. (5):

$$\text{Therefore wt\% O} = [p_{CO}/(\text{wt\% C} \times k_{CO})] = [(1/760)/(0.2 \times 493.853)]$$
$$= \mathbf{1.332 \times 10^{-5}\%}$$

Example 20.6 Find out wt% O in liquid steel to remove C from HM (content 3% C, 1.2% Si, 0.8% Mn, 0.04% S, 0.4% P) at 1600 °C.

Given $\log k_{CO} = [(1056/T) + 2.131]$

$e_c^c = 0.14, e_c^{Mn} = -0.012, e_c^P = 0.051, e_c^S = 0.046, e_c^{Si} = 0.08$

$e_o^c = -0.13, e_o^{Mn} = -0.021, e_o^P = 0.07, e_o^S = -0.133, e_o^{Si} = -0.131.$

Solution

T = 1600 + 273 = 1873 K.
So, log k_{CO} = [(1056/1873) + 2.131] = 2.6948.
Therefore, k_{CO} = 495.22.

$$\text{Again}, \log f_c = e_c^S[\text{wt\% S}] + e_c^C[\text{wt\% C}] + e_c^{Si}[\text{wt\% Si}] + e_c^{Mn}[\text{wt\% Mn}]$$
$$+ e_c^P[\text{wt\% P}]$$
$$= 0.046 \times 0.04 + 0.14 \times 3.0 + 0.08 \times 1.2 + (-0.012) \times 0.8 + 0.051 \times 0.4$$
$$= 0.5286$$

Therefore, f_c = 3.378.

$$\text{Similarly}, \log f_o = e_o^S[\text{wt\% S}] + e_o^C[\text{wt\% C}] + e_o^{Si}[\text{wt\% Si}] + e_o^{Mn}[\text{wt\% Mn}]$$
$$+ e_o^P[\text{wt\% P}]$$
$$= (-0.133) \times 0.04 + (-0.13) \times 3.0 + (-0.131) \times 1.2 + (-0.021) \times 0.8$$
$$+ 0.07 \times 0.4$$
$$= -0.54132$$

Therefore, f_o = 0.2875.
We know that k_{CO} = {p_{co}/h_c × h_o}.
Therefore, h_c × h_o = {p_{co}/k_{CO}} = {1/k_{CO}} {since p_{co} = 1 and h_i = f_i × [wt% i]}.
So, f_c × [wt% C] × f_o × [wt% O] = {1/k_{CO}}.

$$\text{Therefore}, [\text{wt\% O}] = \{1/k_{CO}\} \times \{1/(f_o \times f_c \times [\text{wt\% C}])\}$$
$$= \{1/(495.22 \times 0.2875 \times 3.378 \times 3.0)\}$$
$$= 1/1442.8359 = \mathbf{6.93 \times 10^{-4}}$$

Example 20.7 Calculate the chemical potential of nitrogen in liquid steel at 1600 °C. Steel contents 0.5% C, 0.5% Mn, 0.2% P and 0.01% N.

Given: [wt% N] f_N = k_N $(p_{N2})^{1/2}$, where log k_N = {(−188.1/T) − 1.246}.
e_N^C = 0.25, e_N^N = 0, e_N^P = 0.051, e_N^{Mn} = −0.02.

Solution

$T = 1600 + 273 = 1873$ K.

$\log k_N = \{(-188.1/1873) - 1.246\} = -1.3464$

Therefore, $k_N = 0.045$.

$$\text{Again}, \log f_N = e_N^C \text{wt\%C} + e_N^N \text{wt\%N} + e_N^P \text{Wt\%P} + e_N^{Mn} \text{Wt\%Mn}$$
$$= 0.25 \times 0.5 + 0 \times 0.01 + 0.051 \times 0.2 + (-0.02) \times 0.5$$
$$= 0.1252$$

Therefore, $f_N = 1.334$.

Since $k_N (p_{N2})^{1/2} = [\text{wt\% N}] f_N = 0.01 \times 1.334 = 0.01334$.

Therefore, $(p_{N2})^{1/2} = 0.01334/0.045 = 0.296$.

So, $p_{N2} = 0.0876$.

$$\text{Chemical potential of nitrogen in liquid steel} = \mu_{N2} = \text{RT} \ln p_{N2}$$
$$= 8.314 \times 1873 \times \ln(0.0876)$$
$$= \mathbf{-37.865\,kJ/molofN_2}$$

Example 20.8 Calculate the oxygen potential of liquid steel in contact with a molten slag at 1600 °C. Assume equilibrium partitioning of oxygen between slag and metal.

Given:

(1) $[O]_{wt\%}$ + Fe (l) = FeO (l), $k_1 = 4.35$ at 1600 °C.
(2) O_2 (g) = 2 $[O]_{wt\%}$, $k_2 = 6.89 \times 10^6$ at 1600 °C.

Solution

$T = 1600 + 273 = 1873$ K.

$k_1 = [a_{FeO}/(a_O \cdot a_{Fe})] = (1/h_O)$ [since $a_{Fe} = 1$ and $a_{FeO} = 1$, pure at std state].

Therefore, $h_O = (1/k_1) = (1/4.35) = 0.23$.

Again $k_2 = [(h_O)^2/P_{O2}]$.

Or $P_{O2} = [(h_O)^2/k_2] = [(0.23)^2/(6.89 \times 10^6)] = 7.67 \times 10^{-9}$.

$$\text{Therefore, oxygen potential of liquid steel} = \mu_{O2} = \text{RT} \ln p_{O2}$$
$$= 8.314 \times 1873 \times \ln\left(7.67 \times 10^{-9}\right)$$
$$= -290979.63\,\text{J/mol}$$
$$= \mathbf{-290.98\,kJ/mol}$$

Example 20.9 Calculate and compare thermodynamic efficiency of de-sulphurization of molten steel by pure CaO and molten slag of $CaO–SiO_2–Al_2O_3$. Also find out residual sulphur content in metal if residual oxygen content is 0.001 wt%.

$$\text{Given}: \frac{[w_O]}{[w_S]} = 3 \times 10^{-2}, \text{for } X_{CaO} = 0.6, X_{SiO_2} = 0.1 \text{ and } X_{Al_2O_3} = 0.3 \text{ at } 1650\,^\circ\text{C}.$$

$$C'_S = 7.5 \cdot C_S, \text{slag}, C_S = 0.2.$$

Solution

Assume, wt% of S in slag as 2. i.e. $(w_S) = 2$.

At 1650 °C, for CaO,

$\frac{[w_O]}{[w_S]} = 3 \times 10^{-2}$ (given)

For slag, $C_S = 0.2$.

Now,

$C'_S = \frac{(w_S)\cdot[h_O]}{[h_S]}$ (Eq. 20.155)

$C'_S = \frac{(w_S)\cdot[w_O]}{[w_S]}$ (Taking $[h_O] = [w_O]$ and $[h_S] = [w_S]$ as approximation).

$\frac{C'_S}{(W_S)} = \frac{[W_O]}{[W_S]}$

$\frac{7.5\cdot C_S}{2} = \frac{[w_O]}{[w_S]}$

$\frac{(7.5)\cdot(0.2)}{2} = \frac{[w_O]}{[w_S]}$

$\frac{[w_O]}{[w_S]} = 0.75$

$\frac{[w_O]}{[w_S]}$ is used as *index of thermodynamic efficiency*.

As given, $w_O = 0.001$, then

$\frac{[w_O]}{[w_S]} = 0.75$

$\frac{0.001}{[w_S]} = 0.75$

$[w_S] = \frac{0.001}{0.75}$

$\mathbf{w_S = 1.33 \times 10^{-3}}$ (answer for slag)

If $W_O = 0.001$, then

$\frac{[w_O]}{[w_S]} = 3 \times 10^{-2}$

$[w_S] = \frac{0.001}{3\times10^{-2}}$

$[w_S] = 0.33$ for pure CaO.

The thermodynamic efficiency,

$$\left(\frac{slag}{CaO}\right) = \frac{0.75}{3 \times 10^{-2}} = 25$$

So, the slag is 25 times more effective as a de-sulphurizer as compared to pure lime.

Problems

Problem 20.1 Find out activity of [O] content in liquid steel during reaction of Mn at 1600 °C. Mn content in liquid steel is 0.8%.

Given : $\log k_{Mn} = (12440/T) - 5.33$. [Ans : 0.0609]

Problem 20.2 Estimate the activity of S in liquid steel containing 0.04% S, 0.8% C, 0.1% Si, 0.5% Mn and 0.04% P at 1600 °C in 1 wt% standard state.

Given: $e_S^C = 0.11$, $e_S^{Si} = 0.063$, $e_S^S = -0.028$, $e_S^{Mn} = -0.026$ and $e_S^P = 0.029$.

[Ans: 0.048]

Problem 20.3 Carbon–oxygen reaction in steelmaking:

$[C] + [O] = \{CO\}$

At 0.2% C, and $p_{CO} = 10^{-5}$ atm; find out the [wt% O] at 1600 °C.

Given : $\log k = \left(\frac{1056}{T}\right) + 2.131$ [Ans : 10^{-7}wt% O]

Problem 20.4 Solution of nitrogen in liquid iron may be assumed to obey Sievert's law. Nitrogen content in liquid iron at 1873 K in equilibrium with 1 atm pressure of nitrogen is measured as 0.044 (mass%). What will be the equilibrium nitrogen content in liquid iron (mass%) if nitrogen is reduced to 0.25 atm? [Ans: 0.022%].

References

1. R.G. Ward, *An Introduction to the Physical Chemistry of Iron & Steel Making* (The English Language Book Society, and Edward Arnold (Publishers) Ltd, London, 1962)
2. A.K. Chakrabarti, *Steel Making* (Printice-Hall of India Pvt Ltd, New Delhi, 2007)
3. C. Bodsworth, Physical *Chemistry of Iron and Steel Manufacture* (CBS Publishers & Distributors, Delhi, 1988)
4. E.T. Turkdogan, *Fundametals of Steelmaking* (The Institute of Materials, London, 1996)
5. S. Basu, A.K. Lahiri, Trans. Indian Inst. Met. **66**(5–6), 555 (2013)
6. S.S. Bedarkar, R. Singh, Trans. Indian Inst. Met. **66**(3), 207 (2013)
7. R.H. Tupkary, *An Introduction to Modern Steelmaking*, 5th edn. (Khanna Publishers, Delhi, 1991)
8. A. Ghosh, *Principles of Secondary Processing and Casting of Liquid Steel* (Oxford & IBH Publishing Co, New Delhi, 1990)
9. E.T. Turkdogan, Arch. Eisenhuttenwesen **54**, 4 (1983)
10. S.K. Dutta, A.B. Lele, Trans. Indian Inst. Metal. **56**(1), 19 (2003)
11. N.K. Bharal, *Key Note Lecture at the National Seminar on Recent Advances in Making, Shaping and Applications of Stainless steel* (Vadodara, India, 2001), p. 57 (Privately Collected)
12. *Slag Atlas* Verein Deutscher Eisenhuttenleute (Verlag Stahleisen mBH, Dusseldorf, 1981)

Part VI
Pollution in Iron and Steel Industries

21 Carbon Foot Prints for Iron and Steel Production

The iron and steel industry has one of the largest carbon footprints of any single industrial sector. Ever looming for the steel industry, the movement to monetize a penalty for CO_2 generation is gathering momentum. Production of a tonne of steel generates almost two tonnes of CO_2 emissions that accounting for about 5% of the world's total greenhouse gas (GHG) emissions. Ironmaking is clearly responsible for a massive amount of CO_2 generation. Using the 1.6 tonnes of CO_2 per tonne of hot metal and multiplying by the tonnage of hot metal produced each year gives about 1.64 billion tonnes of CO_2 per year. Ironmaking is responsible for 5–6% of the entire production of CO_2 by all of civilization. The processing steps in steelmaking generate an additional 1–2%.

21.1 Introduction

Carbon emissions are a growing issue on a global scale. Modern society has advanced through industrialization and that has led to better standards of living and prosperity worldwide but also contributed to increased emissions. Steel is a cornerstone and key driver for the world's economy. Steel is at the core of the green economy, in which economic growth and environmental responsibility work hand in hand. Steel is everywhere in our lives. No other material has the same unique combination of strength, formability and versatility. World crude steel production reached 1606 million tonnes (Mt) for the year 2013 [1]. The prevailing process makes steel from iron ore, which is mostly iron oxide, by heating it with carbon; the process forms carbon dioxide (CO_2) as a by-product. Production of a tonne of steel generates almost two tonnes of CO_2 emissions, according to steel industry figures, accounting for as much as 5% of the world's total greenhouse gas (GHG) emissions [2]. The iron and steel industry has one of the largest carbon footprints of any single industrial sector. That is because of the sector's size and the incredibly energy-intensive processes of mining and transporting iron ore, smelting that ore into iron in blast furnaces and then turning the iron into steel. Together they contribute more than 3% of global man-made emissions, says the World Resources Institute [3].

Today, coke-based blast furnaces produce well over 90% of the world's iron. Natural gas (methane) is responsible for about 5%, coal for about 2% (primarily in rotary kilns) and only about

S. K. Dutta: *JPC Bulletin on Iron & Steel*, XIV(5), May 2014, pp 13–21. (Reproduce with permission from *Joint Plant Committee* (JPC), Ministry of Steel, Government of India, Kolkata, India).

S. K. Dutta and Y. B. Chokshi, *Basic Concepts of Iron and Steel Making*,
https://doi.org/10.1007/978-981-15-2437-0_21

1%, or less, is made with charcoal. All these fuels, except for natural gas, share one important characteristic. They are comprised almost totally of carbon and generate lots of carbon dioxide (CO_2) as a by-product. Including the processing step to make the coke from metallurgical coal, approximately 1.6 tonnes of CO_2 are produced for every tonne of iron production. Different sources give figures varying from 1.5 to 2.0 tonnes of CO_2.

Ironmaking is clearly responsible for a massive amount of CO_2 generation. Using the 1.6 tonnes of CO_2 per tonne of hot metal figure and multiplying by the tonnage of hot metal produced each year gives about 1.64 billion tonnes of CO_2 per year. Figures for the total contribution of CO_2 for all of mankind also vary, from 28.2 billion tonnes per year to 31.9 billion tonnes per year [4]. Ironmaking is responsible for 5–6% of the entire production of CO_2 by all of civilization. The processing steps in steelmaking generate an additional 1–2%. Note that each 1% represents 282 million tonnes per year.

The amount of greenhouse gases (GHG) in the atmosphere hits a new record in 2012, the World Meteorological Organization announced. Carbon dioxide alone reached 393.1 parts per million (ppm) in 2012, or 141% of the pre-industrial level. The latest reading follows a 2.2 ppm increase between 2011 and 2012, representing a jump from the steady 2.02 upward climb that characterized the last decade. Emitted through human activities, namely the burning of fossil fuels, CO_2 remains in the atmosphere for hundreds and even thousands of years [5]. The greenhouse gas of most relevance to the world steel industry is carbon dioxide (CO_2). On average, 1.8 tonnes of CO_2 are emitted for every tonne of steel produced. According to the International Energy Agency [6], in 2010, the iron and steel industry accounted for approximately 6.7% of total world CO_2 emissions.

Steel company of Arcelor Mittal (USA) normalized carbon emissions increased 18.6% from 2010 to 2011, according to the company's report [2], released in late 2012. In 2010, the steel giant emitted 1.4 tonnes of CO_2 for every tonne of steel it produced. In 2011, this figure increased to 1.6 tonnes of CO_2 per tonne of steel. The company says that despite the year-on-year increase, its normalized carbon emissions remain lower than the industry average of 1.6 tonnes of CO_2 per tonne of steel in both 2010 and 2011.

Indian steel company of SAIL has reduced load of PM emission from 2.2 kg per tonne crude steel (kg/tcs) in 2007–2008 to 1.01 kg/tcs in 2011–2012, a reduction of 54% during the last 5 years [7]. CO_2 emissions for India's RINL are reported 2.61 t/tcs in 2011–2012.

21.2 Iron and Steel Sector

The production of steel at an integrated iron and steel plant is accomplished using several interrelated processes. The major processes are: (1) coke production; (2) sinter production; (3) hot metal production; (4) raw liquid steel production; (5) ladle metallurgy; (6) continuous casting; (7) hot and cold rolling; and (8) finished product preparation. The operations for secondary steelmaking, where ferrous scrap is recycled by smelting and refining in electric arc furnaces (EAF) include only (4) through (8) above [8]. The interrelation of these operations is shown in a general flow diagram of the iron and steel industry in Fig. 21.1.

The GHG emissions in steelmaking are generated as one of the following: (1) process emissions, in which raw materials and combustion both may contribute to CO_2 emissions; (2) emissions from combustion sources alone; and (3) indirect emissions from consumption of electricity (primarily in EAF and in finishing operations such as rolling mills at both integrated and EAF plants). The major process units at iron and steel facilities include the following: [8]

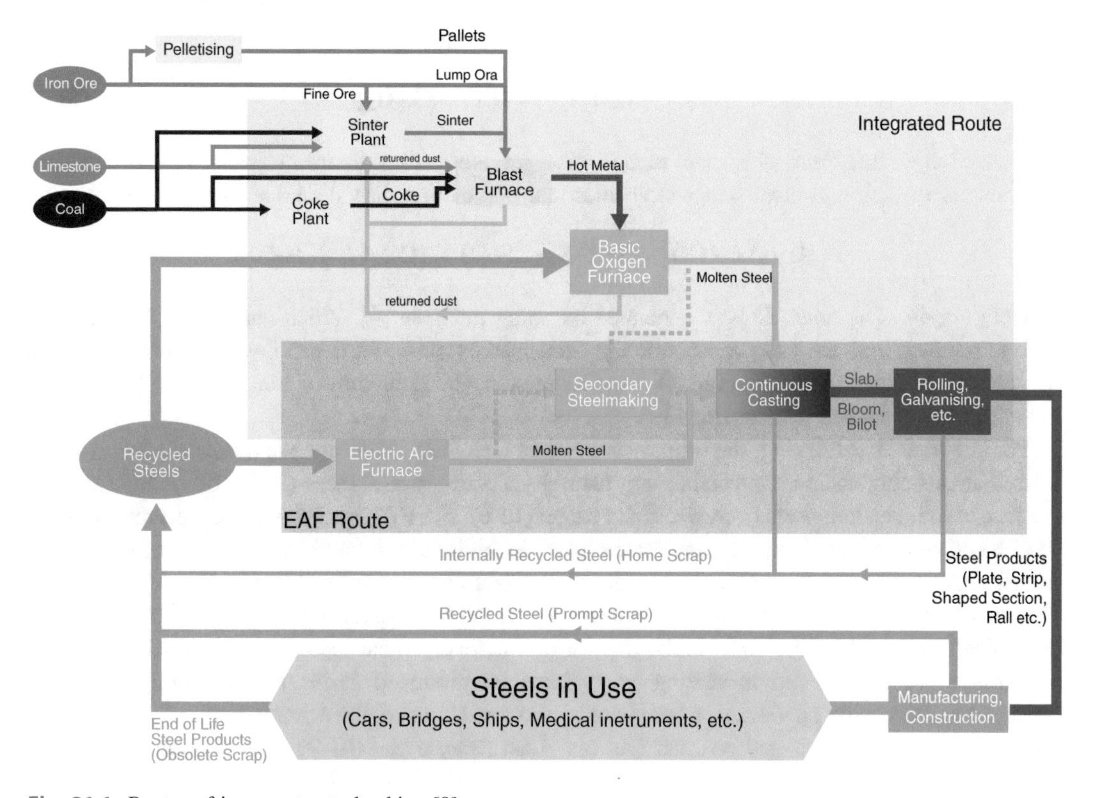

Fig. 21.1 Routes of iron ore to steelmaking [8]

- Sinter plant,
- Non-recovery coke oven battery combustion stack,
- Coke pushing,
- Blast furnace,
- Basic oxygen furnace (BOF) exhaust,
- EAF exhaust.

To reduce iron with coke or charcoal each atom of oxygen in the iron oxide (iron ore) requires one atom of carbon. In a blast furnace, the carbon from the coke or charcoal is first partially oxidized to carbon monoxide (CO) using gaseous oxygen:

$$3\mathrm{C(s)} + 1.5\{\mathrm{O_2}\} = 3\{\mathrm{CO}\} \tag{21.1}$$

This oxygen is provided by the blast air (heated air enriched with additional oxygen, then injected into the blast furnace at the tuyeres). This carbon monoxide diffuses into the highly porous ore and collects an additional oxygen atom from the iron oxide, creating metallic iron (Fe) and forming carbon dioxide:

$$\mathrm{Fe_2O_3(s)} + 3\{\mathrm{CO}\} = 2\mathrm{Fe(s)} + 3\{\mathrm{CO_2}\} \tag{21.2}$$

On the other hand, when natural gas (methane) is used, each molecule of CH_4 is first reformed into one carbon monoxide molecule and two hydrogen molecules:

$$\{CH_4\} + 0.5\{O_2\} = \{CO\} + 2\{H_2\} \tag{21.3}$$

Each of these three molecules will take one oxygen atom from the iron oxide. So the products of the reduction reaction are two water molecules and one carbon dioxide molecule:

$$Fe_2O_3(s) + \{CO\} + 2\{H_2\} = 2Fe(s) + \{CO_2\} + 2\{H_2O\} \tag{21.4}$$

Only one-third as much CO_2 is generated by using methane [4]. When natural gas (methane) is used to produce iron, the CO_2 contribution is decreased by 50–65%, depending on how one account for the CO_2 sources. At the major CO_2 producing step, the reduction of iron from oxide to metal, natural gas represents a 67% savings [9].

For integrated steelmaking, the primary sources of GHG emissions are blast furnace stoves (43%), miscellaneous combustion sources burning natural gas and process gases (30%), other process units (15%) and indirect emissions from electricity usage (12%). For EAF steelmaking, the primary sources of GHG emissions include indirect emissions from electricity usage (50%), combustion of natural gas in miscellaneous combustion units (40%) and steel production in the EAF (10%). For coke facilities, the battery stack is the highest source with over 95% of the GHG emissions for recovery (by-product) coke plants and 99% of the GHG emissions for non-recovery (heat recovery) plants [8].

Furthermore, several carbon-bearing by-products are produced during the iron and steelmaking process. During coking operations, where coal is converted to coke, the primary by-products are coke, coke oven gas (COG), tar and residual fuel oil. Blast furnace gas (BFG) is a by-product produced during the production of crude iron. If recovered, the COG and BFG gaseous by-products are considered fuel gases because they contain methane and other hydrocarbon components. These gasses are typically used for oven under-firing and as a combustion gas for the furnaces and boilers used to provide electricity and steam to the plant [10]. All by-products can be used for internal use or sold off to other companies.

21.3 Estimation of CO_2 Emissions

21.3.1 One Method

21.3.1.1 Coke Production

This guidance provides two methods for calculating the CO_2 emissions from coke manufacture. Facilities should chose between the two based on whether or not the coke they consumed was produced onsite. This is because coke manufacture may entail the use of by-products from industrial activities that took place within the site of coke production. Facilities that consume coke produced offsite should not account for the emissions from these by-products; otherwise, emissions from these by-products might be double-counted [11]. The separate consideration of emissions from onsite and offsite coke manufacture also allows the separation of emissions based on ownership. In other words, if coke is purchased from an entity outside the organizational boundaries of the reporting company. Likewise, if the coke production facility is offsite.

Calculating CO_2 emissions from onsite coke production:

$$E_{CO_2,\text{energy}} = \begin{bmatrix} CC \cdot C_{CC} + \sum_a (PM_a \cdot C_a) + BG \cdot C_{BG} \\ -CO \cdot C_{CO} - COG \cdot C_{COG} - \sum_b (COB_b \cdot C_b) \end{bmatrix} \cdot \frac{44}{12} \quad (21.5)$$

where:

$E_{CO_2,\text{energy}}$ emissions of CO_2 from onsite coke production, tonnes,
CC quantity of coking coal consumed for coke production in onsite integrated iron and steel production facilities, tonnes,
PM_a quantity of other process material *a*, other than those listed as separate terms, such as natural gas and fuel oil, consumed for coke and sinter production in onsite coke production and iron and steel production facilities, tonnes,
BG quantity of blast furnace gas consumed in coke ovens, m_3 (or other units such as tonnes or GJ),
CO quantity of coke produced onsite at iron and steel production facilities, tonnes,
COG quantity of coke oven gas transferred offsite, m^3 (or other units such as tonnes or GJ),
COB_b quantity of coke oven by-product *b*, transferred offsite to other facilities, tonnes,
C_x carbon content of material input or output *x*, tonnes C/(unit for material *x*) [tonnes C/tonne].

Calculating CO_2 emissions from offsite coke production:

$$E_{CO_2,\text{energy}} = \begin{bmatrix} CC \cdot C_{CC} + \sum_a (PM_a \cdot C_a) - NIC \cdot C_{NIC} \\ -COG \cdot C_{COG} - \sum_b (COB_b \cdot C_b) \end{bmatrix} \cdot \frac{44}{12} \quad (21.6)$$

where:

$E_{CO_2,\text{energy}}$ emissions of CO_2 from offsite coke production (tonnes),
CC quantity of coking coal used in non-integrated coke production facilities (tonnes),
PM_a quantity of other process material *a*, other than coking coal, such as natural gas and fuel oil consumed nationally in non-integrated coke production (tonnes),
NIC quantity of coke produced offsite in non-integrated coke production facilities nationally (tonnes),
COG quantity of coke oven gas produced in offsite non-integrated coke production facility (tonnes),
COB_b quantity of coke oven by-product *b*, produced nationally in offsite non-integrated facilities and transferred offsite to other facilities (tonnes),
C_x carbon content of material input or output *x*, tonnes C/(unit for material *x*) [e.g. tonnes C/tonne].

21.3.1.2 Sinter Production

Calculating CO_2 emissions from sinter production [11]:

$$E_{CO_2} = \begin{bmatrix} CBR \cdot C_{CBR} + COG \cdot C_{COG} + BG \cdot C_{BG} \\ + \sum_a (PM_a \cdot C_a) - SOG \cdot C_{SOG} \end{bmatrix} \cdot \frac{44}{12} \quad (21.7)$$

where:

E_{CO_2} emissions of CO_2 from sinter production (tonnes),
CBR quantity of purchased and onsite produced coke breeze used for sinter production (tonnes),
COG quantity of coke oven gas consumed in blast furnace in sinter production (m^3),
BG quantity of blast furnace gas consumed in sinter production (m^3),
PM_a quantity of other process material a, other than those listed as separate terms, such as natural gas and fuel oil, consumed during sinter production (tonnes),
SOG quantity of sinter off-gas transferred offsite either to iron and steel production facilities or other facilities (m^3),
C_x carbon content of material input or output x, tonnes C/(unit for material x) [e.g. tonnes C/tonne].

21.3.1.3 Carbon Combustion

Calculating CO_2 emissions using carbon content data that are expressed on a mass basis:

$$E = A_{f,m} \cdot F_{c,m} \cdot F_{ox} \cdot (44/12) \tag{21.8}$$

where:

E Amount of CO_2 emitted (tonnes),
$A_{f,m}$ Mass of fuel consumed (tonnes),
$F_{c,\ m}$ Carbon content of fuel on a mass basis (tonnes carbon/tonne),
F_{ox} Fraction oxidation factor, (44/12) = Ratio of the molecular weight of carbon to that of CO_2.

Calculating CO_2 emissions from stationary combustion sources using carbon content data expressed on an energy basis [11]:

$$E = A \cdot HV_f \cdot F_{c,h} \cdot F_{ox} \cdot (44/12) \tag{21.9}$$

where:

E Amount of CO_2 emitted (tonnes),
A Mass of fuel consumed (e.g. metric tonnes),
HV_f Heating value of fuel (MJ/kg),
$F_{c,h}$ Carbon content of fuel on a heating value basis (tonnes C/GJ),
F_{ox} Fraction oxidation factor, (44/12) = Ratio of the molecular weight of carbon to that of CO_2.

A small fraction of a fuel's carbon content can escape oxidation and remain as a solid after combustion in the form of ash or soot (for solid fuels) or particulate emissions (for natural gas and other gaseous fuels). This unoxidized fraction is a function of several factors, including fuel type, combustion technology, equipment age and operating practices. This fraction can be assumed to contribute no further to CO_2 emissions, so it is easily corrected for in estimating CO_2 emissions. The stationary combustion CO_2 methods in this tool use an 'oxidation factor' to account for the unoxidized fraction (where 1.0 = complete oxidation). In general, variability in the oxidation factor is low for gaseous and liquid fuels, but can be much larger for solid fuels. The oxidation factor for solid fuels is ranged from 0.88 to 0.99.

21.3.1.4 Lime Production

Two different methods are presented here to allow facilities to calculate the CO_2 emissions from limestone and dolomite production. Calculating CO_2 emissions from lime production [11]:

$$E_{CO_2} = \sum_i (EF_i \cdot M_i \cdot F_i) - M_d \cdot C_d \cdot (1 - F_d) \cdot EF_d \quad (21.10)$$

where:

E_{CO_2} emissions of CO_2 from lime production (tonnes),
EF_i emission factor for carbonate i (tonnes CO_2/tonne carbonate) (e.g. 0.44 for $CaCO_3$ and 0.48 for $CaMg(CO_3)_2$),
M_i weight or mass of carbonate i consumed (tonnes),
F_i fraction calcination achieved for carbonate i (fraction). [e.g. A value of 1.0 (i.e. 100% calcination)],
M_d weight or mass of lime kiln dust (LKD) (tonnes),
C_d weight fraction of original carbonate in the LKD (fraction),
F_d fraction calcination achieved for the LKD (fraction) [e.g. A value of 1.0 (i.e. 100% calcination)],
EF_d emission factor for the uncalcined carbonate in the LKD (tonnes CO_2/tonne carbonate).

21.3.1.5 Iron and Steel Production

The CO_2 emissions from iron and steelmaking can be calculated by the following equation [11]:

$$E_{CO_2} = \begin{bmatrix} PC \cdot C_{PC} + \sum_a (COB_a \cdot C_a) + CI \cdot C_{CI} + L \cdot C_L + D \cdot C_D + CE \cdot C_{CE} \\ + \sum_b (O_b \cdot C_b) + COG \cdot C_{COG} - S \cdot C_S - IP \cdot C_{IP} - BG \cdot C_{BG} \end{bmatrix} \cdot \frac{44}{12} \quad (21.11)$$

where:

E_{CO_2} emissions of CO_2 from iron and steel production (tonnes),
PC quantity of coke consumed in iron and steel production (not including sinter production) (tonnes),
COB_a quantity of onsite coke oven by-product a, consumed in blast furnace (tonnes),
CI quantity of coal directly injected into blast furnace (tonnes),
L quantity of limestone consumed in iron and steel production (tonnes),
D quantity of dolomite consumed in iron and steel production (tonnes),
CE quantity of carbon electrodes consumed in EAFs (tonnes),
O_b quantity of other carbonaceous and process material b, consumed in iron and steel production, such as sinter or waste plastic (tonnes),
COG quantity of coke oven gas consumed in blast furnace in iron and steel production (m^3),
S quantity of steel produced (tonnes),
IP quantity of iron production not converted to steel (tonnes),
BG quantity of blast furnace gas transferred offsite (m^3),
C_x carbon content of material input or output x, tonnes C/(unit for material x) [e.g. tonnes C/tonne].

21.3.1.6 Direct Reduced Iron (DRI) Production

The CO_2 emissions from DRI production stem from the combustion of fuel, coke breeze, metallurgical coke or other carbonaceous materials. The emissions can be calculated [11]:

$$E_{CO_2} = (\mathrm{DRI_{NG}} \cdot C_{NG} + \mathrm{DRI_{BZ}} \cdot C_{BZ} \cdot \mathrm{DRI_{CK}} \cdot C_{CK}) \cdot \frac{44}{12} \tag{21.12}$$

where:

E_{CO_2} emissions of CO_2 (tonnes),
$\mathrm{DRI_{NG}}$ amount of natural gas used in direct reduced iron production (GJ),
$\mathrm{DRI_{BZ}}$ amount of coke breeze used in direct reduced iron production (GJ),
$\mathrm{DRI_{CK}}$ amount of metallurgical coke used in direct reduced iron production (GJ),
C_{NG} carbon content of natural gas (tonne C/GJ),
C_{BZ} carbon content of coke breeze (tonne C/GJ),
C_{CK} carbon content of metallurgical coke (tonne C/GJ).

21.3.2 Another Methods

The iron and steel sectors are divided into two types of facilities: (1) primary iron and steel plants that produce iron and steel from coal and/or purchased coke and iron ore and (2) EAF facilities that produce steel from scrap.

(1) Primary Facilities: The method for calculating GHG emissions from primary iron and steel facilities includes calculating emissions from carbonate flux and adjusting stationary combustion emissions to account for carbon sold offsite in carbon bearing products and by-products. Emissions of CO_2 from use of the carbonate flux are calculated based on the amount of flux used and the stoichiometric ratio of CO_2 to $CaCO_3$ and $MgCO_3$. The CO_2 emissions from carbon inputs are adjusted to account for the amount of carbon that remains in products sold offsite. Equation (21.13) represents the method used to calculate CO_2 emissions from iron and steel production at primary facilities [10].

$$\begin{aligned}\text{Emissions} = {} & [(\text{Flux} \cdot \text{CF}_{\text{Flux}}) - (\text{B–P} \cdot \text{CF}_{\text{B–P}}) - (\text{Iron} \cdot \text{CF}_{\text{I}}) \\ & - (\text{Steel} \cdot \text{CFs})][\text{CO}_2(\text{m.w.})/\text{C}(\text{m.w.})]\end{aligned} \tag{21.13}$$

where:

Flux Mass of flux used,
CF_{Flux} Flux carbon factor (mass C/mass flux),
B–P Mass or volume of by-products sold,
$\text{CF}_{\text{B–P}}$ By-product carbon factor (mass C/(mass or volume of by-products sold),
Iron Mass of iron sold,
CF_{I} Iron–carbon factor (mass C/mass iron sold),

Steel	Mass of steel sold,
CF_S	Steel carbon factor (mass C/mass steel sold),
CO_2 (m.w.)	Molecular weight of CO_2,
C (m.w.)	Molecular weight of carbon.

(2) EAF Facilities: The method for calculating GHG emissions from EAF steel facilities includes calculating emissions from carbonate flux and use of carbon electrodes. Emissions of CO_2 from the use of carbonate flux are calculated based on the amount of flux used and the stoichiometric ratio of CO_2 to $CaCO_3$ and $MgCO_3$. The emissions from the use of electrodes in EAF steel production are estimated based on the number of electrodes used and the carbon content of the electrodes. Equation (21.14) represents the method used to calculate CO_2 emissions from steel production at EAF facilities [10].

$$\text{Emissions} = [(\text{Flux} \cdot \text{CF}_{\text{Flux}}) + (\text{ Electrode} \cdot \text{CF}_{\text{E}})] \times [\text{CO}_2(\text{m.w.})/\text{C}(\text{m.w.})] \tag{21.14}$$

where:

Flux	Mass of flux used,
CF_{Flux}	Flux carbon factor (mass C/mass flux),
Electrode	Mass of carbon electrode used,
CF_E	Electrode carbon factor (mass C/mass of electrode),
CO_2 (m.w.)	Molecular weight of CO_2,
C (m.w.)	Molecular weight of carbon.

21.4 Product and by-Product

In many applications, steel has a very long life and as a result the contribution of modern steels in improving the energy efficiency of buildings, plants, machinery and transportation are much more important in helping mankind reduce its carbon footprint than the emissions associated with the initial steel production [6].

The production of steel results in the generation of by-products that can reduce CO_2 emissions by substituting natural resources in other industries. For example, blast furnace slag is used by the cement industry allowing it to reduce its CO_2 emissions significantly. Steelmaking slags are also used as civil works aggregates, thereby saving natural resources and environmental impact. World Steel Association strongly believes that by-products and natural resources should compete within the same legal framework as they can both serve the same purpose.

21.5 Summaries

Three ways to reduce carbon emissions:

- Reduce energy consumption—limited potential. USA and European countries reduced energy and CO_2 per tonne steel $\sim$50% since 1970s,
- Sequester CO_2 underground—many complications, does not reduce emissions,
- Use energy source with less carbon—great potential.

Roughly half of the CO_2 emitted by human activities is absorbed by the biosphere and ocean. As a result of this, climate is changing, weather is more extreme, temperature is increasing, ice sheets and glaciers are melting, and sea's water levels are rising. To combat these rising levels and their effects, immediate and multilateral action is needed. Limiting climate change will require large and sustained reductions of greenhouse gas emissions; otherwise, we will jeopardize the future of our children, grandchildren and many future generations.

Probable Questions

1. What do you mean by greenhouse gas (GHG) emissions?
2. Discuss the estimation of CO_2 emissions for coke and sinter production.
3. Discuss the estimation of CO_2 emissions for carbon combustion and lime production.
4. Discuss the estimation of CO_2 emissions for iron and steel production.
5. Discuss the estimation of CO_2 emissions for steel production at EAF.

References

1. *World Steel in Figures 2014* (World Steel Association)
2. www.environmentalleader.com/2013/05/10/mit-develops-ghg-free-steel/, 10 May 2013
3. http://knowledge.allianz.com/environment/climate_change/?651/ten-sources-of-greenhouse-gases-gallery
4. R. Hunter, *Direct from Midrex*, 3rd/4th Quarter (2009)
5. www.natureworldnews.com/articles/4791/20131106/greenhouse-gases-reached-new-record-in-2012.htm
6. Steel's contribution to a low carbon future. worldsteel.org, Mar 2013
7. *Annual Report 2012–13* (Ministry of Steel, Government of India)
8. *Available and Emerging Technologies for Reducing Greenhouse Gas Emissions from the Iron and Steel Industry* (U.S. Environmental Protection Agency, Sept 2012)
9. J. Kopfle, *Green Steel Summit* (Midrex Tech, 20–21 May 2010)
10. *Draft Iron & Steel Production—Guidance, Climate Leaders* (U.S. Environmental Protection Agency, Aug 2003)
11. *Iron and Steel Version 2.0 Guidance*: *Calculating Greenhouse Gas Emissions from Iron and Steel Production* (Jan 2008)

Appendix A
Ellingham Diagram

See Fig. A.1.

S. K. Dutta and Y. B. Chokshi, *Basic Concepts of Iron and Steel Making*,
https://doi.org/10.1007/978-981-15-2437-0

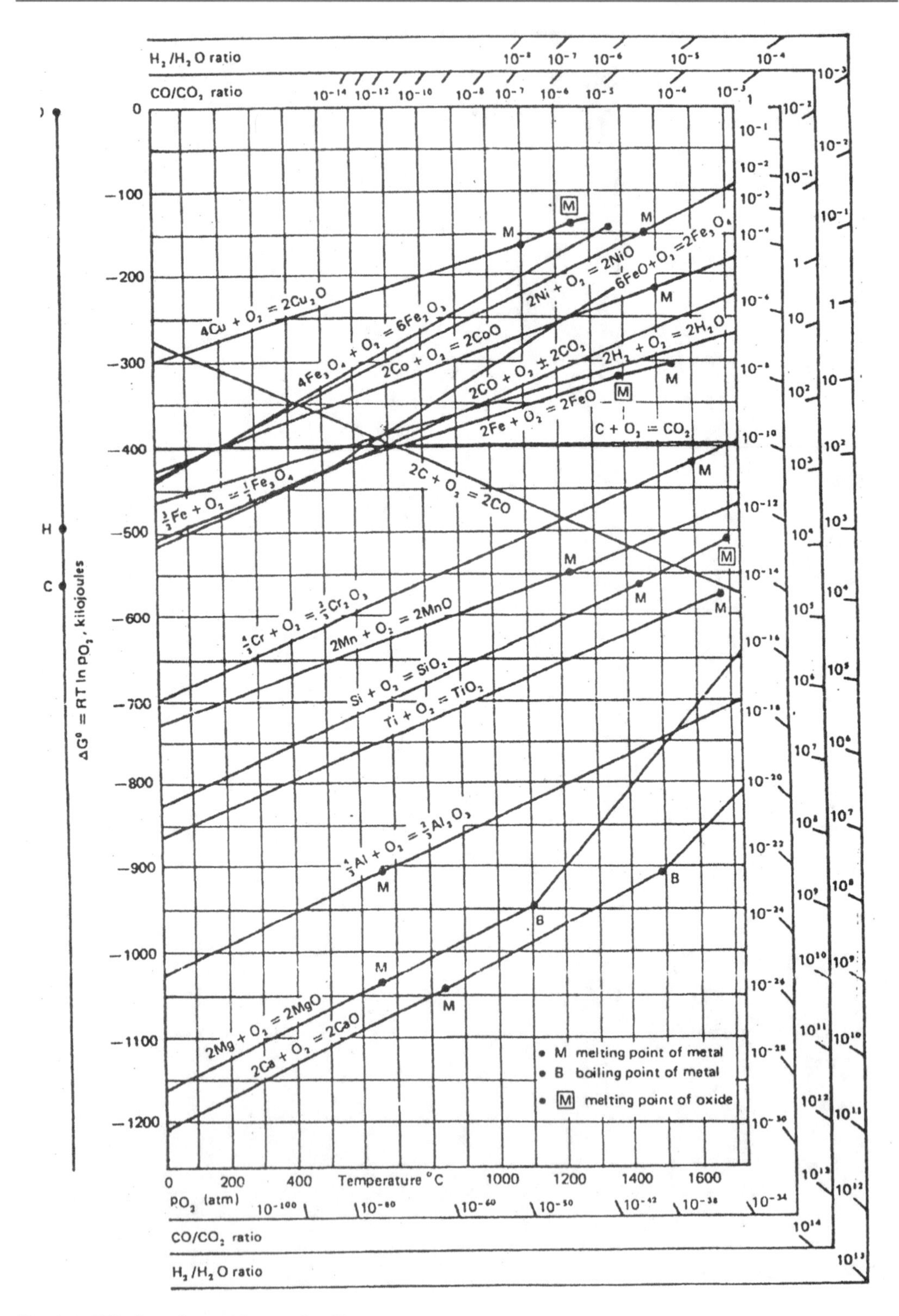

Fig. A.1 Ellingham diagram for metal oxides

Appendix B
Physical Properties of Metals

Metal	Symbol	Atomic weight	Specific gravity	Melting point (°C)	Boiling point (°C)
Aluminium	Al	26.981	2.7	660	2520
Antimony	Sb	121.7	6.68	630.5	1590
Arsenic	As	74.921	5.7	Volatilizes	610 (sublimes)
Barium	Ba	137.3	3.5	710	1,770
Beryllium	Be	9.012	1.8	1,280	2,450
Bismuth	Bi	208.98	9.8	271.3	1,530
Boron	B	10.81	2.3	2,030	2,550 (sublimes)
Cadmium	Cd	112.4	8.65	320.9	767
Calcium	Ca	40.08	1.55	843	1484
Chromium	Cr	51.996	7.19	1,857	2,672
Cobalt	Co	58.933	8.9	1493	2930
Copper	Cu	63.54	8.96	1,083	2,580
Gold	Au	196.97	19.32	1063	2860
Hafnium	Hf	178.4	13.29	2,227	4,600
Iron	Fe	55.84	7.9	1539	2860
Lead	Pb	207.2	11.37	327.4	1,750
Lithium	Li	6.94	0.53	181	1324
Magnesium	Mg	24.305	1.74	651	1,107
Manganese	Mn	54.938	7.3	1,244	2,150
Molybdenum	Mo	95.9	10.22	2,610	5,560
Neptunium	Np	237.05	–	637	–
Nickel	Ni	58.69	8.88	1,445	3,075
Niobium	Nb	92.906	8.57	2,415	3,300
Palladium	Pd	106.42	12.0	1552	2940
Platinum	Pt	195	21.45	1769	4100
Plutonium	Pu	239.06	19.5	639	3,235
Potassium	K	39.09	0.86	63.2	779
Rhodium	Rh	102.91	12.4	1966	3700
Scandium	Sc	44.956	3.0	1538	2870
Silicon	Si	28.085	2.37	1,414	2,287
Silver	Ag	107.87	10.5	960.8	2200
Sodium	Na	22.989	0.97	97.8	883
Tantalum	Ta	180.95	16.6	2,996	6,000

(continued)

S. K. Dutta and Y. B. Chokshi, *Basic Concepts of Iron and Steel Making*,
https://doi.org/10.1007/978-981-15-2437-0

Metal	Symbol	Atomic weight	Specific gravity	Melting point (°C)	Boiling point (°C)
Thorium	Th	232.04	11.5	1,750	4,200
Tin	Sn	118.71	7.29	232	2,270
Titanium	Ti	47.88	4.54	1668	3285
Tungsten	W	183.8	19.32	3,410	5,927
Uranium	U	238.03	19.05	1,132	3,813
Vanadium	V	50.941	6.11	1,900	3,000
Zinc	Zn	65.39	7.14	419.5	907
Zirconium	Zr	91.224	6.45	1852	4400

Source Extraction of nuclear and non-ferrous metals: Sujay Kumar Dutta, Dharmesh R. Lodhari, Springer Nature, Singapore, 2018

Appendix C
Standard Free Energy Change for Some Important Reactions ΔG_T°

Reaction	ΔG_T°, J/mol product[a]	ΔG_{1000}°, kJ/mol	ΔG_{1473}°, kJ/mol
C (s) + ½ O_2 (g) = CO (g)	−111712.8 − 87.65 T	−199.37	−240.82
C (s) + O_2 (g) = CO_2 (g)	−394132.8 − 0.84 T	−394.97	−395.37
Fe (s) + ½ O_2 (g) = FeO (s)	−264889.04 + 65.35 T	−199.54	−168.63
Fe (l) + ½ O_2 (g) = FeO (l)	−232714.08 + 45.31 T	−187.40	−165.97
3FeO (s) + ½ O_2 (g) = Fe_3O_4 (s)	−312210.08 + 125.10 T	−187.11	−127.94
2Fe_3O_4(s) + ½O_2(g) = 3Fe_2O_3(s)	−249450.08 + 140.67 T	−36.26	−14.08
3 Fe (s) + C (s) = Fe_3C (s)	+25104 − 17.02 T	+8.08	+0.0335
H_2 (g) + ½ O_2 (g) = H_2O (g)	−246437.6 + 54.81 T	−191.63	−165.70
C (s) + 2 H_2 (g) = CH_4 (g)	−69119.68 + 51.25 T log T − 65.35 T	+19.28	+73.79

[a]*Source Metallurgical Thermo-chemistry*: O. Kubeschewski and C. B. Alcock, 5th Ed Maxwell Macmillan, 1989

S. K. Dutta and Y. B. Chokshi, *Basic Concepts of Iron and Steel Making*,
https://doi.org/10.1007/978-981-15-2437-0

Appendix D
Free Energy and Enthalpy Values for Some Important Reactions

Reaction	ΔG_T° (J/mol CO/H_2)	ΔG_T° (kJ/mol CO/H_2)		ΔG_{298}° (kJ/mol CO/H_2)
		298 K	1373 K	
$3Fe_2O_3$ (s) + CO (g) = $2Fe_3O_4$ (s) + CO_2 (g)	−32969.92 − 53.85 T	−49.02	−101.52	−52.43
Fe_3O_4 (s) + CO (g) = 3FeO (s) + CO_2 (g)	29790 − 38.28 T	18.38	−18.95	40.46
FeO (s) + CO (g) = Fe (s) + CO_2 (g)	−17530.96 + 21.46 T	−11.13	9.79	−18.54
½CO_2 (g) + ½C (s) = CO (g)	85353.6 − 87.23 T	59.36	−25.69	85.35
1/3 Fe_2O_3 (s) + CO (g) = 2/3 Fe (s) + CO_2 (g)	−8730.63 − 0.18 T	−8.78	−8.96	−9.2
1/3 Fe_2O_3 (s) + C (s) = 2/3 Fe (s) + CO (g)	161976.57 −174.64 T	109.93	−60.34	251.97
3/2 Fe (s) + CO (g) = ½ Fe_3C (s) + ½ CO_2 (g)	−72801.6 + 78.725 T	−49.34	27.42	−73.66
$3Fe_2O_3$ (s) + H_2 (g) = $2Fe_3O_4$ (s) + H_2O (g)	3012.48 − 85.86 T	−22.57	−106.28	−55.31
Fe_3O_4 (s) + H_2 (g) = 3FeO (s) + H_2O (g)	65772.48 − 70.29 T	44.83	−23.71	37.57
FeO (s) + H_2 (g) = Fe (s) + H_2O (g)	18451.44 − 10.54 T	15.31	5.03	−21.42
1/3 Fe_2O_3 (s) + H_2 (g) = 2/3 Fe (s) + H_2O (g)	27251.77 − 32.19 T	17.66	−13.73	119.32
CH_4 (g) + H_2O (g) = CO (g) + 3 H_2 (g)	203844.48 − 51.25 T log T − 77.11 T	143.08	−96.88	206.3
CH_4 (g) + ½ O_2 (g) = CO (g) + 2 H_2 (g)	−42593.12 − 51.25 T log T − 22.3 T	−87.03	−273.54	−35.7
½ CH_4 (g) + ½ CO_2 (g) = CO (g) + H_2 (g)	119913.44 − 25.625 T log T − 54.555 T	84.76	−50.82	123.7
CH_4 (g) + ½ H_2O (g) + 1/4 O_2 (g) = CO (g) + 5/2 H_2 (g)	80625.68 − 51.25 T log T − 49.705 T	28.03	−185.21	85.3
3Fe (s) + CO (g) + H_2 (g) = Fe_3C (s) + H_2O (g)	−109620.8 +125.44 T	−72.24	50.06	−150.185
3/2 Fe (s) + ½ CH_4 (g) = ½ Fe_3C (s) + H_2 (g)	47111.84 − 25.625T log T + 24.165 T	35.42	−23.41	49.98

S. K. Dutta and Y. B. Chokshi, *Basic Concepts of Iron and Steel Making*,
https://doi.org/10.1007/978-981-15-2437-0

Appendix E
Free Energy Values for Some Reactions

S. No.	Reaction	Temperature range (K)	$\Delta G^\circ = \Delta H^\circ - T\Delta S^\circ$		Probable accuracy (±J)
			$-\Delta H^\circ$ (kJ)	$-\Delta S^\circ$ (J/K)	
1	4/3<Al> + {O_2} = 2/3<Al_2O_3>	298–930	1073.61	181.17	42,000
2	4/3<Al> + {O_2} = 2/3<Al_2O_3>	930–2318	1077.38	185.35	42,000
3	4/3<Al> + {O_2} = 2/3<Al_2O_3>	2318–2330	1005.00	153.97	42,000
4	4/3<Al> + {S_2} = 2/3<Al_2S_3>	298	ΔG° = −550.196		
5	<Al_2O_3> + <SiO_2> = <Al_2SiO_5>				
6	Andalusite	298–1600	163.39	−22.59	13,000
7	Cyanite	298–1600	165.69	3.77	13,000
8	Sillimmanite	298–1600	192.25	−10.46	13,000
9	2<Ba> + {O_2} = 2<BaO>	298	1116.29	0.00	0
10	2<Ba> + {S_2} = 2<BaS>	298	ΔG° = −928.848		
11	<C> + {O_2} = {CO_2}	298–2500	394.13	−0.84	4000
12	2<C> + {O_2} = 2{CO}	298–2500	223.43	−175.31	4000
13	<C> + {S_2} = {CS_2}	298–1600	12.97	−7.24	4000
14	2<Ca> + {O_2} = 2<CaO>	298–1124	1266.25	−197.99	42,000
15	2(Ca) + {O_2} = 2<CaO>	1124–1760	1284.91	214.56	42,000
16	2(Ca) + {O_2} = 2(CaO)	ca. 1873	1184.49	179.62	42,000
17	2<Ca>$_{FCC}$ + {S_2} = 2<CaS>	298–673	1083.11	190.87	4000
18	2<Ca>$_{HCP}$ + {S_2} = 2<CaS>	673–1124	1084.07	192.13	4000
19	2(Ca) + {S_2} = 2<CaS>	1124–1760	1102.74	208.70	4000
20	2(Ca) + {S_2} = 2(CaS)	ca. 1873	1026.75	173.76	21,000
21	<CaO> + <SiO_2> = <$CaSiO_3$>$_a$	298–1483	89.12	0.50	4000
22	<CaO> + <SiO_2> = <$CaSiO_3$>$_b$	1483–1813	83.26	−3.43	13,000
23	2<CaO> + <SiO_2> = <Ca_2SiO_4>	298–1500	124.68	0.00	13,000
24	3<CaO> + <SiO_2> = <Ca_3SiO_5>	298–1500	122.80	0.00	42,000
25	3<CaO> + <P_2O_5> = <$Ca_3P_2O_8$>$_a$	298–1830	1695.98	379.07	50,000
26	4<CaO> + <P_2O_5> = <$Ca_4P_2O_9$>	298–1940	1738.79	395.81	50,000
27	2Ce + {S_2} = 2<CeS>	298–2200	1117.13	167.36	42,000
28	2<Co> + {O_2} = 2<CoO>	298–1763	467.77	143.72	42,000
29	2(Co) + {O_2} = 2<CoO>	1763–2200	506.52	164.01	42,000
30	4/3<Cr> + {O_2} = 2/3<Cr_2O_3>	298–1868	746.84	173.22	13,000
31	4/3(Cr) + {O_2} = 2/3<Cr_2O_3>	1868–2500	768.77	184.97	13,000

(continued)

S. K. Dutta and Y. B. Chokshi, *Basic Concepts of Iron and Steel Making*,
https://doi.org/10.1007/978-981-15-2437-0

S. No.	Reaction	Temperature range (K)	$\Delta G° = \Delta H° - T\Delta S°$		Probable accuracy (±J)
			$-\Delta H°$ (kJ)	$-\Delta S°$ (J/K)	
32	$<Cr_2O_3> + <FeO> = <Fe_2Cr_2O_4>$	1173–1700	15.69	−11.92	8000
33	$4<Cu> + \{O_2\} = 2<Cu_2O>$	298–1357	333.46	126.02	4000
34	$4(Cu) + \{O_2\} = 2<Cu_2O>$	1357–1509	385.51	164.35	4000
35	$4(Cu) + \{O_2\} = 2(Cu_2O)$	1509–1573	273.05	89.70	4000
36	$2(1-x)<Fe> + \{O_2\} = 2<Fe_{1-x}O>$	848–1644	525.93	128.41	4000
37	$2(1-x)<Fe> + \{O_2\} = 2(Fe_{1-x}O)$	1644–1808	464.84	91.25	13,000
38	$2(1-x)[Fe] + \{O_2\} = 2(Fe_{1-x}O)$	1808–2000	494.13	107.57	13,000
39	$6<FeO> + \{O_2\} = 2<Fe_3O_4>$	298–1642	624.46	250.20	13,000
40	$4<Fe_3O_4> + \{O_2\} = 6<Fe_2O_3>$	298–1730	498.94	281.37	42,000
41	$2<Fe>_a + \{S_2\} = 2<FeS>_a$	298–412	310.95	130.46	4000
42	$2<Fe>_a + \{S_2\} = 2<FeS>_b$	412–1179	300.49	105.10	4000
43	$2<Fe>_g + \{S_2\} = 2<FeS>_b$	1179–1261	301.83	106.61	4000
44	$2[Fe] + \{S_2\} = 2(FeS)$	ca. 1873	268.11	79.66	13,000
45	$2[Fe] + (Fe_{(1-x)}O) + (SiO_2) = (FeSiO_3)$	1873	$\Delta G° = -11.72$		4000
46	$2x<Fe> + 2<Fe_{(1-x)}O> + <SiO_2> = <Fe_2SiO_4>$	843–1478	40.17	15.48	13,000
47	$2x<Fe> + 2<Fe_{(1-x)}O> + <SiO_2> = (Fe_2SiO_4)$	1478–1644	−34.02	−34.73	13,000
48	$2x<Fe> + 2(Fe_{(1-x)}O) + <SiO_2> = (Fe_2SiO_4)$	1644–1808	27.07	2.51	13,000
49	$2x[Fe] + 2(Fe_{(1-x)}O) + <SiO_2> = (Fe_2SiO_4)$	1808–1986	28.58	3.35	13,000
50	$2x[Fe] + 2(Fe_{(1-x)}O) + (SiO_2) = (Fe_2SiO_4)$	>1986	37.36	7.78	13,000
51	$2\{H_2\} + \{O_2\} = 2\{H_2O\}$	373–2500	493.71	111.92	4000
52	$2\{H_2\} + \{S_2\} = 2\{H_2S\}$	298–1800	180.58	98.78	4000
53	$2<Mg> + \{O_2\} = 2<MgO>$	298–923	1215.03	192.88	42,000
54	$2[Mg] + \{O_2\} = 2<MgO>$	923–1380	1248.51	231.79	42,000
55	$2\{Mg\} + \{O_2\} = 2<MgO>$	1380–2000	1501.03	429.28	42,000
56	$2[Mg] + \{O_2\} = 2(MgO)$	ca. 1873	1061.48	175.31	42,000
57	$2<Mg> + \{S_2\} = 2<MgS>$	298–923	833.87	190.79	42,000
58	$2[Mg] + \{S_2\} = 2<MgS>$	923–1380	851.86	210.46	42,000
59	$2\{Mg\} + \{S_2\} = 2<MgS>$	1380–2000	1124.24	407.94	42,000
60	$2[Mg] + \{S_2\} = 2(MgS)$	ca. 1873	737.64	164.43	42,000
61	$<MgO> + <SiO_2> = <MgSiO_3>$	298–1600	37.24	4.60	4000
62	$2<MgO> + <SiO_2> = <Mg_2SiO_4>$	298–1700	63.26	0.00	13,000
63	$3<MgO> + \{P_2O_5\} = <Mg_3P_2O_8>$	1273–1600	1479.50	395.81	50,000
64	$2<Mn> + \{O_2\} = 2<MnO>$	298–1500	769.44	144.89	13,000
65	$2[Mn] + \{O_2\} = 2<MnO>$	1500–2051	798.31	164.22	13,000
66	$2[Mn] + \{O_2\} = 2(MnO)$	2051–2200	678.64	105.65	13,000
67	$2<Mn>_a + \{S_2\} = 2<MnS>$	298–1000	535.55	128.20	13,000
68	$2<Mn>_b + \{S_2\} = 2<MnS>$	1000–1374	540.03	132.67	13,000
69	$2<Mn>_g + \{S_2\} = 2<MnS>$	1374–1410	544.59	135.98	13,000
70	$2<Mn>_\delta + \{S_2\} = 2<MnS>$	1410–1517	548.19	138.53	13,000
71	$2[Mn] + \{S_2\} = 2<MnS>$	1517–1803	577.48	157.82	13,000

(continued)

S. No.	Reaction	Temperature range (K)	$\Delta G^\circ = \Delta H^\circ - T\Delta S^\circ$		Probable accuracy ($\pm$J)
			$-\Delta H^\circ$ (kJ)	$-\Delta S^\circ$ (J/K)	
72	$2[Mn] + \{S_2\} = 2(MnS)$	1803–2000	525.26	128.87	13,000
73	$<MnO> + <SiO_2> = <MnSiO_3>$	298–1600	29.71	12.55	42,000
74	$3<MnO> + \{P_2O_5\} = <Mn_3P_2O_8>$	298–1320	1455.61	446.01	50,000
75	$<Mo> + \{O_2\} = <MoO_2>$	298–1100	475.30	180.20	42,000
76	$2/5\{P_2\} + \{O_2\} = 2/5<P_2O_5>$	298–631	634.29	231.79	42,000
77	$2/5\{P_2\} + \{O_2\} = 2/5\{P_2O_5\}$	631–1400	619.23	206.56	42,000
78	$2<Pb> + \{O_2\} = 2<PbO>$	298–600	437.23	194.26	8000
79	$2[Pb] + \{O_2\} = 2<PbO>$	600–1150	436.43	193.22	13,000
80	$4<Na> + \{O_2\} = 2<Na_2O>$	298–371	813.79	230.96	42,000
81	$4[Na] + \{O_2\} = 2<Na_2O>$	371–1150	824.25	259.41	42,000
82	$4\{Na\} + \{O_2\} = 2<Na_2O>$	1150–1500	1211.27	585.34	42,000
83	$4<Na> + \{S_2\} = 2<Na_2S>$	298–371	870.27	234.72	42,000
84	$4[Na] + \{S_2\} = 2<Na_2S>$	371–1187	880.73	263.17	42,000
85	$4\{Na\} + \{S_2\} = 2<Na_2S>$	1187–1192	1267.75	589.11	42,000
86	$4\{Na\} + \{S_2\} = 2(Na_2S)$	1192–1600	1257.71	580.74	42,000
87	$<Na_2O> + <SiO_2> = <Na_2SiO_3>$	298–1361	232.42	−5.86	42,000
88	$(Na_2O) + <SiO_2> = <Na_2SiO_3>$	1361–1600	180.25	−44.14	42,000
89	$2<Ni> + \{O_2\} = 2<NiO>$	298–1725	489.11	197.07	13,000
90	$2[Ni] + \{O_2\} = 2<NiO>$	1725–2200	524.26	217.40	13,000
91	$½\{S_2\} + \{O_2\} = \{SO_2\}$	298–2000	296.42	−7.15	13,000
92	$\{S_2\} = 2\{S\}$	298–2200	−323.21	−124.26	42,000
93	$<Si> + \{O_2\} = <SiO_2>$	298–1700	871.53	181.17	13,000
94	$[Si] + \{O_2\} = <SiO_2>$	1700–1973	910.31	204.14	13,000
95	$[Si] + \{O_2\} = (SiO_2)$	1973–2200	901.53	199.70	13,000
96	$2\ Si + \{O_2\} = 2\{SiO\}$	1500–1920	381.58	−31.80	13,000
97	$Ti + \{O_2\} = <TiO_2>$	298–2080	910.02	173.22	42,000
98	$4/3\ V + \{O_2\} = 2/3\ V_2O_3$	298–1995	865.25	162.76	42,000

[a]*Source An Introduction to the Physical Chemistry of Iron and Steel Making*: R. G. Ward, ELBS, 1962, pp. 212–213
NB It should be noted that these values are originally given in cal/mole and the data are modified to J/mol by converting: 1 cal/mole = 4.184 J/mol

Printed in the United States
By Bookmasters